PLANT BIOLOGY

SECOND EDITION

THOMAS L. ROST

MICHAEL G. BARBOUR

C. RALPH STOCKING

TERENCE M. MURPHY

with contributions from
Robert M. Thornton and Michael S. Plotkin

University of California
Davis, California

THOMSON
━━━━━ ✦ ━━━━━ ™
BROOKS/COLE

Australia • Canada • Mexico • Singapore • Spain
United Kingdom • United States

Plant Biology, second edition
Rost, Barbour, Stocking, and Murphy

ACQUISITIONS EDITOR: Peter Adams

DEVELOPMENT EDITORS: Suzannah Alexander/Mary Arbogast

ASSISTANT EDITOR: Kari Hopperstead

EDITORIAL ASSISTANT: Kristin Lenore

TECHNOLOGY PROJECT MANAGER: Travis Metz

MARKETING MANAGER: Ann Caven

MARKETING ASSISTANT: Leyla Jowza

ADVERTISING PROJECT MANAGER: Kelley McAllister

PROJECT MANAGER, EDITORIAL PRODUCTION: Belinda Krohmer

ART DIRECTOR: Lee Friedman

PRINT/MEDIA BUYER: Rebecca Cross

PERMISSIONS EDITOR: Sarah Harkrader

PRODUCTION SERVICE: Graphic World, Inc.

PHOTO RESEARCHER: Kathleen Olson

COVER DESIGNER: Irene Morris

COVER IMAGE: ©Lindsey P. Martin/Corbis

COVER PRINTER: Transcontinental Interglobe

COMPOSITOR: Graphic World, Inc.

PRINTER: Transcontinental Interglobe

For more information about our products, contact us at:
Thomson Learning Academic Resource Center
1-800-423-0563
For permission to use material from this text or product, submit a request online at http://www.thomsonrights.com.
Any additional questions about permissions can be submitted by email to thomsonrights@thomson.com.

Thomson Higher Education
10 Davis Drive
Belmont, CA 94002-3098
USA

Asia (including India)
Thomson Learning
5 Shenton Way
#01-01 UIC Building
Singapore 068808

Australia/New Zealand
Thomson Learning Australia
102 Dodds Street
Southbank, Victoria 3006
Australia

Canada
Thomson Nelson
1120 Birchmount Road
Toronto, Ontario M1K 5G4
Canada

UK/Europe/Middle East/Africa
Thomson Learning
High Holborn House
50/51 Bedford Row
London WC1R 4LR
United Kingdom

Latin America
Thomson Learning
Seneca, 53
Colonia Polanco
11560 Mexico
D.F. Mexico

Spain (includes Portugal)
Thomson Paraninfo
Calle Magallanes, 25
28015 Madrid, Spain

Library of Congress Control Number: 2004115123

Student Edition: ISBN 0-534-38061-1

International Student Edition: ISBN 0-495-01393-5 (Not for sale in the United States)

Cover Image: East Indian Lotus (*Nelumbo nucifera*)

This volume is dedicated to the past authors of this series on botany and plant biology, R. M. Holman, W. W. Robbins, and T. E. Weier, and to the hundreds of instructors and thousands of students who have used the books over the years.

Contents in Brief

Contents

Preface

The last six years have been an exciting time for biology, especially for plant biologists. In this time, the number of gene sequences in databases has ballooned, and it is now possible to compare the sequences of conserved genes from a large number of organisms in all the kingdoms of life. Gene sequences provide marvelous information for determining evolutionary relationships, and our understanding of the tree of life has taken some new twists and turns.

We have seen the completion of the sequencing of several genomes. The human genome was perhaps the most trumpeted, but many others have been completed, including the genome of a small research plant, *Arabidopsis thaliana*, the "white mouse" of plant biology. Rice will be the next plant—and the first agricultural crop—to have its genome sequenced. Scientists are taking a new genome-wide view of the growth and development of plants.

Plant biologists have been among the first to recognize and embrace the power of modern phylogenetic logic in determining the evolutionary progression of the earth's diverse organisms. The major changes in the text—and the ones that we feel put this book in the forefront of introductory plant biology textbooks—deal with the description of the evolutionary origins and relationships of plants. With this new edition, the biodiversity part of the book largely adopts the best-supported, most current phylogenetic view of the organisms covered in our text. We introduce the topic of cladistics and show its application to various kinds of information, including gene sequences. We then describe how cladistics and sequence information have changed and enriched our understanding of plant evolution. This is a subject that excites strong passions among biologists. We show students how to appreciate different types of evidence—including both gene sequences and morphological traits—that describe the past events and directions of evolution.

Genetic modification has become big business, in plant biology more than any other field. When we wrote the first edition, we could only point to one or two genetically modified crops available to consumers. Now over 20 percent of the world's soybean, corn, cotton, and canola plants—and

75 percent of the soybeans grown in the U.S.—have been engineered for new traits. The new traits make foods and fibers cheaper to grow or pest-resistant. New varieties will even be more healthful. At the same time, many people in the world have become concerned about genetically modified plants, the procedures by which they are developed, approved, and distributed, and the possibility of gene "leakage" from the modified plants to nearby wild relatives. This edition deals with the opportunities and concerns in an objective manner.

There also has been much recent excitement in the field of ecology. Climate change continues to hold the attention of scientists and non-scientists alike. Can photosynthesis by plants ease the problems caused by our profligate burning of fossil fuels? Can we manage ecosystems to stem the present rate of biodiversity loss? Satellite sensors have given ecologists new tools for measuring the distribution and health of plant life over large areas.

We have taken these developments into account—as well as many excellent suggestions received from reviewers—in revising the chapters of this textbook. Our goal has been to keep the text topical, improve its readability and usefulness as a guide, and keep its size and language accessible to beginning biology students.

Other changes include:

- The chapter on background chemistry has been reorganized for clarity and new summary tables have been added.

- Information on transcription, translation, the cell cycle, and signaling has been moved to the chapter on development. The information on hormones has been rewritten to focus on various phases and processes of growth and development rather than the hormones themselves.

- The Biotechnology chapter has new information on PCR, genomics, and cloning; also, there are new figures showing recent applications of the technology.

- The chapter on prokaryotes now clearly distinguishes the Archaea and Bacteria.

- Chapters on plant structure have been carefully edited and new images have been added to enrich the already strong illustrations in the first edition.

- Within each biodiversity chapter we discuss organisms in terms of their phylogeny, rather than in terms of their Linnaean relationships.

- Cladograms are used in each chapter as a way to diagrammatically summarize phylogenetic relationships.

- There are many new sidebars throughout the book's chapters, organized into four areas: Plants, People, and the Environment; Economic Botany; Biotechnology; and In Depth (technical points).

- Each chapter ends with a selection of references, carefully chosen for readability, to articles that are available through InfoTrac®, an on-line subscription reference service. More technical references can be found through the Brooks/Cole–Thomson website.

- We added appendixes on useful and practical botanical subjects, including a list of toxic plants and information on choosing and nurturing house plants.

Bearing in mind the differences in preparation and experience of the students who will use this textbook, we have elected to retain the sequence of topics of the first edition. We begin with chapters on the structure, function, reproduction, physiology, and genetics of flowering plants because they are more familiar to readers, making the new information more easily absorbed. We then present the evolutionary survey, with detailed excursions into the Prokaryotes, Protists, and Fungi, as well as into divisions of the Kingdom Plantae. The overall sequence of major subjects in the book builds from metabolism to whole plant function to reproduction within flowering plants, then from simpler to more advanced organisms, concluding with two ecological chapters that tie together the entire book. We recognize that instructors differ in their preferences for chapter sequence, and we have taken great pains to ensure that chapters are independent, so that it is possible to start with survey chapters and proceed to the reproduction, structure and physiology of flowering plants. We hope that instructors and students will find the chapters easy to apprehend in whatever sequence they read them.

This book is the capstone of a chain of plant biology textbooks that is probably longer than that of any other existing biology textbook in the world. The ancestral book, *A Textbook of General Botany*, was co-authored in 1924 by Richard M. Holman of the University of California, Berkeley, and Wilfred W. Robbins in the Botany Division of what was then called the University Farm, located in the small town of Davis. The book was widely adopted through its four editions. After Richard Holman passed away, Professor Robbins invited a young faculty member in the Farm's Botany Division to be co-author of a 1950 book, *Botany: An Introduction to Plant Science*. That faculty member was T. Elliot Weier, a cytologist and pioneer in the use of the electron microscope. After Wilfred Robbins' death, Dr. Weier invited a colleague in the department, C. Ralph Stocking, a plant physiologist, to become co-author of the second and third editions. Michael Barbour, a plant ecologist, joined Weier and Stocking on the fourth (1970) and fifth (1974) editions, and Tom Rost, a developmental anatomist, joined for the sixth edition (1982). Plant physiologist Robert Thornton helped them write two editions of a smaller book called *Botany: A Brief Introduction to Plant Biology* (Rost et al., 1979 and 1984). Terence Murphy, a molecular plant biologist, has become the eighth co-author in a story that has wound through 13 versions/editions, 74 years, hundreds of thousands of books printed, two publishers, and four generations of readers. Welcome to you, the fourth generation!

We thank the many people who provided us with help, suggestions, comments, and criticisms. First are two people who have made significant contributions to the revision of eight chapters: Robert Thornton and Michael Plotkin, who helped with systematics and evolution. Colleagues who provided new photographs include Bo Liu, Ernesto Sandoval, and Tim Metcalf, Section of Plant Biology; and Eduardo Blumwald, Pomology Department, University of California, Davis. Judy Kjelstrom, Paul Gepts, Bruce German, Bruce Hammock, Carl Keen, Janet King, and Richard Roush, University of California, Davis; and Daniel Gladish, Miami University, Hamilton, Ohio, provided the material for sidebars. We also are grateful to our original acquisitions editor at Brooks/Cole, Nedah Rose, whose enthusiastic support got this edition off to a great start, and to our developmental editor, Suzannah Alexander, whose critical insight and care have contributed greatly to the result. The reviewers who provided excellent advice include: M. Stephen Ailstock, Anne Arundel Community College; Susan Alcala, Northwest Vista College; Natalie Barratt, Baldwin-Wallace College; Mary Berbee, University of British Columbia; Susan Bornstein-Forst, Marian College; Richard R. Bounds, Mount Olive College; Judith K. Brown, University of Arizona, Tuscon; Gregory T. Chandler, University of North Carolina, Wilmington; William B. Cook, Midwestern State Univesity; Kenneth J. Curry, University of Southern Mississippi; David B. Czarnecki, Loras College; Lawrence J. Davenport, Samford University; Roger del Moral, University of Washington, Seattle; David Domozych, Skidmore College; Diane S. Doss, South Puget Sound Community College; Frederick B. Essig, University of South Florida; Michael A. Fidanza, Penn State University, Berks; Loretta Graham Fogwill, Red River College; Mike Gipson, Oklahoma Christian University; Joel Hagen, Radford University; Laszlo Hanzely, Northern Illinois University; Stephanie G. Harvey, Georgia Southwestern State University; Brian Hedlund, University of Nevada, Las Vegas; David S. Hibbett, Clark University; A. Scott Holaday, Texas Tech University; Laura Jaquish, Northwestern Michigan College; Walter S. Judd, University of Florida, Gainesville; Geoffrey S. Kennedy, University of Wisconsin, Milwaukee; James W. Kimbrough, University of Florida, Gainesville; Judy Kjelstrom, University of California, Davis;

Robert L. Koenig, Southwest Texas Junior College; Erica F. Kosal, North Carolina Wesleyan College; Thomas G. Lammers, University of Wisconsin, Oshkosh; Ira Levine, University of Southern Maine; Michael Howard Marcovitz, Midland Lutheran College; Andrew G. McCubbin, Washington State University, Pullman; Martina McGloughlin, University of California, Davis; Andrew S. Methven, Eastern Illinois University; James A. Nienow, Valdosta State University; Rachel Pfister, University of Arizona, Tucson; Daniel Potter, University of California, Davis; Edward E. Schilling, The University of Tennessee, Knoxville; Andrea Schwarzbach, Kent State University; Barbara Greene Shipes, Hampton University; Nancy L. Smith-Huerta, Miami University; Teresa Snyder-Leiby, State University of New York, New Paltz; Ann E. Stapleton, University of North Carolina, Wilmington; Dennis W. Stevenson, The New York Botanical Garden; P. Roger Sweets, University of Indianapolis; John Taylor, University of California, Berkeley; Jim Taylor, Ouachita Baptist University; Leslie R. Towill, Arizona State University; Michael R. Twiss, Clarkson University; Garland R. Upchurch, Jr., Texas State University, San Marcos; C. Gerald Van Dyke, North Carolina State University; Joseph M. Wahome, Mississippi Valley State University; Yongbao Wang, University of Miami; Yunqiu Wang, University of Miami; Cherie L. R. Wetzel, City College of San Francisco; and James C. Zech, Sul Ross State University.

About Plant Biology

Visit us on the web at http://biology.brookscole.com/plantbio2 for additional resources, such as flashcards, tutorial quizzes, InfoTrac exercises, further readings, and web links.

1. Plants include nearly half a million species of mosses, ferns, conifers, and flowering plants. In addition, plantlike relatives—certain bacteria, fungi, and algae, which also are included in this book—total another million species. **Plant biology** is the study of organisms classified either as plants or their (sometimes distant) plantlike relatives.

2. Plants and their plantlike relatives are of vital importance to all life on Earth, including humans. Economic wealth is largely dependent on plant products. A major challenge of the 21st century is to attain sustainable use of our plant resources, which means no loss in the carrying capacity of the earth and no loss in the diversity of organisms that coexist on our planet.

3. The **scientific method** is a process of formulating predictions (hypotheses) about the world and then testing those hypotheses. Although it is a powerful and currently widely accepted method used for explaining our environment, the scientific method has some limitations, and it was historically used to justify actions that had negative social and ecological consequences.

4. The phylogenetic (evolutionary) relationships among plants, plantlike relatives, bacteria, fungi, and animals have become better understood but also more complex in the last half-dozen years, largely because of new techniques that analyze the sequence of base-pairs in the DNA of cell nuclei and of cell organelles such as chloroplasts. The division of life into a handful of kingdoms, a commonly accepted starting point for classification—and one used in the previous edition of this book—no longer appears possible. Many familiar names for groups of plants and the familiar hierarchies for placing these names are now known to be inappropriate.

1.1 THE IMPORTANCE OF BOTANICAL KNOWLEDGE

For most of you, this textbook and the class it accompanies represent the only formal education in organismal plant biology you will acquire in college. You are heading toward careers that seem to have little to do with plants, such as engineering, aerospace, medicine, chemistry, climatology, marine science, history, or political science. Our goal was to write a book that would efficiently and interestingly present a modern survey of plant biology. We also wanted to present the many interrelationships between plants and your central field of interest—connections that are important but probably would remain unknown to you without this textbook.

For those of you who are going on to a career in plant biology, such as field botanist, ethnobotanist/ethnoecologist, science writer/editor, laboratory researcher, agricultural economist, teacher, or greenhouse conservator **(Fig. 1.1),** we wanted to write a textbook that gives you enough depth in the field to prepare you well for upper division courses. In your future work, you will be called on to show nonbotanists the relevance of plant biology to their own areas of expertise.

The linkages among different fields of knowledge may be as important as the narrower information within each field. As each area of expertise expands in its own direction, the gulf between areas widens and the importance of those who can see connections is heightened.

1.2 THE IMPORTANCE OF PLANTS

Seen from the perspective of the moon, 385,000 km away, Earth has a blue color **(Fig. 1.2a).** Blue oceans cover more than two thirds of the planet's surface, and a blue atmosphere extends away from the entire globe, a fuzzy halo dimming into black space. Seen from the perspective of 10 km above the surface, however, the earth looks green **(Fig. 1.2b).** The green color is caused by an enormous number of plants that carpet the ground, extend into the air, occupy the soil, and float below the surface of lakes and oceans. This tangle of plant tissue selectively absorbs red and blue wavelengths of sunlight, and either reflects green wavelengths or allows them to pass through. Consequently, sunlight reflected from a forest, a meadow, or a field of corn is green, as is the dappled shade beneath a tree.

The red and blue wavelengths of light are transformed by plant tissue into chemical energy in a process called **photosynthesis.** Technically, a **plant** is an organism that is green and photosynthetic (producing organic sugar from inorganic carbon dioxide, water vapor, and light). In addition, plant cells (the basic units of which all organisms are composed) are surrounded by a rigid wall made of **cellulose,** a molecule rarely found in other organisms. Plants have multicellular bodies usually well suited to life on land because they can control water loss, they have strengthening tissue that keeps them upright, they can somewhat regulate their temperature, and they can reproduce with microscopic, drought-tolerant cells called **spores.**

This technical definition of a plant applies to approximately 300,000 species of trees, shrubs, herbs, grasses, ferns, mosses, and algae. These different types of plants exhibit an incredible diversity in habitat, shape, life history, evolutionary history, ecology, and human use. Most of this textbook is about them. They are formally classified as belonging to the plant kingdom **(Fig. 1.3).**

There are another million species of plantlike organisms that are not classified as belonging to the plant kingdom, which this textbook also addresses. They are certain bacteria, fungi, and algae. Although some are green and photosynthetic, others engulf living food or feed on dead organic remains. They cannot regulate water loss and they do not possess strengthening tissue **(Fig. 1.4).** Many of them grow in aquatic habitats rather than on land. They exhibit enormous morphologic and habitat diversity and have great economic and ecologic importance to humans. They are part of the science of **botany,** or plant biology, which is the study of plants and plantlike organisms. Plants and plantlike organisms provide many **ecologic services.** That is, by merely living they serve the needs of other organisms, including humans. For example, they are sources of food, fabric, shelter, and medicine. They produce atmospheric oxygen and organic nitrogen. They build new land and inhibit its loss

a

b

c

Figure 1.1 Examples of botanical careers. (**a**) Field botanists document the locations of rare plants; they also quantify/classify vegetation as habitat types for other organisms. (**b**) Ethnobotanists/ethnoecologists interview indigenous people about the uses of native plants and the impact of their management techniques on the landscape. (**c**) Laboratory technicians or researchers explore the genetics, metabolism, morphologic development, and ecologic tolerances of selected plant species. (**d**) Science teachers "translate" knowledge about plants to students. (**e**) Conservators collect, propagate, and classify plants—studying and preserving plant diversity and communicating their values to members of the community.

d

e

a

b

Figure 1.2 (**a**) Earth from space. The globe is blue, as seen from the distance of its moon. (**b**) At closer range, Earth is predominantly green because of the prevalence of plants.

from erosion. They control atmospheric temperature. They decompose and cycle essential mineral nutrients.

Civilizations have risen and fallen throughout recorded time depending on their access to and control of plants—especially of trees for lumber to make warships; as fuel to smelt metals, cure pottery, and generate power and heat; and as sources of wealth in the form of spices and industrial products such as rubber and oil. Anthropologists often describe human cultures and history in such terms as Stone Age, Bronze Age, and Iron Age, but truly humans have

Figure 1.3 Examples of organisms in the plant kingdom: (**a**) scarlet oak leaves; (**b**) tree ferns in Hawaii; (**c**) a moss; (**d**) the broad-leaved herb mule ears, a relative of sunflower; and (**e**) the grass wheat.

Figure 1.4 Examples of plantlike organisms: (**a**) photosynthetic Bacteria; (**b**) a mushroom in the kingdom Fungi; and (**c**) algae (protists) floating on the surface of a pond.

always lived in a **Wood Age,** because wood has been the ultimate measure and creator of wealth **(Fig. 1.5).**

Our earliest written myths and legends feature plants. The *Epic of Gilgamesh,* dated to 4,700 years ago and created by storytellers in ancient Mesopotamia (now Iraq), highlights the importance of forests and conveys an understanding of the ecologic consequences of logging. Gilgamesh was ruler of Uruk and wanted to make a name for himself by building a great city. Building required lumber, but the vast cedar forest that covered mountain slopes to the east had never been

entered by humans. A demigod named Humbaba protected the forest for nature and the gods. Nevertheless, Gilgamesh and his companions traveled to the forest with adzes and axes. They briefly lost themselves in their initial contemplation of the forest's beauty and holiness, but soon set to work felling trees. When Humbaba challenged them, they killed him. Then the cedars filled the air with a sad song, and the gods cast curses and promises of fire, flood, and drought on Gilgamesh and on generations of humans to come. These promises later came true when Mesopotamia's growth

Figure 1.5 Conifer forests such as this coast redwood forest have enormous economic value. The wood from a single old-growth redwood tree has a retail value of more than $50,000.

Figure 1.6 Forest decline in central Europe. Beginning in the late 1970s, many square kilometers of forest died because of a complex of factors including ozone from automobile exhaust and acids from coal burning.

stripped it of natural resources, degraded the environment, and contributed to its loss of regional power.

Ironically, just as we have come to recognize the critical, pervasive importance of plants, we also recognize that we are losing natural plant cover to agriculture, urbanization, overgrazing, pollution, and extinction at a faster rate than ever before. We are finding the truth behind four "environmental laws" described in 1961 by plant biologist Barry Commoner: (1) Everything is connected to everything else; (2) everything must go somewhere; (3) nature knows best; and (4) there is no such thing as a free lunch.

In Chapter 27, we write that the term **conservation** once meant the "wise use" consumption of a natural resource at a rate that would result in its sustained, continued existence far into the future. However, the history of our farms, forests, pastures, and fisheries suggests that technological, growth-oriented human cultures have not been able to determine what that ideal level of resource use is. We have consistently overexploited resources beyond the balance point, thus degrading the landscape. In addition, human activities interact with soil, water, and air in unexpected ways, resulting in such potential global catastrophes as acid rain, ozone depletion (see "PLANTS, PEOPLE, AND THE ENVIRONMENT: Unexpected Links among Chlorofluorocarbons, Climate, and Plants" sidebar), climate change, and forest decline (Fig. 1.6).

1.3 THE SCIENTIFIC METHOD

Scientists in every field have a set of methods in common. They observe, ask questions, make educated guesses about possible answers, base predictions on those guesses, and then devise ways to test their predictions. If a predicted result actually occurs, that is evidence that the possible answer could be correct. More formally, the guesses are called **hypotheses** or theories, and the tests and interpretations are called the **scientific method.** The scientific method was codified and encouraged in the 17th century by Rene Descartes and Sir Francis Bacon, among others. It has proven to be a very powerful method for advancing human understanding.

In science, a hypothesis is never seen as absolute truth. It is merely the best approximate explanation, and we can expect it to be modified in the future as evidence from new predictions and tests becomes available. The test-evaluate-refine-retest cycle has been entered hundreds of times for some hypotheses that are currently accepted, but hypotheses about subjects at the very edge of our current understanding may have passed through very few cycles. Sometimes, the new information calls for a complete rejection of the original hypothesis and the creation of a new one, providing the scientific equivalent of a political revolution (Fig. 1.7).

The external world, not internal convictions, should be the testing ground for hypotheses, according to the scientific

PLANTS, PEOPLE, AND THE ENVIRONMENT:

Unexpected Links among Chlorofluorocarbons, Climate, and Plants

In the 1930s, refrigeration became widespread in homes, trains, trucks, and agriculture. The basic component in refrigerators is a gas that contains chlorine or bromine, fluoride, and carbon. The gas absorbs heat when it is allowed to expand. Freon was an early trade name for such gases, but they are more technically called chlorofluorocarbons (CFCs). They are very efficient, nontoxic, and generally nonreactive. They are also used as foam-blowing agents and as cleansers in computer industries. A related compound, methyl bromide, has commonly been used to sterilize soils.

Four decades later, however, scientists began to realize that when CFCs leaked from refrigerators or were emitted into the atmosphere from other industrial sources, they rose high into the stratosphere (25–50 km above the earth's surface) and destroyed ozone there. Ozone is an oxygen molecule that is made up of three atoms instead of the usual two. It is a natural component of the stratosphere, and it is ecologically important because it absorbs ultraviolet radiation from the sun. Chlorine atoms act as catalysts in this destruction of ozone. Because a catalyst is a molecule that enhances a chemical reaction without being used up itself, this means that a single chlorine atom can destroy thousands of molecules of ozone. The concentration of CFCs has reached a point where there is as large as a 20% loss of ozone, depending on the latitude and the season.

Ozone depletion has several serious consequences. One is the likely increase of skin cancers, cataracts, and immune deficiency diseases among humans. Cancer is expected to increase by 15% during the next few decades in the heavily populated north-temperate zones of the world. Ultraviolet radiation, in particular ultraviolet B (UVB), is known to derange plant metabolism. We may expect yields of such sensitive crops as soybeans to be depressed by 25%. The impact on natural vegetation is unknown, but teams of scientists who have erected UVB lamps over grassland and tundra vegetation are studying the long-term effects of UVB (Fig. 1).

The danger of ozone depletion was convincing enough to have led to the most comprehensive international agreement ever established: a protocol signed in Copenhagen in 1992 by nearly all the countries of the world, who agreed to dramatically reduce further production of CFCs and to eliminate all production by 2030. The catalytic nature of CFCs, however, means that their impact on the global environment will linger far into the future, well beyond the year 2030.

The elimination of CFCs will have some negative effects on agriculture because agribusiness has come to depend on industrial chillers and the use of methyl bromide to sterilize soils and to eliminate plant pathogens. Methyl bromide is used extensively in fruit, vegetable, cotton, cocoa, coffee, and grain production. Imaginative and possibly costly substitutes will be needed to replace the CFC-based technology.

One psychological consequence of the CFC situation is a realization that the impacts of technology's byproducts have become global, extending far from their point of creation into what had been considered to be pristine wilderness. Human influence has become so pervasive that no place on the globe is unaffected. As one environmentalist concluded, this is surely the signal of the death of wild nature. We are learning that the first two laws of ecology formulated by Barry Commoner are true: Everything is connected to everything else, and everything must go somewhere.

Michael G. Barbour

Figure 1 Researchers directed by Philip Grime (pictured), University of Sheffield, are investigating the ecologic impact of 5% to 20% more ultraviolet B, greater concentrations of carbon dioxide in the atmosphere, higher soil temperatures, and decreased rainfall on U.K. grasslands.

Websites for further study:

ENFO—Information on the Environment: http://www.enfo.ie/leaflets/bs23.htm

Almanac of Policy Issues:
http://www.policyalmanac.org/environment/archive/ozone.shtml

British Antarctic Survey:
http://www.antarctica.ac.uk/Key_Topics/The_Ozone_Hole/Ozone_explained/

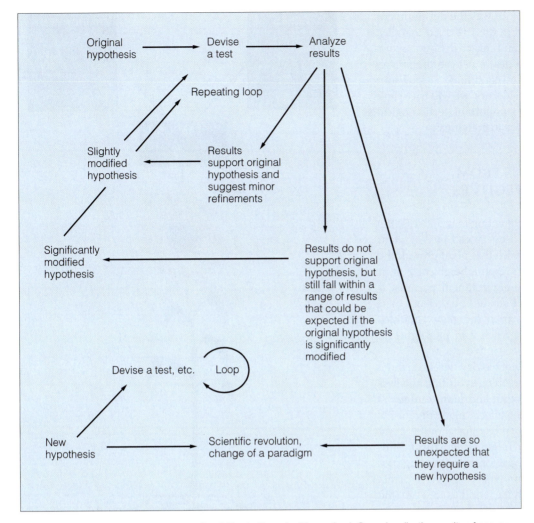

Figure 1.7 Diagram of the sequence of activities in the scientific method. Occasionally, the results of one or more tests are so different from expectations that the underlying assumptions and theories must be radically changed, resulting in a "revolution" or "paradigm shift."

method. That means that those who use the scientific method should be objective and unbiased. In fact, however, personal bias is difficult to remove completely. The questions asked, the range of expected answers, and the tests used to check predictions are all products of the culture that surrounds the scientist. In addition, one's beliefs may subconsciously influence one's science. For example, historian Greg Mitmann studied a group of biologists at the University of Chicago whose ideas about ecology were dominant during the mid-20th century. In essence, their research showed that nature was organized around cooperative behavior and community. Mitmann pointed out that these scientists were socialists and pacifists in their private lives. Did their personal beliefs lead them to interpret their scientific findings in a certain light? (Currently, by contrast, the Chicago School's ideas have been turned upside down; most biologists now hypothesize that nature is organized around competition and individuality.)

Carolyn Merchant, a professor at the University of California, Berkeley, has argued that the scientific method is a peculiarly masculine invention that has fostered exploitation of natural resources. The scientific method, she claims, invites us to become isolated observers who impersonally manipulate nature to test assumptions. The ever-giving Earth Mother image of pre-science cultures has been replaced with an inanimate globe whose treasures must be forcibly extracted. Merchant writes that our scientific subjugation of nature has paralleled a time of subjugation of women, and she implies that one has caused the other.

Technology is the application of information to industrial or commercial objectives. It does not require an understanding of how or why a given process functions. Science and technology are ultimately linked, but they are not the same thing, and each can progress at independent rates. For example, consider the development of antibiotics, the "wonder drugs" of the mid-20th century. Penicillin, the first antibiotic, was widely and effectively prescribed by doctors for the cure of many bacterial infections beginning in the late 1940s. A scientific understanding of how the drug actually worked, however, was not attained for another 20 years. Only then did researchers discover that penicillin interferes with how a bacterial cell builds a wall around itself, and that if the wall is absent, the cell contents burst open and the cell dies. Another example of technology leading science was the development of plant-killing auxins (herbicides) in the 1940s. To this day, plant biologists do not understand exactly how auxins derange plant growth and lead to death, yet the pragmatic, wide use of auxin-based herbicides continues. Throughout this textbook, and especially in the sidebars, we will try to link the science of plant biology with important technological applications, attempting to do justice to both areas.

1.4 STUDYING PLANTS FROM DIFFERENT PERSPECTIVES

Every plant is a product of two interacting components: the genetic (inherited) material, which every cell in that plant carries, plus the environment in which that plant grew. If the genetic potential of the plant is to reach a height of 2 m, it will attain that height only if the environment permits. If certain nutrients are lacking, or if shade is too deep, or if night temperatures are too cold, or if too many animals browse the plant, then it will not reach 2 m in height, no matter how much time goes by.

Two major disciplines within plant biology deal with these two components: **Plant genetics** (including **evolution** and **systematics**) is the study of plant heredity, and **plant ecology** is the study of how the environment affects plant organisms **(Fig. 1.8).** Knowledge of genetics is a prerequisite for the study of plant evolution and classification (**plant systematics**). Similarly, knowledge of ecology is essential for reconstructing past climates and landscapes and for understanding how plants have come to be distributed around the world as they are (the disciplines of **paleoecology** and **biogeography**).

One of the many ways genes and environment become integrated is through plant metabolism. **Metabolism** is the process by which plants perform photosynthesis, transport materials internally, construct unique molecules, and use hormones to affect their behavior. The study of metabolism includes the disciplines of **plant physiology** and **plant molecular biology.**

The combination of genes, environment, and metabolism works together to produce an individual plant. The plant can then be studied in several basic ways. The study of how a plant develops from a single cell into diverse tissues and organs and an array of outer surfaces and shapes is the discipline of **plant morphology.** The study of a plant's internal structure, its tissues and cell types, is called **plant anatomy.** Plants also can be studied according to their taxonomic classification: Microbiology is the study of bacteria, mycology is the study of fungi, phycology is the study of algae, bryology is the study of mosses, and so on (Fig. 1.8).

We have chosen to write about each of these disciplines in a certain order. We begin in Chapters 2 through 15 with a multiple focus on physiology, morphology, and anatomy. In these chapters, you will find answers to the following questions:

- How do the billions of cells, dozens of tissues, and several major plant parts cooperate to achieve a complex, functioning organism?

Figure 1.8 Interrelationships among several plant biology disciplines.

- Why do roots grow downward but stems upward?
- How is it possible for leaves and flowers to track the sun's path during the day, maximizing the amount of light they receive?
- How can a species regulate where its seeds germinate, thus increasing the chances of survival of offspring?
- How does a plant repair injuries? How does a plant tell time and season so that its many life cycle events occur at the appropriate times?
- How do plants acquire and transport energy, carbohydrates, water, and nutrients?
- How do plants solve the problems of reproduction, drought, competition, and a changing environment while being rooted to a single location?

In Chapters 16 and 17, we describe the rapidly growing science of plant genetics and its technological implications for breeding new crops with increased yields and tolerances to diseases and weeds. These chapters answer the following questions:

- Can our understanding of genetics reach a point where we are capable of designing and engineering new species by moving bits of genetic material around or by creating the genetic material ourselves?

- Can we create genetic bank accounts for all rare and endangered species, as insurance against extinction?

In Chapters 18 through 25 we focus on the taxonomic diversity of plants and plantlike organisms, highlighting the unique way each group's morphology, anatomy, physiology, and reproduction are suited to its particular range of habitats. We also try to reconstruct the evolutionary history of each group. These chapters address the following questions:

- To which existing or extinct group is a plant most closely related?

- When did a group begin to evolve in a separate line?

- Is the group still actively evolving, or has it stagnated in some way?

- How has the group adapted to its range of habitats?

1.5 A BRIEF SURVEY OF PLANT CLASSIFICATION

Kingdoms, Domains, Divisions or Phyla, and Clades

As recently as 10 years ago, most biologists accepted a five-kingdom classification first developed in detail by the plant ecologist Robert Whittaker in 1969. The names of the five kingdoms were Monera, Fungi, Protista, Plantae, and Animalia. Each **kingdom** was presumed to be a **monophyletic** group of species; that is, all species in a kingdom were related to a single common ancestral species. The kingdom **Monera** included bacteria; Fungi included molds, mildews, and mushrooms; Protista included a great variety of simple organisms, some of which were photosynthetic and largely aquatic organisms informally called algae; Plantae included more complex photosynthetic organisms that typically grew on land; and Animalia included typically motile, multicellular, nonphotosynthetic organisms distributed in habitats throughout the world.

The recent use of molecular biology techniques, and a process called **cladistics,** has largely shown that the five-kingdom approach did not recognize natural (evolutionarily related) groups. By comparing the sequence of base pairs in the genetic material from one group of organisms with that from another, the overlap (the percent similarity of two strands of DNA) can be quantified. It no longer is possible to claim that the five kingdoms are monophyletic. For example, the Monera must be split into two groups, each of which does appear to be monophyletic: the domain Bacteria (which contains most of the familiar bacteria) and the domain

Archaea (which superficially resemble bacteria but differ in habitat, metabolism, and cell structure). *Domain* is a neutral term, which is generally applied to groups of organisms as large as a kingdom or larger that are monophyletic.

Domain **Eukarya** includes what we commonly recognize as the plant, animal, and fungal kingdoms. The **Protista** is not a monophyletic group, but there is no agreement as to where to place the various groups of organisms within it. At the moment, *protist* is an informal name that can be applied to the species of algae, slime molds, and protozoans that were formerly in the kingdom.

Each kingdom can be divided into divisions or phyla, the members of each having originated from a common ancestor. In the kingdom Plantae, for example, the division (phylum) Tracheophyta includes all plants with vascular tissue. Division names end in *-phyta.* Below divisions are classes (ending in *-opsida*), orders (ending in *-ales*), families (ending in *-aceae*), genera, and species (Fig. 1.9). This hierarchy is part of the Linnaean system of taxonomy.

Taxonomy is the science of naming and categorizing organisms. Most taxonomists believe that a classification system should be both maximally useful to people and accurately reflective of what we know about the relationships among the organisms. Although the Linnaean system is easy to use, it has been found that evolutionary groups do not always correspond exactly with Linnaean categories. More importantly, the Linnaean system promotes the false idea that two groups of equal rank are biologically and evolutionarily equivalent units. Those who attempt to compare two equivalent Linnaean groups often are misled.

This textbook handles taxonomy in a new and exciting way. We have downplayed the traditional Linnaean system, and instead we emphasize evolutionary groups. These groups, known to biologists as **clades,** are uncovered by methods explained in later chapters. Clades are assembled in a branching diagram of relationships that is called a **cladogram** or a **phylogenetic tree.** Relationships in a cladogram emphasize shared features inherited from a common ancestor.

Linnaeus, in contrast, based his system on the notion that all species were created as they are now, and never changed. He grouped organisms on the basis of arbitrary similarities and attempted to make his system as easy to use as possible. Because we know now that many of the similarities he identified are coincidental and do not represent features shared because of descent from a common ancestor, it is not surprising that his taxonomic system fails to meet the needs of modern biologists.

A taxonomy based on evolutionary groups does not use hierarchical, nested categories that denote rank. Rather, it relies on one overriding principle: Only evolutionary groups (clades) should be named. These evolutionary groups consist of a common ancestor and all the descendants of that ancestor. The relationship of one clade to any other clade is determined by looking at a phylogenetic tree, and not by comparing arbitrary ranks.

Although phylogenetic trees may bring new names and ideas about plant groups, we believe that all readers will soon discover their benefits. Cladograms contain much

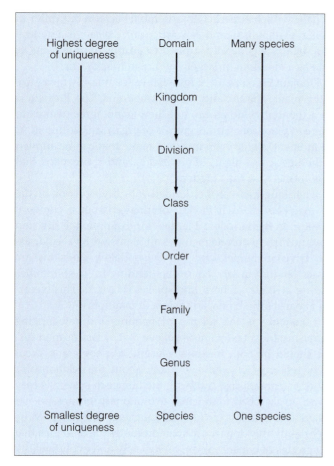

Highest degree of uniqueness	Domain	Many species
	Kingdom	
	Division	
	Class	
	Order	
	Family	
	Genus	
Smallest degree of uniqueness	Species	One species

Figure 1.9 Relationships among domains, kingdoms, divisions, and smaller classification units such as classes, orders, families, genera, and species.

more information than the rigid categories of the old system; they promote a different way of thinking about biological diversity; they move the focus from correctly naming to understanding the evolution of the organisms under study; and, as Dobzhansky famously said about evolution in general, they provide "a framework in which everything else begins to make sense."

Plants have the best-supported and most thoroughly detailed phylogenies of all organisms. There are several areas of uncertainty (noted in the text), but also many areas where we have great confidence. Undoubtedly, future discoveries will require that some cladograms be modified. We believe that thinking phylogenetically is more useful than taxonomic "pigeon-holing," even if particular trees are someday amended.

The Domains Bacteria and Archaea Include Photosynthetic Bacteria

Organisms in the domain Bacteria have **prokaryotic** cells. Such cells lack an organized nucleus with a bounding membrane. They have no specialized compartments (**organelles**) within the cell sap (**cytoplasm**), their cell volume is very small,

and the genetic material (DNA) in their chromosome is organized into a closed circle rather than in a line. Sexual reproduction is unknown, although these organisms do exchange bits of genetic material occasionally across cellular bridges.

The domain **Archaea** also is composed of prokaryotic organisms, but they occupy very different habitats and have cell structure and chemistry that are different from that of **Bacteria.** The differences are so profound that one can conclude Archaea organisms are as different from Bacteria as prokaryotes are from the domain Eukarya. The Archaea are divided into three groups on the basis of habitat: bacteria of sulfur-rich anaerobic hot springs and deep ocean hydrothermal vents; bacteria of anaerobic swamps and the guts of termites; and bacteria of extremely saline waters. In this last group, the **extreme halophiles,** are photosynthetic organisms that harvest light with a unique pigment called bacteriorhodopsin.

Bacteria is also the informal name given to species in the two domains. Bacteria exist as single cells or as filaments of cells. The earliest forms of life, as revealed by fossils 3 to 4 billion years old, apparently were bacteria. Bacteria are microscopic, but their numbers and weight (biomass), collectively for the earth, are very large. Bacteria occur in virtually every habitat on Earth, including the atmosphere, soil, water, and stressful places that are hot, cold, oxygen-deficient, and even within the surface layer of rocks in Antarctica.

Bacteria are enormously beneficial as decomposer organisms, which are able to break down complex organic material into simpler nutrient molecules that can then be taken up by other plants. Bacteria also have detrimental effects as **pathogens** (disease-causing agents). Human diseases caused by bacteria include botulism, bubonic plague, cholera, diphtheria, gonorrhea, leprosy, meningitis, syphilis, tetanus, tuberculosis, and typhoid fever. Other bacteria cause diseases of plants.

One group of Bacteria, the **Cyanobacteria** or **blue–green algae,** has the pigment chlorophyll in its cells. These bacteria are green and are capable of conducting photosynthesis. Another characteristic that they share with algae and species in the kingdom Plantae is that their cells are surrounded by a rigid wall. Many *Cyanobacteria* species also have the unique capacity for **nitrogen fixation**—that is, to incorporate inorganic nitrogen gas from the atmosphere into organic nitrogen in the form of ammonium.

The Kingdom Fungi Contains Decomposers and Pathogens

Fungi are nonphotosynthetic organisms with **eukaryotic** cells—that is, their cells have a membrane-bounded nucleus, special organelles within the cytoplasm, linear DNA, and much larger cell volumes than prokaryotes. Fungi include molds, mildews, mushrooms, rusts, and smuts. They are typically microscopic and filamentous, and their cells are surrounded by a rigid wall made of chitin, a substance more commonly found in some species of the kingdom Animalia than in algae and members of the kingdom Plantae. They reproduce sexually in a variety of complex life cycles, and with spores. Fungi are widely distributed throughout the world, but they are mainly terrestrial.

Fungi are ecologically important because many of them are decomposers, similar to bacteria. Others form intimate associations with the roots of plants, improving the absorbing capacity of the plant's root system. Some fungi, such as mushrooms and morels, are important foods for animals and humans. The decomposing action of yeast fungi creates flavored cheeses, leavened bread, and alcoholic drinks. Fungi such as *Penicillium* have been used commercially to produce antibiotic drugs. Other fungi are pathogens, invading the tissue of animals or plants, causing illness, and reducing crop yields by billions of dollars annually.

Protists Include the Grasses of the Sea

Protists are simple, often microscopic, and usually aquatic organisms. Their cells are eukaryotic, and they reproduce both sexually and asexually. This nonhomogeneous group of species contains some photosynthetic organisms, called algae, and some nonphotosynthetic organisms called slime molds, foraminiferans, and protozoans. It is a catch-all group, containing such widely divergent groups as those related to the ancestral forms of the plant, animal, and fungal kingdoms. As biologists learn more, the various protist groups will no doubt be arranged in different clades in the near future.

Algae occur as single cells or as clusters, filaments, sheets, or three-dimensional packets of cells. Every cell in most multicellular algal bodies can carry out photosynthesis and obtain water and nutrients directly from its liquid environment. Algae are important because great numbers of them float in the uppermost layers of all oceans and lakes. There small animals graze on them, which in turn provide food for larger fish and ultimately for humans. These microscopic algae are called **phytoplankters** and are commonly referred to as "grasses of the sea" because they form the base of a vast natural **food chain,** which transfers energy from sunlight to all organisms within a pond, lake, or ocean. Phytoplankters also produce half of all the oxygen in the atmosphere as a byproduct of photosynthesis.

The Plant Kingdom Contains Complex Plants Adapted to Life on Land

The **plant kingdom** contains mosses, ferns, pine trees, oak trees, shrubs, vines, grasses, and broad-leaved herbs—all the organisms we informally call plants. Organisms in this kingdom are adapted to life on land, and some of them are among the more recent forms of life to evolve and appear in the fossil record. Plants that produce seeds, cones, and flowers are relatively large and abundant and give the landscapes of the planet their characteristic appearance, making up most of the biomass of forests, meadows, shrub lands, deserts, marshes, woodlands, and grasslands.

Members of the kingdom **Plantae** share certain unique biochemical traits. They have eukaryotic cells with walls made of cellulose, they accumulate starch as a carbohydrate storage product, and they have special types of chlorophylls and other pigments. Only green algae have these same traits, which is why they are also part of the plant kingdom.

Plants have more complex bodies than bacteria, fungi, or protists. This complexity is visible from cell to cell and from region to region within the plant's body. Some cells and tissues are specialized to transport fluids, to store reserves, to perform photosynthesis, or to add strength. Different parts or regions of the plant form such unlike structures as leaves, stems, roots, flowers, and seeds.

Plants have major ecologic and economic importance. They form the base of terrestrial food chains, they are the principal human crops, and they provide building materials, clothing, cordage, medicines, and beverages. Our terrestrial ecosystems are dependent on organisms in the kingdom Plantae.

1.6 A CHALLENGE FOR THE 21ST CENTURY

Chapters 26 and 27, the final chapters of this textbook, focus on the discipline of ecology. They take us through past and present landscapes, describing the distribution of plants and their elegant solutions to environmental stresses. How do plants meet the challenges of a changing environment for continued existence? They must partition time and energy, within an individual life span, in such a way that their kind continues for another generation.

What can plants tell us by their presence, vigor, or abundance about the past, present, and future of their habitat? If, as Barry Commoner says, "Nature knows best," can we ever understand the environment well enough to permit us to restore endangered vegetation and degraded landscapes? Can we continue to increase our human population while retaining natural biological diversity and developing a sustainable use of the world's forests, grasslands, and cropland? As a species, we have not yet been successful in achieving this objective. It will remain a major challenge well into the 21st century, a century in which we expect readers such as you to contribute new insights for books yet to be written and wise actions yet to be taken. We wish you every success.

KEY TERMS

algae	cladistics
Animalia	cladogram
Archaea	conservation
Bacteria	Cyanobacteria
biogeography	cytoplasm
blue–green algae	ecologic services
botany	Eukarya
cellulose	eukaryotic
clade	evolution

extreme halophiles

food chain

Fungi

hypothesis

kingdom

metabolism

monophyletic

organelle

paleoecology

pathogens

photosynthesis

phylogenetic tree

phytoplankter

plant

plant anatomy

plant biology

plant ecology

plant genetics

plant molecular biology

plant morphology

plant physiology

plant systematics

Plantae

prokaryotic

Protista

scientific method

spore

SUMMARY

1. Plant biology is the study of plants, of organisms formally classified in the kingdom Plantae, and of an equal number of plantlike organisms formally classified in the domain Bacteria, the kingdom Fungi, and protists.

2. These half-million organisms share at least some of the following traits: They are photosynthetic, nonmotile, and have cells surrounded by a rigid wall; they reproduce by spores; and they have relatively simple unicellular or multicellular bodies lacking obvious digestive, nervous, and muscular systems.

3. Plants and plantlike organisms not only pervade nearly every habitat on Earth, they are also of central importance to every ecosystem and to the human population. They are at the base of natural food chains. They produce oxygen, incorporate nitrogen, and stabilize land surfaces. They provide food, fabric, pharmaceuticals, and structural products for humans.

4. Plant biologists use the scientific method to test hypotheses about plant behavior. The scientific method involves formulating a hypothesis, developing a test of that hypothesis, and then incorporating the results to formulate a modified, new, more accurate hypothesis. The method is an endless loop, because hypotheses can only be disproved, not proved. Each hypothesis can only be considered as an approximation of the truth, not a complete explanation. Technology is based on scientific study, but it can proceed, to a certain degree, independently of science.

5. Botanists study plant form and development (morphology), plant metabolism (physiology), plant genetics and evolution, plant structure (anatomy), and plant ecology.

6. Plantlike organisms include Cyanobacteria in the domain Bacteria. These organisms are prokaryotic (lack a cell nucleus). Other plantlike organisms are molds in the kingdom Fungi. Although the latter do possess rigid cell walls and reproduce with spores (as do plants), they are nonphotosyn-

thetic. Bacteria and molds are ecologically important as decomposers. They also are important pathogens (disease-causing agents).

7. Algae are simple, photosynthetic, and generally aquatic protists. Some algae share important biochemical traits with plants and are in the plant kingdom, but their bodies lack the complex adaptations to life on land exhibited by other members of the plant kingdom.

8. A major challenge of the 21st century is to achieve sustainable use of plant natural resources in the face of an increasing human population and increasing pollution from human technology.

Questions

1. Why does the color green in nature typically signify the presence of plants?

2. List any three ecologic services provided by plants and plantlike organisms. List any three ways in which they are economically important to humans. Why can it be said that humans have lived in a "Wood Age" through most of their history?

3. What is the difference between science and technology? How does the story of our use of chlorofluorocarbons illustrate the difference?

4. What is the subject matter of the disciplines of genetics, ecology, physiology, morphology, and anatomy?

5. In what way are organisms in the domain Bacteria different from protists, animals, fungi, and plants? How are fungi and algae different from organisms in the plant kingdom?

6. Give some examples of organisms in the plant kingdom and of plantlike organisms in the Bacteria, protists, and fungi. Is there any single trait that all half-million species share?

7. Why is sustainable use of natural resources an important challenge for humans in the 21st century?

 ### InfoTrac® College Edition

http://infotrac.thomsonlearning.com

Plants and Climate

Seaborg, D. 2004. The greenhouse diet: as if climate change isn't bad enough: elevated CO_2 may deplete the food you eat of protein, with devastating ecological consequences. *Earth Island Journal* 18:39. (Keywords: "Seaborg" and "greenhouse")

Anon. 2003. Rising temperatures spur biological chaos (global warming effects on seasonality). *USA Today (Magazine)* June 131:14. (Keywords: "biological" and "chaos")

Scientific Method

Guttman, B.S. 2004. The real method of scientific discovery: scientists don't sit around in their labs trying to establish generalizations. Instead they engage in mystery-solving essentially like that of detective work, and it often involves a creative, imaginative leap. *Skeptical Inquirer* 28:45. (Keywords: "Guttman" and "mystery")

Classification of Living Organisms

Flam, F. 1996. New findings overturn traditional 'kingdoms of life.' *Knight Ridder/Tribune News Service* Jan. 15, p. 115. (Keywords: "Flam" and "kingdoms")

The Chemistry of Life

Visit us on the web at http://biology.brookscole.com/plantbio2
for additional resources, such as flashcards, tutorial quizzes,
InfoTrac exercises, further readings, and web links.

1. Living organisms are made from chemical compounds, and everything they do must obey all the laws of chemistry and physics. To understand organisms, one must have a basic appreciation of the principles of chemistry.

2. Most of the substances in organisms are based on carbon-containing compounds. Carbon atoms can form complex chains held together by stable covalent bonds. The large number of possible structures provides a basis for the complex activities of organisms.

3. The chemical properties of water make it uniquely suited as a milieu for living beings.

4. The functional chemical units of living organisms—carbohydrates, lipids, proteins, and nucleic acids—are large polymeric molecules formed from simpler monomers—amino acids, sugars, fatty acids, and nucleotides. The information that directs the assembly of the polymers is encoded in the structure of the nucleic acid polymers.

2.1 CHEMISTRY AND PLANTS

There are many ways to study plants. Poets, painters, anthropologists, and agriculturalists all have their own ideas about the importance and usefulness of the earth's flora (see "PLANTS, PEOPLE, AND THE ENVIRONMENT: Plants as Pharmacists" sidebar on page 21). In recent years, biologists have made notable strides in understanding how plants function. Much of this understanding has come from deducing the chemical basis of life, particularly plant life. To fathom the most important discoveries about plant life, one must grasp the basic concepts of chemistry. What are plants made of? Why are the stems of some plants (such as grasses) soft, whereas those of others (such as trees) are hard? What substances do plants need to grow? How is it possible to manipulate the characteristics of plants through biotechnology? The answers to all these questions are based in chemistry. This chapter provides an overview of the concepts needed to understand plant cell structure and function.

2.2 THE UNITS OF MATTER

Every type of matter—including all the components of living cells and organisms and all the nonliving materials on which living things depend—is built from very small units. **Atoms (Fig. 2.1)** and **molecules** are the smallest particles that retain the chemical characteristics of their type of matter.

Matter made from only one type of atom is called an **element.** There are 92 elements found in nature. Although physicists have produced several more elements, they are unstable and short-lived. Some of the most prominent elements in living organisms are listed in **Table 2.1,** together with the letter

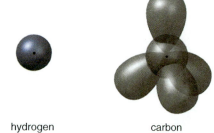

hydrogen   carbon

Figure 2.1 The basic structures of the hydrogen and carbon atoms. Hydrogen has one proton in its nucleus, surrounded by one orbital with one electron. Carbon has six protons and six electrons in its nucleus, surrounded by one inner orbital with two electrons and four outer orbitals with one electron each.

symbols that chemists use to identify them. (The Periodic Table showing all the elements may be found in the appendix.)

Matter made from molecules is called a **compound.** Molecules are made of two or more atoms. There are many thousands of kinds of molecules, which differ in size, shape, and behavior. The largest molecules in a plant cell carry hereditary information; they are slender threads greater than 1 mm in length when extended. Molecules of table sugar (sucrose) are closer to the average size; about 300,000 molecules of sucrose placed end to end would span a millimeter. A plant builds all of its own molecules by rearranging the parts of simpler molecules taken from the environment.

Chemists indicate the composition of a molecule by using the elemental symbols and subscripts to show the kind and number of atoms present. Thus, the symbol CH_4 represents a molecule that has one carbon atom (the subscript 1 is assumed if no number is supplied) and four hydrogen atoms. The five atoms in this molecule are held together by interactions among their component parts.

Table 2.1 The Twelve Most Common Elements in Living Organisms		
Element	Symbol	Number of Protons
Hydrogen	H	1
Carbon	C	6
Nitrogen	N	7
Oxygen	O	8
Sodium	Na	11
Magnesium	Mg	12
Phosphorus	P	15
Sulfur	S	16
Chlorine	Cl	17
Potassium	K	19
Calcium	Ca	20
Iron	Fe	26

The elements are listed in order of size (number of protons).

Molecules Are Made of Atoms

A single molecule is formed from atoms arranged in specific positions relative to one another. The atoms in turn are formed from three types of particles: **protons,** each of which has 1 unit of mass and 1 unit of positive electrical charge; **neutrons,** which have 1 unit of mass but no electrical charge; and **electrons,** which have 0.0005 units of mass and 1 unit of negative electrical charge. The protons and neutrons form the atomic **nucleus,** a small kernel at the center of an atom. If an atom were magnified to the size of a house, the nucleus would be about as large as a pinhead. Electrons move around the nucleus in **orbitals,** ill-defined regions in the space outside the nucleus. The size of the atom, as measured by how many can be packed into a given volume, is determined by the distance of the outer electrons from the nucleus. The weight of the atom is determined mainly by the number of protons and neutrons, because each is about 1,840 times heavier than an electron.

The number of protons in the nucleus, known as the atomic number, determines the chemical characteristics of an atom and its identity. Although neutrons affect the weight of an atom, they influence its chemical characteristics only slightly. It is not uncommon in nature for an element to consist of atoms that differ in their number of neutrons. Such atoms are called isotopes of the element. For instance, the atoms of the element carbon may have six, seven, or eight neutrons. Because every atom of carbon has six protons—and because neutrons weigh as much as protons—the relative weights of these three isotopes are 12, 13, and 14; therefore, these isotopes are identified as ^{12}C, ^{13}C, and ^{14}C. The balance in the number of neutrons and protons seems to be a factor in holding the nucleus together. Nuclei with about the same number of neutrons and protons, such as ^{12}C and ^{13}C, tend to be stable. Nuclei with the number of neutrons much different from the number of protons may be unstable. Regardless of differences in stability, atoms with the same atomic number represent the same element and can take the same places within molecules.

Unstable atoms spontaneously decompose, a process known as radioactive decay. Atoms of ^{14}C are radioactive, giving off energetic electrons called beta (β) particles as they decay into a more stable state. Radioactive atoms tend to be rare, especially among the elements that make up living organisms; therefore, radioactivity is seldom a factor in the chemistry of life. However, high concentrations of radioactive material can damage cells because the high-energy particles of radioactive decay can destroy the complex molecules they hit. On the other hand, low concentrations of radioactive atoms are useful in biological studies because their decay products are like spotlights, enabling researchers to track the fate of these atoms in chemical reactions and to trace their movements through cells and tissues.

Electrical Forces Attach Electrons to Nuclei

Electrical forces attach the electrons to the nuclei to form atoms and molecules. Each electron carries a unit of negative electrical charge, and each proton carries the same amount of positive electrical charge. Two particles attract one another if they carry opposite charges, which is why electrons are attracted to the nuclei. Particles repel one another if they carry charges of the same sign, so two or more electrons avoid one another as they move around the atom. An atom or molecule that has equal numbers of electrons and protons is said to be electrically neutral because any force that its electrons exert on a distant object is countered by an opposite force exerted by the protons.

Although nuclei take up fairly definite positions in a molecule, electrons are more difficult to locate. The orbitals that they occupy are not fixed lines, like the orbits of planets around the sun. Rather, they are fuzzy areas around the nuclei where a probability exists that an electron might be found at any particular time. In some parts of the orbital, the probability is greater (which means the electron spends more time there); in other parts, the probability is smaller (the electron spends less time). There is a general rule that applies to all orbitals. An orbital may have zero, one, or two electrons—no more. Once two electrons have occupied the orbital closest to the nucleus (where the attractive forces are greatest), no other electrons can join them. Other electrons must move to orbitals farther away. An atom or molecule is most stable when it has exactly two electrons in each orbital.

The potential energy of an electron depends on the position of the orbital it occupies. We can think of energy as the capacity to do work, which means to exert a force over a distance. It takes work, and thus energy, to move an electron away from the nucleus against the attractive electrical force. That energy is stored in the electron's position (thus, *potential* energy) and given up if the electron falls back toward the nucleus. Thus, an electron in an orbital close to the nucleus has less potential energy than one in an orbital farther from the nucleus. Orbitals in which electrons have the same potential energy are said to be in the same shell. Electrons can change shells by losing or gaining energy. As an electron drops into an orbital of a shell closer to the nucleus, it gives up energy in the form of a photon (a unit of light or heat radiation), which leaves the atom and can sometimes be detected as a flash. Conversely, for an electron to move into an orbital of a shell farther from the nucleus, it must absorb a photon from the environment. Such changes in electron energy are at the heart of photosynthesis (see Chapter 10). An electron can also leave its home orbital and take up residence in the orbital of another atom, or it can move to an orbital shared by its atom and another atom. These movements lead to the formation of ionic and covalent bonds, respectively, and they result in the formation of new molecules.

Ionic Bonds Consist of Electrical Forces

It is possible for atoms and molecules to have more electrons than protons or vice versa. When they do, the atoms or molecules are called ions. The excess charge of an ion is denoted with a superscript. For example, K^+ indicates the potassium ion, with one more proton than electrons; SO_4^{2-} indicates the sulfate ion, with two more electrons than protons. Ions with a net positive charge are called **cations;** those with a net negative charge are called **anions.** Ions are formed when an electron moves from one originally neutral atom to another. For instance, the transfer of an electron from a sodium atom to a chlorine atom (Fig. 2.2) results in the formation of a sodium cation (Na^+) and a chlorine anion (called chloride [Cl^-]). The loss of an electron is called **oxidation;** the gain of an electron is called **reduction.** In this transfer, the sodium is oxidized and the chlorine is reduced (that is, the chlorine atom's charge is reduced from 0 to -1). Because the oxidation and reduction occur together, they often are referred to as an **oxidation–reduction (redox)** reaction.

Ions with the same sign (both positive or both negative) repel each other, whereas ions with opposite signs attract each other. When two ions of opposite charge are held close together by electrical forces, chemists often say they are joined by an **ionic bond.** A compound containing anions and cations held together by ionic bonds is called a **salt.** As one example, sodium and chloride ions, connected by ionic bonds, form sodium chloride crystals, also known as table salt. The inorganic nutrients in fertilizer—for instance, calcium phosphate or potassium nitrate—also are salts.

Covalent Bonds Consist of Shared Electrons

The strongest type of chemical bond is a **covalent bond.** Most types of molecules are formed from covalently bonded atoms. Although some of the orbitals in such a molecule surround a single nucleus and are said to be nonbonding orbitals, other orbitals are distributed between two or more nuclei. These are said to be bonding orbitals, and the electrons occupying them are bonding electrons, shared by the nuclei. The nuclei that share the bonding electrons are said to be joined by a covalent bond. The covalent bond is strong because the electrons in the bonding orbital are in relatively stable positions—that is, their attraction to the nuclei is strong and their mutual repulsion is minimized. It would take a lot of energy to separate the nuclei and move the electrons into different orbitals. Covalent bonds within a molecule can be shown in different ways (Fig. 2.3).

It is important to know how orbitals form around the nuclei of hydrogen (H), carbon (C), nitrogen (N), and oxygen (O), because these make up the bulk of the molecules of the cell. H is the simplest nucleus, consisting of just one proton. The charge on the proton is so weak that only one orbital forms around the nucleus. C, N, and O nuclei have six, seven, or eight protons, respectively, and they exert a much stronger attraction for electrons. Each of these nuclei is the center for five orbitals. The first orbital is in a spherical region very close to the nucleus. The two electrons in this orbital do not participate in covalent bonds. The four remaining orbitals are oblong and tend to orient themselves so that they are as far apart as possible, because the negative charges of the electrons repel one another. They are said to point to the vertices of a tetrahedron (Fig. 2.4).

When atoms come together to make molecules, bonding orbitals are formed. You can think of them as representing an overlap, or hybrid, of two orbitals from adjacent atoms (although they have a shape of their own). If possible, these orbitals form in such a way that each contains exactly two electrons.

Carbon has six electrons. Two of the electrons are in the inner, nonbonding orbital; four are available for sharing in the outer, bonding orbitals. Thus, all four of the outer orbitals can participate in the formation of bonding orbitals. This is seen most simply with CH_4, or methane. The C nucleus with its inner electrons forms the center of the molecule. Each of the outer, oblong orbitals forms a bonding orbital with an H nucleus embedded in one end and the C nucleus in the other (Fig. 2.4). Each of these orbitals contains two electrons. If we think of an H nucleus as getting a half-share in the electrons of a bonding orbital, then one unit of negative charge is available to balance its unit of positive charge. Similarly, there are six units of negative charge available to balance the six protons in the C nucleus: two in the inner

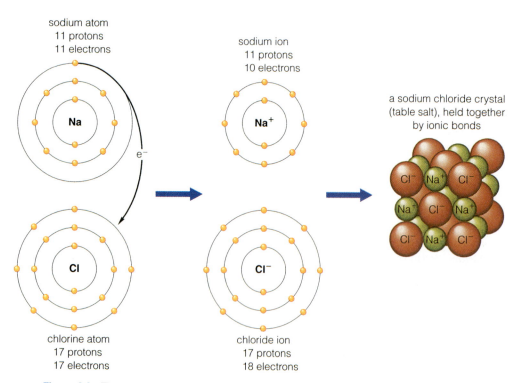

sodium atom
11 protons
11 electrons

sodium ion
11 protons
10 electrons

a sodium chloride crystal (table salt), held together by ionic bonds

Na

Na^+

e^-

chlorine atom
17 protons
17 electrons

chloride ion
17 protons
18 electrons

Cl

Cl^-

Figure 2.2 The oxidation of sodium and reduction of chlorine to form ions. In crystals, ions are held together by the mutual attraction of opposite electrical charges.

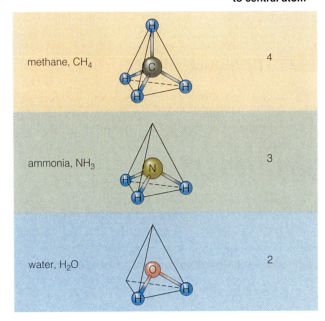

orbital and one from each of the four bonding orbitals (again, a half-share of the two electrons in each orbital). Overall, the molecule is neutral. This illustrates the general rule that each C nucleus forms four covalent bonds. Of course, C can form bonds with other nuclei besides H, such as N and O. Most importantly, it can form bonds with other C nuclei, to produce chains of carbon atoms. In a **hydrocarbon compound,** each C nucleus is bound to one or more other C nuclei and to hydrogen nuclei **(Fig. 2.5).**

Nitrogen has one more proton than carbon, and thus has one more electron than carbon. The extra electron fills up one of the outer orbitals, so that each N nucleus can participate in only three bonds. A good example is NH_3 (ammonia; Fig. 2.4). Similar considerations apply to the behavior of oxygen, as illustrated by H_2O (water). Oxygen has two more protons than carbon, and thus attracts two more electrons. These fill up two of the outer orbitals, so that an O nucleus can form only two covalent bonds with hydrogen atoms.

It is possible for two nuclei to form two or three bonding orbitals, and thus to share four or six electrons. The sharing of four electrons is known as a double bond; sharing six electrons is a triple bond. Chemists represent double bonds by two lines—for example, C=C. Ethylene, a very short hydrocarbon chain with two carbon and four hydrogen atoms, contains a double bond (Fig. 2.5). The two C nuclei are connected by two bonding orbitals: one occupying an oval region in the center, and the other consisting of two sausage-shaped regions on either side. The double bond resists being twisted, so that the six nuclei of this molecule sit rigidly in a single plane. In longer hydrocarbon chains, occasional double bonds

Figure 2.4 The three-dimensional structures of methane, ammonia, and water and their relationship to a tetrahedron.

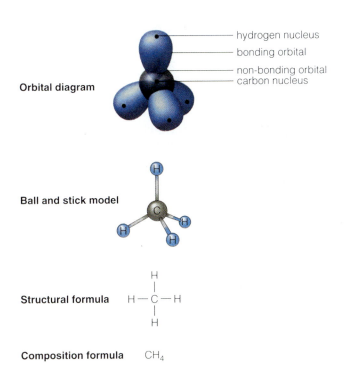

Figure 2.3 Several ways to symbolize a molecule of methane, which is held together by covalent bonds. Methane, a major component of natural gas, is produced when certain bacteria decompose organic matter in the absence of oxygen.

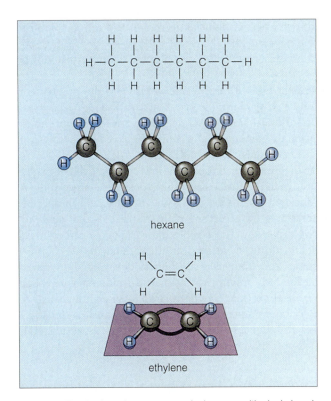

Figure 2.5 Two hydrocarbon compounds: hexane, with single bonds between the carbon atoms, and ethylene, which contains a double bond. Because atoms on either end of a single bond can rotate relative to each other, a molecule such as hexane is very flexible. In contrast, all six atoms of ethylene are confined to a single plane.

produce kinks in otherwise flexible molecules. Although triple bonds are rare in the molecules found in cells, double bonds such as C=C, C=O, and C=N are common.

THE MILIEU OF LIFE

Water is the environment of life—even for terrestrial organisms—at the cellular and molecular levels. The physical and chemical properties of water are essential to cellular reactions and processes. Perhaps the three most important characteristics of water are its ability to dissolve a great many compounds, to remain a liquid over a wide range of temperatures, and to form weak hydrogen bonds with itself and with other molecules. These characteristics derive from the arrangement of its atoms.

Water Owes Its Unique Properties to Its Polarity

The amount and distribution of electrical charge in a molecule are important in predicting its behavior. Some molecules have their electrons spaced evenly throughout the orbitals; these molecules are called **nonpolar.** Other molecules have local regions of positive, negative, or both charges; these are said to be **polar.**

Polarity is established within a molecule because electrons that are shared between two unlike nuclei may spend more of their time closer to one nucleus than the other. A measure called **electronegativity** expresses the tendency of a nucleus to attract electrons. Nuclei of oxygen and nitrogen are more electronegative than those of carbon, sulfur (S), hydrogen, and phosphorus (P). Because of these differences, a molecule tends to be more negative near N and O nuclei and more positive near C, H, P, and S nuclei. Water is the quintessential example of a polar molecule **(Fig. 2.6).** The high electronegativity of the O nucleus attracts electrons—not only the electrons in the nonbonding orbitals, but also the electrons in the bonding orbitals. As a result, there is a deficiency of electrons near the H nuclei. This means that there is a slight negative charge at the O nucleus and slight positive charges at the two H nuclei. This property of water is extremely important in determining the structures of the molecules that make up a living organism. These molecules—proteins, carbohydrates, lipids, and nucleic acids (introduced later in this chapter)—assume three-dimensional shapes, determined in part by their interaction with water. Their three-dimensional shapes are critical for their functions. This is why all life exists in a watery environment. Although some organisms can survive being dried and rehydrated, as far as we know no life can function without water.

Hydrogen Bonds Form between Polar Molecules

A strong polarity in an orbital involving H leads to a new type of bond. Again, water provides the best example. The H of one water molecule can be attracted to the electrons around the O of a neighboring water molecule because of their opposite charges. In fact, the H can bounce between

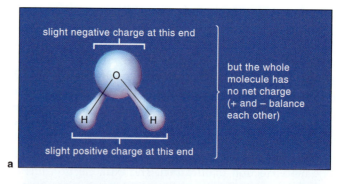

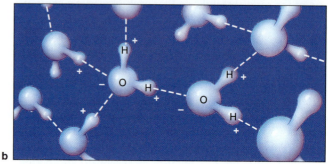

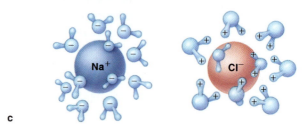

Figure 2.6 Properties of water. The polarity of water (shown in **a**) is the reason water can (**b**) form hydrogen bonds and (**c**) dissolve substances, such as the salt ions shown here.

the electronegative Os of the two water molecules. This is a stable situation; therefore, it is said that a **hydrogen bond** has formed between the molecules. A hydrogen bond is represented by a dotted line: for example, O···H. Each of the two Hs in a water molecule can participate in a hydrogen bond. Even though this bond is only about 1/16 as strong as a covalent bond, the large number of hydrogen bonds among water molecules makes water more stable (less volatile) than we may otherwise expect. This means that a water molecule can evaporate (leave the liquid state and move into the gaseous state) only by simultaneously breaking all the hydrogen bonds that connect it to other molecules in the liquid. Consequently, water can absorb a large amount of heat as it evaporates, and it releases the same amount of heat as it condenses. The absorption of heat removes heat that otherwise would increase the water's temperature; the condensation provides heat to keep the water warm. Thus, evaporation and condensation tend to stabilize the temperature of liquid water. This fact is important to living systems, which are damaged by large swings in temperature.

Other molecules besides water can form hydrogen bonds. This happens when an H nucleus that is sharing electrons with O or N comes close to another O or N, with the three nuclei approximately in a straight line, such as: O-H···N. Many biological molecules that contain O-H or N-H groups participate in hydrogen bonds with water by donating their H to the bond; other molecules with groups such as C=O participate by attracting an H from a water molecule. Biological molecules may also form hydrogen bonds with each other. Examples can be found later in this chapter when carbohydrates (Fig. 2.10a) and nucleic acids (Fig. 2.16) are introduced.

Water Dissolves Polar and Ionic Substances

The polarity of water and its ability to form hydrogen bonds make it an excellent **solvent** for many other substances. A pure substance in the solid form consists of many molecules of the same kind packed regularly (a crystal) or irregularly (an amorphous solid such as glass) and held together with chemical (but not covalent) bonds. On contact with water, the molecules at the surface of the solid can leave to become surrounded by and bound to water molecules. The mass is said to *dissolve;* the molecules that become intermixed in the water (the **solvent**) are the **solute** molecules. This homogeneous mixture is called a **solution.** The more hydrogen bonds that can form between the solute and the water, the more easily soluble is the solute. Polar, nonionic molecules, such as ammonia (NH_3, Fig. 2.4), are very soluble for that reason. Ions also are soluble, but for a different reason. Water molecules surround each ion with their oppositely charged poles pointing at the ion (Fig. 2.6c). Most biological molecules either form hydrogen bonds with water or are ionic, and thus they dissolve in water. This allows the molecules to move, mix, and react chemically with one another; it is one of the principal reasons all life exists in a water solution.

Nonpolar molecules such as hydrocarbons (in which the C and H nuclei have similar electronegativities and the electrons are evenly distributed) cannot form hydrogen bonds with water or anything else. If tossed into water, these molecules force the water to form a sort of cage around them—an energetically unfavorable (unstable) situation. Over time, the nonpolar molecules will tend to move together and stick together to minimize the amount of water that is used to form cages: this tendency is called a **hydrophobic bond** (because in sticking together the nonpolar molecules act as though they had a fear of or an aversion to water). In contrast, nonpolar solutes will dissolve in nonpolar solvents (such as oil or kerosene), but polar solutes will not. In general, we can predict that polar or **hydrophilic** molecules will dissolve in polar solvents (water), and nonpolar or **hydrophobic** molecules will dissolve in nonpolar solvents (oil).

Acids Donate—and Bases Accept—Hydrogen Nuclei

In water solutions, both within and outside living cells, there is a rapid exchange of H nuclei among solutes and the water solvent. Molecules that contain an H nucleus bonded to a strongly electronegative atom (the same molecules that might donate an H to a hydrogen bond) can lose the H nucleus entirely. The H is not really lost, of course; in general, it is transferred to a molecule of the surrounding solvent (water) forming a hydronium ion. The original molecule retains the two electrons in the bonding orbital and thus acquires an extra unit of negative charge. Molecules that donate an H nucleus are called **acids.** Molecules that accept an H nucleus are called **bases**, for example:

H-Cl (hydrochloric acid) + H-O-H ⇔ H-O-H (hydronium) + Cl⁻ (chloride)
                                                    H
      (acid)   (base)   (acid)   (base)

Water can be both an acid and a base, as follows:

H-O-H + H-O-H ⇔ H-O-H (hydronium) + O-H⁻ (hydroxyl)
                          H
   (acid)   (base)   (acid)   (base)

These are reversible reactions: The hydronium and chloride can react together to form hydrochloric acid and water, and the hydronium and hydroxyl can react to form two molecules of water. In these examples (considering the reactions in both directions), hydrochloric acid, water, and hydronium are the acids, and chloride, hydroxyl, and water are the bases.

The concentration of hydronium ions determines the acidity of the solution. More hydronium means a more acidic solution. Chemists express the hydronium concentration by a measure known as pH (Fig. 2.7). A change of 1 unit on the pH scale is a 10-fold change in hydronium concentration. The lower the pH, the greater the hydronium concentration. Therefore, a solution with pH of 4 has 10 times as much hydronium as a solution with pH of 5. Acids and bases can be defined as substances that decrease and increase, respectively, the pH of a solution, and solutions can be classified as acidic or basic according to their pH. The acidic juice squeezed from a lemon has a pH of about 2 to 3. Pure water, which is neither acidic nor basic, has a pH of 7. Household bleach, a basic solution, has a pH of about 12.

Instead of talking about hydronium, chemists often abbreviate the concept by referring to the H⁺ ion itself. From this viewpoint, pH refers to the H⁺ concentration, although free H⁺ never occurs in water solution. Most of the figures in this book will indicate hydronium by the symbol H⁺.

2.4 THE SUBSTANCE OF LIFE

Living Organisms Are Made of Chemicals

All plants (and other organisms) are made of chemical compounds. These compounds influence the shapes that plant cells and their subunits take and determine their functions. Compounds that contain primarily carbon are called **organic** compounds, because they were first associated with living organisms. As the techniques of laboratory carbon chemistry have advanced, it has become useful to distinguish the molecules that actually occur in cells—they often are referred to as **bio-organic.** Bio-organic molecules are based on a carbon skeleton and generally include oxygen and hydrogen. They may often contain nitrogen, phosphorus, and/or sulfur in

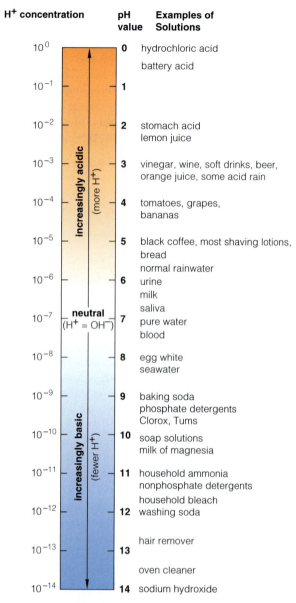

Figure 2.7 The pH scale. Acidic solutions (pH <7) are associated with a sour taste. Basic solutions (pH >7) often are bitter and have a slippery feel, rather like soap.

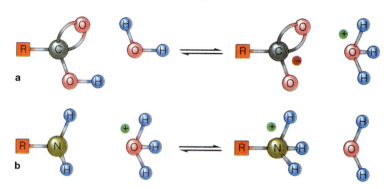

Figure 2.8 Acid–base relations in functional groups. (**a**) The release of an H⁺ nucleus by the acidic group, R-COOH. Note that the functional group ends up with a negative charge. (**b**) In accepting an H⁺ nucleus, the basic amino group, R-NH₂, acquires a positive charge.

their structures. Other elements that may be associated with bio-organic molecules (generally attached with ionic bonds) include iron, calcium, potassium, and magnesium, and less frequently sodium, boron, zinc, manganese, molybdenum, chlorine, and copper.

Biological molecules often are large and complex, but they are easier to understand when we use the concept of **functional groups**. A functional group is a small part of a larger molecule, but one that can participate in chemical reactions. Some functional groups are acids and bases (Fig. 2.8). For instance, the functional groups carboxyl (R-COOH) and phosphoryl (R-OPO₃H₂) are acids and can donate H⁺ nuclei to water (R stands for the remainder of the molecule and is not part of the functional group). The amino group (R-NH₂) acts as a base and accepts H⁺ nuclei from hydronium. Other functional groups may participate in hydrogen bonds, hydrophobic bonds, or oxidation–reduction reactions.

Most of the organic compounds found in living systems can be classified into four families: carbohydrates, lipids, proteins, and nucleic acids (Table 2.2). Together these molecules act as structural and fuel molecules, speed up chemical reactions, and serve as libraries of genetic information. The following sections discuss these compounds.

Table 2.2 The Most Important Types of Biomolecules

Class	Components	Polymer	Role
Carbohydrate	Simple sugars	Starch, glycogen, cellulose	Energy, storage, structure (for example, cell wall)
Protein	Amino acids	Polypeptide chain	Catalysis, structure, movement
Lipid	Fatty acids, glycerol, phosphate, or sugar	Phospholipid, glycolipid, triglyceride	Membrane structure, energy storage
Nucleic acid	Base (adenine [A], guanine [G], cytosine [C], thymine [T], or uracil [U]), sugar, phosphate	DNA, RNA	Information storage, protein synthesis

PLANTS, PEOPLE, AND THE ENVIRONMENT:

Plants as Pharmacists

Plants are superb chemists. They synthesize a great variety of chemicals beyond those needed to perform the basic functions of cells. For many years these chemicals were known as secondary compounds because they were not needed for the functioning of all cells. The term suggested that the compounds had no significance in the life of the plant. Now it is recognized that these compounds often play important roles, either as components of specialized cells or in the interactions between plants and other organisms with which they associate. For instance, some of the chemicals are attractants, promoting the transfer of pollen from one plant to another of its species or the dispersal of seed. Some of the chemicals are antibiotics or toxins, restricting the ability of pathogens and herbivores to feed off the plant.

The subject of ethnobotany, the use of plant extracts by people of many cultures, recently has become popular. An understanding of these uses may help us control disease bacteria that are becoming resistant to current antibiotics. They may also help cure conditions like AIDS and give us new ways of alleviating pain, anxiety, or neurologic disorders. The adjacent table lists several plants that produce compounds that affect the human nervous system. A few of these are in common and accepted use; a few are in common use but are legally discouraged; and some are only used medicinally.

Many of the chemicals in plants are used to prevent or to fight cancer. A large class of compounds, called antioxidants, prevents damage that seems to lead to cancer. These compounds include ascorbic acid (vitamin C), alpha-tocopherol (vitamin E), and beta-carotene (which is converted into vitamin A), all of which are accumulated in many fruits and vegetables. These work by reacting with (and thus taking out of circulation) chemicals with unpaired electrons (free radicals) that would otherwise combine with, oxidize, and inactivate DNA, RNA, proteins, or membrane lipids. They are accumulated in plants to protect the plants' own cells, but they probably do the same job in human bodies. Some plants have special compounds that may perform the same function. Broccoli contains sulforaphane, garlic and onions have allyl sulfides, and tea has catechins. Many of these chemicals have been found to prevent cancer in experimental animals. The evidence that they prevent human cancers is less well established and is a subject of current research.

Chemicals used to fight cancer are generally toxic to cancer cells. Because cancer cells reproduce rapidly, the most effective compounds are those that interfere with cell division. Such compounds include vincristine and vinblastine, which come from the periwinkle plant (*Vinca rosea*), and Taxol from yew trees (*Taxus brevifolia*). Psoralens from celery, parsley, and citrus leaves make cells sensitive to ultraviolet light; this treatment allows physicians to kill cells in localized regions by irradiating them with an ultraviolet laser. Gossypol (from cottonseed), which has been touted as a potential male contraceptive, also exhibits anticancer activity, possibly by stimulating the formation of free radicals that oxidize membrane lipids in cancer cells.

Even though chemists are becoming more proficient at designing and synthesizing complex molecules, the use of plant-derived chemicals will continue and expand. The specificity of biological catalysts in plant cells (enzymes; see Chapter 8) is the reason. There are many molecular structures, formed within plant cells, that are impossible for even the most skilled chemists to duplicate.

Figure 1 Stems of a male *Ephedra* plant. *Ephedra equisetina* (common name Ma Huang) is the source of ephedrine, a stimulant, and pseudoephedrine, found in over-the-counter allergy medicines. Although the Food and Drug Administration does not regulate "dietary supplements," such as natural ephedra, a 2003 report indicated that the use of ephedra, particularly by athletes, may be associated with serious risks, including heart attacks, strokes, hypertension, respiratory depression, and psychiatric problems.

Plant Sources of Psychoactive and Neuroactive Agents		
Source	Chemical	Effect, Use
Atropa belladonna (belladonna)	Atropine	Pupil dilation, heart rate acceleration
Cannabis sativum (hemp)	Tetrahydrocannabinol	Nausea reduction, appetite stimulation
Coffea arabica (coffee)	Caffeine	Stimulant, diuretic
Erythroxylon coca (coca)	Cocaine	Stimulant, anesthetic
Lophophora williamsii (peyote)	Mescaline	Hallucinogen
Nicotiana tabacum (tobacco)	Nicotine	Stimulant
Papaver somniferum (poppy)	Morphine, codeine	Anesthetic, narcotic
Thea sinensis (tea)	Theophylline	Stimulant, diuretic
Theobroma cacao (cocoa)	Theobromine	Stimulant, diuretic

Websites for further study:

American Chemical Society—Chocolate:
http://www.chocolate.org/choclove/neuroactive.html

Psychnet-UK: Substance Related Disorders: http://www.psychnet-uk.com/clinical_psychology/clinical_psychology_substance_related_disorders3.htm

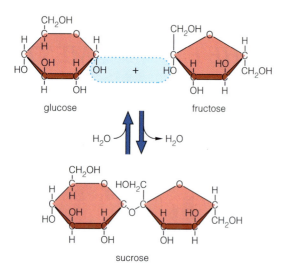

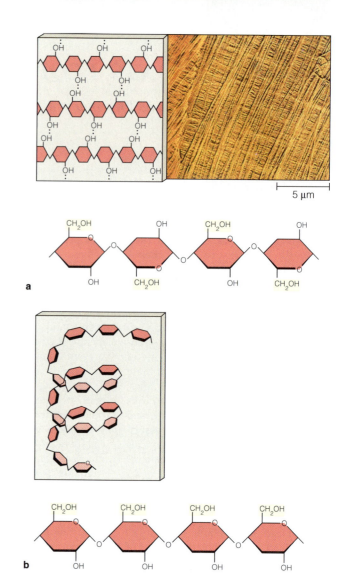

Figure 2.9 Structual formulas of two simple sugars, glucose and fructose, the most common sugar monomers in plants, and of the oligosaccharide sucrose, also known as common table sugar. The formation of the bond in sucrose between the glucose and fructose is formally a dehydration reaction (releases H_2O). The bond can be broken by hydrolysis (adding H_2O).

Figure 2.10 Models of the polysaccharides (a) cellulose and (b) starch. In both molecules, the monomer is the sugar glucose. The two molecules differ in the ways the adjacent glucose monomers are connected. This allows the cellulose molecule to form straight chains, whereas the starch molecule forms helices in solution. The frame in the *top right* is an electron micrograph of cellulose cables.

Carbohydrates Include Sugars and Polysaccharides

Carbohydrates include simple sugars, oligosaccharides, and polysaccharides. Simple sugars are formed from carbon atoms with associated oxygen and hydrogen atoms in the proportions $C_nH_{2n}O_n$ (n = 3–7). The carbon atoms in a simple sugar often are arranged in a ring, with each carbon covalently bonded to two other carbon atoms, one H and one O **(Fig. 2.9)**. **Oligosaccharides** are small chains of two or more simple sugars. **Polysaccharides** are long chains of simple sugars **(Fig. 2.10)**. Some polysaccharides—for instance, **cellulose** and **pectin**—contribute to structure, especially in the cell wall. Cellulose forms cables that are very strong and can keep a cell from bursting under pressure (see Chapter 3); pectin acts as a glue to hold cells together in tissues (see Chapter 4).

Other polysaccharides are used for storage of energy. **Starch,** for instance, a polymer of the simple sugar glucose, is made and stored in green plants. Glycogen, another polymer of glucose that is generally larger than starch and more highly branched, is made and stored in some algae and in animals. These compounds, when broken down, provide energy to run chemical reactions and other activities of an organism.

Energy in this context has different constraints than it did when energy levels of electrons were discussed in Section 2.2. When a molecule absorbs a photon and an electron gains potential energy, or it emits a photon and an electron loses potential energy, total energy is conserved. In an organism, when molecules participate in a chemical reaction and their potential energy is converted to other forms, such as the potential energy of new molecules or kinetic energy (energy of motion) or heat, the total amount of energy is conserved—it is the same before and after the reaction. But useful energy, known as **free energy,** is not conserved. A portion is lost during every reaction. This is because a portion is converted to heat and other forms that are disordered and lost to the environment. (Chemists say that *entropy* has increased.) To say that polysaccharides store energy means that these compounds are rich in free energy. But the free energy is dissipated as these compounds are used. This is one reason organisms need a continual supply of energy.

Lipids Are Insoluble in Water

The name **lipid** applies to any oil-soluble (nonpolar) substance in the cell; however, two classes of lipids—the **phospholipids** and **glycolipids,** which associate to form thin sheets **(Fig. 2.11)**—play special roles in plant cells. These lipids do have water-soluble (polar, hydrophilic) functional groups, which may be phosphate groups (in phospholipids) or sugars (in glycolipids). They also have water-insoluble (nonpolar, hydrophobic) components, which are the hydrocarbon ends of fatty acids. A fatty acid is a molecule with a hydrocarbon chain attached to a carboxylic acid. The carbon atoms of the hydrocarbon chain

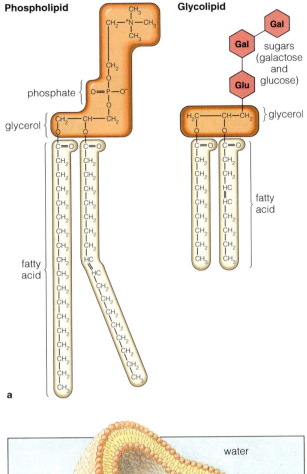

Phospholipid   **Glycolipid**

phosphate

glycerol

fatty acid

sugars (galactose and glucose)

glycerol

fatty acid

a

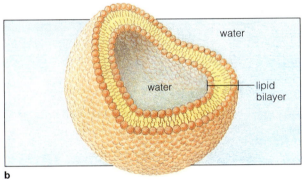

water

water ——— lipid bilayer

b

Figure 2.11 Phospholipids, glycolipids, and membranes. (**a**) Models of a phospholipid and a glycolipid. Both have hydrophobic hydrocarbon tails and a hydrophilic head group. (**b**) Phospholipids in solution form a bilayer complex that keeps water away from the tails and in contact with the heads. The bilayer is flexible, and the individual molecules can move relative to one another, as long as they stay in the plane of the bilayer.

may all be connected by single covalent bonds, in which case the chain is flexible but generally straight, or they may have one or more double bonds, which insert a bend in the chain. A hydrocarbon chain with one or more double bonds is said to be **unsaturated;** a hydrocarbon chain with all single bonds is **saturated** (with H atoms). The phosphate or sugars and the fatty acids are connected by a three-carbon molecule called glycerol (Fig. 2.11a). By forming sheets two molecules in thickness, phospholipids and glycolipids are able to bury their hydrophobic hydrocarbons in the middle, away from the solvent water, and expose their hydrophilic components to the solvent. These sheets provide the structural basis for the **membranes** that are found throughout cells (Fig. 2.11b).

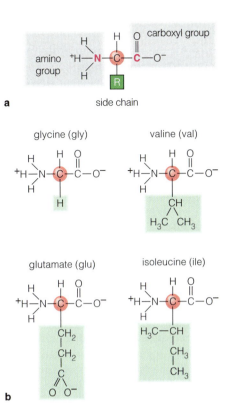

amino group

carboxyl group

a   side chain

glycine (gly)   valine (val)

glutamate (glu)   isoleucine (ile)

b

Figure 2.12 Amino acid structure. (**a**) The general formula for amino acids. (**b**) Four of the 20 possible amino acids found in proteins. Notice that these molecules differ only in their side chains.

Sterols are another type of lipid that contributes to membrane structure. Sterols are large, multiringed hydrocarbons that dissolve in the hydrophobic part of a membrane and keep it flexible, preventing it from developing cracks. In animals, the primary sterol is cholesterol; in plants (including the vegetables and fruits that you eat), there is no cholesterol; a related molecule, ergosterol, has the same function.

Other lipids, the **triglycerides,** are used as energy storage compounds because the oxidation of their components releases a great deal of useful free energy. Triglycerides have three fatty acid molecules attached to a glycerol connector. They are insoluble in water and form discrete lipid bodies inside plant cells. Extracted from plant organs, they form oils (liquid at room temperature) or fats (solid at room temperature), depending on the lengths of the fatty acid hydrocarbons and whether they have double bonds. (Fatty acids in oils are shorter, have more double bonds, or both.) Most salad oils are triglycerides extracted from seeds of maize, soybean, sunflower, sesame, cotton, peanut, or other plants. Three commercially important fats from plants are palm "oil," coconut "oil," and cocoa butter.

Proteins Have Diverse Shapes and Functions

Proteins are large molecules formed by stringing together between one hundred and several hundred amino acids into a long, unbranched chain. The **amino acids** are the **monomers** (individual units) that form the protein **polymer** (multiunit molecule). Each amino acid has a **backbone** containing an amino group, a central carbon, and a carboxyl (carboxylic acid) group **(Fig. 2.12).** The amino and carboxylic acid groups are responsible for the name **amino acid.** Twenty different types of amino acids are found in proteins. They all have the same backbone, but each has a different **side chain** attached to the central carbon.

The amino acids can be linked together at their backbones: an amino group of one amino acid attached to the carboxyl group of a second amino acid, the amino group of the second amino acid attached to the carboxyl group of a third amino acid, and so on **(Fig. 2.13).** The bonds linking the amino acids are called **peptide** bonds; therefore the chain is called a **polypeptide** chain. Any one of the 20 different types of amino acids may be placed in any of the positions of the chain. Thus, there are 20^{100} different possible types of chains that are 100 amino acids in length, and even more possibilities exist with longer chains. Every particular type of protein has its own specific arrangement of amino acids, sometimes called its **primary structure.** This arrangement promotes the folding of the protein into a very specific three-dimensional shape, with different parts of the protein held together by hydrogen bonds between parts of the backbone **(secondary structure)** and by ionic, hydrogen, and hydrophobic bonds between the side chains of the different amino acids **(tertiary structure; Fig. 2.14).** Individual protein molecules may associate with each other or with other types of molecules to form larger structural complexes **(quaternary structures).**

The functional properties of the protein depend on its three-dimensional shape. Large protein complexes give form to the cell, direct movement within the cell, or provide a scaffold for chemical reactions. All **enzymes,** which catalyze the chemical reactions of the cell (see Chapter 8 for a discussion of catalysis), are proteins. Proteins are high-energy compounds; therefore, some proteins may be used for storage of energy (for example, in seeds).

Heating a solution of protein to a temperature of 50°C or more breaks the relatively weak ionic, hydrogen, and hydrophobic bonds between the side chains and unwraps the protein; this changes its three-dimensional shape, even though it does not alter the order of the amino acids in the polypeptide chain. Protein molecules that have been heated or otherwise treated to change their three-dimensional shape lose their function. These protein molecules are said to be **denatured.** Denatured proteins often form large masses and become insoluble. The proteins in a cooked egg white are a good example.

Cells depend on functioning proteins to live. The function of a protein depends on its three-dimensional shape. And a protein's shape depends on the arrangement of amino acids in its polypeptide chain. Thus, the information that guides the arrangement of amino acids into a polypeptide chain is a key element of life. This information is carried by other biological molecules, as is explained in the following section.

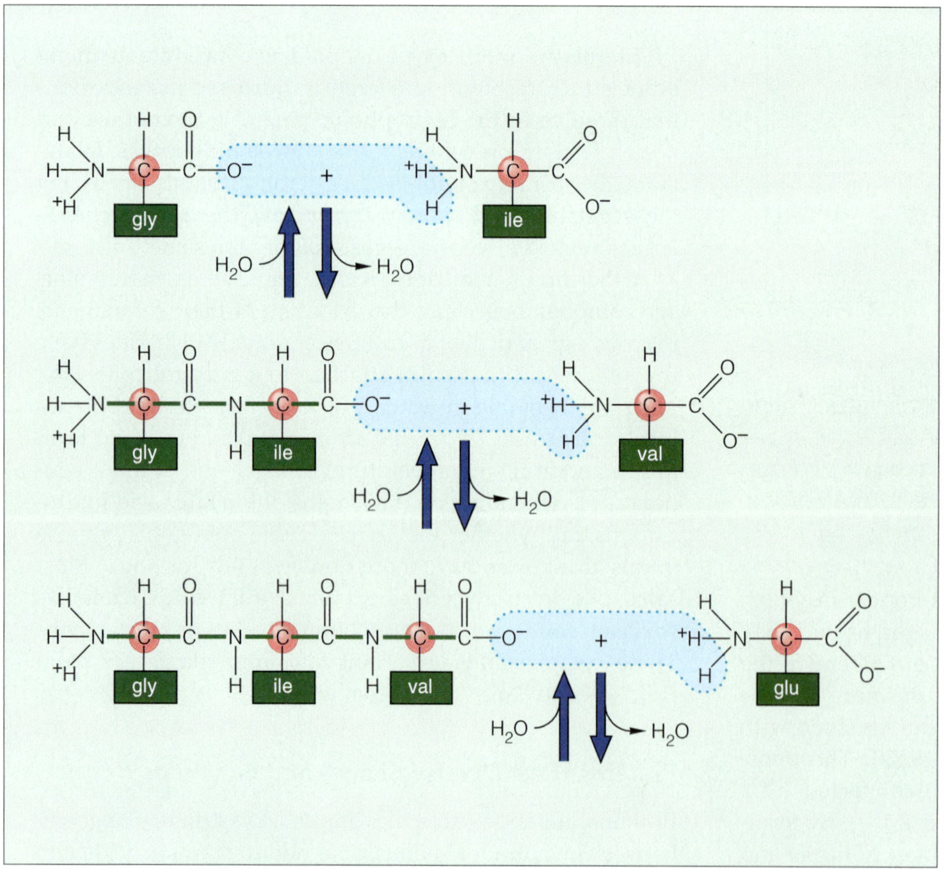

Nucleic Acids Store and Transmit Information

There are two types of **nucleic acid: deoxyribonucleic acid (DNA)** and **ribonucleic acid (RNA).** Both DNA and RNA are polymers—that is, long unbranched chains of nucleotide monomers. Each nucleotide is formed from a simple five-carbon sugar attached to a phosphate group and to a one- or two-ringed molecule called a base. (The word *base* in this context has a different meaning from that used earlier in this chapter. There, base referred to any compound that accepted H^+. Here, it refers to a few specific molecules.) The sugars give the nucleic acids their names: the sugar of DNA nucleotides is deoxyribose; the sugar of RNA nucleotides is ribose. Chains of nucleotides are formed by attaching the phosphate group of one nucleotide to the sugar of another **(Fig. 2.15).**

Figure 2.13 Peptide bonding in protein formation. Different amino acids can polymerize through peptide bonds to form a polypeptide chain. The length of the chain, and the order of the amino acids, can vary according to the type of protein. The formation of a peptide bond is formally a dehydration reaction (releases H_2O). A peptide bond can be broken by hydrolysis (adding H_2O).

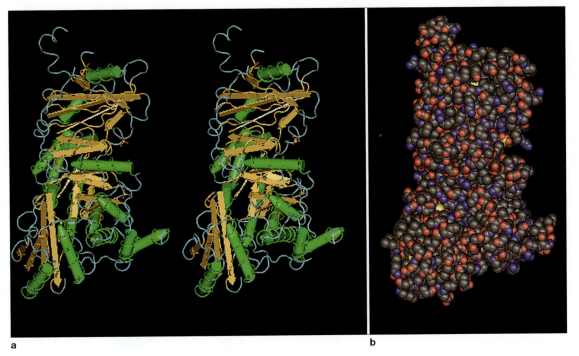

a   b

Figure 2.14 Models of protein structure. This protein is a subunit of rubisco, a key enzyme in photosynthesis (see Chapter 10). (**a**) A stereo diagram showing the twists of the protein's backbone. The green cylinders indicate regions where the backbone forms a helix (called an *alpha-helix*). The orange ribbons show regions where several quasi-parallel stretches of backbone are connected (called a *beta-pleated sheet*). Hold the page about 12 inches from your face and stare at the drawing while focusing at an imaginary point in the distance. The two images should seem to converge to form a three-dimensional image in the middle. (**b**) The same protein described with a "space-filling" model, which shows the individual atoms.

Different nucleotides are distinguished by their different bases. There are four types of nucleotides in DNA, which have bases called A (adenine), T (thymine), G (guanine), or C (cytosine). There are four corresponding types of nucleotides in RNA, with A, U (uracil), G, or C bases. The nucleotide bases can bind together with hydrogen bonds to form complementary pairs. A binds only to T (or U); G binds only to C (Fig. 2.16).

DNA is formed from two chains of nucleotides wound around each other in a formation called a **double helix** (Fig. 2.17). At each position, the base of one chain is hydrogen-bonded to the corresponding base of the other chain. Therefore, the two chains are called complementary, just as the individual bases are complementary. DNA stores **genetic information** in its sequence of nucleotide bases. Thus, a sequence of bases such as ATGCCC has a different meaning from another sequence, such as AAGTTA. Consequently, the order of the bases in DNA is critical to its function, as the order of amino acids is critical to the function of proteins. Although there are not as many types of bases as amino acids, the potential variation for DNA is much greater than for proteins

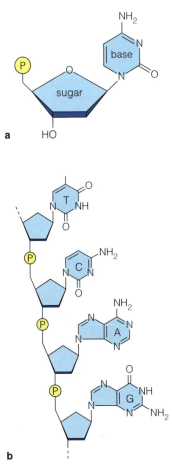

Figure 2.15 Structures of a nucleic acid polymer. (**a**) The basic nucleotide subunit has a phosphate (P), sugar, and base. DNA and RNA nucleotides are distinguished by their sugars (the one shown here is deoxyribose from DNA). (**b**) To form a nucleotide chain, nucleotides are connected through their sugars and phosphates.

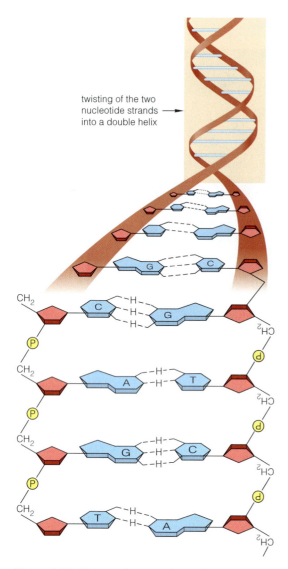

twisting of the two nucleotide strands into a double helix →

Figure 2.16 The complementary bases A and T and also G and C can form base pair complexes, connected by two and three hydrogen bonds, respectively.

T.M. Murphy

Figure 2.17 A model of DNA, the genetic material, showing two chains of nucleotides wound tightly around each other in a helix. Each base (part of a nucleotide) on a chain binds to a complementary base on the opposite chain.

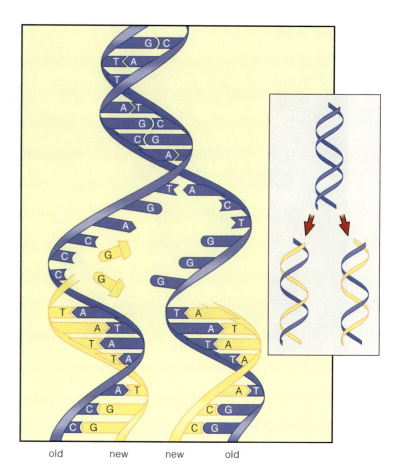

old   new   new   old

Figure 2.18 The replication of a DNA molecule. Shown is a DNA molecule that has been partially unzipped by the breaking of the hydrogen bonds between the bases. Each strand serves as a template for creating a new, complementary DNA strand. The template strand and the new strand stay together after replication is complete.

because DNA molecules are so much longer. A single DNA molecule generally has several million nucleotides.

Genetic information is used to specify the order of amino acids in proteins. A set of three bases on a DNA molecule specifies one amino acid according to a **genetic code.** Therefore, a section of DNA containing 100 sets of three bases, all in order, can specify the amino acid sequence of a protein of 100 amino acids. Such a section of DNA (together with extra bases controlling when the genetic information is expressed to form proteins) is a **gene.**

RNA is made from a single strand of nucleotides. There are various types of RNA. Some are extended, whereas others may loop back on themselves and be wound into different shapes. The different types of RNA molecules have different functions, but almost all are involved in some way in the use of genetic information to synthesize proteins. How DNA and RNA contribute to the process of synthesizing proteins is described in Chapter 15.

DNA REPLICATION Reproduction is one of the most important and complex characteristics of living organisms. It is axiomatic that when a cell or organism reproduces, it must pass its genetic information to its progeny. This means that new copies of the information, and thus new copies of DNA base sequences, must be synthesized. The original DNA chains serve as *templates* to guide the formation of the new DNA. When a cell produces a copy of DNA identical to an original DNA double helix, we say that the original has been *replicated.*

The basic process of **DNA replication** involves the separation of the two original complementary strands of DNA. Each strand can serve as a template for the assembly of a new complementary strand (Fig. 2.18). An enzyme attaches the nucleotides together one by one, choosing at each step a nucleotide with a base that is complementary to the opposite base on the template. The result is two identical double helical molecules of DNA, each with one original and one new strand.

The actual synthesis of a new DNA molecule is more complicated, with synthesis starting in the middle of the molecule and working outward in both directions. Additional enzymes are needed to separate the strands and to promote unwinding and rewinding of the double helices, as well as to weld together separate, adjacent segments of newly synthesized strands. It is amazing that the process works at all, given the millions of nucleotides that must be joined in proper sequences to produce one new full copy of genetic information. It is essential that it does work, however, because DNA forms the molecular foundation for all organisms and most of their functions.

KEY TERMS

acids	hydrocarbon compound
amino acids	hydrogen bond
anions	hydrophilic
atoms	hydrophobic
bases	hydrophobic bond
carbohydrates	ionic bond
cations	lipid
cellulose	membranes
compound	molecules
covalent bond	monomers
deoxyribonucleic acid (DNA)	neutrons
	nonpolar
DNA replication	nucleic acid
double helix	nucleus
electronegativity	oligosaccharides
electrons	orbitals
elements	oxidation–reduction (redox)
enzymes	pectin
free energy	peptide bond
functional groups	phospholipids
gene	polar
genetic code	polymer
glycolipids	polypeptide chain
polysaccharides	secondary structure
primary structure	side chain
proteins	starch
protons	sterols
quaternary structures	tertiary structure
ribonucleic acid (RNA)	triglycerides
salt	

SUMMARY

1. All matter is made up of simple units called molecules. Molecules are made of atoms. An atom contains a nucleus (with protons and neutrons) and electrons, which move around the nucleus in orbitals.

2. Matter made from one type of atom is an element; matter made of more than one type of atom is a compound. The number of protons in the nucleus defines the type of atom; the number of neutrons contributes to the weight of the atom and the stability of the nucleus. Unstable nuclei are radioactive.

3. Protons have a positive charge. Electrons have a negative charge. An atom or molecule with the same number of protons and electrons is electrically neutral.

4. An atom or molecule with more protons than electrons has a net positive charge and is called a cation. An atom or molecule with fewer protons than electrons has a net negative charge and is called an anion.

5. The transfer of an electron from one atom or molecule to another oxidizes the first and reduces the second.

6. The attractive force that holds together a positive ion and a negative ion is called an ionic bond.

7. Electrons that are shared between two atoms in a bonding orbital form a covalent bond between the two atoms and hold them together. Hydrogen atoms participate in one covalent bond, oxygen atoms in two, nitrogen atoms in three, and carbon atoms in four.

8. Some atomic nuclei in a molecule pull electrons more strongly than others, creating a polarity of electric charge in the molecule. Water is an example of a polar compound, with a partial negative charge at the oxygen atoms and a partial positive charge near the hydrogen atoms. Electrical forces result in hydrogen bonds being formed between two water molecules or other molecules where a hydrogen nucleus can be positioned between two strongly electronegative nuclei (O or N).

9. Polar molecules tend to dissolve in water and are called hydrophilic. Nonpolar molecules tend not to dissolve in water and are called hydrophobic.

10. Molecules that donate hydrogen nuclei are called acids; molecules that accept the hydrogen nuclei are called bases. Excess hydrogen nuclei in water are attached to the water molecules to form hydronium ions. The pH scale measures the concentration of hydronium ions in a solution.

11. A protein is a linear chain of amino acids connected by peptide bonds. Each type of protein has a specific order of amino acids and a specific three-dimensional structure. A protein that loses its three-dimensional structure cannot function.

12. Carbohydrates are simple sugars and polymers of sugars. Some contribute to structure, especially of the cell wall, and others are a form of energy storage.

13. Lipids are hydrophobic compounds made from hydrocarbons and other molecules. Phospholipids and glycolipids form sheets, called membranes, that separate compartments in a cell. Triglycerides serve as forms of energy storage.

14. Nucleic acids include DNA and RNA. Both are long, linear chains of nucleotides. DNA, a double-stranded molecule, stores genetic information in the sequence of its nucleotide bases. RNA, a single-stranded molecule, is an intermediate in the use of genetic information to make functional proteins.

15. The presence of two complementary chains in DNA and their use as templates form the chemical basis for the replication of genetic information.

Questions

1. How many protons and how many electrons are present in a molecule of (a) methane (CH_4); (b) ammonia (NH_3); and (c) water (H_2O)?

2. When an atom of ^{14}C undergoes radioactive decay, its nucleus emits an electron. Because the electron carries one negative charge, the nucleus has gained one positive charge. One of the neutrons in the nucleus has turned into a proton! Is the atom still an atom of carbon? Explain your answer.

3. Sketch a molecule of methane showing the positions of the nuclei of the five atoms and the general positions of the bonding and nonbonding electrons. Illustrate that the total number of electrons in the molecule equals the total number of protons in the five nuclei.

4. Sketch a molecule of water showing the positions of the nuclei of the three atoms and the general positions of the bonding and nonbonding electrons. Illustrate that the total number of electrons in the molecule equals the total number of protons in the three nuclei. Do all the outer orbitals of oxygen form chemical bonds? Show how the tendency of electrons to move to the larger nucleus gives the molecule a polarity.

5. Sketch the following hydrocarbon molecules showing the positions of the C and H atoms (as well as you can on a flat piece of paper) and the numbers of covalent bonds between adjacent atoms. Remember that all carbon atoms should have four bonds and that some carbon atoms may have double bonds between them. Also, some carbon chains can form rings.

a. ethane (C_2H_6)

b. butane (C_4H_{10})

c. hexane (C_6H_{14})

d. cyclohexane (C_6H_{12})

e. ethylene (C_2H_4)

f. benzene (C_6H_6)

6. The pH of pure water is 7. What will the pH be if (a) you increase the hydrogen (hydronium) ion concentration 1,000-fold; or (b) you add a compound that forms complexes with 99% of the hydrogen (hydronium) ions?

7. Explain why salad oil and vinegar separate after you shake them up together.

8. Rearrange the following list so that the largest items are at the bottom and the smallest at the top:

carbon atom

water

neutron

protein

proton

DNA

amino acid

electron

sugar

triglyceride

9. Match the monomers and polymers in the following lists:

amino acid	DNA
fatty acid	cellulose
nucleotide	enzyme
sugar	RNA
	phospholipid
	starch
	protein

10. (a) Assume that one strand of DNA contains bases in the following order: AATGCTACGTTAA. Write the order of the bases in the complementary strand of the DNA molecule. (b) Assume that a strand of DNA contains bases in the following order: TTTGCACTAAAA. Write the order of the bases in a strand of RNA that would be formed using this DNA as a template.

 InfoTrac® College Edition

http://infotrac.thomsonlearning.com

Water and hydrogen bonds

Pennisi, E. 1991. Split hydrogen bond allows water to flow. *Science News* 140:359. (Keywords: "hydrogen bond" and "water")

Phytochemicals

Bjerklie, D. 2003. What you need to know about…fruits & vegetables. The news isn't that fruits and vegetables are good for you. It's that they are so good for you they could save your life. *Time* 162(16):50 (October 20). (Keywords: "Bjerklie" and "vegetables")

Yanick, P., Jr. 2003. Successful treatment of carcinogenesis with phytochemicals. *Townsend Letter for Doctors and Patients* June, p. 56. (Keywords: "carcinogenesis" and "phytochemicals")

Webb, D. 2003. Phytonutrients: the hidden keys to disease prevention, good health. *Environmental Nutrition* 26:1. (Keywords: "phytonutrients" and "hidden keys")

The Plant Cell and the Cell Cycle

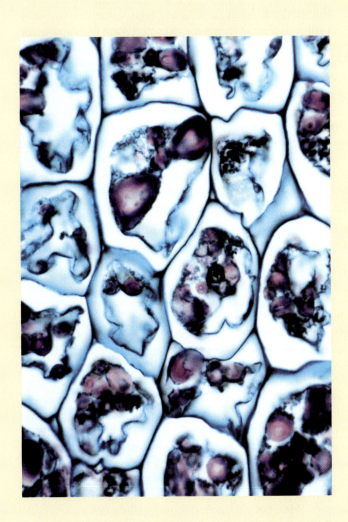

1. Every plant is constructed from small compartments called cells. Each cell is a living individual, possessing the basic characteristics of life, including movement, metabolism, and the ability to reproduce. Some cells in a plant develop specialized capabilities that contribute to the life of the whole organism.

2. Plant cells contain organelles with specialized functions. The nucleus, ribosomes, and endomembrane system participate in the synthesis of proteins; the plastids and mitochondria capture and convert energy into useful forms; the cytoskeleton directs the movement of other components around the cell. Learning the anatomy of a cell helps one understand its activities.

3. Cells reproduce by dividing. Cell division is the most complicated process that any cell can undergo. Specific genes and proteins cooperate to regulate the timing of the events in cell division.

3.1 CELLS AND MICROSCOPY

Cells Are the Basic Units of Plant Structure and Function

In the late 1600s, an English experimentalist named Robert Hooke used his improved version of a microscope to look at shavings of cork tissue (the dead outer bark of an oak tree). He described "little boxes or cells distinct from one another…that perfectly enclosed air." Later, Nehemiah Grew, an English clergyman, recognized that leaves were formed from collections of cells filled not with air, but with fluid and green inclusions (Fig. 3.1). It took many years for the ubiquity of cells to be realized, but in 1838 the Belgian botanist Matthias Schleiden and zoologist Theodor Schwann proposed that all plants and animals are composed of cells. Later, in 1858, Rudolf Virchow suggested that cells possess a characteristic of life ascribed by earlier observers only to organisms—that is, cells reproduce themselves, and all cells arise by reproduction from previous cells. This set of propositions, now known as the **cell theory,** is one of the key principles of biology.

Almost all cells have certain similarities in structure, because they share the same activities and the same problems. Most cells grow—that is, they get larger, and they divide to form new cells. In mature plants, many cells stop growing, but even these continue to synthesize new components. All these cells must accumulate chemicals that they need for the synthesis of new components. They must find sources of energy that promote the chemical reactions needed for synthesis. They must store and interpret the genetic instructions that direct the synthesis of these components at the right times and places. They must get rid of worn-out components and exclude toxins from sensitive reactions. Cells must control their own size, which means controlling the amount of water that moves into or out of

them. All these functions are important to all cells, those of Archaea, Bacteria, animals, and protists, as well as plants. But these various types of organisms do differ somewhat in their cell structures. Archaea and bacteria (prokaryotes), for instance, have cells that appear simpler than those of animals, fungi, protists, and plants (eukaryotes). Prokaryotes are important in the evolution and ecology of plants, and their cells are described in Chapter 19. This chapter focuses on plant cells, their structures, and their methods of carrying out essential functions.

With few exceptions, each cell in the plant body plays a role in the health and activities of the whole plant. To be effective, some cells have specialized structures or chemicals. Certain cells are specialized for rapid growth and cell division. Other cells have a protective function. Cells on the outer layer of a stem, for instance, secrete water-impermeable chemicals, such as waxes; these keep water vapor from diffusing out of the plant, thus keeping the interior moist. Still other cells have a structural role, such as stiffening large organs so that they can support their own weight. Some cells are responsible for the transport of compounds from one part of the plant to another. Certain cells play key roles in sexual reproduction. In each case, the cell forms specialized structures that allow it to accomplish its mission in the life of the plant.

The specialized structures within cells are called **organelles** ("little organs," by analogy with the organs contained in the body of a multicellular organism). These are associated with some of the general and specialized functions that the cells must perform. The next section explains one of the most effective methods for studying organelles.

Microscopes Allow One to See Small, Otherwise Invisible Objects

Plant cells have been studied with a light microscope ever since Robert Hooke looked at cork. In fact, the development of the microscope was the technical breakthrough that led to the discovery of cells, and improvements in microscopy continue to contribute to our understanding of cell structure and function. Light microscopes use lenses, which bend light rays so that the object looks larger (Fig. 3.2). With a good compound microscope (one that has many lenses arranged in series), you can easily see cells that are 20 to 200 micrometers (μm) in diameter, and you should be able to see components as small as 1 μm in diameter (1 μm is 10^{-6} meter or one millionth of a meter or about four hundred thousandths of an inch) (Fig. 3.2c).

A light microscope has the advantage of being usable with live specimens, but it also has limitations. One involves *contrast.* Many of the organelles of a cell do not absorb light well; therefore, light rays coming from them look the same as rays from adjacent parts of the cell. This means that you cannot tell that the organelle is there. Microscopists partially solve this problem by staining the cells. Certain stains color particular organelles, thus increasing their contrast. However, even in stained samples the scattering of light from other parts of the sample tends to wash out the image, reducing the contrast. The thicker the slice of tissue, the more serious the scattering problem. One solution to this problem is to cut a thin slice of the sample, but the soft substance of a living cell cannot withstand the chemical treatment needed for making very thin slices; therefore, this treatment kills the specimen.

Even if the contrast of a sample in a light microscope is good, the microscope's *resolution*—its ability to distinguish separate objects—is limited by several factors. Because different colors of light are affected differently by lenses, it is impossible to focus an image perfectly when it is illuminated with white light (which contains rays of all colors). Furthermore, the resolving power of a microscope is limited to one half of the wavelength of the light being used. Because the shortest wavelength seen by the human eye is about 0.4 μm, the absolutely smallest object that can be resolved in a light microscope is about 0.2 μm in diameter.

New techniques of microscopy have minimized many of these limitations. Contrast can be dramatically increased by **confocal microscopy.** In this system, the illumination (a laser) and the detecting lens are both focused on one point in the sample at a time, scanning across the sample to assemble a whole picture. Because only one point is illuminated, there is no reduction in contrast from scattered light from other parts of the sample. Even in a relatively thick sample, the focal point of illumination can be very exact, which means that the light can be focused on different levels of a sample. If one takes separate pictures of different levels, three-dimensional pictures of a cell can be assembled.

Resolution can be improved by using **transmission electron microscopy** (Fig. 3.2d). Instead of light, electron microscopes use beams of electrons. Quantum theory tells us that electrons, although normally thought of as particles, also behave like light waves, with wavelengths about 1 million

Figure 3.1 Plant cells through the microscope. (**a**) A drawing of cell walls from the cork tissue of an oak (*Quercus* sp.) tree, published in 1665 by Robert Hooke in his *Micrographia*. (**b**) A light micrograph of leaf tissue from the aquatic plant *Elodea*, showing how the tissue is divided into cells.

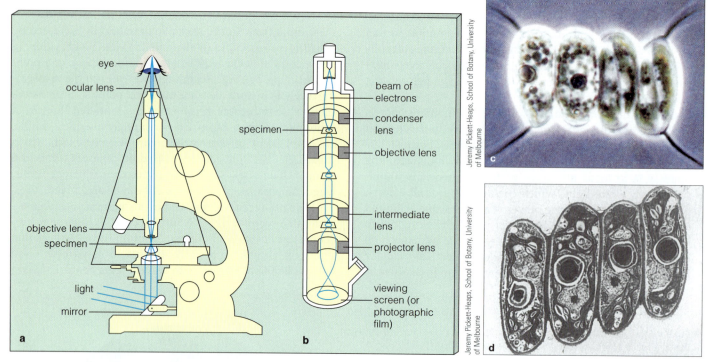

Figure 3.2 A comparison of the light microscope (**a**), the transmission electron microscope (**b**), and the images they produce, (**c**) and (**d**), respectively. A light or electron beam is focused on the sample with glass or magnetic lenses, respectively. From each part of the sample the beam radiates toward the objective lens, forming a larger image on the other side. A series of lenses remagnifies this image, which eventually is focused on the eye (light microscope) or a photographic film (electron microscope). Both micrographs show the same organism—the green alga *Scenedesmus*—at the same size. Notice that the electron micrograph (**d**) has the better resolution, but is black and white; only light micrographs (**c**) can show natural color. Some electron micrographs have color added later to highlight important elements of the picture.

times shorter than those of visible light. Electron beams, having a negative charge, are bent by magnets; therefore, in a transmission electron microscope, magnets serve as lenses. Because the human eye cannot see electrons, the final image is made visible by using the electrons to excite a fluorescent plate or to expose photographic film. These electron microscopes have limitations as well. Electron beams cannot pass through air or through a whole cell. Therefore, the sample must be sliced ultrathin and examined in a vacuum. This technique clearly cannot be used while the samples are alive and functioning. Nevertheless, transmission electron microscopy has been responsible for the discovery of most of the smaller organelles in the cell.

A second type of electron microscopy, called **scanning electron microscopy**, works on an entirely different principle. A very narrow electron beam is scanned across a sample in a series of lines. The electrons that bounce off the sample are collected and used to modulate a similar beam that is forming a picture in a television picture tube. The result is a television picture of the sample (see example in Fig. 4.12). The resolution of the picture can be very high, although only the surface of the sample is shown. Like transmission electron microscopy, scanning electron microscopy must be conducted in a vacuum. In the past, this technique has required complex and careful preparative techniques that kill the

cells. Recent versions of scanning electron microscopes, however, operate in a relatively low vacuum; therefore, live plant cells and insects can be viewed without killing them. Movements and other functions that require living cells can now be studied.

Throughout human history, advances in technology have led to discoveries in basic science. This has also been true with the microscope. From the first discovery of cells, the continuing development of microscopy has led to a long series of breakthroughs in our understanding of plant cell structure and function.

3.2   THE PLANT CELL

Living cells are found throughout the plant body. They make up the internal, photosynthetic cells of the leaf that convert light energy to chemical energy. They make up the pith and cortex of the stem and the cortex of the root. They make up the bulk of fleshy fruits. These cells all have similar organelles (**Fig. 3.3**), those needed for general growth and maintenance of cell function. Some also have specialized organelles for specific functions. The next section describes the components found in a generalized living plant cell.

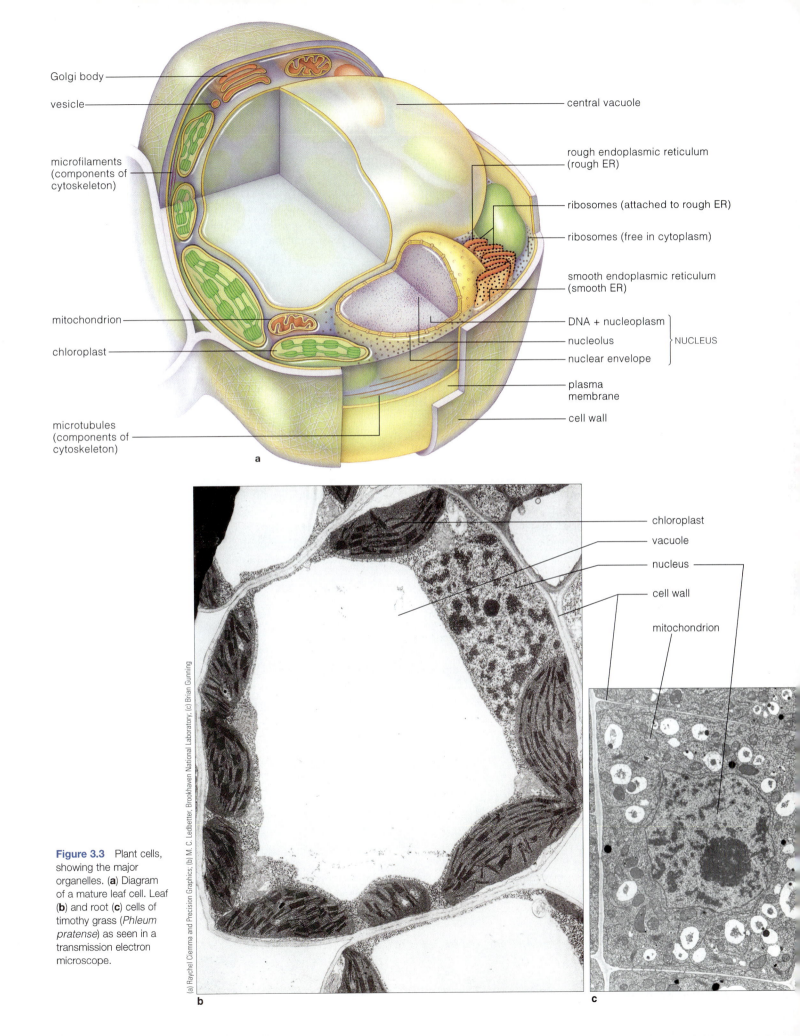

Golgi body

vesicle

microfilaments
(components of
cytoskeleton)

mitochondrion

chloroplast

microtubules
(components of
cytoskeleton)

central vacuole

rough endoplasmic reticulum
(rough ER)

ribosomes (attached to rough ER)

ribosomes (free in cytoplasm)

smooth endoplasmic reticulum
(smooth ER)

DNA + nucleoplasm

nucleolus ⎫ NUCLEUS

nuclear envelope ⎭

plasma
membrane

cell wall

a

chloroplast

vacuole

nucleus

cell wall

mitochondrion

b

c

Figure 3.3 Plant cells, showing the major organelles. (**a**) Diagram of a mature leaf cell. Leaf (**b**) and root (**c**) cells of timothy grass (*Phleum pratense*) as seen in a transmission electron microscope.

(a) Raychel Ciemma and Precision Graphics; (b) M. C. Ledbetter, Brookhaven National Laboratory; (c) Brian Gunning

The Plasma Membrane Controls Movement of Material into and out of the Cell

A thin membrane, the **plasma membrane,** surrounds each cell. Membranes are composed of approximately half phospholipid and half protein, with a small amount of sterols, another form of lipid **(Fig. 3.4).** The phospholipids and sterols provide a flexible, continuous, hydrophobic (water-excluding) sheet two molecules in thickness (called the **phospholipid bilayer**). This separates the aqueous solution inside the cell (called the **cytoplasm**) from that outside the cell. The phospholipid bilayer prevents ions, amino acids, proteins, carbohydrates, nucleic acids, and other water-soluble compounds inside the cell from leaking out and also prevents those outside the cell from diffusing in. This means that for one of these compounds to move in or out, there must be a special pathway or carrier. These pathways are provided by special proteins in the bilayer.

Each type of protein in a membrane performs a different function. Some types provide the special pathways by which compounds can move into or out of a cell in a highly regulated manner. For instance, some proteins function as **ion pumps;** one important type (the **proton pump**) pumps H^+ ions from the inside to the outside of the cell (see Chapter 11 for more details). Another pumps Ca^{+2} ions to the outside of the cell. The characteristic of a pump is that it can move ions from lower to higher concentration by using cellular energy in the form of adenosine triphosphate (ATP) (see p. 130 and Chapters 9 and 10). Some proteins form **channels** for certain substances (such as K^+ ions, sucrose, or even water) to diffuse across the membrane (only from higher to lower concentration), either alone or in combination with another substance (such as H^+). The combination of phospholipids (which form a relatively impermeable sheet) and proteins (which pass specific materials through the sheet) allows the plasma membrane to control transport into and out of the cytoplasm. It is thus a *selectively permeable* membrane.

An unusual property of plant cells is the existence of connections, the **plasmodesmata** (singular, *plasmodesma*) **(Fig. 3.5),** between the plasma membranes of adjacent cells. At a plasmodesma, the plasma membrane on one cell extends outward through the cell wall (see the next section), forming a tube and then connecting to the plasma membrane of the next cell. The continuous cytoplasm in a set of cells connected by plasmodesmata is sometimes called the **symplast.** A plasmodesma forms a passageway that may allow materials to move from the cytoplasm of one cell to the next. It is a striking idea that plant cells are so interconnected by plasmodesmata that a plant might be considered a single super-large cell, partially divided into the compartments we call cells. However, a plasmodesma is not simply an open tube. Inside are another membrane tube and proteins that control the transport of materials (especially large molecules) through the plasmodesma. Small molecules such as sugars and amino acids seem to pass easily. Larger molecules (such as proteins) and organelles generally cannot pass, although recently it has been discovered that plant viruses can open plasmodesmata so that the virus particles can spread their infection from cell to cell.

Does the plasma membrane define the limits of the cell? Many traditional cell biologists consider the plasma mem-

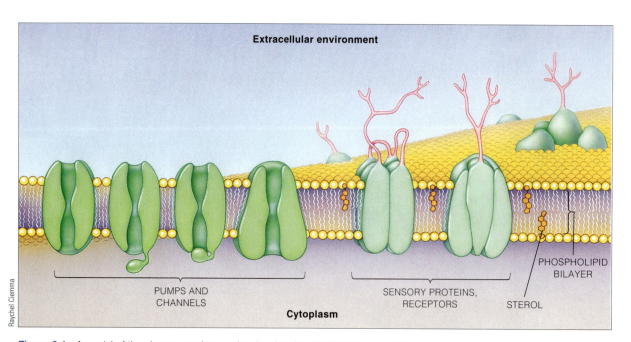

Extracellular environment

PUMPS AND CHANNELS

SENSORY PROTEINS, RECEPTORS

STEROL

PHOSPHOLIPID BILAYER

Cytoplasm

Raychel Ciemma

Figure 3.4   A model of the plasma membrane, showing the phospholipid bilayer, sterols, and various types of proteins floating in the bilayer.

brane to be the outer limit of the cell. According to this view, everything outside the plasma membrane is extracellular—that is, something that may have come from the cell but is not really part of it. Others point to recent research that shows there is considerable metabolic activity outside the plasma membrane. They consider the space outside the cell, next to the plasma membrane and within the fibrils of the cell wall, to be part of the cell. In either case, there is a continuity of the space around adjacent cells. This space is called the **apoplast.** It is clear that the apoplast is an important space in a plant, but it is an open question whether it is part of the plant's cells.

The Cell Wall Limits Cell Expansion

A plant cell (or any cell) that is surrounded only by a plasma membrane and placed in pure water will expand until it bursts. This is because a cell contains a relatively high concentration of solutes, held in by the plasma membrane. The solutes decrease the effective concentration of the water inside the cell. Water, like other chemicals, tends to move from a region of high concentration to one of low concentration. The plasma membrane prevents solutes from moving out, but it does allow water to move in. The flow of water from a relatively dilute solution to a more concentrated solution across a selectively permeable membrane is called **osmosis.** Although the inflow of water dilutes the solution inside the cell, that solution never becomes as dilute as pure water. Thus, the inflow continues until the plasma membrane, which is not infinitely expandable, ruptures.

For plant cells, this problem of rupturing in a dilute solution, such as rainwater, is solved by a **cell wall (Fig. 3.6).** The cell wall is a relatively rigid structure that surrounds the cell

just outside the plasma membrane. It is made of **microfibrils** formed from a polysaccharide, **cellulose.** Cellulose is a linear, unbranched chain of the sugar glucose (see Fig. 2.10). Two of these chains can line up side by side connected by hydrogen bonds. About 40 of these chains form a microfibril, which is flexible but strong, like a cable or a nylon cord. Among the microfibrils are other, less highly organized polysaccharides (hemicelluloses and pectins) and proteins.

The microfibrils and other components form a porous network, so that water and solutes easily penetrate to the plasma membrane. If, however, an inflow of water expands the cell, it forces the plasma membrane against the cell wall. The plasma membrane cannot penetrate the cell wall; therefore, the pressure from the water tends to expand the cell wall. The cell wall is tough but elastic; therefore, the more it expands, the more it resists further expansion. At some point the resistance of the cell wall exactly balances the pressure of osmosis, stopping the flow of water into the cell. This keeps the plasma membrane, and consequently the cell, from expanding further **(Fig. 3.7).**

When the cell stops expanding, the osmotic forces pulling water into the cell do not disappear. They are simply balanced by the pressure exerted by the cell wall. The internal hydrostatic pressure in a cell can be very high. Typical cell walls maintain pressures of 0.75 megapascal (7.5 times atmospheric pressure at sea level). Cells with such internal pressure (**turgor** pressure) become stiff and incompressible. This means that they can hold heavy weights. Thus, even plant cells with a thin cell wall can form and support large plant organs, so long as the solution in the apoplast is sufficiently dilute that a large turgor pressure is generated. When the supply of water is cut off, a leaf wilts. This occurs because water is lost from the cell and turgor pressure decreases. Thus, in addition to preventing the cell from bursting when it is surrounded by a dilute solution, the cell wall also is responsible for maintaining the high turgor pressure that gives a plant organ much of its strength.

Osmosis works in both directions. If the concentration of solutes outside the cell is greater than that inside—such as is

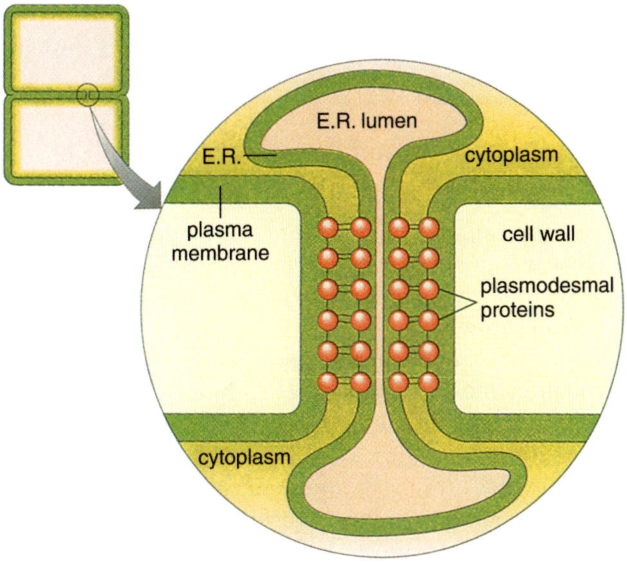

E.R. lumen

E.R.

cytoplasm

plasma membrane

cell wall

plasmodesmal proteins

cytoplasm

Figure 3.5 Model of a plasmodesma, showing the endoplasmic reticulum and proteins that are thought to control the flow of materials through the channel.

Figure 3.6 A small section of cell wall, as seen in a transmission electron microscope. The filaments are cellulose.

PROTOPLAST   SOLUTION

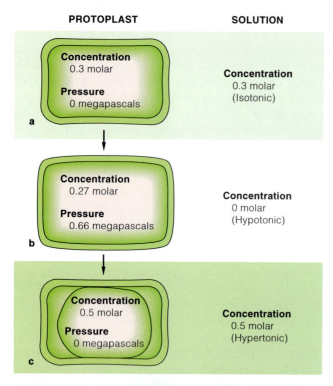

Concentration
0.3 molar

Pressure
0 megapascals

a

Concentration
0.3 molar
(Isotonic)

Concentration
0.27 molar

Pressure
0.66 megapascals

b

Concentration
0 molar
(Hypotonic)

Concentration
0.5 molar

Pressure
0 megapascals

c

Concentration
0.5 molar
(Hypertonic)

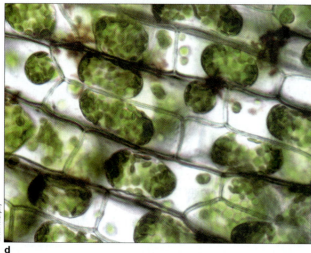

T. M. Murphy

d

Figure 3.7 The effect of osmosis on cell size. (**a**) The cell in this example is assumed to have an initial internal concentration of solutes, which we will designate as 0.3 molar. The cell is in a solution of the same concentration. The cell protoplast (volume inside the plasma membrane) is exactly the size of the cell wall in its resting (unstretched) state. (**b**) The cell is transferred to pure water. Water moves into the protoplast by osmosis, and the protoplast expands, pushing against the cell wall. The wall exerts a back pressure on the protoplast, which inhibits further influx of water. Also, the expansion of the protoplast dilutes the solutes inside the protoplast. Water stops entering the cell when the pressure exerted by the cell wall equals the osmotic pressure forcing water into the cell. (**c**) The cell is transferred to a solution of 0.5 molar. Water moves out of the protoplast by osmosis, and the protoplast shrinks until the concentration of solutes in the protoplast equals 0.5 molar. The plasma membrane pulls away from the cell wall. This effect is called *plasmolysis*. (**d**) Plasmolyzed *Elodea* cells. Compare these cells to the ones in Fig. 3.1b.

found in salt flats—then water will flow out of the cytoplasm. The volume of the **protoplast** (the space inside the plasma membrane) will shrink, and the plasma membrane will pull away from the cell wall (Fig. 3.7). This effect is called **plasmolysis.** A plasmolyzed cell has no turgor pressure. Thus, an accumulation of salt in the soil can cause a plant to wilt.

The cell wall that forms while the cell is growing is the *primary cell wall.* After cells stop growing, some deposit an additional cell wall layer between the primary cell wall and the plasma membrane—a *secondary cell wall* (see Fig. 4.9). The secondary cell wall generally contains cellulose microfibrils and a strong, water-impermeable substance called **lignin.** Lignin is formed from subunits (derived from an amino acid) that are crosslinked with many covalent bonds, forming a three-dimensional network. The crosslinks make these cell walls especially rigid and much more able to resist stretching or compression. Secondary cell walls provide much of the strength of wood (see Chapter 5). Other specialized types of cell walls have **cutin** covering them or **suberin** embedded in them. Cutin and suberin are waxy compounds, made from modified fatty acids, that are especially impermeable to water. Cutinized cell walls are found on the surface of leaves and other organs that are exposed to the air. They retard the evaporation of water from the cells into the air. They also form a barrier to potential pathogens such as bacteria, fungi, and viruses. The adaptations of specialized plant cells often involve chemical changes to their cell walls.

3.4   THE ORGANELLES OF PROTEIN SYNTHESIS AND TRANSPORT

Much of the cell is made of protein, and most cell activities depend on specialized proteins. Plant cells have specialized organelles that are responsible for producing proteins and for making sure that each is placed in its proper position in the cell.

The Nucleus Stores and Expresses Genetic Information

One of the larger and more prominent organelles is the **nucleus.** It is ovoid or irregular in shape and up to 25 μm in diameter. It stains densely with many of the stains used for light or electron microscopy. It is surrounded by a double membrane: the **nuclear envelope.** Filaments of a protein called lamin line the inner surface of the envelope and stabilize its structure. The inner and outer membranes of the nuclear envelope connect to form pores, through which molecules may pass (**Fig. 3.8**). In the nucleoplasm (the portion inside the nuclear envelope) are found the **chromosomes,** which contain DNA and protein. In each chromosome, one long, double-helical molecule of DNA is wound around special proteins (*histones*) to form a chain of nucleosomes. Other

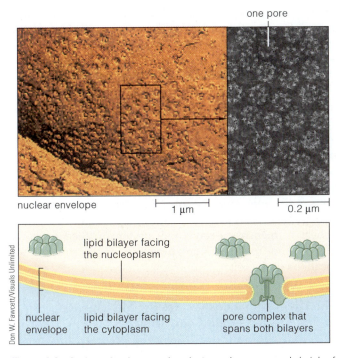

one pore

nuclear envelope

1 μm   0.2 μm

Don W. Fawcett/Visuals Unlimited

lipid bilayer facing
the nucleoplasm

nuclear
envelope

lipid bilayer facing
the cytoplasm

pore complex that
spans both bilayers

Figure 3.8 Surface view by scanning electron microscopy and sketch of the nucleus, showing the nucleolus and the nuclear envelope with pores.

proteins form scaffolds, which hold loops and helices of the chain of nucleosomes in place **(Fig. 3.9).**

The DNA molecules in the chromosomes store genetic information in their sequences of nucleotides. This information is used to direct the synthesis of proteins. The first step in the process is transcription—the DNA is used as a template to direct the synthesis of RNA (see Chapter 15 for a more detailed explanation). Much of this RNA, for unknown reasons, stays in the nucleus or is rapidly broken down. However, a small fraction of the RNA, messenger RNA (mRNA), carries the genetic information of its DNA template out of the nucleus and into the cytoplasm.

Also in the nucleoplasm are **nucleoli** (singular, *nucleolus*), seen in the light or electron microscope as densely staining regions. These are accumulations of RNA–protein complexes. The RNA–protein complexes are ribosomes (see the next section), and nucleoli are the sites where the ribosomes are being assembled. At the centers of the nucleoli are DNA templates, which guide the synthesis of ribosomal RNA.

Ribosomes and Associated Components Synthesize Proteins

In the cytoplasm are the **ribosomes,** which are small dense bodies formed from ribosomal RNA and special proteins. In combination with other molecules, ribosomes synthesize proteins. The active ribosomes are found clustered together

Figure 3.9 The organization of DNA in nucleoplasm.

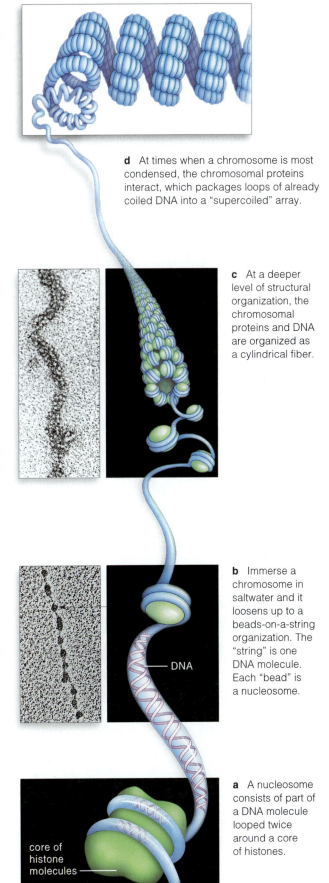

d At times when a chromosome is most condensed, the chromosomal proteins interact, which packages loops of already coiled DNA into a "supercoiled" array.

c At a deeper level of structural organization, the chromosomal proteins and DNA are organized as a cylindrical fiber.

b Immerse a chromosome in saltwater and it loosens up to a beads-on-a-string organization. The "string" is one DNA molecule. Each "bead" is a nucleosome.

DNA

a A nucleosome consists of part of a DNA molecule looped twice around a core of histones.

core of histone molecules

in **polyribosomes (Fig. 3.10).** The ribosomes in a polyribosome are held together because each is attached to the same mRNA. Because the mRNA carries the information for the particular type of protein to be synthesized, all ribosomes in one polyribosome are making the same type of protein.

Although a polyribosome looks like a fixed object in an electron micrograph or a diagram, in a living cell, the ribosomes move rapidly along the mRNA, reading its base sequence and adding amino acids to a growing protein chain. At the end of the mRNA, the ribosomes fall off, releasing the completed protein into the cytoplasm.

The Endoplasmic Reticulum Packages Proteins

In a plant cell, the **endoplasmic reticulum (ER)** is generally a branched, tubular membrane, often near the periphery of the cell. In three dimensions, it is a closed structure (or several closed structures); therefore, solute molecules cannot move from the cytoplasm into the inside (the **lumen**) except by passing through the membrane.

There are many places in the cell in addition to the cytoplasm that need new proteins, including the membranes (plasma membrane, nuclear envelope, and other membranes to be described), certain membrane-surrounded spaces (vacuoles, vesicles), and the apoplast. One of the functions of the ER is to serve as the site where proteins are synthesized and packaged for transport to these locations. Ribosomes in a polyribosome attach to the surface of the ER. In an electron micrograph, this forms a structure known as **rough ER** (Fig. 3.10), as opposed to the **smooth ER,** which does not have ribosomes attached. As the ribosomes synthesize a protein, the protein is injected through the membrane into the lumen. Carbohydrates often are attached to proteins in the ER. This helps protect the proteins from breakdown by destructive enzymes. The proteins in the lumen may be considered to be packaged—that is, separated from the cytoplasm by a membrane. A small sphere of membrane-containing proteins (called a *vesicle*) may bud off from the ER and can carry the proteins to other locations in the cell.

The Golgi Apparatus Guides the Movement of Proteins to Certain Compartments

A **Golgi apparatus** looks like a stack of membranous, flattened bladders, called **cisternae,** which often are swollen toward the edges. (It is also called a **dictyosome.**) Although it is difficult to see in an electron micrograph, there is a direction to the organelle, with the cisternae on one side differing from those on the other.

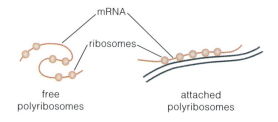

free
polyribosomes

attached
polyribosomes

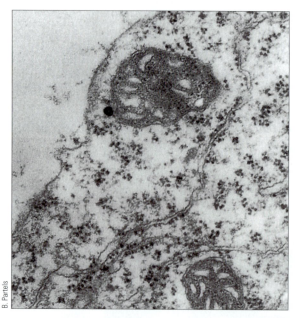

Figure 3.10 Polyribosomes as observed by transmission electron microscopy. Some of the polyribosomes in this cell from a wheat root tip were free in the cytoplasm and some were attached to the endoplasmic reticulum (ER), forming rough ER.

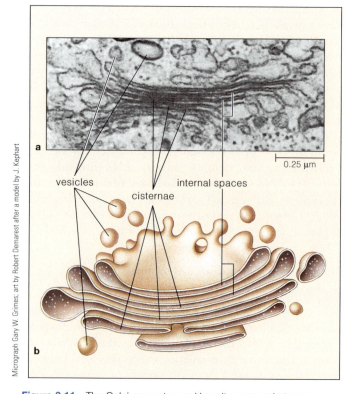

Figure 3.11 The Golgi apparatus and how it moves substances through a cell. (**a**) Transmission electron micrograph. Note the cisternae and associated vesicles. (**b**) A model of a Golgi apparatus.

The Golgi apparatus directs the flow of proteins and other substances from the ER to their destinations in the cell. For example, cell wall proteins, hemicellulose, and pectin in ER packages pass through the cisternae of a Golgi apparatus and then move to the plasma membrane inside a small membranous sphere. When the sphere joins the plasma membrane, its membrane becomes part of the plasma membrane. Its protein, hemicellulose, and pectin contents are then released to the outside of the cell (Fig. 3.11).

The trafficking between the Golgi apparatus, the ER, and other organelles of the cell is rapid and continuous. Even though the organelles usually appear to be separate in electron micrographs and can be isolated from one another by biochemical techniques, they can be considered parts of a complex network called the **endomembrane system.**

3.5 THE ORGANELLES OF ENERGY METABOLISM

The synthesis, packaging, and transport of proteins—and many other functions—require free energy. Plant cells have two specialized organelles that provide this energy: plastids and mitochondria.

Plastids Convert Light Energy to Chemical Energy

Plastids are complex organelles found in every living plant cell. One cell may have 20 to 50 plastids, each 2 to 10 μm in diameter. Characteristically, they are surrounded by a double membrane. They contain DNA and ribosomes—a full protein-synthesizing system similar but not identical to the one in the nucleus and cytoplasm. Some of the proteins of the plastid are made by this system; others are made in the cytoplasm and are transported into the plastid across its outer membranes.

Dividing plant cells always contain some small plastids, called **proplastids (Fig. 3.12c).** These have a few short internal membranes and some crystalline associations of membranous material, called *prolamellar bodies.* As cells mature, their plastids develop and acquire special characteristics. In this process the components of prolamellar bodies apparently are reorganized and combined with new lipids and proteins to form more extensive internal membranes.

Many cells in leaves contain plastids called **chloroplasts** (*chloro-* is derived from a Greek word for a yellow-green color; **Fig. 3.12a–b**). Chloroplasts contain an elaborate array of membranes, the **thylakoids.** Incorporated in the thylakoid membranes are proteins that bind the green compound **chlorophyll.** It is this compound that gives green plant tissues their color. Surrounding the thylakoids is a thick solution of enzymes, the **stroma.** Together, the proteins in the thylakoids and the stromal enzymes perform photosynthesis (see Chapter 10), during which light energy is converted to chemical energy. These plastids can also store carbohydrates, the products of photosynthesis, in the form of starch grains.

In roots and some nongreen tissues in stems, the plastids are **leukoplasts** (*leuko-* is derived from the Greek word for "white"; **Fig. 3.12d**). In these plastids, the thylakoids are missing, although there may be some less organized membranes. These plastids also are able to store carbohydrates (imported from photosynthetic tissue) in the form of starch because they have the enzymes for starch synthesis. Under a light microscope, one often sees white, refractile, shiny particles: starch grains. When leukoplasts contain large granules of starch, they often are called **amyloplasts** (*amylo-* is derived from the Greek word for "starch"; **Fig. 3.12e**).

In certain colored tissues (for instance, tomato fruits and carrot roots), the plastids accumulate high concentrations of specialized lipids—carotenes and xanthophylls. The orange-to-red color of these lipids gives the plastids their name, **chromoplasts** (*chromo-* meaning "color"; **Fig. 3.12f**).

Mitochondria Make Useful Forms of Chemical Energy

In plant cells, as in the cells of all eukaryotes, there are **mitochondria (Fig. 3.13).** Mitochondria are made of two membrane sacs, one within the other. The inner membrane forms folds or fingerlike projections called **cristae.** These folds increase the surface area available for chemical reactions taking place on the membrane. In the center of the organelle, inside the inner membrane, is a viscous solution of enzymes, the **matrix.** Like plastids, mitochondria contain DNA and ribosomes, and they are able to synthesize some types of proteins.

Mitochondria are best known as the sites of oxidative respiration, the places where organic molecules are broken down into CO_2 and H_2O (see Chapter 9). During this process, some of the energy that is released is used to synthesize the energy-rich molecule ATP. ATP powers many of the important chemical reactions in the cell. Mitochondria are the source of most of the ATP in any cell that is not actively photosynthesizing.

3.6 OTHER CELLULAR STRUCTURES

Cells have several types of organelles in addition to those already described in this chapter. Some of these organelles seem rather passive, serving mainly as storage compartments. Others are hard to see (even by microscopy), but they are very active, moving many of the other organelles around the inside of the cell in a continuous dance.

Vacuoles Store Substances

In each living plant cell, a large compartment, bounded by a single membrane, makes up a large fraction of the cell volume (Fig. 3.3). This is the **vacuole.** The membrane surrounding it is called the tonoplast. Like the plasma membrane, the tonoplast has a number of embedded protein pumps and channels that control the flow of ions and organic molecules into and out of the vacuole.

Figure 3.12 Plastids. (**a**) Model of a chloroplast. (**b**) A maize (*Zea mays*) leaf chloroplast, showing dense thylakoid membranes. (**c**) A maize proplastid, with only a few internal (prolamellar) membranes. (**d**) A small leukoplast from an inner white leaf of endive (*Cichorium endiva*). (**e**) An amyloplast from a bean (*Phaseolus vulgaris*) seedling, showing large starch grains. (**f**) A chromoplast from a mature red pepper (*Capsicum* sp.). The dark circles are globules of red and orange xanthophylls.

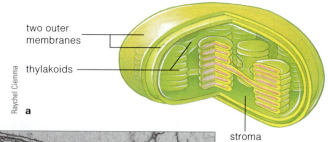

two outer membranes

thylakoids

stroma

Raychel Ciemma

a

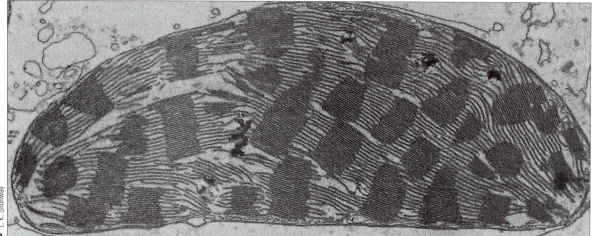

L. K. Shunway

b

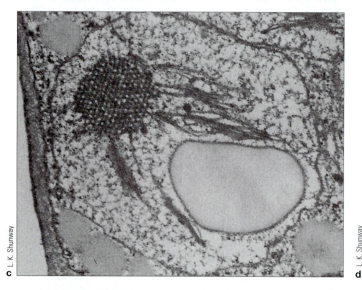

L. K. Shunway

c

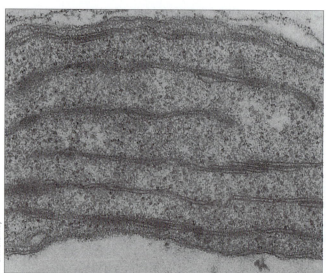

L. K. Shunway

d

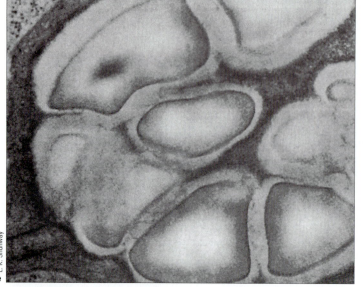

L. K. Shunway

e

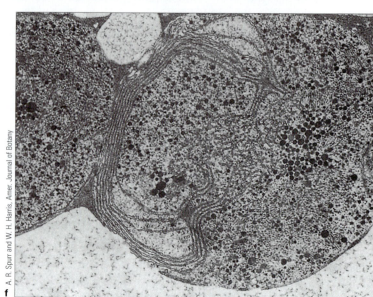

A. R. Spurr and W. H. Harris, Amer. Journal of Botany

f

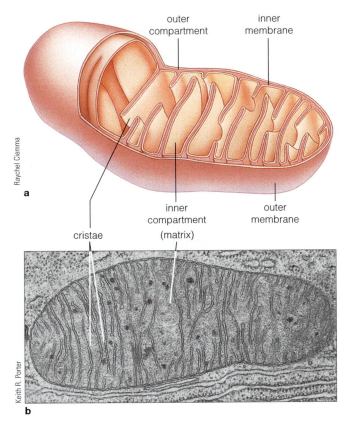

Figure 3.13 Mitochondria. (a) Model of a typical mitochondrion. (b) Transmission electron micrograph of a mitochondrion from the young leaf of a safflower (*Carthamus tintorius*) seedling.

outer compartment
inner membrane
inner compartment (matrix)
outer membrane
cristae

Raychel Ciemma

Keith R. Porter

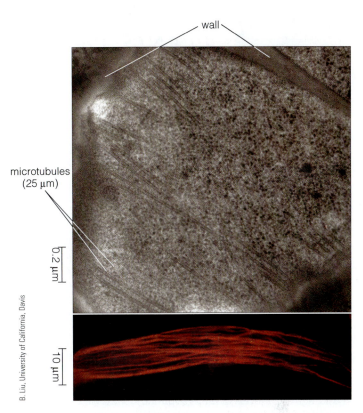

wall
microtubules (25 µm)
0.2 µm
10 µm

B. Liu, University of California, Davis

Figure 3.14 (a) Microtubules in a root cell of mouse-ear cress (*Arabidopsis thaliana*). (b) Bundles of microfilaments in a pollen tube of *Tradescantia virginiana*. These have been complexed with an anti-actin antibody attached to a fluorescent compound that glows red when irradiated with ultraviolet light.

Depending on the cell, the vacuole may have any of several functions. It may accumulate ions (or other substances), increasing the osmotic pull of water into the cell and increasing the turgor pressure inside the cell. It may store nutrients such as sucrose (common sugar). (Our commercial sugar comes from extracting sucrose from the vacuoles of sugarcane stem or sugar beet root storage cells. Other nutritious chemicals may be stored in vacuoles. See "PLANTS, PEOPLE, AND THE ENVIRONMENT: Foods and Health" sidebar) The vacuole may also accumulate compounds that are toxic to herbivores. Finally, it may serve as a dump for wastes that the cell cannot keep and cannot excrete. Sometimes substances accumulate in the vacuole to such a high concentration that they form crystals.

Other Organelles Transport and Store Substances and Compartmentalize Reactions

The cytoplasm has many types of small, round bodies surrounded by a single membrane (Fig. 3.3), as is the vacuole. These include **vesicles**—both the small spheres of membrane that bud off the ER and carry proteins (ER vesicles) and those that travel from the Golgi apparatus to the plasma membrane (Golgi vesicles). There are also similar but more permanent organelles in the cell. In some specialized cells, protein bodies and lipid bodies store proteins or lipids. **Peroxisomes** and **glyoxysomes** serve as compartments for enzy-

matic reactions that need to be separated from the cytoplasm (because some products of those reactions would be toxic to enzymes in the cytoplasm). **Lysosomes** contain enzymes that break down proteins, carbohydrates, and nucleic acids. The lysosomes may have some function in removing wastes within the living cell. In addition, when they break, they release their enzymes, which dissolve the cell. In the life of a plant, there are several times when it is important to get rid of certain cells; the lysosomes accomplish this.

The Cytoskeleton Controls Form and Movement within the Cell

In high-resolution electron micrographs, a collection of long, filamentous structures within the cytoplasm can sometimes be seen. These are evidence of a higher order of organization of the organelles within the cell. They form the **cytoskeleton**, which sometimes keeps the organelles in particular places and sometimes directs their movement around the cell (the movement is called **cyclosis** or cytoplasmic streaming).

There are several different types of structures in the cytoskeleton. One type is relatively thick (0.024 µm in diameter): the **microtubule (Fig. 3.14a).** It is assembled from protein subunits called *tubulin*. Tubulins assemble in a helical manner to form a hollow cylinder. Once assembled, microtubules are fairly rigid, although they can lengthen by the addition of tubulin molecules to one end (or shorten by the

PLANTS, PEOPLE, AND THE ENVIRONMENT:

Foods and Health

Food is one of life's great delights. Food nourishes, comforts, educates, protects, and provides pleasure every day of our lives. The chemicals in foods play a major role in determining how healthy we are. Plants are a major source of chemicals for energy, synthesis of body tissues, and control of environmental toxins.

Modern science and technology have provided unparalleled values to consumers in the breadth of choices in delicious, safe, and nutritious foods. These values have been driven by scientific knowledge at all levels of the agricultural enterprise from genetic improvements in production agriculture, to mathematic control of food process engineering, from molecular understanding of food safety, to statistical precision in the analysis of consumer sensation. One of science's great achievements of the 20th century was building the quantitative knowledge of the nutritional requirements for humans and animals and practically eliminating primary nutrient deficiencies in the developed world.

The successes in health that were achieved by defining the essential nutrients has generated new challenges, both with respect to a better understanding of mechanisms important to chronic disease processes, and the expansion in food choices. At a time when the opportunities to produce a wide variety of foods to improve health would seem unprecedented, more and more individuals lacking suffi-cient knowledge of themselves, or their foods, are choosing diets that increase their risk for metabolic imbalances and disease rather than making optimal choices. The prevalence of diet-related diseases such as obesity, type 2 diabetes, food allergies and intolerances, and cardiovascular disease is increasing throughout the world, high-lighting our discouraging lack of scientific knowledge of the precise molecular and mechanistic links between food and health.

During the last decade, there has been a paradigm shift in how food and diet are viewed with respect to human health and well-being. As an example, the Food and Nutrition Board of the Institute of Medicine (IOM), working in cooperation with scientists from Canada, recently developed Dietary Reference Intakes (DRIs) for the essential nutrients. In marked contrast to the previous Recommended Dietary Allowances (RDAs), the new DRI values for nutrients are defined as targets for the "individual," rather than the general population. The philosophy underlying the development of the new DRI values is to identify the amount of essential nutrients that an individual needs to consume to reduce his or her risk for chronic disease. Notably, for several nutrients, the new DRI values for certain essential nutrients have been set at levels that cannot be easily met from nonfortified foods.

The recent surge of interest by the U.S. Department of Agriculture in chemical constituents of plants (phytochemicals) il-lustrates the growing awareness of the potential health benefits associated with consuming certain phytochemicals. Importantly, the phytochemical profile, and content, of foods can be markedly influenced by agricultural practices, as well as by food processing. Public health officials are increasingly looking to diet as a means to reduce many common chronic diseases. This interest in dietary solutions is occurring in part as a consequence of recent steep increases in the cost of conventional medicine, and in part because of an increasing awareness that through diet one can significantly reduce the risk for many common diseases. Thus, at the national and international level, the line between food and medicine is becoming increasingly blurred, and the need for research on nutrition is becoming ever more important.

Adapted from "Foods for Health" by Bruce German, Bruce Hammock, Carl Keen, and Janet King, University of California, Davis.

 Websites for further study:

Ohio State University, Extension Fact Sheet: Phytochemicals—Vitamins of the Future?: http://ohioline.osu.edu/hyg-fact/5000/5050.html

Quackwatch.org: Antioxidants and Other Phytochemicals: www.quackwatch.org/03HealthPromotion/antioxidants.html

---

loss of tubulins from the same end); therefore, the length of a microtubule can change rapidly.

Certain *motor proteins* can move along microtubules. These proteins—one type called kinesins, another dyneins—move by stepping along the microtubule strand, making and breaking connections with adjacent tubulin subunits. The motor proteins may be attached to other organelles, such as vesicles. In this way, microtubules guide the move-ment of organelles around the cytoplasm. The power generated by motor proteins comes from the breakdown of the high-energy molecule ATP (see Chapter 8 for a description of this high-energy molecule).

Microtubules are key organelles in cell division (see p. 44). They also form the basis of *cilia* and *flagella*, motile organs that move cells by beating back and forth. Cilia and flagella are never found in the cells of flowering plants, but they are important to some algae and to male gametes of lower plants.

Another type of structure in the cytoskeleton is the microfilament (Fig. 3.14b). The **microfilament,** thinner (about 0.007 μm in diameter) and more flexible than the microtubule, is made of protein subunits called *actin* that fit together in a long, helical strand. Microfilaments often are found in bundles. Like microtubules, they can serve as guides for the movement of organelles. However, myosin, rather than kinesins or dyneins, is the motor protein that moves along a microfilament. The energy for this movement comes from the breakdown of ATP.

Many types of specialized proteins connect microtubules and microfilaments to other organelles. These connections are thought to coordinate many of the processes of the cell. For instance, they might direct vesicles to move to the plasma membrane and deposit new wall material when triggered by the presence of hormones. This coordination is most important, and most spectacular, during the complex process of cell reproduction. As you read the next section, notice how microtubules appear to direct several of the events of cell division.

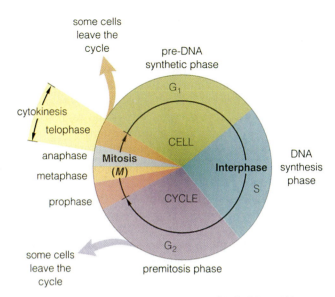

Figure 3.15 The cell cycle, consisting of the G1, S, G2, and M phases. Approximately 50% of new daughter cells leave the cycle at either G1 or G2 and begin to differentiate; the remaining cells will cycle again.

3.7 THE CELL CYCLE

What Are the Phases of the Cell Cycle?

In higher plants, cells form aggregates—tissues and organs—that have specific structures and functions (see Chapter 4). The production of new cells allows the plant body to grow by forming new organs and by increasing the number of cells in an existing tissue or organ. The following section examines the beginning of the tissue- and organ-forming process, describing how cells divide and what mechanisms control the process.

The process of division occurs in special regions of the plant called **meristems.** (Meristems are found at the tips of roots and shoots and in some other regions of a plant.) When cells divide, they must progress through a series of steps called the **cell cycle.** One cell cycle is the interval of time between the formation of a cell and its division to form two new cells.

The cell cycle has four phases: G1 ("gap"), S ("synthesis" of DNA), G2, and M ("mitosis") (Fig. 3.15). Traditionally, the G1, S, and G2 phases of the cycle are grouped together and called **interphase.** The remaining M phase is actually composed of two parts: **karyokinesis** (nuclear division) and **cytokinesis** (cytoplasmic division).

Specific Metabolic Events Occur in Each Cell Cycle Phase

Cells in each phase of the cell cycle are unique in structure and molecular composition. During G1, the cell prepares itself metabolically for DNA synthesis. These preparations include both the accumulation and synthesis of specific enzymes to control DNA synthesis and the production of the DNA subunits so that a supply is on hand when synthesis begins.

The second phase of the cell cycle is the S phase. During this portion of the cycle, the cell duplicates its DNA molecules (see Chapter 2). The time needed for a cell to progress through the phases of the cell cycle depends on the amount of DNA per nucleus. Plants with more DNA in their nuclei have longer cell cycle times than plants with less DNA. The time for a complete cycle ranges from a few hours to days.

Unless cells are metabolically active, they cannot progress through the phases of the cell cycle. Cells must be able to synthesize enzymes and other proteins needed for specific cell cycle events. DNA polymerase, the key protein in the synthesis of new DNA, is an example of a cell cycle–specific protein. The amounts of this protein and of histones—proteins that form the nucleosomes around which DNA winds—increase at the end of the G1 phase and during the S phase.

3.8 REGULATION OF THE CELL CYCLE

The Principal Control Point Hypothesis Identifies the Control Points in the Cell Cycle

The high-energy molecule ATP is essential for the synthesis of DNA and proteins, and for other processes. Cycling cells in meristems require a constant supply of carbohydrates to make ATP. If root tips are removed from seedlings and placed into sterile culture, their cells will progress through the cell cycle only if carbohydrate (sucrose) is added to the culture medium. Without sucrose, the cells stop cycling, but only at specific places in the cell cycle. Jack Van't Hof and his collaborators at Brookhaven National Laboratory on Long Island observed that such cells became arrested in G1 and G2 phases. The ratio of cells arrested in G1 and G2 in such experiments varies among different species. For example, in peas (*Pisum sativum*) 50% stop in G1 and 50% stop in G2, whereas in sunflowers (*Helianthus annuus*) 90% stop in G1 and 10% stop in G2. Cell cycle progression resumes after sucrose is added back to the culture medium.

These experiments led Van't Hof to develop the Principal Control Point Hypothesis, which proposes that control points exist in the plant cell cycle at G1 and G2 phases (Fig. 3.16). If a cell progresses past the G1 or G2 control point, it then automatically progresses through DNA synthesis (S phase) or cell division (M phase), respectively. Metabolically, this means that during the G1 phase a cell synthesizes certain proteins in preparation for DNA synthesis in the S phase. If the cell forms these critical macromolecules, it progresses through the S phase; but if it does not, the cell is arrested in G1. The same interpretation would apply to the G2 control point, except that at that time a dividing cell would need proteins that act during the M phase.

Microtubules Set the Plane of Cell Division

Microtubules (see p. 42) are intimately involved in many aspects of the cell cycle, from the regulation of cell wall formation and cell shape to the movement of chromosomes and control of the plane of cell division (Fig. 3.17). During G1 and S of the cell cycle, the microtubules are located around the periphery of the cell, next to the plasma membrane and cell wall. These microtubules are involved in the deposition of cellulose in the cell wall. The cell will elongate at right angles to the axis of the microtubules and cellulose, a process described in more detail in Chapter 15.

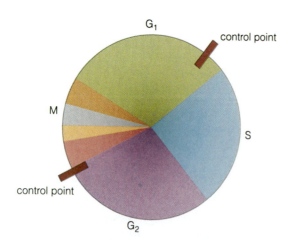

Figure 3.16 Cell cycle control points. Two control points located in G1 and G2 regulate the cell cycle in plant cells and other eukaryotes. Consequently, a plant cell population stressed by starvation will show cells arrested in G1 and G2 because the cells lack critical proteins needed for DNA synthesis (S phase) and mitosis (M phase). The G1 control points are regulated by the activity of a gene called *cdc2*.

During G2 the microtubules move into a new location and form a band surrounding the nucleus, but pressed close to the cell wall (Figs. 3.17a–c and 3.18). This organization is called the **preprophase band (PPB)** of microtubules. Its formation precedes mitosis by hours. The orientation of the PPB somehow marks the position of the new cell wall that will form at the end of mitosis. The new cell wall defines the *plane of cell division.*

Several factors may influence the position of the PPB. Within tissues, gradients of chemicals such as plant hormones regulate the start of differentiation events such as mitosis. The hormone gradients may induce the movement of the microtubules to the PPB as a first step in the overall induction of cell division.

Mitosis Occurs in Stages and Is Followed by Cytokinesis

The purpose of mitosis is to separate the DNA, which was doubled in amount during the S phase, in such a way as to produce two complete sets of genetic information. The process of mitosis is generally divided into four phases to make it easier to understand.

PROPHASE   During G1, S, and G2 (interphase) phases, the DNA molecules are long and apparently tangled in the nucleus. At the start of **prophase,** the DNA molecules thicken by coiling on themselves several times to form dense chromosomes (Figs. 3.9d and 3.19a,b). The nucleolus gradually disappears.

It becomes apparent during late prophase that each chromosome is composed of two intertwined DNA molecules connected by a constricted region called a **centromere** (Fig. 3.19b,c). The nuclear membrane breaks down during late prophase, but the nucleoplasm does not mix with the general cytoplasm; there is a clear zone between them, devoid of organelles.

METAPHASE   In **metaphase,** the chromosomes now arrange themselves on the equatorial plane of the cell, usually with their **centromeres** aligned (Fig. 3.19c). This position may be in the middle of the cell, or it may be off center, resulting in an unequal cell division. The spindle fibers, composed of bundles of microtubules, can now be seen. The spindle extends from the poles near the ends of the cell to their attachment point on each chromosome, the **kinetochore,** where motor proteins connect the centromere to the microtubules. All of the spindle fibers collectively are called the *spindle apparatus.*

The chromosomes are now distinct bodies of two closely associated halves, each half known as a **chromatid.** In plants such as corn (*Zea mays*), which have been intensively studied, each chromosome can be recognized (on the basis of its shape) and numbered. There are 20 chromosomes in corn, but only 10 different types that can be distinguished by their size and form. The 20 chromosomes of corn may be

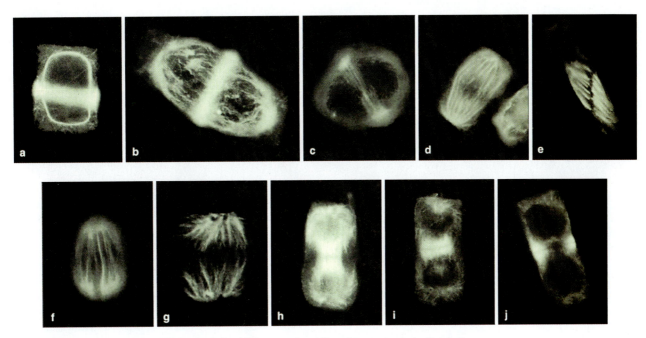

Figure 3.17 The role of microtubules in mitosis. Onion (*Allium cepa*) root tip cells were treated with tubulin antibodies tagged with a fluorescent dye and photographed by fluorescent microscopy. Thus, the photographs show only the microtubules. (**a–c**) Early prophase showing the preprophase band of microtubules plus some microtubules along the outside of the cell just inside the cell wall. (**d,e**) The spindle apparatus in early metaphase. (**f**) Metaphase chromosomes are located in the center of the cells. (**g**) Anaphase chromosomes migrate back toward the poles, and the spindle apparatus microtubules will be dispersed. (**h,i**) The dense accumulation of microtubules in the center of the cell shows the early formation of the cell plate. (**j**) Late telophase showing microtubules only at the periphery of the cell plate. (Used by permission of Dr. S. Wick. *Jour. Cell Biol.* 89:685–690 (1981). ©The Rockefeller University Press, NY.)

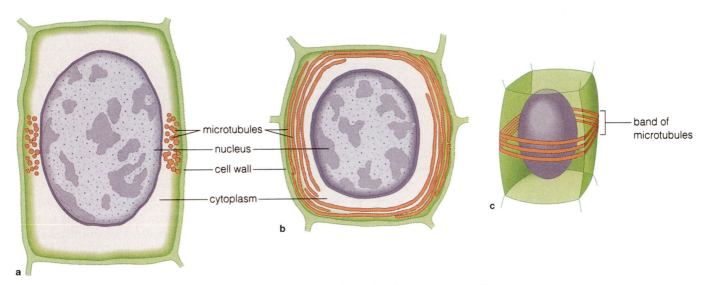

Figure 3.18 Formation of the preprophase band of microtubules in meristematic cells of a young leaf of tobacco (*Nicotiana tabacum*). (**a**) Section at right angles to the plane of the future cell plate shows a cross section of microtubules. (**b**) Section in the plane of the future cell plate shows microtubules encircling the nucleus. (**c**) Three-dimensional drawing of **a** and **b**. (Redrawn after K. Esau and J. Cronshaw. *Protoplasma* 65:1. ©1966 by Springer-Verlag. Reprinted with permission of the publisher.)

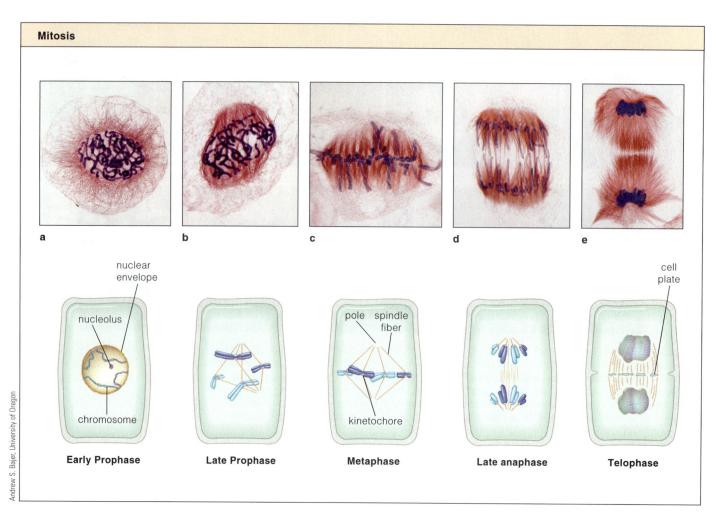

a   b   c   d   e

nuclear envelope

cell plate

nucleolus

pole spindle fiber

chromosome

kinetochore

Early Prophase   **Late Prophase**   **Metaphase**   **Late anaphase**   **Telophase**

Andrew S. Bajer, University of Oregon

Figure 3.19   The stages of mitosis, shown in light micrographs of cells from the endosperm of the blood lily (*Haemanthus katherinae*); proteins (including microtubules) are red, and chromosomes are blue. These cells are unusual in that they do not have cell walls. The bottom row diagrams mitosis in the more typical meristem cell, with a cell wall. Cells from the G2 phase, with duplicated chromosomes, form the starting material for mitosis. (**a**) Early prophase: The chromosomes start to become thicker and shorter by coiling tightly. The nuclear envelope is still present. (**b**) Late prophase: The nuclear envelope and nucleolus have disappeared. On each chromosome, a kinetochore becomes visible. Spindle fibers are starting to appear. (**c**) Metaphase. The coiled chromatids are distinct. The spindle is organized, and the kinetochores line up in the middle of the spindle. (**d**) Late anaphase: The chromatids have moved to opposite poles of the cell. (**e**) Telophase: Chromatids have aggregated at the opposite poles of the cell. The nuclear envelopes have not yet formed. A cell plate has started to form.

arranged in 10 pairs. Geneticists have mapped corn chromosomes, showing the relative positions of many genes along them. Because each gene is replicated, when the chromosomes split longitudinally, each chromatid contains a full set of genes.

ANAPHASE   The chromosomes do not remain long in the equatorial plane. The individual chromatids of each pair soon separate from each other and move to opposite poles of the cell (Fig. 3.19d). This period is called **anaphase.**

The mechanism of chromosome movement and the functional role of the spindle fiber microtubules are not fully understood. There are two major mechanisms to explain chromosome movement. The *assembly–disassembly mecha-*

nism contends that microtubules move chromosomes by losing tubulin molecules from one end of the spindle. The removal of tubulin subunits from both poles of the spindle would shorten the spindle and pull the chromosomes apart. The *motor protein mechanism* contends that the chromosomes slide over microtubules of the spindle. Motor proteins at the kinetochore provide the sliding force.

TELOPHASE   When the divided chromosomes have reached the opposite poles of the cell, **telophase** begins. The chromosomes aggregate and begin to uncoil (Fig. 3.19e) into long, thin chromatin strands. The nuclear envelope and the nucleolus re-form, and the cells return once again to G1 of the cell cycle.

CYTOKINESIS Cytokinesis begins before telophase is finished. A new cell wall, called the **cell plate,** starts to form between the separated nuclei (Fig. 3.19e). This process involves the **phragmoplast** (Fig. 3.17i), consisting of a band of microtubules that re-forms perpendicular to the cell plate and many small membrane-bound vesicles. The small vesicles contain cell wall material, which is deposited first in the center of the cell and then outward until the new wall attaches to the side walls. The point of attachment, you will recall, was previously "marked" by the PPB (Fig. 3.18). Some of the new cells that are formed by mitosis and cytokinesis will divide again in plant meristems, and others will begin to differentiate to form specialized cells.

KEY TERMS

amyloplasts	metaphase
anaphase	microfibrils
apoplast	microfilament
cell cycle	microtubule
cell plate	mitochondria
cell theory	nuclear envelope
cellulose	nucleoli
cell wall	nucleus
centromere	organelles
channels	osmosis
chlorophyll	peroxisomes
chloroplasts	phospholipid bilayer
chromatid	phragmoplast
chromoplasts	plasma membrane
chromosomes	plasmodesmata
cisternae	plasmolysis
confocal microscopy	plastids
cristae	polyribosomes
cutin	preprophase band (PPB)
cyclosis	proplastids
cytokinesis	prophase
cytoplasm	proton pumps
cytoskeleton	protoplast
dictyosome	ribosomes
endoplasmic reticulum (ER)	scanning electron microscopy
glyoxysomes	stroma
Golgi apparatus	suberin
interphase	symplast
ion pumps	telophase
kinetochore	thylakoids
leukoplasts	transmission electron microscopy
lignin	turgor
lysosomes	vacuole
matrix	vesicles
meristems	

SUMMARY

1. All organisms are formed from one or more cells. All cells demonstrate the basic functions of life, including controlling their size, accumulating nutrients, extracting energy from the environment, reproducing, and withstanding or tolerating environmental stresses. Cells are produced only by reproduction from previously existing cells.

2. Because cells are very small, an essential tool for their study is microscopy. Light microscopes can be used to visualize live samples, but contrast and resolution are major problems in light microscopy. These limitations can be minimized or circumvented by confocal microscopy, transmission electron microscopy, and scanning electron microscopy.

3. The plasma membrane surrounds the cytoplasm of the cell. It is formed from a lipid bilayer and proteins. The lipid bilayer prevents the diffusion of water-soluble molecules into or out of the cell. Proteins in the plasma membrane control the passage of molecules through the plasma membrane, among other functions.

4. The cell wall, formed from carbohydrates and proteins, prevents excess water from entering the cell through osmosis and provides the turgor pressure that gives most plant organs a firm structure.

5. The nucleus contains chromosomes made of DNA and protein. DNA serves as a template for mRNA synthesis; mRNA directs protein synthesis. The nucleolus (or plural, nucleoli) inside the nucleus is responsible for the synthesis of ribosomal RNA. Ribosomal RNA plus specific proteins from ribosomes leave the nucleus and, together with mRNA in the cytoplasm, form the polyribosomes, where proteins are actually synthesized.

6. The endoplasmic reticulum (ER) is a tubular compartment extending around the periphery of the cell. Polyribosomes that are synthesizing proteins destined for export from the cell or for import into vesicles attach to the surface of the ER and inject their protein products into its lumen. Proteins leave the ER inside vesicles. The Golgi apparatus directs the flow of proteins and other substances in vesicles from the ER to their destinations in the cell.

7. Plastids are bounded by a double membrane and have their own DNA- and protein-synthesizing systems. There are several types of plastids, including chloroplasts, which enclose thylakoid membranes and catalyze the reactions of photosynthesis; amyloplasts, which store starch; and chromoplasts, which contain high concentrations of colored carotenoids and xanthophylls.

8. Mitochondria also are bounded by a double membrane and have their own DNA- and protein-synthesizing systems. They catalyze the oxidation of organic compounds to CO_2 and H_2O and use a portion of the energy released to synthesize ATP.

9. Among the specialized compartments bounded by a single membrane are the vacuole and some specialized vesicles. Vacuoles store salts, sugars, defensive compounds toxic to predators, and wastes. Some vesicles store energy-rich lipids or

proteins (liposomes, protein bodies); some compartmentalize enzymatic reactions (glyoxysomes, peroxisomes); and some control the release of self-destructive enzymes (lysosomes).

10. The cytoskeleton, which is formed from protein filaments, directs the organization and movement of organelles within the cell. Motor proteins (dynein, kinesin, myosin) move along filaments, dragging other organelles with them.

11. Cells reproduce by passing through a series of events called the cell cycle. The four phases of the cell cycle are: G1, the pre-DNA synthesis phase; S, the DNA synthesis phase; G2, the premitotic phase; and M, the cell division phase.

12. Higher plant cells and fission yeast have control points at G1 and G2 phases. If critical metabolic events do not occur at these points, DNA synthesis and mitosis will not occur.

13. Microtubules have several roles in plant cell division, including control of cell plate location and regulation of chromosome movement. The PPB of microtubules forms in the G2 phase and marks the plane of cell division.

14. The stages of mitosis are prophase, metaphase, anaphase, and telophase.

Mitotic Stage	Summary of Events
Prophase	Chromosomes coil; the nuclear envelope and the nucleolus break down.
Metaphase	Chromosomes move to the equatorial plane; the spindle apparatus forms and attaches to the chromosomes at the kinetochores.
Anaphase	Chromosomes move to the poles by some mechanism involving the spindle fibers.
Telophase	Chromosomes uncoil; the nuclear envelope and the nucleolus re-form.

15. During cytokinesis, the cell plate, containing cell wall materials, forms first in the middle of the cell and then moves to the periphery, dividing the cytoplasm in two.

Questions

1. Explain why it is necessary to use a microscope to see a cell. Describe the difference between the image of a cell seen at high magnification with low resolution and one seen at the same magnification but with high resolution.

2. Assume a cell is a cube with 10-μm sides. What is the volume of the cell in cubic micrometers? What is the volume in cubic meters? How many of these cells would fill a teaspoon (approximately 5 cm³)?

3. List the organelles that are needed to synthesize a protein and move it to the outside of a cell.

4. Which of the following organelles is (are) absolutely essential for the synthesis of a protein?

nuclear envelope

ribosome

microtubule

endoplasmic reticulum

Golgi apparatus

Which of the above will be needed for treatment of the protein after it is made?

5. Almost every living plant cell has mitochondria. Explain why the presence of mitochondria is essential for a cell's well-being.

6. In what organ(s) of the plant would you find the most chloroplasts?

7. Assume that a cell must reproduce itself exactly. List (as completely as you can) the events that must occur in G1, S, G2, and M phases for the reproduction to be successful.

8. List the phases of mitosis and describe what happens in each. If you add a chemical that blocks the formation of microtubules to a cell that is starting mitosis, mitosis eventually stops. In what phase will it stop?

 InfoTrac® College Edition

http://infotrac.thomsonlearning.com

Microscopy

Boguslavsky, J. 2003. Optical slices of life: confocal microscopy enables researchers to obtain high-resolution, serial optical sections from live specimens. *The Scientist* 17:39. (Keywords: "confocal," "optical," and "live")

Wright-Smith, C., Smith, C.M. 2001. Atomic force microscopy. *The Scientist* 15:23. (Keywords: "Smith," "atomic," and "force")

Cell Walls

Jung, H.G., Engels, F.M. 2002. Alfalfa stem tissues: cell wall deposition, composition, and degradability. *Crop Science* 42:524. (Keywords: "alfalfa" and "deposition")

Strauss, E. 1998. When walls can talk, plant biologists listen. *Science* 282:28. (Keywords: "plant," "walls," and "signals")

Zambryski, P. 1995. Plasmodesmata: plant channels for molecules on the move. *Science* 270:1943. (Keywords: "plasmodesmata" and "Zambryski")

Cell Division

Chapman, M.J., Dolan, M.F., Margulis, L. 2000. Centrioles and kinetosomes: Form, function, and evolution. *Quarterly Review of Biology* 75:409. (Keywords: "centrioles" and "kinetosomes")

Ivanov, V.B., Dubrovsky, J.G. 1997. Estimation of the cell-cycle duration in the root apical meristem: A model of linkage between cell-cycle duration, rate of cell production, and rate of root growth. *International Journal of Plant Sciences* 158:757. (Keywords: "cell-cycle" and "root")

4

The Organization of the Plant Body: Cells, Tissues, and Meristems

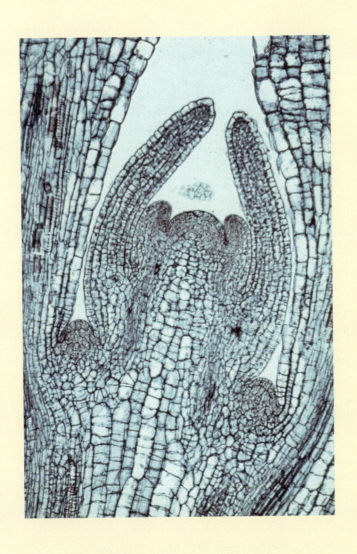

ORGANIZATION OF THE PLANT BODY

PLANT CELLS AND TISSUES

There Are Three Types of Simple Tissues: Parenchyma, Collenchyma, and Sclerenchyma

■ *PLANTS, PEOPLE, AND THE ENVIRONMENT:* **Plant Anatomists as Detectives?**

Complex Tissues Make Up the Plant's Vascular System and Outer Covering

■ *ECONOMIC BOTANY:* **Music from Plants**

Secretory Tissues Produce and Secrete Materials

MERISTEMS: WHERE CELLS DIVIDE

What Is a Meristem?

What Are the Categories of Meristems, and How Do They Differ?

SUMMARY

Visit us on the web at http://biology.brookscole.com/plantbio2 for additional resources, such as flashcards, tutorial quizzes, InfoTrac exercises, further readings, and web links.

1. The plant body is composed of individual cells that are organized into aggregates of cells called tissues. The cells of each tissue function as a unit.

2. Simple tissues are composed of cells that are all of the same type. Complex tissues are composed of more than one cell type. Tissues may function as structural supports, protective coverings, or transporters of water and nutrients.

3. Meristems are the sites of cell division and differentiation in the plant body. A hierarchy of meristems exists in the plant body, each with a specific role in plant development.

4.1 ORGANIZATION OF THE PLANT BODY

The next time you are outside, notice the amazing variation in the forms that plants take. Despite the big differences among them, they all have evolved a basically common mechanism for development and a similar internal body plan. In this and the next few chapters, the internal structures of vascular plants are examined, including structures for ferns, cone-bearing plants (gymnosperms) such as pine trees, and flowering plants (angiosperms) like rose bushes and grasses.

The plant body of most vascular plants consists of an aboveground part, the **shoot system,** which includes stems, leaves, buds, flowers, and fruit, and a belowground part, the **root system,** composed of main roots and branches **(Fig. 4.1).** This plant body is constructed from millions of tiny cells, each having a characteristic shape and function. This chapter examines several different cell types, tissues (aggregates of cells), and their origins from unique parts of the plant body called *meristems.*

4.2 PLANT CELLS AND TISSUES

Around each plant cell is a cell wall. Living cells filled with water exert force (turgor pressure) against their walls, making each cell a rigid box. Plant cells are glued to each other by a material called *pectin,* and collectively they form a strong, yet flexible, plant body.

Plants consist of many different types of cells that are organized into aggregates called tissues. Tissues are derived from specialized groups of dividing cells called meristems. **Meristems** are the source of cells and tissues; therefore, they are not strictly speaking tissues themselves. The organs of the plant—leaves, stems, roots, and flower parts—are composed of tissues arranged in different patterns. Tissues in the plant body are made up of living and dead cells. The dead cells, often with thick, strong cell walls, are retained as strengthening cells. Knock on your wood table right now and you will see firsthand how strong and hard these dead cells are.

Table 4.1 Vascular Plant Tissues and Cell Types

Simple Tissues	Cell Types
Parenchyma tissue	Parenchyma cell
Collenchyma tissue	Collenchyma cell
Sclerenchyma tissue	Fiber
	Sclereid

Complex Tissues	Cell Types
Xylem	Vessel member
	Tracheid
	Fiber
	Parenchyma cell
Phloem	Sieve-tube member
	Sieve cell
	Companion cell
	Albuminous cell
	Fiber
	Sclereid
	Parenchyma cell
Epidermis	Guard cell
	Epidermal cell
	Subsidiary cell
	Trichome (hair)
Periderm	Phellem (cork) cell
	Phelloderm cell
Secretory structures	Trichome
	Laticifer

This table is limited to examples from this chapter. There are many other examples.

The main tissues of plants may be grouped into three systems (Fig. 4.1). The **ground tissue system** is the most extensive, at least in leaves (mesophyll) and young green stems (pith and cortex). The **vascular tissue system** contains two types of conducting tissues that distribute water and solutes (xylem) and sugars (phloem) through the plant body. The **dermal tissue system** covers and protects the plant surface (epidermis and periderm).

Some of the tissues are composed mostly of a single cell type; these are called simple tissues. Tissues made from aggregates of different cell types are called complex tissues. Tissues, simple or complex, act together as a unit to accomplish a collective function and are derived from meristems. Table 4.1 lists the plant tissues described in this chapter and their cell types.

There Are Three Types of Simple Tissues: Parenchyma, Collenchyma, and Sclerenchyma

The simple tissues are made of cells that are the workhorse cells of the plant body. They conduct photosynthesis, load things in and out of the vascular system, hold up the weight

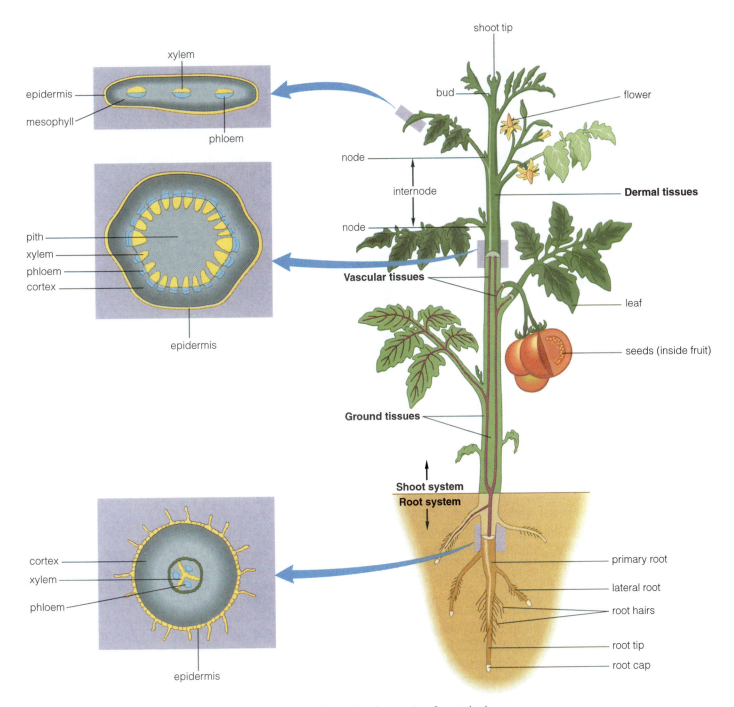

Figure 4.1 The plant body, shown here as a tomato plant, consists of the shoot system (leaves, buds, stems, flowers, and fruits) and the root system (roots). Each organ is made up of cells organized into tissue systems: dermal, vascular, and ground. One way the vegetative organs (leaves, stems, and roots) differ from each other is in the distribution of the tissues.

of the plant, store things, and generally perform the important business and housekeeping chores needed to keep the plant body healthy and functioning.

PARENCHYMA Parenchyma cells are usually somewhat spherical or elongated, but they may have diverse shapes **(Fig. 4.2).** They usually have a thin primary cell wall, but they may have a secondary wall, which is sometimes ligni-

fied (Fig. 4.2b). Lignin is a polymer that is embedded between the cell wall cellulose (see Chapter 3). It renders the wall impermeable to water, so that water movement occurs only through openings in the cell wall called **pits.** Lignin is also quite hard and strong, making lignified cells rigid and supportive even when the cells are dead.

Living parenchyma cells found in all plant organs perform the basic metabolic functions of cells: respiration, pho-

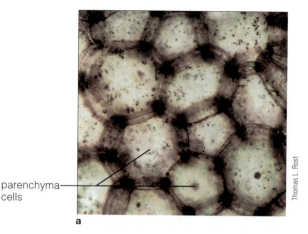

parenchyma cells

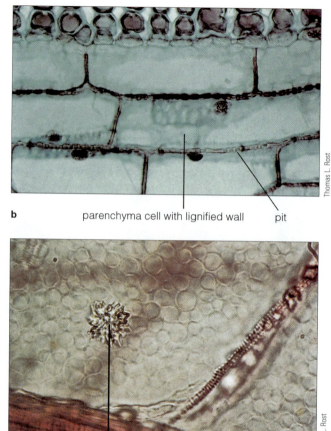

b   parenchyma cell with lignified wall   pit

Thomas L. Rost

Figure 4.2 Types of parenchyma cells.
(**a**) Pith parenchyma cells from impatiens (*Impatiens* sp.) stem. ×280. (**b**) Pine (*Pinus* sp.) leaf parenchyma cells with lignified wall. ×200.
(**c**) Leaf parenchyma cells with calcium oxalate crystal. ×400.

c   crystal

Thomas L. Rost

tosynthesis, storage, and secretion. Parenchyma cells usually live for 1 to 2 years but have been known to live hundreds of years in some plants, such as cactus.

Crystals of many different shapes and sizes, usually made of calcium oxalate, are commonly found inside the vacuoles of parenchyma cells (Fig. 4.2c). They may be involved in regulating the pH of cells by crystallizing oxalic acid (which is a way of taking it out of solution).

Parenchyma cells may occur as aggregates forming **parenchyma tissue;** the cortex and pith of stems, the cortex of roots, and the mesophyll of leaves are composed of parenchyma tissue **(Fig. 4.3a–c)**. The **cortex** is the region between the plant's epidermal and vascular tissues in most stems and roots. The **pith** is usually composed of storage parenchyma cells and lies at the center of many stems, inside the cylinder of vascular tissues. In some stems, such as corn **(Fig. 4.3d),** the vascular tissue is dispersed in bundles. In this case, the parenchyma tissue making up the bulk of the stem is simply called *ground tissue,* and the terms *cortex* and *pith* are not used. The **mesophyll** makes up the bulk of most leaves and is the site of most photosynthesis and water storage in leaves (Fig. 4.3c). Sometimes the air spaces are large, especially in the stems and leaves of plants that grow in wet places.

Parenchyma cells are unique in that mature ones can be developmentally reprogrammed to form into different cell types, especially after wounding. For instance, within several hours after a *Coleus* stem is wounded, the parenchyma cells immediately around the wound start to divide. After about 2 days, some of these cells differentiate into xylem cells, which can transport water around the wound **(Fig. 4.4).**

Transfer cells are modified parenchyma cells that have many cell wall ingrowths **(Fig. 4.5)**. This enables these cells to improve the transport of water and minerals over short distances between themselves and attached cells. Transfer cells are found at the ends of files of vascular cells, where they help load and unload sugars and other substances.

COLLENCHYMA Collenchyma is a tissue specialized to support young stems and leaf petioles (petioles are the part of the leaf that attach the leaf blade to the stem). Its cells often are the outermost cells of the cortex, being just inside the epidermis in young stems and the petioles of leaves. Collenchyma cells are elongated, often contain chloroplasts, and are living at maturity. The walls of collenchyma cells are composed of alternating layers of pectin and cellulose. In the most common type of collenchyma, the cell walls are thickened at the corners **(Fig. 4.6)**. These thickenings are quite flexible and will stretch without snapping back (they are said to be plastic). Collenchyma is, therefore, an ideal strengthen-

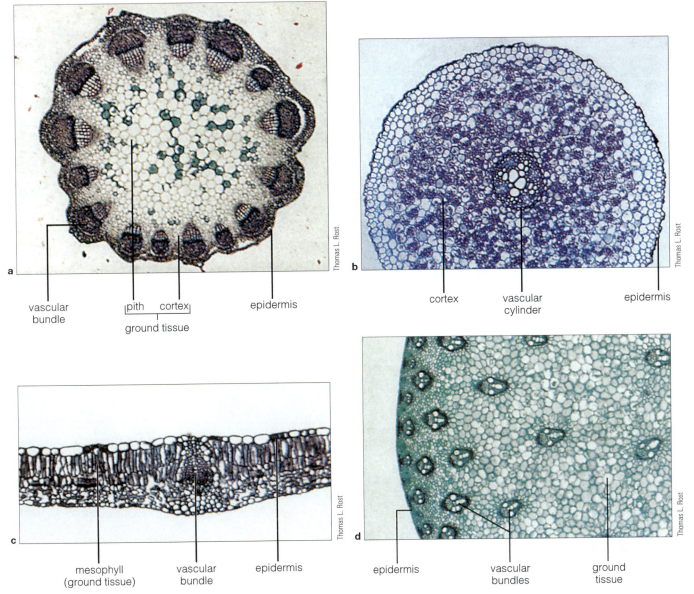

vascular
bundle

pith cortex

ground tissue

epidermis

a

cortex

vascular
cylinder

epidermis

b

mesophyll
(ground tissue)

vascular
bundle

epidermis

c

epidermis

vascular
bundles

ground
tissue

d

Figure 4.3 Cross sections of vegetative organs showing the distribution of parenchyma in the ground tissue (cortex and pith in the stem cortex in the root, mesophyll in the leaf). The epidermis, vascular bundle in the stem and leaf, and vascular cylinder in the root (xylem and phloem) also are labeled for each organ. (**a**) Clover (*Trifolium* sp.) is a typical stem with cortex and pith. ×35. (**b**) Buttercup root (*Ranunculus* sp.) has cortex. ×60. (**c**) Lilac leaf (*Syringa vulgaris*) has vascular bundles embedded in mesophyll. ×75. (**d**) Corn stem (*Zea mays*) shows vascular bundles scattered in the ground tissue. ×16.

Figure 4.4 Redifferentiation of parenchyma cells. This *Coleus blumei* stem shows a severed vascular bundle and the regeneration of xylem cells around the wound. ×17.

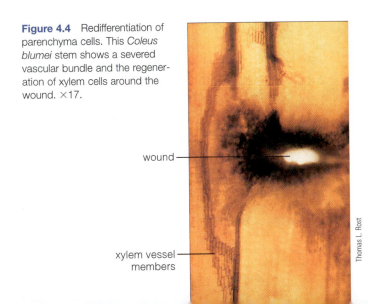

wound

xylem vessel
members

ing tissue in young stems because it also allows for tissue growth. Collenchyma cells occur as aggregates (collenchyma tissue), forming a cylinder surrounding a stem, or as strands, such as make up the ridges of a celery stalk.

SCLERENCHYMA Sclerenchyma tissue is composed of cells with rigid cell walls and they function to support the weight of a plant organ. There are two types of sclerenchyma cells: **fibers** and **sclereids.** These cells tend to have thick, lignified secondary cell walls. They are dead at maturity.

Fibers can occur in aggregates forming a continuous cylinder around stems, they may connect end to end to form multicellular strands acting like strengthening cables exactly

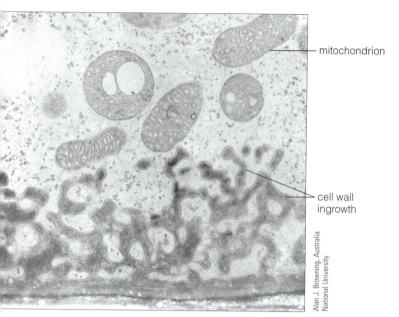

Figure 4.5 Transfer cell from *Funaria* sporophyte (see Chapter 22). Note the many ingrowths of the cell wall. ×2900. (Used with permission from Gunning, B.E.S., Transfer cells and their roles in transport of solutes in plants. *Science Progress Oxford* 977(64):539–568. Photo taken by Dr. Alan J. Browning, Australia National University.)

mitochondrion

cell wall ingrowth

Alan J. Browning, Australia National University

Figure 4.6 Section of marigold stem (*Calendula officinalis*) showing pink-stained collenchyma cells with thickened corners. ×170.

collenchyma cell

Thomas L. Rost

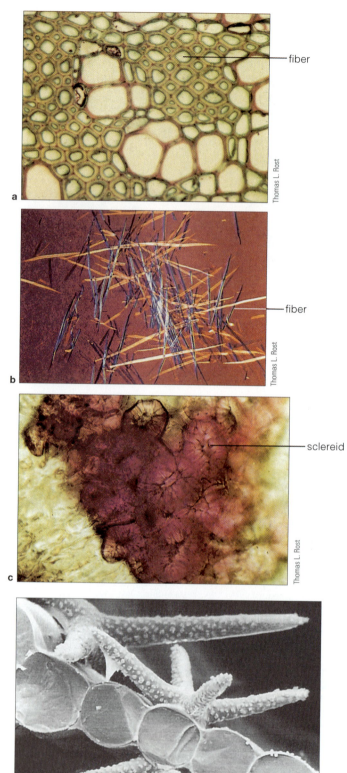

fiber

fiber

sclereid

Thomas L. Rost

Lesley Sunell

Figure 4.7 Sclerenchyma. (a) Cross section of geranium (*Pelargonium* sp.) stem showing clusters of green-stained fibers. ×200. (b) Photograph taken with polarized light, showing a maceration of cells from the wood from tulip tree (*Liriodendron tulipifera*). The long thin cells are fibers. ×100. (c) Stone cells are a type of sclereid found in pear (*Pyrus* sp.) fruit. ×200. (d) Star-shaped sclereid in water lily (*Nymphia* sp.) leaf. ×380.

like re-bar in concrete, or they sometimes appear as individual cells or small groups of cells as a component of vascular tissues. They are long, narrow cells with thick, pitted cell walls and tapered ends **(Fig. 4.7a,b).** Fibers are sometimes elastic and can be stretched to a degree, but they will snap back to their original lengths.

Sclereids sometimes occur as sheets (an example being the hard outer layer of some seed coats), but they usually occur in small clusters or as solitary cells. Sclereids have many striking shapes, from elaborately branched cells, to star-shaped cells, to the simple stone cells that give a gritty texture to pear fruits **(Fig. 4.7c,d).** Sclereid cell walls often are thicker than the walls of fibers.

PLANTS, PEOPLE, AND THE ENVIRONMENT:

Plant Anatomists as Detectives?

FBI agents, police, lawyers, medical and veterinary doctors, agricultural extension specialists, and members of the public have all had occasion to contact me over the past several years concerning the identification of plant material. Usually this material was chewed, digested, burned, dried, or otherwise distorted, so that the identity of the plant or plant organ was no longer apparent. Yet, every plant species contains cell types with special structures, shapes, sizes, and staining reactions—characteristics that make it possible to identify the plant. Regardless of the kind of case, similar microtechnique tools and methods are used, and the application of basic knowledge of plant anatomy is required.

Plant anatomists have been involved in many legal cases, but mostly we are asked for opinions dealing with medical or agricultural problems. For example, agricultural extension specialists need help in identifying root specimens pulled from clogged sewers. Veterinarians who suspect that sick animals have eaten poisonous plants will call plant anatomists. Anatomists have also worked with anthropologists interested in ancient basketry or the identification of wood and charcoal pieces found in ancient fire pits.

Probably the most famous legal case involving a plant anatomist was the Lindbergh kidnapping. In 1932, the infant son of celebrated aviator Charles Lindbergh and his wife, Anne Morrow Lindbergh, a noted writer and poet, was kidnapped out of a second-story nursery. The kidnapper sent a ransom note, but the baby was later found dead. The only evidence left at the crime scene was a crude wooden ladder leaning against the second-story window. The police and FBI gave the ladder to the Forest Products Laboratory in Wisconsin for analysis. A plant anatomist there, Arthur Koehler, was able to specifically identify the wood used. After a man named Bruno Richard Hauptmann was arrested for the crime, police found that several boards were missing from the floor of the attic in his house. Pieces of this wood were sent to scientists at the Forest Products United States Department of Agriculture (USDA) Laboratory in Wisconsin, and they positively identified the wood as a match to the ladder. This testimony was among those used to convict Hauptmann, who was later executed.

Most cases are less poignant. In 1984, I was contacted by the Sheriff's office in Calaveras County, California (the part of California's gold country made famous by Mark Twain's story of the jumping frog contest). In this instance, a man was suspected of growing marijuana (*Cannabis sativa*) in an elaborate hydroponic setup in his attic. The suspect was warned of an impending raid just before the sheriff arrived. He hid the stems of the plants outside his house and tried to burn the stem stumps and roots in his fireplace. The sheriff arrested him and took the partially burned material as evidence. The suspected grower reportedly told a sheriff's deputy that he would get off because it was not possible to prove that he was burning marijuana in his fireplace. I examined the material taken as evidence and compared it with known specimens of marijuana from herbarium sheets and from identified marijuana stems. The charred evidence and the known specimens showed similar wood anatomy. I testified in a pretrial hearing that the partially burned material was most likely marijuana. The defendant accepted a plea bargain, and the case was never brought to trial.

Several of my cases have come from the San Diego Zoo. One of them involved a group of Hanuman langur monkeys, a rare Asian species. For 2 years the monkeys had been fed mostly *Acacia* leaves. Suddenly, three monkeys died after bouts of weight loss, diarrhea, and vomiting. An autopsy revealed intestinal lesions and plugging with masses of fibrous plant material. I examined some of this material, together with samples of *Acacia* leaf browse and other plant materials within reach of the monkey enclosure. Microscopic examination of this material revealed partially digested vascular strands with attached thick-walled cells and small epidermal fragments. By making polarized light images and comparing them with known specimens, I was able to identify the material as *Acacia* leaf vascular bundles. The keepers changed the monkeys' feed.

The point of these stories is that the study of plant anatomy has uses far beyond just knowing what is inside the plant. Basic information and a few simple tools—such as a razor blade, some common dyes, and a microscope—can go a long way toward solving criminal and medical puzzles. So study hard; you never know when your knowledge of plant anatomy might come in handy.

Thomas L. Rost, Section of Plant Biology, University of California, Davis, CA 95616

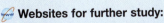 **Websites for further study:**

Charleslindbergh.com—Kidnapping:
http://www.charleslindbergh.com/kidnap/index.asp

Forensic Botany:
http://www.dal.ca/~jvandomm/zvandommelen/xgrad/vandommelenst.html

Complex Tissues Make Up the Plant's Vascular System and Outer Covering

THE VASCULAR SYSTEM: XYLEM   The vascular system consists of an interconnected network of cells that traverse the entire body of the plant (Fig. 4.1). All cells of the plant require minerals and water, which are absorbed by the roots and transported by the xylem. Sugars are manufactured in the leaves and transported by the phloem.

The **xylem** is a complex tissue made up of different kinds of cells that work together to transport water and dissolved minerals. The cell types found in xylem are: the water-conducting cells—**tracheids** and **vessel members** (the latter join together end to end to make **vessels**); fibers, for strength and support; and parenchyma cells, which help load minerals in and out of the vessel members and tracheids.

In leaves and young stems, the xylem is found in discrete bundles called **vascular bundles** (Figs. 4.3a and 4.8a), and in young roots, the xylem occurs in groups of cells at or near the center of the root known as the **vascular cylinder** (Fig. 4.3b). Xylem in those locations is called **primary xylem.** It is formed in the root and shoot apex very early in organ development (see discussion later in this chapter). Xylem that forms later in the development of stems and roots is organized in cylinder patterns and is called **secondary xylem** (Fig. 4.8b). Usually, leaves have only primary xylem.

Parenchyma cells are the only living cells found in xylem. They usually have a thin primary cell wall, but in secondary xylem they often have a lignified secondary wall. Fibers function to support xylem tissue and hold it rigid, rather like a steel rod would be used to hold up a plastic water pipe.

Vessel members and tracheids have many structural and functional characteristics in common; therefore, the term **tracheary element** is used to refer to them generally. Tracheary elements are not living at maturity. Before the cells die, their protoplast degenerates, and the cell wall becomes thickened with cellulose and lignin.

The secondary cell wall, deposited after the cell stops enlarging, forms in one of several patterns: annular (ring shaped), spiral, scalariform (ladderlike), reticulate (forming a network), or pitted (Fig. 4.9). These patterns relate to the timing of their formation. Tracheary elements that form and function in primary xylem of organs that are still elongating have annular or spiral secondary cell walls. Such cells are able to stretch as the surrounding cells elongate, and they can still function in water transport. Vessel members that form after organ elongation has ended (in late-forming primary or secondary xylem) tend to have pitted cell walls. These cells are quite rigid and very strong, but they are unable to stretch.

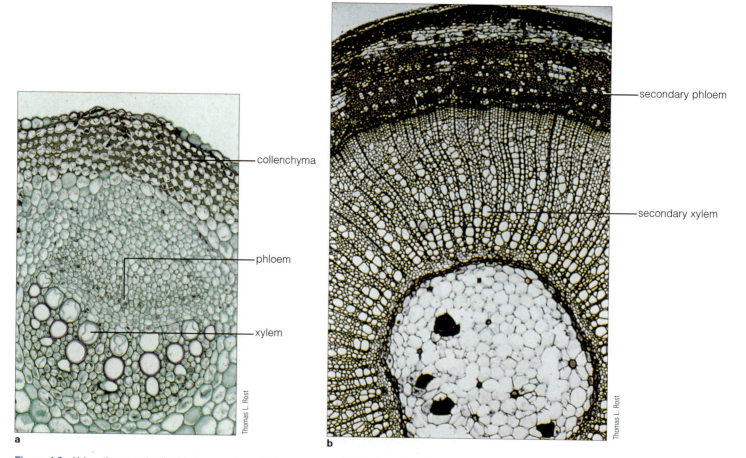

Figure 4.8   Xylem tissue and cells. (**a**) Cross section of primary vascular bundle of a sunflower (*Helianthus annuus*). ×72. Primary xylem occurs in young stems and forms from the primary meristem procambium. (**b**) Cross section of a 1-year-old basswood stem (*Tilia americana*) showing a ring of secondary xylem. ×69.

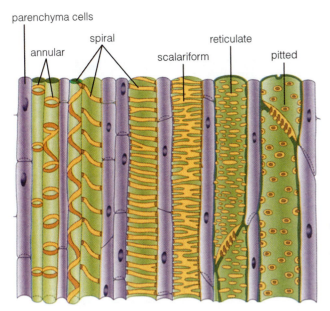

parenchyma cells
spiral
annular
reticulate
scalariform
pitted

Because lignified secondary cell walls are impermeable to water, the only way that water can be exchanged between cells is through tiny openings called pits. A pit is not actually a hole in the cell wall; the primary cell wall remains intact, rather like a loose membrane across the pit opening. There are two types of pits: **simple pits** and **bordered pits.** Simple pits (Fig. 4.10a) are openings in the secondary walls of fibers and lignified parenchyma cells. Bordered pits occur in tracheids, vessel members, and some fibers (Fig. 4.10b–d). These pits have an expanded border of secondary wall that extends over a small pit chamber.

Figure 4.9 The different patterns of secondary cell walls in tracheary elements. Vessel members that form in the early primary xylem have secondary wall thickening as rings or spirals. As the organ grows, these cells can stretch. Note the two cell files labeled *annular*. The one on the left has stretched and the one on the right has not. The xylem vessel members and tracheids that form after growth is finished tend to have pitted secondary cell walls. Intermediate forms of secondary cell walls, scalariform and reticulate, also can be observed.

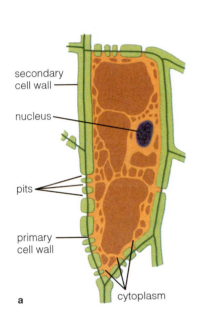

secondary cell wall
nucleus
pits
primary cell wall
cytoplasm

a

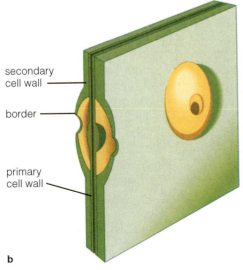

secondary cell wall
border
primary cell wall

b

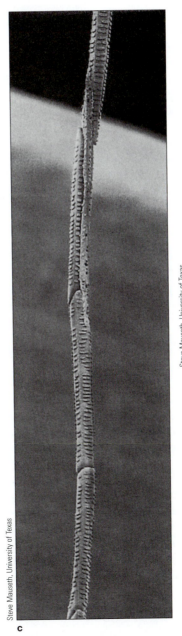

Steve Mauseth, University of Texas

c

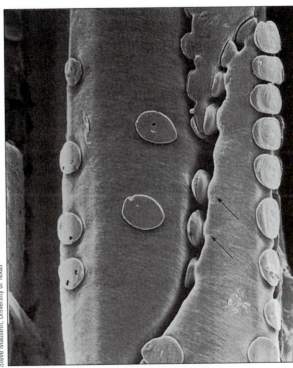

Steve Mauseth, University of Texas

d

Figure 4.10 Pits and their locations. (**a**) Drawing of a parenchyma cell with a thick secondary cell wall, showing the location of simple pits. Note that the primary cell wall (*black line*) remains across the pit opening. (**b**) Drawing of a bordered pit with a raised border. This type of pit is found in the walls of most tracheary elements. (**c,d**) These are plastic casts made by injecting resin into xylem tissue. After the resin hardened, the walls were digested away. The hardened resin indicates the pathway of water and shows how the bordered pits apparently fill with water. ×232 (**c**); ×1400 (**d**).

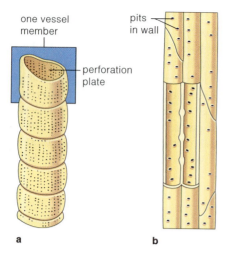

one vessel member

pits in wall

perforation plate

a   b

Figure 4.11 Tracheary elements compared. (**a**) Vessel members join end to end, but they digest out the end walls (simple perforation plate) forming a tube called a vessel. (**b**) Tracheids join end to end and along their sides and are connected by bordered pits.

Tracheids and vessel members have some structural and functional characteristics in common, but there also are differences (**Fig. 4.11**). A **vessel** member is a cell with an oblique, pointed, or transverse end. The ends of mature vessel members are partially or completely digested away during their development to form a perforation plate. There are several different types of perforation plates; simple and scalariform are the most common types (Figs. 4.11a, **4.12**).

A vessel is a series of vessel members connected end to end (Fig. 4.11a). Vessels often are several centimeters in length, and in some vines and trees, they may be many meters in length. Because a vessel has a specific length, the terminal cell of a vessel will have a closed end wall containing bordered pits. Vessels connect to other vessels end to end and laterally through pits.

A tracheid is an elongated cell with more or less pointed ends (Fig. 4.11b). Tracheids are joined at overlapping ends through bordered pits, and they do not have perforation plates. Both tracheids and vessel members may be present in a single flowering plant. Most cone-bearing plants (gymnosperms), however, have only tracheids. Chapter 11 discusses the mechanism of water transport through the xylem and the advantages and disadvantages of tracheids and vessel members.

THE VASCULAR SYSTEM: PHLOEM   Phloem is the tissue that transports sugar through the plant. **Primary phloem** occurs in vascular bundles near the primary xylem in young stems and leaves and in the vascular cylinder in roots. **Secondary phloem** occurs outside the secondary xylem in older stems and roots, usually in plants that live longer than 1 year. In flowering plants (angiosperms), the phloem is made up of several different types of cells (**Fig. 4.13**): sieve-tube members, companion cells, parenchyma, and sometimes fibers and/or sclereids. **Sieve-tube members** are cells that join at their ends to form long sieve tubes. **Sieve tubes** are the con-

perforation plate

Sonia Cook, University of California, Davis

Figure 4.12   Scalariform perforation plate is nicely shown in this scanning electron microscope view of a group of xylem cells from tulip tree wood (*Liriodendron tuliperfera*). ×300.

ducting elements of the phloem (**Fig. 4.14a,b**), which transport sugars produced by photosynthesis in the leaves to other plant parts.

A young sieve-tube member contains a nucleus, plastids, mitochondria, and Golgi stacks. As the cell matures, its nucleus disintegrates, the plastids and mitochondria shrink, and the cytoplasm becomes reduced to a thin peripheral layer. The central part of the cell becomes occupied by a mass of dense material. This mass, which was called slime in the early literature, can be seen with the light microscope. It is currently called **P-protein** (**Fig. 4.15a**) because it is actually composed of a complex of proteins that may be involved in moving materials through the sieve tubes.

When the sieve-tube member is mature, one or more **companion cells** lie connected to it by plasmodesmata (Fig. 4.13). Because companion cells have a nucleus and a full complement of organelles, it is thought that they regulate the metabolism of their adjacent sieve-tube member. Companion cells

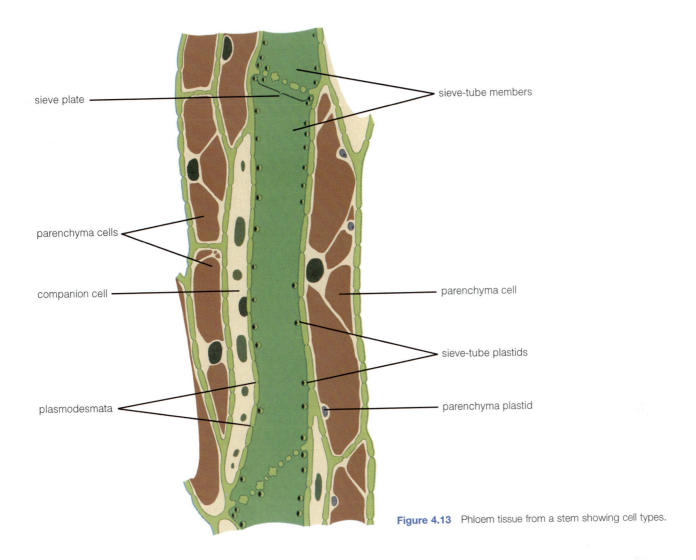

sieve plate

sieve-tube members

parenchyma cells

parenchyma cell

companion cell

sieve-tube plastids

plasmodesmata

parenchyma plastid

Figure 4.13 Phloem tissue from a stem showing cell types.

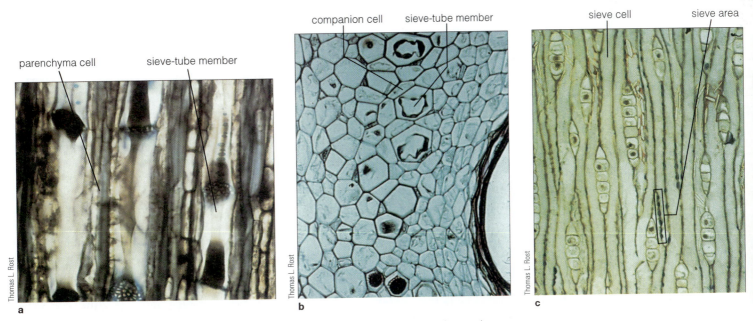

parenchyma cell

sieve-tube member

companion cell

sieve-tube member

sieve cell

sieve area

Thomas L. Rost

a

Thomas L. Rost

b

Thomas L. Rost

c

Figure 4.14 Phloem. (**a**) Elm (*Ulmus* sp.) tangential longitudinal section (TLS) showing sieve-tube members and parenchyma cells in secondary phloem. ×375. (**b**) Primary phloem in cucumber (*Cucurbita pepo*) stem shown in cross section. ×130. (**c**) Pine (*Pinus* sp.) secondary phloem in TLS showing sieve cells. Note that the callose in the sieve areas is stained blue in this photograph. ×228.

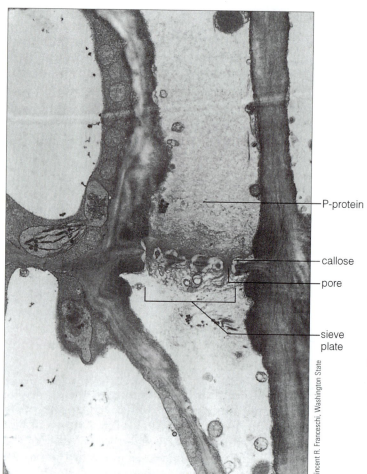

P-protein

callose

pore

sieve plate

Vincent R. Franceschi, Washington State

a

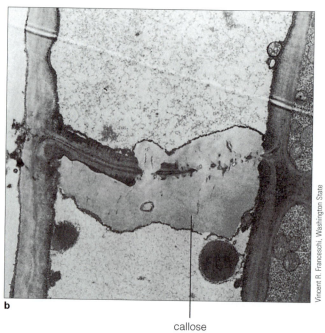

b

callose

Vincent R. Franceschi, Washington State

Figure 4.15 P-protein and callose. (**a**) Beet (*Beta vulgaris*) petiole sieve-tube members with P-protein and callose surrounding the pores at the sieve plate. ×10,600. (**b**) Beet petiole 1 hour after being stressed by exposure to cold temperature. Note the massive amount of callose that now plugs the pores. ×10,200.

also are known to play an important role in the mechanism of loading and unloading the phloem.

The walls of mature sieve-tube members contain aggregates of small pores called **sieve areas.** One or more sieve areas on the end wall of a sieve-tube member is called a **sieve plate** (Figs. 4.13 and 4.15a). The end walls of two connecting sieve-tube members are thickened, and strands of cytoplasm and P-protein pass through pores adjoining them. Sieve-tube members live and function from 1 to 3 years, except in a few trees such as palms, which live much longer.

In many studies of the structure of mature sieve-tube members, a carbohydrate known as **callose** was seen to surround the margins of the pores in the sieve areas (Fig. 4.15a). In some instances, protein also is collected at the **sieve plate.** It has been demonstrated in other instances that callose can form rapidly in response to aging, wounding, and other stresses (Fig. 4.15b), and the result is to limit the loss of cell sap from injured cells.

In gymnosperms and ferns, **sieve cells** rather than sieve-tube members are the conducting elements in the phloem. These cells are quite long, with tapered ends (Fig. 4.14c). They have sieve areas but no sieve plates at their ends. Sieve cells apparently function similarly to sieve tubes and usually lack nuclei at maturity. Adjacent **albuminous cells** are short, living cells that act as companion cells to these sieve cells.

Phloem fibers are usually long, tapered cells with ligni-fied cell walls. Mature fibers from the ramie plant (*Boehmeria nivea*) have been reported to be up to 55 cm in length. Such fibers are valued for their commercial uses, especially for rope and fabric. Phloem parenchyma cells are usually living cells that function in phloem loading and unloading.

THE OUTER COVERING OF THE PLANT: EPIDERMIS The epidermis is the outer covering of the plant. It is a complex tissue composed of epidermal cells, guard cells, and tri-chomes (hairs) of various types. The epidermis is usually one layer of cells, but it may be as many as five or six layers in the leaves of some succulent plants and in the aerial roots of certain orchids. The epidermis protects the inner tissues from drying and from infection by some pathogens. It also regulates the movement of water and gases out of and into the plant.

Epidermal cells are the main cell type making up the epidermis. These cells are living, lack chloroplasts, are usu-ally somewhat elongated, and often have walls with irregu-lar contours (Fig. 4.16). The outer walls of epidermal cells often are thicker than the inner and side walls. The outer epidermal wall is coated with a waxy substance (cutin) forming an impermeable layer called the cuticle (Fig. 4.17). All parts of the plant, except the tip of the shoot apex and the root cap, have a cuticle. In roots, the cuticle often is very thin

Richard H. Falk, Virginia Commonwealth University

a

Richard H. Falk, Virginia Commonwealth University

b

— hair

— guard cell

— epidermal cell

Figure 4.16 Epidermis and epidermal cells. (**a**) Upper leaf surface of Shepard's purse (*Capsella bursa-pastoris*). ×70. (**b**) Higher magnification showing epidermal cells, star-shaped hairs, and stomata. ×350.

Figure 4.17 Photograph of pincushion tree (*Hakea* sp.) leaf shown in cross section. Note the thick cuticle and the small channels that cross the cuticle. ×184.

cuticle

Thomas L. Rost

(and has been reported to be totally lacking on the surface of root hairs).

Young stems, leaves, flower parts, and even some roots in exceptional instances have specialized epidermal cells called **guard cells.** Between each pair of guard cells is a small opening, or pore, through which gases enter and leave the underlying tissues. Two guard cells and the pore constitute one **stoma** (plural, **stomata; Fig. 4.18**). Guard cells differ from other epidermal cells by their crescent shape and because they contain chloroplasts. Another type of epidermal cell, the **subsidiary cell,** forms in close association with guard cells and functions in stomatal opening and closing. The physiologic role of stomata and the mechanism of their opening and closing are discussed in Chapter 11.

guard cell

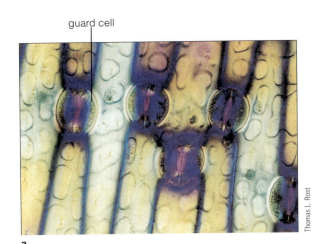

Thomas L. Rost

a

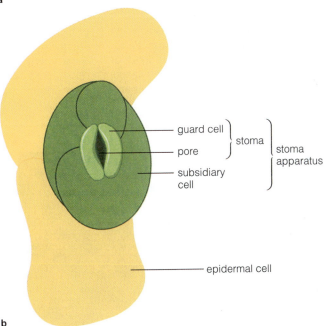

guard cell ⎱ stoma
pore ⎰
⎱ stoma apparatus
subsidiary cell

epidermal cell

b

Figure 4.18 Stomata. (**a**) Surface view of *Iris* sp. leaf as seen through polarized light showing several stomata composed of two guard cells surrounding a pore. ×415. (**b**) Diagram of stomatal apparatus consisting of two guard cells and surrounding subsidiary cells such as would be found on leaves in a plant such as *Sedum*.

ECONOMIC BOTANY:
Music from Plants

The direct relation between plants and music is obvious if one thinks about it even a little. The earliest musical instruments were probably drums made of grooved or hollowed logs, and to this day drums are usually made of wood cylinders, typically with skins or synthetic materials stretched over them. The bodies of ancient stringed instruments, such as lyres and harps, were made of wood. Tambourines, pan pipes, whistles, recorders, early flutes, shawms, accordions, organs, harpsichords, and pianos were, and still are, made largely of wood. Clarinets, English horns, oboes, bassoons, violins, violas, cellos, contrabasses ("stand-up bass"), xylophones, and guitars (electric and acoustic) are made of various different kinds of wood carefully chosen for bodies, bells, mouthpieces, backs, sides, fronts, bridges, and fingerboards carefully chosen for their acoustic and other physical properties.

But wood is not the only material that contributes to music making. Many of you have made a sort of whistle using your hands and a blade of grass. Paper, most of which is made of various kinds of plant fibers, can be used to make a kazoo. The oldest types of traditional flutes and whistles were made of various types of cane, a cousin of bamboo. For 2,000 to 3,000 years the preferred type of cane has traditionally been *Arundo donax* (giant cane). The hollow stems from this species have always been the best for making reeds for woodwind instruments, such as clarinets **(Fig. 1)**, saxophones, oboes, and bassoons, and for other not-so-obvious examples, such as the Chinese sheng (perhaps the oldest reed instrument), Turkish zurna, Egyptian mizmar, Vietnamese ding tac ta (played by inhaling), Scot and Irish bagpipes, hornpipes, krummhorns, concertinas, bandoneons, small organs, some harmonicas, melodicas, and many others.

Anyone who has played the clarinet or other reed instrument for any length of time knows that getting a reed that produces just the right sound quality from the instrument is a challenge. Players who cut their own reeds can tell you that the raw material is highly variable in its playing properties. Commercially, clarinet reeds are made from quartered pieces of *A. donax* internodes that are milled into the classic reed shape **(Fig. 2)**. In the past, some commercially made reeds often were

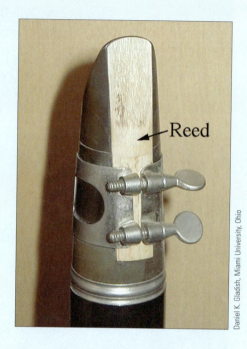

Figure 1 A clarinet "reed" is a piece of *Arundo donax* stem that has been carved or machine milled to an appropriate shape and clamped to the mouthpiece of the instrument. Air from the lungs of the player causes the reed to vibrate, which creates a tone. The sound qualities and pitch of the tone can be modulated by the muscles of the player's mouth and the appropriate placement of fingers over the holes and keys on the clarinet body.

Figure 4.19 Glandular hair in tobacco (*Nicotiana tabacum*). ×22. The thickened bulblike end of this trichome secretes sticky material.

Trichomes are epidermal outgrowths and may be a single cell or multicellular **(Fig. 4.19)**. In roots, for example, root hairs are extensions of single epidermal cells that increase the root surface area in contact with soil water. In some leaves, elaborate multicellular trichomes may form (Fig. 4.16b). Trichomes are discussed again in the following chapters.

THE OUTER COVERING OF THE PLANT: PERIDERM The periderm is a protective layer that forms in older stems and roots after those organs expand and the epidermis splits and is lost. It is a secondary tissue. This tissue is several cell layers deep and is composed of **phellem** (cork) cells on the outside, a layer of dividing cells (**phellogen** or **cork cambium**), and the phelloderm toward the inside **(Fig. 4.20)**. Phellem cells are dead at maturity and have a waxy substance (suberin) embedded in their cell walls.

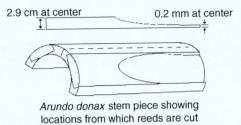

2.9 cm at center   0.2 mm at center

Arundo donax stem piece showing locations from which reeds are cut

Figure 2   Traditionally, internodes of *Arundo donax* were quartered and then carved into the characteristic clarinet reed shape. This process is now done commercially by milling machines.

discarded because they did not sound quite right when used with a particular instrument. When I started the project that is described below, a considerable traditional lore had developed among players and people who manufactured reeds as to how to select and process them.

In 1992, as part of an effort to improve their products, Rico International, an American reed-making company, asked me to research whether there were aspects of the anatomy of *A. donax* plants that predisposed them to be especially good or bad for clarinet reeds. We asked several skilled symphonic performers to play a number of clarinet reeds and to identify the best and the worst among them. Each reed was randomly given a numeric code identifier, and the reeds were sent to me for analysis. I had no idea which reeds played well.

Thin sections (slices) of each reed were carefully cut. The sections were treated with stains to increase the contrast of the cells and tissues in the reed, and photographs were then taken of each section using a precision photomicroscope. Twenty-three anatomic characteristics (for example, the ratio of vascular to ground tissue, the size of vascular bundles, the shape of vascular bundles, the proportion of vascular bundles that was bundle sheath tissue, the length and diameter of parenchyma cells, the thickness of parenchyma cell walls, and so on) were collected for each reed. The combined data were analyzed statistically.

I found that the coded reeds "clustered" into two groups when several characteristics of the vascular system were considered in combination. The statistical conclusion for any particular clarinet reed matched the professional players' assessments 92% of the time for good reeds and 85% of the time for bad reeds. My analysis showed that reeds made better music when their vascular bundles were more or less lined up, uniformly distributed through the tissue, and had thick, fibrous bundle sheaths that completely surrounded the vascular bundles (**Fig. 3**). Because my study was done for a private company, it was never published, but another group of scientists in Australia (which included a clarinet player) did a similar study. Using similar methods, they came to the same conclusions that I did (Kolesik et al. 1998. *Annals of Botany* 81:151–155). It was satisfying to know that other scientists had confirmed

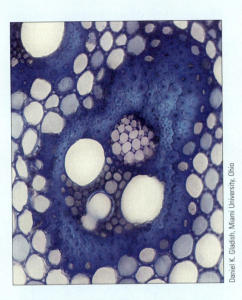

Daniel K. Gladish, Miami University, Ohio

Figure 3   The best sounding reeds have vascular bundles with thick bundle sheaths of fiber cells that completely surround each vascular bundle. The more of the vascular bundles that look like this the better.

my results. Clarinet reed manufacturers can now use these data to help them select the best canes for new reeds. It is clear to me that science can be used to help artists get better results.

Dr. Daniel K. Gladish, Botany Department, Miami University, Hamilton, OH

Websites for further study:

Miami University Department of Botany—Development of Root Systems: http://www.cas.muohio.edu/botany/bot/dkg.html

Alaska Science Forum—Secrets of the Woodwind Reed: http://www.gi.alaska.edu/ScienceForum/ASF10/1066.html

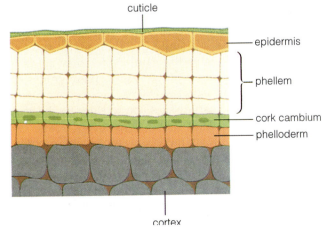

Figure 4.20   Diagram of periderm showing the layers: cork cambium, phelloderm, and phellem.

cuticle

epidermis

phellem

cork cambium

phelloderm

cortex

Phelloderm cells live longer than phellem cells and are parenchymalike.

Secretory Tissues Produce and Secrete Materials

Secretory structures occur primarily in leaves and stems. These may be composed of single secretory cells or complex multicellular structures (Fig. 4.19). Some trichomes, for example, may secrete materials out of the plant to attract insect pollinators. They may also form inside the plant body and secrete materials within the plant; cells called laticifers **(Fig. 4.21)**, for example, secrete latex, which discourages herbivores from eating the plant. They also form complex ducts inside wood of trees. These interesting structures are discussed again with leaves and stems in Chapters 5 and 6.

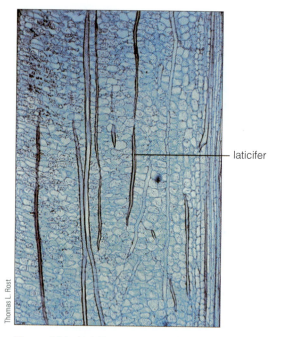

laticifer

Figure 4.21 Laticifer cells in spurge (*Euphorbia* sp.) stems, shown here in longitudinal section. ×20.

Table 4.2 contains a summary of the structure and function of the common cell types found in seed plants.

4.3   MERISTEMS: WHERE CELLS DIVIDE

We now know the tissues and cell types making up the vascular plant, but how do these tissues and cells come to be? Do plant cells specialize and group into tissues in the same manner that animal cells do? The answer is, largely, no. The growth patterns of animals and plants differ in one very significant way. Animal cells divide mostly during the embryo stage of development. After the animal reaches adult size, cells divide only in the bone marrow (to produce new blood cells), in the intestinal lining (to repair wounded tissues), and in the epithelial layers (to form new skin, nails, and hair). In vascular plants, an entirely different process exists: Cell division continues through the whole lifetime of the plant, but it occurs in special regions of the plant body called meristems (derived from the Greek word meaning "to divide").

What Is a Meristem?

A **meristem** is a site in the plant body where new cells form and the complex processes of growth and differentiation are initiated. **Growth** means the irreversible increase in size that comes from cell division and cell enlargement. **Cell differentiation** refers to the changes that a cell undergoes structurally and biochemically so that it can perform a specialized function. Because cells and tissues are derived

from meristems, we do not consider meristems themselves to be tissues, although they are sometimes referred to as meristematic cells and tissues.

There are different categories of meristems, each with a specific function. Shoot and root apical meristems are at the tips of branches and roots **(Fig. 4.22)**; they are the ultimate sources of all cells in the plant. **Primary meristems,** the next level of meristems, originate in apical meristems and produce, or more correctly, differentiate into the primary tissues. The **secondary meristems** produce the secondary tissues. These categories of meristems allow vascular plants to grow very large and to a great age. The next section examines some details of the location and function of each meristem.

What Are the Categories of Meristems, and How Do They Differ?

ROOT AND SHOOT APICAL MERISTEMS   The vascular plant body is polar, meaning that it has a shoot end and a root end. At the tip of each branch is a **shoot apical meristem** (SAM), and at the tip of each root is a **root apical meristem** (RAM; Figs. 4.22 and **4.23**). These two apical meristems are the sites of the formation of new cells by cell division. Theoretically, apical meristems could operate forever. This does not occur, however, because some factor will always limit the size of a plant. Whether it is because of a scarcity of nutrients, structural limitations, or heredity, eventually a plant ceases to grow. A branch, for example, can carry only a certain weight before it breaks. Also, each plant and plant organ (leaves, stems, and roots) has a system for genetic regulation of growth; every species seems to have an optimum size.

PRIMARY MERISTEMS   A shoot tip and a root tip of a representative plant are shown in Figures 4.22 and 4.23. If you made a very thin longitudinal section through them (this would be done by embedding a shoot or root tip in plastic resin or paraffin and cutting a thin section), you would see that the cells of the SAM and RAM and those just basal to them (toward the more mature cells) are small, with relatively dense protoplasts (Fig. 4.23). Cells with these characteristics usually are capable of dividing, and thus are referred to as **meristematic cells.**

Cells immediately basal to the SAM are ordered into distinct files of cells (Figs. 4.22 and 4.23). These newly ordered cells are still meristematic (they can divide); they are in a sense the embryonic stages of the tissues. These groups of cells are called the **primary meristems,** and they have two roles: to form the primary tissues and to elongate the root and shoot.

There are three primary meristems: protoderm, procambium, and ground meristem (Fig. 4.22). The cells of the **protoderm** differentiate into the epidermis. The **procambium** cells differentiate into the cells of the primary xylem and primary phloem. The **ground meristem** differentiates into the cells of the pith and cortex of stems and roots and the mesophyll of leaves.

Table 4.2 Summary of Plant Cell Types

Cell Type		Characteristics	Location	Function
Parenchyma		Living at maturity; usually more or less spherical or elongated in shape; usually has primary cell walls only	Cortex and pith of roots and stems; mesophyll of leaves; xylem and phloem	Site of basic metabolic cell functions
Transfer cell		Living at maturity; modified parenchyma cell ingrowths	Xylem, phloem, secretory structures	Involved in short distance transport containing cell wall
Collenchyma		Living at maturity; elongate with plastic cell walls	Leaf petioles, young stems	Support
Sclereid		Dead at maturity; elongated, star-shaped, or stone-shaped with thick, lignified secondary cell walls	Cortex, pith, mesophyll	Strength
Fiber		Dead at maturity; long, thin cell with thick, lignified secondary cell walls	Xylem, phloem, cortex	Most important strengthening and supportive cell type
Vessel member		Dead at maturity; lignified secondary cell walls that may be spiral, ringed, or pitted; perforation plates in open end walls; members connect together to form vessels	Xylem	Transport of water and dissolved minerals
Tracheid		Dead at maturity; similar to vessel member, except with narrower diameter and closed pitted ends; do not form vessels	Xylem	Transport of water
Sieve-tube member		Living at maturity, but does not retain cell nucleus; elongate cell with aggregates of pores (sieve plates) on end walls and side walls; cells join end to end to form sieve tubes	Phloem	Transport of sugars
Sieve cell		Living at maturity; similar to sieve-tube members, does not develop sieve plates or join to form sieve tubes	Phloem in gymnosperms and ferns	Transport of sugars
Companion cell		Living at maturity	Connected to sieve tube	Metabolic regulation of sieve-tube members
Albuminous cell		Living at maturity; counterpart to companion cells, found in gymnosperms and ferns	Phloem in conifers and ferns	Metabolic regulation of sieve cells
Guard cells		Living at maturity; occur in pairs separated by a pore to form a stoma	Leaves, young stems	Gas exchange
Epidermal cell		Living at maturity; contains no chlorophyll; outer wall impregnated with cutin	Epidermis	Inhibit evaporation of water
Subsidiary cell		Living at maturity; type of epidermal cell, in contact with guard cells	Epidermis	Regulate guard cell opening

Continued

Table 4.2 Summary of Plant Cell Types (continued)

Cell Type		Characteristics	Location	Function
Trichome		Living at maturity; elongated epidermal structure composed of one or more cells	Epidermis	Absorption, secretion, storage, and more
Phellem cell		Dead at maturity; also called cork cell; cell walls impregnated with suberin (a wax)	Outer layers of periderm	Replacement of epidermis in old stems and roots
Phelloderm cell		Most living at maturity	Inner layers of periderm	Replacement of epidermis in old stems and roots
Secretory cell		Living at maturity; several different types; may be solitary or part of a multicellular secretory structure	Most tissues and organs	Secretion of different secretory products

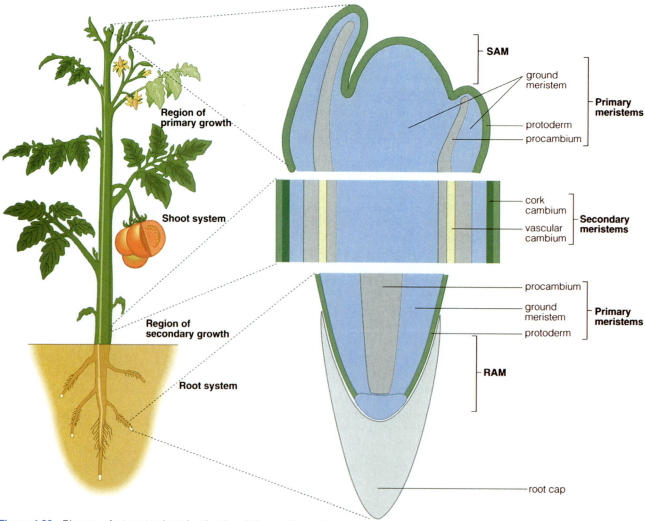

Figure 4.22   Diagram of a tomato plant showing the relative positions of the root apical meristem (RAM) and shoot apical meristem (SAM), the primary meristems (protoderm, ground meristem, and procambium), and the secondary meristems (vascular cambium and cork cambium) in both the shoot and root systems.

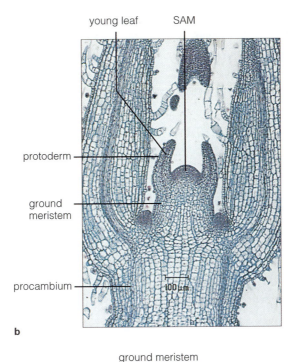

young leaf   SAM

protoderm

ground
meristem

procambium

100 μm

David R. Frazier/Photo Researchers, Inc.

a

b

Figure 4.23 Shoot and root tips. (**a**) Photograph of a living plant showing the position of the shoot apex. The shoot apical meristem (SAM) is surrounded by several small leaves. (**b**) Median longitudinal section through the SAM of a *Coleus blumei* plant. ×49. (**c**) Photograph of a living root tip. The root apical meristem (RAM) is located at the very tip of the root, just inside the root cap. (**d**) Median longitudinal section through SAM of an *Arabidopsis thaliana* root tip showing the root cap and RAM. ×426.

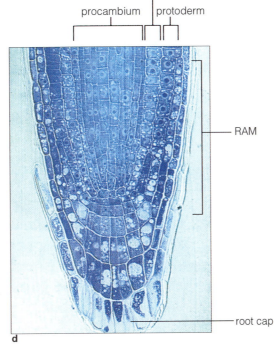

ground meristem

procambium   protoderm

RAM

root cap

G. I. Bernard/NHPA

c

d

and growth in a lateral direction, thereby increasing the thickness and circumference of stems and roots. The wood in trees, for example, is really secondary growth resulting from the activity of secondary meristems. Not all plants have secondary meristems. There are thousands of species that grow only one season and usually lack secondary growth. Leaves also usually lack secondary growth.

The bodies of many plants have two secondary meristems: the vascular cambium and the cork cambium (Fig. 4.22). **Vascular cambium** differentiates into secondary xylem and secondary phloem, and the **cork cambium** differentiates into the periderm **(Fig. 4.24)**. These are discussed at length in Chapters 5 and 7 on stems and roots.

The primary meristems near the tips of the roots and shoots are the site of most elongation. They produce new cells, which then enlarge primarily by elongation. However, in many plants, the branch or root continues to increase in girth as well. This increase in girth requires lateral growth, which involves the formation and activity of the next category of meristems, called secondary meristems.

SECONDARY MERISTEMS The secondary meristems are responsible for cell division, initiation of cell differentiation,

OTHER MERISTEMS There are several other meristems. In stems, for example, an intercalary meristem occurs within the stem to regulate its elongation. As leaves develop, there are leaf-specific meristems that regulate leaf shape. The intercalary meristem at the base of grass leaves allows the leaf to continue to grow after being grazed or mowed. Other meristems are involved in forming buds and roots in unusual places, such as at the base of trees, and also in the repair of wounds. Some of these other meristems are discussed in the following chapters.

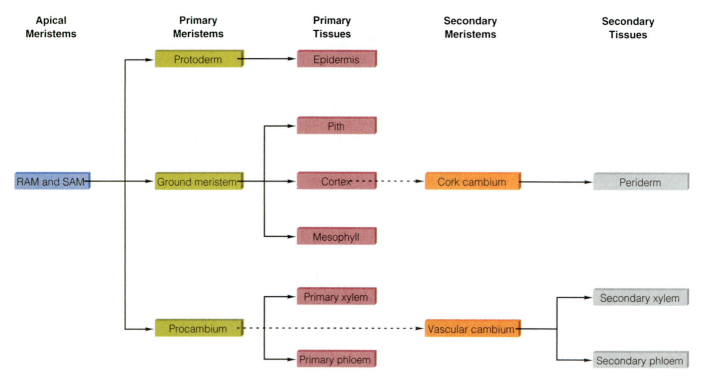

Figure 4.24 Summary of meristems and the tissues they generate.

KEY TERMS

albuminous cells

bordered pits

callose

cell differentiation

companion cells

cork cambium

cortex

dermal tissue system

epidermal cells

epidermis

fibers

ground meristem

ground tissue system

growth

guard cells

meristem

mesophyll

parenchyma tissue

periderm

phellem

phelloderm

phellogen

pith

pits

P-protein

primary meristems

primary phloem

primary xylem

procambium

protoderm

root apical meristem (RAM)

root system

sclereids

sclerenchyma

secondary meristems

secondary phloem

secondary xylem

shoot apical meristem (SAM)

shoot system

sieve areas

sieve cells

sieve plate

sieve tubes

sieve-tube members

simple pits

stoma (stomata)

subsidiary cell

tracheary element

tracheids

trichomes

vascular bundles

vascular cambium

vascular cylinder

vascular tissue system

vessel members

SUMMARY

1. The vascular plant body is organized into a shoot system (stems, leaves, buds, flowers, and fruits) and a root system. Flowering plants (angiosperms), cone-bearing plants (gymnosperms), and ferns are all vascular plants.

2. Cells in the plant body are organized into the ground, vascular, and dermal tissue systems. There are several different cell types making up the plant body. These cells are organized into aggregates called tissues.

3. The simple tissues are parenchyma, collenchyma, and sclerenchyma. (Refer to Tables 4.1 and 4.2 and Figure 4.24 for summaries of cell types, tissues, and meristems.)

4. The complex tissues are epidermis, xylem, phloem, and periderm.

5. Secretory structures are specialized to secrete various substances within and outside the plant body.

6. Each cell originates from a cell that was once meristematic. The process whereby a cell changes into a mature cell is called **differentiation.**

7. Meristems are the sites of cell division, cell elongation, and the beginning of cell differentiation.

8. Apical meristems occur at the tips of stems and roots and are the ultimate source of all cells in the shoot and root systems.

9. Primary meristems (protoderm, procambium, and ground meristem) form the primary tissues (epidermis, primary xylem and phloem, and pith and cortex).

10. Secondary meristems (vascular and cork cambium) form the secondary tissues (secondary xylem and secondary phloem, and phellem and phelloderm, respectively).

11. The positional information theory states that the differentiation of all cells in the plant body is regulated by a gradient of at least two substances that trigger gene expression patterns.

Questions

1. What are the tissues found in the plant body? How are they organized in each vegetative organ?

2. What are the features, cell types, and functions of each of the tissues listed below?

epidermis   parenchyma
periderm   collenchyma
xylem   sclerenchyma (fibers and sclereids)
phloem

3. Define the following xylem terms:

vessel member   tracheid
vessel   bordered pit

4. Define the following phloem terms:

sieve-tube member   companion cell
sieve tube   sieve plate

5. Describe the function of the following meristems:

root and shoot apical meristems

primary meristems: protoderm, ground meristem, and procambium

secondary meristems: vascular cambium and cork cambium (or phellogen)

6. Each tissue in the plant body has a different function. Describe these functions and discuss how the tissues and cells communicate with each other.

InfoTrac® College Edition

http://infotrac.thomsonlearning.com

Meristems

Pizzolato, T.D., Sundberg, M.D. 2002. Initiation of the vascular system in the shoot of Zea mays L. (Poaceae): II. The procambial leaf traces. *International Journal of Plant Sciences* 163:353. (Keywords: "vascular" and "Zea")

Forensic Botany

Lane, M.A., Anderson, L.C., Barkley, T.M., Bock, J.H., Gifford, E.M., Hall, D.W., Norris, D.O., Rost, T.L., Stern, W.L. 1990. Forensic botany. *BioScience* 40:34. (Keywords: "forensic" and "botany")

The Shoot System I: The Stem

Visit us on the web at http://biology.brookscole.com/plantbio2 for additional resources, such as flashcards, tutorial quizzes, InfoTrac exercises, further readings, and web links.

1. The shoot system is composed of the stem and its lateral appendages: leaves, buds, and flowers. Leaves are arranged in different patterns (phyllotaxis): alternate, opposite, whorled, and spiral.

2. Stems provide support to the leaves, buds, and flowers. They conduct water and nutrients and produce new cells in meristems (shoot apical meristem [SAM] and primary and secondary meristems).

3. Dicot stems and monocot stems are usually different. Dicot stems tend to have vascular bundles distributed in a ring, whereas they tend to be scattered in monocot stems.

4. Stems are composed of the following: epidermis, cortex and pith, xylem and phloem, and periderm.

5. Secondary xylem is formed by the division of cells in the vascular cambium and is called wood. The bark is composed of all of the tissues outside the vascular cambium, including the periderm (formed from cork cambium) and the secondary phloem.

6. Several different types of modified stems (rhizomes, spines, and others) have important functions.

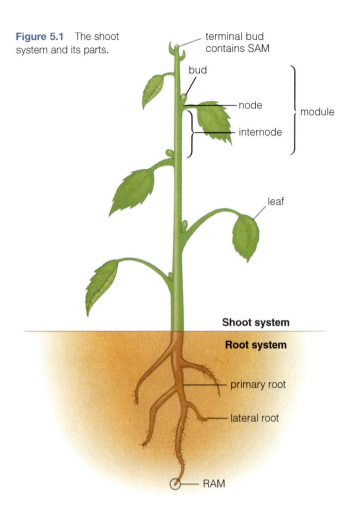

Figure 5.1 The shoot system and its parts.

terminal bud contains SAM
bud
node
internode
module
leaf
Shoot system
Root system
primary root
lateral root
RAM

5.1 THE FUNCTIONS AND ORGANIZATION OF THE SHOOT SYSTEM

The shoot system of a typical flowering plant consists of the stem and the attached leaves, buds, flowers, and fruits. The leaves are displayed in a way that maximizes their exposure to sunlight. Flowers and fruits are located on stems in positions that allow for pollination and the dispersal of fruits and seeds. Internally, stems provide pathways for the movement of water and dissolved minerals from the roots into the leaves and for food synthesized in leaves to move into roots. Some stems are modified for different functions such as the storage of water and various food substances.

The stem is actually composed of repeated units called modules. A module is a segment of stem—an internode—plus the leaf and bud attached to the stem **(Fig. 5.1).** The point of attachment is called a node.

The SAMs and primary meristems of the stem are located in buds at the ends of the branches and just above the nodes. Farther down the stem in some plants are secondary meristems, which make secondary tissues. These meristems produce new cells and act as sites for the start of cell elongation and differentiation.

Flowering plants are divided into two groups: dicotyledonous plants (dicots) and monocotyledonous plants (monocots). One of the bases for this division is that the dicots, such as peas (*Pisum* sp.) and oaks (*Quercus* sp.), produce embryos with two cotyledons (seed leaves), whereas monocots, such as corn (*Zea mays*) and onions (*Allium cepa*),

produce embryos with only one cotyledon (see Chapter 14). The stems of these two groups also have major differences in the distribution of their tissues **(Fig. 5.2)** and in the operation of their meristems. As we examine primary and secondary growth in stems, we will contrast the anatomy of dicots and monocots.

5.2 PRIMARY GROWTH AND STEM ANATOMY

Primary Tissues of Dicot Stems Develop from the Primary Meristems

The SAM is composed of dividing cells. It is responsible for the initiation of new leaves and buds and for making the three primary meristems **(Fig. 5.3)**. The three primary meristems—protoderm, ground meristem, and procambium—and their tissue products are discussed in the following sections.

PROTODERM TO EPIDERMIS The outermost layer of cells in the shoot tip is the *protoderm.* This layer is called a primary meristem because its cells are still dividing and because primary meristems differentiate into primary tissues. When the cells of the protoderm stop dividing and mature, they are called *epidermis* **(Fig. 5.4)**. Epidermal cells,

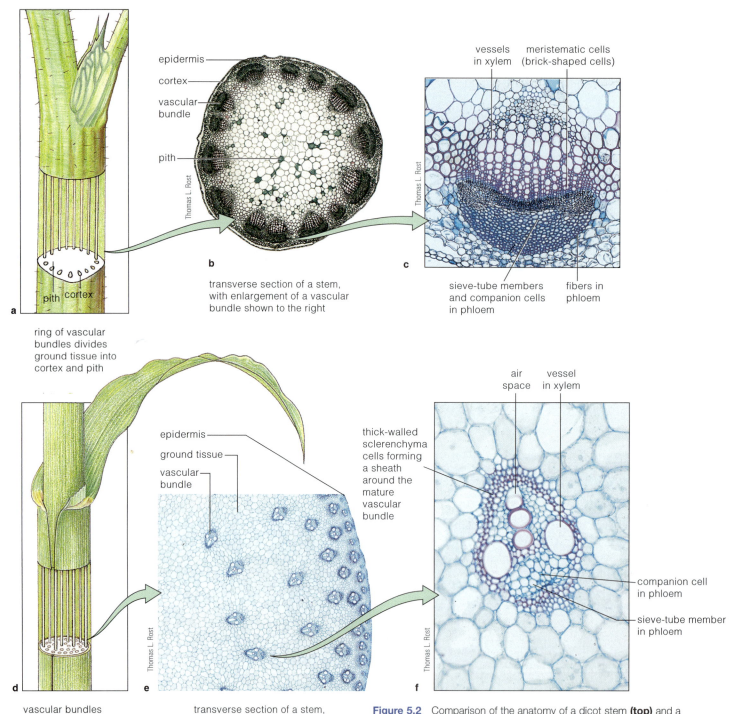

epidermis

cortex

vascular bundle

pith

b

transverse section of a stem, with enlargement of a vascular bundle shown to the right

vessels in xylem

meristematic cells (brick-shaped cells)

c

sieve-tube members and companion cells in phloem

fibers in phloem

a

pith cortex

ring of vascular bundles divides ground tissue into cortex and pith

epidermis

ground tissue

vascular bundle

air space

vessel in xylem

thick-walled sclerenchyma cells forming a sheath around the mature vascular bundle

companion cell in phloem

sieve-tube member in phloem

d

e

vascular bundles distributed through ground tissue

transverse section of a stem, with enlargement of a vascular bundle shown to the right

f

Figure 5.2 Comparison of the anatomy of a dicot stem **(top)** and a monocot stem **(bottom)**. (a) Typically, dicot stems have a pith surrounded by a cylinder of vascular bundles, as shown in a clover (*Trifolium* sp.) stem (**b**). ×37. (**c**) These bundles usually have primary phloem toward the outside and primary xylem toward the inside. ×266. (**d**) Typically, monocot stems have scattered vascular bundles, as shown in a portion of a corn stem (*Zea mays*) (**e**). ×14. (**f**) The primary phloem in these bundles is usually positioned toward the outside. ×270.

guard cells, different kinds of trichomes or hairs, and a cuticle make up the epidermis (see Chapter 4).

GROUND MERISTEM TO PITH AND CORTEX In the very center of the shoot tip and just inside the protoderm is the *ground meristem* (Figs. 5.3 and 5.4). These cells slowly lose their ability to divide, and they differentiate primarily into parenchyma cells of the cortex (the cylinder of cells just inside the epidermis) and the pith (the core of cells in the center of the stem) (Fig. 5.2b). Along with their different positions in the stem, the cortex and pith have different functions. The parenchyma cells nearest the outside of the cortex sometimes contain chloroplasts for photosynthesis. Sometimes, the parenchyma cells of the cortex or pith store starch. The pith region of the stem in some plants may

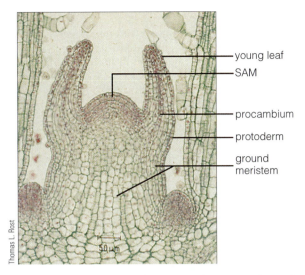

young leaf
SAM
procambium
protoderm
ground meristem

Thomas L. Rost

50 µm

Figure 5.3 Shoot apex of *Coleus blumei*, a common houseplant. ×114.

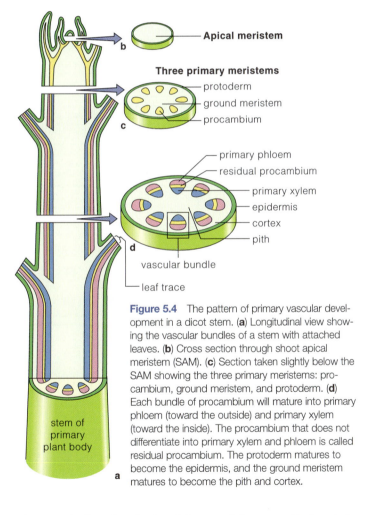

Apical meristem

b

Three primary meristems

protoderm
ground meristem
procambium

c

primary phloem
residual procambium
primary xylem
epidermis
cortex
pith

d

vascular bundle

leaf trace

stem of primary plant body

a

Figure 5.4 The pattern of primary vascular development in a dicot stem. (**a**) Longitudinal view showing the vascular bundles of a stem with attached leaves. (**b**) Cross section through shoot apical meristem (SAM). (**c**) Section taken slightly below the SAM showing the three primary meristems: procambium, ground meristem, and protoderm. (**d**) Each bundle of procambium will mature into primary phloem (toward the outside) and primary xylem (toward the inside). The procambium that does not differentiate into primary xylem and phloem is called residual procambium. The protoderm matures to become the epidermis, and the ground meristem matures to become the pith and cortex.

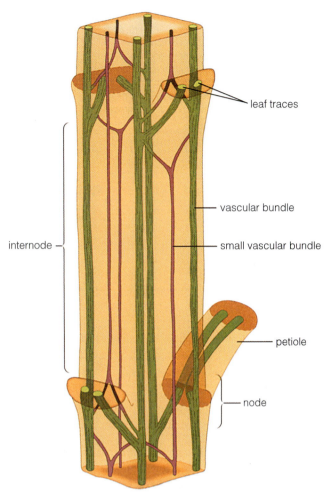

leaf traces

vascular bundle

small vascular bundle

internode

petiole

node

Figure 5.5 Three-dimensional view of the entire primary vascular system in a *Coleus* stem. In this stem there are two leaves per node, and two leaf traces pass into each leaf. A single large vascular bundle is found in the corners with a smaller bundle between them.

become hollow by the breakdown of the centrally located parenchyma cells.

PROCAMBIUM TO PRIMARY XYLEM AND PHLOEM The *procambium* tends to form as small bundles (Figs. 5.3 and 5.4) of relatively thin, long cells with dense cytoplasm. The bundles are usually arranged in a ring just inside the outer cylinder of

ground meristem and below the SAM. Procambium cells divide, and then at some position down the axis they stop dividing and differentiate into primary xylem and primary phloem. Each bundle of procambium becomes a vascular bundle, with primary xylem cells toward the inside of the stem and primary phloem cells toward the outside (Fig. 5.2). In plants exhibiting secondary growth, some procambium between the primary xylem and phloem remains undifferentiated; such cells are called residual procambium (Fig. 5.4d).

The Distribution of the Primary Vascular Bundles Depends on the Position of Leaves

Vascular bundles in dicot stems are distributed in a vascular cylinder. The vascular bundles that network into the attached leaves are called leaf traces (Fig. 5.4a). The organization of primary vascular bundles in stems depends on the number and distribution of leaves and on the number of traces that branch into the leaves (Fig. 5.5) and also into buds. The number of vascular bundles in the vascular cylinder and the number of leaf traces differ by species and are dependent on the number and arrangement of leaves.

Figure 5.6 The four basic patterns of leaf arrangement (phyllotaxis). (**a**) Alternate has one leaf per node and a 180-degree angle of divergence between leaves. (**b**) Opposite has two leaves per node and a 90-degree angle between sets of leaves. (**c**) Whorled usually has three to five leaves per node; in the case of three leaves per node, the angle of divergence is 60 degrees. (**d**) Spiral has one leaf per node and 137.5 degrees between leaves.

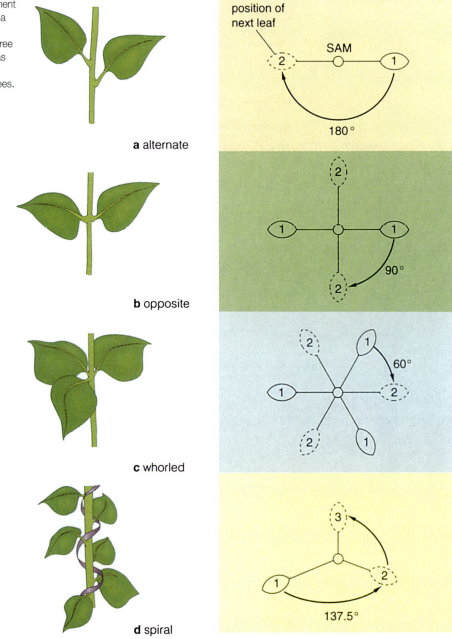

a alternate

b opposite

c whorled

d spiral

The arrangement of leaves on the axis of the stem is called *phyllotaxis* (*phyllo-* is derived from the Greek for "leaf," and *-taxis* is derived from the Greek for "arrangement"). There are four basic patterns of phyllotaxis (Fig. 5.6). Plants with alternate phyllotaxis have only one leaf per node, and the leaves are positioned 180 degrees from each other. The angle separating one leaf or set of leaves from another is called the *angle of divergence*. Opposite phyllotaxis means that the shoot has two leaves per node; the angle of divergence between leaves in successive sets is 90 degrees. In whorled phyllotaxis there are three or more leaves per node, and the angle of divergence between successive sets of leaves depends on the leaf number per set. The angle is 60 degrees in plants with three leaves per node (Fig. 5.6c). Plants with spiral phyllotaxis have one leaf per

node, and the angle of divergence between leaves is 137.5 degrees (Fig. 5.6d).

Primary Growth Differs in Monocot and Dicot Stems

Monocot stems differ in several ways from the patterns described for dicot stems. One main difference is that the vascular bundles tend to be scattered throughout the stem instead of being in a ring (Fig. 5.2). The terms *pith* and *cortex* usually are not used when the bundles are scattered; instead, the term *ground tissue* is used for all the parenchyma tissue surrounding the vascular bundles. There are some exceptions; wheat (*Triticum* sp.) stems, for example, are hollow in the stem internodes (Fig. 5.7), rather than having ground tissue there.

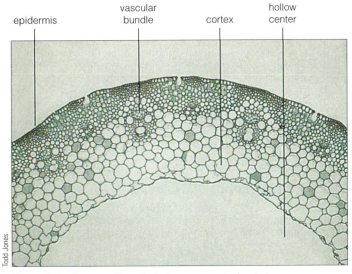

Figure 5.7 Cross section of a hollow monocot wheat stem (*Triticum* sp.). ×38.

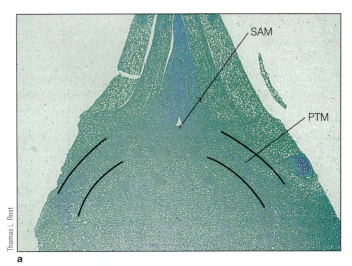

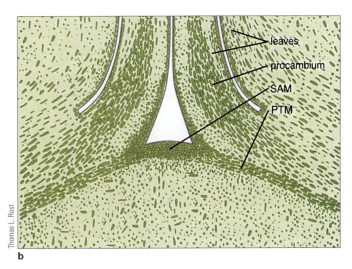

Figure 5.8 (a) Longitudinal section of *Iris* shoot apex to show the primary thickening meristem (PTM). ×108. (b) Diagram of a similar apex to more clearly show the "umbrella-shaped" PTM.

If you were to compare the shapes of a monocot stem and a dicot stem, one of the first things you would notice is that a monocot stem is about the same diameter near its apex as at its base. This shape is primarily because of the activity of the **primary thickening meristem** (PTM), which is absent in dicot stems. The PTM is unique in contributing to both elongation and lateral growth, a characteristic resulting from its umbrella-like shape **(Fig. 5.8).** The SAM and the primary meristems also are present in these tips.

5.3 SECONDARY GROWTH AND THE ANATOMY OF WOOD

Most monocots and many dicots show little or no secondary growth. These are herbaceous (nonwoody) plants, which normally complete their life cycle in one growing season. By contrast, many other dicots, such as oaks (*Quercus* sp.) and maples (*Acer* sp.), and gymnosperms, such as pines (*Pinus* sp.) and firs (*Abies* sp.), show secondary growth starting in their first year of growth. In some plants, this continues for many, even hundreds, of years; these are called woody plants.

Secondary Xylem and Phloem Develop from Vascular Cambium

Woody plants develop thicker, more massive stems because of the growth of secondary xylem and phloem from their secondary meristems. The first step in making secondary xylem and phloem is to form the vascular cambium (plural, cambia). Development of the vascular cambium involves coordinated cell division in the *residual procambium* inside the vascular bundles and the parenchyma cells between the bundles **(Fig. 5.9).** The signal for this cell division is probably given by a plant hormone. When the residual procam-

bium cells divide, they are referred to as **fascicular cambium** (fasciculus is derived from the Latin word for "bundle"). Next, the cells between the vascular bundles divide; these are referred to as the **interfascicular cambium** ("between the bundles"). The parenchyma cells near the bundles divide first, and the adjoining cells divide until a complete ring forms (Figs. 5.9 and **5.10**). Once the cylinder of dividing cells—the fascicular cambium and the interfascicular cambium—is complete, it is called **vascular cambium** (Fig. 5.9).

The vascular cambium is an interesting meristem in that it is only one or two cells in thickness but divides in two directions. The cells formed to the outside become secondary phloem, and the cells formed to the inside become secondary xylem (Figs. 5.9 and 5.10). **Figure 5.11** summarizes the successive divisions that contribute to thickening of the secondary xylem and phloem. Vascular cambium cells produce more xylem cells than phloem cells.

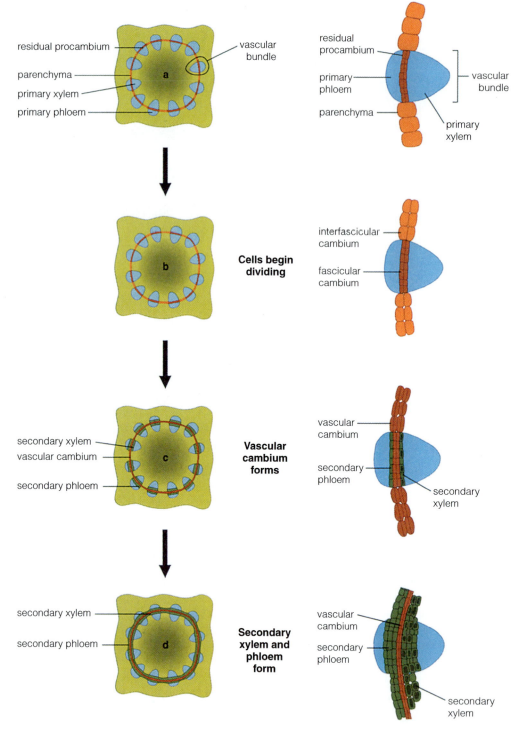

Figure 5.9 Formation of the vascular cambium. (**a**) At the completion of primary growth, some residual procambium cells remain between the primary xylem and primary phloem, and parenchyma cells occur between the vascular bundles. (**b**) After the cells begin dividing, the residual procambium is then called the fascicular cambium, and the cells between the bundles are called the interfascicular cambium. (**c**) When the fascicular and interfascicular cambia are all dividing and become connected as a ring, they are called the vascular cambium. (**d**) Secondary xylem (inside) and secondary phloem (outside) form from the vascular cambium.

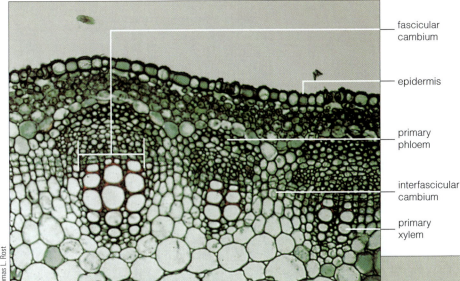

fascicular cambium

epidermis

primary phloem

interfascicular cambium

primary xylem

Thomas L. Rost

a

Figure 5.10 Cross sections of an alfalfa (*Medicago sativa*) stem, showing (**a**) the location of the fascicular and interfascicular cambia and (**b**) some secondary vascular tissue. ×241 (**a**); ×246 (**b**).

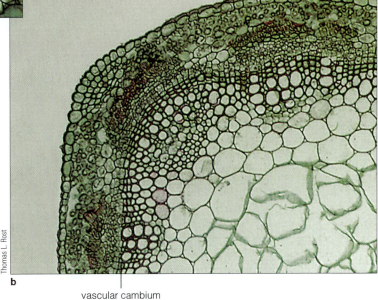

Thomas L. Rost

b

vascular cambium

Some vascular cambium cells, called **fusiform initials,** form into cells of the **axial system;** vessel members are examples **(Figs. 5.12 and 5.13).** The cells of the **ray system** are formed from vascular cambium cells called **ray initials.** The rays are composed of only two different cell types: ray parenchyma cells and ray tracheids. The cells of the axial system function in the longitudinal movement of water and minerals; the cells of the ray system transport water and minerals radially.

Wood Is Composed of Secondary Xylem

The *secondary xylem* is collectively called *wood*. Examining thin sections of wood under a microscope reveals its many distinctive characteristics. Typically, these sections are made

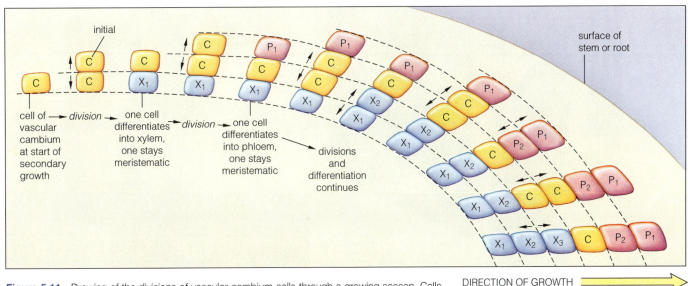

Figure 5.11 Drawing of the divisions of vascular cambium cells through a growing season. Cells labeled *C* are the initial cells; their derivative cells become secondary phloem to the outside (P₁, P₂, and so on) and secondary xylem to the inside (X₁, X₂, and so on). The result is that the stem gets wider and the vascular cambium keeps increasing in circumference and moving outward.

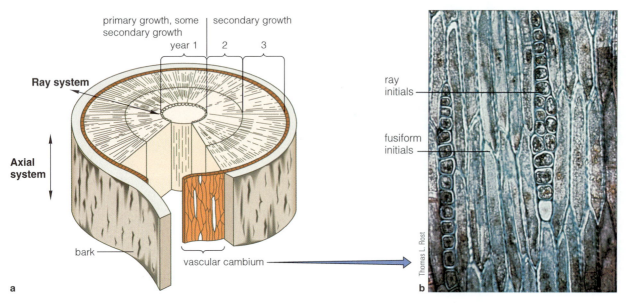

Figure 5.12 (**a**) Drawing of a 4-year-old woody stem showing the growth increments as annual rings. The lines radiating from the center are the rays; all the rays together make up the ray system. All the other cells in the wood make up the axial system. Bark refers to all the tissue from the vascular cambium to the outside. (**b**) The vascular cambium has two types of initial cells, the ray initials make the rays, and the fusiform initials make the cells of the axial system. ×275.

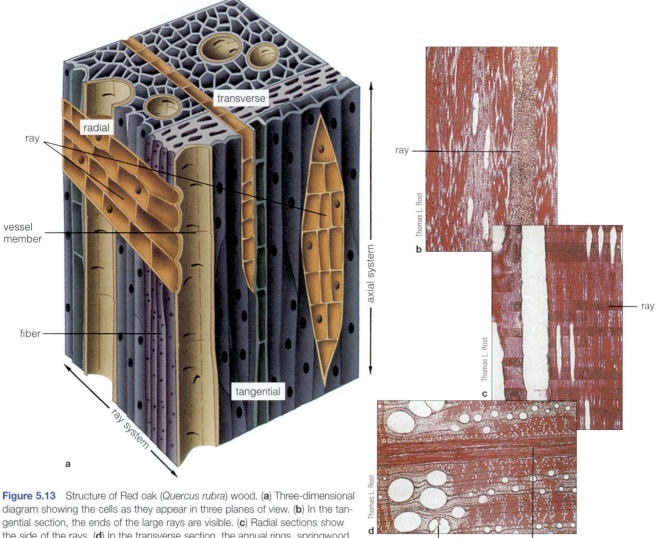

Figure 5.13 Structure of Red oak (*Quercus rubra*) wood. (**a**) Three-dimensional diagram showing the cells as they appear in three planes of view. (**b**) In the tangential section, the ends of the large rays are visible. (**c**) Radial sections show the side of the rays. (**d**) In the transverse section, the annual rings, springwood and summerwood, are obvious. ×20 (**b–d**).

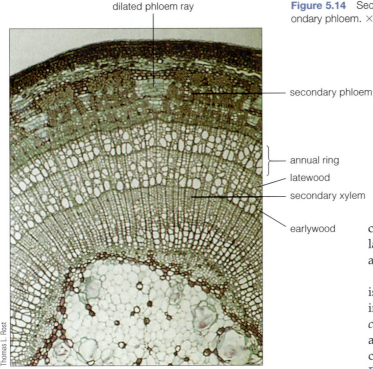

dilated phloem ray

secondary phloem

annual ring

latewood

secondary xylem

earlywood

Thomas L. Rost

Figure 5.14 Section of a 3-year-old basswood (*Tilia americana*) stem including the secondary phloem. ×23.

in three different planes of view: tangential, radial, and transverse (Fig. 5.13).

Tangential sections of wood (Fig. 5.13b) show an end view of the rays. Wood cut in this plane makes interesting grain patterns used in furniture. Radial sections show the rays from a side view (Fig. 5.13c). Transverse sections of wood show an end view of the cells of the axial system (Fig. 5.13d).

Annual rings are concentric rings of cells in the secondary xylem (Figs. 5.12 and 5.14). In trees growing in temperate climates, one annual ring forms each growing season. This means that it is usually possible to determine the age of a tree by counting the growth rings. Trees growing in tropical rain forests tend to have irregular growth rings because the climate is always warm and growth happens all during the year. On the basis of the annual ring counts, the oldest known trees—the redwoods (*Sequoia sempervirens*) and bristlecone pines (*Pinus longaeva*)—can be thousands of

years old. In addition, the growth rate of a tree during each growing season can be determined from the thickness of the annual rings. Ecologists use this information to speculate about climate conditions in the past. The study of tree rings is called dendrochronology.

Each annual ring has two components (Fig. 5.14). The cells in the inner part of an annual ring tend to be larger in diameter because they form in the springtime during the first growth spurt of the new season and are called the **springwood** or **earlywood.** The cells that form later in the growing season tend to have smaller diameters and are called **summerwood** or **latewood.**

One of the structural features used to categorize wood is the distribution of large-diameter vessel members seen in transverse view. In some trees, such as Red oak (*Quercus rubra*) (Fig. 5.13), the large-diameter vessel members are located primarily in the springwood. This pattern is called **ring porous.** In other trees, such as elm (*Ulmus* sp.; Fig. 5.15), the large-diameter vessel members are quite uniformly distributed throughout both the springwood and summerwood, although the cells in summerwood tend to be slightly smaller in diameter. This pattern is called **diffuse porous.**

If you have ever cut down a tree, you may have noticed that the wood in the center of the tree is often darker than the wood near the periphery (Fig. 5.16). The lighter wood near the periphery is called **sapwood.** The secondary xylem cells in this part of the stem are the functioning xylem cells. This is where most of the actual transport of water and dissolved minerals takes place. The darker wood in the center is called **heartwood.** These cells often are filled with resin and other materials that block them, so that they no longer transport. In some woody plants, heartwood vessel members are blocked by structures called tyloses. This is true of white oak (*Quercus alba*), the wood used to make wine barrels (see "ECONOMIC BOTANY: How Do You Make a Barrel?"). In some wood, parenchyma cells lie adjacent to vessel members. A tylose forms when the cell wall of the parenchyma cell actually grows through a pit and into the

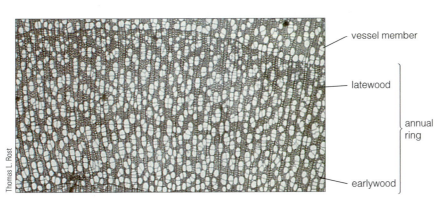

vessel member

latewood

annual ring

earlywood

Thomas L. Rost

Figure 5.15 Transverse section of diffuse porous wood in elm (*Ulmus* sp.). ×57.

ECONOMIC BOTANY:

How Do You Make a Barrel?

The highest quality red wines, as well as chardonnay (a white wine), are carefully aged in oak barrels for several months. Vintners are experts—artists, actually—at selecting the best barrel and at timing the aging process to produce the best-tasting wine. Did you ever wonder where these barrels come from and how they are made? I found out when I visited the Mendocino Cooperage, a barrel-making factory at northern California's Fetzer Winery in Hopland.

No wood is better suited for aging wine than oak—specifically, white oak (*Quercus alba*). If you look at the cellular structure of any species of oak (Fig. 5.13, Red oak), you will see a multitude of wide- and short-vessel members, surrounded by fibers with thick cell walls, and large, multicellular rays (see Chapter 4 if you do not recall these cell types). This combination makes for a wood that is very dense and strong, yet flexible enough to make the rounded barrels.

Two other features make white oak the wood of choice. One is that the vessel members are plugged by tyloses, the bubble-shaped ingrowths from surrounding parenchyma cells. Tyloses inhibit pathogen movement in inactive wood, and

they make white oak wood good for barrels because the tyloses also make the wood leak-resistant (Fig. 5.17). Nothing, including water, can move up a vessel when it is plugged with tyloses; therefore, nothing can leak out of these barrels, either. The other feature is that the cells making up the wood contain many complex chemicals phenolics (complex polymers of phenol units; phenols are six-carbon ring molecules), carbohydrates, and other carbon compounds that add desirable aromas and tastes to aged wine. These aromas and tastes have become traditional and are expected by wine drinkers.

The best American barrels come from white oak trees grown in the cold climates of Minnesota and Iowa. The wood from these old trees (100–200 years old) has tight grain (small annual rings) because the trees grow very slowly and uniformly. The wood is clear and light colored.

Making barrels is a rare and highly valued craft. There are few master coopers (barrel makers) in the United States. At the beginning of the coopering process, white oak wood is sawn into boards about 1 meter in length, a little more than 2.5 cm in thickness, and 7.5 to 12.5 cm in width. The boards are stacked so that air can cir-

culate among them, and they are aged outside for 2 years. The boards are then examined one at a time for imperfections, and the perfect ones are planed and trimmed into barrel staves (**Fig. 1**). The staves are tapered toward their ends, and their edges are beveled so that they will fit snugly together when bent.

Each barrel is made of the same number of staves. These are collected together in a steel ring, and a hoop is pushed snugly part way up the forming barrel (**Fig. 2**). The barrel is then held over a small open flame (Fig. 2) to heat the wood and release whatever water is still left in the cells of the wood. At the same time, a long cable is wrapped around the bottom of the forming barrel (see the bottom of **Fig. 3**). As the barrel continues to heat, the cooper wets the outside. The barrel slowly turns. As the cable pulls tighter, the staves bend and end up tightly pressed together (Fig. 3). After a certain amount of time, the cooper pulls the barrel off the fire and adds a second hoop on the other end. The new barrel is then placed into a machine to even all the staves, and the remaining steel hoops are pounded into position. No screws or nails are used at any step in the process.

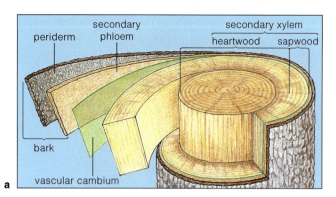

Thomas L. Rost

Figure 5.16 (**a**) Diagram showing the sapwood, heartwood, and other components of a woody stem. (**b**) Slice of a woody stem of mulberry branch (*Morus* sp.) showing the coloration difference between sapwood and heartwood. The dark area from the center to the lower right is an embedded branch that started to grow when this stem was small. This embedded branch will make a "knot" in a board made from this part of the stem.

The barrel ends are also made of oak-wood pieces. These are fired briefly, sawn to the correct length, and fitted together with wood pegs. Rush leaves are placed between the boards to seal them together. The covers are then pressed together and planed smooth, and an edge is cut all around the cover. The cooper rubs a flour-and-water paste into the groove at the lip of the barrel and hammers the cover into place very cleverly. The barrel is complete **(Fig. 4)**. Barrels are then pressure tested for leaks with water and air. The ends of the staves are examined for water seepage and are plugged if necessary.

New barrels impart a very specific flavor and aroma to the wine. Chardonnay, as an example, will age in oak barrels for 7 to 8 months. Red wines may be aged longer. A typical barrel loses its characteristic flavor after being used twice. After that, the barrel may be used for several more years to store less vintage wines. A typical wine barrel has a working life of about 8 years, after which you may find it cut in half and growing tomatoes in someone's garden. Making good barrels is really an art form. I will look at them with new eyes from now on.

Figures 1–4 Images of wine barrel making.

Websites for further study:

History of the Wine Barrel:
http://www.eresonant.com/pages/history/history-barrels.html

Wine Institute: http://www.wineinstitute.org/wilib/brrls2.htm

Bouchard Cooperages—Research News:
http://www.bouchardcooperages.com/website/research/research_news_intro.html

vessel member. Tyloses look like bubbles; they can fill a vessel member and completely block it (Fig. 5.17 a and b).

Take another look at Figure 5.16b. Note the dark region labeled branch in the lower right of the image. This is a branch, or knot. If a branch forms early in the life of a tree, it will be connected at a growth ring near the middle of the stem axis. As the stem continues to enlarge, it gets wider as it forms more and more growth rings. This new growth embeds the base of old branches. When these are cut through to make boards, they will be the knots that I'm sure you have seen many times.

Gymnosperm Wood Differs from Angiosperm Wood

The wood described so far in this chapter is angiosperm (flowering plants) wood: hardwoods such as oak (*Quercus*), elm (*Ulmus*), and walnut (*Juglans*). Gymnosperm (seed plants that do not form flowers; see Chapter 24) wood includes softwoods such as pine (*Pinus*), fir (*Abies*), and redwood (*Sequoia*). It has a simpler structure than angiosperm wood. Angiosperm wood is composed of several different types of cells, which may provide interesting and highly valued grain patterns; gymnosperm wood is composed of only a few cell types, mostly tracheids in the axial system, and

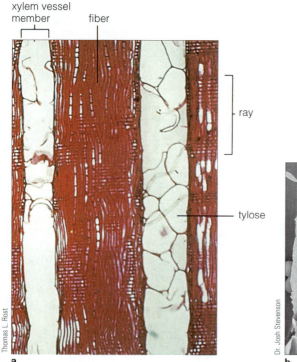

xylem vessel member | fiber | ray | tylose

Thomas L. Rost

a

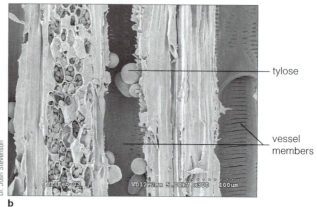

tylose | vessel members

Dr. Josh Stevenson

b

very simple rays. A popular wood for building is the redwood (*Sequoia sempervirens*). This wood (shown in tangential, radial, and transverse views in **Fig. 5.18**) is quite simple in structure. For example, in Figure 5.18d, the annual rings of a redwood tree are apparent, and, indeed, they are all tracheids. Annual rings can be discerned because the springwood cells are a little wider than the summerwood cells.

Some gymnosperm wood have resin ducts **(Fig. 5.19)**, which are secretory structures that produce and transport

Figure 5.17 (**a**) Radial section of wood of white oak (*Quercus alba*) showing tyloses plugging the large vessel to the left in the photograph. ×48. (**b**) Longitudinal view of grape wood (*Vitus vinifera*) showing tyloses inside a vessel member. Line scale100 μm. (Courtesy Dr. Josh Stevenson, Department of Biology and Chemistry, Texas A&M International University).

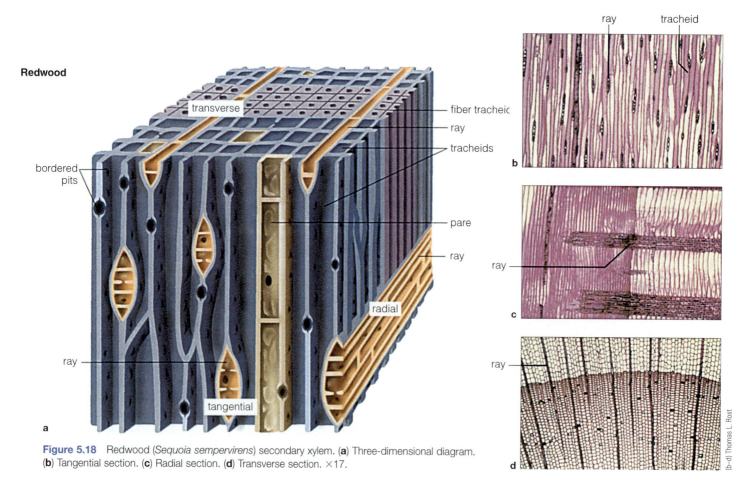

Redwood

transverse | fiber tracheid | ray | tracheids | bordered pits | pare | ray | radial | ray | tangential

ray | tracheid | ray | ray

a
b
c
d

(b-d) Thomas L. Rost

Figure 5.18 Redwood (*Sequoia sempervirens*) secondary xylem. (**a**) Three-dimensional diagram. (**b**) Tangential section. (**c**) Radial section. (**d**) Transverse section. ×17.

resin. The resin is synthesized and secreted by a lining of *epithelial cells.* Turpentine is one commercial product made from it. Resin has several names. It is called *sap* when it flows through the resin ducts all the way to the outside of the stem; but when it hardens, it is called *rosin.* Fossilized rosin is the *amber* popular for jewelry. Insects can get stuck in sap, which is why trapped insect bodies are sometimes found in amber.

Bark Is Composed of Secondary Phloem and Periderm

The protective covering over the wood of a tree is commonly called *bark.* In botanical terms, the bark is everything between the vascular cambium and the outside of the woody stem. The actual composition of the bark varies a little, depending on the age of the tree. The bark of a 1- or 2-year-old tree includes the secondary phloem,

maybe a few cells of the cortex, and one or two increments of **periderm.** In a very old tree, the bark would include the layers of secondary phloem plus several layers of periderm (Fig. 5.16).

SECONDARY PHLOEM Secondary phloem (Fig. 5.14) forms to the outside of the vascular cambium. The types of cells in the secondary phloem are sieve-tube members, companion cells, phloem parenchyma, phloem fibers, sclereids in the axial system, and ray parenchyma in the ray system. When secondary phloem is examined in radial, tangential, and transverse sections, its cellular orientation is similar to that in wood (xylem). A difference in transverse section is that secondary phloem does not develop in annual ring increments. You cannot count phloem rings to determine tree age (Fig. 5.14). Phloem rays are composed of phloem ray parenchyma cells. In some trees, such as basswood (*Tilia americana*) (Fig. 5.14), wedge-shaped areas of the secondary phloem, called dilated rays, occur, which are composed mostly of secondary phloem ray parenchyma cells.

PERIDERM Usually, stems of plants that live more than 1 year will lose their epidermis sometime during the first or second year of growth. As the stem increases in diameter because of secondary growth from the vascular cambium, the cells of the epidermis stretch but cannot keep up with the increasing circumference. Eventually, the epidermis cracks, dries up, and flakes away from the stem.

At the same time, some cells (usually in the outer cortex) start to divide all around the periphery of the stem **(Fig. 5.20).** This new layer of dividing cells is the **cork cambium** (or **phellogen**). The cork cambium divides in two directions, to form **phellem cells** (also called cork cells) to the outside and **phelloderm cells** to the inside.

These three layers—the phellem, the cork cambium, and the phelloderm—make up the **periderm** (Fig. 5.20c). Phellem cells form in regular rows, have waxy (contain a waxy substance called suberin) cell walls, and are usually dead by the time the periderm is fully functional. The phelloderm cells also form in regular rows, but these cells tend to live longer and look like parenchyma cells. The function of the periderm is to serve as an impermeable layer, inhibiting water evaporation from the protected cell layers and protecting against insect and pathogen invasion.

The cells near the surface of the stems of young woody trees often are living. These cells require oxygen to function. Consequently, a structure called a *lenticel* is present in the bark of young branches of different woody plants **(Fig. 5.21).** **Lenticels** are specialized regions of the periderm consisting of loosely packaged parenchyma cells. Their purpose is to provide a place where gases can be exchanged.

At the start of each new growing season, a new increment of periderm is initiated. This means that in most trees an entirely new cork cambium is generated each spring. This new cork cambium is initiated from secondary phloem parenchyma cells, which are triggered to divide much like the outer cortex cells that divided to form the first increment

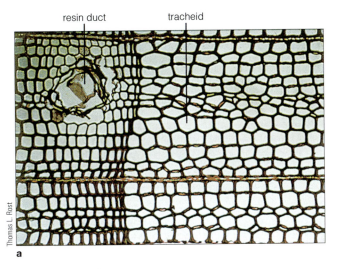

resin duct · tracheid

Thomas L. Rost

a

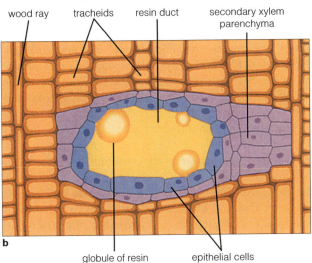

wood ray · tracheids · resin duct · secondary xylem parenchyma

globule of resin · epithelial cells

b

Figure 5.19 (**a**) Pine wood (*Pinus* sp.) transverse section showing a resin duct. ×118. (**b**) Diagram of a pine wood resin duct showing the secretory cells and globules of resin.

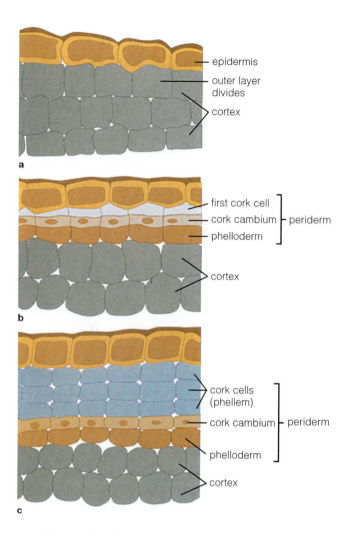

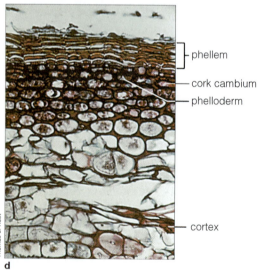

Figure 5.20 The cork cambium. (**a**) Cross section diagram showing a portion of the epidermis and outer cortex of a mature stem. (**b**) The outer cortex layer divides to become the "first" cork cambium. (**c**) The cork cambium makes cork cells (phellem) to the outside and the phelloderm to the inside, these three layers make the periderm. Note how these cells form in regular rows. (**d**) Microscopic view of cork cambium and periderm layers in elder (*Sambucus* sp.) stem. ×200.

(Figure 5.20 a labels: epidermis; outer layer divides; cortex)

(Figure 5.20 b labels: first cork cell; cork cambium — periderm; phelloderm; cortex)

(Figure 5.20 c labels: cork cells (phellem); cork cambium — periderm; phelloderm; cortex)

(Figure 5.20 d labels: phellem; cork cambium; phelloderm; cortex)

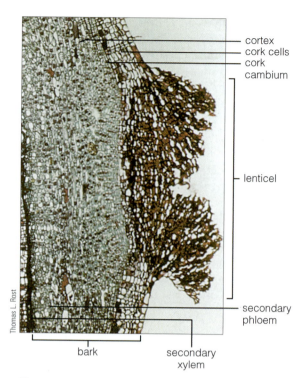

Figure 5.21 Transverse section of a woody stem of elder (*Sambucus* sp.). ×83.

(Figure 5.21 labels: cortex; cork cells; cork cambium; lenticel; secondary phloem; bark; secondary xylem)

Figure 5.22 Bark of pine (*Pinus* sp.) showing the layers of the bark.

of cork cambium. The consequence of this is that one year's accumulation of periderm stacks on another, forming layers **(Fig. 5.22).**

There are always exceptions to the rule, and cork oak (*Quercus suber*; **Fig. 5.23a**) is one. Instead of new cambium being generated yearly, a single cork cambium lasts the entire life of the tree. Interestingly, it divides only to the outside to form cork cells. Sheets of cork are harvested every 8 to 10 years for the manufacturing of corks for wine bottles and other things. Natural cork is increasingly rare; consequently, many wine makers are currently using artificial corks for wine bottles.

In bark, the innermost periderm layer is the current active layer. *Girdling*, the removal of a continuous strip of bark around the circumference of a tree, is a sure way to kill it. That is because the nutrient-transporting secondary phloem tissue will be severed.

The external appearance of bark varies among species of trees. Most species can be recognized by their external bark

a   b   c

Figure 5.23 External views of the bark of (**a**) cork oak (*Quercus suber*), (**b**) birch (*Betula papyifera*), and (**c**) *Eucalyptus* sp. showing their different appearance.

texture (Fig. 5.23). There are several main patterns of bark. The ring bark of, for example, paper birch trees (*Betula papyrifera*), forms in continuous rings **(Fig. 5.23b).** Scale bark, which is characteristic of pine trees, forms as small overlapping scales (Fig. 5.22). Shag bark, such as that found in *Eucalyptus*, has long overlapping thin sheets **(Fig. 5.23c).** There also are intermediate forms that are useful for tree identification.

Buds Are Compressed Branches Waiting to Elongate

The buds at the ends of woody branches are actually short compressed branches. These buds are covered with several hard, modified leaves called bud scales, which keep the inside of the bud fresh and moist. The bud at the end of a branch is called the terminal bud; those in the axils (at the base of the petiole) of leaves on the side of a branch are called lateral buds **(Fig. 5.24).** Some buds produce flower parts instead of leaves; these are called flower buds. In some fruit trees, such as almonds and apricots (both *Prunus* sp.), the first buds to open in the spring are flower buds.

When a bud elongates, as in the early spring, the bud scales open and finally fall off, leaving small scars called bud-scale scars. When a leaf falls, it leaves a leaf scar (Figs. 5.24 and 5.25). The vascular bundles that pass through the leaf into the stem also break off when the leaf falls; these scars are called bundle scars. The structure of the leaf scar and the number and distribution pattern of the bundle scars are characteristic of the plant and can be used to identify plants in the winter when all their leaves have fallen (Fig. 5.25).

Some Monocot Stems Have Secondary Growth

Most monocots lack a vascular cambium and do not form secondary xylem and phloem. Such plants cannot support large vertical shoot systems. Other monocots such as different types of lilies (*Lilium* sp.) can become very large plants by producing elaborate underground branches (rhizomes).

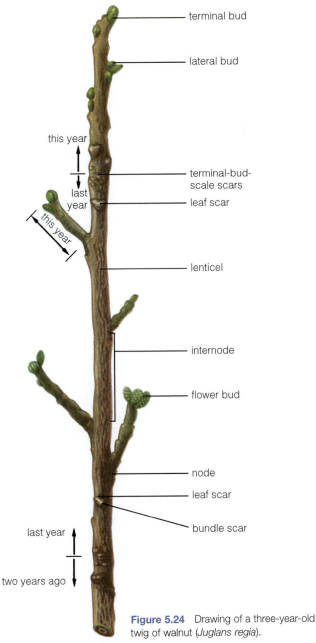

Figure 5.24 Drawing of a three-year-old twig of walnut (*Juglans regia*).

terminal bud
lateral bud
this year
last year
this year
terminal-bud-scale scars
leaf scar
lenticel
internode
flower bud
node
leaf scar
bundle scar
last year
two years ago

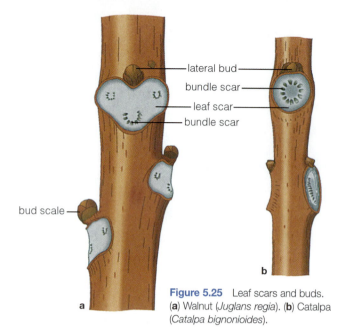

lateral bud
bundle scar
leaf scar
bundle scar

bud scale

b

a

Figure 5.25 Leaf scars and buds. (a) Walnut (*Juglans regia*). (b) Catalpa (*Catalpa bignonioides*).

Treelike monocots **(Fig. 5.26)** such as coconut trees (*Cocos* sp.) and Joshua trees (*Yucca brevifolia*) can reach enormous size. There are actually three different types of monocot trees: (1) palms, such as the coconut tree, which are unbranched and do not have true secondary growth (Fig. 5.26a); (2) pandans (*Pandanus* sp.), which have branched stems but also lack true secondary growth (Fig. 5.26b); and (3) tree lilies (such as *Yucca*) and others (Fig. 5.26c), which have branches, a cambium, and true secondary growth.

Palm stems have a large number of vascular bundles, most of which are leaf traces. The leaves are large, and the leaf base wraps around the stem. Each leaf has many leaf traces. Some thickening does occur at the bases of palms, but this comes from basal adventitious roots (look at the very bottom of the palm tree in Fig. 5.26a). Some thickening throughout the stem results from the division and enlargement of parenchyma cells; this is called *diffuse secondary growth*. It is not true secondary growth because a cambium is lacking.

Several monocot plants, including *Yucca*, *Agave* (century plant), and *Dracaena* (dragon's blood tree), have true secondary growth from a cambium. The stems of these plants are tapered—that is, thin at the top and thick at the base. This architecture results from the pattern of cells the cambium produces. The monocot cambium is unique in forming mostly parenchyma cells with secondary vascular bundles embedded in this ground tissue **(Fig. 5.27).** The secondary vascular bundles are different from primary vascular bundles in that the xylem surrounds the phloem.

Some older monocot stems have a corklike layer that replaces the epidermis. This layer, called *storied cork,* is not considered to be periderm because no cork cambium is present. Storied cork forms by the division of parenchyma cells along the stem periphery.

a

Thomas L. Rost

b

D. Cavagnaro, Visuals Unlimited

c

Thomas L. Rost

Figure 5.26 Treelike monocot plants. (a) Palms are unbranched and lack true secondary growth. (b) *Pandanus* sp. is branched and lacks true secondary growth. (c) The Joshua tree (*Yucca brevifolia*) has a branched stem and true secondary growth.

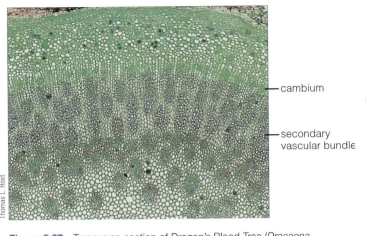

cambium

secondary
vascular bundle

Thomas L. Rost

Figure 5.27 Transverse section of Dragon's Blood Tree (*Dracaena draco*). The cambium layer produces secondary vascular bundles and parenchyma cells. ×12.

| 5.4 | STEM MODIFICATIONS FOR SPECIAL FUNCTIONS |

Stems of both dicots and monocots may be adapted for functions other than support, transport, and the production of new growth. They may, for instance, serve as protective devices or as attachment organs for vines, carry on photosynthesis, or store food or water.

Rhizomes are underground stems **(Fig. 5.28a).** They are usually light colored and burrow into the ground just below the surface. In some plants, such as Bermuda grass (*Cynodon dactylon*), the rhizomes can be quite deep, about 40 cm. Because rhizomes are modified stems, they have internodes and nodes. Small scalelike leaves sometimes form at the nodes, but they do not grow or become photosynthetic. The buds in the axils of these scale leaves elongate, producing new branches that extend to the soil surface and form new plants.

Tubers are the enlarged terminal portions of underground rhizomes. The potato plant (*Solanum tuberosum*) has three types of stems: (1) upright leafy stems; (2) underground rhizomes; and (3) swollen ends of rhizomes, the tubers **(Figs. 5.28b and 5.29a).** The structure of a tuber shows it to be a stem. The eyes of a tuber are actually lateral buds formed in the axil of small scale leaves at a node. The internodes of the tuber are short, and the tuber body is filled with parenchyma cells containing starch, a storage form of sugar.

Corms and bulbs are shoot structures modified for storage of food. A **corm** is a short, thickened underground stem with thin, papery leaves. *Gladiolus* sp. corms are good examples. The central portion of the corm accumulates stored food, which is used at the time of flowering **(Fig. 5.28c).** New corms can form from the lateral buds on the main corm. A **bulb** differs from a corm in that the food is stored in specialized fleshy leaves **(Fig. 5.28d).** The stem portion is small and has at least one terminal bud (to produce a new, upright leafy stem) and a lateral bud (to produce a new bulb). Food stored in the bulb is used up during the initial growth spurt

rhizome

a

b

c

(a–d) Thomas L. Rost

d

Figure 5.28 Specialized stems. (**a**) Rhizome of *Canna* lily. (**b**) Potato tuber (*Solanum tuberosum*). (**c**) top view of corm in *Gladiolus*. (**d**) Split view of a daffodil (*Narcissus*) bulb showing short stem and thick fleshy leaves.

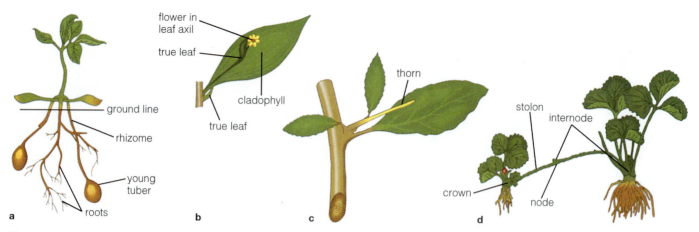

Figure 5.29 (**a**) A young potato plant (*Solanum tuberosum*) showing development of young tubers developing at the ends of rhizomes. (**b**) Cladophyll in butcher's broom (*Ruscus aculeatus*). (**c**) Thorn growing from the axil of a *Pyracantha* sp. leaf. (**d**) Stolon in strawberry (*Fragaria* sp.).

each spring. A table onion (*Allium cepa*) is an example of a bulb.

Cladophylls (also called *cladodes*) are flattened photosynthetic stems that function as and resemble leaves. They are not actually leaves in a developmental sense because they develop from buds in the axils of small, scalelike leaves. Cladophylls may bear flowers, fruits, and small leaves. Butcher's broom (*Ruscus aculeatus*; **Fig. 5.29b**), asparagus, and some cacti are examples of plants with cladophylls.

Thorns originate from the axils of leaves (**Fig. 5.29c**) and help to protect the plant from predators. Some thorns actually have leaves growing on them. From developmental evidence, we know that prickles and spines (though outwardly similar structures) are not modified stems. Spines are actually modified leaves; prickles, such as those on rose stems, are really modified clusters of epidermal hairs.

Stolons, also called *runners*, are aboveground horizontal stems. Strawberry plants (*Fragaria* sp.) send out stolons (**Fig. 5.29d**). At each node of the stolon, a small leaf will form, and in its axil a root and a bud will sprout to initiate a new strawberry plant. The stolon helps the plant spread. Some successful weed plants, such as Bermuda grass, also can spread by stolons.

5.5 ECONOMIC VALUE OF WOODY STEMS

Trees are, without doubt, among the most valuable things on earth. Our forests are home to countless plants and animals. They also are a source of the raw material for many useful products. They purify our air, keep our soil from washing away, and affect our weather patterns. The implications of the loss of our forests are too terrible to consider. We must be extremely careful to protect and manage this natural resource even as we grow and harvest woody plants for thousands of uses.

Trees are grown for wood, for sugar (from maple trees, *Acer saccharum*), and for secondary products such as turpentine (from the pine, *Pinus palustris*), natural rubber (from the Brazilian rubber tree, *Hevea brasiliensis*), and chewing gum (from the sapodilla tree, *Achras sapota*). These products are harvested by making cuts in the woody stem and collecting the sap that runs out (**Fig. 5.30**). The raw secondary product is then processed to make commercially important materials. This is a truly renewable resource because trees can live and continue to produce for many years.

In contrast, when trees are used for lumber, they are destroyed. In the manufacturing of lumber, a tree is removed from the forest with large machines and trucked or floated down a river to a sawmill. The log is moved to the mill and oriented so that tangential cuts are made along its

George Hunter/Tony Stone Images

Figure 5.30 Maple syrup being harvested from a sugar maple tree (*Acer saccharum*).

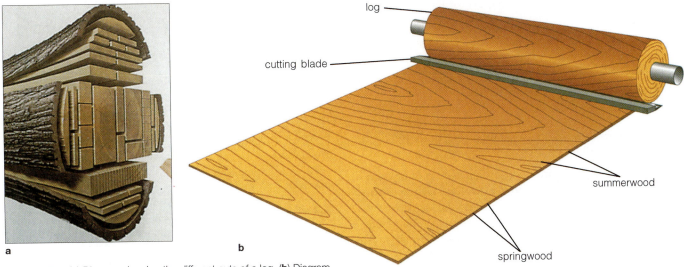

label: log
label: cutting blade
label: summerwood
label: springwood

Figure 5.31 (**a**) Diagram showing the different cuts of a log. (**b**) Diagram showing a sheet of plywood being peeled from a slowly revolving log.

length to make boards (Fig. 5.31a). Different types of wood grain can be obtained, depending on the type of cut made and on the quality and type of wood used.

Sheets of plywood are made by steaming the log and then placing it against a large lathe that literally peels thin sheets of wood from the slowly revolving log (Fig. 5.31b). The sheets are then cut to size, and multiple sheets are glued together to make plywood of the desired thickness.

For other industrial uses, wood is not sawed into boards or cut into sheets, but broken down by machines into individual fibers, tracheids, and vessel members. This pulp is then processed to make paper and cardboard products. In paper making, the lignin in the secondary cell walls is removed. In the past, such waste products have caused environmental pollution. Recently, innovative products using these wastes have been developed, including soil conditioners and adhesives.

Recycling paper products is now a major activity, and this important practice means that our natural tree resources may be preserved for many generations to come. Processing recycled paper is not without its hazards; it releases vast quantities of toxic inks and chemicals used in printing. Intelligent processing, however, has resulted in the discovery of ways to detoxify and reuse these materials.

KEY TERMS

axial system
bulb
cladophylls
cork cambium
corm
diffuse porous
earlywood
fascicular cambium

fusiform initials
heartwood
interfascicular cambium
latewood
lenticels
periderm
phellem cells
phelloderm cells

phellogen
primary thickening meristem
ray initials
ray system
rhizomes
ring porous
sapwood

secondary phloem
secondary xylem
springwood
stolons
summerwood
thorns
tubers
vascular cambium

SUMMARY

1. The functions of stems are: (1) to provide the axis for the attachment of the leaves, buds, and flowers; (2) to provide pathways for movement of water and minerals from the roots and food from the leaves; and (3) to produce new cells, tissues, leaves, and buds.

2. The SAM forms the three primary meristems: protoderm, procambium, and ground meristem.

3. Bundles of procambium form in a ring pattern (in dicot stems) between the two parts of the ground meristem (which will become the pith and cortex). In monocot stems, the bundles are scattered throughout the ground tissue, and the terms pith and cortex are usually not used. Procambium cells will become primary xylem and phloem.

4. Leaves are arranged on the stem according to four basic patterns (phyllotaxis): alternate, opposite, whorled, and spiral.

5. The vascular cambium in dicot stems forms from the fascicular cambium (by division of residual procambium cells within the vascular bundles) and from the interfascicular cambium (by division of parenchyma cells between vascular bundles).

6. Cells in wood are arranged into axial and ray systems. The cells of the axial system come from fusiform initials in

the vascular cambium; those of the ray system come from ray initials.

7. One year's increment of growth in wood equals one annual ring. Cells in earlywood (springwood) have relatively large diameters and form in the spring. Latewood (summerwood) cells form later in the summer and have smaller diameters. Ring porous wood has large vessel members only in the earlywood; diffuse porous wood has large vessel members throughout the annual ring, but the cells in latewood are smaller.

8. Angiosperm wood (hardwood) is more complex anatomically than gymnosperm wood (softwood). Gymnosperm wood is composed mostly of tracheids and simple rays, whereas angiosperm wood is composed of vessel members, tracheids, fibers, parenchyma cells, and more complex rays.

9. The bark consists of all tissues outside of the vascular cambium—secondary phloem, layer or layers of periderm. Bark occurs in different patterns.

10. The periderm forms from the cork cambium by divisions of outer cortex cells. Cork cells form to the outside of the cork cambium, and phelloderm cells form to the inside. During subsequent years, new layers of periderm are added in the secondary phloem.

11. Different kinds of buds occur on woody stems: terminal buds, lateral buds, and flower buds. Bud scale scars and leaf scars form after bud scales and leaves fall from the stem.

12. Monocot stems have scattered vascular bundles and usually do not have secondary growth. Palm trees increase their thickness by diffuse division of parenchyma cells. Monocot trees, such as *Yucca* and *Dracaena,* have a cambium and true secondary growth with a vascular cambium.

13. Several different modified stems occur in flowering plants, including rhizomes, corms, bulbs, tubers, cladophylls, thorns, and stolons.

14. Trees are ecologically essential and provide many products for humans. It is imperative that they be conserved.

Questions

1. Be able to diagram and describe the primary meristems that develop into the primary tissues.

Table of Primary Meristems and Primary Tissues

Primary Meristem	Primary Tissue
Protoderm	Epidermis
Ground meristem	Cortex and pith
Procambium	Xylem and phloem

2. Given an illustration of a section of a stem, be able to identify the tissues and cell types present. Also, be able to make a diagram (cross section or longitudinal section) of a stem and label all of the tissues and cell types.

3. Name and describe the four types of phyllotaxis. List one plant of each type found in your area.

4. Describe the initiation of vascular cambium and the formation of secondary xylem and phloem in a stem.

5. Describe the initiation and activity of the cork cambium forming the periderm.

6. How do the stems of monocots and dicots differ anatomically?

 InfoTrac® College Edition

http://infotrac.thomsonlearning.com

Wood

Kaib, M. 1999. Enlightenment in burnt forests (Laboratory of Tree-Ring Research). *Whole Earth* Winter, p. 20. (Keywords: "burnt" and "tree-ring")

Layton, J. 2001. Grains of truth: Wood fragments harbour tales of fortune, history and justice. Translating them is all in a day's work for this timber sleuth (champions of science: Jugo Ilic). *Ecos* April-June, p. 31. (Keywords: "Layton" and "timber")

Scuderi, L.A. 1993. A 2000-year tree ring record of annual temperatures in the Sierra Nevada Mountains. *Science* 259:1433. (Keywords: "Sierra" and "tree-ring")

Bark

Swain, R.B. 1996. The beauties of winter bark. *Horticulture, The Magazine of American Gardening* 74:44. (Keywords: "winter" and "bark")

The Shoot System II: The Form and Structure of Leaves

Visit us on the web at http://biology.brookscole.com/plantbio2 for additional resources, such as flashcards, tutorial quizzes, InfoTrac exercises, further readings, and web links.

1. Typically, the most important functions of leaves are photosynthesis (creation of chemical energy, in the form of carbohydrate, from sunlight and carbon dioxide) and transpiration (evaporation of water from the leaf surface). Some leaves are modified for specialized functions such as water storage, protection, and others.

2. A leaf usually consists of a leaf blade and a petiole. Leaves with one leaf blade are called *simple leaves*; leaves with many leaflets are called *compound leaves*.

3. Leaves are composed of epidermis, mesophyll, xylem, and phloem tissues. A waxy cuticle covers the epidermis; the mesophyll is where photosynthesis usually occurs. The xylem conducts water, and the phloem transports sugars.

4. Leaf formation is initiated at the shoot apical meristem.

5. Leaf fall in the autumn involves an active process of cell division and cell breakdown at the abscission layer in the petiole.

6.1   THE FUNCTIONS OF LEAVES

Green plants, algae, and a few species of bacteria use sunlight as an energy source. Other inhabitants of the earth obtain energy by consuming—directly or indirectly—what the green plants produce. **Photosynthesis** is the process used by plants to synthesize sugars and release oxygen; the chloroplasts in plant cells trap light energy and convert it to chemical energy.

Leaves provide large surfaces for the absorption of light and carbon dioxide. Both are the raw materials for photosynthesis. In general, most leaves are thin; therefore, no cells lie far from the surface. This architecture makes it easy for leaf cells and chloroplasts to absorb light and to imbibe carbon dioxide gas. At the same time, however, it promotes the loss of water from plants, because any opening that will permit carbon dioxide to pass also will allow water to evaporate. Evaporation, also called **transpiration** in plants, helps cool the leaf and acts as the driving force for water transport (see Chapter 11); however, excessive evaporation places the plant in danger of dehydration. To a great extent, leaf form (morphology) and anatomy are a compromise between capturing light and carbon dioxide and conserving water. Water-saving structures (for example, a thick waxy cuticle to inhibit evaporation) become important for plants that occupy hot, dry regions, such as rocky cliffs and deserts, where water loss is a great hazard.

This chapter discusses the form and anatomy of leaves, describes their development, and notes the modifications that help plants deal with environmental challenges.

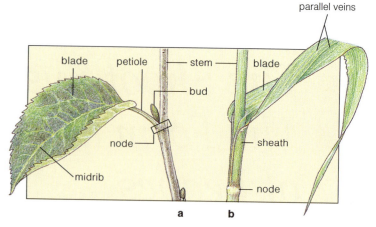

Figure 6.1 Typical dicot (**a**) and monocot (**b**) leaves are compared. Dicot leaves usually have netted venation and a midrib; monocot leaves usually have parallel veins. (Art by D. & V. Hennings)

Leaves Are Shaped to Capture Light

The leaves of dicot and monocot plants (defined in Chapter 5) have similar functions, but their basic designs differ. Monocot plants often have leaves with a strap-shaped leaf blade, their vascular bundles commonly run parallel through the leaf, and their leaf bases usually wrap around the stem (Fig. 6.1b). Dicot leaves typically are composed of a flat, thin portion (the blade), with the vascular bundles in a netted pattern, and a narrow portion (the petiole) (Fig. 6.1a). The **blade** is the part that absorbs most of the light; the **petiole** is designed to hold the blade away from the stem. Not only does this give the blade greater exposure to sunlight, but it also allows the blade to move in the air so that gas exchange can take place easily.

THE LEAF BLADE   In most dicot and monocot plants, the cells of the leaf blade contain many chloroplasts and perform most of the photosynthesis in the plant. The blade provides a broad, flat surface for capturing light and carbon dioxide. Leaves with a single blade (Fig. 6.2a) are called **simple leaves.** The form of the blade varies widely among different plants, though. For instance, the leaf edges or margins may be entire (smooth), toothed, or lobed. Some leaves are so deeply divided that the blade forms several separate units, or leaflets (Fig. 6.2b); leaves having this design are called **compound leaves.** There are two types of compound leaves: A *palmately compound leaf* has its leaflets diverging from a single point, and a *pinnately compound leaf* has leaflets arranged along an axis. These variations in leaves are useful in classifying plants.

Quite often in nature there is a practical reason for differences in structures that have similar functions. For compound leaves, the question naturally arises of how a plant might benefit by having its blade divided into leaflets. Perhaps the most likely answer is that the spaces between leaflets allow better airflow over the leaf surface. This may help cool the leaf (because of increased evaporation) while improving carbon dioxide uptake. However,

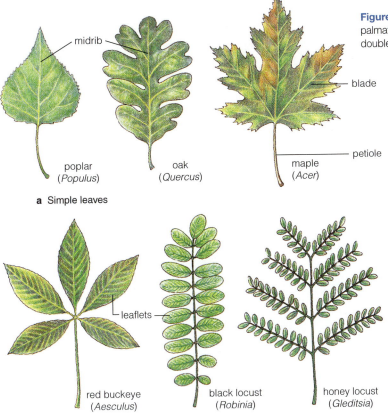

midrib

blade

petiole

poplar
(*Populus*)

oak
(*Quercus*)

maple
(*Acer*)

a Simple leaves

leaflets

red buckeye
(*Aesculus*)

black locust
(*Robinia*)

honey locust
(*Gleditsia*)

b Compound leaves

extending the leaf blade away from the stem, the petiole reduces the extent to which the blade is shaded by other leaves. It also allows the blade to move in response to air currents. This is important because a layer of stagnant air, the boundary layer, forms at the leaf surface. During photosynthesis, the boundary layer becomes depleted of carbon dioxide. Leaf movements as a result of air currents help bring fresh air containing carbon dioxide to the leaf surface and cool the leaf by increasing evaporation. These movements are passive reactions to air movement and are not under any kind of mechanical regulation.

Petioles are variable in shape. They may be long or short, cylindrical or flat in cross section. They are usually attached to the base of the leaf blade (Fig. 6.1a). Leaves that lack a petiole are called *sessile*.

THE SHEATH In most monocot leaves, the leaf base typically wraps entirely around the stem to form the **sheath (Fig. 6.3).** In grasses (for example, corn) the sheath extends almost the entire length of the stem internode (Fig. 6.1b). If you were to carefully examine a leaf of corn or of several other grasses, you would see a small flap of tissue, a **ligule,** extending upward from the sheath (Fig. 6.3b,c). The function of the ligule is not really understood, but it apparently keeps water and dirt from sifting down between the stem and leaf sheath. In some grass species, such as barley, two additional flaps of leaf tissue, *auricles,* extend around the stem at the juncture of the sheath and blade (Fig. 6.3c).

Another interesting characteristic of grass leaves is that they grow from the base of the leaf sheath. Anyone who has had to cut the lawn in the summer has experienced this phe-

there may be other adaptive or developmental reasons for these differences.

THE PETIOLE Dicot leaf blades are not directly attached to the stem. Instead, a petiole serves that purpose. The petiole is the narrow base of most dicot leaves. Its function is to improve photosynthetic efficiency in several ways. By

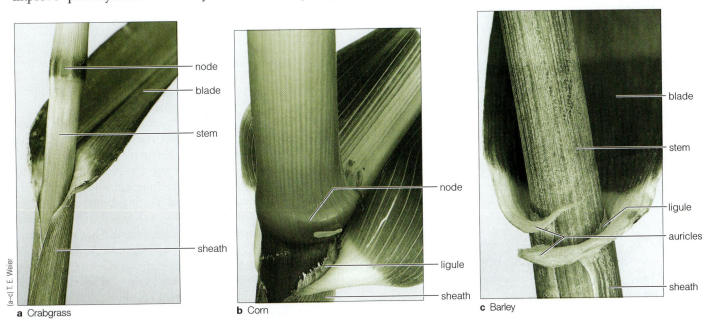

(a–c) T. E. Weier

node

blade

stem

sheath

node

ligule

sheath

blade

stem

ligule

auricles

sheath

a Crabgrass

b Corn

c Barley

Figure 6.3 Examples of monocot leaves. (**a**) Crabgrass (*Digitaria sanguinalis*). (**b**) Corn (*Zea mays*), showing ligule. (**c**) Barley (*Hordeum vulgare*).

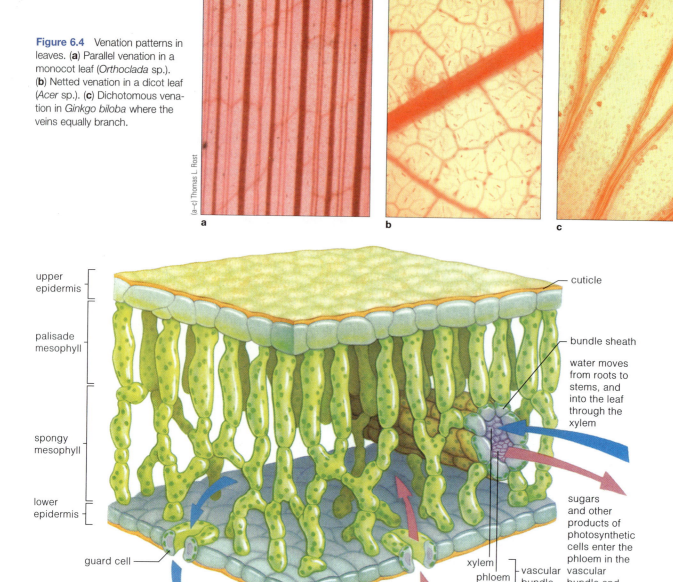

Figure 6.4 Venation patterns in leaves. (**a**) Parallel venation in a monocot leaf (*Orthoclada* sp.). (**b**) Netted venation in a dicot leaf (*Acer* sp.). (**c**) Dichotomous venation in *Ginkgo biloba* where the veins equally branch.

(a–c) Thomas L. Rost

upper epidermis

palisade mesophyll

spongy mesophyll

lower epidermis

guard cell

cuticle

bundle sheath

water moves from roots to stems, and into the leaf through the xylem

xylem
phloem
} vascular bundle (vein)

sugars and other products of photosynthetic cells enter the phloem in the vascular bundle and *depart* from the leaf

Oxygen and water vapor *depart* from the leaf through stomata.

a

carbon dioxide from the air *enters* the leaf through stomata

palisade mesophyll

spongy mesophyll

b

Figure 6.5 Tissues of a typical leaf blade. (**a**) Diagram of the upper and lower epidermis, palisade and spongy mesophyll, and a vascular bundle (vein). (Art by Raychel Ciemma) (**b**) Leaf blade of pennyroyal (*Mentha pulegium*). ×350. (Photo by Lesley Sunnell, UC Santa Cruz.)

nomenon. Have you ever wondered how the grass blades keep growing week after week? The reason for this is the **intercalary meristem** of the leaf, a unique feature at the base of grass leaves. The function of this particular meristem is to make new cells at the base of the grass leaf, allowing for continued growth of the mature leaf. The intercalary meristem is not active for the entire life of the leaf, and it will eventually stop dividing when the leaf reaches a certain age or length.

LEAF VEINS   The internal connections of the leaf to the rest of the plant are through veins. Leaf **veins** are actually vascular bundles. Vascular bundles (see Chapter 4) are composed of xylem (to transport water) and phloem (to transport sugars). The pattern they make in leaves often is quite elaborate (**Fig. 6.4**). Leaves differ from one another in many ways, including the arrangement of veins. Monocot leaves, for example, have **parallel venation,** meaning that there are several major veins that run parallel from the base to the tip of the leaf, and minor veins run perpendicular to them (Figs. 6.1b and 6.4a). In contrast, many dicot leaves have **netted venation** (Fig. 6.4b). In this pattern, a major vein, the *midvein* (also called the midrib) usually runs up the middle of the leaf, and lateral veins branch out from it. The precise pattern of veins differs among species of plants and is sometimes used as an identifying characteristic. Another type common in ferns and some gymnosperms is called **open dichotomous venation;** in this pattern, the veins have Y-branches with no small veins interconnecting them (Fig. 6.4c).

The Arrangement of Leaf Cells Depends on Their Functions

A closer look at the cells and tissues of the leaf blade **(Fig. 6.5)** reveals anatomical structures that aid in photosynthesis, transport of food and water, and the transpiration of gases. The outside surface of the leaf is covered by an epidermis, the ground tissue is chloroplast-filled mesophyll, and the vascular tissue is in the form of vascular bundles (veins). The petiole has specialized structures that enable it to support the leaf blade and conduct food, water, and minerals. The petiole also is the point at which leaves are shed in the autumn, at the end of the growing season.

THE EPIDERMIS The epidermis covers the entire surface of the blade, petiole, and leaf sheath; it is continuous with the epidermis on the surface of the stem. In most leaves, the epidermis is a single layer of cells that consists of epidermal cells, guard cells, subsidiary cells, and trichomes. Epidermal cells are flattened in a cross-sectional view of a leaf (Fig. 6.5). In surface view, epidermal cells often are irregular and sometimes look like puzzle pieces **(Fig. 6.6a).** The outer cell walls of epidermal cells are sometimes thickened and are covered by a waxy layer, the cuticle (Fig. 6.5a). The waxy cuticle inhibits the evaporation of water through the outer epidermal cell wall.

Because the cuticle blocks most evaporation, special openings are needed in the epidermis for the controlled exchange of gases. These openings exist as small pores, each lying between two guard cells (Fig. 6.6). Two guard cells plus the pore form a **stoma** (plural, **stomata**). In most species of plants, the guard cells are surrounded by specialized epidermal cells called subsidiary cells **(Fig. 6.6b).** These cells typically occur in a recognizable pattern, and they may play a role in the mechanism to open and close the stomatal pore. The term **stomatal apparatus** describes the guard cells and their subsidiary cells. The stoma permits the entry of carbon dioxide needed for photosynthesis and the loss of water vapor by transpiration. Transpiration (see Chapter 11) cools the leaf surface by evaporation and, in the process, pulls water up from the roots. There are hundreds or even thousands of stomata per square centimeter of leaf area. Usually, there are more stomata on the bottom (abaxial surface) of the leaf than on the top **(Table 6.1),** probably to minimize water loss caused by direct sunlight on the top of the leaf. Stomata also occur in the epidermis of young stems and some flower parts most likely to ensure water movement to these plant parts.

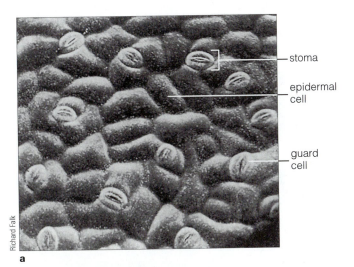

Richard Falk

a

— stoma
— epidermal cell
— guard cell

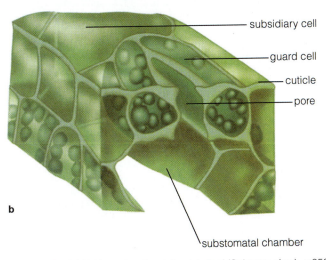

— subsidiary cell
— guard cell
— cuticle
— pore

b

— substomatal chamber

Figure 6.6 (**a**) Epidermal surface of potato leaf (*Solanum nigra*). ×350. (**b**) Diagram of a single stomatal apparatus.

Table 6.1 Average Numbers of Stomata per Square Centimeter in Different Plant Species		
Plant	Upper Epidermis	Lower Epidermis
Alfalfa (*Medicago*)	16,900	13,800
Apple (*Malus*)	0	28,400
Bean (*Phaseolus*)	4,000	28,100
Cabbage (*Brassica*)	14,100	22,600
Corn (*Zea*)	5,200	6,800
English oak (*Quercus*)	0	45,000
Oat (*Avena*)	2,500	2,300
Potato (*Solanum*)	5,100	16,100
Tomato (*Lycopersicon*)	1,200	13,000

PLANTS, PEOPLE, AND THE ENVIRONMENT:

Insect-eating Plants

Insectivorous plants have special mechanisms for snaring and digesting their prey. Some of these involve leaves with elaborate trichomes. The Venus flytrap (*Dionaea muscipula*), for example, has specialized leaves that look as if they are hinged along the central vein separating the two lobes of the blade **(Fig. 1a)**. On the surface of the lobes are two kinds of trichomes: trigger hairs and secretory hairs. Trigger hairs are long cells with a multicellular base. When an insect moves two or more of these hairs, the trap is triggered, closing the leaf blade and entrapping the insect **(Fig. 1b)**. The insect's struggles push its body against several short secretory hairs with multicellular heads. These hairs secrete digestive enzymes that break down the soft parts of the insect's body into nutrients that the leaf then absorbs.

Sundews (*Drosera capillaris*) capture insects by producing a thick, sticky mucilage from "tentacles" **(Fig. 1c)**, which are multicellular structures much more complex than simple trichomes. When an insect lands on the leaf, it sticks to the mucilage, and the leaf curls to trap the insect. Secreted enzymes then digest the insect body. Pitcher plants secrete substances from glands located along the lip of the pitcher-shaped leaves **(Fig. 1d)**.

An insect that lands to feed and happens to crawl to the inside of the "pitcher" is likely to slip down the side. Downward-pointed sharp hairs impede the insect's progress as it tries to crawl out **(Fig. 1e)**. Eventually the insect tires and falls into the digestive fluid at the bottom of the pitcher. The amino acids, lipids, sugars, and minerals, especially nitrogen, released from breaking down the insect's body are used by the plant for growth. Insectivorous plants can survive without capturing and digesting insects, but this highly developed specialization gives them an advantage in nutrient-poor marshes and bogs.

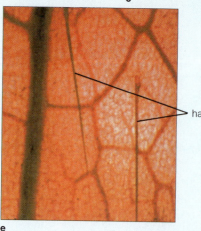

hair

Figure 1 Insectivorous plants that use trichomes on their leaves to attract, capture, or digest insects. (**a**) Venus fly trap (*Dionaea muscipula*). (**b**) Venus fly trap, which has captured a fly by closing leaf lobes. (**c**) Sundew (*Drosera capillaris*) captures insects by trapping them in mucilage secreted by "tentacles." (**d**) Pitcher plant (*Sarracenia purpurea*) entices insects to land by secreting sweet substances, only to block, with sharp hairs (**e**), the egress of those insects that wander into the leaf cup.

(a–e) Thomas L. Rost

Websites for further study:

UCLA Mathias Botanical Garden: http://www.botgard.ucla.edu/html/botanytextbooks/lifeforms/insectivorousplants/fulltextonly.html

Darwin—Insectivorous Plants:
http://pages.britishlibrary.net/charles.darwin3/insectivorous/insect_fm.htm

International Carnivorous Plant Society: http://www.carnivorousplants.org/

Several different types of **trichomes** occur as part of epidermal leaf surfaces **(Fig. 6.7).** Some trichomes are secretory structures. These usually have a stalk and a multicellular head that does the secreting (Fig. 6.7b). The material secreted often is designed to attract pollinators to flowers. The leaves of some desert plants such as saltbush (*Atriplex* sp.) have several short hairs. The end cell has a whitish cuticle, and the cells are filled with water. The hairs not only store water but also reflect sunlight and insulate the leaf against the extreme desert heat. Other leaves, such as those of the olive tree (*Olea europea*), have a mat of branched hairs that act as heat insulators. Leaves modified to eat insects as a food (mainly nitrogen) source also have specialized trichomes.

THE MESOPHYLL The most important function of a leaf is to use sunlight to manufacture carbohydrates. To this end, most of the leaf blade is made up of parenchyma cells containing chloroplasts. Collectively, this is a tissue, called **mesophyll.** The mesophyll in a typical dicot leaf is organized into two distinct regions (Fig. 6.5). The **palisade mesophyll** is usually on the upper (abaxial) surface of the leaf. The individual cells, called **palisade parenchyma cells,** are packed together and are shaped like columns oriented at right angles to the leaf surface. These cells are tightly packed together to more efficiently absorb sunlight. The **spongy mesophyll** (usually located on the bottom or abaxial portion of the leaf) is made of **spongy parenchyma cells,** which are usually irregularly shaped with abundant air spaces between them to allow for more efficient air exchange from their cell surfaces.

Cells of the mesophyll have thin, moist cell walls that allow rapid inward diffusion of carbon dioxide and outward diffusion of oxygen and water vapor (Fig. 6.5). Just under the stomata there often is an air space called a **substomatal chamber** (Fig. 6.6). The cell walls exposed to this space provide the main evaporative surfaces for gas exchange.

The veins consist of vascular bundles that form a network throughout the leaf for transport of water and nutrients. The xylem conducts water and dissolved minerals from roots to leaves, and the phloem transports carbohydrates made through photosynthesis in the mesophyll from leaves to the rest of the plant.

In a typical dicot leaf, a large-diameter midrib (the central vascular bundle) runs the length of the leaf (Figs. 6.1 and 6.2). In cross section, the xylem is in the upper part of the bundle **(Fig. 6.8).** A simple way to remember this is that in the stem the xylem is toward the middle so if a vascular bundle diverged from the stem into the leaf the xylem would end up toward the top of the leaf. Together with the vessel members and tracheids that transport water, xylem often has fibers around the outside of the bundle, which provide mechanical support, and parenchyma cells between the vessel members. The phloem makes up the lower part of the bundle. Phloem in flowering plants contains sieve-tube members (STMs, which conduct sugars), companion cells

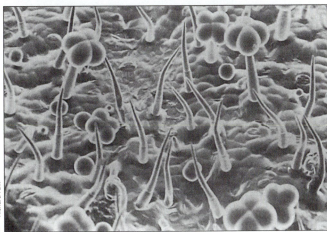

Figure 6.7 Trichomes. (**a**) Branched trichome on the leaf of *Aleurites* sp. ×210. (**b**) Pointed unicellular hairs from *Croton* sp. ×120.

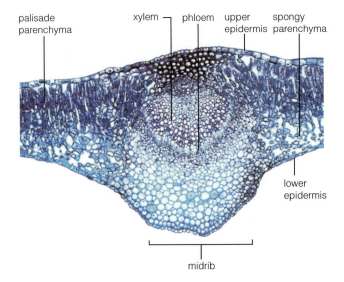

palisade parenchyma · xylem · phloem · upper epidermis · spongy parenchyma · lower epidermis · midrib

Figure 6.8 A lilac leaf (*Syringa vulgaris*) in cross section showing the midrib. ×93.

(which work with the STMs), fibers for support, and parenchyma cells (which load the sugars into the STMs). Most of the smaller vascular bundles have the same basic structure, with the smallest having only a few cells. If you look carefully at Figure 6.4b, you will see that many leaf veins end blindly in the leaf mesophyll. **Figure 6.9** shows what a leaf vein ending looks like; often, the very tip of the vein is quite simple with only one or two tracheary elements surrounded by bundle sheath cells. Sometimes a single layer of cells called the bundle sheath surrounds the vascular bundles (Figs. 6.5a and 6.9). The cells in the bundle sheath work to load sugars into the phloem and to unload water and minerals out of the xylem.

The main vascular system difference between many dicot and monocot leaves is that monocot leaves have several main bundles running parallel along the length of the leaf (Fig. 6.1b) rather than a single midrib with branches, as found in typical dicot leaves. In cross section, leaf vascular bundles have xylem at the top of the bundle and phloem at the bottom. In some grasses, a special type of photosynthesis called C_4 photosynthesis occurs. This type of photosynthesis can operate at lower carbon dioxide concentrations and greater temperature than regular photosynthesis. The C_4 photosynthesis requires more steps to convert carbon dioxide into sugars, and it involves a special role for the bundle sheath cells (see Chapter 10 for a description). Leaves of C_4 plants have enlarged bundle sheath cells because of this **(Fig. 6.10).**

Where Do New Leaves Come From?

Leaves originate from meristems. If you push away the older leaves of a bud, you will see several crescent-shaped bumps on the flanks of the very tip of the shoot apical meristem (SAM) (see Chapter 4). These are the new leaves being formed **(Fig. 6.11a).** In the early stages of development they are called **leaf primordia** (singular, **primordium**). The initiation of a new leaf involves a complicated set of events. Among them is the movement of some kind of chemical signal from the vascular bundles or perhaps from the already initiated leaf primordia. The location of a new leaf depends on the plant's phyllotaxis (see Chapter 5). The signal will trigger the cells at the initiation site to start dividing. These cells become the leaf primordium **(Fig. 6.11b).** The shape of the new leaf will be determined by how the cells in the primordium divide and enlarge.

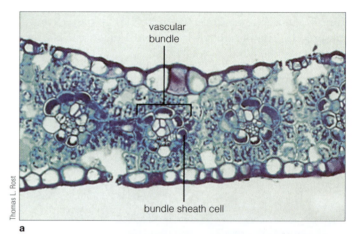

vascular bundle

bundle sheath cell

Thomas L. Rost

a

bundle sheath cell

Thomas L. Rost

b

Figure 6.10 C_4 leaf anatomy. (**a**) Corn leaf (*Zea mays*) cross section. Note the large bundle sheath. ×90. (**b**) Paradermal section (section cut parallel to the epidermis) through the leaf at the level of the vascular bundles shows the large bundle sheath cells. ×90.

vein ending
tracheid

bundle-sheath
parenchyma

spongy
parenchyma

mesophyll cell

intercellular
space

Figure 6.9 Diagram of a longitudinal section through a vein ending showing the mesophyll cells, bundle sheath cells, and tracheids.

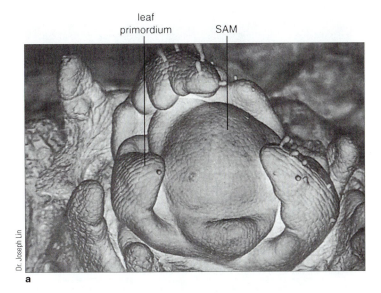

leaf primordium
SAM

Dr. Joseph Lin

a

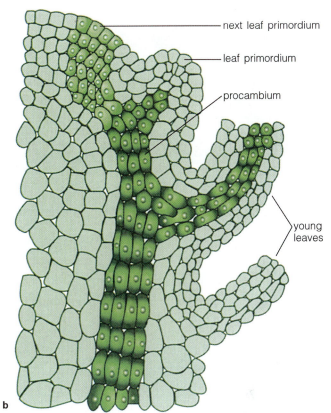

next leaf primordium

leaf primordium

procambium

young leaves

b

Figure 6.11 The development of leaf primordia. (**a**) Scanning electron micrograph of shoot apex of summer adonis (*Adonis aestivalis*) showing newly developing leaf primordia. SAM, Shoot apical meristem. ×80. (**b**) Diagram of a longitudinal section of the shoot apex showing leaf primordia in different stages of development. The *line* at the top indicates the position where the next leaf primordium will be initiated. The *shaded cells* are procambium.

6.2 LEAF FORM AND SPECIALIZED LEAVES

Leaf Shape May Depend on the Plant's Age and the Environment

Leaves that form during different stages of a plant's life cycle can have different shapes. The first leaves that form are part of the embryo found in seeds. These first *seed leaves*, or *cotyledons*, are slightly flattened and often oval-shaped (**Fig. 6.12**). The cotyledons are mostly storage organs, which wither and die during seedling growth; but, in some plants, they enlarge and conduct photosynthesis. In bean plants, the first true leaves are simple, and later stage leaves are compound (Fig. 6.12).

The phenomenon of different leaf shapes on a single plant is called **heterophylly.** One kind of heterophylly is related to the age of the plant. In ivy (*Hedera helix*), age-dependent changes in leaf form are related to the reproductive maturity of the plant. Leaves of juvenile ivy plants have three lobes, the plant is a vine, and juvenile plants do not flower. An adult ivy plant is upright, the leaves are not lobed, and the plant can produce flowers (**Fig. 6.13**).

Another kind of heterophylly is induced by the environment. In this case, the environment that the shoot apex is exposed to during leaf development influences the form of the leaf. For example, in certain marsh plants, leaves that develop underwater are thin and have very deep lobes; these are called *water leaves* (**Fig. 6.14**). When the shoot tip extends above the water in the summertime, thicker leaves with reduced lobing are formed; these are called *air leaves*.

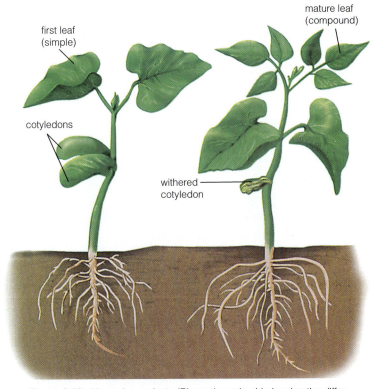

first leaf (simple)

mature leaf (compound)

cotyledons

withered cotyledon

Figure 6.12 Young bean plants (*Phaseolus vulgaris*) showing the different leaf shapes that form at different stages in the plant's life cycle: cotyledons, first leaves are simple with smooth margin, and mature leaves are compound.

Thomas L. Rost

a

b

Figure 6.13
Age-dependent
leaves of ivy (*Hedera
helix*). (**a**) Juvenile.
(**b**) Adult.

Thomas L. Rost

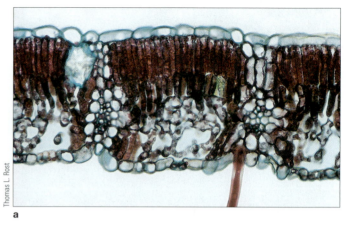

Thomas L. Rost

a

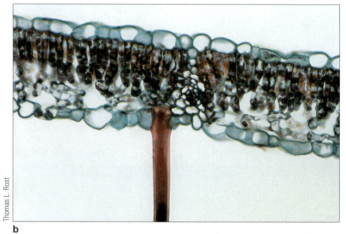

Thomas L. Rost

b

Figure 6.15   Sun and shade leaves. Sun leaves grow in full sun and tend to be thicker (**a**); shade leaves tend to be thinner (**b**) and have a larger surface area. ×244.

One last example can be seen on several tree species. Leaves that form on bottom branches, where they are mostly in the shade, tend to be thin and relatively large in surface area; these are *shade leaves.* The leaves that develop near the top of the same tree and in more direct sunlight, however,

tend to be thicker and smaller. These are *sun leaves* (**Fig. 6.15**). In each instance of heterophylly, the different leaf designs are adaptive to their environments.

Figure 6.14   Effect of environmental conditions on the development of leaves. Buttercup (*Ranunculus aquatilis*) leaves that form when the shoot tip is under water are thin and more deeply lobed; those that form in the air are thicker and less lobed. (After John G. Torrey, *Development in Flowering Plants*, by permission of Macmillan Publishing Company, © 1967 by John G. Torrey.)

6.3   ADAPTATIONS FOR ENVIRONMENTAL EXTREMES

The concept of evolution is discussed in Chapter 18, but the fundamental idea is that some individuals in a plant population will be better suited to the environment in which they grow because of some special characteristic, and therefore potentially produce more offspring and are more "fit." For example, of the plants growing in a dry climate, those with a thick cuticle to protect against water evaporation might be more successful—that is, more of their progeny will survive. Over time, the individual plants best suited for a particular environment will survive to maturity and will propagate new plants that also will be adapted for that particular environment. When researchers have compared plants from different environments, they have found that the species sharing a common environment tend to share specialized structural characteristics.

Plants that grow successfully in dry climates are called **xerophytes.** These plants tend to have leaves that are designed to conserve water, store water, and insulate against

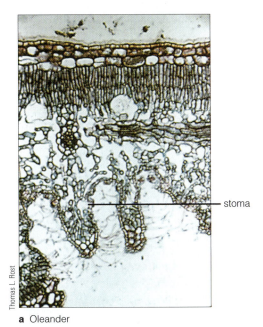

a Oleander

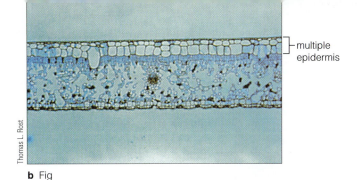

b Fig

multiple epidermis

mesophyll cells

c Jade

Figure 6.16 Xerophytic adaptations help plants survive dry and hot environments. **(a)** Oleander (*Nerium oleander*) with sunken stomata. ×225. **(b)** Fig (*Ficus* sp.) leaf with multiple epidermis that stores water and insulates against solar radiation. ×85. **(c)** Jade plant (*Crassula argentea*) mesophyll cells are tightly packed with few air spaces to inhibit evaporation. ×35.

stoma

heat. The oleander (*Nerium oleander*) **(Fig. 6.16a)** and fig (*Ficus* sp.; **Fig. 6.16b**) are good examples. The leaves have sunken stomata and a thick cuticle, to inhibit water loss, and sometimes a multiple layer of epidermis. The jade plant (*Crassula argentea*) has thick leaves with few air spaces in the mesophyll **(Fig. 6.16c).**

Another anatomical feature common to xerophytic plants is an abundance of fibers in leaves. These long, rigid cells help to support the leaf and keep it from losing its shape when it dries. These fibers often have commercial value.

Plants that grow in moist environments are called **hydrophytes.** The leaves of such plants tend to be thin, have a thin cuticle, and often are deeply lobed (Fig. 6.14). Because these leaves always have an abundance of water, they lack characteristics to conserve water.

Leaves from plants in moderate climates are called **mesophytes.** The typical dicot leaves discussed in this chapter are mesophytes (Fig. 6.5).

6.4 MODIFICATIONS FOR SPECIAL FUNCTIONS

Spines are one of a handful of leaf modifications that help plants protect against predators. Spines of various species of cactus are actually modified leaves **(Fig. 6.17a).** They are made of cells with hard cell walls. In the black locust, the spines are stipules, which are projections that form at the base of some leaves. Spines are pointed and quite dangerous to potential predators.

Figure 6.17 Leaf modifications. **(a)** Cactus (*Opuntia* sp.) spines. **(b)** Tendrils from leaflets of Trumpet flower (*Bignonia capreolata*). **(c)** Plantlets along the leaf margin of air-plant (*Kalanchoe pinnata*). (Photos by (a) Thomas L. Rost; (b) William J. Weber/Visuals Unlimited; (c) Jerome Wexler/Photo Researchers)

a

tendril

b

plantlets

c

ECONOMIC BOTANY:

Commercially Important Fibers

People have used fibers from plants for at least 10,000 years. Cotton fibers are known to have been used in Mexico almost 9,000 years ago. Cotton fibers, however, are not really *fibers* in terms of their cell type; they are epidermal hairs that grow from the cotton seed coat. The kind of fibers discussed here are elongated sclerenchyma fibers joined into long strands by their overlapping ends (see Chapter 4). Bundles of these fibers can be long and very strong. Archaeological evidence has revealed that Native Americans were using *Agave* (sisal) to make cords about 10,000 years ago, and Stone Age Europeans were using flax plants (*Linum usitatissimum*) for linen at about the same time.

Where do fibers come from? They tend to form in strands around vascular bundles and within the ground tissues of stems and leaves. Their purpose in the plant is to strengthen stems and leaves so they do not break when moved by the wind. They act like reinforcing bars used in concrete structures. Ancient peoples figured out how to extract these strands and use them for fabrics and ropes.

There are basically two types of commercially important fibers: hard fibers and soft fibers (see the **Table**). Hard fibers occur in leaves of monocotyledonous plants such as *Agave* **(Fig. 1).** Their long, strap-shaped leaves contain thick bundles of fibers, often surrounding the vascular bundles. These fibers are very stiff and often have thick lignified cell walls (see Chapter 4). Fiber strands are extracted from the leaves, usually by crushing or beating the leaves and then scraping them to expose the fiber bundles. Once exposed, they are pulled and dried. These fibers (called sisal) are woven into coarse fabric or twisted and braided into rope.

Another example is Manila hemp (*Musa textilis*), which is used to make the huge ropes that tie down ships, because it resists decay and breakdown by salt water. Some Manila hemp also is used to make lightweight fabric and tea bags.

Soft fibers (sometimes called *bast fibers*) are found mostly in the stems of dicotyledonous plants, where they occur just outside the phloem as part of vascular bundles. Individual soft fiber cells are longer than those of hard fibers. They tend to be rather soft and flexible and are sometimes unlignified. These properties make such fibers valuable for clothing fabric, string, rope, and even cigarette paper and money.

Probably the most famous soft fiber is linen from the flax plant, which currently accounts for about 2% of the world's textiles **(Fig. 2).** The plant, usually about 1 meter (more than 3 feet) in height, is pulled from the soil and bundled to dry. The fibers are extracted by a process called retting. This involves placing the dried plants in a watery bacterial slurry and allowing the bacteria to decay the ground tissues away, leaving the fiber bundles. These are then combed out, removed, and dried **(Fig. 3).** A newer way to extract them is to pass the undried plants through crushers, comb the bundles free, and dry them. The dried fibers are then spun into yarn, dyed, and woven into fabric. The elegant feel and texture of linen have made it popular for a long time. Its only problem is that it wrinkles easily. Currently, linen fibers are chemically treated, and the yarn is woven with cotton, rayon, or both. This makes contemporary fabrics that are more wrinkle resistant but retain some of the attractive texture of the original.

Ramie (*Boehmeria nivea*) has the longest known fiber cells. They can be up to 550 mm (almost 2 feet) in length and are stronger than cotton fibers. Mercerized ramie, made by treating the fiber bundles with alkali, is among the strongest natural fibers used in clothing. The next time you see a very exotic looking fabric, look at the label, you will probably find it is part linen or ramie.

Commercially Important Fibers			
Species	Common Name	Uses	Fiber Length, mm
Hard fibers			
Agave sisalana	Sisal	Rope, coarse fabric	0.8–8
Musa textilis	Manila hemp	Marine rope, cloth	2–12
Phormium tenax	New Zealand hemp	Rope	2–12
Soft fibers			
Linum usitatissimum	Flax (linen)	Cloth, paper money	9–70
Cannabis sativa	Hemp	Canvas, rope	5–55
Boehmeria nivea	Ramie	Fabric	50–550
Corchorus capsularis	Jute	Burlap, carpet backing	0.8–6

Figure 1 *Agave sisaliana* plants (foreground) are grown as a crop in various places in the world; shown here in Malaysia. The large tree in the background is a baobab (*Adonsonia digatata*). The swollen trunk of this tree stores water.

Figure 2 Flax plant, *Linum usitatissimum*.

Figure 3 Flax fibers used to make linen are being processed.

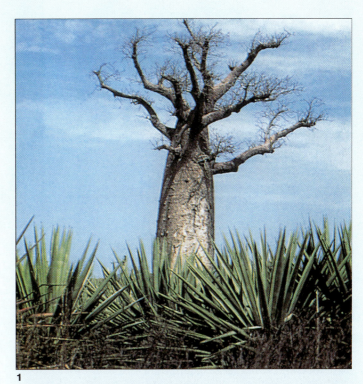

1

(1) Tim Metcalf; (2–3) Masters of Linen/USA

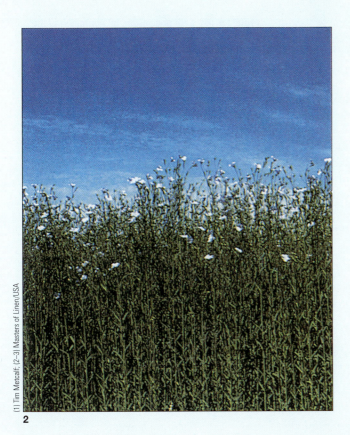

2

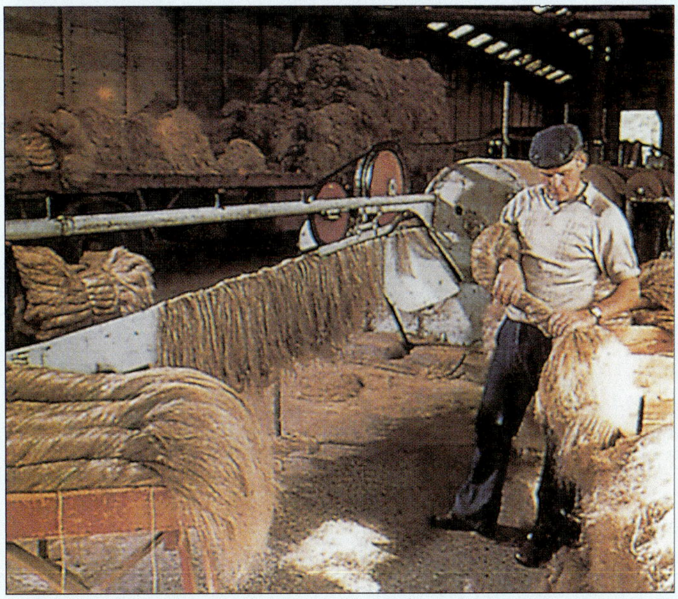

3

Websites for further study:

Flax and Linen: http://www.rootsweb.com/~belghist/Flanders/Pages/flaxLinen.htm

Scherzer—Manila Hemp:
http://www.univie.ac.at/Voelkerkunde/apsis/aufi/aurel/novara/nov1319.htm

Coir—Coconut Fiber: http://www.bio.ilstu.edu/armstrong/syllabi/coir/coir.htm

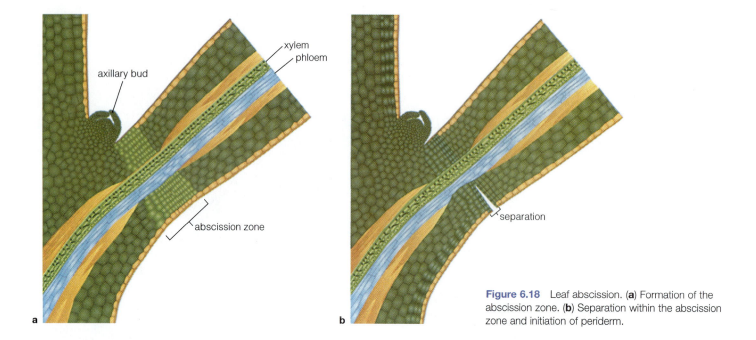

axillary bud

xylem
phloem

abscission zone

separation

a

b

Figure 6.18 Leaf abscission. (**a**) Formation of the abscission zone. (**b**) Separation within the abscission zone and initiation of periderm.

Tendrils are modified leaflets that wrap around things and support the shoot. In some plants, the entire leaf becomes a tendril (Fig. 6.17b).

Bulbs have very thick leaves sometimes called bulb scales. Bulbs are modified branches with a short, thick stem and short, thick storage leaves. These leaves, or bulb scales, store food and water (see Fig. 5.28d).

Some leaves produce plantlets. In the air-plant (*Kalanchoe pinnata*; Fig. 6.17c), the leaves have notches along their margins. In the bottom of each notch, a special meristem develops that produces a new plantlet. After these little plantlets grow for a short time, they fall off the leaf and root into the soil. Similar mechanisms occur on other plants. This means of reproducing is a form of vegetative, or asexual, reproduction (see Chapter 12).

Leaf Abscission Is the Seasonal Removal of Leaves

The separation of plant parts from the parent plant is a normal, continually occurring process. Leaves fall, fruits drop, flower parts wither and fall away, and even branch tips or whole branches may separate as a normal part of a plant's life. A familiar example is the fall of leaves from certain woody dicot plants in the autumn. In practically all cases, separation, or **abscission,** is the result of differentiation in a specialized region known as the **abscission zone** at the base of the petiole (Fig. 6.18a). Parenchyma cells making up the abscission zone are usually smaller than cells in adjacent tissues, and they may lack lignin in their cell walls. Xylem and phloem cells tend to be shorter in the vascular bundles at the base of the petiole, and fibers often are absent in the abscission zone. These anatomical features make this zone an area of structural weakness (Fig. 6.18b). As the zone gradually weakens, the vascular bundle cells become plugged, and the

leaf finally falls. The leaf scar seals over with waxy materials so that pathogens cannot enter after the leaf falls off.

Environmental cues, such as cold temperature or short days, are important controls for abscission. These cues induce hormonal changes that, in turn, affect the formation of the abscission zone. The leaves move much of their nutrients back to the stem; then they lose color, fall to the ground, and decompose, returning whatever nutrients they still retain back to the soil to be recycled.

KEY TERMS

abscission	palisade parenchyma cells
abscission zone	parallel venation
blade	petiole
bulbs	photosynthesis
compound leaves	sheath
heterophylly	simple leaves
hydrophytes	spines
intercalary meristem	spongy mesophyll
leaf primordial (primordium)	spongy parenchyma cells
ligule	stoma (stomata)
mesophyll	stomatal apparatus
mesophytes	substomatal chamber
netted venation	tendrils
open dichotomous venation	transpiration
palisade mesophyll	trichomes
	xerophytes

SUMMARY

1. The main functions of typical leaves are photosynthesis and transpiration.

2. Leaves of flowering plants are generally flat and have thin blades that are attached to the stem by petioles or sheaths. Veins (vascular bundles) transport food and water and strengthen the leaf blade. Leaf blades may be simple or compound; leaf margins may be entire, toothed, or lobed.

3. A cross section through the blade of a typical leaf shows the following tissues: upper epidermis, palisade and spongy mesophyll, vascular bundles, and lower epidermis. Generally, a waxy cuticle coats the epidermis. Guard cells of the epidermis form stomata that control gas exchange.

4. Mesophyll tissue is composed of palisade and spongy parenchyma cells and veins. Chloroplasts found in mesophyll cells convert sunlight to energy stored in carbohydrates during photosynthesis.

5. Modified leaves such as spines, bulbs, and tendrils serve many special functions, such as protecting against herbivores and holding on to a substrate.

6. Young leaves are produced at the SAM.

7. Environmental factors such as strong light intensity, nutrient deficiency, and too much or too little water can affect leaf form and anatomy. *Heterophylly* refers to different leaf forms induced by environmental changes—for example, air and water leaves.

8. The formation of a definite abscission zone across the petiole is responsible for leaf fall. Environmental cues and hormonal changes affect the formation of the abscission zone.

Questions

1. What are some functions of leaves?

2. What are the main distinctions between monocot and dicot leaves?

3. Describe the parts of a simple and a compound leaf.

4. Describe the general anatomical characteristics of a leaf. Make a cross-sectional diagram and label the three tissue systems and the tissues in each.

5. What are the specific functions of the epidermis, mesophyll, xylem, and phloem in leaves?

6. Some leaves are modified to live in extreme environments. Identify one characteristic of each of the following leaf types: mesophyte, hydrophyte, and xerophyte.

7. Leaves fall off many plants in the autumn. Describe the process of leaf abscission.

8. Describe the functions of these modified leaves: tendrils, bulb scales, and spines.

 InfoTrac® College Edition

http://infotrac.thomsonlearning.com

Leaves

Adler, T. 1996. Mapping out a leaf's mountains and valleys. *Science News* 149:325. (Keywords: "Adler" and "leaf")

Baker-Brosh, K.F., Peet, R.K. 1997. The ecological significance of lobed and toothed leaves in temperate forest trees. *Ecology* 78:1250–1255. (Keywords: "toothed" and "leaves")

Marcotrigiano, M. 2001. Genetic mosaics and the analysis of leaf development. *International Journal of Plant Sciences* 162:513. (Keywords: "mosaics" and "leaf")

Nadeau, J.A., Sack, F.D. 2002. Control of stomatal distribution on the Arabidopsis leaf surface. *Science* 296:1697. (Keywords: "stomatal" and "distribution")

The Root System

Visit us on the web at http://biology.brookscole.com/plantbio2
for additional resources, such as flashcards, tutorial quizzes,
InfoTrac exercises, further readings, and web links.

1. The principal functions of roots are absorption of water and nutrients, conduction of absorbed materials into the plant body, and anchorage of the plant in the soil. Many roots have relationships with bacteria and fungi in the rhizosphere (soil zone near the root).

2. The root is initiated in the embryo as the radicle. It penetrates into the soil and branches.

3. The root tip is composed of the root cap, the root apical meristem, the region of cell elongation, and the region of cell maturation.

4. Roots are composed of the following tissues: epidermis, cortex, endodermis, pericycle, xylem, and phloem.

5. The endodermis regulates ion movement into the xylem. The Casparian strip embedded in the cell wall inhibits mineral movement through the wall. The pericycle is the site of lateral root initiation and contributes to vascular cambium and cork cambium formation.

7.1 THE FUNCTIONS OF ROOTS

Although plant biologists have studied plants for hundreds of years, they have largely ignored the root system and its functions. This is probably because roots are underground, where they cannot be seen. Though they are hidden from view, roots play a critical role in the everyday activities of plants. The main functions of roots are anchorage of the plant body in the soil (or to a surface, in the case of some vines), absorption of water and minerals from the soil, storage of foods, and conduction of food and water from the soil and from storage reserves into the shoot. Other root functions also are described in this chapter.

The root system becomes more complex as the plant grows from a single root in a young seedling to a massive system of branched roots, often weighing several tons in large trees. During all stages in a plant's growth cycle, there is a balance between the

shoot system and the root system **(Fig. 7.1).** The root system must be able to supply the shoot with sufficient water and mineral nutrients, and the shoot system must manufacture enough food to maintain the root system.

The contact zone between the root surface and the soil is called the rhizosphere **(Fig. 7.2).** This region, only a few millimeters in thickness, is an interesting and unique zone. Its complex chemistry includes organic material, gases, and nutrients from the roots. Consequently, the bacteria and fungi near roots often are richer and more diverse than in soil farther away. Soil fungi and bacteria form important symbiotic (mutually beneficial) relationships with roots. These also are discussed later in this chapter.

Every time you try to pull weeds you are reminded of the *anchorage* function of roots. Plants with unusual roots provide anchorage in atypical ways. In ivy (*Hedera helix*) and the climbing fig (*Ficus pumila;* see "PLANTS, PEOPLE, AND THE ENVIRONMENT: The Climbing Fig and Its Adhesive Pads"), clusters of roots develop from the stems to form an unusual adhesive pad allowing these vines to cling to vertical surfaces. Parasitic plants such as dodder (*Cuscuta* sp.) sink specialized roots into their host and then tap into its water and nutrient supply.

All plants need water. In herbaceous (nonwoody) plants, water accounts for about 90% of the plant's weight. Water is needed for all root processes and for every metabolic reaction. Plants also need the dissolved salts and minerals, such as potassium, sulfur, phosphorus, calcium, and magnesium, contained in the soil water. For these reasons, plants have developed an elaborate system in roots for the *absorption* and *conduction* of water.

shoot system

1 m

root system

Figure 7.1 Root and shoot systems of a tree at least 10 years old. Note that the majority of the roots lie within the upper 1 m of soil and that the total volume of roots is equal to or greater than the shoot branches.

PLANTS, PEOPLE, AND THE ENVIRONMENT:

The Climbing Fig and Its Adhesive Pads

Everybody has seen vines growing on the sides of buildings and fences. Have you ever wondered how they manage to climb so high and stick to different surfaces?

Climbing vines have developed three different growth strategies—twining, winding, and clinging. *Twining* vines such as wisteria (*Wisteria sinensis*) encircle poles and branches by twisting their stems just like winding a rope around a pole. *Winding* vines such as clematis (*Clematis hybrida*) support themselves by tendrils. Tendrils, remember, are modified leaves that can twist around poles and other things and attach to them. A third type are called *clinging* vines; English ivy (*Hedera helix*) is probably the best-known example.

Another common plant that uses this strategy is the climbing fig (*Ficus pumila*). Clinging vines climb surfaces by means of clusters of adventitious roots that emerge from internodes and form into an interesting unique and sticky structure called an adhesive pad.

Ficus pumila, the climbing fig, is commonly used as an ornamental to cover walls and fences **(Fig. 1).** It is an aggressive plant, and it sprouts shoots that rapidly cover almost any surface including glass windows. Perhaps the first report of the clinging vine mechanism in climbing fig was by Charles Darwin (1875) in his book, *The Movements and Habits of Climbing Plants,* where he notes that roots on the stems secreted a "viscid fluid," possibly "India-rubber," that adheres to any surface.

Adhesive Pads Hold the Climbing Fig to Any Surface

The adhesive pad in *F. pumila* develops from a cluster of adventitious roots that are initiated just basal to a node near the tips of shoots **(Figs. 2 and 3).** Adventitious roots are initiated in pairs on either side of a vascular bundle at the second to third internodes of young stems. After emerging through the cortex and epidermis of the stem, the adventitious root clusters elongate until they are a few millimeters in length (Fig. 2). The roots then stop elongating, root hairs form on them, and they secrete a very sticky cement substance. The clusters of adventitious roots and their root hairs all stick together forming the adhesive pad (Fig. 3). If the adventitious roots do not touch a substrate, they usually dry up; if they touch moist soil, they tend to branch and change to a terrestrial form. If you ever see a climbing fig, try pulling one away from the wall; you will see how strong they stick.

Websites for further study:

Ecology and Evolutionary Biology Conservatory:
http://florawww.eeb.uconn.edu/acc_num/198500793.html

Hawaiian Ecosystems at Risk—Plants of Hawaii:
http://www.hear.org/starr/hiplants/reports/html/ficus_pumila.htm

Figure 1 Climbing fig plant (*Ficus pumila*) growing on a privacy wall in Davis, CA.

Figure 2 Tip of a climbing fig branch showing the position of an adhesive pad (*red circle*).

Figure 3 Close-up view of an adhesive pad taken through a glass plate. Note that some of the adventitious roots have merged and some are separate. Hairs on the surfaces of these roots secrete a sticky substance that glues the pad to any surface.

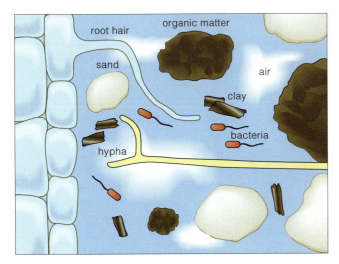

Figure 7.2 The soil area immediately around a root is called the rhizosphere. This area, shown near the root tip, is rich in soil microorganisms, bacteria and fungi (hypha), and in nutrients from the root body and from sloughed-off root cap cells. (Redrawn with permission from Epstein, E., and Bloom, A.J. (2004). *Mineral Nutrition of Plants: Principles and Perspectives, 2nd Ed.* Sinauer Associates, Sunderland, MA, p. 415.)

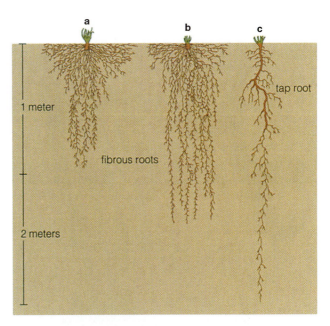

Figure 7.3 Fibrous and tap root systems. (**a**) Shallow, spreading, fibrous root system. (**b**) Fibrous root system penetrating the soil evenly up to 1 m in this example. (**c**) Tap root system in which main primary root penetrates soil 2 m or more.

All roots, even slender ones with a primary function of absorption, may temporarily store small amounts of food. For example, when sugar moves into roots more rapidly than it can be used by growing cells, it may be converted to starch and stored. During slow growth periods, rather large quantities of starch are stored in woody roots of orchard trees. This food constitutes a reserve that is used when flowering and active growth are resumed in spring. Carrots (*Daucus carota*) and beets (*Beta vulgarus*) are common root crops.

7.2 TYPES OF ROOT SYSTEMS

There are two basic types of root systems: fibrous and tap. They are distinguished by the way they develop and by their appearance. Many grasses and small garden plants, for instance, when pulled up bring with them a massive clump of soil. This happens because the **fibrous root system** of these plants consists of several main roots that branch to form a dense mass of roots (Fig. 7.3a,b). A typical annual grass such as corn (*Zea mays*) or rye (*Elymus cereale*) will build an immense fibrous root system in one growing season. A single rye plant 50 cm in height, with 80 tillers (shoot branches), may have a root system surface area of about 210 m², compared with only about 5 m² for its aboveground shoot system.

Plants with a large storage root, such as carrots (*Daucus carota*), have a **tap root system,** consisting of one main root from which lateral roots branch (Fig. 7.3c). Some desert plants have a rapidly growing tap root system that enables them to penetrate the soil quickly to reach deep sources of water.

Differences in the Design of Root Systems Help Plants Compete for Water and Minerals

Plants that grow in close proximity compete for water, mineral nutrients, and light energy. As discussed in Chapter 26, plants reduce the effects of competition by using different parts of the environment, including the soil. This is one reason why root systems of different species also occupy different depths in the soil.

When plants growing in the prairie in the middle United States are carefully excavated, three general categories of roots are evident, on the basis of how deep the roots grow (Fig. 7.3). Some grassland species, such as blue grama (*Bouteloua gracilis*), possess a shallow root system, most of the roots being within the top 15 cm of soil. Other species, such as buffalo grass (*Buchloe dactyloides*), have evenly distributed roots as deep as 1.5 m. Still others, such as locoweed (*Crotalaria sagittalis*), have a tap root system, which lacks width but runs deep. By using different depths of the soil, these plants reduce competition for moisture and dissolved minerals.

Plants Have Different Types of Roots

When seeds germinate, the embryonic root or **radicle** (see Chapter 14) extends by the division and elongation of cells to form the primary root (Fig. 7.4a). Tap root systems develop from one primary root, which then forms lateral roots. Further branching results in succeeding orders of roots. Fibrous root systems develop in a slightly different way. The embryos of most grasses have a single radicle, but in addition several other embryonic roots form just above the radicle; these are called **seminal roots.** The seminal roots emerge soon after the radicle, and all of these roots branch, making a fibrous root system (Fig. 7.4b).

Roots called **adventitious roots** originate on leaves and stems. There are several common examples of adventitious roots. In a young corn plant, soon after germination **prop roots** develop on the stem just above the soil (Fig. 7.5a). Prop roots absorb water and minerals, but they also support the plant in the soil.

lateral roots

seminal roots

tap root (primary root)

primary roots

a

b

Figure 7.4 **(a)** Seedling of pea (*Pisum sativum*) with a tap root and several lateral roots. **(b)** Seedling of wheat (*Triticum aestivum*). Seminal roots emerge from the hypocotyl (these are adventitious roots because they do not emerge from another root) and create the fibrous root system.

Banyan (*Ficus bengalensis*) trees grow in the salty mud of tropical lagoons and tidal marshes. Branches of these trees form adventitious roots (also called aerial roots because they are exposed to air). These extend down from branches into the soil, where they enlarge and actually hold up the large branches **(Fig. 7.5b).** These roots absorb water and nutrients, but their most important function is to prop up the stem. In India, merchants once held open-air bazaars among the prop roots and expansive branches of the banyan.

Mangrove trees (*Rhizophora mangle*) are native to low tidal shores and marshes in tropical and subtropical regions. In the mangrove **(Fig. 7.5c),** small adventitious roots called **pneumatophores** stick up from the mud. These roots absorb oxygen and increase its availability to the submerged roots. There are many other examples of adventitious roots.

Pieces of stem, such as a cane from a blackberry plant or a branch of willow, can be induced to make roots from their cut ends simply by placing them in moist soil. Leaves from *Begonia* and several other plants also can be rooted simply by soaking them in water. Many commercially important ornamentals are reproduced by root propagation from the leaves or stems.

leaf

node
stem

prop roots

lateral root

J. W. Perry Photo Library

a

b

aerial roots

c

Figure 7.5 Types of adventitious roots. **(a)** Corn (*Zea mays*) stem showing prop roots, which emerge from the stem just above the soil. These roots help support the shoot system. **(b)** Banyan (*Ficus bengalensis*) tree with an extensive aerial root system. **(c)** Extensive adventitious root system of mangrove (*Rizophora mangle*) growing in the tidal zone of Australia's tropical coast. Note the many air roots (pneumatophores) sticking up from the mud.

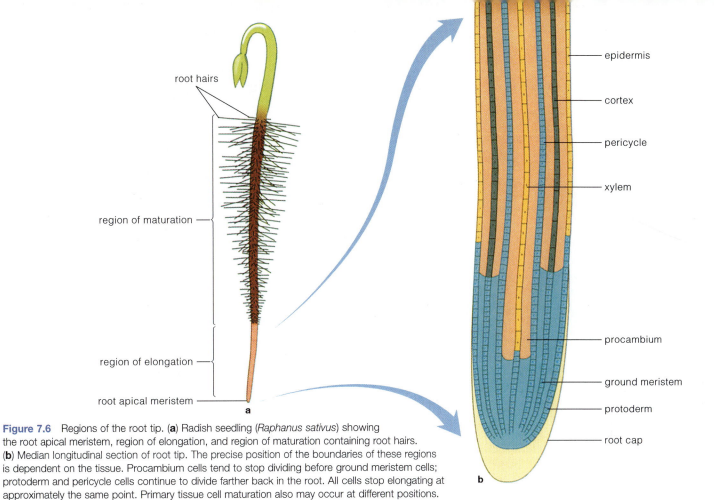

root hairs

region of maturation

region of elongation

root apical meristem

a

epidermis

cortex

pericycle

xylem

procambium

ground meristem

protoderm

root cap

b

Figure 7.6 Regions of the root tip. (**a**) Radish seedling (*Raphanus sativus*) showing the root apical meristem, region of elongation, and region of maturation containing root hairs. (**b**) Median longitudinal section of root tip. The precise position of the boundaries of these regions is dependent on the tissue. Procambium cells tend to stop dividing before ground meristem cells; protoderm and pericycle cells continue to divide farther back in the root. All cells stop elongating at approximately the same point. Primary tissue cell maturation also may occur at different positions.

7.3 THE DEVELOPMENT OF ROOTS

The tips of functional roots are thin (Fig. 7.4a) and usually white. If you were to dig up the roots of a big tree, you would see many very large roots, each of which could be followed through its branches to a thin, white tip. These tiny root tips are important parts of the root system because it is here, in just a few millimeters, where many of the important functions of roots take place.

The Root Tip Is Organized into Regions and Is Protected by a Root Cap

In a longitudinal section of a root tip viewed through a microscope, it is apparent that the cells are organized into three regions: the **root apical meristem** (RAM), the **region of elongation,** and the **region of maturation (Fig. 7.6a).** The developmental events that take place in the cells of each region are somewhat specific, but the regions do overlap **(Fig. 7.6b).**

The root cap at the tip of the root apex protects the RAM, a group of small, regularly shaped cells, most of which are dividing. These cells are organized into two

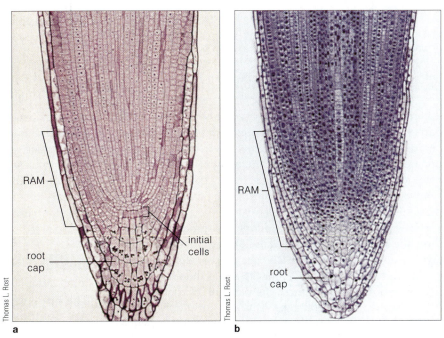

RAM

root cap

initial cells

a

RAM

root cap

b

Thomas L. Rost

Figure 7.7 The patterns of root apical meristem (RAM) divisions. (**a**) Longitudinal section of the root tip of flax (*Linum grandiflorum*). Notice that all cell files connect directly to specific tiers of cells just above the root cap. ×220. (**b**) Longitudinal section of onion (*Allium cepa*) root tip. In this type of root apex, the cell files terminate at a group of cells without any apparent organization. ×89.

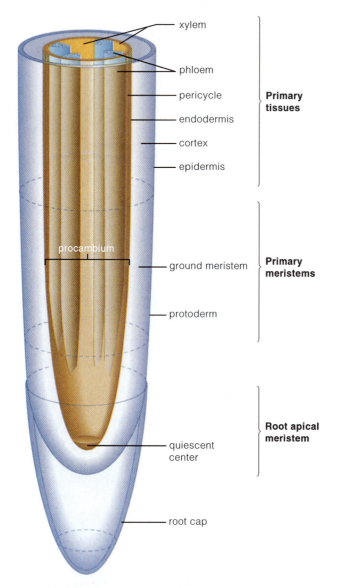

Figure 7.8 Three-dimensional diagram of a root in longitudinal view to show the relative positions of the primary meristems and primary tissues. The quiescent center is a small group of cells that are metabolically quiescent.

xylem
phloem
pericycle
endodermis
cortex
epidermis

Primary tissues

procambium

ground meristem
protoderm

Primary meristems

quiescent center

Root apical meristem

root cap

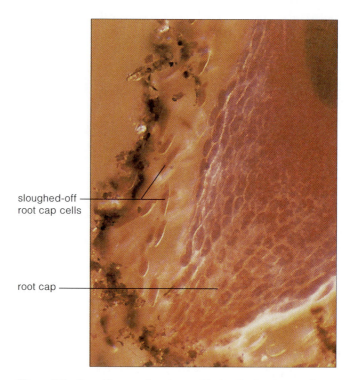

sloughed-off root cap cells

root cap

Figure 7.9 Corn (*Zea mays*) root cap with sloughed root cap cells (border cells) in the soil. ×60. (Photo by Professor Margaret McMully, McMaster University, Carlton, Ontario, Canada)

different patterns (Fig. 7.7a,b). A small, centrally located part of the RAM is called the quiescent center (QC) (Fig. 7.8), because its cells divide at an extremely slow rate. The function of the QC is not exactly known, but it seems to be activated during times of acute stress. It may be a site for the synthesis of plant hormones important for controlling root development.

Cells just apical to the QC divide and produce cells to form the root cap (Figs. 7.7 and 7.8), the thimble-shaped layer of cells that protect the RAM as the root elongates and pushes through the soil. The root cap also is the site of gravity perception, which controls the direction of root growth (see Chapter 15). Root cap cells are constantly being sloughed off at the very tip, but new cells are added by the apical meristem. The sloughed-off cells (also called border cells) can remain alive in the soil for a time, where they provide nutrients for soil bacteria and fungi in the rhizosphere (Figs. 7.2 and 7.9).

Just basal to the meristem region, toward the body of the root, is the region of elongation. Careful examination of a longitudinal section of a root shows that the boundary where cells stop dividing and start elongating is different for each tissue. The region of maturation is the site of root hair formation and the maturation of other cell types. The precise position of cell maturation in different cell files is variable; cells in some files mature close to the tip, and others mature farther back (Fig. 7.6b).

The Root Apical Meristem Forms Three Primary Meristems

As mentioned in Chapter 4, cells change their structure according to their position, and in roots the process of change begins in the RAM. The RAM differentiates into three primary meristems: the **protoderm, ground meristem,** and **procambium** (Fig. 7.8). These then go on to become the primary tissues of the root, as described in the next section and summarized in Table 7.1.

7.4 THE STRUCTURE OF ROOTS

The Epidermis, Cortex, and Vascular Cylinder Are Composed of Specialized Tissues

EPIDERMIS Protoderm cells differentiate into the epidermis, which in roots is composed primarily of long epidermal cells. Some cells of the protoderm develop into **root hairs** (Fig. 7.10) by the extension of epidermal cell walls into the

Table 7.1 Summary of Tissues and Meristems in Roots

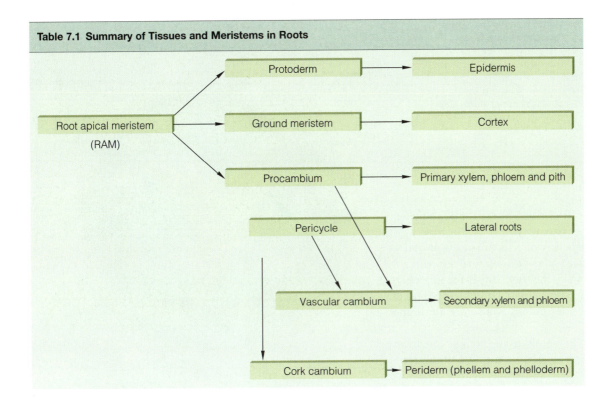

Root apical meristem (RAM)
- → Protoderm → Epidermis
- → Ground meristem → Cortex
- → Procambium → Primary xylem, phloem and pith

Pericycle → Lateral roots

Vascular cambium → Secondary xylem and phloem

Cork cambium → Periderm (phellem and phelloderm)

surrounding soil. Root hairs may be quite long, but they are only one cell. The cell walls of root hairs are thin and composed principally of cellulose and pectic substances. Root hairs tend to be sticky; therefore, soil particles cling to them (Fig. 7.10b). In most plants, the life of any one root hair is short; it functions for only a few days or weeks. New hairs are constantly forming at the apical end of the root hair zone, whereas those at the basal end are dying. Thus, as the root advances through the soil, fresh, actively growing root hairs are constantly coming into contact with new soil particles. In the rye plant, root hairs develop at an average rate in excess of 100 million per day.

Although nearly all ordinary land plants possess root hairs, a few plants, such as some gymnosperm trees (for example, firs), apparently lack them. Also, many aquatic plants have no root hairs. Moreover, land plants (for example, corn) that normally develop root hairs when the root system grows in the soil may develop no root hairs when the roots grow in water. In plants devoid of root hairs, absorption is accomplished entirely through the epidermal cells.

The epidermis in roots is usually one cell layer in thickness; but in the aerial roots of certain plants, such as orchids, a multilayered epidermis develops that stores and possibly absorbs water from the moist air.

J. W. Perry Photo Library

a

labels: seed, primary root, root hairs, root apex

b

labels: nucleus, vacuole, cytoplasm, soil particle, new root hair, new root hair

Figure 7.10 The development of root hairs. (**a**) Radish seedling (*Raphanus sativa*). (**b**) Stages in the development of root hairs. Note that the external epidermal cell wall protrudes and that the cell cytoplasm and nucleus move into the root hair near the tip. Root hairs are in close contact with soil particles and increase the water absorptive surface of the root.

Figure 7.11 A dicot root cross section from the area where maximum absorption occurs. The epidermis contains root hairs. The cortex has abundant air spaces between parenchyma cells. The endodermis is bounded by a Casparian strip. The pericycle is one cell layer in thickness. The primary xylem is distributed into three protoxylem points, with metaxylem in the middle of the root. The primary phloem alternates with primary xylem. Residual procambium occurs between the primary xylem and primary phloem.

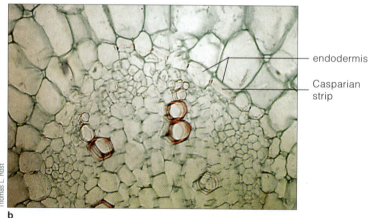

root hair
endodermis
cortex
epidermis
primary phloem
intercellular air spaces
protoxylem vessel member
metaxylem vessel member
pericycle

CORTEX The root cortex is derived from ground meristem and is composed chiefly of parenchyma cells. The innermost layer of the cortex, a single row of cells called the **endodermis (Figs. 7.11 and 7.12a,b),** plays a special role in controlling mineral accumulation by the root. This is the role of the **Casparian strip,** a waxy material embedded in the upper, lower (transverse), and side (radial) walls of endodermal cells (Fig. 7.12).

Water and dissolved minerals from the soil can move from cell to cell by two paths: They can travel through the porous walls of the cortex and epidermis, or they can move

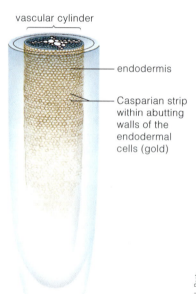

vascular cylinder

endodermis

Casparian strip within abutting walls of the endodermal cells (gold)

a

Figure 7.12 Control of nutrient movement into the xylem is a function of the endodermis. (**a,b**) The endodermis is a single layer of inner cortex cells that have a waxy strip (Casparian strip) embedded into their transverse and radial cell walls. (Art by Raychel Ciemma) (**c**) The strip keeps water from moving indiscriminately through the cell walls and into the vascular cylinder. (**d**) The Casparian strip makes water move through the endodermal cells, and in this way the plasma membrane can selectively control the uptake of nutrients. (Art by Leonard Morgan)

endodermis

Casparian strip

Thomas L. Rost

b

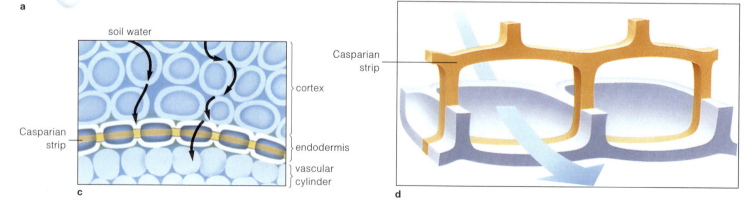

soil water

Casparian strip

cortex

endodermis

vascular cylinder

c

Casparian strip

d

through the living cells **(Fig. 7.12c,d).** Movement through the cell wall is free movement without any constraints. Movement into a living cell, however, is regulated because it involves crossing the plasma membrane. Some minerals can move across the membrane by diffusion. (This involves a gradient in which the concentration of a mineral on one side of a semipermeable membrane will try to equalize on the other side.) Some minerals (for example, potassium) can be moved across membranes through special proteins embedded in the membrane. These proteins can actually pump minerals into a living cell, even against a diffusion gradient.

The function of the endodermis is to guarantee that the minerals that finally reach the vascular cylinder can do so only by first passing across at least one plasma membrane. One reason this is important in roots is that it provides a mechanism to increase the concentration of needed minerals through pumps in the endodermis cell membrane.

In the roots of many plants, an **exodermis** containing Casparian strips occurs at the outer layer of the cortex, just inside the epidermis. This layer is present in many grass roots and in the aerial roots of orchids **(Fig. 7.13).** The exodermis apparently also functions to regulate ion absorption and accumulation.

VASCULAR CYLINDER The entire central cylinder of roots is composed of vascular tissue that differentiates from the procambium cells **(Fig. 7.14).** In the roots of dicot plants, the primary xylem usually consists of a central core of xylem elements organized into two or more radiating points (Fig. 7.11). In most monocot roots, the very center of the root is composed of parenchyma cells with the primary xylem and phloem forming in a ring **(Fig. 7.15).** The first xylem element to mature, the protoxylem, develops at the outer points of the xylem (Figs. 7.11 and 7.14b). Metaxylem, the last primary xylem to mature, differentiates in the center of the vascular cylinder (Fig. 7.14b,c).

The protoxylem is capable of transporting water while the root is elongating, which requires both the strength to withstand the forces that move water and still be flexible enough to stretch as the root elongates. This dual ability comes from a secondary cell wall in the shape of annular rings or spirals (see Fig. 4.9).

Metaxylem cells mature in regions of the root where elongation has been completed. Because they are no longer required to elongate, they form thick secondary cell walls with pits through which lateral exchange of water and minerals may take place. Protoxylem cells often become crushed after the metaxylem develops, but by then these cells are not needed.

In roots of monocots such as asparagus (*Asparagus officinale*), a central region of parenchyma cells forms (Fig. 7.15). This region is sometimes called a pith (this refers to the location of ground tissue in the center of stems, which is formed from ground meristem); but in roots it is part of the vascular cylinder and originates from procambium. Xylem (both protoxylem and metaxylem) of roots consists of several other cell types, including vessel elements, tracheids, parenchyma, and fibers.

Phloem cells form in the areas between the protoxylem arms (Figs. 7.11 and 7.14). The protophloem is actually the first part of the vascular system to become functional. These cells form at the periphery of the phloem and function primarily during root elongation. Metaphloem develops toward the inside and functions during the plant's adult life. Phloem of roots may consist of parenchyma, fibers, sieve-tube members, and companion cells.

The outer boundary of the vascular cylinder is the **pericycle** (Figs. 7.11 and 7.14c). This tissue is unique in that it remains capable of dividing for a long time. It has three functions in dicot roots: (1) It is the site where the development of lateral roots is initiated; (2) it contributes to the formation of vascular cambium; and (3) it contributes to the

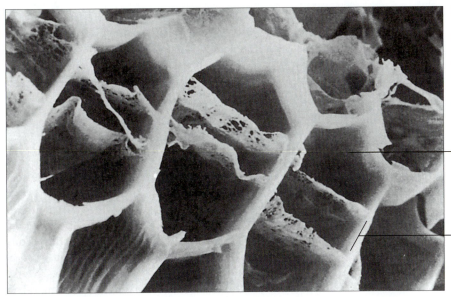

Figure 7.13 The Casparian strip in exodermis cells of an aerial root of *Epidendrum* orchid. The plasma membrane has been pulled away from the cell wall by soaking the roots in a salt solution. This is a scanning electron microscope image showing that the membrane is attached at the Casparian strips. ×800. (Zankowski et al., *Lindleyana* 2 (1987): 1–7. Used with permission of the authors.)

exodermal cell

Casparian strip

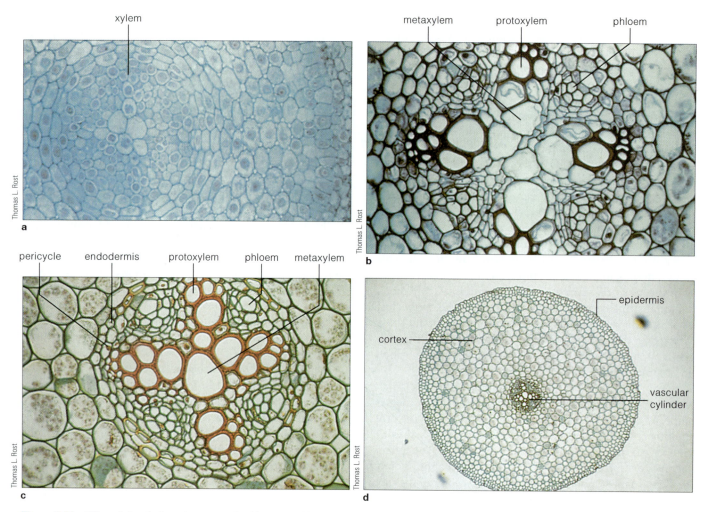

xylem

metaxylem   protoxylem   phloem

a

b

pericycle   endodermis   protoxylem   phloem   metaxylem

epidermis

cortex

vascular
cylinder

c

d

Figure 7.14 Differentiation in the primary growth of buttercup (*Ranunculus* sp.) roots. (**a**) Immature region, where no secondary walls have yet formed in the xylem. ×170. (**b**) The cells of the protoxylem have developed secondary walls (shown here stained red with safranin). This particular root has four protoxylem points with phloem between. ×190. (**c**) Fully mature root with all primary tissues differentiated. Note that the endodermal cells adjacent to the protoxylem points are lacking secondary walls. ×170. (**d**) The same section as (**c**), but showing all tissues. ×190.

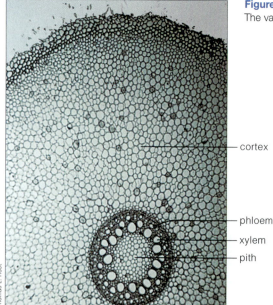

cortex

phloem
xylem
pith

Figure 7.15 *Asparagus officinale* root in transverse section showing all tissues and root regions. The vascular cylinder (stele) has a pith of all parenchyma cells. ×50.

formation of the cork cambium. In monocot roots, no secondary growth occurs; therefore, pericycle is involved only in lateral root initiation. The pericycle usually is one cell layer in thickness, but, in some roots, it has multiple layers.

Lateral Roots Are Initiated in the Pericycle

The initiation of lateral roots at particular locations is controlled by chemical growth regulators in the root that cause pericycle cells to begin dividing at specific sites **(Fig. 7.16a).** The **lateral root primordia** that result **(Fig. 7.16d)** continue to form new cells, which, in turn, elongate. Endodermal cells outside the primordium also divide for a short time, contributing cells to the tip of the new lateral root. As it expands **(Fig. 7.16b),** the lateral root pushes its way through and destroys the cortical cells and the outer epidermis. The breakdown of these cortical cells is thought to be at least

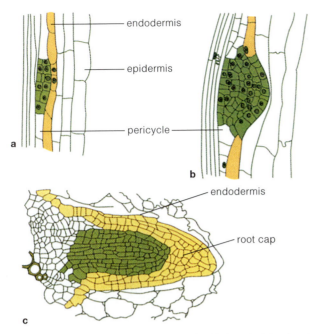

Figure 7.16 Lateral root formation. (**a**) Initiation of a lateral root of carrot (*Daucus carota*) by division of cells in the pericycle. (**b**) Formation of the lateral root primordium. (**c**) The young root pushing through the cortex. (**d**) Cleared pea (*Pisum sativum*) root showing root primordia. (**e**) Cleared pea root showing vascular connections of the main axis of the root and the lateral roots.

Maude Hinchee

partly the result of digestive enzymes released from the lateral root primordium. As the lateral root emerges, its cells become organized into a root cap and RAM (**Fig. 7.16c**,d). The wound formed by lateral root emergence is quickly healed by the secretion of mucilage and waxy substances by adjacent cortical cells. The vascular system of the main root axis and the lateral roots is connected (**Fig. 7.16e**).

The Vascular Cambium and the Cork Cambium Partially Form from the Pericycle

In long-lived dicot plants, the older regions of roots form secondary vascular tissues by activating a secondary meristem, the vascular cambium. Secondary growth is initiated by the division of pericycle cells and also some leftover or **residual procambium** cells located between the arcs of xylem and phloem (**Fig. 7.17a**). Residual procambium cells are actually

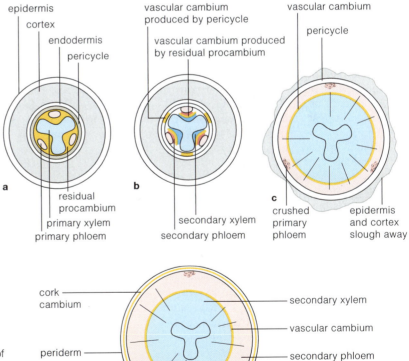

Figure 7.17 Secondary growth in a root. (**a**) At the completion of primary growth, an arc of residual procambium cells remains between the primary xylem and primary phloem. The pericycle is a complete cylinder. (**b**) The residual procambium starts to divide and joins with the pericycle cells outside the xylem arms to form a continuous cylinder of vascular cambium. The pericycle just outside the protoxylem points divides to form at least two cell layers. The inner layer joins with the residual procambium to form the vascular cambium; the outer layer stays part of the pericycle cylinder. (**c**) The vascular cambium forms secondary xylem internally and secondary phloem externally. The primary phloem is being pushed outward. (**d**) A cylinder of vascular cambium produces secondary xylem to the inside and secondary phloem to the outside. The primary xylem remains in the center of the root, the primary phloem has been crushed, and the pericycle that remains will form the cork cambium. The cork cambium forms the periderm after the epidermis and cortex die. The term bark refers to everything outside the vascular cambium.

PLANTS, PEOPLE, AND THE ENVIRONMENT:
Myths and Popular Uses of Roots

Over the millennia, through trial and error, ancient peoples developed traditional uses—medical and otherwise—for berries, seeds, leaves, stems, and roots. Much of this plant lore has little validity in medical fact, but some traditional plants have proven to be effective remedies.

During the Middle Ages in Europe, most people believed that the way a plant looked (its shape or color), how it smelled, or some other characteristic provided clues about its uses for people. This was called the Doctrine of Signatures (see "PLANTS, PEOPLE, AND THE ENVIRONMENT: Doctrine of Signatures" in Chapter 14). For example, the roots of the mandrake (*Mandragora officinarum*) are thick and fleshy and tend to be irregularly branched. With a little imagination, at least if you lived in the Middle Ages, the shape of a person could be seen in a root carefully extracted from the soil. People in medieval times assigned considerable importance to the manlike appearance of these roots; they believed that such roots could bring good fortune **(Fig. 1)**. In some cases, the roots were even ground up and eaten as a love potion. Mandrake root does not seem to have much real medicinal value. It does not really give anyone good luck, and it certainly does not work as a love potion; but, in some cases, an infusion made from mandrake root may help to control a cough.

Another root that looks vaguely like a person is ginseng (*Panax quinquefolium*). In Asia, ginseng root has been considered a cure-all for many ailments. Ginseng also grows in the United States, but it has only recently become popular. Currently, you can find it in most health food stores. Most likely, the subtle, supposedly restorative power associated with ginseng is psychological; however, some recent evidence suggests that chemicals in ginseng roots do act to calm some people.

A root with more widely accepted utility comes from cassava (*Manihot esculenta*), a shrubby South American plant. The whitish latex that exudes from a cut plant contains hydrocyanic acid, a deadly poison used by Brazilian Indians to make poisoned arrows. The interesting thing, however, is that when the latex is removed, the rich, starchy pulp is a good food source. Native people ground up cassava root, placed it in a sack, and hung it from a tree; when all the juice dripped out, the meal that remained could be made into cakes and eaten. It was called *farina*. Now cassava is grown commercially. You are probably familiar with the starch-rich pellets manufactured from cassava root, called tapioca.

Native Americans also have a rich plant lore that includes roots. The roots of several plants were used as a source of soap. One of these, soap weed (*Yucca* sp.), is currently used as a component of popular shampoos. Yucca root juice also was used as glue to fasten feathers to arrows. One type of soap root was used to induce vomiting as a step in ritual purification ceremonies; another type was thrown in streams to stupefy fish so that they could be harvested without using fishing gear. Roots are obviously important to the plant, but many also have uses for people.

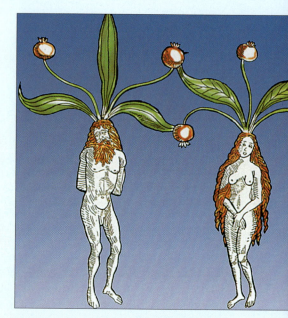

Figure 1 Shown are colored drawings of mandrake plants made from a black and white original from an old German Herbal. Herbals were among the first books ever made. They contained drawings of plants and sometimes discussed how they could be used. (From Lewis, W.H., Elvin-Lewis, M.P.F. 1977. *Medical botany: Plants affecting man's health.* New York: Wiley-Interscience, by permission of Wiley-Interscience. [Original art of "male and female" mandrake (*Mandragora officinarum*) from Peter Schöffer. *The German Herbarius,* Mencz. 1485.])

 Websites for further study:

Cassava Information Network: http://www.cassava.org/

Shaman Shop:
http://www.shamanshop.net/store/proddetail.cfm/ItemID/5281.0/CategoryID/3000.0/SubCatID/0.0/file.htm

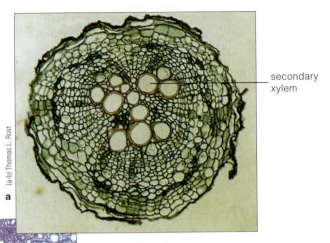

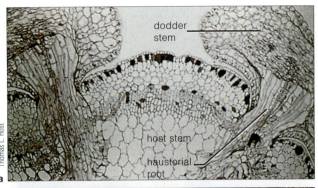

Figure 7.18 (**a**) Alfalfa (*Medicago sativa*) root transverse section with some secondary tissues and periderm on its periphery. ×125. (**b**) Woody root of oak (*Quercus* sp.). ×77.

Figure 7.19 (**a**) Haustorial roots of a plant parasite, dodder (*Cuscuta* sp.), infecting the stem of a host plant. ×53. (**b**) External view of dodder on tomato plants.

procambial cells that did not develop into primary xylem or primary phloem. They are now induced to divide, and they form secondary xylem to the inside and phloem to the outside **(Fig. 7.17b).** After a time, the crescent-shaped region of dividing cells joins with the pericycle, which also begins to divide, forming at least two layers of pericycle cells. The inner layer joins to the residual procambium to form an intact ring of vascular cambium **(Fig. 7.17c).** The outer layer remains as pericycle.

The secondary xylem or wood in roots with several years of secondary growth looks very much like that of woody stems. The only difference is that in young roots primary xylem occupies the middle of the root **(Fig. 7.18a,b),** whereas in young stems pith occupies the middle. Continued growth expands the root and finally causes the splitting, sloughing off, and destruction of the cortex and epidermis (Fig. 7.17c). The pressure created by this expansion apparently stimulates the remaining ring of pericycle cells to divide again. This last function of the pericycle converts it into the cork cambium, which forms the periderm (see Fig. 4.20). The bark in woody roots (which includes all cells from the vascular cambium outward) appears similar to that in stems, but it may be thinner and smoother on its outer surface **(Fig. 7.17d).**

Interestingly, there is only one monocot plant, the dragon's blood tree (*Dracaena draco*), known to have secondary growth in roots. Even very tall, treelike monocots like palms lack secondary growth in roots.

Some Roots Have Special Functions

Roots of a great many plants do not have the general characteristics common to most roots. This chapter has already discussed examples of adventitious roots, which arise from nonroot origins, and the uniquely shaped roots of clinging vines. Other plants, especially those forming partnerships with microorganisms, have specialized root structures. Some important examples are haustorial roots, root nodules, mycorrhizae, and contractile roots.

Parasitic plants such as dodder (*Cuscuta* sp.) anchor themselves by sinking **haustorial roots** into the vascular tissue of a host stem, thus tapping the host's water and nutrient supply **(Fig. 7.19).**

Although nitrogen is one of the most important elements needed by plants, most plants cannot use atmospheric nitrogen (N_2) directly. Certain legumes, such as peas (*Pisum sativum*) and soybeans (*Glycine max*), are capable of fixing nitrogen—that is, changing N_2 that diffuses into the soil into NH_4^+ (ammonium ion), which is usable by the plant. Nitrogen fixation is the result of an unusual relationship between the bacterium *Rhizobium* and the roots of legumes (see Chapter 19, Fig. 19.15). Root cells are infected by the passage of a thin infection thread of the bacteria into root hair cells and on through to the cortical cells. The bacteria then divide and stimulate the cortical cells to divide, thereby forming a **root nodule (Fig. 7.20).** The bacteria are the actual agents for fixing the nitrogen.

Mycorrhizae are short, forked roots common to as many as 90% of seed plants. These specialized root structures represent an association of roots with a soil-borne fungus (see Chapter 20).

Two types of mycorrhizal roots may occur, distinguished by whether the fungus penetrates into the root cells. **Ectotrophic mycorrhizae** are found in roots of such trees as

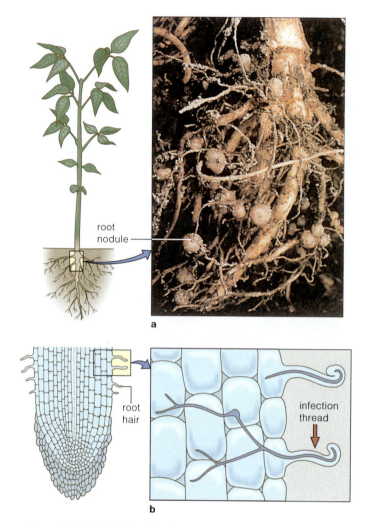

Figure 7.20 Bacterial nodules in legume roots. **(a)** Legume plants form bacterial nodules on roots. (Photo by Adrian P. Davies/Bruce Coleman Ltd.) **(b)** Bacteria enter the plant by passing through a tiny infection thread that penetrates root hairs. Once inside the host, the bacteria penetrate to the cortex of the root forming a swollen mass of cells filled with bacteria. The bacteria, called *Rhizobium* sp., fix nitrogen, which then passes up the plant body in a usable form.

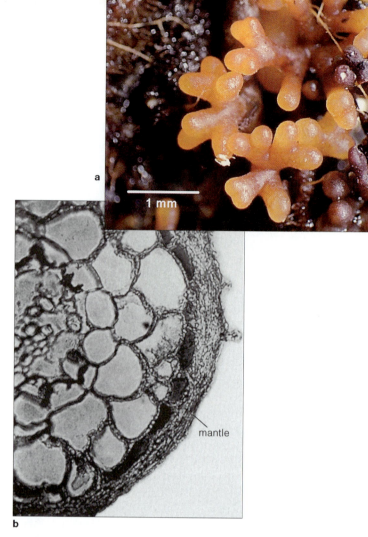

Figure 7.21 Ectotrophic mycorrhizae. **(a)** The typical Y-branched form in pine roots (*Pinus pinaster*). (line scale = 1 mm) (Photo by J. Parlade, IRTA-Departament de Proteccio Vegetal, Barcelona, Spain; reproduced by permission.) **(b)** Transverse section through an infected mycorrhizal root showing the mantle of fungal hyphae, and the hyphae growing between the cortical cell walls. ×240.

pine (*Pinus*), birch (*Betula*), willow (*Salix*), and oak (*Quercus*) trees. This type causes a drastic change in the root shape **(Fig. 7.21),** but the fungus does not penetrate the root cells. Instead, the fungus penetrates between the cell walls of the cortex, and it forms a covering sheath (or *mantle*) of fungal hyphae around the entire root. These mycorrhizal roots are about 0.5 cm in length, lack a root cap, and have a simple vascular cylinder. **Endotrophic mycorrhizae** do not form a mantle over the root, and the fungus actually enters the cortex cells **(Fig. 7.22).** Mycorrhizae make roots more efficient in mineral absorption, but they are apparently not absolutely essential for the growth of the usual host plants (see Chapter 11). This is known because plants that are artificially fed adequate nutrients can grow without mycorrhizae. Mycorrhizae also may be beneficial to their host plants by secreting hormones or antibiotic agents that reduce the potential of plant disease.

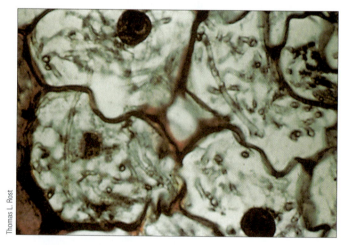

Figure 7.22 Photograph of *Coralloriza* sp. rhizome with endotrophic mycorrhizae, showing fungal hyphae actually inside the host cells. (Note: A rhizome is actually a stem that grows underground. This example is being used, however, because it shows the nature of the infection process very well, and it looks the same in root cells.) ×735.

Figure 7.23 Contractile roots of water hyacinth (*Hyacinthus orientalis*). Notice the wrinkled surface at the base of these roots where the contraction occurs.

Roots of the dandelion (*Taraxacum officinale*), water hyacinth (*Hyacintha orientalis*), and some other plants are capable of contracting, which keeps aboveground parts near the soil surface. This contraction is caused by the radial expansion (or, in some instances, the collapse) of cells in the root cortex (Fig. 7.23).

KEY TERMS

adventitious roots	prop roots
Casparian strip	protoderm
ectotrophic mycorrhizae	radicle
endodermis	region of elongation
exodermis	region of maturation
fibrous root system	residual procambium
ground meristem	root apical meristem
haustorial root	root hairs
lateral root primordial	root nodule
pericycle	seminal roots
pneumatophores	tap root system
procambium	

SUMMARY

1. The principal functions of roots are absorption of water and nutrients, conduction of absorbed materials and food, and anchorage of the plant in the soil.

2. The primary root develops from the radicle in the embryo. It generally penetrates the soil to some depth; if it dominates, a tap root system results, with a main root axis and branches.

3. Fibrous root systems are formed by seminal roots arising in the embryo in addition to the radicle. Grasses are good examples of plants with fibrous roots.

4. Adventitious roots may arise on stems and leaves.

5. The root tip is composed of (a) the root cap, which protects (b) the meristematic region; (c) a region of elongation; and (d) a region of maturation, characterized externally by root hairs and internally by the formation of primary vascular tissues.

6. The epidermis forms as the outer tissue of the root. Water is absorbed through the epidermal cells and the root hairs. The next layer is the cortex. Its cells mainly store nutrients. The endodermis is the innermost cell layer of the cortex. The Casparian strip is a waxy substance found in the radial and transverse walls. Water cannot move across the Casparian strip. Therefore, all water with dissolved nutrients must pass through the protoplasts of endodermal cells. An exodermis is present just inside the epidermis in many roots; it may also have Casparian strips and function in ion absorption and regulation.

7. In cross section, the primary xylem in dicot roots is star-shaped, with protoxylem at the points. Primary phloem arises between the arms of primary xylem.

8. Roots of certain grasses usually have central parenchyma and many protoxylem points.

9. The pericycle, a row of cells internal to the endodermis, represents the outermost row of cells of the vascular cylinder. Cells of pericycle may eventually initiate the differentiation of the vascular cambium and the cork cambium in dicot roots. The pericycle initiates lateral roots in both dicots and monocots.

10. A vascular cambium originates in dicot roots from procambium cells between primary xylem and phloem and from pericycle cells exterior to the radiating points of primary xylem. The vascular cambium forms secondary xylem internally and secondary phloem externally. The resulting increase in diameter stretches and tears the endodermis, cortex, and epidermis. A cork cambium develops from the pericycle and forms the periderm.

11. Haustorial roots from parasitic plants penetrate into the host. Bacterial nodules occur in roots of nitrogen-fixing legumes. Mycorrhizae are roots infected by beneficial fungi. Contractile roots pull the shoot tight to the soil surface.

Questions

1. Be able to identify in a diagram or photograph the following root structures and tissues:

root cap	cortex
root apical meristem (RAM)	endodermis
region of elongation	Casparian strip
region of maturation	pericycle
root hair	vascular cylinder
epidermis	(xylem and phloem)

2. State how each item listed in Question #1 contributes to the function of the root.

3. Discuss the tissues and cells involved in mineral uptake and transport in a root.

4. Describe the differences between primary and secondary tissues in a root. Where are these located? Be able to make a labeled diagram to show both primary and secondary xylem and phloem.

5. Make a diagram to show the position of cork cambium and vascular cambium in a dicot root.

6. Describe the structure and function of the root cap.

7. Describe two symbiotic associations involving roots (root nodules and mycorrhizae). What are the microorganisms involved, and how does the association alter the structure of the root? How do both partners benefit from the association?

8. Describe two modified roots.

InfoTrac® College Edition

http://infotrac.thomsonlearning.com

Roots

Comis, D. 1998. Miracle plants withstand flood and drought. *World and I* 13:158. (Keywords: "Comis" and "flood")

Sanford, R.L. Jr. 1987. Apogeotropic roots in an Amazon rain forest. *Science* 235:1062. (Keywords: "Apogeotropic" and "roots")

Stone-Palmquist, M.E., Mauseth, J.D. 2002. The structure of enlarged storage roots in cacti. *International Journal of Plant Sciences* 163:89. (Keywords: "storage" and "cacti")

Watson, G.W., Hennen, G. 1989. Journey to the bottom of a tree (what life is like for a tree root). *American Forests* 95:26. (Keywords: "Watson" and "Hennen")

Mycorrhizae

Amaranthus, M. 2003. Forest primeval and the urban landscape: bridging the gap with mycorrhizae. *Arbor Age* 23:8. (Keywords: "mycorrhizae" and "forest")

Chasan, R. 1998. Notes from underground (Arbuscular mycorrhizal fungi and their ecological relationships with plants). *BioScience* 48:6. (Keywords: "Chasan" and "underground")

Concepts of Metabolism

Visit us on the web at http://biology.brookscole.com/plantbio2 for additional resources, such as flashcards, tutorial quizzes, InfoTrac exercises, further readings, and web links.

1. Almost every chemical reaction that takes place in a cell is catalyzed (accelerated) by a specific enzyme. Each type of enzyme is formed from a unique kind of protein molecule. Some types of enzymes also require special nonprotein cofactors to work.

2. Chemical reactions in a cell form a network of interrelated metabolic pathways. The cell controls the synthesis and breakdown of chemicals by regulating the activities of enzymes that catalyze key reactions in the network.

3. Every reaction—and every set of interrelated (coupled) reactions—that proceeds forward spontaneously (with or without enzymatic catalysis) loses free energy. A reaction that gains free energy can be forced to proceed by coupling it to another reaction that loses even more free energy.

4. Two key reactions that lose free energy—the hydrolysis of adenosine triphosphate and the oxidation of nicotinamide adenine dinucleotide—power many of the activities of a cell.

8.1 SWIMMING UPSTREAM

Living organisms have a problem. They consist of highly ordered forms of matter in a universe that favors an ever greater state of disorder. Furthermore, they need a constant supply of ordered, complex molecules to stay alive. Similar to salmon swimming upstream, life itself moves against the universe's chaotic tendencies. That it can do so stems from the ability of plants to supply not only themselves but almost all other living organisms with the basic molecules needed for life. To overcome the universal current of disorder, plants use the sun's energy to synthesize carbohydrates, hydrocarbons, and other complex molecules from much simpler molecules. They then reverse the process and use the energy stored in the new molecules to drive various life-sustaining processes. But synthesizing and destroying complex molecules poses another problem: how to orchestrate thousands of chemical reactions so that they fit together in the complex process called metabolism. This chapter describes some of the elegant solutions cells have devised to control the rates of chemical reactions, as well as to surmount the energy hurdles to building complex molecules and structures.

8.2 CHEMICAL REACTIONS

Living cells are made of thousands of different chemical compounds. Some of these are small, simple compounds, such as water, sugars, and amino acids; others, such as proteins and nucleic acids, are large, complex molecules. Many of these compounds, especially the complex ones, are used to construct the organelles seen inside a cell.

Some compounds are the building blocks of more complex molecules. Still others have specialized functions in the life of the cell, such as defending against pathogens or herbivores.

Virtually every plant cell synthesizes all the different complex compounds that it contains. These compounds are formed from very simple chemicals containing carbon, hydrogen, oxygen, and nitrogen, as well as smaller amounts of other elements such as sulfur and phosphorus. Water (H_2O), carbon dioxide (CO_2), ammonium (NH_4^+) or nitrate (NO_3^-) ion, sulfate (SO_4^{2-}), and phosphate (PO_4^{3-}) represent the major basic stocks for the production of all the other molecules in the cell. Water, which makes up more than 90% of the weight of the protoplast (the portion of the cell inside the cell wall), is primarily a solvent, but it also is the source of most of the hydrogen atoms and some of the oxygen atoms in organic molecules. Carbon dioxide is the primary source of carbon and is a major source of oxygen. Ammonium and nitrate are the primary sources of nitrogen in proteins and nucleic acids. Sulfate is the source of the sulfur found in some amino acids, and inorganic phosphate is incorporated into nucleotides.

The rearrangement of atoms from their initial positions in certain molecules to new positions in other molecules is called a *chemical reaction*. In cells of biological organisms, the reactions often are called *biochemical reactions*. It is clear that there must be thousands of biochemical reactions in plant cells to produce the thousands of bio-organic molecules found there.

Chemical Reactions Must Overcome Energy Barriers

Although every chemical reaction is different, there are concepts that describe reactions in general. Because a chemical reaction involves the rearrangement of atoms, it must also involve the breaking and re-formation of some of the covalent bonds holding the atoms together (see Chapter 2). The covalent bonds represent shared pairs of electrons in orbitals that are reasonably stable. To break an existing bond, it is necessary to stretch or bend it. This destabilizes the electrons, enabling them to move to other, even more stable places. Just as it takes energy to stretch or bend a spring, it takes energy to stretch or bend a covalent bond. And just as the energy used to stretch a spring is not lost—it is stored as **potential energy**—the energy used to stretch a bond also is stored as potential energy. The energy can be recovered as the stretched bond either springs back (and the electrons move back to their original positions) or breaks (and the electrons move to other stable positions).

Figure 8.1 shows the potential energy relationships of the various phases of the following reaction:

$$A + B \rightarrow C + D$$

where A and B, the **substrates** (or reactants) of the reaction, collide and rearrange their covalent bonds to form C and D, the products of the reaction. The two molecules, A and B, have a certain amount of potential energy inherent in their separate covalent bonds. As they collide, their bonds bend

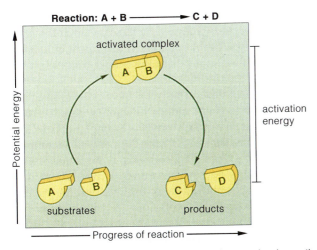

Figure 8.1 The changes in potential energy of two molecules as they exchange atoms in a chemical reaction.

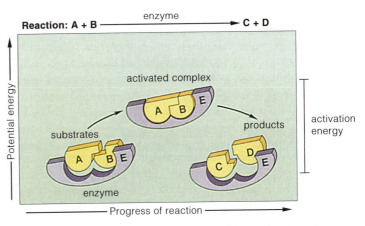

Figure 8.2 The changes in potential energy of two molecules as they exchange atoms in an enzyme-catalyzed reaction. Note that the activation energy is less in this reaction than it was in the uncatalyzed reaction diagrammed in Figure 8.1.

(that is, the electron orbitals move from their most stable positions), and their potential energy goes up. This distorted, high-energy state is called an *activated complex.* As the activated complex forms, electrons can move to new orbitals, and bonds can thus shift. For instance, the orbitals holding an atom to the other atoms of A can be broken, and new orbitals can form to attach the atom to the atoms of B. At this step, C and D form. As the new bonds shift into their most stable positions, the potential energy of C and D decreases to its normal value. For A and B to react to form C and D, they must attain an increase in potential energy, called the **activation energy.** The activation energy is an energy barrier that must be scaled for the reaction to proceed.

The law of conservation of energy says that energy cannot be created or destroyed in normal chemical reactions. Where does the energy come from to bend the bonds of A and B and form the activated state? That increasing the temperature speeds the reaction suggests that it comes from the energy of motion—the **kinetic energy**—of the two substrates. The average amount of kinetic energy of a mixture depends on the temperature. The greater the temperature, the faster the molecules in the mixture are moving. In a solution such as the cytoplasm, molecules are closely packed and travel for only short distances before bouncing against other molecules. This means that the molecules of water and other solutes continually collide with the substrate molecules; thus, the kinetic energy of the substrate molecules matches the kinetic energy of the other molecules in the mixture.

Although the temperature determines the average energy of all the molecules, including the substrates, individual molecules have more or less than the average.

At a normal temperature, for instance, 20°C, only a few A and B pairs will come together with enough kinetic energy to provide the potential energy needed to form the activation complex. This limits the rate at which the reaction can proceed. If the temperature increases, the reaction goes faster because a greater proportion of the A and B pairs will have enough kinetic energy.

A Catalyst Speeds a Reaction by Decreasing the Energy Barrier

The rate of a chemical reaction is important. Each of the thousands of reactions in a cell must proceed at a reasonable rate for the cell to function. There are a few ways to speed up a reaction. Increasing the concentrations of the substrates speeds the reaction by increasing the probability that pairs of substrate molecules will meet. Increasing the temperature increases the rate at which molecules move, and thus the frequency with which pairs of substrate molecules will meet. As mentioned earlier, increasing the temperature also speeds the reaction by increasing the probability that substrate molecules will have enough kinetic energy to form an activated complex.

Although cells can increase the concentrations of specific molecules, they can do so only up to a certain point. Increasing the temperature is not a good strategy for speeding biochemical reactions in a cell because many of a cell's components are harmed by excessive heat. Instead, cells rely on **catalysts** to speed up reactions. Catalysts are not substrates or products of the reaction. They are neither used up nor formed as the reaction proceeds, although they do interact closely with the substrates. A catalyst can be a simple substance, such as the metallic element palladium, or it can be a complex biomolecule. A catalyst is thought to work by forming a temporary complex with one or both substrates, attaching to it (or them) with reversible bonds (bonds that will break after the reaction is completed). The formation of the complex distorts the bonds of the substrate, so that further bond bending or stretching requires less energy.

A diagram showing the effect of a catalyst would look similar to the original energy diagram (compare **Fig. 8.2** with Fig. 8.1). Substrates plus catalyst have a certain amount of potential energy. When they come together to form an activated complex (which may be quite different in form from the uncatalyzed activated complex described in Fig. 8.1), they have a high potential energy. As the products are formed and the new bonds relax, the potential energy of the trio decreases again. The difference between the catalyzed and the original, uncatalyzed reaction is that the amount of activation energy is less. Therefore, at a particular temperature, the group of substrates plus catalyst is more likely than the substrates alone to have enough energy to form an activated complex.

Enzymes Catalyze Specific Reactions

One of the features that characterizes all living cells is the presence of biological catalysts called enzymes. Enzymes catalyze the thousands of reactions in a cell. Enzymes differ from simple catalysts, such as palladium, in their specificity. Each enzyme works on only one set of substrates (or, at most, a few related substrates); each enzyme catalyzes one type of reaction. That means that a cell must have thousands of enzymes to catalyze all the reactions it needs.

With few exceptions, enzymes are made of protein molecules. As described in Chapter 2, proteins are large polymeric molecules with complex three-dimensional shapes. Each different type of protein has a different shape. In particular, enzyme proteins have **active sites,** which are grooves or crevices in their surfaces, into which one or both of the substrates can fit **(Fig. 8.3).** The shape of the active site is almost complementary to that of the substrate—that is, the active site almost fits the substrate as a glove fits a hand. The substrate is held in the active site by hydrogen bonds and other forces that are weak individually but very effective as a group. As the substrate is held in the active site, its bonds are distorted in any of a number of ways. Sometimes pulling the substrate into a slightly misfitting groove distorts the substrate's shape; sometimes the active site changes shape once the substrate is present; sometimes electric charges in the active site push and pull the electrons of the substrate; sometimes functional groups (side chains of amino acids) in the active site react (temporarily) with the substrate. In any case, the distortion of the substrate makes it susceptible to the particular reaction catalyzed by the enzyme. If the reaction involves two substrates, the active site might bind both, bringing them close together in the proper orientation for the reaction to occur.

Some enzymes depend only on their protein structure for their activity; others cannot function without certain nonprotein substances called *cofactors*. Examples of cofactors include metal ions, such as Fe^{2+} and Fe^{3+}, as well as complex organic molecules without metal ions. Some cofactors are bound to enzymes by covalent bonds. Others are loosely bound and may be removed easily from the enzyme protein. Often, but not always, the cofactors are able to accept or donate electrons in oxidation–reduction reactions. If a cofactor (sometimes called a **coenzyme**) is loosely bound to its enzyme and is capable of accepting and donating electrons (or hydrogen atoms: electrons plus protons), it may serve as a carrier of electrons (or hydrogen atoms) from one reaction to another. Plants require trace amounts of certain metals that serve as cofactors (for instance, iron, copper, and molybdenum), but they make all the bio-organic cofactors they use. Many of the bio-organic cofactors that are common in plants—such as riboflavin, thiamin, niacin, and pantothenic acid—cannot be synthesized by humans. We obtain these compounds from the plants we eat.

The results of some plant enzyme activities are easy to detect. The darkening of an apple fruit after it has been cut (or bitten) results from the action of the enzyme polyphenol oxidase on chemicals released from the cells. The softening of a tomato fruit as it ripens is caused by the action of several enzymes on the polysaccharides of the cell walls (including cellulase, which breaks down cellulose, and polygalacturonase, which cleaves the chains of pectin). Papain, an enzyme from papaya fruit, digests proteins in the fruit as it ripens. It also can be extracted from the fruit and used to tenderize meat quickly before it is cooked.

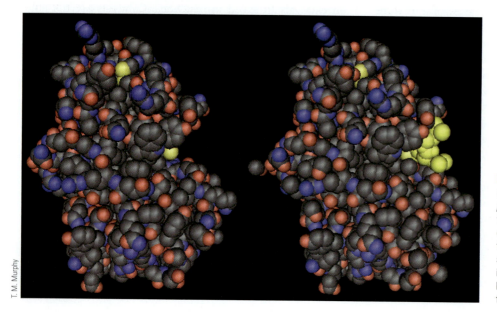

Figure 8.3 Three-dimensional model of an enzyme, papain, from papaya (*Carica papaya*) fruit. The image at the left shows the active site (notch at the right of the molecule). The yellow atom is a sulfur, essential for enzyme activity. The image at the right shows a substrate molecule (yellow-green) in the active site. Papain hydrolyzes the bonds between amino acids in a polypeptide chain. It is used as a meat tenderizer.

T. M. Murphy

The Rates of Reactions Can Be Controlled by the Synthesis, Activation, and Inhibition of Enzymes

In a normal cell, the temperature is too low for most essential biochemical reactions to occur at reasonable rates. For cells to function, all the reactions must occur fairly rapidly and in a balanced fashion. In this way, the components of the cells can be synthesized in the proper proportions, and no one material is formed (using up all the substrates) at the expense of all the others. The synthesis of the various organelles must be balanced, and synthetic reactions must be coordinated with the reactions that breakdown molecules into their component parts.

Cells have several ways of controlling the rates of individual reactions. One method is to synthesize the enzyme that catalyzes a particular reaction when, and only when, that reaction is needed. The formation and germination of a seed provides a good example. A seed contains an embryo with associated tissues that store nutrients used for the growth of the embryo when the seed germinates. The nutrients include starch, and starch is formed in the cells from simple carbohydrate subunits imported in the form of sucrose. Two enzymes (at least) are required to break down sucrose and use the products to form starch; these enzymes increase in amount as the seed is formed. Later, they disappear. When the seed germinates, the starch must be broken down into subunits, and the enzymes that accomplish this are synthesized during the early stages of germination. The synthesis of enzymes is a key aspect of development and is discussed in more detail in Chapter 15.

The control of reaction rates by the synthesis and breakdown of enzymes is rather slow and crude. A much quicker and finer control occurs through the regulation of the catalytic activity of already existing enzymes. Such regulation is possible because the catalytic activity of an enzyme depends on its three-dimensional shape, and the shape depends in turn on weak bonds (such as hydrogen bonds) that can easily be broken and re-formed. Not all enzymes can be regulated in this way; those that can generally have a special site on their surface that will bind a regulatory molecule. In some cases, a special enzyme attaches a phosphate group to that site covalently. In other cases, some other molecule (there are many possibilities) binds to the site through weak bonds, such as hydrogen or ionic bonds. Binding to the regulatory molecule changes the shape of the enzyme, and that change affects the shape of the active site **(Fig. 8.4).** The overall result is either (1) to lower the affinity of the active site for the substrate or its catalytic efficiency once the substrate has been bound (in which case the enzyme is inhibited), or (2) to increase the affinity of the active site for the substrate or its catalytic efficiency (in which case the enzyme is activated). The regulatory step used depends on the enzyme. It may involve a normal compound found in the cell (perhaps the substrate or product of another chemical reaction). For the regulated enzyme, this compound serves as a signal to turn off or on the reaction it catalyzes. Later in this chapter we show how regulation of an enzyme in this way can maintain a steady concentration of a compound.

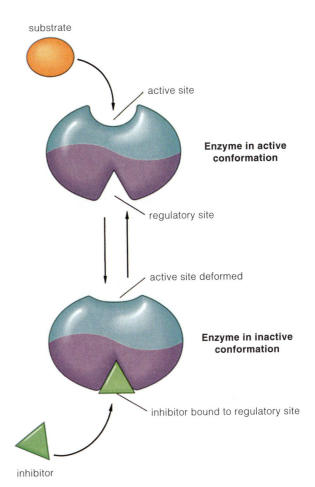

Figure 8.4 A regulatory enzyme has two sites: an active (catalytic) site, and a site at which a regulatory molecule may bind. In the presence of the regulatory molecule (inhibitor), the enzyme changes its shape. The substrate binding in the active site may also contribute to (or oppose) the shape change.

8.3 METABOLIC PATHWAYS

Most bio-organic molecules produced in a cell are formed not from a single chemical reaction but from a sequence of reactions, in which the product of one reaction becomes the substrate for the next. A series of such linked reactions in a cell is called a **metabolic pathway.**

Reactions Are Linked When One Reaction's Product Is Another's Substrate

It is not possible to make complex molecules in a single step; it takes a series of simple steps. A single step might be the addition of the components of water to the molecule, or a reduction (addition of one or two electrons), or perhaps the removal of a CO_2 group. A series of such steps, each in the correct order, can lead to quite complex compounds, as seen in Chapters 9 and 10.

In a metabolic pathway **(Fig. 8.5),** the product of the first reaction is the substrate of the second reaction, the product of the second reaction is the substrate of the third, and so

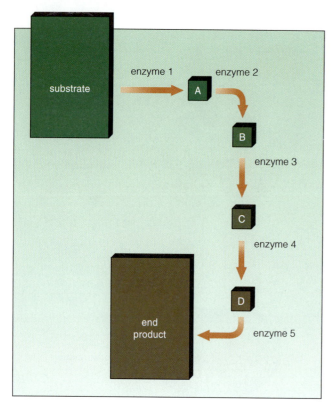

Figure 8.5 A metabolic pathway. Each letter represents a different compound. Each reaction is catalyzed by a different enzyme.

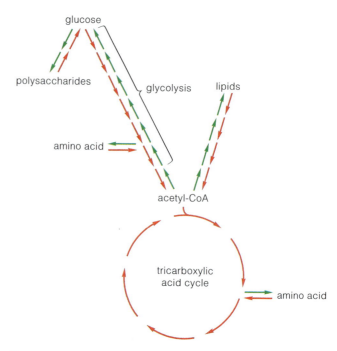

Figure 8.6 A simplified picture of some interacting metabolic pathways in a typical plant cell. Note that the anabolic reaction (*green arrows*) synthesizing substances and the catabolic reactions (*red arrows*) breaking them down generally involve different enzymes. *Glycolysis* and the *tricarboxylic acid cycle* are pathways described in Chapter 9.

forth. Each reaction is catalyzed by a separate enzyme. The products of reactions, other than the final product of a series, are called **intermediates.** For this reason, the collection of all the metabolic pathways in cells is called **intermediary metabolism.** The enzymes that catalyze the reactions are linked: Each enzyme depends on the ones before it to provide substrates. Without the substrates, the enzymes cannot function. In the same way, each enzyme depends on those after it to complete the formation of the pathway's product. Without the enzymes for the later reactions, intermediates will accumulate, and no final product will be formed.

In Cells, Linked Reactions Form a Complex Network

Although Figure 8.5 shows a simple linear pathway, few, if any, pathways in a cell are this simple. In some cases, two or more compounds are needed to form a product. Frequently, one intermediate is used as a substrate by several separate enzymes to produce different products. In this case, the intermediate is called a *branch point* because, diagrammatically, the pathways appear to branch off in different directions (Fig. 8.6). Some intermediates are key compounds in the cell because they serve as substrates for many different reactions.

Figure 8.6 gives a rough idea of the complexity of the metabolic reactions in a typical plant cell. This diagram shows several metabolic pathways linked together, with each arrow representing a separate reaction with a separate enzyme. Each pair of adjacent reactions is connected by an intermediate, although only a few key intermediates are shown.

The reactions that produce subunits of functional or structural molecules (such as proteins or cellulose) are some-

times called **anabolic** reactions. Reactions that break damaged or unwanted molecules down into their component parts are called **catabolic** reactions. Figure 8.6 indicates these two classes of reactions, with arrows pointed in opposite directions to denote that the synthesis and breakdown of materials generally (but not always) involve different enzymes. In some cells (such as those in meristems), anabolic reactions predominate; in others (such as the cells in ripening fruit), catabolic reactions are most active. In most cells of plants and other organisms, both sets of reactions occur; there is a constant synthesis and breakdown—called a *turnover*—of cell components.

The Concentration of a Compound Often Controls Its Production in a Cell

Cells cannot afford to allow metabolic pathways to run amok. If pathways operated without any control, materials and energy would be wasted in the formation of unneeded compounds or the breakdown of necessary ones. Certain intermediates might accumulate to toxic levels; others vital to the cell might become scarce. That reactions are organized into pathways immediately suggests an efficient way of controlling the formation of the end product of a pathway. Because each reaction depends on the ones before it to provide substrates, the cell need only shut off the activity of the first enzyme in the pathway to restrict the formation of the final product. A control mechanism often found in cells is that the first enzyme (*branch-point enzyme*) of a pathway has a regulatory site, and the final product of that pathway serves as its inhibitor (Fig. 8.7). This mechanism keeps the

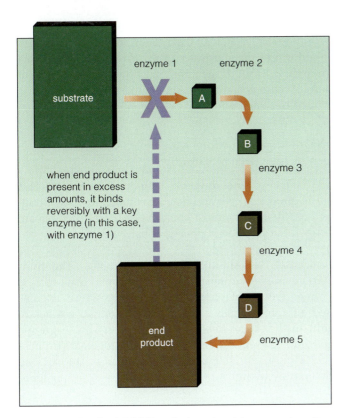

Figure 8.7 Feedback inhibition of a branch-point enzyme stops the formation of the final product of the pathway. The *dotted line* indicates that the final product of the pathway inhibits the catalytic activity of enzyme 1.

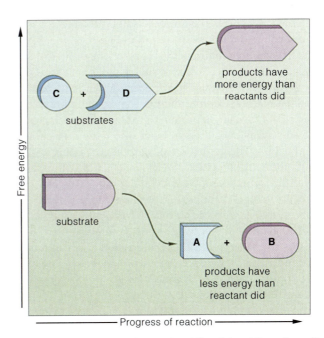

Figure 8.8 Free energy changes in uphill and downhill reactions. Only the downhill reaction will go forward spontaneously. Uphill reactions require an input of free energy from some other source.

concentration of the product constant. As the product accumulates and its concentration increases, the product inhibits the branch-point enzyme; therefore, its synthesis slows down. As the cells use up the product, its concentration decreases, the inhibition of the branch-point enzyme is released, and synthesis resumes. This type of control mechanism is called **feedback inhibition.** It is just one example of the ways in which the control of enzyme activity can be used to orchestrate metabolic reactions.

<table>
<tr><td>**8.4**</td><td>**THE ROLE OF FREE ENERGY IN METABOLISM**</td></tr>
</table>

Free Energy Determines the Direction a Reversible Reaction Will Take

Most chemical and biochemical reactions are reversible at least in theory, so that:

$$A + B \leftrightarrow C + D$$

This means that the substrates and products of a reaction can be interconverted. Under arbitrary conditions, a reversible reaction generally favors one direction over the other. Therefore, the rate of the forward reaction, $A + B \rightarrow C + D$, often differs initially from that of the reverse reaction, $A + B \leftarrow C + D$. As the forward and reverse reactions proceed, the overall reaction reaches a state of *equilibrium.* The rates of the forward and reverse reactions become equal, and the concentrations of the

substrates and products become constant. Enzymes are thought to reduce the *time* needed to reach equilibrium but not to change the *ratio* of substrates and products at equilibrium.

What determines the direction in which a reaction proceeds? There are several factors: (a) the initial concentrations of substrates and products in the reaction's mixture; (b) the relative stability of the bonds in the substrates and products; (c) the number of substrates and products that participate in the reaction; and (d) the temperature.

Chemists and biochemists group all the above factors into one useful concept: the **free energy.** Every chemical or set of chemicals can be assigned a free energy value. That is, the greater the concentration of substrates in a reaction mixture, the greater their free energy. The more stable the bonds of the substrates, the lower their free energy. The more independent molecules included in the substrates, the lower their free energy. All these statements apply also to the *products* of a reaction.

Every chemical reaction that can be imagined is associated with a change in free energy, which is defined as the difference between the free energies of the products and substrates. The reason that this concept is important is that *every reaction that proceeds forward spontaneously (with or without enzymatic catalysis) loses free energy.* Every reaction that shows no change in free energy is at equilibrium. Every reaction that is associated with an increase in free energy will not occur spontaneously and will need free energy from some source to proceed. These concepts are diagrammed in **Figure 8.8.** This figure shows why spontaneous reactions, those associated with a decrease in free energy, are sometimes called *downhill reactions* and why reactions that do not proceed spontaneously may be called *uphill reactions.* Note that free energy includes potential energy but also includes other factors; thus, the graph in Figure 8.8 is not the same as—and cannot be compared with—the graph in Figure 8.1.

The free energy requirements for reactions provide a difficulty for cells because many of the anabolic reactions (on which the growth and survival of a cell depend) are associated with an increase in free energy. The combination of 100 amino acids into a single protein represents a large increase in free energy, as does the synthesis of a large molecule of RNA or DNA. The reduction of carbon-containing molecules to form the hydrocarbon chains in lipid molecules involves the formation of bonds that are inherently less stable in an oxygen atmosphere—again representing an increase in free energy. Movement, whether it be the whipping of flagella or the separation of chromosomes during mitosis, depends on uphill chemical reactions. The accumulation of nutrients by pumping ions from an area of low concentration to an area of high concentration (see Chapter 11) also depends on reactions that require the input of free energy. Some of the great discoveries in biology have been the ways in which cells provide sources of free energy to force anabolic reactions forward, thus obtaining the materials they need to power such uphill processes as growth.

In Coupled Reactions, a Downhill Reaction Can Push an Uphill Reaction

When biologists try to unravel the complexities of metabolism, one of the ways in which they simplify the biochemical reactions is to think of partial reactions. In oxidation–reduction reactions, for example, one compound is oxidized (loses electrons), and another is reduced (gains electrons). The oxidation can be written as a separate, but partial, reaction, and so can the reduction. Neither reaction occurs by itself, however, because electrons must be conserved. Because they must occur together, the reactions are said to be *coupled*.

Another example is the transfer of a phosphate group. One compound might lose a phosphate group through a **hydrolysis reaction** (the addition of the components of water, H^+ and $OH-$, breaks the bond between the phosphate and the rest of the compound); another compound might gain a phosphate group through a **condensation reaction** (the opposite of hydrolysis, producing water). These two partial reactions can be coupled so that they occur together. In this case, neither the hydrolysis nor the condensation actually occurs; instead, the phosphate group is transferred from the first compound to the second without ever being independent.

The coupling of two reactions means that the change in free energy of the overall reaction equals the sum of the changes in free energy of the partial reactions. This is important because if one of the partial reactions is a strong uphill reaction but the other is an even stronger downhill reaction, the overall reaction will be downhill, and both reactions will proceed spontaneously. It is not unreasonable to suggest that the downhill partial reaction has forced the uphill partial reaction to run forward. If the uphill reaction involves

the synthesis of DNA, then the cell has provided free energy (in the form of the substrates of the downhill partial reaction) to make DNA. If the uphill reaction involves the separation of chromosomes in anaphase, then the cell has provided free energy for that process.

Two Downhill Reactions Run Much of the Cell's Machinery

There are a number of downhill, spontaneous reactions that occur in almost all cells and that are coupled to many uphill reactions. Two of the most common of these reactions are described in the following paragraphs.

HYDROLYSIS OF ADENOSINE TRIPHOSPHATE The compound **adenosine triphosphate** (ATP) is found in all cells. It is formed from the nucleotide adenosine—which is composed of the complex organic base adenine, covalently bound to the sugar ribose—plus three phosphate groups (Fig. 8.9a). The three phosphates hooked together (triphosphate) form a relatively unstable assemblage. They can be split apart by hydrolysis (addition of water) to form adeno-

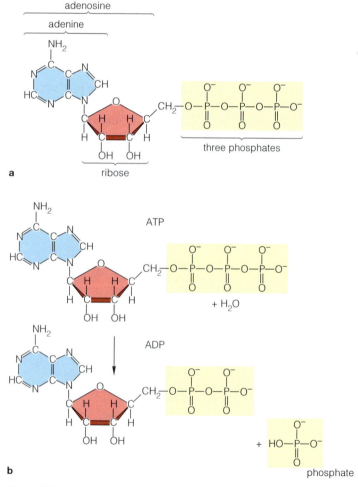

Figure 8.9 Adenosine triphosphate (ATP), the main energy carrier in cells. (**a**) The structural formula for ATP. (**b**) The hydrolysis of ATP, a strongly downhill reaction.

sine diphosphate (ADP) and an independent phosphate group **(Fig. 8.9b)**. This hydrolysis reaction is spontaneous and is associated with a release of free energy of 7 to 12 kcal per mole of ATP, just enough to run any of several uphill reactions in a cell. For instance, it provides energy to pump protons across a membrane, to force the addition of an amino acid to a growing protein, and to extend a cellulose chain by one glucose unit. It also powers the motor proteins that attach to microtubules: these move cilia and flagella and the chromosomes in mitosis. ATP is the quintessential *high-energy compound* or *energy carrier*. That it is formed in the mitochondrion (by a condensation of ADP with phosphate—formally the reverse of the reaction shown in Fig. 8.9b) is the main reason that the mitochondrion is called the "power-house of the cell."

OXIDATION OF NICOTINAMIDE ADENINE DINUCLEOTIDE AND NICOTINAMIDE ADENINE DINUCLEOTIDE PHOSPHATE The coenzymes **nicotinamide adenine dinucleotide (NADH)** and **nicotinamide adenine dinucleotide phosphate (NADPH)** are essential molecules in eukaryotic cells. The reactive part of these molecules is their nicotinamide functional group. (Our bodies cannot synthesize nicotinamide; we must obtain it in food, as the vitamin called *niacin*. Plant cells, however, can synthesize niacin and use it to make NADH and NADPH.) NADH is the reduced form of the molecule, which can readily donate electrons to other molecules; the oxidized form, remaining after the electrons have gone, is denoted as NAD^+. The oxidation of NADH by oxygen gas **(Fig. 8.10)** is associ-

ated with the release of 52 kcal per mole of free energy! This is enough free energy to power the formation of several molecules of ATP. In fact, in the mitochondrion, the oxidation of NADH is coupled to the formation of two to three molecules of ATP. The oxidation of NADPH (a parallel reaction to that of NADH) is used to synthesize fatty acids and some amino acids.

Of course, the energy locked in the bonds of ATP, NADH, and NADPH does not come without cost. The reactions that form these compounds are uphill reactions, and they occur only when they are coupled to other downhill reactions. The metabolic pathways that provide the energy for the synthesis of these compounds are described in Chapters 9 (which discusses cellular respiration) and 10 (which discusses photosynthesis).

KEY TERMS

activation energy	intermediary metabolism
active sites	intermediates
adenosine triphosphate (ATP)	kinetic energy
	metabolic pathway
anabolic	nicotinamide adenine dinucleotide (NADH)
catabolic	
catalysts	nicotinamide adenine dinucleotide phosphate (NADPH)
coenzyme	
condensation reaction	
feedback inhibition	potential energy
free energy	substrates
hydrolysis reaction	

SUMMARY

1. The amount of potential energy needed to force molecules to react together limits the rate of a chemical reaction. Chemical reactions can be speeded up by increasing the concentration of the substrates, increasing the temperature, or adding a catalyst.

2. Enzymes are catalysts made from protein. Most enzymes catalyze a single reaction from a specific substrate or set of substrates. In cells, the rate of a reaction is determined by the amount of enzyme and by the activity of the enzyme that is present.

3. Enzyme-catalyzed reactions can be linked together in metabolic pathways, in which the product of one reaction is the substrate for the next, to produce complex biochemicals. In a cell, the network of metabolic pathways is complex, with a few key compounds being converted into many other compounds.

4. Feedback inhibition of the first enzyme in a pathway by the final product of the pathway enables the cell to control the concentration of that product.

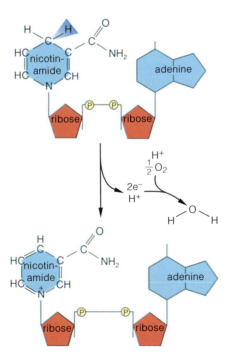

Figure 8.10 The oxidation of nicotinamide adenine dinucleotide (NADH) to NAD$^+$ by oxygen. Notice that the NADH loses one H (hydrogen) and one chemical bond between the H and C (carbon), which represents two electrons. The electrons may be transferred to a series of compounds before they reach oxygen.

5. The change in free energy of a reaction defines the direction in which it will proceed. The change in free energy is influenced by the initial concentrations of substrates and products in the reaction mixture, the relative stability of the bonds in the substrates and products, the number of independent substrates and products involved in the reaction, and the temperature. A spontaneous reaction always loses free energy.

6. If two partial reactions are coupled, the free energy change of the overall reaction is the sum of their separate free energy changes. A downhill reaction can thus force an otherwise uphill reaction to proceed.

7. Two reactions, the hydrolysis of ATP and the oxidation of NADH, are coupled to—and drive—many of the important functions of a cell.

Questions

1. Name the sources of the elements carbon, hydrogen, oxygen, nitrogen, sulfur, and phosphorus that plants use for growth. Why are enzymes needed to convert these sources to bio-organic compounds?

2. The diagram below shows the energy relations for two chemical reactions in which chemical compounds A and B are converted to C and D, and F and G are converted to H and I. Which reaction will proceed faster? Explain.

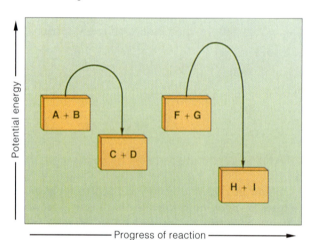

3. Would increasing the temperature by 5°C change the diagram in Question #2? What would happen to the reactions?

4. Would adding an enzyme that catalyzes the first reaction (A + B → C + D) change the diagram in Question #2? How?

5. Adding a weak acid to a solution containing an enzyme usually inactivates the enzyme, often irreversibly. Boiling the solution does the same thing. Suggest an explanation for these observations.

6. Give an example from your own experience (and other than the one in the text) of the action of the enzyme polyphenol oxidase. Do the same for the enzyme polygalacturonase.

7. Assume that three enzymes (E1, E2, and E3) form a metabolic pathway, such as:

$$A \quad \xrightarrow{E1} \quad B \quad \xrightarrow{E2} \quad C \quad \xrightarrow{E3} \quad D$$

Which factors will influence the rate of formation of D?

a. the amount of E1

b. the amount of E2

c. the amount of E3

d. the temperature

e. all of the above

8. How are catabolic reactions energetically coupled to anabolic reactions in a cell?

9. In the metabolic pathway described in Question #7, are the reactions catalyzed by E1 and E2 coupled reactions? Explain your answer.

10. List five uphill reactions that occur in a living plant cell.

11. Maltose is formed from two glucose molecules bound together with a covalent bond. The enzyme maltase breaks the two glucoses apart by catalyzing the reaction:

$$maltose + H_2O \rightarrow 2 \text{ glucose}$$

A solution of maltose alone produces glucose very slowly, but when maltase is added to the solution, glucose appears rapidly. Which of the following statements best explains this effect?

a. The free energy of glucose is much less than that of maltose.

b. The free energy of glucose is much more than that of maltose.

c. Maltase makes the reaction irreversible.

d. Maltase decreases the activation energy for the forward reaction.

e. Maltase increases the activation energy for the reverse reaction.

12. No one ever detects the formation of maltose when maltase is added to a solution of glucose. Which of the statements listed as possible answers for Question #11 best explains why?

13. In the presence of ATP, the enzyme luciferase (an enzyme found in the abdomens of fireflies) and its cofactor luciferin produce light. Light is energy. Energy cannot be created; therefore, the light energy must have come from somewhere. Where does the light energy come from?

a. from the breakdown of the luciferase and luciferin

b. from the hydrolysis of ATP

c. from the heat energy in the water molecules

 InfoTrac® College Edition

http://infotrac.thomsonlearning.com

Enzymes

Carnegie Institution. 1998. Constructing "designer" plant enzymes. *M2 Presswire* Nov. 13. (Keywords: "Carnegie" and "plant enzymes")

LaBell, F. 1994. Efficiency from enzymes for fruit juice, bread. *Prepared Foods* 163:83. (Keywords: "enzymes" and "fruit")

Plant Metabolism

Brown, K. 2003. Working weeds: A German company develops a way to peek into plant metabolism. *Scientific American* 288:34. (Keywords: "weeds" and "metabolism")

Comis, D. 1992. Resetting a plant's thermostat. *Agricultural Research* 40:14. (Keywords: "plant" and "thermostat")

Respiration

Visit us on the web at http://biology.brookscole.com/plantbio2
for additional resources, such as flashcards, tutorial quizzes,
InfoTrac exercises, further readings, and web links.

1. All living cells require energy to drive the reactions of life. Respiration releases energy trapped as chemical bond energy in organic food molecules. Some of this energy is transferred to adenosine diphosphate (ADP) and nicotinamide adenine dinucleotide phosphate ($NADP^+$) in the formation of adenosine triphosphate (ATP) and reduced nicotinamide adenine dinucleotide (NADPH). These energy carriers may then be moved in the cell and coupled to energy requiring reactions.

2. When a six-carbon-atom sugar molecule is respired in a cell, the sugar is broken down into pyruvate in a series of enzymatic reactions called glycolysis. In the absence of molecular oxygen, the pyruvate may be enzymatically broken down to form CO_2 and ethyl alcohol, a process called alcoholic fermentation.

3. In the presence of molecular oxygen, the pyruvate is broken down in *aerobic respiration,* an integrated series of enzymatic reactions called the *tricarboxylic acid cycle* and the *electron transport chain.* During these reactions, CO_2 and water are formed and some energy is trapped in the energy carriers ATP and NADPH.

4. Enzymes of glycolysis and alcoholic respiration are in the cell cytoplasm; enzymes involved in aerobic respiration are in the mitochondrial matrix or the inner membrane of the mitochondrial envelope.

5. The chemiosmotic theory of ATP formation proposes that the proton gradient that develops across the mitochondrial envelope during the terminal oxidation reactions drives ATP synthesis from ADP and inorganic phosphorus.

6. Some cells have developed alternate pathways of respiration. The pentose phosphate pathway transfers energy from glucose to $NADP^+$ (forming NADPH) and produces ribose-5-phosphate, necessary for the synthesis of nucleic acids. The glyoxylate pathway converts fat into intermediate compounds and ultimately into sucrose.

7. Environmental factors such as cell hydration, temperature, oxygen supply, and food availability affect the rate of respiration of a plant, as do factors such as cell type and the age and species of the plant.

9.1 THE RELEASE OF ENERGY FROM FOOD

To stay alive, every living cell must obtain energy in a usable form, which it does by oxidizing food molecules. Without energy from the oxidation of food, cell membranes cannot maintain differential permeability, roots will not accumulate solutes, and cytoplasm will not flow. Cells cannot synthesize amino acids, proteins, or fats without the energy and intermediate carbon compounds produced by respiration. This chapter examines respiration in detail to explore why it is so fundamental to life.

Early biologists applied the term respiration to the exchange of gases between an organism and its environment. Even today, many people consider respiration and breathing in animals as synonyms. Breathing is merely the visible indication that chemical reactions are taking place in animal cells. We define **respiration** as the oxidation of organic molecules within cells accompanied by the release of usable energy.

Digestion Converts Complex Food into Simpler Molecules

Before large insoluble food molecules can be respired, they must be broken down into smaller soluble components. Cells cannot oxidize even a simple soluble food molecule such as sucrose until it is broken into simpler sugars. **Digestion,** the breaking apart of complex foods (carbohydrates, fats, and proteins) into simpler compounds (sugars, fatty acids, glycerin, and amino acids), occurs easily in the presence of water and specific enzymes. Large amounts of energy are not released, because the transfer of electrons through oxidation–reduction reactions is not involved. Cells use water molecules in digestion; therefore, the process is a hydrolysis reaction. Green plants differ from animals in that they normally do not ingest complex foods but synthesize them. Plants do digest foods, however, although unlike animals, they have no special organs for digestion. Digestion occurs in any cell that stores complex food, even temporarily. During photosynthesis, starch often is stored as insoluble starch granules in the chloroplasts. When the time comes for this temporary starch reserve to be used by the cells or to be transported out of the cells, it must be changed back to soluble sugar. This chemical transformation of the insoluble starch into soluble glucose is one example of digestion (Fig. 9.1).

Respiration Is an Oxidation–Reduction Process

Once the large food molecules have been digested, respiration results in the transfer of energy from food molecules to energy-carrier molecules in the cell (see Chapter 8). Respiration is an oxidation–reduction process because it involves the removal of electrons from electron donor molecules to electron acceptor molecules; nevertheless, many steps do *not* involve the use of molecular oxygen. Indeed, many organisms, such as yeast and some bacteria, obtain their energy from food without any reactions that involve molecular oxygen. Respiration that does not involve molecular oxygen is called **anaerobic respiration.** In the more common type of respiration, **aerobic respiration,** molecular oxygen plays a major role. Because of this difference in molecular oxygen usage, many biochemists restrict the definition of respiration to designate the final series of reactions that occur in aerobic respiration. However, this chapter uses the term *respiration* in a broader sense to include all of the reactions involved in the oxidation of food. The cellular organelles most involved in aerobic respiration are the mitochondria (see Chapter 3). Mitochondria produce most of the energy and the interme-

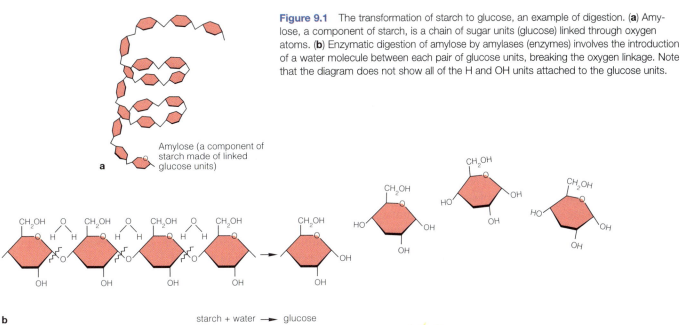

Figure 9.1 The transformation of starch to glucose, an example of digestion. (**a**) Amylose, a component of starch, is a chain of sugar units (glucose) linked through oxygen atoms. (**b**) Enzymatic digestion of amylose by amylases (enzymes) involves the introduction of a water molecule between each pair of glucose units, breaking the oxygen linkage. Note that the diagram does not show all of the H and OH units attached to the glucose units.

Amylose (a component of starch made of linked glucose units)

a

b

starch + water ⟶ glucose

diate carbon compounds used as molecular building blocks in the cell. Environmental factors also play a role, for they can alter the rate and effectiveness of respiration.

Respiration Is an Integrated Series of Reactions

Respiration is more than a simple oxidation reaction; it is a series of chemical reactions that supplies the energy for most cellular processes (**Fig. 9.2**). Although the reactions of respiration might seem overwhelming if we listed them all, they fall into patterns. In particular, some consume energy as food molecules are prepared for breakdown, whereas others result in the transfer of energy to molecules that act as energy carriers in the cell. It is more important for you to have an understanding of the process of respiration as a whole and its roles in the life of a plant than it is for you to worry about mastering each step in detail.

When we consider the total energy cycle in the living plant, we find that stored chemical energy is moved from one part of the plant to another in the form of potential energy in food molecules. At the cellular level, some of the potential energy in the food is trapped during respiration in the energy carriers ATP and NADH or NADPH, which may move throughout the cell.

Respiration also performs another important task in the life of the cell. During the many steps in the respiratory breakdown of food, essential intermediate compounds are formed. Living cells use many of these intermediate compounds as carbon-containing building blocks from which they synthesize compounds such as proteins, fats, nucleic acids, and hormones. Respiration, through the oxidation of food, has several functions: (1) It supplies energy in a form available to do work in the cell, and (2) it produces intermediate carbon compounds essential for the continued growth and metabolism of cells.

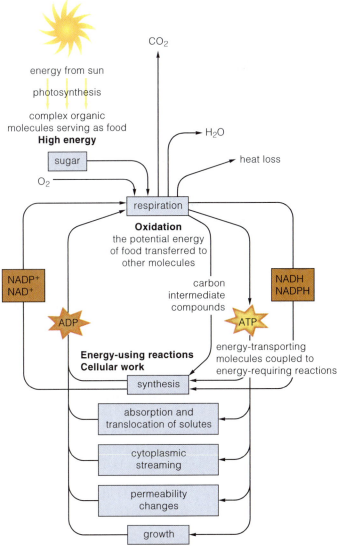

Figure 9.2 An overview of metabolism. Energy from sunlight is trapped in food during photosynthesis. Energy is released from food during respiration and is used to perform work in the cell.

The Transfer of Energy Occurs through Coupled Reactions

To use the energy stored in food molecules, cells couple energy-yielding reactions with energy-consuming reactions. Such mechanisms transfer energy from sugar to the energy carriers ATP and NADH or NADPH (Fig. 9.2). In some reactions, NAD^+ is reduced, and in others, $NADP^+$ is reduced. Whether NAD^+ or $NADP^+$ is reduced by accepting electrons and hydrogen depends on the enzymes catalyzing the particular reaction. When 1 mole, 1 g molecular weight (180 g), of glucose is burned, CO_2, H_2O, and 686 kcal of energy are released. This amount of energy released at once as heat would destroy living cells. Instead, energy transformations in the cells occur slowly and in small steps through a sequence of many reactions. Specific enzymes catalyze each of these small reactions, allowing them to proceed at moderate rates at low temperatures. About 40% of the energy released is trapped in the bond energy of the ATP and NADPH formed. Because these molecules are in solution in the cell, they may diffuse or be carried by cytoplasmic streaming throughout a cell. Both ATP and NADPH eventually yield energy that performs work in the cell.

Because the total amounts of ADP and $NADP^+$ in a cell are quite small, a cell can never accumulate large amounts of ATP or NADPH. Consequently, cellular energy is stored as carbohydrates, fats, and protein. A cell at rest and not using energy rapidly has converted most of its ADP to ATP. At cell rest, the ATP is not used to drive work reactions; consequently, ATP is not hydrolyzed to ADP and inorganic phosphate (Pi). Some of the reactions of respiration require ADP, and because the concentration of ADP in a resting cell is very low, respiration is very slow. If such a cell is stimulated to do work—for example, to increase synthetic reactions, the rate of cytoplasmic streaming, or the amount of salt uptake—the rate of respiration will increase. When work is being done, energy liberated from the hydrolysis of ATP to ADP and Pi helps drive these work reactions. The ADP and Pi liberated become available to the respiratory pathways, where they are synthesized again into ATP trapping more energy and increasing the rate of respiration. This is one example of many control mechanisms that regulate the rates of various metabolic reactions in the cell.

9.2 THE REACTIONS OF RESPIRATION

If you look at the complete oxidation of a simple six-carbon sugar in the presence of molecular oxygen, only CO_2, H_2O, and heat (energy) are the final products. This overall process is written as:

$$C_6H_{12}O_6 + 6\,O_2 \rightarrow 6\,CO_2 + 6\,H_2O + 686\,kcal$$
$$\text{sugar} + \text{oxygen} \rightarrow \text{carbon dioxide} + \text{water} + \text{energy}$$

The equation tells us nothing about the way the reaction occurs, the intermediate steps, or other possible products.

It took many years for scientists to discover the many steps involved in respiration. To study respiration, we separate it into phases, each of which includes reactions that produce key intermediate compounds. In the first phase, **glycolysis,** for every molecule of the six-carbon sugar (glucose) that is broken down and oxidized, two molecules of pyruvate (a three-carbon organic compound) are produced. No molecular oxygen takes part in the reactions, and the energy change is small. Most of the energy in the chemical bonds of the sugar being broken down still remains in the pyruvate formed. The pyruvate is then oxidized in either an aerobic or anaerobic pathway. In anaerobic respiration, the pyruvic acid is only partially oxidized. Alcohol and CO_2 are formed and only a small amount of usable energy is released. In aerobic respiration, the pyruvate moves into the mitochondria. There, in the presence of molecular oxygen, it is oxidized to carbon dioxide and water. A large amount of energy is trapped in ATP and NADPH.

Glycolysis Is the First Phase of Respiration

There are three major steps in the process of glycolysis (Fig. 9.3):

1. *Phosphorylation:* preparation of the six-carbon sugar for reaction by the addition of phosphate from ATP.

2. *Sugar cleavage:* splitting of the sugar phosphate into two three-carbon-atom fragments.

3. *Pyruvate formation:* oxidation of the fragments to form pyruvic acid with the formation of NADH and ATP.

The six-carbon sugar, the most common sugar respired in the cell, is glucose. Although glucose is energy-rich, it is stable and does not react readily with oxygen at life-sustaining temperatures, and it is not readily broken down into intermediate products. Before a cell can break down glucose and release its stored energy, glucose must react enzymatically with ATP, which donates a phosphate group and energy. This reaction occurs at two points in glycolysis so that two ATP molecules provide energy and phosphate in the phosphorylation of the sugar molecule (Fig. 9.3). Note that ATP used in these reactions is replaced during later steps in respiration. The phosphorylated sugar also undergoes a series of internal rearrangements that result in the formation of another six-carbon sugar phosphate, fructose 1,6-bisphosphate, which is split in the next reactions.

Another enzyme catalyzes the splitting of fructose 1,6-bisphosphate into two different three-carbon sugars, dihydroxyacetone phosphate and glyceraldehyde 3-phosphate. These three-carbon sugar phosphates are in equilibrium and may be converted enzymatically into one another. Glyceraldehyde 3-phosphate is oxidized to pyruvate (Fig. 9.3). As it is broken down, the equilibrium shifts so that more dihydroxyacetone phosphate is converted into glyceraldehyde 3-phosphate. Thus, both three-carbon sugar phosphates are actually available for pyruvate formation. Although molecular oxygen does not take part, the three-carbon sugar phosphate is oxidized by the transfer of two of its electrons and

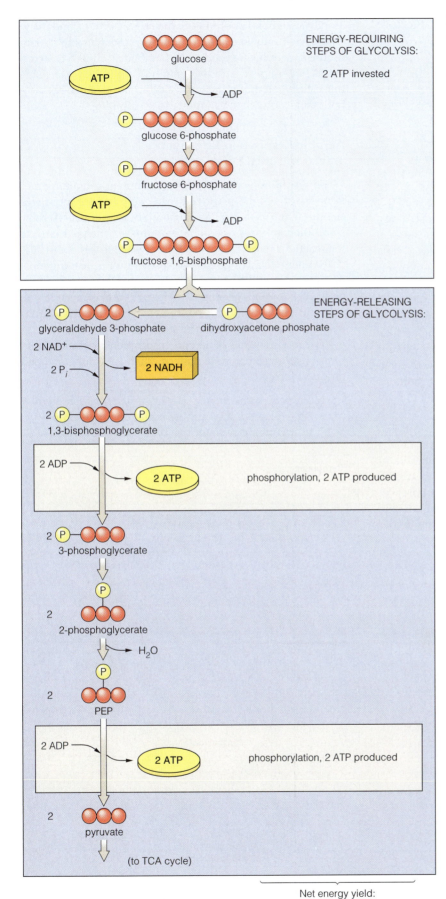

ENERGY-REQUIRING
STEPS OF GLYCOLYSIS:

2 ATP invested

ATP → ADP

glucose

glucose 6-phosphate

fructose 6-phosphate

ATP → ADP

fructose 1,6-bisphosphate

ENERGY-RELEASING
STEPS OF GLYCOLYSIS:

2 glyceraldehyde 3-phosphate ← dihydroxyacetone phosphate

2 NAD⁺

2 P$_i$ → 2 NADH

2 1,3-bisphosphoglycerate

2 ADP → 2 ATP   phosphorylation, 2 ATP produced

2 3-phosphoglycerate

2 2-phosphoglycerate

H$_2$O

2 PEP

2 ADP → 2 ATP   phosphorylation, 2 ATP produced

2 pyruvate

(to TCA cycle)

Net energy yield:
2 ATP
2 NADH

Figure 9.3 Three major steps in the process of glycolysis. ADP, Adenosine diphosphate; ATP, adenosine triphosphate; TCA, tricarboxylic acid; NADH, reduced nicotinamide adenine dinucleotide; PEP, phosphoenolpyruvate; Pi, inorganic phosphate.

hydrogen to a hydrogen acceptor, NAD^+. This effectively transfers some of the energy originally in the chemical bonds of the three-carbon sugar to the newly formed NADH. Also, during the oxidation of each three-carbon sugar phosphate molecule to pyruvate, energy is used to form two molecules of ATP from two molecules of ADP plus two phosphates. The oxidation of the two three-carbon sugars forms a total of four ATP and two NADH molecules. However, because two molecules of ATP are used to phosphorylate a glucose molecule in preparation for its breakdown (Fig. 9.3), glycolysis results in a net yield of two ATP and two NADH molecules. Glycolysis benefits the cell because usable intermediate compounds are formed, and although only a small amount (approximately 20%) of the energy of the glucose molecule is trapped in ATP and NADH, only about 3% has been lost as heat.

The fate of the pyruvate that results from glycolysis depends on the availability of molecular oxygen. If molecular oxygen is absent, anaerobic respiration (fermentation) may occur in most plant cells, forming ethyl alcohol and CO_2. If molecular oxygen is present, aerobic respiration may occur, and the pyruvate diffuses into the mitochondria and passes through two more phases of respiration: (1) the **tricarboxylic acid (TCA) cycle,** sometimes called the **Krebs cycle** after the Nobel Laureate physiologist Hans Krebs, whose research contributed a great deal to our knowledge of respiration; and (2) the **terminal electron transport chain.**

Anaerobic Respiration Transforms Only a Small Amount of Energy

Normally, higher plants cannot live long in the absence of molecular oxygen. Under anaerobic conditions, their cells have the following characteristics: (1) They do not oxidize food completely enough to yield adequate energy for their life processes; (2) they may produce poisonous products; and (3) they may not synthesize some necessary intermediate compounds.

Anaerobic respiration takes place in two steps (Fig. 9.4):

1. One CO_2 molecule is enzymatically split off from each pyruvate. This leaves a two-carbon compound, acetaldehyde.

2. NADH, formed during glycolysis, reduces acetaldehyde to alcohol. NAD^+ is thereby released to take part in glycolysis again.

Thus, during anaerobic respiration, the energy trapped in NADH is used to form alcohol (alcoholic fermentation). It is significant in the life of the cell that for each mole of glucose fermented to alcohol a net of only two moles of ATP are available to do work in the cell. This is about 14 kcal or less than 3% of the energy available in one mole of glucose. About 16% of the energy is lost as heat, whereas almost 84% is still locked in the two molecules of alcohol formed, and thus is unavailable to the plant. In addition, the alcohol itself may be toxic to the plant. We can conclude that anaerobic respiration is not an efficient way of using the energy in food.

Some fruits, notably apples, may be held for long periods in an atmosphere containing small amounts of oxygen and continue to give off carbon dioxide. Yeast may live actively in an atmosphere with small amounts of oxygen and produce relatively large amounts of carbon dioxide and alcohol from the pyruvate formed during glycolysis. However, yeast has a limited tolerance for alcohol. When the alcohol concentration of the medium in which yeast is living reaches about 12%, the yeast cells are killed. Consequently, wines and other naturally fermented alcoholic beverages do not have an alcohol content greater than about 12%. Yeast grows much more vigorously under aerobic conditions than in the absence of oxygen. Indeed, relatively few organisms grow only under strictly anaerobic conditions. Some fungi, bacteria, and many animal cells may live and grow slowly under anaerobic conditions by producing products other than alcohol. When oxygen is not supplied rapidly enough to vigorously exercising muscles in your body, the muscle cells may produce not alcohol, but lactic acid, which may cause cramping.

Aerobic Respiration Effectively Transforms the Energy in Food

When respiration occurs in an environment containing molecular oxygen, the plant cell oxidizes pyruvate more efficiently than it can under anaerobic conditions. Aerobic respiration releases adequate amounts of energy that fuel the reactions required for life. Glycolysis releases about 20% of the energy contained in a glucose molecule, leaving most of the energy still locked in the bonds of the two pyruvate molecules formed. During aerobic oxidation of pyruvate to CO_2 and H_2O, 40% to 50% of the energy in the pyruvate is trapped in a potentially useful form as ATP and NADH.

The reaction steps of aerobic respiration are divided into three parts according to the intermediate molecules and end products formed (Fig. 9.5):

1. *Entry of carbon into the TCA cycle of respiration.* Pyruvate releases CO_2, and the remaining two-carbon-atom fragments enter the TCA cycle—so named

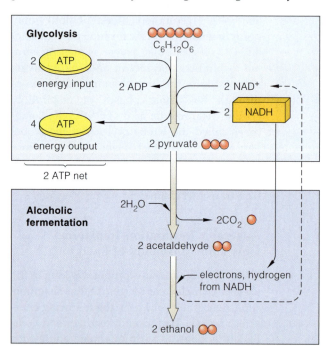

Figure 9.4 The reactions of alcoholic fermentation (anaerobic respiration). In the absence of molecular oxygen, pyruvate from glycolysis is converted to carbon dioxide and alcohol (ethanol). ATP, Adenosine triphosphate.

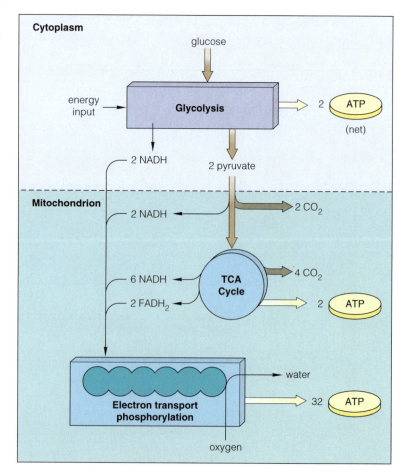

Figure 9.5 Overview of the complete oxidation of glucose during aerobic respiration. ATP, Adenosine triphosphate; TCA, tricarboxylic acid.

ECONOMIC BOTANY:

How to Make a Better Beer

Beer making is a complex art, one that has developed over centuries. Although the techniques were originally perfected by serendipity and trial and error, they currently are based on important scientific principles. The major steps are:

- germinating the grain (barley, *Hordeum vulgare* L.)
- roasting the germinated grain to produce malt
- extracting the malt sugars; boiling the wort
- adding hops during boiling to extract their essential oils
- inoculating the wort with yeast and incubating it to allow fermentation
- bottling

The success of these steps depends greatly on the enzymes of respiration in yeast. The methods of culturing pure yeast strains and some methods of studying enzymes were developed at research laboratories of the Carlsberg brewery in Copenhagen.

Moistening the grain to start germination activates enzymes that break down starch, stored in the endosperm, into sugars. The major enzymes are α-amylase and maltase. α-Amylase hydrolyzes starch, cutting the chain of glucose molecules into single glucose and maltose (two glucoses connected). Maltase hydrolyzes maltose, separating the two glucoses. At the same time, seedlings start to grow, using the glucose as an energy source. Once the hydrolysis is almost finished, roasting kills the seedlings and inactivates the enzymes, leaving the glucose available to support yeast fermentation. Some over-roasting, which oxidizes a portion of the glucose, provides caramel flavor and color and some sweetness, because the modified glucose cannot be used up by the yeast.

The roasted malt is ground into a powder and boiled with water to dissolve the sugars and to kill unwanted microorganisms. (Amateur brewers may buy a concentrated malt extract, which they add to boiling water.) Hop flowers or extracts of the flowers may be added to provide bitter flavor, aroma, and some protection against infection by unwanted microorganisms. Then the liquid, called wort, is cooled, and a culture of yeast (*Saccaromyces carlsbergensis*) is added ("pitched") into the liquid. The mixture is incubated for several days at 4°C (lagers) or 20°C (ales) until the fermentation is complete. It is during this time that enzymes of glycolysis in the yeast break down the glucose. Because the liquid is not stirred (or air is prevented from entering), oxygen is used up, and most of the glucose is metabolized by alcoholic fermentation, which produces ethanol and CO_2. Sometimes much of the CO_2 escapes; therefore, for proper fizziness, more CO_2 must be added after the beer is bottled. In commercial beers, this often is done by injecting purified CO_2 gas into pasteurized beer as the beer is bottled, but it can also be done by adding a little sugar to unpasteurized beer just before the bottle is sealed. In the latter method, the yeast will produce a little more CO_2 (and ethanol), and there will be a layer of yeast at the bottom of the bottle.

What gives different beers their unique taste? It is a combination of the four ingredients: water, malt, yeast, and hops. The malt is critical—how much, what kind, and how much it has been roasted. The light Pilsner lagers are made from lightly roasted malt; the brown ales and stouts include some varieties of malt that are roasted at greater temperatures for longer times. The amounts and varieties of hops also are major factors. Most American lagers are very lightly hopped; pale ales, in contrast, are more bitter and aromatic.

Beers, like all foods, should be appreciated for the art and science that go into their creation and for the subtle nuances of taste that they can provide to a meal.

 Websites for further study:

The Beer History Library: http://www.beerhistory.com/library/

---

because several three-carboxyl (COOH) acids play prominent roles.

2. *The TCA cycle.* The two-carbon-atom fragments release CO_2 and become oxidized. ATP and NADH are formed.

3. *Electron transport and terminal oxidation.* Oxygen accepts electrons and H^+ from the reduced coenzymes. ATP and water are formed.

ENTRY OF CARBON INTO THE TRICARBOXYLIC ACID CYCLE
One of the most complex series of reactions in respiration oxidizes pyruvate, thereby forming the carbon molecules that will enter the TCA cycle. In these reactions, several vitamins, particularly those of the vitamin B complex (niacin, thiamin, pantothenic acid), serve as coenzymes or parts of coenzymes (see Chapter 8). In the final stages, each molecule of pyruvate oxidized forms a molecule of CO_2, with the transfer of two electrons and two hydrogen atoms to NAD^+

to form NADH. About 53 kcal of energy are transferred from a mole of pyruvate to NAD^+ when NADH is formed.

The remaining part of pyruvate is now a two-carbon-atom fragment called an acetyl group. It complexes with coenzyme A (CoA) to produce the reactive substance, acetyl-CoA (Fig. 9.6), the molecule that enters the TCA cycle. In the presence of a specific enzyme, acetyl-CoA transfers (donates) its acetyl group to other acceptor molecules.

In this way, each acetyl group, which still contains a high percentage of the energy originally present in the glucose molecule, may be transferred from one series of reactions to another in the cell. This transfer is analogous to the transfer of electrons and hydrogen by the coenzymes NADH and NADPH. Although pyruvate is the usual donor of acetyl groups to CoA, it is not the only donor. The breakdown of fat and protein (amino acids) forms acetyl groups that also may be donated to CoA and enter the TCA cycle.

THE TRICARBOXYLIC ACID CYCLE OF RESPIRATION In the TCA cycle, acetyl-CoA donates its acetyl group to an organic acid acceptor molecule, oxaloacetate, found in mitochondria. The union of oxaloacetate and the acetyl group produces citrate, which is gradually broken down in a cyclic series of reactions in which seven other organic acids including a new molecule of the acetyl acceptor, oxaloacetate, are formed (Fig. 9.6). During these reactions, pairs of electrons (and hydrogen atoms) are transferred to the electron carriers NAD^+, $NADP^+$, and FAD (flavin adenine dinucleotide). All of the atoms brought in with the acetyl group are gradually removed, and the carbon and oxygen are released as molecules of CO_2. Because the reactions form new oxaloacetate molecules, capable of accepting other acetyl groups, the cycle begins again.

In one turn of the cycle, some of the potential energy in one acetyl group is lost as heat to the environment, but about 66% is trapped in one molecule of ATP, three molecules of NADH, and one molecule of $FADH_2$. The ATP is free to carry its energy to other parts of the cell where work is being done. The energy in the NADH and the $FADH_2$ generally is used in the synthesis of more ATP during the final stages of respiration.

The TCA cycle provides the following benefits to the cell:

1. It produces intermediate compounds that may act as starting materials for synthetic pathways of other molecules such as proteins and lipids.

2. It transfers electrons to energy carriers NAD^+ and FAD.

3. It produces the energy carrier ATP from ADP and inorganic phosphate.

Although all the energy originally associated with the reduced carbon atoms in the sugar molecule (food) is released or transferred by the time the TCA cycle is completed, only about 10% is directly trapped in the ATP formed during glycolysis and the TCA cycle. About 56% is first trapped in reduced nucleotides (see Chapter 8). How does the cell obtain the energy trapped in these reduced nucleotides (NADH, NADPH, $FADH_2$) to do cellular work? It uses an electron transport chain.

THE ELECTRON TRANSPORT CHAIN During electron transport, the final series of reactions in aerobic respiration, cells do not obtain energy directly from the reduced nucleotides. Instead, they oxidize the nucleotides and trap part of their energy in ATP. In a stepwise series of oxidations, electrons

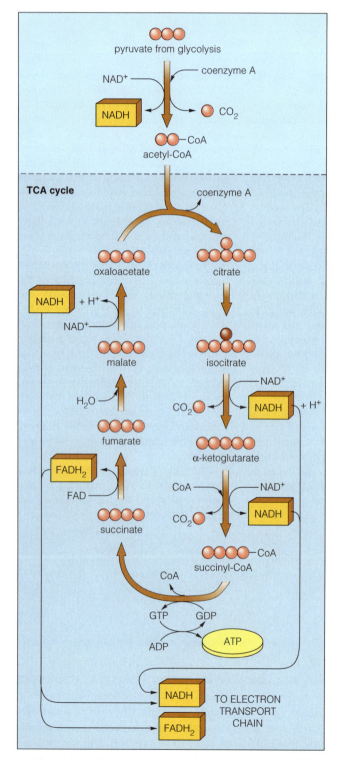

Figure 9.6 Several steps in the tricarboxylic acid (TCA) cycle leading to the production of reduced nucleotides. ADP, Adenosine diphosphate; ATP, adenosine triphosphate; CoA, coenzyme A; FAD, flavin adenine dinucleotide; GDP, guanosine diphosphate; GTP, guanosine triphosphate.

are transferred from the reduced nucleotides through an electron transport chain of carriers that are alternately reduced and oxidized (Fig. 9.5). These electron carriers include a flavoprotein, coenzyme Q, and several cytochromes. **Cytochromes** contain iron atoms that are reduced to the ferrous (Fe^{2+}) form and then give up their electrons, becoming oxidized to the ferric (Fe^{3+}) form. The last step transfers a pair of electrons to an oxygen atom. Two hydrogen ions from the cellular environment then combine with the oxygen, forming water. If free oxygen were not available, this last transfer of electrons could not take place. The flow of electrons would cease, and the entire aerobic respiratory sequence would cease.

During electron transport, three molecules of ATP are synthesized from ADP and inorganic phosphate for each NADH or NADPH molecule oxidized, and two molecules of ATP are synthesized for each $FADH_2$ oxidized. This process, which couples oxidation to phosphorylation, is called **oxidative phosphorylation.** Complete aerobic respiration of a glucose molecule thus involves a stepwise breakdown of the sugar molecule, uses six molecules of oxygen, and forms six CO_2 molecules and six H_2O molecules. The process transfers about 40% of the potential energy in the glucose to some 36 molecules of ATP, which can do work in the cell. About 60% of the energy is lost as heat during the various steps of respiration.

Some Plant Cells Have Alternate Pathways of Respiration

Although aerobic respiration is the most common method plant cells use to oxidize foods, alternate pathways exist that also provide energy and organic building blocks. Two common alternate pathways are the pentose phosphate pathway (PPP) and the glyoxylate cycle.

THE PENTOSE PHOSPHATE PATHWAY The PPP occurs in the cytoplasm of some tissues. Biochemically, the PPP starts in the same way as glycolysis—that is, with the phosphorylation of a glucose molecule to produce glucose 6-phosphate. But instead of producing fructose 6-phosphate, the next steps in the PPP are the oxidation of the glucose 6-phosphate to form two molecules of NADPH, one molecule of CO_2, and one molecule of ribulose 5-phosphate from which ribose 5-phosphate may be synthesized (Fig. 9.7). The rest of the cycle consists of a series of sugar transformations, and glucose may be regenerated in the process. Young growing plant tissues appear to use the TCA cycle as the predominant pathway of glucose oxidation, whereas aerial parts of the plant and older tissues seem to use PPP as well.

The PPP has two important consequences.

1. The pathway transfers energy in glucose to NADP, by forming NADPH, which is used as an energy source in synthetic reactions.

2. The pathway produces the five-carbon sugar phosphate, ribose 5-phosphate, which is necessary for the synthesis of some essential compounds such as nucleic acids.

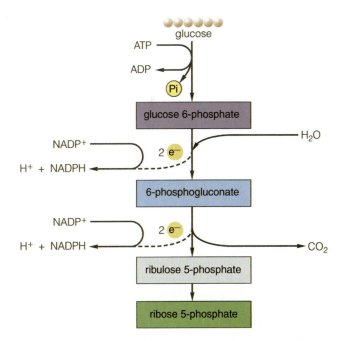

Figure 9.7 The first steps of the pentose phosphate pathway. ADP, Adenosine diphosphate; ATP, adenosine triphosphate.

THE GLYOXYLATE CYCLE When fat-storing seeds germinate, the developing embryo requires energy and carbon compounds for growth. Because fats and oils do not dissolve in water, they are not transported from storage cells in the seed to cells in the growing seedling. Instead, the fats must be broken down to their constituent long-chain fatty acids and glycerol, which are then further metabolized. Although glycerol may be converted into dihydroxyacetone phosphate, thus entering the glycolytic sequence, only a relatively small amount of energy or carbon units may be derived from it. Fatty acids, however, yield large amounts of acetyl-CoA when they are oxidized.

The disappearance of reserve fat and oil in fat-storing seeds during germination usually is accompanied by a significant increase in the sucrose content of the seeds including the embryo. An important part of this process involves a series of reactions called the glyoxylate cycle. The glyoxylate cycle is important because it is a cellular mechanism that helps to convert the immobile products of fat breakdown into soluble sugar. These water-soluble sugar molecules can be transported to sites in the plant requiring soluble food.

9.3 RESPIRATION AND CELL STRUCTURE

One cannot understand the entire process of respiration without knowledge of the role that cell structure plays in the process. Biochemists and plant physiologists isolate and purify the various cell organelles and membranes and by studying them determine how cellular structures function in cell metabolism. Some enzymes are confined, during the life of the cell, to specific locations in the cell. Two experiments

Figure 9.8 (**a**) A plant mitochondrion, showing the cristae, the double nature of the mitochondrial envelope, and the site of reactions in aerobic respiration. (**b**) Section through cristae. (**c**) Detail of membranes, showing location of electron transport system and adenosine triphosphate (ATP) synthesis. TCA, Tricarboxylic acid. (b–c Art by Raychel Ciemma)

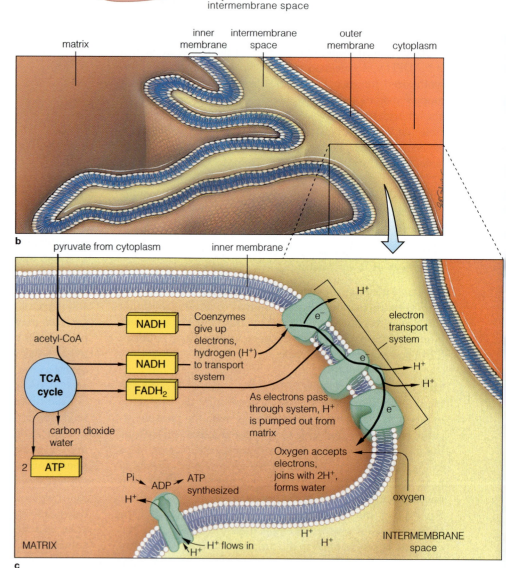

inner membrane site of electron transport ATP synthesis

outer membrane

matrix site of most of the TCA cycle (except for succinate oxidation and citrate transformation to isocitrate)

a

cristae

intermembrane space

matrix / inner membrane / intermembrane space / outer membrane / cytoplasm

b

pyruvate from cytoplasm / inner membrane

acetyl-CoA

NADH

NADH

FADH$_2$

Coenzymes give up electrons, hydrogen (H$^+$) to transport system

TCA cycle

carbon dioxide water

2 **ATP**

Pi, ADP, ATP synthesized

H$^+$

MATRIX

e$^-$

H$^+$

e

H$^+$

H$^+$

e$^-$

electron transport system

As electrons pass through system, H$^+$ is pumped out from matrix

Oxygen accepts electrons, joins with 2H$^+$, forms water

oxygen

H$^+$ flows in

H$^+$

H$^+$

H$^+$

INTERMEMBRANE space

c

show this to be true for some enzymes catalyzing respiration reactions.

The first experiment demonstrates that when plant tissue is ground until all cells and organelles are broken apart, but the enzymes are preserved, glycolysis and alcohol fermentation will occur in the cell free preparation, but complete aerobic respiration will not. This means that the enzymes of glycolysis and fermentation are soluble enzymes and do not depend on intact organelles or membranes for their action.

The second experiment shows that if mitochondria are isolated from plant tissue, and pyruvate, ADP, Mg^{2+}, and inorganic phosphate are added, the pyruvate will break down, and the entire TCA cycle will occur, as will electron transport. ATP will be formed, molecular oxygen will be absorbed, and H$_2$O and CO$_2$ will be produced. The results of this experiment, in light of the first, means that at least some steps in aerobic respiration require intact mitochondria.

In fact, biochemists have gone on to show that all but one of the enzymes of the TCA cycle are present in the mitochondrial matrix, and one is located in the inner mitochondrial membrane. The electron transport chain enzymes and the ATP synthesizing enzymes are also part of the inner mitochondrial membrane (**Fig. 9.8**). From such findings, researchers have been able to determine where in a cell specific reactions occur (**Table 9.1**).

Experimental findings indicate that a relationship exists between mitochondrial structure and respiration and that the membrane structure in mitochondria is important for ATP synthesis and electron transport. A mitochondrion, you will recall, consists of a matrix surrounded by a double membrane envelope made up of an outer and an inner membrane. These membranes separate the mitochondrion from the rest of the cytoplasm (Fig. 9.8). The outer membrane is permeable to water and to small neutral molecules such as urea but not to hydrogen ions or simple sugars.

Reactions in the Inner Mitochondrial Membrane Include Adenosine Triphosphate Synthesis and Electron Transport

The final stages of respiration, electron transport and oxidative phosphorylation, occur in the inner membrane. The electron carriers NAD$^+$ and FAD, which were reduced to NADH and FADH$_2$ during the oxidation of organic acids in the TCA cycle in the matrix, come in contact with the inner mem-

Table 9.1 Intracellular Location of Some Respiratory Reactions

Reactions Catalyzed	Location of Reactions
Glycolysis	Cell cytoplasm
Alcoholic fermentation	Cell cytoplasm
Tricarboxylic acid cycle	Mitochondrial matrix (most reactions)
Citrate to isocitrate and succinic oxidation	Inner mitochondrial membrane
Electron transport	Inner mitochondrial membrane
Adenosine triphosphate synthesis	Inner mitochondrial membrane

brane, where they become oxidized (lose electrons) to other electron carriers. This reaction regenerates NAD^+ and FAD.

The electron carriers making up the electron transport chain are present in aggregated clusters in the inner membrane. Each cluster appears to consist of a fixed number of molecules of each carrier. This close organization of electron carriers facilitates the rapid exchange of electrons from a reduced carrier to the next carrier in the chain as soon as it is oxidized. Electrons flow from high-energy levels in the NADH or $FADH_2$ along the carrier chain down an energy gradient and finally are transferred to oxygen. Hydrogen ions are taken up from the surrounding matrix, and water is formed. Each reaction step releases energy: Some is lost as heat, but some is used to drive H^+ (protons) out of the mitochondrial matrix through the inner mitochondrial membrane and into the surrounding cytoplasm. *Because the inner membrane prevents the free diffusion of H^+ back into the matrix, an energy gradient is established across the membrane.* This gradient has two components:

1. an osmotic component—because there is a difference of H^+ concentration across the membrane

2. an electrical component—because there is an accumulation of positive charges, H^+, between the inner and outer mitochondrial membrane or within the cytoplasm

This proton gradient is a form of free energy (see Chapter 8), which is believed to be involved in ATP synthesis.

The Chemiosmotic Theory Explains the Synthesis of Adenosine Triphosphate in Mitochondria

In 1961, Peter Mitchell, a British biochemist, hypothesized that the proton gradient that is developed across the inner mitochondrial membrane during electron transport drives ATP synthesis. Although the results of many experiments on oxidative phosphorylation are best explained by this theory, the details of the process are still not well understood. However, the **chemiosmotic theory** (so named because the action is both chemical and osmotic) is now generally accepted as one of the best explanations of the mechanism of ATP syn-

thesis in mitochondria, as well as ATP synthesis in chloroplasts during photosynthesis. Energy harvest from the proton gradient across the inner mitochondrial membrane is brought about by a complex of enzymes known as **ATP synthetase,** which is embedded in and extends across the inner mitochondrial membrane (Fig. 9.8). It is believed that special channels in the ATP synthetase allow protons to move down the chemiosmotic gradient into the matrix, discharging the gradient. Some of the energy released by this flow of protons is transferred to ATP when ADP and inorganic phosphate are combined during ATP synthesis. Exactly how this is accomplished is unclear.

9.4 EFFECTS OF ENVIRONMENTAL FACTORS ON RESPIRATION

Conditions inside and outside the cell affect the rate of respiration, as well as the rates of other cellular activities, such as photosynthesis, absorption of water and mineral salts, cell division, and growth. The rate of respiration usually is expressed as the quantity of CO_2 released or O_2 absorbed per unit of cell weight per unit of time. Many factors that affect respiration rate act indirectly—for example, soil flooding decreases the availability of oxygen to respiring cells, whereas shading decreases photosynthesis, and therefore reduces the amount of food available to respiring cells. Respiration rate also depends on cell type and age. Cells of different ages or from different kinds of plants may respond differently to the same environmental factor. For example, at low oxygen concentrations, CO_2 is released at a much greater rate from cells in rice seedlings than from cells in wheat seedlings, undoubtedly because rice seedlings have a greater capacity for fermentation than wheat seedlings. This is because rice seedlings are adapted to grow under water, where O_2 concentration is low.

A few examples of the effects of environmental factors, age, and species of plant on respiration are discussed in the following section.

Cell Hydration May Indirectly Affect the Rate of Respiration

The water content of active protoplasm may be as high as 90%, and small fluctuations in water content have little effect on the rate of respiration. However, in some plants, starch is hydrolyzed to sugar during water stress, for example, when a plant wilts. In these plants, the rate of respiration may increase during wilting as a result of an increase in the amount of respirable food.

One of the most dramatic effects of the influence of cell water content on respiration occurs when seeds mature and begin to dry. When the water content of seeds decreases to between 16% and 17%, there is a sharp decrease in the rate of CO_2 released from their cells. The water content of mature seeds frequently becomes less than 10% of the seed weight. Growth ceases, mineral salts are not absorbed, and cell divi-

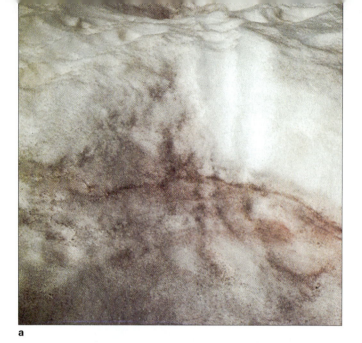

a   b

Figure 9.9 Adaptations to temperature extremes. (**a**) Cyanobacteria growing in a snowbank. (**b**) Cyanobacteria growing in hot springs.

sion stops, as do many, if not all, reactions that synthesize new molecules. Respiration goes on at a very slow rate; therefore, cells consume little oxygen, little carbon dioxide is given off, and only a very little amount of heat is released. Energy is needed only to maintain a little-understood steady state in the quiescent protoplasm. In this dormant state, seeds may be kept in large storage bins such as grain elevators for long periods.

If even a little water is added to viable seeds, it is imbibed. The seeds then swell, respiration increases rapidly, and if the seeds have an adequate oxygen supply, growth begins and the seeds germinate. However, if confined too closely, as grain in a grain elevator, too high a moisture content in the grain may cause a rapid rate of respiration that can increase the temperature inside the mass and kill the grain.

Living Cells Are Very Sensitive to Fluctuations in Temperature

Temperature has a marked effect on most biological reactions, especially those that are controlled by enzymes. In the temperature range from near freezing (0°C/32°F) to about 30°C (86°F), the rate of respiration approximately doubles for every increase of 10°C (18°F) in temperature. At temperatures greater than 35 to 40°C (95 to 113°F), protein molecules, which make up enzymes, progressively unfold (become denatured), resulting in the loss of enzyme activity. The longer a cell is subjected to a high temperature, the greater the loss of enzyme activity. At first an increase in temperature of cells to 35 or 40°C may cause the respiration rate to increase, but soon the rate decreases and eventually respiration ceases.

Nevertheless, over the long periods needed for adaptation to occur, certain organisms have evolved characteristics enabling them to survive in otherwise hostile environments.

For instance, some species of algae and bacteria are adapted to respire and grow under temperature extremes that would kill most other organisms. Particularly impressive is the presence of some of these organisms in hot springs and streams where temperatures may exceed 60°C (140°F). Others grow in snow where the temperature remains near freezing (**Fig. 9.9**).

In some plants such as the potato, a decrease in temperature causes the hydrolysis of starch to sugar. Because cells cannot respire starch directly—but can respire sugar—instead of decreasing the rate of respiration as you might expect, a decrease in temperature actually may increase the rate of respiration in potato cells. It has been demonstrated that more sugar accumulates in potato tubers stored at 0°C than in those stored at 4.5°C. At the same time, the respiration of potatoes stored at 0°C is markedly faster than it is in potatoes stored at 4.5°C, undoubtedly because of the greater sugar concentration. This accumulation of sugar in potatoes stored at or near freezing is of considerable importance to people who process potatoes, because high sugar levels frequently cause undesirable browning in heat-processed potato products.

Factors That Affect Photosynthesis May Indirectly Affect Respiration

Photosynthesizing cells produce their own food in the light. Under normal conditions, they must also make enough food to supply the needs of all the other living cells in the plant. All living cells of a plant kept in the dark continue to use food even though photosynthesis has stopped. The stored food reserves, particularly starch and sugars, rapidly become depleted. If this condition continues, the plant will eventually starve to death. In fact, any factor such as light, temperature, and CO_2 level in the air that limits photosynthesis and thus reduces food availability must indirectly influence respiration.

ECONOMIC BOTANY:

Control of Respiration after Fruit Harvest

An understanding of the process of respiration—and of means to control and modify it—is critically important to the producers, handlers, and shippers of fruit and vegetables. High rates of respiration are associated with a complex series of biochemical changes during the ripening and maturation of fruit. If the rate of respiration is reduced, the rate of maturation is subsequently delayed, and storage life is extended.

One of the most effective methods for reducing the rate of respiration is to cool the produce. Most long-range shipping of fruit is done in refrigerated railcars and trucks. Because all living cells respire—and because respiration produces heat—the amount of refrigeration required to ship and store fruit is considerable. For example, 40,000 boxes of apples at 0°C (32°F) produce enough heat to melt 2.8 tons of ice in 24 hours; at 4°C (39°F), the respiratory heat loss would melt 4.8 tons of ice; and at 16°C (61°F), 19 tons of ice would melt. Control of this heat production represents a considerable cost in marketing the product.

Fruits and vegetables vary greatly in their susceptibility to low temperature and the optimum conditions they require for successful storage. Bananas, tomatoes, and summer squash (all of tropical origin) suffer chilling injury when stored at temperatures less than 10 to 13°C (50–55°F) but greater than freezing. (Chilling injury, which shows up when the produce is taken out of cold storage, can include internal browning, scalding, and pitting.) Cold weather fruits and vegetables such as pears, some apples, and onions, by contrast, tolerate storage at near-freezing temperatures for extended periods. The shipper's objective is to store the produce at the lowest economical temperature that still maintains the quality of the particular fruit or vegetable being shipped.

Grocery stores have tasty, crisp apples from Washington and Oregon year-round, even though the apple harvest lasts only a few weeks. If the apples are kept at room temperature after harvest, their rate of respiration rapidly increases, and production of ethylene (a gaseous plant hormone; see Chapter 15) increases. Ethylene stimulates the aging process and decreases fruit quality. Within a few weeks, the apples become soft, mealy, and undesirable.

Cold storage significantly prolongs the storage life of apples. Also, because oxygen is necessary for aerobic respiration, if the oxygen level in the air in the storage room is reduced from about 21% to about 1% to 3%, storage life is increased (particularly if the CO_2 level is increased from the usual 0.03% to 1–3%, storage life is increased at the same time). Reducing the oxygen level to zero would not be desirable because this stimulates anaerobic respiration and may result in a more rapid deterioration of the apples. A large fraction of Washington and Oregon apples are stored under refrigeration in rooms with modified atmospheres. These apples can be stored for as long as 9 months and still be acceptable for consumption.

In addition to the other changes that accompany respiration, there is a loss of the sugar respired. If large amounts of living material are involved, a considerable weight of sugar may be lost. As an example of this weight loss, let us look at sugar beets. At harvest, the tops of the sugar beets are removed, and the roots are mechanically dug up and loaded into trucks for transport to the sugar processing plant. At the plant, beets often are stored in large piles on a cement apron. If the pile loses 1% of its weight in 20 days (a reasonable figure), a 100-ton pile of sugar beets would lose 1 ton of its weight. Of course, part of the weight loss is water, but a significant portion is sugar. No practical means of reducing the rate of respiration in harvested sugar beets exists, but the faster the beets are processed after harvest, the smaller the amount of sugar that will be lost.

> **Websites for further study:**
>
> Fruit Conservation and Processing:
> http://technofruits2001.cirad.fr/en/mndmang_en.html
>
> International Institute of Tropical Agriculture:
> http://www.iita.org/info/trn_mat/irg64/irg643.html

Oxygen Gas Must Be Available for Aerobic Respiration

Cells carrying out aerobic respiration must have a continual supply of oxygen diffusing into the cells. They produce CO_2, which diffuses across the plasma membrane and into the surrounding environment. Most plant cells can continue for a time to oxidize foods even in the absence of gaseous oxygen by switching to anaerobic respiration. However, higher plants require molecular oxygen. Rarely does the concentration of oxygen in the atmosphere deviate enough from the normal 21% to appreciably affect the rate of respiration. However, underground stems, seeds, and roots may be in an oxygen-poor environment, because microorganisms and the plant parts themselves may use the oxygen in the soil atmosphere faster than it is replaced from the air. Under these conditions, respiration in the cells of these underground organs may decrease. Similar conditions of low oxygen level and high carbon dioxide concentration may occur in the internal cells of bulky plant organs such as large fleshy fruit. However, in general, the diffusion of O_2 through the intercellular spaces is rapid enough so that aerobic respiration occurs even within bulky plant parts.

Although the absence of oxygen is detrimental to the cells of most plants, modified atmospheres made of low concentrations of oxygen and greater levels of inert gases, coupled with low temperatures, effectively prolong storage life and improve the quality of many fruit and vegetable products. Shippers and handlers of fresh produce have developed controlled atmosphere containers and storage bins to take advantage of these effects (see sidebar "ECONOMIC BOTANY: Control of Respiration after Fruit Harvest").

KEY TERMS

adenosine triphosphate (ATP) synthetase

aerobic respiration

alcoholic fermentation

anaerobic respiration

chemiosmotic theory

cytochromes

digestion

glycolysis

oxidative phosphorylation

respiration

terminal electron transport chain

tricarboxylic acid cycle

SUMMARY

1. The two major functions of respiration are (a) the transformation of chemical bond energy stored in food through the production of ATP and reduced nucleotides and (b) the production of intermediate products that are used in the synthetic reactions of the cell.

2. Energy-yielding oxidative reactions may be coupled to energy-requiring reactions.

3. Phosphorylation prepares sugar for oxidation. During phosphorylation, an organic phosphorus donor, ATP, transfers phosphorus to the sugar molecule. Sugar cleavage then occurs, resulting in the production of two three-carbon sugar phosphate intermediates. These, in turn, are oxidized to pyruvic acid. This entire sequence of reactions is glycolysis.

4. In the absence of molecular oxygen, pyruvate usually is metabolized to alcohol and carbon dioxide in the final stages of anaerobic respiration in plant cells. Only small amounts of energy are released in this process.

5. In the presence of molecular oxygen, pyruvate becomes further oxidized to carbon dioxide and water through the TCA cycle. During aerobic respiration, large amounts of energy are stored in ATP and reduced nucleotides, which are used when work is done in the cell. Reduced nucleotides are oxidized in the electron transport chain, and ATP is produced.

6. Respiration is carried out through an integration of reactions going on in the cytoplasm (glycolysis) and mitochondria (TCA cycle and electron transport chain).

7. According to the chemiosmotic theory, the synthesis of ATP from ADP and inorganic phosphate during electron transport is driven by the proton gradient that develops across the inner mitochondrial membrane.

8. Some cells exhibit alternate pathways of respiration. One alternate pathway, the PPP, transfers energy from glucose to NADP, forming NADPH, and produces ribose 5-phosphate, which is necessary for the synthesis of nucleic acids. Another pathway, the glyoxylate cycle, which occurs in fat-storing seeds, converts fat into intermediate compounds and ultimately into sucrose, which may then be transported to growing cells.

9. Cell hydration, temperature, oxygen supply, food availability, the plant type, and plant age all affect the rate of respiration.

Questions

1. Define digestion.

2. What two major roles does respiration play in the life of a plant?

3. Why does respiration increase when a cell is stimulated to do work such as salt absorption or cytoplasmic streaming?

4. Name and describe the three major steps in glycolysis.

5. How would you determine whether a plant was carrying out aerobic or anaerobic respiration?

6. Why is anaerobic respiration not an effective way of using food?

7. Describe an experiment showing that some of the enzymes involved in aerobic respiration are soluble and others are located in the mitochondrial membranes.

8. What is the chemiosmotic theory of ATP synthesis?

9. Explain how you would expect the rate of respiration in root cells to change when a plant kept in the dark for a period is moved into the light.

10. By means of a graph, show how you would expect the rate of respiration to change if the temperature were increased from 10 to 20 to 30 to 40 to 50 to 60°C.

 InfoTrac® College Edition

http://infotrac.thomsonlearning.com

Plant Respiration

Dayton, L. 2001. Philodendrons like it hot and heavy (the respiration of *Philodendron selloum*). *Science* 292:186. (Keywords: "Philodendron" and "respiration")

Fermentation

Rofe, J. 2001. RAVE on (review of recent research in enology and viticulture). *Wines & Vines* 82:40. (Keywords: "fermentation" and "RAVE")

Brewing

Stewart, G. 2000. A brewer's delight (history of brewing techniques). *Chemistry and Industry* Nov. 6, p. 706. (Keywords: "brewer's" and "delight")

Photosynthesis

Visit us on the web at http://biology.brookscole.com/plantbio2 for additional resources, such as flashcards, tutorial quizzes, InfoTrac exercises, further readings, and web links.

1. Photosynthesis is the primary energy-storing process on which almost all life, both plant and animal, depends. The energy from sunlight is stored as chemical energy in organic compounds through a series of light- and temperature-sensitive reactions. Carbon dioxide and water are the raw materials, and the products are sugar and oxygen.

2. Chlorophyll in green plants absorbs light energy, which activates electrons in special chlorophyll *a* molecules. These electrons move along a chain of electron carriers, and some of their energy is stored in the production of adenosine triphosphate (ATP) or reduced nicotinamide adenine dinucleotide phosphate (NADPH). These reactions take place in association with thylakoid membranes.

3. Temperature-sensitive enzymatic reactions in the stroma use the ATP and NADPH to reduce CO_2 and to produce sugar.

4. The C_3 cycle is the major path of carbon assimilation in green plants. In this cycle, two three-carbon-atom molecules of phosphoglyceric acid are formed, hence the name C_3.

5. Some plants have adapted to extreme environmental conditions by evolving variations on the carbon assimilation pathway. Some plants use crassulacean acid metabolism. Crassulacean acid metabolism plants absorb CO_2 through open stomata at night and form organic acids. During the day, when stomata are closed in these plants, CO_2 (released in the leaves from the organic acid formed) is converted into carbohydrate by the C_3 cycle.

6. Other plants have evolved another variation of carbon assimilation, the C_4 pathway. These plants effectively concentrate CO_2 in their leaf mesophyll chloroplasts by forming four-carbon organic acids. The organic acids are transported into the bundle sheath chloroplasts and lose the CO_2, which is used in the C_3 cycle there.

7. In the light and at high temperatures, chloroplasts take up O_2 and release CO_2 (photorespiration). Photorespiration is reduced in C_4 plants.

8. Usually only about 0.3% to 0.5% of light energy that strikes a leaf is stored during photosynthesis, but under ideal conditions this may increase several-fold. Photosynthesis may be limited by CO_2 concentration, light, temperature, minerals, and other environmental and hereditary factors.

10.1 THE HARNESSING OF LIGHT ENERGY BY PLANTS

With few exceptions, all living cells require a continuous supply of energy, which comes directly or indirectly from the sun. Although terrestrial green plants use large amounts of energy directly from the sun in both transpiration and photosynthesis, only in photosynthesis is light energy stored as chemical energy for future use. Billions of years ago,

Cyanobacteria (also known as blue–green algae) created an oxygen-rich atmosphere through their photosynthetic activity. Since that time, photosynthetic organisms have continued to support life by being the original source of energy for other organisms. The great importance of photosynthesis is twofold: the liberation of oxygen as an end product and the transformation of low-energy compounds (carbon dioxide and water) into high-energy compounds (sugars). Perhaps someday humans will use other sources of energy to drive the energy-requiring steps in the production of food and fiber. Currently, however, except for a few species of bacteria (see "IN DEPTH: Chemosynthesis" sidebar), all life is dependent on the energy-storing reactions of photosynthesis. Although the subject of this chapter is primarily "photosynthesis in green plants," you should keep in mind that the photosynthesis carried out by aquatic life—including photosynthetic bacteria and red, green, yellow, golden, and brown algae—liberates at least as much oxygen per day as that produced by terrestrial green plants.

10.2 DEVELOPING A GENERAL EQUATION FOR PHOTOSYNTHESIS

Like respiration and other complex processes occurring in living cells, photosynthesis consists of many reaction steps. It is easier to approach photosynthesis, or any biochemical process, by looking at an overview. Indeed, this is how scientific knowledge about photosynthesis evolved; therefore, this section traces the discoveries that led to a general understanding of photosynthesis. Later sections detail the steps of specific reactions.

Early Observations Showed the Roles of Raw Materials and Products

Until the early 17th century, scholars believed that plants derived the bulk of their substance from soil humus. A simple experiment performed by Flemish physician and chemist Joannes van Helmont disproved this idea. He planted a 2.27-kg (5-lb) willow (*Salix*) branch in 90.7 kg (200 lb) of carefully dried soil and supplied rainwater to the plant as needed. In 5 years, it grew to a weight of 67.7 kg (169 lb), but according to van Helmont's measurements, the soil had lost only 57 g (2 oz). Consequently, he reasoned that the plant substance must have come from water. This was a logical deduction, though not entirely correct. Almost two centuries elapsed before van Helmont's findings were finally explained correctly.

Our knowledge of photosynthesis begins with the observations of a religious reformer, philosopher, and spare-time naturalist, Joseph Priestley. In 1772, Priestley reported that a sprig of mint could restore confined air that had been made impure by a burning candle. The plant changed the air so that a mouse was able to live in it. The experiment was not always successful, probably because Priestley (who did not know about the role of light) did not always provide ade-

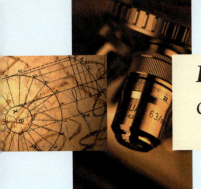

IN DEPTH:

Chemosynthesis

In the 1970s, scientists discovered vents deep in the Pacific Ocean associated with the upwelling of hot sea water. Rich animal populations including large clams, crabs, mussels, and worms are clustered around these hydrothermal vents. Some of these animals live deep within the vents. It was shown that organic material produced by photosynthetic organisms near the ocean surface did not provide adequate food to sustain the life of these organisms. Instead, they obtain their food from dense microbial mats consisting primarily of bacteria. Many of these bacteria use reduced inorganic compounds, such as H_2S, that are in high concentration in the vents. These microorganisms oxidize the reduced inorganic compounds and use the energy liberated to assimilate CO_2 into organic compounds in a process known as **chemosynthesis.** Chemosynthe-

sis is similar to photosynthesis in several respects, but chemical energy rather than light energy drives the synthetic reactions. In addition to obtaining organic matter from the microbial mass, some of the animal species in the vents live in symbiotic association with chemosynthetic microorganisms.

In 1996, the bacteria in a cave in southern Romania were studied using radioactive ^{14}C bicarbonate to determine their source of carbon. It was found that radioactive carbon was incorporated into

microbial lipids. Because light was not available to these bacteria, chemosynthetic fixation must have occurred. Organisms that can produce their own food from inorganic materials using reduced inorganic compounds instead of light as a source of energy are **chemoautotrophic organisms.** Just as chemoautotrophic bacteria serve as a food base for the marine life in the hydrothermal vents in the ocean, the chemoautotrophic bacteria in the cave in Romania serve as a food base for a variety of cave-dwelling organisms.

 Websites for further study:

NOAA—Chemosynthesis:
http://www.pmel.noaa.gov/vents/nemo/explorer/concepts/chemosynthesis.html

Office of Naval Research—Hydrothermal Vents:
http://www.onr.navy.mil/focus/ocean/habitats/vents2.htm

The Movile Cave Project: http://www.geocities.com/rainforest/vines/5771/

quate illumination for his plants. In 1780, a Geneva pastor, Jean Senebier, published his own research and pointed out another important part of the process: that "fixed air," carbon dioxide, was required. Thus, in the new terminology of French chemist Antoine Lavoisier, it could be said that green plants in the light use carbon dioxide and produce oxygen.

But what was the fate of the carbon dioxide? A Dutch physician, Jan Ingen-Housz, answered this question in 1796 when he found that carbon went into the nutrition of the plant. In 1804, 32 years after Priestley's early observations, the final part of the overall reaction of photosynthesis was explained by the Swiss botanist and physicist Nicolas de Saussure, who observed that water was involved in the process. Now the experiment performed by van Helmont almost 200 years earlier could be explained:

$$\text{carbon dioxide + water} \xrightarrow[\text{green plants}]{\text{light energy}} \text{oxygen + organic matter}$$

Almost 50 years elapsed before scientists identified carbohydrates as the organic matter formed during photosynthesis in most plants. Between 1862 and 1864 a German plant physiologist, Julius von Sachs, observed that starch grains occur in the chloroplasts of higher plants, and that if leaves

containing starch are kept in darkness for some time, the starch disappears. If these leaves are exposed to light, starch reappears in the chloroplasts. von Sachs was the first to connect the appearance of starch, a carbohydrate, with both the fixation of carbon in chloroplasts and the presence of light.

It is easy to demonstrate in the laboratory that starch forms during photosynthesis; however, it is much more difficult to show that sugar forms *before* starch does. Proof that sugar is the first carbohydrate produced by photosynthesis had to await the availability of radioactive carbon (^{14}C). This development is discussed in a later section in this chapter.

Comparative Studies Showed That Several Molecules May Reduce CO_2 in the Light

By glancing at the overall equation for photosynthesis,

$$6CO_2 + 6H_2O \xrightarrow{\text{light}} C_6H_{12}O_6 + 6O_2$$

you might conclude that the carbon dioxide molecule splits, liberating the oxygen molecule. Indeed, most scientists believed that the reaction proceeded in this manner until the early 1930s when Cornelis van Niel, working at Stanford University, compared photosynthesis in a number of different groups of photosynthetic bacteria. The green and purple sulfur bacteria use hydrogen sulfide instead of water to reduce carbon dioxide, for example, and van Niel found that sulfur instead of oxygen is liberated, as follows:

$$6CO_2 + 12H_2S \xrightarrow{\text{light}} C_6H_{12}O_6 + 6H_2O + 12S$$

The sulfur can come only from the hydrogen sulfide. Because the hydrogen sulfide serves the same role in these bacteria as water does in higher plants, van Niel reasoned that the oxygen evolved by higher plants comes from water, not from carbon dioxide. (Experiments using radioactive tracers have since verified van Niel's insight.)

After comparing similar reactions in other organisms, van Niel concluded that a general equation for photosynthesis should be written as:

$$6CO_2 + 12H_2A \xrightarrow{\text{light}} C_6H_{12}O_6 + 6H_2O + 12A$$
carbon dioxide + hydrogen donor ⟶ carbohydrate + water + A

H_2A can be H_2O, H_2S, H_2, or any other molecule capable of donating an electron, and the reaction requires an input of energy. When H_2A gives up its electron, it is oxidized to A.

10.3 LIGHT REACTIONS AND ENZYMATIC REACTIONS (PREVIOUSLY CALLED DARK REACTIONS)

Although the general equation for photosynthesis identifies the reactants and products, it tells us nothing about the individual reactions that, taken together, make up this complex process. To supply food and fiber to the increasing world population, we need to be able to increase crop yields. We need to know, among other things, the specific reactions of photosynthesis. Research spanning 100 years has shown that photosynthesis involves both light absorption and enzymatic reactions.

Between 1883 and 1885, a German physiologist, T.W. Engelmann, in a remarkably simple experiment, demonstrated which colors of light are used in photosynthesis. The spectrum of visible light varies from violet to red, as can be observed when white light is broken into its components by passing it through a prism. Engelmann placed together on a microscope slide a living filament of a green alga and some bacteria that would migrate toward high concentrations of dissolved oxygen. He reasoned that the bacteria would cluster near regions of the alga generating the most oxygen from photosynthesis. When he placed a filament of the green alga, *Spirogyra*, in a spectrum produced by passing light through a prism, he found that the bacteria migrated to the sections of the algal filament exposed to the red and blue light. This demonstrated that red and blue light were trapped by the photosynthetic organelles of the alga, and that they supplied energy to drive photosynthesis and liberate oxygen **(Fig. 10.1).**

At approximately the same time, J. Reinke, another German scientist, was studying the effect of changing the intensity (quantity) of light on photosynthesis. Reinke observed that the rate of photosynthesis increased proportionally to an increase in the intensity of light, but only at low-to-mod-

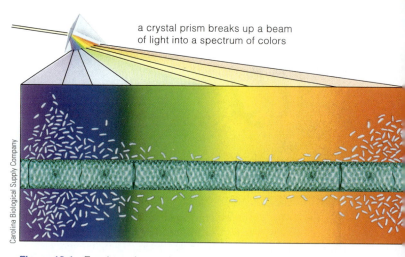
a crystal prism breaks up a beam of light into a spectrum of colors

Carolina Biological Supply Company

Figure 10.1 Engelmann's experiment, demonstrating that red and blue light are effective in photosynthesis. The bacteria migrate to that portion of the algal filament where oxygen is produced. (Art by Raychel Ciemma.)

erate light intensities. At greater light intensities, the rate of photosynthesis was unaffected by changing the light intensity. This indicated that the reaction was then proceeding at maximum rate and was no longer affected by increasing light intensity; it had become light-saturated.

A further study and a more comprehensive interpretation of this phenomenon was conducted in 1905 by F.F. Blackman, a British plant physiologist. Blackman realized that the rate of light reactions is relatively unaffected by a change in temperature, but that the rate of some light-insensitive reactions approximately doubles with every 10°C increase in temperature, over a temperature range of about 10 to 25°C. He found that when photosynthesis was proceeding rapidly under adequate levels of CO_2 and high light intensities, the rate of photosynthesis more than doubled with a temperature increase of 10°C. However, if the light intensity was low, an increase in temperature had little effect on the rate of photosynthesis. Blackman reasoned that photosynthesis may be divided into two general parts: (1) photochemical reactions, *light reactions*, which are insensitive to temperature changes, and (2) temperature-sensitive reactions, previously called *dark reactions*.

The temperature-sensitive (enzymatic) reactions do not directly depend on light. They can occur either in the light or in the dark. However, in the chloroplast, the activities of several of the enzymes are affected by products of light reactions, such as changes in pH and NADPH concentration. In addition, there are special chloroplast proteins known as thioredoxins that regulate the activities of some dark reaction enzymes. Thus, the rate of the dark reactions depends indirectly on the presence of light.

10.4 CHLOROPLASTS: SITES OF PHOTOSYNTHESIS

Because many of the individual reaction steps of photosynthesis are dependent on the specific cellular structure in which they occur, before we can examine photosynthesis in detail, we need to understand the cellular site where it takes place. Early studies with intact plants showed that oxygen is liberated and starch is formed in chloroplasts and that photosynthesis consists of both light and enzymatic reactions. It does not necessarily follow, however, that all of the reaction steps of photosynthesis take place in chloroplasts. To find out, scientists isolated intact chloroplasts and parts of chloroplasts from the cell and studied the role they play in the complex process of photosynthesis. One of the earliest successful attempts to do this was by Robin Hill in Cambridge, England. In 1932, he demonstrated that chloroplasts isolated from the cell could still trap light energy and liberate oxygen. Then, in 1954, Daniel Arnon, at the University of California, and others proved that isolated chloroplasts could convert light energy to chemical energy and use this energy to reduce CO_2.

Chloroplast Structure Is Important in Trapping Light Energy

Not all chloroplasts have the same shape, but they do have a universally similar structural organization that is critically important for photosynthesis. Chloroplasts, seen with the light microscope in living cells, appear to be homogeneously green; but in an electron micrograph, the double-membrane envelope and the complex internal membranes are apparent **(Fig. 10.2).** There are two types of internal membranes: those

forming the **grana** (singular, *granum*) and those interconnecting the grana, the **stroma lamellae.** Together, these two membrane types constitute the **thylakoids** in the chloroplasts.

All biological membranes have high concentrations of both lipids and proteins. Biochemists believe that the chlorophylls and other pigments that are located in the grana and stroma lamellae are in close contact with lipids and with some of the photosynthetic enzyme systems, forming definite protein, lipid, and pigment patterns **(Fig. 10.3).** Such a structural arrangement would improve the cell's ability to trap light energy and transform it into chemical energy during photosynthesis. Light energy initiates a series of reactions in which electrons from chlorophyll flow to other compounds (electron acceptors) in the internal membranes.

Experiments Reveal a Division of Labor in Chloroplasts

To determine what role chloroplasts play in photosynthesis, researchers need to study them independently of the rest of the plant cell. To isolate intact, functional chloroplasts, researchers carefully cut or grind cells in a dilute, buffered sugar solution or other solution of appropriate osmotic strength. They then separate the chloroplasts that remain

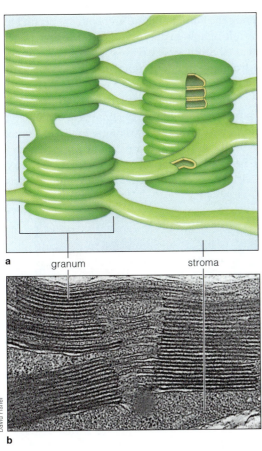

Figure 10.2 Chloroplast membrane systems. (**a**) Diagram of thylakoid membrane system of chloroplast. (**b**) Electron micrograph of chloroplast, showing membranes. (Art by Raychel Ciemma.)

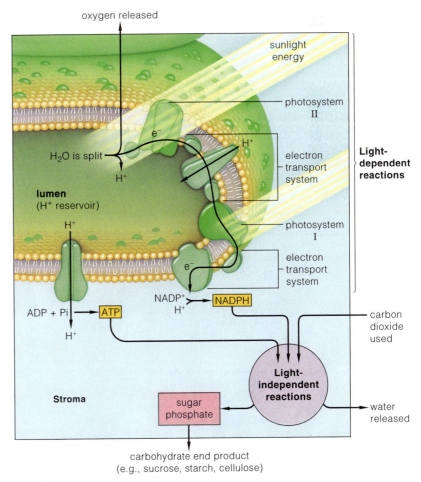

Figure 10.3 Diagram of a section of chloroplast granum showing where reactions take place. ADP, Adenosine diphosphate.

intact in the solution from the rest of the cell contents by placing the ground material in a centrifuge tube and spinning it at moderate speeds. Large and heavy particles—nuclei, starch grains, and cell wall fragments—collect at the bottom of the tube. The liquid containing the chloroplasts and other, smaller cellular components is then poured into another tube and centrifuged at a slightly greater speed. The chloroplasts, free of most of the cellular debris, separate out at the bottom of the tube. Electron micrographs of these plastids show some to be complete, with internal membranes embedded in stroma and surrounded by the outer envelope. The others have lost their envelope and stroma and consist only of a system of membranes that still contain all the light-absorbing pigments (Fig. 10.4).

Research has shown that the intact chloroplasts will carry out the complete process of photosynthesis. The broken plastids will carry out only part of the reactions of photosynthesis, but they will liberate oxygen, as Hill demonstrated. A simple laboratory experiment to verify Hill's results uses dichlorophenol indophenol, which is a blue dye in its oxidized form (DCIP) and is colorless in its reduced form (DCIPH$_2$). If DCIP is mixed with isolated broken chloroplasts (thylakoid fragments) in the light, oxygen is liberated as DCIP loses color. This reaction will not, however, take place in the dark.

$$2 \text{ DCIP (oxidized, blue)} + 2H_2O \xrightarrow[\text{thylakoids}]{\text{light}} 2 \text{ DCIPH}_2 \text{ (reduced, colorless)} + O_2$$

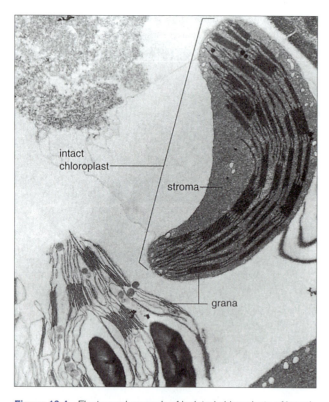

Figure 10.4 Electron micrograph of isolated chloroplasts of broad bean (*Vicia faba*), showing one intact chloroplast and chloroplasts that have lost their outer envelope and stroma. ×2,500.

Experiments with isolated chloroplasts and with isolated thylakoids reveal a division of labor within the chloroplast. The green thylakoids capture light, liberate oxygen from water, form ATP from adenosine diphosphate (ADP) and phosphate, and reduce NADP$^+$ to NADPH. The colorless stroma contains water-soluble enzymes, captures carbon dioxide, and uses the energy from ATP and NADPH in sugar synthesis.

10.5 CONVERTING LIGHT ENERGY TO CHEMICAL ENERGY

Knowledge of the details involved in the conversion of light energy to chemical energy during photosynthesis is not necessary to appreciate the significance of the process. However, a consideration of the major steps in the reaction, as we now know them, will help us to understand how the details of such a complex reaction are being unraveled. Many of the individual steps in the overall reaction of photosynthesis are the subject of current research. The more scientists know about the process, the closer humans come to producing sugar directly from the raw materials of light, carbon dioxide, and water.

Light Has the Characteristics of Both Waves and Particles

First, let us consider the nature of light. Physicists have two models describing the nature of light, and both are needed to understand the role of light in photosynthesis. The first model interprets light as electromagnetic waves. Light is only a small part of the **electromagnetic energy spectrum** that comes to us from the sun or outer space. The longest waves, radio waves, may be hundreds or thousands of meters in length, whereas gamma rays have wavelengths a fraction of a nanometer (1 nanometer [nm] = 10^{-9} m; 1,000,000 nm = 1 mm) in length. When white light passes through a prism or through water droplets, it is resolved into its separate component colors, forming the visible spectrum (Fig. 10.5). White light is

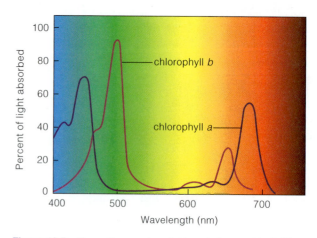

Figure 10.5 Absorption spectra of chlorophylls *a* and *b* at different wavelengths of light. Graph shows the fraction of received light that is absorbed when the pigment is exposed to various wavelengths of light. The relation between wavelength and color is also shown.

composed of wavelengths ranging from red (wavelengths of 640–740 nm) to violet (wavelengths of 400–425 nm). We cannot see wavelengths longer than red—that is, the infrared and radio waves—or shorter than violet—that is, the ultraviolet, X-rays, and gamma rays. The visible spectrum is only a small part of the complete electromagnetic spectrum, and only a part of the visible spectrum provides the energy for photosynthesis.

The second model of light states that in addition to acting as if it travels in waves, light also acts as if it were composed of discrete units or packets of energy called **photons.** Each photon contains an amount of energy that is inversely proportional to the wavelength of light characteristic for that photon. Therefore, the short wavelength of blue light has more energy per photon than does the longer wavelength of red light. When light is absorbed by a pigment (as in the case of photosynthesis by chloroplast pigments), *only one photon is absorbed by one pigment molecule at a time. The energy of the photon is absorbed by an electron of the pigment molecule, giving this electron more energy.*

Reactions in Thylakoid Membranes Transform Light Energy into Chemical Energy

ABSORPTION OF LIGHT ENERGY BY PLANT PIGMENTS
Chlorophyll appears green because it absorbs some of the blue and red wavelengths of white light, leaving proportionally more green light to be transmitted or reflected and seen. *It is the absorbed light that is used in photosynthesis.* Scientists can measure the amount of any specific wavelength of light that is absorbed by a pigment by using a **spectrophotometer,** an instrument that measures and then plots the percentage of light absorbed for each wavelength of the entire visible spectrum. The resulting graph is an **absorption spectrum.**

Vascular plants contain two major types of chlorophyll: *a* and *b*. The absorption spectra of chlorophylls *a* and *b* in solution (Fig. 10.5) show that they absorb much of the red, blue, indigo, and violet light; these are the wavelengths that are used most in photosynthesis. Part of the red and most of the yellow, orange, and green light are scarcely absorbed at all unless the chlorophyll solution is very concentrated. If, instead of a chlorophyll solution, a very thin green leaf is placed in the spectrophotometer, the absorption spectrum is quite similar, although not identical, to those from the chlorophyll solutions. The difference can be attributed in part to the yellow pigments in the leaf, which absorb blue light. Also, chlorophyll in the leaf is in close association with proteins and lipids in the chloroplasts; chlorophyll in solution is not.

What happens when the chlorophyll molecules absorb light energy? How is light energy converted into chemical bond energy during photosynthesis? Recall from Chapter 2 that molecules are composed of atoms having positively charged nuclei and one or more electrons spinning around the nuclei. Energy is required to move an electron away from the positively charged nucleus. When a photon is

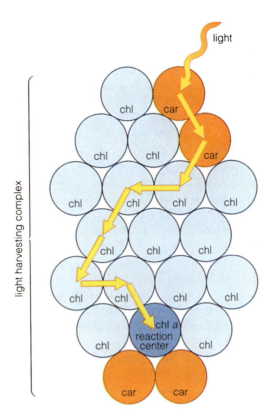

Figure 10.6 Accessory pigments in the light-harvesting complex absorb light energy and transfer it to the reaction center, where a special chlorophyll *a* molecule is excited and loses an electron with high energy. car, Carotene; chl, chlorophyll.

absorbed or trapped by a chlorophyll molecule, the energy of the photon causes an electron from one of its atoms to be moved into a higher energy state. The chlorophyll molecule now is in an excited state. This condition is unstable, and the electron tends to move rapidly, usually within a billionth of a second, back to its original energy level. As it does this, the absorbed energy can be transferred to an adjacent pigment molecule by a process called resonance. Eventually the energy can be transferred to a special chlorophyll *a* molecule (reaction center; Fig. 10.6), which through a complex series of steps drives electrons from water to reduce $NADP^+$. The formation of high-energy NADPH represents the conversion of light energy to chemical energy. The NADPH subsequently reduces CO_2 in enzymatic reactions leading to sugar formation.

THE TWO PHOTOSYSTEMS Two photosystems are involved in the trapping of light energy during photosynthesis. In the 1950s, Robert Emerson at the University of Illinois found that red light of about 680 nm or longer wavelength is inefficient in photosynthesis, but that adding light of shorter wavelength increases the efficiency of the long wavelengths of light. This observation led to the realization that there are two light reactions and two pigment systems, designated as **photosystems I and II,** in thylakoids of plants that evolve oxygen. Each photosystem contains several protein molecules complexed with pigment molecules, chlorophylls *a* and *b*. Photosystem II also contains carotene (a reddish orange pigment).

In addition, each photosystem has a complex of electron acceptor/donor molecules.

Chlorophyll *a* and chlorophyll *b* are fat soluble and are found in the thylakoids together with other fat-soluble pigments such as **carotenoids,** which are red, yellow, or purple accessory pigments. These pigments act as light traps and are grouped together in functional units, **light-harvesting complexes.** The energy absorbed by these pigments is transferred to a specific chlorophyll *a* molecule in the reaction center. This chlorophyll *a* molecule becomes excited and loses an electron (Fig. 10.6). The reaction centers of photosystem I and II are called P_{700} and P_{680}, respectively, because they absorb light maximally at those wavelengths. The chlorophyll *a* molecules in the reaction centers have unique light absorption properties, in part because of their association in a chlorophyll–protein complex. In addition to having P_{700} as its reaction center, photosystem I has a greater proportion of chlorophyll *a* than chlorophyll *b* in its light-harvesting complex; it also is sensitive to longer wavelength light. In contrast, photosystem II, which is sensitive to shorter wavelength light, contains P_{680} and has almost equal amounts of chlorophylls *a* and *b* in its light-harvesting complex.

REDUCED NICOTINAMIDE ADENINE DINUCLEOTIDE PHOSPHATE FORMATION The light reactions of photosynthesis are oxidation–reduction processes involving the transfer of large amounts of energy in many relatively small steps. These energy transfers are initiated with the absorption of light energy and involve the flow of electrons from reducing agents (electron donors) to oxidizing agents (electron acceptors) located in the thylakoids. Each carrier differs from the others in its ability to gain or release electrons. In addition, each carrier is located in the thylakoid membrane in a precise order so it forms a chain with other molecules along which electrons flow. The relative tendency of some of the electron carriers in the thylakoids to accept or release electrons is expressed as volts on the vertical scale in **Figure 10.7.** A carrier high on the scale will spontaneously transfer an electron to a carrier lower on the scale, without the input of more energy into the system. Each time an electron (and energy) is transferred from one molecule to the next, some energy is lost as heat, but the rest of the energy is then in the electron acceptor. For electron carriers higher on the scale to gain electrons from a carrier lower on the scale, however, there must be an input of light energy.

The absorption of light energy and subsequent transport of electrons to $NADP^+$ has two important consequences: Water molecules are split, releasing H^+, an electron, and oxygen, and NADPH is formed. Because electrons cannot move along the electron transport chain until the chlorophyll *a* molecule at reaction center I loses an electron, we start our discussion of electron flow with reactions occurring

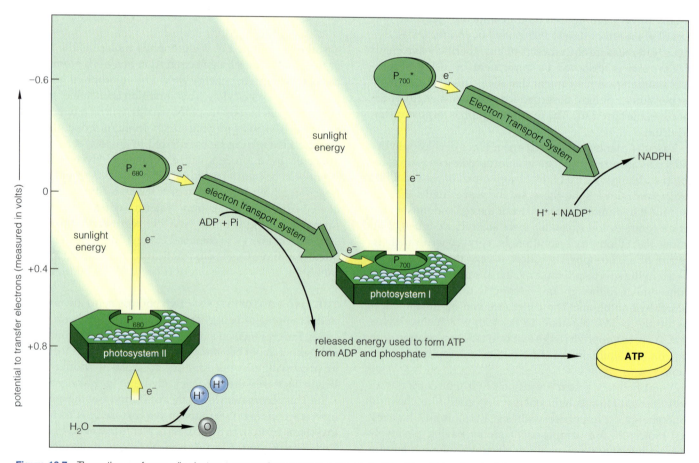

Figure 10.7 The pathway of noncyclic electron transport from water to reduced nicotinamide adenine dinucleotide phosphate (NADPH), with the associated adenosine triphosphate (ATP) synthesis. Pi, Inorganic phosphate.

in reaction center I. As the steps in this electron flow process are described, trace them in Figure 10.7. Observe the energy change as light energy is converted into electrical energy (electron flow) and then to the energy of chemical bonds in the NADPH molecule.

When a photon is absorbed by the light-harvesting complex associated with photosystem I, the energy is transferred to the reaction center P_{700}. An electron in this special chlorophyll *a* molecule gains energy and moves into a higher energy level, symbolized by P_{700}^*. (Note that the position of P_{700}^* higher on the scale in Figure 10.7 indicates that the reaction center has gained energy, not that it has physically moved in the thylakoid membrane.) In this high energy, excited state, the P_{700}^* has a tendency to lose an electron to an electron acceptor near it in the carrier. The carriers are located in the thylakoid membrane in a precise order; therefore, they form a chain with the other molecules along which electrons flow.

Because the electron lost by the chlorophyll *a* is in a higher energy state from the absorption of light energy, no further input of energy is needed for the electron to move down an electron transport chain. The electron is transferred quickly from one carrier molecule to another, illustrated as moving down the electron chain and finally arriving at an $NADP^+$ molecule. At this point, $NADP^+$ is reduced to $NADP^-$ (a process requiring two electrons). It picks up a proton (H^+) from the stroma and becomes the energy-rich NADPH, which is now available for further enzymatic reactions of photosynthesis.

Meanwhile, the chlorophyll *a* at reaction center I is left as a positive ion. It holds its remaining electrons very strongly in this low-energy state. It cannot accept another photon until the lost electron is replaced. In this condition, photosystem I must await events in photosystem II. A photon strikes photosystem II and drives an electron into a higher energy level, producing an excited state, P_{680}^*. The electron lost from P_{680}^* moves to an electron acceptor and down the electron transport chain. Finally it joins the chlorophyll *a* ion of photosystem I. Now the chlorophyll *a* molecule in the reaction center of photosystem I can absorb another photon of light. But now photosystem II has a positive chlorophyll *a* ion, which has a powerful tendency to replace its lost electron. In this highly oxidized state, P_{680} molecules gain electrons from water with the accompanying formation of oxygen and hydrogen ions, a reaction known as *photolysis*. The exact steps in this oxygen liberation process are only poorly known, but manganese, chloride, and at least one enzyme appear to be involved.

This reaction probably takes place near the inner surface of the thylakoid membrane, and the hydrogen ions produced contribute to the establishment of a proton gradient across the thylakoid membrane. In addition, during the flow of electrons to photosystem I, hydrogen ions are picked up from the stromal side of the thylakoids and released into the space within the thylakoid.

If we step back from the individual steps described and view the light and electron transport reactions as a single unit, it appears that two photons have pushed two electrons

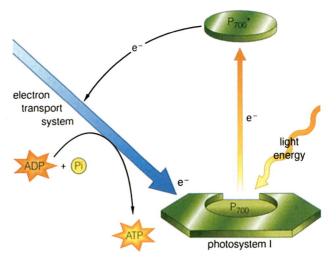

Figure 10.8 The pathway of cyclic electron transport, with the production of adenosine triphosphate (ATP) from adenosine diphosphate (ADP) and inorganic phosphate (Pi).

from water to $NADP^+$. The pathway transfers energy from light to storage in the chemical bonds of NADPH and establishes a proton gradient across the thylakoid membranes. The enzyme complex ATP synthetase catalyzes the formation of ATP in response to a proton gradient across the thylakoid membrane, in a manner similar to ATP formation in respiration in mitochondria. This pathway is **noncyclic electron transport** because the electrons from light-excited P_{700}^* flow to $NADP^+$ and do not cycle. Noncyclic electron transport involves both photosystems.

In contrast, **cyclic electron transport (Fig. 10.8)** involves only photosystem I, and electrons from the light-excited P_{700}^* do not flow to $NADP^+$. Instead, they pass down the electron transport chain and back to the chlorophyll *a* ion in P_{700}, completing the cycle. No NADPH is produced, but a proton is removed from the stroma and released into the space within the thylakoids. ATP may be formed as a result of this proton gradient, as detailed in the next section.

ADENOSINE TRIPHOSPHATE SYNTHESIS During electron flow, hydrogen ions (protons) are moved from the stroma and added to the fluid inside the thylakoids. The result is a proton gradient—*a difference in H^+ concentration across the thylakoid membrane.* Scientists believe that this proton gradient provides the necessary energy to form ATP from ADP and inorganic phosphate (Pi). According to this idea, the enzyme ATP synthetase, which occurs in the thylakoid membrane, forms ATP from ADP and Pi in the presence of the H^+ gradient. In the process, H^+ is removed from the space within the thylakoid and is released to the stroma (Fig. 10.3). Consequently, the H^+ gradient across the membrane is decreased.

This light-driven production of ATP in chloroplasts is called **photophosphorylation.** There are two types of photophosphorylation: cyclic and noncyclic photophosphorylation. In **cyclic photophosphorylation** (Fig. 10.7), electrons flow from light-excited chlorophyll molecules to electron acceptors and cyclically back to chlorophyll. Only photosystem I takes part in this process. No oxygen is liberated, and no $NADP^+$ is reduced, because it does not receive electrons. A proton gradient is formed during cyclic electron flow, and light energy is converted into chemical energy in the ATP molecules. However, because NADPH is not formed, cyclic photophosphorylation alone does not produce the molecules needed to bring about CO_2 reduction and sugar formation. Its significance in photosynthesis is limited to producing the H^+ gradient that leads to energy conservation in ATP production.

Noncyclic photophosphorylation (Fig. 10.7) produces both ATP and NADPH. In this series of reactions, electrons from excited chlorophyll are trapped in $NADP^+$ in the formation of NADPH and do not cycle back to chlorophyll. Note that photosystems I and II are involved in noncyclic photophosphorylation, and both ATP and NADPH are formed. Energy stored in the bonds of these molecules drives the CO_2 reduction reactions of photosynthesis.

10.6 THE REDUCTION OF CO_2 TO SUGAR: THE CARBON CYCLE OF PHOTOSYNTHESIS

The preceding section dealt with the transformation of electromagnetic energy (light) into the chemical bonds of NADPH and ATP. This section considers how energy stored in NADPH and ATP is used to reduce CO_2 to sugar.

Enzymes Catalyze Many Light-independent Reactions in Photosynthesis

The process of carbon dioxide reduction to carbohydrates requires many enzymic reactions. All the enzymes that directly participate in photosynthesis occur in the chloroplasts. Many of them are water soluble and are found in the stroma.

One of the most extensively studied enzymes of photosynthesis, and the enzyme that probably occurs in greatest concentration in many leaf cells, is **ribulose bisphosphate carboxylase/oxygenase (rubisco).** This is one of the most important enzymes in plants because it catalyzes the first step in the carbon cycle of photosynthesis:

$$\text{carbon dioxide + ribulose bisphosphate} \xrightarrow{\text{rubisco}} \text{2 phosphoglyceric acid}$$

In this reaction, CO_2 combines with the five-carbon sugar ribulose bisphosphate (RuBP) in the plastid stroma to produce two molecules of the three-carbon acid, phosphoglyceric acid (PGA). This is a spontaneous reaction involving little change in energy.

The Photosynthetic Carbon Reduction Cycle Was Identified by Use of Radioactive Carbon Dioxide and Paper Chromatography

Although early plant physiologists analyzing the carbohydrate content of photosynthesizing leaves easily recognized that both starch and sugars usually accumulate during pho-tosynthesis, they could not discover the mechanism for CO_2 reduction and sugar synthesis without learning the sequence in which carbon compounds form. The development of two sensitive analytical techniques enabled scientists to isolate the carbon compounds formed during enzymatic reactions. One method used radioactive carbon (^{14}C) in carbon dioxide to trace each intermediate product of the carbon reduction cycle. The other, two-dimensional paper chromatography, permitted investigators to easily and accurately separate minute amounts of different organic compounds from one another. Using these techniques together, investigators could follow carbon in the series of reactions composing the carbon reduction cycle of photosynthesis.

In the 1950s, a group of scientists at the University of California, under the direction of Melvin Calvin, was particularly successful in using this technique to work out the early carbon pathway of photosynthesis. Calvin was awarded a Nobel Prize for this achievement.

The most successful method finally adopted by this group was to expose cells of a green alga, *Chlorella*, to radioactive $^{14}CO_2$ in the light and then kill the cells in boiling alcohol after short time intervals. The alcohol extract contained essentially all of the radioactive-labeled compounds formed in the cells. By varying the time to which the plant was exposed to $^{14}CO_2$, researchers determined the order in which compounds were synthesized during exposure to $^{14}CO_2$. Photosynthesis is rapid, and the first products formed are quickly changed into other substances. Thus, Calvin used short exposures to $^{14}CO_2$, in the range of 5 to 10 seconds.

If photosynthesis proceeds for an hour or so in an atmosphere containing $^{14}CO_2$, most of the labeled carbon is in carbohydrates (sugar or starch). If photosynthesis is stopped after only a few seconds, however, most of the labeled carbon is found in PGA. PGA formed during photosynthesis is thus an intermediate between carbon dioxide and sugar. By stopping photosynthesis at increasingly longer intervals (from a few seconds to several minutes), investigators were able to trace other intermediates and to construct the entire pathway.

The C_3 Pathway Is the Major Path of Carbon Reduction and Assimilation in Plants

The photosynthetic carbon reduction cycle of photosynthesis often is called the Calvin cycle (in honor of its discoverer) or the C_3 cycle of photosynthesis (because the first product, PGA, contains three carbons). This cycle occurs widely in plants that photosynthesize. However, it is reasonable to assume that during the evolution of plants in a wide variety of ecological situations, adaptations and modifications in the metabolic process have occurred. As we consider the C_3 pathway, we should note that memorizing the sequence of all these reactions and intermediate compounds is not particularly useful, but understanding the roles played by this complex series of reactions in the metabolism of the plant is

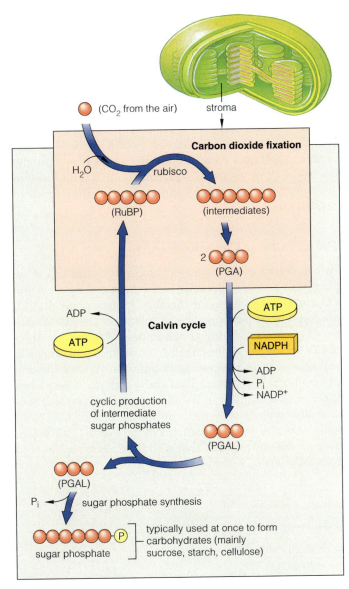

(CO₂ from the air)

stroma

Carbon dioxide fixation

H_2O rubisco

(RuBP)   (intermediates)

2 (PGA)

ADP

Calvin cycle

ATP

ATP

NADPH

ADP
P_i
$NADP^+$

cyclic production
of intermediate
sugar phosphates

(PGAL)

(PGAL)

P_i sugar phosphate synthesis

sugar phosphate

typically used at once to form
carbohydrates (mainly
sucrose, starch, cellulose)

Figure 10.9 Some major steps in the C_3 pathway of photosynthesis. Carbon dioxide enters the cycle when the enzyme rubisco combines CO_2 with ribulose bisphosphate (RuBP) to produce two molecules of phosphoglyceric acid (PGA). Carbon atoms of the key molecules are shown in *red*. All of the intermediates have one or two phosphate groups attached. For simplicity, only the phosphates on the resulting sugar phosphate are shown. ADP, Adenosine diphosphate; ATP, adenosine triphosphate; PGAL, phosphoglyceraldehyde; Pi, inorganic phosphate.

sugar phosphate, fructose 1,6-bisphosphate. The plant by this process has essentially reduced one CO_2.

4. As these reactions continue, some of the fructose bisphosphate may be transformed into other carbohydrates, including starch (by reactions not part of the C_3 cycle).

5. RuBP is regenerated. This, in turn, accepts more CO_2. Thus, a cycle of carbon compounds exists, with CO_2 from the air and electrons from water entering the cycle, and various sugars being produced.

Although carbon assimilation occurs in most photosynthetic organisms by the steps outlined above, some variations do occur. Two of these variations, found in some succulent plants and in some plants of tropical origin, are described in Section 10.8.

10.7 **PHOTORESPIRATION**

Because respiration releases CO_2, if respiration in leaves occurs in the light, then the measurement of rates of photosynthesis by measuring CO_2 uptake can be difficult. Scientists wondered for a long time whether respiration proceeds in green cells at the same rate in the light as it does in the dark. If respiration does occur during photosynthesis, it would release the CO_2 just fixed by photosynthesis. Studies show that light actually does stimulate the loss of fixed CO_2, a process called photorespiration, but the degree of stimulation depends on the type of photosynthesis in each plant. Photorespiration differs from aerobic respiration in that it yields no energized energy carriers and does not occur in the dark. In addition, photorespiration involves interactions among three subcellular organelles: chloroplasts, peroxisomes, and mitochondria.

Photorespiration starts when rubisco acts on RuBP. The activity of the enzyme rubisco is determined by the ratio of O_2 to CO_2 in the cell. When CO_2 concentration in the leaf is high, rubisco catalyzes the addition of CO_2 to RuBP, as in the C_3 cycle of photosynthesis. When the concentration of O_2 is high and that of CO_2 is low, as may occur on hot afternoons, rubisco catalyzes instead the addition of O_2 to RuBP. One of the end products of this reaction is a compound that is oxidized to CO_2, but without the formation of ATP or NADPH. Under some conditions (such as hot, sunny days), 50% of the carbon reduced during photosynthesis may be reoxidized to CO_2 during photorespiration in C_3 plants. This is a severe loss of energy.

C_3 plants have very high rates of photorespiration (particularly on hot, bright days), but some other plants called "C_4 plants" show little or no photorespiration. Mesophyll cells in C_4 plants have a high capacity to trap CO_2 and to present rubisco (in special cells called "bundle sheath" cells; see the following section) with a high concentration of CO_2 in the form of organic acids. The high level of CO_2 that occurs in bundle sheath cells when the organic acids lose their CO_2, coupled with lower levels of O_2 that may occur at high tem-

important. The key points to note in the C_3 carbon reduction cycle **(Fig. 10.9)** are the following:

1. CO_2 enters the cycle when it combines with RuBP that is produced in the stroma. Two molecules of PGA are produced.

2. Energy is stored in NADPH and ATP when these molecules are formed by light and electron-transport reactions in the thylakoids, and it is transferred into stored energy in the phosphoglyceraldehyde (PGAL) formed in the carbon reduction step of the cycle.

3. PGAL may be converted enzymatically into another three-carbon sugar phosphate, dihydroxyacetone phosphate. These two molecules combine to form a

peratures, suppresses photorespiration. Consequently, C_4 plants may produce two or three times as much sugar as C_3 plants during the hot, bright days of summer. This is part of the reason why C_4 plants such as sugar cane (*Saccharum officinarum*) and corn (*Zea mays*) are so productive (fix high amounts of CO_2). Under milder conditions, when photorespiration is less likely to occur, the C_3 plants are more efficient than C_4 plants, in part because they expend less energy to capture CO_2.

10.8 ENVIRONMENTAL STRESS AND PHOTOSYNTHESIS

Plants have adapted in various ways to extremely dry conditions existing in desert areas. To survive, plants must either endure recurrent drought or avoid drought by such means as carrying out the active part of their life cycle rapidly during brief rainy periods. Some plants, including some succulents, have developed methods of storing and conserving water. The parenchyma tissue in succulent plants is highly developed, vacuoles are large, and intercellular spaces are reduced. When moisture is available, these succulents absorb and store large amounts of water, and during periods of drought they resist the loss of water to the environment.

Some Succulents Trap CO_2 at Night

In contrast to most mesophytes, which have stomata open during the day and closed at night, many succulents have their stoma closed during the day and open at night. This adaptation reduces water loss during the day, when water stress is high. It could, however, be disadvantageous for photosynthesis by reducing CO_2 uptake in the daylight, which is when photosynthesis can occur. These succulent plants minimized this disadvantage by evolving a particular type of carbon metabolism called **crassulacean acid metabolism** (CAM). Botanists first observed it in plants belonging to the Crassulaceae family, but they now know that it is also common among 10 other large families, including Cactaceae, Euphorbiaceae, Liliaceae, and Orchidaceae.

The major features unique to CAM are the following:

1. During the night stomata are open, and the leaves rapidly absorb CO_2. During the day stomata are closed, preventing or greatly reducing CO_2 absorption and water loss.

2. At night, when CO_2 is rapidly absorbed, the enzyme phosphoenolpyruvate (PEP) carboxylase initiates the fixation of CO_2. Generally, malate, a four-carbon compound, is produced.

3. The total amount of organic acids rapidly increases in the leaf-cell vacuoles at night.

4. Leaf acidity rapidly decreases during the following day, as the organic acids are decarboxylated and CO_2 is released into the leaf mesophyll.

5. Even though stomata are closed during the day, the C_3 cycle of photosynthesis usually takes place and converts the internally released CO_2 into carbohydrate.

These features give CAM plants an effective mechanism for trapping CO_2 at night, releasing it internally during the day, and using it for photosynthesis when light is available but stomata are closed (which reduces or prevents CO_2 entrance from outside air). CAM plants use the released CO_2 in the usual way, combining it with RuBP to yield two molecules of PGA and to complete the C_3 cycle of photosynthesis.

The C_4 Pathway Concentrates CO_2

We can determine how efficiently a plant absorbs CO_2 if we place it in a closed container in the light and measure the CO_2 content of the air in the chamber. At some point, the CO_2 produced by respiration will just balance or compensate for the CO_2 absorbed during photosynthesis. The concentration of CO_2 remaining in the chamber under these conditions is known as the CO_2 **compensation point,** and it varies among different plants. If a bean plant and a corn plant are placed together in a chamber in the light, the corn plant will successfully compete with the bean for the limited CO_2. Both will eventually die of starvation, but the bean will die before the corn does **(Fig. 10.10).** That is because plants such as corn have very low CO_2 compensation points.

Figure 10.10 Corn (*Zea mays*), a C_4 plant (*right*), with its low CO_2 compensation point is able to survive at a lower CO_2 concentration than bean (*Phaseolus vulgaris*), a C_3 plant (*left*), when they are grown together in a closed chamber in light for 10 days.

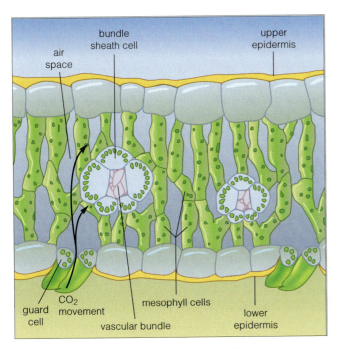

Figure 10.11 Photosynthesis in corn (*Zea mays*). A section through a leaf shows the concentric arrangement of bundle sheath and mesophyll cells. Compare this diagram with Figure 6.10a. (Art by Raychel Ciemma.)

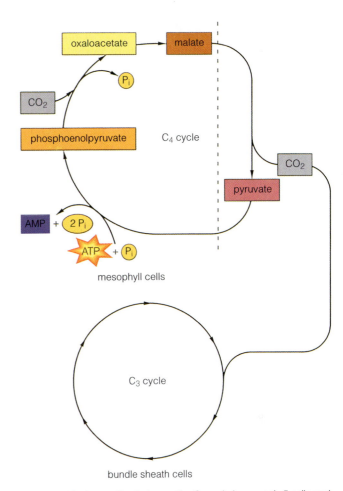

Figure 10.12 Interaction between the C_4 cycle in mesophyll cells and the C_3 cycle in bundle sheath cells.

In general, such plants seem to be adapted to grow in habitats with high light intensities and high temperatures, and they use water more efficiently. They also have certain structural features, such as the arrangement of large parenchyma cells in a "bundle sheath" around veins (Fig. 10.11). Often, though not always, the chloroplasts of the bundle sheath parenchyma cells lack or have greatly reduced grana and frequently store starch. In contrast, the mesophyll cells between veins contain chloroplasts that have typical grana but little or no starch.

Plants that have these specialized bundle sheath and mesophyll cells rely on a special type of photosynthesis. Instead of CO_2 combining with RuBP (as it does as a first step in the C_3 cycle under the influence of the enzyme rubisco), another enzyme, PEP carboxylase, catalyzes the union of CO_2 with a three-carbon acid, PEP, to form the four-carbon acid oxaloacetate (Fig. 10.12). Thus, this pathway is called the C_4 cycle. Phosphoenolpyruvic carboxylase has a strong affinity for CO_2. Later, two other four-carbon compounds, malate or aspartate, may form from oxaloacetate. These four-carbon acids diffuse from mesophyll to bundle sheath plastids, where they release CO_2 and become a three-carbon acid. This acid returns to the mesophyll, where it picks up another CO_2. The CO_2 released in the bundle sheath plastid then enters the normal C_3 cycle to form sugars.

This pathway, discovered in Hawaii in 1965 by H.P. Kortschak, C.E. Hartt, and G.O. Burr and extensively studied by M.D. Hatch and C.R. Slack in Australia, also is known as the Hatch–Slack cycle. It differs from the C_3 or (Calvin) cycle in that it ensures an efficient absorption of CO_2 and results in a low CO_2 compensation point.

Note that photosynthesis is similar in CAM and C_4 plants, in that CO_2 is first incorporated into organic acids and then is liberated to enter the C_3 carbon cycle. Such plants differ, however, in that CAM plants trap CO_2 at night in the same mesophyll cells that liberate and use it in the C_3 cycle during the day. C_4 plants trap CO_2 in mesophyll cells during the day and then transport the organic acids into bundle sheath cells, where the C_3 cycle takes place (also during the day). C_3, CAM, and C_4 plants all use the C_3 pathway to produce carbohydrate from CO_2.

10.9 FACTORS AFFECTING PRODUCTIVITY

The photosynthetic leaf is a highly evolved unit with many features that promote an efficient capture of light and carbon dioxide. Nevertheless, only about 0.3% to 0.5% of the light energy that strikes a leaf is stored in photosynthesis. The yield may be increased by a factor of 10 under ideal conditions. In a hungry world, there is much to be gained by exploring and controlling the conditions that limit plant productivity, the amount of living tissue produced per unit of time by a plant or population of plants.

Greater Productivity Can Be Bred into Plants

Plant productivity is determined partly by the environment and partly by the hereditary traits of the plant. One avenue toward greater productivity is to breed more efficient plants. This approach has already been quite successful in the case of cereal grains. Norman Borlaug received the Nobel Prize in 1970 for developing high-yielding wheat (*Triticum*) strains so productive as to constitute a *green revolution* in tropical countries. Unfortunately, these strains require high levels of fertilizer application—an expensive and pollution-creating practice. There also are potential genetic problems associated with these grains.

Photorespiration is one example of a hereditary trait that reduces plant productivity, and the C_4 system offers a compensating hereditary advantage. Now that research has uncovered these traits, perhaps breeding programs or the use of recombinant DNA technology will lead to new C_4 varieties, as well as to C_3 plants that are less prone to photorespiration.

Fluctuations in the Environment Alter the Rate of Photosynthesis

In addition to heredity, the environment also has many general influences on photosynthesis and plant productivity. To a certain degree, some of the environmental factors are under the control of the plant grower. Crop yields can be influenced by modifying such factors, particularly in controlled growth chambers or in greenhouses. Factors generally most easily controlled are water and the mineral elements of the soil. The control of temperature, light (intensity, quality, and duration), and carbon dioxide usually require special equipment.

TEMPERATURE Plants are capable of photosynthesis over a wide temperature range. Plants native to Arctic regions may photosynthesize at temperatures below 0°C. Some lichens from the Antarctic carry out photosynthesis at temperatures of −18°C and have a photosynthetic temperature optimum near 0°C. Cyanobacteria in the water of hot springs may carry on photosynthesis at a temperature as high as 75°C. C_4 species generally have greater temperature optima for photosynthesis than do C_3 species.

Most plants function best between temperatures of 10°C and 25°C. If there is adequate light intensity and a normal supply of carbon dioxide, the rate of photosynthesis of most common land plants increases with an increase in temperature up to about 25°C; above this range there is a continuous decrease in the rate as the temperature increases. The longer the exposure to a given high temperature, the greater the decrease in photosynthetic rate. Under conditions of low light intensity, an increase in temperature beyond a certain minimum will not produce an increase in photosynthesis. These conditions may occur in the winter in greenhouses. If the temperature is increased too much, the plants will suffer, because although the rate of photosynthesis has not changed, respiration has increased as a result of higher temperature.

LIGHT Both light intensity and wavelength affect the rate of photosynthesis. Under moderate temperature conditions and with sufficient carbon dioxide, carbohydrates produced by a given area of leaf surface increase with increasing light intensity, up to an optimum, after which their production decreases. It is not the intensity of light falling on the leaf surface that affects photosynthesis as much as it is the intensity to which the chloroplasts are exposed. Surface hairs, thick cuticle, thick epidermis, and other structural leaf features diminish the light intensity that reaches the chloroplasts.

Intense light appears to retard the rate of photosynthesis. Many plants living in deserts and other places where the light is very bright often have structural adaptations that tend to diminish the intensity of light reaching the chloroplasts. The usual light intensity in arid and semiarid regions is considerably greater than the optimum for photosynthesis in many plants, especially crop plants introduced by humans. In these regions, the light intensity is probably nearer the optimum for photosynthesis on days when the sky is overcast than on clear, sunny days. Leaves on the surface of the plant canopy receive more intense light than those beneath, which are shaded. Therefore, some of the leaves receive light of optimum intensity, whereas others may receive light either greater than or less than the optimum.

A special situation exists in a dense stand of plants such as a forest, where the vertical distribution of light varies dramatically—from full sunlight on the upper leaves of a canopy to very dim light on the forest floor. Plants growing below a leaf canopy receive full sunlight for relatively short periods, when the sun is directly overhead and pulses of light penetrate through holes in the canopy. When breezes move the canopy of leaves, these brief exposures to direct sunlight may become very short indeed. One may wonder whether plants can efficiently use light from these *sunflecks,* as they are called. Surprisingly, recent research shows that sunflecks may indeed contribute a majority of light energy used by understory vines, shrubs, and herbs. Research in Hawaii indicates that as much as 60% of the daily carbon gain by seedlings of such understory trees as *Euphorbia forbesii* and *Claoxylon sandwicense* comes from sunflecks.

Under many natural conditions, the quality of light striking green plants varies widely. Such variations may determine seedling survival and growth on forest floors. The leaves of the overstory canopy absorb predominantly red and blue light. Thus, lower leaves, screened from direct sunlight by upper leaves, receive light rich in green wavelengths, which are less efficiently used. Plants growing in deep water, such as marine algae, are subjected to light in the shorter (blue–green) wavelengths. Many such plants have developed accessory pigment systems adapted to absorb blue–green light and use it in photosynthesis. The ability of plants to adapt to the particular quality and quantity of light received is important for survival.

CARBON DIOXIDE Earth's atmosphere contains about 78% nitrogen and about 21% oxygen. The remaining small percentage is composed of traces of other gases, including approximately 0.037% carbon dioxide. Land plants absorb atmospheric CO_2 through their stomata. Once it enters the intercellular spaces of the leaf, the CO_2 dissolves in the water within the walls of the palisade and spongy parenchyma cells. During photosynthesis, carbon dioxide is removed from solution in the chloroplasts. A diffusion gradient for carbon dioxide is set up between the outside atmosphere and the chloroplasts, resulting in the diffusion of CO_2 from the atmosphere to the chloroplasts. At the same time, oxygen liberated during photosynthesis is used in respiration, or it diffuses outward in aqueous solution through the cell walls, to intercellular spaces, and from there through stomata to the air around the leaf.

If all the chlorophyll-bearing cells of the earth's plants are constantly taking carbon dioxide from the atmosphere during daylight, the quantity of this gas used must be enormous, and natural processes must continually replenish the supply. A hectare (2.5 acres) of corn (*Zea mays*) may contain 24,700 plants. During a growing season of 100 days, it may accumulate 6,200 kg (6.83 tons) of carbon, equivalent to 22,700 kg (25.0 tons) of CO_2. All of this carbon is derived from the CO_2 of the atmosphere.

The current atmospheric supply of carbon dioxide would be used up in about 22 to 30 years if it were not constantly renewed. However, it is not the possible depletion of atmospheric CO_2, but rather the continual increase in the level of CO_2, that is contributing to the threat of global warming, with all of its catastrophic consequences.

Atmospheric CO_2 immediately around leaves limits the rate of photosynthesis for C_3 plants. This may be particularly true in greenhouses kept closed in the winter. Under these conditions, the CO_2 may be reduced to much less than the average level in the air (0.034%). On the other hand, scientists have determined experimentally that (at normal temperatures and light intensities) an artificial increase in CO_2—up to a concentration of 0.6%—may increase the rate of photosynthesis, but only for a limited period. It appears that this high level of CO_2 is injurious to some plants after 10 to 15 days of exposure. Such information is of importance to greenhouse managers.

WATER   Most of the water absorbed by plant roots is lost as transpiration. A significant amount of the rest is retained by the highly hydrated living cells, but only 1% or less is actually used in photosynthesis. However, the rate of photosynthesis may be changed by small differences in the water content of the chlorophyll-bearing cells. Drought reduces the rate of photosynthesis in some plants, such as long pod bean (*Vicia faba*), because the turgor of guard cells is reduced and because stomata tend to close when plants are deprived of water. Then the decreased diffusion of CO_2 into the leaf limits the rate of photosynthesis. Other plants, such as the bean (*Phaseolus vulgaris*), are able to compensate for mild water deficit by increasing the solute concentration of their guard cells. This prevents a loss in turgor. Consequently, the stomata remain open, and the rate of photosynthesis is less affected, but the rate of water loss is increased.

MINERAL NUTRIENTS   Obviously, all of the essential elements for normal plant growth and development are required for an actively photosynthesizing plant. However, several elements are specifically required for the development of photosynthetic systems. For instance, magnesium and nitrogen are part of the chlorophyll molecule, and iron is necessary for its synthesis; the water-splitting, oxygen-evolving system requires manganese, chloride, and calcium; some electron carriers contain iron, and all proteins (including enzymes) contain nitrogen. Consequently, poor soils can result in plants with poorly developed photosynthetic capacities. In these cases, yields can be greatly increased by effective fertilizer programs.

KEY TERMS

absorption spectrum	light-harvesting complexes
carboxylase/oxygenase (rubisco)	noncyclic electron transport
carotenoids	noncyclic photophosphorylation
compensation point	P_{680}
crassulacean acid metabolism (CAM)	P_{700}
cyclic electron transport	photons
cyclic photophosphorylation	photophosphorylation
	photosystems I and II
electromagnetic energy spectrum	ribulose bisphosphate
	stroma lamellae
grana	thylakoids

SUMMARY

1. Photosynthesis is the primary energy-storing process of life, in which light energy is stored as chemical energy in organic compounds.

2. Carbon dioxide and water are the raw materials. The products of photosynthesis are sugar and oxygen.

3. Light reactions and electron transport occur in thylakoid membranes. Thylakoids of green plants that evolve oxygen have two photosystems: photosystems I and II.

4. Light energy absorbed by light-harvesting pigment complexes in the two photosystems is funneled to special chlorophyll *a* molecules in the reaction centers. The chlorophyll *a* molecule becomes excited as one of its electrons is driven into a higher energy level.

5. The electron that has gained energy escapes from the chlorophyll *a* molecule and passes down an electron transport chain.

6. NADP$^+$ acts as the final electron acceptor and forms NADPH, thereby storing some of the energy acquired by the absorption of light by the pigments.

7. During electron transport, a proton (H^+) gradient is formed across the thylakoid membrane. This proton gradient drives the synthesis of ATP.

8. The compounds NADPH and ATP are the first molecules in which light energy absorbed by chloroplasts is stored as chemical energy.

9. Electrons move from water to replace those lost by chlorophyll in the reaction center P_{680}; H^+ and O_2 are formed.

10. Enzymes catalyzing the fixation of inorganic carbon are found in the stroma.

11. In the stroma, the H^+ and electrons of NADPH are transferred to organic compounds and are used in a series of reactions that reduce CO_2 and produce molecules of sugar. ATP is used as an energy source.

12. Two major pathways of carbon fixation are the C_3 and C_4 pathways. In C_3 plants, photorespiration may release up to 50% of the previously captured CO_2. This major loss in captured CO_2 does not happen in C_4 plants, which are therefore more productive at high temperatures.

13. In CAM plants, CO_2 is captured at night when stomata are open. The resulting organic acid releases CO_2 to photosynthesizing cells during the day, when stomata are closed.

14. Photosynthesis may be limited by CO_2, light, temperature, water, minerals, and the plant's hereditary efficiency.

15. The atmosphere's concentration of CO_2 has been increasing for the last two centuries because of the removal of forests and the burning of fossil fuel. As a result, global temperatures are expected to increase.

Questions

1. Write the general equation for photosynthesis in words.

2. Describe an experiment to demonstrate which colors of light are used in photosynthesis.

3. When light is absorbed by chlorophyll in a leaf, what happens to an electron in the reaction center?

4. What is the significance of the proton gradient that is established across the thylakoid membrane during electron transport in the thylakoid?

5. What are the first two molecules that store energy from light absorbed by a leaf during photosynthesis?

6. Outline the methods used by scientists to determine which intermediate compounds are formed during the synthesis of sugar from CO_2.

7. Name and briefly distinguish among the three major carbon cycles of photosynthesis.

8. Under what environmental conditions would CAM metabolism and C_4 metabolism be advantageous to a plant? Why?

9. A person was growing plants in a greenhouse in the winter. The greenhouse was closed and heated to keep it warm. The plants grew very slowly. Offer some reasonable explanation for this effect.

 InfoTrac® College Edition

http://infotrac.thomsonlearning.com

History of Photosynthesis

Beale, N., Beale, E. 2001. Sunlight at Southall Green (fictional conversation on the discovery of the sun's role in photosynthesis). *Perspectives in Biology and Medicine* 44:333. (Keywords: "Southall Green")

Environmental Stress and Photosynthesis

Ben-Ari, E.T. 1998. CAM photosynthesis: Not just for desert plants (crassulacean acid metabolism). *BioScience* 48:989. (Keywords: "CAM" and "photosynthesis")

Foyer, C.H., Noctor, G. 1999. Leaves in the dark see the light (protection of chloroplasts from high light intensity). *Science* 284:599. (Keywords: "Foyer" and "dark")

Wood, M. 2002. Robust plants' secret? Rubisco activase! *Agricultural Research* 50:8. (Keywords: "Rubisco" and "activase")

Absorption and Transport Systems

Visit us on the web at http://biology.brookscole.com/plantbio2
for additional resources, such as flashcards, tutorial quizzes,
InfoTrac exercises, further readings, and web links.

1. Water flows into, through, and out of a plant in response to a gradient of forces. The various forces that move water into, through, and out of the plant are combined in the concept of *water potential*. As water flows through the plant, it is moving "downhill" from regions of higher to lower water potential.

2. To replace water lost from leaves by transpiration, water is pulled through the xylem. Plants control the rate of water loss by opening and closing their stomata.

3. Mineral elements in solution in the soil are taken up by root cells through an active, energy-requiring process. A fraction of the minerals are secreted into the xylem and rise to the shoot in the transpirational stream.

4. The products of photosynthesis are transported in the phloem from the leaves (where they are produced) to other parts of the plant (where they are used or stored). An osmotic pump provides the pressure that pushes sap through the phloem sieve tubes.

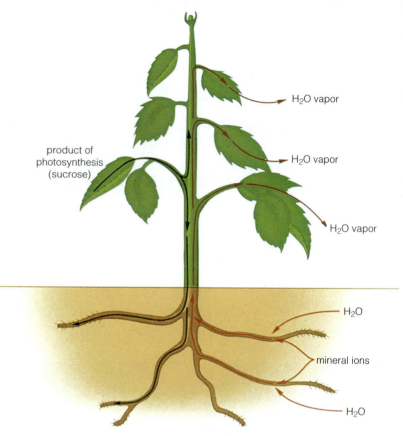

Figure 11.1 The conduction of water, mineral elements, and organic nutrients through a plant.

11.1 TRANSPORT AND LIFE

Is there just a heartbeat between life and death? Often we characterize life in animals by movement, and in higher animals, it is specifically the movement of a beating heart that circulates oxygen, nutrients, and hormones around the body in the blood. Plants clearly lack a heart and blood, yet they have the same general need as animals for transporting substances from one organ to another. Although the mechanisms by which plant transport systems work differ from those of animals, the processes are just as effective and impressive.

The need for transport in plants derives from their complex anatomy and their photosynthetic lifestyle. Plants obtain their energy and carbon from photosynthesis. Photosynthesis requires a constant source of carbon dioxide (CO_2), which comes from the air into the leaf (through open stomata) and then spreads into the air spaces between the mesophyll cells. However, the existence of such open pathways between the photosynthetic cells inside the leaves and the air outside means that water vapor can move out of the leaf, into the drier air, at the same time as CO_2 is moving in. That water must be replaced. Plant cells need a supply of water for maintaining structures, photosynthesis, and growth, and they die if they become dehydrated. The replacement water comes from the soil through the roots. Thus, there must be an effective transport system to get the water from the soil into the roots and on up to the leaves.

Growth of the plant requires mineral nutrients, such as nitrogen, phosphorus, potassium, and iron, as well as carbon. Because emerging leaves and growing stems may be a considerable distance from the soil (the source of minerals), plants must also have a system for transporting minerals to meristematic regions.

Photosynthesis produces carbohydrates that provide energy and carbon skeletons (covalently linked carbons) for the synthesis of other organic molecules. Energy and carbon are needed in all parts of the plant, especially in meristematic regions of stems and roots, and also in flowers, seeds, and fruits. Consequently, there must be a means of transporting carbohydrates from photosynthetic organs to living cells throughout the plant.

The four sections of this chapter describe the mechanisms by which water, mineral nutrients, and carbohydrates are transported from one part of a plant to another (Fig. 11.1).

11.2 TRANSPIRATION AND WATER FLOW

Water is the most abundant compound in a living cell. Without water solutes cannot move from place to place, and enzymes cannot acquire the three-dimensional shape that they need for catalytic activity. Water is a substrate or reactant for many biochemical reactions, and it provides strength and structure to herbaceous plant organs through the turgor pressure it exerts (see Chapter 3). As much as 85% to 95% of the weight of a growing herbaceous plant is water. Because water molecules are connected to each other by hydrogen bonds (see Chapter 2), the water in a plant forms a continuous network of molecules. This network extends into every apical bud, leaf, and root cell; it permeates, with few exceptions, every cell wall and much of the intercellular space. Because the network is continuous, a loss of water from one area affects the entire system.

Although the cuticle that covers the stems and leaves of terrestrial plants is relatively hydrophobic (and thus mostly impermeable to the diffusion of water), stomata, lenticels, and cracks in the cuticle allow a loss of water vapor from the interior of the plant. This loss is called transpiration. The amount of water transpired by plants is considerable. In one study, a single corn (*Zea mays*) plant in Kansas transpired 196 liters (L) of water between May 5 and September 8. One hectare (2.5 acres) of such plants (at 14,800 plants, a low density) would transpire more than 3 million L (800,000 gallons) of water during the season, an amount equivalent to a sheet of water 28 cm in depth over the entire hectare.

A Variety of Factors Affect the Flow of Water in Air, Cells, and Soil

To understand how water moves within a plant and why it is transpired, we must first find out why water moves at all. There are five major forces that move water from place to place: diffusion, osmosis, capillary forces, hydrostatic pressure, and gravity.

The description of these forces begins with molecules, which, far from being static, are in constant motion. In a gas, the individual molecules move independently in random directions. Overall, however, the net flow is from regions of higher concentration to regions of lower concentration. This principle applies to every element and compound in air; and, in particular, it applies to water vapor, which flows from areas of higher humidity to areas of lower humidity. The net flow of molecules from regions of higher to lower concentration is called diffusion. Diffusion is a major force directing the flow of water in the gas phase (Fig. 11.2a).

Liquids are quite different from gases, but liquid water and solute molecules also are in constant motion and also diffuse from regions of higher to lower concentration. To witness diffusion, place a drop of dye in a glass of water. You will see the dye disperse throughout the water. What you may not see is that the water also diffuses into the region of dye. This process becomes very important when the dye (or any other solute) is confined by a **differentially permeable** membrane (such as the plasma membrane) that allows water but not solutes to flow across it. Solutes displace water; therefore, liquid with a higher concentration of solute has a lower concentration of water. The diffusion of water across a selectively permeable membrane from a dilute solution (less solute, more water) to a more concentrated solution (more solute, less water) is called **osmosis**.

A device that uses osmosis to power a flow of water out of a chamber is called an **osmotic pump** (Fig. 11.2b). An osmotic pump is one that works by pressure generated through osmosis. It is easy to set up an example in the laboratory. One ties a bag formed from a dialysis membrane around a tube.

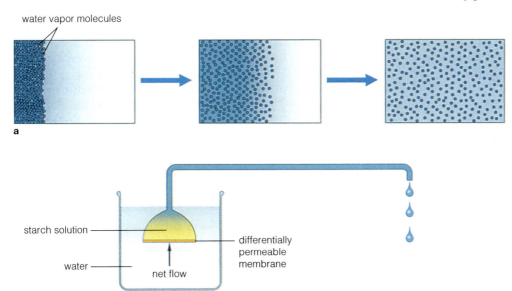

water vapor molecules

a

starch solution

water

net flow

differentially permeable membrane

b

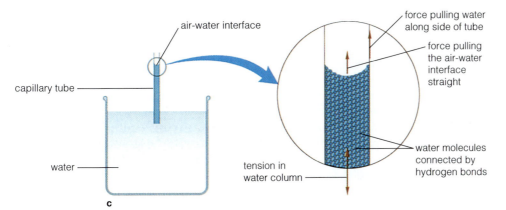

air-water interface

capillary tube

water

force pulling water along side of tube

force pulling the air-water interface straight

water molecules connected by hydrogen bonds

tension in water column

c

Figure 11.2 Some of the major forces that move water. (**a**) Diffusion of water vapor molecules gives a net movement from high to low concentration until the concentration at all points is equal. (**b**) An osmotic pump pulls water from a dilute solution to a more concentrated solution. (**c**) Capillary forces pull water into a narrow tube or any other narrow spaces.

A dialysis membrane is a thin sheet made of modified cellulose fibers. It has pores large enough to pass water molecules but small enough to retain large solute molecules, such as proteins or starch. Inside the bag is a solution of large solute molecules; outside the bag is pure water. Osmosis forces water into the bag. As the volume of water in the bag increases, pressure builds up inside the bag. Because there is an outlet through the capillary tubing, the pressure forces solution through this outlet. Later, this chapter (see Section 11.4) examines how osmotic pumps function in a plant.

Osmosis represents a potent force moving water into cells. As explained in Chapter 3, so long as the apoplast solution—the solution outside of the plasma membrane—is more dilute than cytoplasm, water will tend to flow into cells from the apoplast solution. As the cell wall expands, it exerts a hydrostatic pressure that opposes the flow (see Fig. 3.7). Hydrostatic pressure in cells is called **turgor pressure;** it is important because it stiffens the cells and the tissue they constitute.

Outside the cells, water is pulled into the small spaces between the hydrophilic cellulose microfibrils of the cell wall and is tightly held there. This is because the water molecules form hydrogen bonds with each other and also with the carbohydrates of the walls (see Chapter 2). Water molecules are **cohesive:** They stick together. They also are **adhesive:** They stick to hydrophilic molecules such as carbohydrates. The hydrogen bonds between water molecules and the cellulose surface tend to drag the water along so that it covers as much surface as possible; the hydrogen bonds between the water molecules at the interface between water and air tend to minimize the area of the interface by pulling it level. The liquid is pulled along behind the interface by the hydrogen bonds that connect the individual molecules. Together these forces can generate a great tension that pulls water into the smallest spaces. We can visualize the forces by putting a glass tube with a narrow bore into water. The water is pulled up the tube until enough water has risen that its weight balances the pull (Fig. 11.2c). Because such a tube is called a *capillary tube,* the forces pulling water into it are called **capillary forces.** Capillary forces produce a tension in the water like that in a stretched rubber band. The strength of the bonds holding water molecules together is surprising. In one type of experiment, investigators were able to exert a tension of 1,000 atmospheres (atm; a stretching force of about 15,000 pounds per square inch) on a column of water before it broke. The maximum tension that can develop in a capillary tube depends on the cross-sectional area of the bore, with the smallest bores producing the greatest tensions. Once the capillary bore is filled, there is no more tension; but if water is removed (for instance, by evaporation), tension will be re-established.

Soil particles also are hydrophilic, and they have small spaces that function like capillary tube bores. For this reason, water is pulled into the soil and held there by capillary forces, just as it is pulled into the cell walls. The strength of these forces depends on how much water is present. If the soil is very wet, the spaces will be filled and the tension holding the water will be weak; if the soil is dry, the tension will be stronger. For a plant to pull water from the soil, the root cells must generate an attractive force greater than the tension holding the water in the soil.

Water also moves in response to gravity. It takes force to move water upward. With small herbs, plant physiologists need not be concerned with this effect because from the bottom of the root to the top of the shoot there is little change in height. However, gravity can be a significant factor in tall trees. To move water to the top of a 33-m elm tree takes a pressure of 0.67 megapascal (MPa) (6.7 atm); for a 100-m redwood tree, it takes a pressure of about 2 MPa (20 atm).

Water Potential Defines the Direction of Water Flow

To describe how a combination of forces determines the direction of water flow, plant physiologists use the concept of **water potential.** Water potential takes into account the many forces that move water and combines them to determine when and where water will move through a plant. If we know the relative water potentials in two regions, we know the direction that water will flow: *Water always tends to flow from a region of high water potential to a region of low water potential.* This is true even if we are thinking of quite different phases, such as liquid water in a solution and water vapor in air. For instance, if the water potential of a solution is greater than the water potential of the water vapor immediately above it, the solution will evaporate. If the water potential of the soil around a root is less than the water potential of the root cells, water will flow out of the root into the soil, and vice versa. It is possible to calculate the water potential of a particular plant tissue (or air or soil) from physical measurements. This is particularly useful to agriculturalists, who must estimate the water needs of their plants and the availability of water in their soil.

Transpiration Pulls Water through the Plant

Under most conditions, the flow of water through a plant is powered by the loss of water from the leaves. Water is *pulled* up the plant by transpiration, not pushed by pumping from the roots. In transpiration, the primary event is the diffusion of water vapor from the humid air inside the leaf to the drier air outside the leaf. The loss of water from the leaf generates a strong attractive force that pulls water into the leaf from the vascular system, up the vascular system from the roots to the shoot, and, eventually, into the roots from the soil. There are many steps in the pathway by which the water moves and many factors that influence and control the rate of movement, as discussed in the following sections.

DIFFUSION OF WATER VAPOR THROUGH THE STOMATA The intercellular air spaces in the leaves are close to being *in equilibrium* with the solution in the cellulose fibrils of the cell walls. This means that they are nearly saturated with water vapor, whereas the bulk air outside the leaves is generally

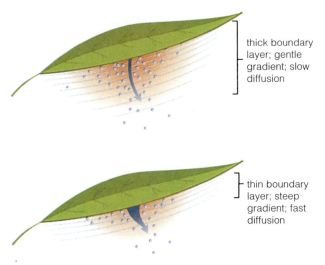

thick boundary layer; gentle gradient; slow diffusion

thin boundary layer; steep gradient; fast diffusion

Figure 11.3 The diffusion of water vapor out of stomata through the boundary layer. Note that the thinner the boundary layer, the steeper the water vapor gradient.

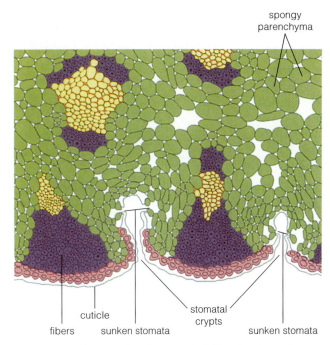

spongy parenchyma

cuticle

fibers   sunken stomata

stomatal crypts

sunken stomata

Figure 11.4 Cross section of a yucca leaf. Note the stomatal crypts that provide a layer of unstirred air over the "sunken" stomata; the thick epidermal cuticle, which limits evaporation from the leaf surface; and the large bundles of fibers, which keep the leaf from drooping if the parenchymal cells wilt.

quite dry. The difference means that there is strong tendency for diffusion of water vapor out of the leaf. This diffusion can occur if there is a pathway with reasonably low resistance. Most of the leaf is covered with the epidermal cuticle, which has a high resistance to water diffusion. However, stomata have a low resistance when open, and water vapor diffuses out through them. This is the route by which most water is lost from a plant.

Water molecules that leave the leaf first pass through the boundary layer, an unstirred layer of air close to the leaf, and then enter the bulk air. The rate of diffusion out of the stomata depends in part on the steepness of the gradient of water vapor concentration **(Fig. 11.3).** All else being equal, a thick boundary layer has a more gentle gradient, and a thin boundary layer has a steeper gradient. Thus, transpiration across a thick boundary layer is slower than across a thin one. Wind stirs up the air close to the leaf and makes the boundary layer thinner. This is why plants transpire much faster on a windy day than on a still one. Anatomical features of a leaf may slow the rate of diffusion by stabilizing a relatively thick boundary layer. For instance, a dense layer of trichomes on the surface of a leaf tends to preserve a boundary layer of relatively motionless air. **Stomatal crypts (Fig. 11.4;** see also Fig. 6.16a), also called sunken stomata, depressions of the leaf surface into which the stomata open, form an effective boundary layer because the air in the crypts is quite still.

Temperature has a major effect on the saturation value of air. Warm air holds much more water than cool air. Thus, the concentration of water vapor inside a warm leaf is much greater than inside a cool leaf, and the gradient between the leaf air and the bulk outside air is generally steeper when the temperature is high (unless the bulk air is saturated). For this reason, plants tend to lose water faster when the temperature is high.

FLOW OF WATER WITHIN LEAVES The loss of water vapor from the intercellular spaces of a leaf decreases the relative humidity of those spaces. This allows water to evaporate from the surrounding cell walls. (Note that this evaporation occurs inside the leaf, not outside.) The removal of water from the cell walls partially dries them, producing capillary forces that attract water from adjacent areas in the leaf **(Fig. 11.5).**

Some of the replacement water will come from the inside of the leaf cells across the plasma membrane. As water leaves the cells, they become smaller. This will decrease turgor pressure (as the cell wall springs back from a more extended to a less extended state) and concentrate the solutes, increasing the net osmotic effect. If much water is removed, the cell may plasmolyze (see Fig. 3.7), and the turgor pressure decreases to zero. The cells become flaccid and lose their ability to support the leaf. We recognize this when the plant wilts.

If the plant is well watered, the water lost from cell walls and from inside the cells will be replaced by water from the xylem. This water flows out of tracheids through the pits in the lignified secondary cell walls and into the fibrous primary cell walls of the mesophyll cells. The spaces between the fibrils of the primary cell walls are very small, but the distances are short—most cells in a leaf are within two to six cell lengths of a small vein—so the resistance to flow is fairly low. This means that the replacement of the cell wall water is rapid, so long as water can be pulled out of the xylem.

FLOW OF WATER THROUGH THE XYLEM The flow of water out of xylem tracheids has an interesting effect: It pulls on the rest of the water in the tracheid, and it pulls on the walls

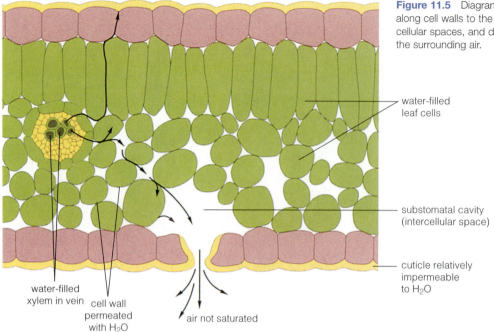

Figure 11.5 Diagram of the flow of water from minor veins along cell walls to the leaf cells, evaporation into the leaf intercellular spaces, and diffusion of water vapor out of a stoma into the surrounding air.

water-filled leaf cells

substomatal cavity (intercellular space)

cuticle relatively impermeable to H₂O

water-filled xylem in vein

cell wall permeated with H₂O

air not saturated

which, in turn, induces a tension in them. If water continues to flow from the leaf tracheid into the leaf cell walls, there will be a constant stream of water flowing up the xylem, powered by a gradient of tension.

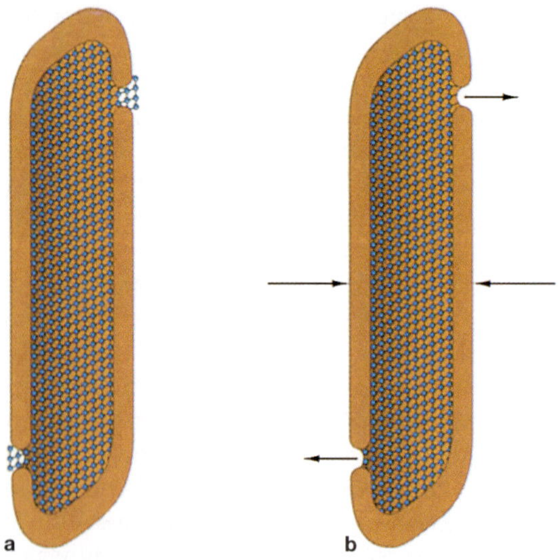

The smaller tracheids, which are separated by pits, provide fairly high resistance to the water flow, and thus require a relatively steep gradient of tension to maintain an adequate flow rate. Vessels, with their larger diameters and longer lengths uninterrupted by pits, pose a much lower resistance to water flow. On the other hand, vessels are more easily inactivated by air bubbles. Air bubbles may form in xylem when gases dissolved in the xylem fluid come out of solution under the influence of high tension or freezing temperatures. A bubble in xylem under tension acts like a cut in a stretched rubber band: It breaks the tension. That is because once formed, the bubble expands almost instantly to fill the tracheid or vessel. However, it stops at the end of the tracheid or vessel because it cannot pass through the pores in the pits. Because there are many small tracheids, a bubble in one tracheid may make that tracheid useless but will have no effect on the overall flow. Because there are fewer vessels, a bubble in a vessel may block a substantial fraction of the water flow. Conifers have only tracheids, no vessels. This is thought to provide an advantage in dry, cold climates (where these trees often are the dominant vegetation), because these are the conditions most likely to produce bubbles in the xylem.

of the tracheid **(Fig. 11.6).** Pull one water molecule out of the central space of the tracheid, and the force is transferred by hydrogen bonds to a network of molecules. The water that is removed from the tracheid cannot be replaced by air because the tiny pores and pits leading into the central space are too small for air bubbles to pass through. The water that is removed cannot collapse the walls, because they are strong and rigid. Thus, the loss of water results in a hydrostatic tension on the rest of the water in the tracheid. The tension pulls water from adjacent tracheids and vessels,

SYMPLASTIC AND APOPLASTIC FLOW THROUGH ROOTS
Eventually, the loss of water throughout the xylem decreases the water potential in the xylem of a growing primary root. This pulls in water from the apoplast of the stele (central region inside the endodermis; see Chapter 7) of that root, as the xylem is directly connected to (and is really part of) the apoplast. In turn, this water is replaced by water flowing into the stele from the root cortex and into the cortex from the soil.

The path of water flow through the cortex of the root and into the stele involves both the apoplast and the symplast (the interconnected cytoplasms of adjacent cells; **Fig. 11.7**). Because there is no cuticle over the epidermis of a primary root, water may flow in between the cells of the epidermis directly into the apoplast of the cortex and all the way to the endodermis. However, it cannot cross the endodermis in the

a   b

Figure 11.6 How capillary forces can convert the loss of water (for instance, by evaporation) into a tension within a tracheid. Diagram shows pits in the wall of a tracheid (exaggerated size); *dots* are water molecules; forces of adhesion and cohesion are shown as *short lines.* (**a**) Before evaporation, there is little tension. (**b**) After evaporation, there is high tension. *Arrows* indicate capillary forces pulling water into the tracheid pits and tension in the water pulling the tracheid wall inward.

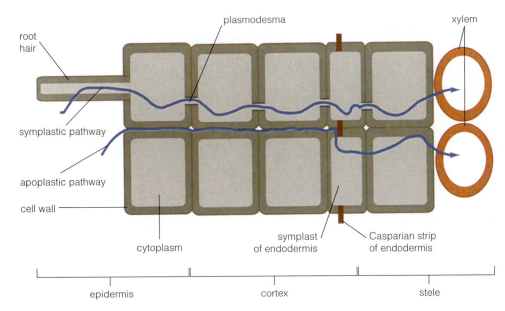

root hair

plasmodesma

xylem

symplastic pathway

apoplastic pathway

cell wall

cytoplasm

symplast of endodermis

Casparian strip of endodermis

epidermis

cortex

stele

Figure 11.7 Apoplastic and symplastic pathways of water transport through the epidermis and cortex of the root. Note that water must flow through the symplast of the endodermis to enter the stele.

replace the water lost from the leaves. This is called *temporary wilt*, because such a plant will recover in a few minutes if the loss of water can be stopped. *Permanent wilt* occurs if the osmotic forces pulling water into cells are not as great as the attractive forces holding water to the soil particles.

apoplast because of the **Casparian strip** (the suberized cell walls that separate the apoplast of the cortex from the apoplast of the stele). To move farther into the root, the water must enter the symplast by crossing the plasma membrane of an endodermal cell. Water may also cross the plasma membrane of cells at the root hairs or in the cortex. Once it does this, it can flow from cell to cell through the symplast via the plasmodesmata. It can cross the endodermis in the symplast, then enter the apoplast, and flow into the xylem.

Note that water must pass through at least two plasma membranes to reach the root xylem from the soil. There is a significant resistance to the flow of water through these membranes, although it is not easy to measure. One indication of this resistance is that transpiration—and therefore water flow—increases if a plant's roots are placed in boiling water long enough to kill the root cells and destroy the cell membranes.

FLOW THROUGH SOIL   The flow of water from the soil into the roots reduces the amount of water in the region of soil around the roots. Most of the water in soil is in small capillary spaces between the soil particles and is bound to the hydrophilic particles by hydrogen bonds. The removal of water near the roots increases the capillary forces that hold water in the soil particles. This results in a gradient of capillary forces between the soil particles near the roots and those farther away. Water flows along the surfaces of the particles and through the capillary spaces in response to this gradient, finally reaching the roots.

Because the capillary spaces are small and the distances may be long, there can be considerable resistance to the flow of water through the soil. This resistance, together with the resistance to flow across the root membranes and through the xylem, is important because it limits the rate at which water can reach the leaves. Even a plant in well-watered soil will wilt in a sudden hot, dry breeze, because water cannot move into the roots and up the xylem quickly enough to

CONTROL OF WATER FLOW   The daily cycle of transpiration is very striking. For most plants, transpiration is very slow at night; it increases starting some minutes after the sun comes up; it peaks in the middle of the day; and it decreases to its night level over the afternoon. By testing plants in experimental growth chambers, researchers have shown that the rate of transpiration is directly related to the intensity of light impinging on the leaves. Other important environmental factors are temperature, relative humidity of the bulk air, and wind speed.

Microscopic examination of the leaf surfaces shows that light affects the opening of the stomata. In dim or no light, the stomata of most plants are closed; as the light intensity increases, the stomata open up to some maximum value. The mechanism by which light controls stomatal aperture has been the subject of investigation for many years, and the following paragraphs provide a detailed (although not complete) description of many steps.

The primary sensing organs of the stomata are the guard cells. Under illumination, the concentration of solutes in the vacuoles of the guard cells increases. How does the solute concentration increase? First, starch, a storage carbohydrate in the chloroplasts of the guard cells, is converted into malic acid (**Fig. 11.8**). Second, the proton pump in the guard cell plasma membrane is activated (see Chapter 3). The proton pump moves H^+, some of which comes from malic acid, across the plasma membrane. (After malic acid loses an H^+, it is called *malate* ion.) This increases the electrical gradient and the pH gradient across the plasma membrane. Potassium ions (K^+) flow into the cell through a channel in response to the charge difference, and chloride ions (Cl^-), in association with H^+ ions, flow into the cell through another channel in response to the H^+ concentration difference.

The accumulation of malate, K^+, and Cl^- increases the osmotic effect drawing water into the guard cells. The signals that turn on the enzymes that form malate and activate the proton pump in the plasma membrane include both red and blue light, but these seem to act in different ways.

Figure 11.8 Events leading to opening of a stoma. **(a)** The production of malate and the influx of K⁺ and Cl⁻ powered by the electrical and pH gradients produced by the proton pump increase the concentration of osmotically active solutes in the guard cells, which pulls water into the cells by osmosis. ADP, Adenosine diphosphate; ATP, adenosine triphosphate; Cl, chloride; K, potassium; Pi, inorganic phosphate. **(b)** Electron micrograph of two guard cells in a bean (*Phaseolus vulgaris*) leaf.

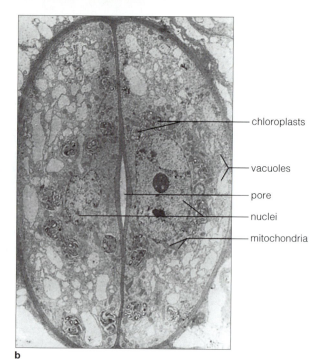

As just stated, the increased solute concentration increases the force drawing water into the guard cells. As the extra volume expands the cell walls, turgor pressure increases until it balances the force drawing water into the cells. Most cells expand in one, two, or sometimes three dimensions as their internal pressure increases; in contrast, guard cells bend away from each other, thus opening the stoma between them (Fig. 11.9). This is because they have specialized cell walls: first, an arrangement of cellulose

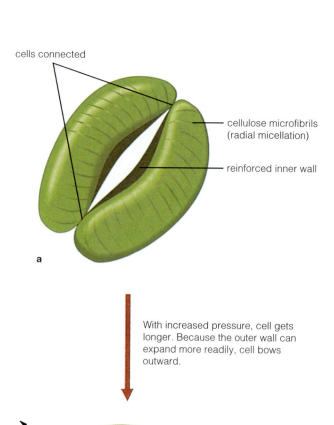

With increased pressure, cell gets longer. Because the outer wall can expand more readily, cell bows outward.

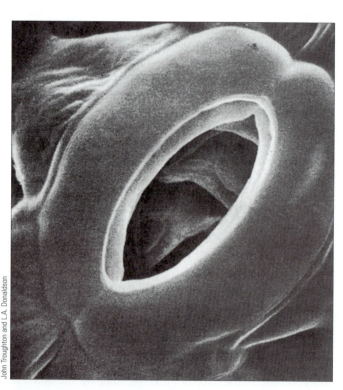

Figure 11.9 How radial micellation and reinforcement of guard cell walls force an expanding cell to bow outward.

microfibrils wrapped around the long axis of the cells (radial micellation); second, a heavier and less extensible wall adjacent to the stoma. The radial micellation is like a girdle, limiting the direction of cell expansion. The cells get longer, not thicker, when they expand. The less extensible central wall means that the cells bow outward as they get longer.

Darkness reverses the process. As light decreases, there is a reduction in the charge difference and the pH difference across the plasma membrane; there also is an opening of channels that conduct K^+ and Cl^- out of the cells. Other stimuli that have the same effect are an increase in the CO_2 concentration inside the leaf and a loss of water to the point where the leaf wilts. The wilting stimulates mesophyll cells to produce abscisic acid (see Chapter 15), a plant hormone, which diffuses to the guard cells and specifically stimulates these cells to release K^+ and Cl^-. As these ions leave the guard cells, the osmotic forces cease to balance turgor pressure, and water leaves the cells. The guard cells contract, and the elastic properties of the cell walls pull the cells together, effectively closing the stoma.

Although the opening and closing of stomata is the primary mechanism by which plants regulate the loss of water from their leaves (and thus also the flow of water through their roots and stems), there are other ways that they can moderate the rate of water loss when the stomata are open. The unstirred boundary layer of air close to the leaf is the key to such control. The larger the unstirred layer, the slower the loss of water. As already mentioned, anatomical adaptations can make the boundary layer thicker and stabilize it. A large concentration of leaf hairs (trichomes) tends to stabilize the air around them. Stomatal crypts, indentations of the leaf surface into which the stomata open, are especially effective. These adaptations are most often found in plants that live in deserts and other dry or windy areas, where the loss of water could be extreme. Some leaves tend to curl up as they dry out, effectively forming a crypt containing unstirred air.

11.3 MINERAL UPTAKE AND TRANSPORT

One of the roles of water in plants is to dissolve mineral elements and then transport them. These elements occur in natural and synthetic fertilizers (see "ECONOMIC BOTANY: Fertilizer" sidebar). Although the need of plants for fertilizer has been known since the development of agriculture, it has been only in the last century that the important components in fertilizer have been identified. Although organic fertilizers (for instance, fish extracts) often are effective, plants do not need to take up organic compounds (such as proteins, vitamins, or carbohydrates) to grow. They synthesize all these compounds themselves. What they need are elements that are substrates or catalysts for the synthetic reactions (Table 11.1). The specific elements that are required have been determined by growing the plants with their roots in aerated solutions (hydroponically) containing different combinations of elements (Fig. 11.10).

Table 11.1 Roles of Mineral Elements in Plants

Element	Primary Roles
Potassium (K)	Osmotic solute, activation of some enzymes
Nitrogen (N)	Structure of amino acids and nucleic acid bases
Phosphorus (P)	Structure of phospholipids, nucleic acids, adenosine triphosphate
Sulfur (S)	Structure of some amino acids
Calcium (Ca)	Structure of cell walls, transmission of developmental signals
Magnesium (Mg)	Structure of chlorophyll, activation of some enzymes
Iron (Fe)	Structure of heme in respiratory, photosynthetic enzymes
Manganese (Mn)	Activation of photosynthetic enzyme
Chloride (Cl)	Activation of photosynthetic enzyme, osmotic solute
Boron (B), cobalt (Co), copper (Cu), zinc (Zn)	Activation of some enzymes

Certain elements are required in fairly high amounts: potassium (K), calcium (Ca), nitrogen (N), phosphorus (P), magnesium (Mg), iron (Fe), and sulfur (S). Also needed are carbon (C), hydrogen (H), and oxygen (O), but carbon and oxygen come from the air; oxygen also comes from water, as does hydrogen, and we have already considered the transport of this compound. (The classical mnemonic for remembering these elements is: C. HOPKiNS CaFe—Mighty good.) These elements form the common compounds synthesized by plant cells. For instance, carbon, hydrogen, and oxygen are essential parts of all protein, carbohydrate, and nucleic acid molecules. Nitrogen is found in both proteins and nucleic acids, sulfur is found in proteins, and phosphorus is found in nucleic acids. Calcium is found in the cell wall and helps hold the other components of the wall together. Magnesium is a component of chlorophyll, and iron is part of some photosynthetic and respiratory enzymes; both atoms help to activate certain other enzymes.

Other elements are required in smaller amounts: manganese (Mn), boron (B), molybdenum (Mo), copper (Cu), zinc (Zn), and chlorine (Cl). These compounds are needed as coenzymes for certain enzymes with important synthetic functions.

Mineral Nutrients Are Solutes in the Soil Solution

SOIL TYPES Most of the elements that plants need exist in the soil. **Soil** is the part of the earth's crust that has been changed by contact with the biotic and abiotic parts of the environment. Soil is a weathered, superficial layer typically

ECONOMIC BOTANY:

Fertilizer

Good soils provide most of the inorganic nutrients needed by a plant for its growth, but these nutrients eventually can be depleted—absorbed by plants and removed in the harvest, or leached (washed) away by water percolating through the soil. Adding fertilizers replenishes the soil. *Complete fertilizers* contain the three elements whose lack is most likely to limit plant growth: nitrogen (N), phosphorus (P), and potassium (K). Containers of commercial fertilizers have a three-number code that tells how much of each nutrient is found in the preparation. One "all-purpose plant food," for instance, is labeled 12-10-12, which means that it contains 12% nitrogen, 10% phosphorus (measured as P_2O_5), and 12% potassium (measured as soluble potash, K_2O). The label generally also tells the form(s) of nitrogen in the preparation. These may include nitrate (NO_3^-), which is soluble and absorbed quickly by plant roots; ammonium ion (NH_4^+, often called *ammonical nitrogen*), which may bind to the soil particles and is released slowly to the plant as nitrate through bacterial action; and organic nitrogen (urea or methylenediurea), which is available to the plant even more slowly as it is broken down by bacteria. Various fertilizers have different proportions of these nutrients, which are sometimes recommended (by the manufacturers) for specific kinds of plants.

Some fertilizers also contain trace elements such as boron, copper, manganese, molybdenum, zinc, and especially iron. These additions may be important for houseplants, growing in containers with a limited amount of soil, and for plants growing in nutrient-poor soils, such as sandy soils and overly leached acidic soils in high rainfall areas. Some fertilizers are formulated to acidify soils, which release bound trace elements such as zinc and cobalt. Plants such as camellias (*Camellia japonica*), sweet gums (*Liquidambar styraciflua*), and cranberries (*Vaccinium macrocarpon*) need acid soils to take up iron and other trace elements.

Organic fertilizers provide nitrogen, phosphorus, and potassium from digested animal or plant matter—one liquid fish emulsion is labeled 5-1-1. These are generally good fertilizers, with slow-release forms of nitrogen and possibly with trace elements (generally unspecified on the label); but they do not provide anything beyond what inorganic fertilizers do. For legumes (peas [*Pisum sativum*] and beans [*Phaseolus vulgaris*]), you can purchase an inoculum of *Rhizobium,* which will promote the formation of nitrogen-fixing nodules. Incomplete fertilizers often are cheap and effective. Potassium nitrate (KNO_3), for example, gives a quick shot of two of the major nutrients, as does ammonium phosphate ($(NH_4)_3PO_4$). Gypsum (calcium sulfate, $CaSO_4$) is a good soil acidifier; so is sul-

fur, which soil bacteria oxidize to acidic sulfur oxides, including sulfuric acid (sulfate). Ferric sulfate ($Fe_2(SO_4)_3$) provides iron in an acidic environment. If you use these or any fertilizers, be careful to read the labels or consult a gardening handbook, because it is possible to overdose plants.

What should you feed your plants? There are a few common signs of nutrient stress that can help you decide. Yellowing of the lower leaves (Fig. 11.10d,e) suggests that the plant is deficient in nitrogen or possibly sulfur; a high-nitrogen complete fertilizer or some ammonium sulfate ($(NH_4)_2SO_4$) would help. A dark purplish color (Fig. 11.10b) suggests a lack of phosphorus; the plants could use a high-phosphorus complete fertilizer or bone meal (calcium phosphate, $Ca_3(PO_4)$). Light yellow or white leaves at the shoot tip (Fig. 11.10g) indicate a lack of iron; iron sulfide (FeS) or an acidic iron chelate (for example, Fe-EDTA) is indicated. If the tips of the leaves are dying, it may be a sign that the plants are taking up and accumulating too many salts. This may come from overfertilizing or from minerals in the water. Give potted plants distilled water or enough tap water to leach high concentrations of minerals out through the bottom of the pot.

Websites for further study:

UFL—Turfgrass Science: http://turf.ufl.edu/residential/fertrecommendations.htm

Make Your Own Complete Fertilizer: http://www.uaf.edu/coop-ext/publications/freepubs/HGA-00131.pdf

---

only 1 to 3 m (3–10 feet) in thickness, made up of physically and chemically modified mineral material associated with organic matter in various stages of decomposition (see "ECONOMIC BOTANY: Soils" sidebar).

Each environment creates its own unique soil. Soils differ in depth, texture, chemistry, and sequence of layers. Soils

are formally named, described, and classified on the basis of these differences. The basic soil classification unit is the soil type. Soil types are grouped into soil series, families, and orders. In the entire world there are only 11 soil orders; representatives of the orders, including more than 14,000 soil series and probably more than 50,000 soil types, occur

Figure 11.10 Tomato plants grown in nutrient culture solutions to show visual symptoms of mineral deficiencies. (**a**) Comparison of whole plant in complete solution and one in solution lacking nitrogen. (**b**) Solution lacking phosphorus. (**c**) Solution lacking potassium. (**d**) Solution lacking nitrogen. (**e**) Solution lacking sulfur. (**f**) Solution lacking calcium. (**g**) Solution lacking iron. (**h**) Solution lacking magnesium.

ECONOMIC BOTANY:

Soils

Soils vary greatly in their physical properties and chemical compositions, as well as in the organisms that live in them. The constituents of soils depend on their parent material and the processes by which they were formed. Different soil constituents include: (1) sand and silt (granular particles of quartz, feldspar, or basalt, with sand particles being much larger than silt); (2) clay (microscopic particles of complex minerals such as mica); and (3) humus (decayed organic matter). Soils occur in layers or *horizons.* The topsoil (A horizon) generally has the smallest particles and the most organic material. Subsoil constitutes the B horizon; and parent material (C horizon) consists of underlying rock fragments that extend to bedrock **(Fig. 1).**

In judging the capacity of a soil to support plant growth, several properties are important, including water-holding capacity, aeration, cation-exchange capacity, salinity and toxicity, and biota.

Water-holding capacity is the amount of water that stays after the soil is allowed to drain by gravity, less the amount that is held to soil particles by forces stronger than those the plant can exert. This is the maximum amount of water available to plants. Well-balanced soils with considerable humus have the best water-holding capacity. Sandy soils tend to allow water to drain away. High-clay soils may become compacted, with little space for water beyond that held tightly to the clay particles.

Aeration is the amount of air in the spaces between soil particles. Root growth and function require respiration, and thus oxygen. The best soils have about equal amounts of air and water in their interparticulate spaces. Sand, which promotes drainage, also promotes aeration. Swampy soils often become anaerobic.

Cation-exchange capacity describes the quantity of cations that can bind to the soil material but be released in the presence of acid generated by the respiration of roots and microorganisms. Clays and humus are the soil constituents that contribute cation-exchange capacity.

The pH of a soil indicates where the soil solution stands on the acid–base scale. The amount of acidic and basic parent material also is important. Neutral soils are good at supporting root function. In basic soils, iron and other elements may be bound into compounds that make them unavailable to plants; in acidic soils, these elements are free, but they may leach from the soil, and thus be depleted. In high-rainfall areas such as tropical rain forests, the soils are acidic and generally nutrient poor because of leaching. The growth of plants depends on the nutrients that are held in decaying organic material and that are deliv-

within the United States. The distribution of particular types of plants often is correlated with the presence of particular soil types. Rhododendrons, for instance, grow only in acid (low pH) soils.

Plant cells, like cells of other organisms, take up mineral elements only when the elements are in solution. In general, that means that the elements are taken up as simple or complex ions—for instance, potassium as K^+, calcium as Ca^{2+}, phosphorus as $H_2PO_4^-$, and nitrogen as NH_4^+ or NO_3^-. These ions are found in the soil solution. They enter the soil solution from the dissolution of crystals in rock and soil particles or from the decomposition of organic matter in the soil.

SOIL FORMATION The process of dissolving elements from rock starts with acidic rain **(Fig. 11.11).** Rain becomes acidic by dissolving CO_2 and, to a lesser extent, sulfur and nitrogen oxide gases such as SO_2 and N_2O_5. In solution, these combine with water to form acids: H_2CO_3, H_2SO_3, and HNO_3. These acids dissociate (lose H^+), increasing the concentration of H^+ and making the raindrop slightly acidic. When this rain falls on rock, it can rapidly or gradually dissolve many of the rock's crystals. Limestone ($CaCO_3$) dissolves

relatively rapidly through the reaction $CaCO_3 + H^+ \rightarrow Ca^{2+} + HCO_3^-$. Other minerals—such as hematite (Fe_2O_3) and feldspar ($K(AlSi_3O_8)$), sulfides such as chalcopyrite ($CuFeS_2$), and magnesium and calcium phosphate rocks—dissolve more slowly.

The rate at which crystals dissolve depends on the amount of crystal surface area in contact with water. Many processes increase the surface area. The dissolving of crystals itself forms cracks. Other crystals on the sides of these cracks come into contact with water. Freezing and thawing of water in a crack breaks off pieces of rock and forms new fissures. This starts the process of soil formation. Water and wind erosion pulverize rock particles; and the smaller the particles, the greater the total surface area. Once there is a small amount of soil solution, lichens and small plants can start to grow. These also accelerate the processes of soil formation and dissolution of minerals. Their rhizoids and roots enlarge the cracks through turgor pressure and emit respiratory CO_2, which forms H_2CO_3, and thus more acid.

Young soils, such as those formed as described earlier, can be a good source of minerals, depending on the composition of the parent rock and the size of the soil particles

ered by wind and water from other regions.

Salinity and *toxicity* measure the presence of high concentrations of elements that interfere with root growth or function. High concentrations of sodium (Na^+) and other elements in saline soils are generally caused by poor drainage (for example, in a "dry lake" with no outflowing river) or by the presence of seawater. Toxic concentrations of some ions may reflect the composition of the parent rock. For example, serpentine soils have toxic concentrations of Mg^{2+} and nickel and low concentrations of Ca^{2+} because they are formed from a magnesium silicate rock containing crystals of heavy metals.

Biota, the organisms living in the soil, may be extremely important in determining the nutrient value of the soil. Nitrogen-fixing bacteria and other bacteria contribute to the nitrogen cycle. Mycorrhizal fungi promote the uptake of phosphate and perhaps other nutrients. Worms, insects, and small mammals break apart compacted soil and improve its texture. All these organisms contribute humus to the soil composition. On the negative side, herbivores and pathogens may kill roots, thus limiting nutrient uptake into the plant.

Although different plants may prefer different types of soils, a good start for a garden can generally be made with a loam, which contains a mixture of clay, silt, and sand particles with a considerable amount of organic matter. The soil should drain well, yet not dry too readily. It should have good air spaces for root respiration. The solution that drains from the soil should be chemically neutral (pH 7). If the soil in your garden is not loam, you often can improve it by tilling in organic soil amendments, for instance, ground bark, peat moss, or compost. This is a good idea each year in any case, because organic components break down through bacterial and fungal metabolism. If your soil is too acid (pH less than 7), you can correct the pH by adding lime (calcium oxide). If your soil is too alkaline (pH greater than 7), you can acidify it with gypsum (calcium sulfate). Both lime and gypsum provide calcium, which improves roots' ability to accumulate nutrients and exclude toxic ions. But before using either lime or gypsum, consult a gardening expert or handbook for guidelines.

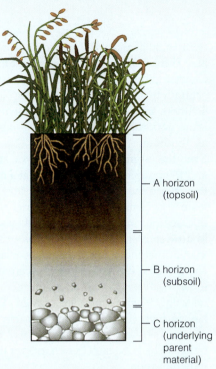

Figure 1 The soil of grassland. The A horizon is alkaline, dark, and rich in humus. The B horizon consists of clay and calcium compounds. The composition of the C horizon depends on the underlying rock.

A horizon (topsoil)

B horizon (subsoil)

C horizon (underlying parent material)

Websites for further study:

Smithsonian Soils Exhibit: http://www.soils.org/smithsonian/

Natural Resources Conservation Service—Soils: http://soils.usda.gov/

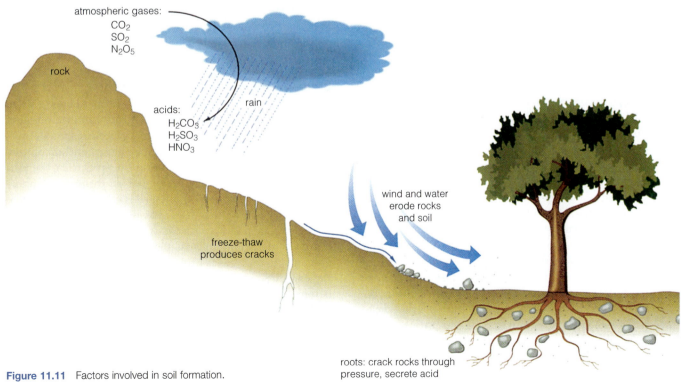

atmospheric gases:
CO_2
SO_2
N_2O_5

rock

acids:
H_2CO_3
H_2SO_3
HNO_3

rain

wind and water erode rocks and soil

freeze-thaw produces cracks

roots: crack rocks through pressure, secrete acid

Figure 11.11 Factors involved in soil formation.

(with smaller particles providing a richer source of minerals). However, the best soils are not those with the greatest concentrations of minerals in their soil solutions. A high concentration of ions increases the osmotic effect of the soil solution, thus limiting the movement of water into plants. More importantly, high concentrations of certain ions, such as aluminum (Al^{3+}) and sodium (Na^+), are toxic to plants. Even the essential ion Mg^{2+} is toxic at high concentrations, and the nutrient $H_2BO_4^-$ is toxic at a concentration no more than twice the optimal concentration. It is better to have a lower concentration of the nutrients, with a source that releases ions into the solution as they are taken up by plants. Mature soils contain clay particles and decomposed organic matter that have fixed negative charges. These negative charges bind electrostatically to cations (positively charged ions), decreasing their concentrations in the solution but releasing them as they are needed. These soils are said to have a high cation exchange capacity. As roots grow, they respire CO_2, which acidifies the soil, releasing H^+. The H^+ binds to the fixed negative charges of the soil in exchange for mineral cations.

NITROGEN FIXATION Nitrogen is a special case. This element is needed in relatively large amounts by plants, but few soils contain much of it. From a global viewpoint, the major storehouse of nitrogen is nitrogen gas (N_2) in the atmosphere, but plants cannot use that form of nitrogen. N_2 must first be converted to NH_4^+ or NO_3^- by a reaction known as nitrogen fixation. Lightning and meteors can oxidize N_2 to NO_3^-, but most nitrogen fixation by far is catalyzed by enzymes in certain bacteria, which reduce N_2 to NH_4^+. Some of these are free-living bacteria in the soil. Some have developed mutualistic associations with specific plants, so that they donate a part of the nitrogen they fix and receive carbohydrates and other favorable living conditions in the plant. The best known example is the bacterium *Rhizobium*, which lives in root cells of legumes (beans, clover, and so forth; see Chapter 19). However, more examples recently have been discovered. Alder trees (*Alnus* sp.), for instance, form associations with bacteria of the actinomycete type.

Although nitrogen can be taken up in either of two forms, NH_4^+ or NO_3^-, it is unusual to find soils that have large quantities of either ion. NH_4^+ (in equilibrium with NH_3) is volatile. NH_4^+ is converted to NO_3^- by some soil bacteria during a process known as nitrification, but NO_3^- is very soluble and is easily leached from the soil. This is why the application of nitrogenous fertilizers is so effective in increasing crop yield. The nitrification of NH_4^+ and fixation of N_2 are just two examples of the chemical reactions that interconvert nitrogen-containing molecules. NO_3^-, once it is taken up by plants, can be converted to NH_4^+, a process known as *nitrate reduction*. NO_3^- in the soil also can be converted by certain bacteria to N_2; this is denitrification. Collectively, these reactions are known as the nitrogen cycle, in recognition that nitrogen moves back and forth between the abiotic and biotic components of the environment (see Fig. 27.4).

Minerals Are Actively Accumulated by Root Cells

All plant cells, particularly those in meristematic regions, require a source of minerals. Young root cells can absorb minerals directly from the soil solution, but the minerals must be transported through the plant to shoot cells, often over long distances. Because the minerals are in water solution, they are in part transported passively in the stream of water that is pulled through the plant by transpiration. However, there are some active processes that contribute to the uptake and transport of mineral ions. Active processes are ones coupled to strongly downhill reactions, such as the hydrolysis of adenosine triphosphate (ATP) or the oxidation of nicotinamide adenine dinucleotide phosphate (NADPH) (see Chapter 8).

MAINTAINING A SUPPLY OF MINERALS If a plant is actively taking up minerals, it will deplete the supply in the area immediately around the roots. There are three processes that replenish the supply of minerals.

The first process is the bulk flow of water in response to transpiration. Water in the soil is pulled toward the roots by the gradient of attractive forces created by the removal of water around the roots. Ions in solution are swept along with the water, although local concentrations of charged organic molecules might withdraw ions from the stream (or release ions to it).

The second process is diffusion, the tendency of materials in solution to move down their concentration gradients. Even if the bulk flow of water were to stop, there would be a net movement of ions toward the roots to replace those taken up by the roots.

The third process depends on the plant, not the solution. This process is growth: Roots continue to grow, more or less rapidly, throughout the lifetime of the plant. The rate of growth can be surprisingly rapid. The uptake of ions occurs just behind the root tip, so that as a root grows it moves to new regions of soil, where it comes in contact with a new supply of ions.

UPTAKE OF MINERALS INTO ROOT CELLS The next step in the travels of a mineral ion into a plant is its transport across the plasma membrane into a root cell. As mentioned earlier, this occurs just behind the growing root tip, in the region where primary tissues (for instance, epidermis, endodermis, and xylem) have differentiated, but where secondary growth has not begun. On entering the epidermis, the ions can move along the symplast—that is, through the plasmodesmata toward parenchymal cells in the center (stele) of the root. Ions may travel as far as the endodermis through the apoplastic pathway. Just as water molecules must cross a plasma membrane to cross the endodermis, so must mineral ions. In fact, this is probably the primary importance of the endodermis. Forcing ions to cross a plasma membrane before they enter the vascular system allows a plant to exclude toxic ions and to concentrate needed nutrients that are present at low concentrations in the soil solution.

A great deal of research has been done in the last few years to understand how ions cross the plasma membrane.

In general, this involves pumps and channels. Root cells have the ability to accumulate ions against their concentration gradient—that is, to pull them into the cell, even though their concentration in the soil solution is less than their concentration in the cytoplasm. This requires energy (ATP), and thus active metabolism. It is one of the most important reasons why the growth of plants requires live, healthy roots.

ATP-generated energy is used to accumulate ions indirectly (Fig. 11.12). The process starts with the plasma membrane proton pump, which uses ATP and pumps H⁺ (protons)

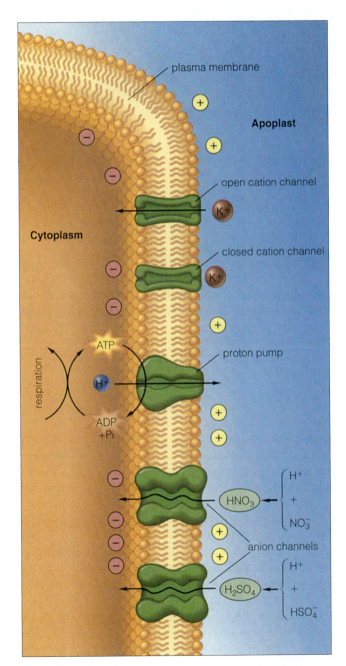

Figure 11.12 Pumps and channels in the plasma membrane. The adenosine triphosphate (ATP)–utilizing proton pump, which forces H⁺ out of the cell, provides the energy charge across the plasma membrane. Channels allow ions to flow into the cytoplasm in response to electrical charge and pH gradients. ADP, Adenosine diphosphate.

from the cytoplasm to the apoplast. The pumping of the protons generates a membrane potential difference, with the inside of the membrane negative (by more than 100 millivolts [mV]) relative to the outside of the membrane. Such a potential difference tends to pull cations, such as K⁺, into the cell. A potential difference of approximately 60 mV will support a concentration difference of 10-fold—that is, K⁺ will tend to be pulled into the cell until its concentration in the cytoplasm is 10 times greater than that immediately outside the plasma membrane. The relation between membrane potential and concentration difference is logarithmic; therefore, a potential difference of approximately 120 mV will support a concentration difference of 100-fold. The potential difference does not ensure that the cations will in fact enter the cell, however. To enter, the cations must pass through specific channels in the plasma membrane, and these channels must be open. The need for specific channels is the molecular mechanism by which the root excludes toxic ions such as Al³⁺.

The plasma membrane proton pump also generates a proton gradient across the plasma membrane, with a greater concentration of H⁺ at the outside. At certain channels, protons form complexes with such anions (negatively charged ions) as NO₃⁻ and HSO₄⁻ (Fig. 11.12). The protons neutralize the negative charges of the anions (which would tend to keep the anions out of the cell), and in fact tend to pull the anions into the cell through those channels by diffusion (because the concentration of H⁺, and thus of the complexes, is greater outside the cell).

MYCORRHIZAE It is a remarkable fact that some roots in soil do not live alone but are closely associated with filaments of fungi, as mycorrhizae (see Fig. 7.21). There are different types of mycorrhizae, depending on the plant and fungus. In some cases, the fungi form a sheath around the root and grow into the spaces between the cortical cells; in others, the fungi penetrate the cortical cell walls. One might be tempted to think of the fungus as a parasite on the plant, obtaining food without providing anything to the plant in exchange; however, plants with mycorrhizae often grow better than plants without mycorrhizae. Pine trees (*Pinus* sp.) in many infertile soils grow only in the presence of mycorrhizal fungi. It turns out that mycorrhizal fungi have high-affinity systems for taking up phosphate, systems lacking in the root cells. The fungi provide phosphate to the root, and the root provides carbon- and energy-rich nutrients to the fungi. This is a classic example of a mutualistic association, one that operates to the mutual advantage of both participants.

Ions Are Transported from the Root to the Shoot

Once mineral ions have entered the root cells, they can function in metabolic reactions in those cells. However, if they are to promote growth in the shoot, they must first be transported there. This requires, first, that the ions move from the stelar cells into the apoplast of the stele. Little is known about the mechanisms involved in this process. One hypothesis suggests that there are special mechanisms by which

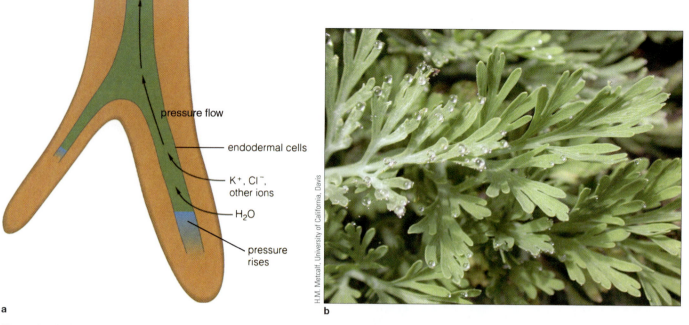

a
b

H.M. Metcalf, University of California, Davis

Figure 11.13 Root pressure. (**a**) Root pressure is generated by an osmotic pump. Solutes accumulate in the stele. Water is pulled in and forced up the xylem. Note that the endodermal cells provide the differentially permeable membrane needed for osmosis. (**b**) Guttation on a California poppy (*Eschscholzia californica*) leaf.

pressure flow
endodermal cells
K^+, Cl^-, other ions
H_2O
pressure rises

ions are transported across the plasma membrane of stelar cells out into the apoplast. If this is true, these mechanisms are regulated in a quite different way from those in cortical cells (which take up the ions). The mechanisms are probably selective (requiring special channels) and possibly active. If they are active, they require ATP, and they can pump ions into the apoplast even if those ions are more concentrated in the apoplast than in the cytoplasm. An alternate hypothesis suggests that ions are accumulated in the vacuoles of developing tracheary elements, probably to a high concentration. Then the ions are released into the apoplast and xylem stream when the xylem elements become mature and their cytoplasm breaks down (see Chapter 4). Currently, there is not enough evidence to reject either of these hypotheses.

Ions secreted into the apoplast can be swept into and through the xylem in the transpiration stream. This process takes the ions to whatever region of the plant has stomata open and transpiration occurring.

Ions that have been transported to the shoot may be taken up into the shoot cells. The uptake requires processes similar to those in root cells, except that the concentrations of ions are probably greater (because they accumulate and the solvent water evaporates). If the salt concentrations in the xylem stream are high, the rate of evaporation is high, and the production of new leaves is slow, then salts may be secreted from the leaves and appear on the surface as crystals. This secretion may occur from specialized trichomes with salt gland cells that actively accumulate and secrete the salts. Even if the salt concentrations in the soil solution are low, salts will tend to build up in the leaves. Eventually, they may reach toxic concentrations. The dead tips of the older leaves of slow-growing houseplants are a sign that salt has accumulated to a toxic level in those leaves. To slow this process, one should water these plants infrequently, but thoroughly. Allow excess water to drain through the pot, carrying away ions that have been accumulated in the soil.

Fertilize the plants infrequently and only as long as they show signs of growth.

Root Pressure Is the Result of an Osmotic Pump

As described earlier, mineral ions are taken up into root cells, passed through the symplast into cells of the stele, and then secreted into the apoplast of the stele. The concentration of ions in the stelar apoplast and in the flowing xylem sap depends in part on the rate of water uptake. If transpiration is occurring, the concentration of ions in the xylem sap may be quite low. But if no transpiration is occurring, ions can accumulate in the apoplast of the stele.

The accumulation of ions in the stele has an osmotic effect. If the soil is saturated with water, the water concentration of the soil solution will be greater than that in the xylem. In this case, water tends to enter the root and stele, building up pressure (**root pressure**) in the xylem and forcing the xylem sap up into the shoot (**Fig. 11.13a**). The flow of xylem sap, when caused by the accumulation of osmotically active salts in the stele, is an example of an osmotic pump. In some grasses and small herbs, water is forced out special openings in the leaves called **hydathodes.** You may see droplets of **guttation** water on the tips of grass leaves on a cool, moist, still morning (**Fig. 11.13b**). This is water forced out of hydathodes by root pressure and should not be confused with dew (water vapor condensed on a cold surface).

11.4 PHLOEM TRANSPORT

Osmosis Can Pump Solutions

Of all the nutrients that must be transported through the plant, the most important is carbohydrate, the product of photosynthesis. Carbohydrate is a source of carbon for the synthesis of all other organic molecules. It also is a source of energy, which is converted into a more useful form (ATP) when the carbohydrate is metabolized in respiration (see Chapter 9). Transport of carbohydrate is sometimes called translocation.

Carbohydrate is synthesized by photosynthesis in the chloroplasts of mature leaf cells. It may be stored temporarily as starch in the chloroplasts. Later, it may be exported from the leaf in the form of sucrose (common table sugar) or, less commonly, other sugars.

The carbohydrate is transported through the phloem. This can be demonstrated by *labeling* the carbohydrate with radioactive CO_2. (Such CO_2 contains the carbon isotope ^{14}C, whose nucleus decays spontaneously, producing high-energy radiation that can be tracked.) When the petioles are sampled shortly after the labeling, sectioned, and tested, the radioactivity is found in the sieve tubes. Using this same labeling technique, it is possible to measure the rate of carbohydrate transport. This rate (1–2 cm or 0.4 to 0.8 inches per minute) is faster than diffusion or transport from individual cell to cell, but it is not as fast as the rate at which water is pulled through the xylem. This suggests that the mechanism of translocation is different from diffusion, or cell-to-cell transport, or transpiration.

One of the most interesting observations about phloem transport is its ability to change direction. Generally, sucrose is transported from leaves, which are net producers of carbohydrate, to roots, which use carbohydrate for growth or storage. Sometimes (for instance, when roots are sprouting new shoots) sucrose is transported up from the roots to the shoots.

The mechanism by which sucrose moves through the phloem was the subject of much research and many arguments by scientists in the 1960s and 1970s. The accumulated evidence, however, currently supports the idea that the sucrose flows through sieve tubes as one component in a bulk flow of solution. The flow is directed by a gradient of hydrostatic pressure and is powered by an osmotic pump.

Plants Use Osmotic Pumps to Transport Sucrose through Sieve Tubes

The phloem can be seen as a dynamic osmotic pump, with a source of solute at one end and a sink at the other. This chapter has already discussed one example of an osmotic pump in the section "Root Pressure Is the Result of an Osmotic Pump." The difference in concentration of ions between the apoplast of the stele (where it was greater) and the apoplast of the cortex and the soil solution (lower) forced water into the stele. The resulting buildup of hydrostatic pressure pushed the xylem sap up the xylem and into the shoot.

In the phloem, sucrose is the main osmotically active solute. It is pumped from photosynthetically active leaf parenchymal cells into sieve tubes of the minor veins **(Fig. 11.14)**. The mechanism of pumping probably involves a carrier (channel) that transports sucrose together with a proton across a plasma membrane. The pH and electric charge differences across the membrane provide the pumping energy. The exact pathway from parenchymal cell to sieve tube is still uncertain.

The accumulation of sucrose in a sieve tube pulls water into the sieve tube from the apoplast by osmosis. This increases hydrostatic pressure inside the sieve tube at the source (for example, photosynthetic tissue or tissue releasing

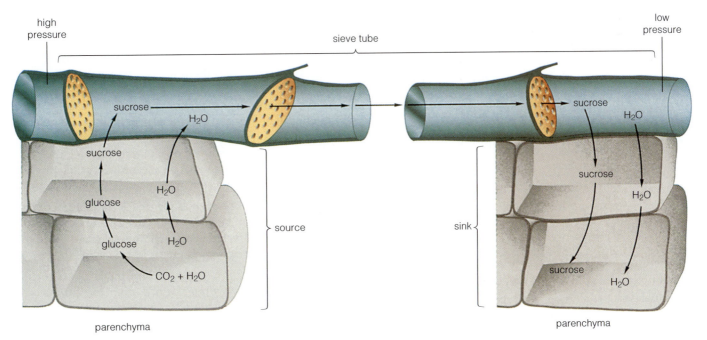

Figure 11.14 The mass-flow mechanism of phloem transport. Sucrose is actively transported into the sieve tubes at the food source region of the plant (leaves or storage organs) and removed at the sink regions (regions of growth or storage). Water follows by osmosis, increasing the pressure in the sieve tubes at the source region and decreasing the pressure at the sink region. The sieve-tube contents flow *en masse* from high- to low-pressure regions.

stored food). The pressure (which can reach more than 20 times atmospheric pressure) starts a flow of solution that will travel to any attached sieve tube in which the pressure is less.

The pressure gradient along the sieve tube depends on the gradient established by differences in sucrose concentration. Because the flow of solution along the sieve tube carries sucrose along with it, you would expect the concentration of sucrose at the source sieve-tube member to decrease eventually and the concentration at the sink sieve tube element to increase. This would eliminate the concentration gradient, and the flow would stop. The loss of the concentration gradient is prevented by two mechanisms: the continual pumping in of sucrose at the source and the removal of sucrose at the sink.

These two processes explain how the direction of phloem transport can change. If an organ is a source, it pumps sucrose into the phloem, the hydrostatic pressure increases, and phloem sap flows out of it. If an organ is a sink, it removes sucrose from the phloem and decreases the hydrostatic pressure, so that more phloem sap containing sucrose flows toward it. Growing tissues, such as shoot and root meristems and expanding fruits, are always sinks. The parenchyma cells of stems and roots may, however, be sinks or sources at different times. A storage root, such as a carrot root, is a sink during the first growing season. It removes sucrose from the phloem and stores starch and sugar in its parenchyma cells. After a winter (and once the shoot starts to bolt or flower), the carbohydrate in the root is converted to sucrose and pumped into the sieve tubes. At that point the root becomes a source.

The osmotic pump is actively regulated. The mesophyll cells of a young dicot leaf start to photosynthesize early, producing carbohydrate, but the leaf also is growing and using carbohydrate. On balance, a growing leaf is a net user of carbohydrate; therefore, it imports sucrose from the phloem system. Once mature, though, the leaf exports sucrose to the phloem system. One might imagine that the transition is gradual as the leaf's rate of growth tapers off. Instead, measurements show that it is abrupt. This suggests that an off/on switch regulates the direction of sucrose transport at the sieve-tube plasma membrane, but the nature of that switch remains a mystery.

KEY TERMS

adhesion	osmosis
capillary forces	osmotic pump
Casparian strip	root pressure
cohesion	soil
differentially permeable	stomatal crypts
guttation	turgor pressure
hydathodes	water potential

SUMMARY

1. Plants lose water to the atmosphere by transpiration, which involves the diffusion of water vapor from air spaces inside the plant through stomata, lenticels, or cracks in the cuticle to the outside air. Plants transpire a large amount of water.

2. The movement of water depends on diffusion, osmosis, capillary forces, hydrostatic pressure, and gravity. The concept of water potential allows one to predict how water will flow in response to a combination of these forces.

3. Under most conditions, water is pulled through a plant. As a first step, water vapor diffuses outward through a relatively unstirred boundary layer of air into the bulk atmosphere. Water evaporates from the cell walls, replacing what was lost. Water is pulled by tension up the xylem and into the cell walls. Finally, water moves from the soil to the xylem of young, growing roots through symplastic and apoplastic pathways.

4. Transpiration is controlled by the size of the stomatal opening, which depends on the guard cells. Under certain conditions (in the light, with low CO_2 concentration), guard cells accumulate solutes. A resulting influx of water increases their turgor pressure and stretches them in a way that opens the stoma. In the dark, in high CO_2, or under drought conditions, the process is reversed.

5. The essential minerals for plants are C, H, O, K, N, P, S, Ca, Mg, and Fe, and, in smaller amounts, Mn, B, Co, Cu, Zn, and Cl.

6. Mineral elements are taken up by roots from the soil solution. The minerals reach the soil solution by dissolving from rock crystals into acidified rainwater or through the decomposition of organic matter.

7. Minerals are taken up into root cells actively and specifically. This means that the processes transporting minerals use ATP and channels in the membrane that pass only the desired mineral ions. The hydrolysis of ATP is used to pump protons out of the cell. Part of the energy released is saved in the proton and electrical gradients across the membrane. The proton and electrical gradients are used to force ions into the cell, where they can be concentrated to 100 times or more of their concentration in the outside solution.

8. In the absence of transpiration, the accumulation of ions in the root stele pulls water into the stele by osmosis and increases the hydrostatic pressure. This so-called root pressure forces water up to and out of leaves, where it appears as guttation.

9. Sucrose, a carbohydrate formed from the products of photosynthesis, is transported from sources (photosynthetic or source tissue) to sinks (growing or expanding tissue) in the sieve tubes of the phloem.

10. The pressure that forces sucrose and other compounds through the sieve tubes comes from a gradient of hydrosta-

tic (turgor) pressure, which is high near sources and low near sinks. The gradient is produced and maintained by the pumping of sucrose into the sieve tube at the source and its removal at the sink.

Questions

1. Match the forces in the left column with their effects in the right column:

osmosis	creates a tension in the water within tracheids
gravity	tends to push water out of a cell
capillary forces	moves water vapor molecules out of a leaf
turgor	opposes the transpirational flow of water up a tree
diffusion	tends to pull water into a cell

2. Predict how each of the following changes will affect the tendency of water to move into or out of a leaf cell.

a. placing the leaf in a closed, humid chamber

b. doubling the concentration of solutes inside the cell

c. soaking a piece of leaf containing the cell in a concentrated solution of sucrose

3. Explain why the rate of transpiration increases when:

a. dawn breaks and the sun comes up

b. the weather becomes very windy

4. Is the circumference of a tree's trunk greater, less, or the same during the day (when the tree is transpiring rapidly) compared with during the night (when there is little transpiration)? Hint: Consider what happens to a single xylem vessel during transpiration.

5. List the seven mineral elements needed in the greatest amounts for optimal growth of a plant. How does each contribute to the structure or function of the plant?

6. What role does acidic rain (in moderate amounts) play in making minerals available to a plant?

7. Can a K^+ ion travel all the way from the soil to the xylem of a root by either the symplastic or the apoplastic pathway? Explain your answer.

8. Why is respiration an important process in roots? Explain how biochemical energy is used to power the uptake of mineral ions into the cytoplasm of a plant cell.

9. Explain the difference between a root in association with *Rhizobium* and with a mycorrhiza.

10. What causes the guttation water seen on the tips of grass leaves early in the morning? Why is guttation water not seen in the afternoon?

11. Does sucrose exported from photosynthesizing leaves in the phloem always travel down the stem to the roots? Explain your answer.

12. Which is greater: the hydrostatic pressure in a sieve tube of a photosynthesizing leaf, or the hydrostatic pressure in a sieve tube of a developing flower?

InfoTrac® College Edition

http://infotrac.thomsonlearning.com

Water Transport

Brown, K. 2001. Xylem may direct water where it's needed. *Science* 291:571. (Keywords: "xylem" and "direct")

Sperry, J.S. 2003. Evolution of water transport and xylem structure (plant-water relationships). *International Journal of Plant Sciences* 164:S115. (Keywords: "xylem" and "relationships")

Mineral Nutrition

Ehleringer, J.R., Schulze, E.D., Ziegler, H., Lange, O.L., Farquhar, G.D., Cowar, I.R. 1985. Xylem-tapping mistletoes: Water or nutrient parasites? *Science* 227:1479. (Keywords: "xylem" and "mistletoes")

Hays, S.M. 1993. Secrets of soil nutrient uptake (ferric-chelate reductase). *Agricultural Research* 41:18. (Keyword: "ferric-chelate")

Life Cycles: Meiosis and the Alternation of Generations

Visit us on the web at http://biology.brookscole.com/plantbio2
for additional resources, such as flashcards, tutorial quizzes,
InfoTrac exercises, further readings, and web links.

1. Life perpetuates itself through reproduction, which is the transfer of genetic information from one generation to the next. This transfer is our definition of *life cycle.* Reproduction can be asexual or sexual.

2. Asexual reproduction requires a cell division known as *mitosis.* Asexual reproduction offers many advantages over sexual reproduction, one of which is that it requires only a single parent. A significant disadvantage of asexual reproduction is the loss of genetic diversity and the likelihood of extinction when the environment changes.

3. Sexual reproduction involves the union of two cells, called *gametes,* which are usually produced by two different individuals. Another kind of cell division, known as *meiosis,* ultimately is necessary to produce gametes.

4. Every species in the kingdom Plantae has both diploid and haploid phases—that is, plants whose cells are all diploid or all haploid. These phases are called generations, and they alternate with each other over time.

5. The fossil record reveals that the most recent groups to evolve have sporic life cycles in which the gametophyte generation is relatively small and the sporophyte generation is dominant in terms of size, complexity, and longevity.

12.1 LIFE CYCLES TRANSFER GENETIC INFORMATION

A basic characteristic of life is that it perpetuates itself. The process can be sexual or asexual. In asexual reproduction, each generation is genetically identical to the last. Asexual reproduction occurs in unicellular and multicellular organisms. For example, a single-celled alga floating near the surface of a lake can divide asexually to produce two single-celled offspring **(Fig. 12.1a).** This individual cell divides to produce two new cells, then each of those cells will divide to produce others, and so on for generations of cells. Similarly, a cell within the root tip of a corn plant can reproduce asexually, first to generate two identical offspring cells and then eventually a whole tissue, layer, or region of the root, consisting of thousands of identical cells **(Fig. 12.1b).** The succulent air plant (*Kalanchoe pinnata*) is able asexually to produce miniature plantlets along leaf edges, each of which can fall off, take root, and become a new plant identical to the parent.

Asexual reproduction requires only a single parent cell or parent organism, and all of the progeny will be identical to that parent. The collection of identical individuals is called a **clone.** Many plants in nature produce clones: strawberries, aspens, and coast redwoods are only a few examples **(Fig. 12.2).** Strawberry plants (*Fragaria* sp.) produce horizontal aboveground stems (stolons or runners) that periodically take root at the nodes and produce new

leaves, flowers, and fruits there. If the runners are severed, the plants remain alive and capable of a fully independent life. When a redwood (*Sequoia sempervirens*) trunk is killed by fire or removed by timber harvest, dormant buds buried under the bark at the base of the stump are stimulated to begin growth. Within decades, a cluster of young redwood trees, all sharing the same root system but having separate trunks, will exist. In time, the parental stump will decompose and disappear, leaving a circular clone of equal-aged offspring. Trembling aspen (*Populus tremuloides*) trees produce special roots that grow horizontally under the soil instead of down. These roots periodically give rise to stems (which in time become mature trees) some distance from the parent tree. An entire grove of aspen, occupying several acres of ground, can be a single

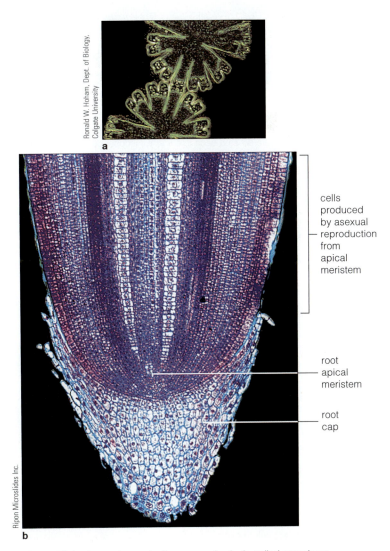

Ronald W. Hoham, Dept. of Biology, Colgate University

a

Ripon Microslides Inc.

b

cells produced by asexual reproduction from apical meristem

root apical meristem

root cap

Figure 12.1 Asexual reproduction occurs in single-celled organisms and in the tissues of multicellular organisms. (**a**) *Micrasterias,* a single-celled freshwater green alga, has just undergone cell division to produce two cells. (**b**) Longitudinal section of a corn plant (*Zea mays*) root tip. Cell division in the apical meristem produces many cells and several different tissue types behind (above) the tip.

Thomas L. Rost

Figure 12.2 Asexual reproduction produces clones of many genetically identical plants. All these trembling aspens (*Populus tremuloides*) in the Sierra Nevada of California are probably connected underground, but if the connections are severed each tree can live independently. These individuals belong to a single clone.

clone of hundreds of genetically identical trees, seemingly independent but actually all connected to each other below ground.

Sexual reproduction, in contrast, causes each generation to be genetically different. For example, when a bean plant produces pods and seeds, each seed produces a plant that is slightly different from the parent: perhaps taller, maybe more frost tolerant, possibly with different petal colors, or capable of flowering a few days sooner than the parent.

Both kinds of reproduction transfer genetic information from parent to offspring, which is the definition of **life cycle** as used throughout this textbook.

Asexual Reproduction Transfers Unchanged Genetic Information through Mitosis

Asexual reproduction requires a particular kind of cell division called **mitosis (Fig. 12.3).** Briefly, mitosis is preceded by a copying process in which enzymes duplicate every chromosome. The two copies of each chromosome, called *sister chromatids,* are joined together at one point. In the first part of mitosis (*prophase*), each chromosome condenses from a threadlike form to a compact rodlike form that can be moved without breaking. A *spindle apparatus* forms too, for moving chromosomes, and spindle fibers link sister chromatids of each chromosome to opposite poles of the spindle.

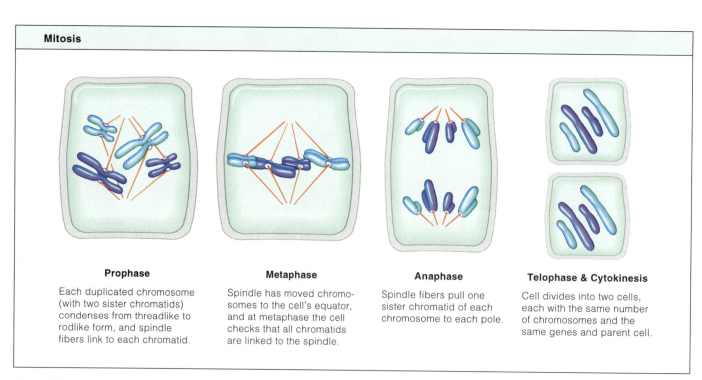

Mitosis

Prophase

Each duplicated chromosome (with two sister chromatids) condenses from threadlike to rodlike form, and spindle fibers link to each chromatid.

Metaphase

Spindle has moved chromosomes to the cell's equator, and at metaphase the cell checks that all chromatids are linked to the spindle.

Anaphase

Spindle fibers pull one sister chromatid of each chromosome to each pole.

Telophase & Cytokinesis

Cell divides into two cells, each with the same number of chromosomes and the same genes and parent cell.

Figure 12.3 A review of mitosis. Dark blue and light blue chromosomes came from different parents that mated to produce the organism with these cells. The original cell has four chromosomes, each of which was duplicated before mitosis to give the chromosome two sister chromatids. Mitosis separates the sister chromatids to produce two cells, each genetically identical to the parent cell.

Spindle fibers pull the chromosomes to the cell's equator. Then, in a stage called *metaphase,* the cell pauses to check whether all the chromatids are linked correctly. In *anaphase,* the spindle pulls sister chromatids to opposite poles of the cell. Next, in *telophase,* the chromosomes uncoil and new nuclear envelopes form (not shown in Fig. 12.3). Finally, in cytokinesis, the cell divides into two cells. Each offspring cell retains the exact same complement of chromosomes (and genes) as that of the parent.

There are several advantages to asexual reproduction. First, only a single parent is required. Therefore, any isolated individual can produce offspring and populate a new part of the species range. Second, asexual reproduction produces offspring that may be just as successful in the habitat as the parent was, unless the parent used up too many resources. Because the parent lived to reproduce, so too should the next generation (providing the environment stays unchanged). Third, asexual reproduction generates offspring faster than sexual reproduction; therefore, an invading species can dominate the landscape quickly. Fourth, asexual reproduction costs less in terms of metabolic energy than sexual reproduction. This is because sexual reproduction requires a plant to invest in reproductive tissue (flowers) even though successful seed formation, dispersal, and germination might not occur in any given year. Asexual reproduction, in contrast, always works.

The only disadvantage to asexual reproduction is that genetic diversity remains fixed. New plants are genetically identical to parents. If the environment changes, all the organisms are equally susceptible to any new stress, such as a new disease, a drought, or a migratory invasion of herbivores. In contrast, sexual reproduction results in genetic diversity, thus giving the maximum probability for continued existence of a species over long periods. The geologic history of earth has consistent themes of environmental change in which species that could not adapt became extinct. Genetic diversity is essential for adaptation.

Sexual Reproduction Transfers New Information through Meiosis

Sexual reproduction requires the union of two cells called **gametes.** The gametes must find each other and join to create a single offspring cell. Sexual reproduction thus poses two problems. One problem is to find a way to bring gametes together, and the other is to reduce the number of chromosomes. If gametes with the normal number of chromosomes fuse, then the resulting offspring cell will have twice the normal number. A repetition of this over several generations would create cells with an unmanageable number of chromosomes.

The solution to the number problem is a type of cell division called **meiosis,** which reduces the number of chromosomes by half. To make sense of meiosis, we need the concept of a chromosome set. The information to build a body is divided between several different kinds of chromosomes. A **chromosome set** consists of one chromosome of each kind. The number in a set varies among species. In humans, there are 23 chromosomes in a set. In a cotton plant (*Gossypium hirsutum*), there are 10 chromosomes in a set. Any cell with just one set of chromosomes is said to be **haploid,** symbolized as the **1n** state. Gametes are haploid, with one set of chromosomes. When two haploid gametes fuse, they create a cell (the **zygote**) that has two sets. A cell with two chromosome sets is **diploid,** symbolized as **2n.** The two copies of a given chromosome in a 2n cell are said to be **homologous.** They make up a pair, each carrying the same information. Thus, a 1n gamete of a cotton plant has 10 *unpaired* chromosomes, and a 2n zygote of a cotton plant has 10 chromosome *pairs.*

Now let us see how meiosis makes haploid cells from a diploid cell **(Fig. 12.4).** Before entering meiosis, the 2n cell duplicates its DNA so each chromosome has two connected copies (*sister chromatids*). Then meiosis carries out two rounds of cell division. The first division, *meiosis I,* opens with *prophase I.* Here each homologous pair of chromosomes comes together in a unique process called **synapsis.** Nothing like it occurs in mitosis. Synapsis is important for two reasons. First, the pairing makes it easy for the cell to divide in a way that produces haploid cells (see later). But before that happens, synapsis allows homologous chromatids to trade segments by a process called **crossing over.**

To show the consequence of crossing over, Figure 12.4 uses two colors for chromosomes from the organism's two parents. Just one crossover is shown, at the lower left in prophase I. Normally, several crossovers occur at random points along each homologous pair. The result is that synapsis and crossing over generate genetic diversity by giving chromosomes new combinations of parental genes. Another event in prophase I is that a spindle forms, and its fibers link one chromosome of each homologous pair to each of the cell's poles.

With prophase complete, the spindle moves chromosomes to the cell's equator. At that point the cell pauses to check whether all the chromosomes are linked properly. The pause for checking is *metaphase I.* Next, in *anaphase I,* the spindle pulls each chromosome with its two sister chromatids to one of the poles. Accurate linkage in prophase I assures that each pole will get a full set of chromosomes. Finally, *telophase I* creates new nuclear envelopes (not shown in Fig. 12.4), and **cytokinesis** divides the cell. Overall, meiosis I converts the original 2n cell to two 1n cells with different combinations of parental genes.

Meiosis II, the second round of division, is simply a mitotic division that separates the sister chromatids and converts the two haploid cells to four. There is no synapsis in prophase II, and hence, no further reduction or crossing over. But as the colors show in Figure 12.4, each of the four haploid cells has a different combination of parental chromosome segments.

Figure 12.4 The two divisions of meiosis. Different colors are used to distinguish chromosomes from the two parents.

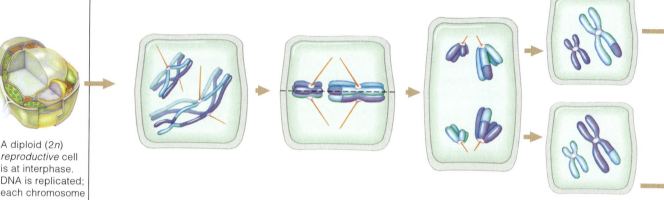

Meiosis I

A diploid (*2n*) *reproductive* cell is at interphase. DNA is replicated; each chromosome has two sister chromatids before meiosis begins.

Prophase I

Chromosomes condense. Synapsis joins homologous chromosomes into pairs that exchange parts (crossing over). One crossover is shown at lower left; normally many occur. After crossing over, homologous chromosomes stay in pairs but peel apart as shown here. Chromosomes link to spindle.

Metaphase I

Further condensed and still joined at crossover sites, chromosome pairs have been moved to the cell's equator. At metaphase, the cell checks for missing links between chromosomes and spindle.

Anaphase I

Spindle pulls one chromosome of each homologous pair to each pole of the cell.

Telophase & Cytokinesis I

Cells divide into two haploid cells, each with a different combination of parental genes. Each chromosome still has two sister chromatids.

12.2 ALTERNATION BETWEEN DIPLOID AND HAPLOID GENERATIONS

As discussed in the preceding section, the sexual life cycle consists of an alternation between haploid cells and diploid cells. This is true of all sexual organisms, whether they are human, animal, plant, fungus, or protist. But organisms vary in the way they handle the diploid and haploid cells. In humans and in animals, *haploid* cells do not multiply by mitosis, but simply become gametes. In contrast, *diploid* animal cells—starting with the zygote—divide mitotically to make diploid bodies.

Matters are quite different in plants. They divide both haploid and diploid cells by mitosis, to make two kinds of multicellular bodies. Plant biologists refer to the creation of both diploid and haploid bodies as an **alternation of diploid and haploid generations. Figure 12.5** shows how a typical flowering plant, a cherry tree, carries out the alternation of diploid and haploid generations. This diagram illustrates some fundamental ideas and terms that are essential to understanding the life of a plant.

The cherry tree, like all large plant bodies, is the diploid part of the life cycle. A diploid plant body is called a **sporophyte** because it makes reproductive units called spores. More specifically, a **spore** is a one-celled reproductive unit that can develop into a new plant without mating with another organism. Actually, the cherry tree makes two kinds of spores in dif-

ferent parts of the flower. One kind of spore will develop into a male haploid plant that makes gametes called *sperm* cells, whereas the other kind will develop into a female haploid plant that makes a gamete called an *egg*. Because haploid plants make gametes, they are called **gametophytes.**

To see how male spores and gametophytes are made, look at the top left of Figure 12.5. Just inside the petals of each flower, there are a number of slender stalks with swollen yellow tips. The tips are **anthers.** When an anther is sectioned, one can see that it contains four cavities, which are **pollen sacs.** Initially, all the cells that make up the pollen sacs are diploid and divide by mitosis, but eventually an inner group of cells divides once by meiosis, each producing four haploid cells. These haploid cells are the male spores.

As discussed earlier, meiosis produces cells with new combinations of genes. The spores made in the anther were built to disseminate those new gene combinations; therefore, they are called **meiospores** to emphasize their meiotic origin and their recombinant nature. This contrasts with **mitospores** that some organisms make to disseminate copies of the parent's own gene combination. Mitospores are made by mitotic division of the parent's cells without intervening meiosis. Within the pollen sac, each male meiospore divides by mitosis to become the tiniest possible gametophyte plant—a body with just two haploid cells. Then its outer wall hardens, making the gametophyte into a micro-

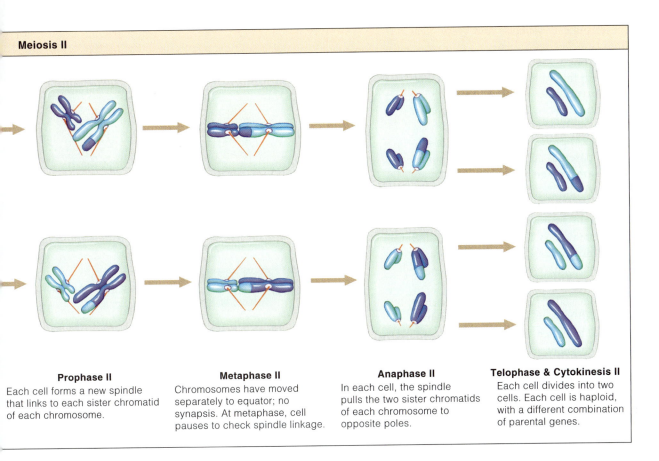

Prophase II
Each cell forms a new spindle that links to each sister chromatid of each chromosome.

Metaphase II
Chromosomes have moved separately to equator; no synapsis. At metaphase, cell pauses to check spindle linkage.

Anaphase II
In each cell, the spindle pulls the two sister chromatids of each chromosome to opposite poles.

Telophase & Cytokinesis II
Each cell divides into two cells. Each cell is haploid, with a different combination of parental genes.

scopic **pollen grain.** The sacs split open, releasing thousands of pollen grains to passing air currents or animals. Later, on reaching the female part of a flower, the pollen grain grows a slender extension called a **pollen tube,** which contains two sperm nuclei (Fig. 12.5, *bottom left*). With the formation of the pollen tube, the male gametophyte has reached maturity.

To see how female structures are made, examine the top right of Figure 12.5. In the center of the flower is a vase-shaped **pistil** with a swollen base called an **ovary.** Chambers in the ovary are lined with microscopic bulges of tissue, each called an **ovule.** At first, all the cells of each ovule are diploid, but eventually an innermost single cell undergoes meiosis and produces four recombinant haploid cells. Three of the four cells degenerate, whereas the fourth matures into a large female spore. This, too, is a meiospore. It is not released, but remains inside the ovule where it divides by mitosis, and the resulting cells divide twice more by mitosis to make a tiny, seven-celled female gametophyte. One gametophyte cell is an **egg.** To unite pollen with eggs, the first step is **pollination**—the transfer of pollen from the anther to the tip of a pistil. There, the pollen grain forms a pollen tube that grows down to an ovule, carrying two haploid sperm nuclei. When the pollen tube reaches an egg, one sperm fuses with the egg to make a zygote, which divides mitotically to form an **embryo** within a seed coat. The embryo is simply a small sporophyte that will grow to become a cherry tree when the seed germinates.

Plants Vary in the Details of Their Life Cycles

Not all plants reproduce like the cherry tree. Later in this textbook, Chapters 21–24 summarize how conifers, ferns, mosses, and algae carry out their life cycles. There are many life cycle similarities among flowering plants and the rest of the kingdom Plantae, but there also are differences. Fortunately, the most important differences can all be brought together in a generalized life cycle (Fig. 12.6). By mastering the vocabulary of Figure 12.6, you will be able to simplify and more easily understand the variations.

Before discussing the generalized life cycle, what is meant by the informal term *plant* needs to be clarified. This clarification requires two more informal terms: embryophytes and green algae. **Embryophytes** are plants that shelter their offspring as embryos within the parental body. They include all the land plants that we know in everyday life. **Green algae,** in contrast, are simpler photosynthetic organisms that do not form embryos. They include green pond scum and the green film that grows on swimming pools and aquarium glass. Despite their many differences, green algae and embryophytes are related to each other more closely than to any other form of life (see Chapter 21). Consequently, many systematists now include both embryophytes and green algae under the informal heading *plants.* This chapter also follows that practice and uses the terms *embryophytes* and *green algae* when discussing differences between them.

To work our way through the generalized life cycle (Fig. 12.6), let us begin with the multicellular diploid plant, the sporophyte. All embryophytes produce sporophytes, ranging from giant trees down to tiny lumps that are smaller than a pinhead. Some green algae also make sporophytes, but many do not. Many sporophytes can reproduce asexually from fragments of a parental body, as when rhizomes spread horizontally and are broken up to produce many separate plants. Each part can regenerate missing parts, resulting in complete organisms. Some green algae reproduce the sporophyte asexually by making **mitosporangium,** a structure in which parental cells divide mitotically and then differentiate into mitospores. Each mitospore is released to grow a new sporophyte with the same genetic constitution as the parent plant.

At particular times, some cells of the sporophyte launch a sexual cycle when they differentiate to form a structure called a **meiosporangium,** in which one or more cells will divide by meiosis. In green algae, a meiosporangium is a single cell. In embryophytes, it is a multicellular structure. In the cherry tree, each pollen sac is a meiosporangium, and each ovule contains a meiosporangium. Eventually, meiospores germinate to develop into gametophytes by means of mitotic divisions and growth. Like sporophytes, the gametophytes may reproduce asexually from fragments of the parental body or, in the case of green algae, by producing mitosporangia and mitospores. Each fragment or mitospore has the potential to develop into a new gametophyte with the same genetic constitution as the parent gametophyte.

When environmental conditions are appropriate, gametophytes take another step in the sexual life cycle by producing structures called **gametangia** (singular, *gametangium*) in which haploid cells differentiate into gametes. In green algae, a gametangium is a single enlarged cell. Embryophytes make more complex gametangia, which contain many cells. However, even there the gametangia are microscopically small. They are particularly small in flowering plants, and they were not shown in our cherry tree life cycle. Some gametangia look like sporangia, and, in some green algae, the gametes look just like mitospores. But there is always one consistent difference between gametes and mitospores: Gametes cannot produce new individuals by themselves. Gametes must fuse in pairs to produce a new organism.

The most complex life cycles have two kinds of gametes that differ in function and appearance. This is true of all flowering plants. As seen in the cherry tree, one gamete (the egg) is large and immobile; the other gamete (the sperm) is small and mobile. In such cases, the two gametes can be given descriptive names and called *male* and *female*. In certain algae, however, the two gametes are equally mobile and equal in size. Even though we cannot visually distinguish between such gametes, there must be profound metabolic and genetic differences, because only certain ones will pair.

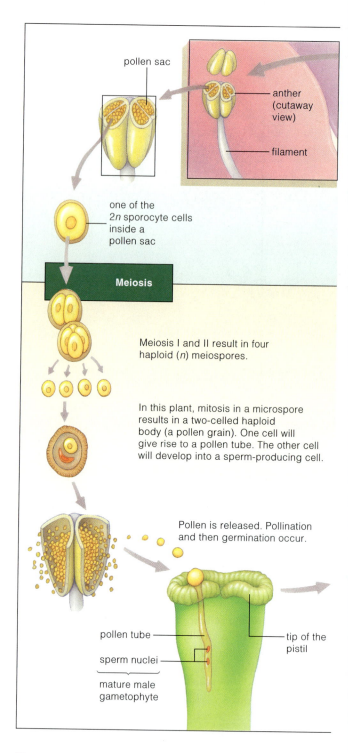

Figure 12.5 Life cycle of cherry (*Prunus*). Start with the flower, which is part of the sporophyte. Look to the flower's top left and see how meiosis in pollen sacs leads to male gametophytes. At bottom left, the pollen tube of the gametophyte has two gametes called sperm. Then at the top right of the flower, it is shown how meiosis in ovules leads to the embryo sac (female gametophyte). Inside the embryo sac is a gamete called an egg. When egg and sperm meet, they fuse to produce a diploid zygote cell (not shown). Mitotic divisions and growth convert the zygote cell into an embryo sporophyte within a seed. On seed germination, the embryo grows into a cherry tree. (Art by Raychel Ciemma.)

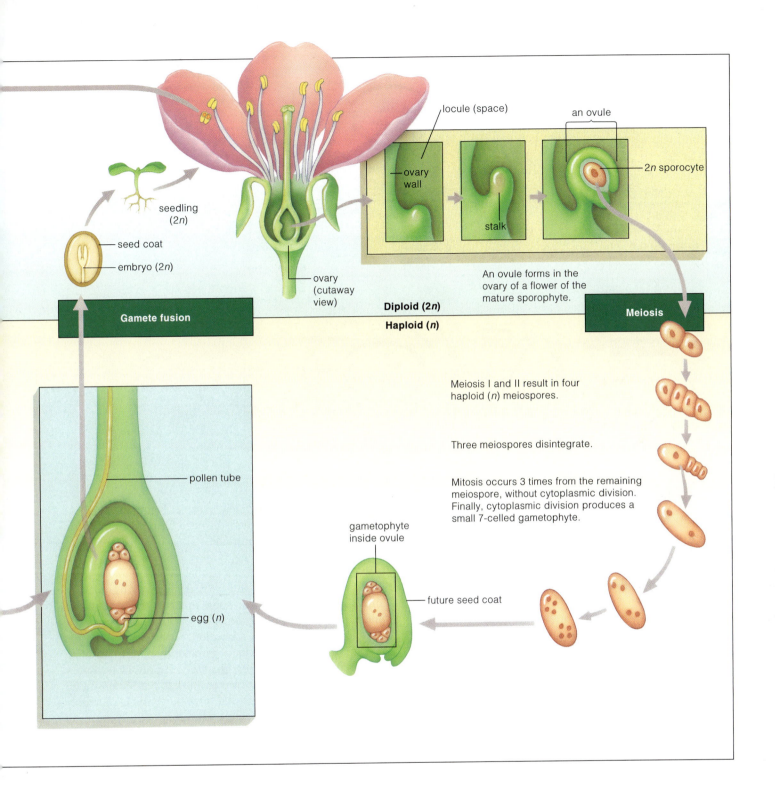

locule (space)

an ovule

ovary
wall

2n sporocyte

stalk

An ovule forms in the
ovary of a flower of the
mature sporophyte.

seedling
(2n)

seed coat

embryo (2n)

ovary
(cutaway
view)

Diploid (2n)

Haploid (n)

Gamete fusion

Meiosis

Meiosis I and II result in four
haploid (n) meiospores.

Three meiospores disintegrate.

Mitosis occurs 3 times from the remaining
meiospore, without cytoplasmic division.
Finally, cytoplasmic division produces a
small 7-celled gametophyte.

pollen tube

gametophyte
inside ovule

future seed coat

egg (n)

In such cases, they are called *mating* types, labeling one kind of gamete *plus* and the other kind *minus*. Just as two male sperm gametes do not fuse, neither will two plus or two minus gametes fuse.

When two compatible gametes meet, the cytoplasmic contents fuse **(plasmogamy),** and soon the nuclei fuse as well **(karyogamy)** to make a diploid zygote cell. In embryophytes, growth and mitotic divisions convert the zygote into a sheltered embryo and later into a mature sporophyte. In seed plants, the embryo phase is prolonged, with the embryo packaged within a protective seed coat and held in an arrested state of metabolism. Embryos of some species can remain dormant but alive within their seeds for decades or even centuries, until the appropriate conditions for germination occur. By contrast, many green algae lack a sporophyte stage, and simply put the zygote through meiosis to start the next gametophyte generation.

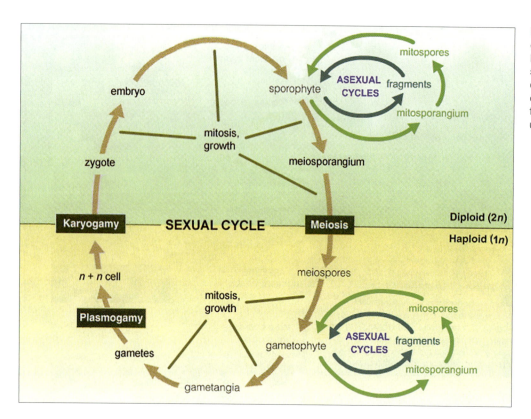

Figure 12.6 Generalized life cycle of organisms in the plant kingdom. No single species has all the steps shown here. The terms used in this diagram are basic, and many synonyms exist for variations. Mastery of these basic terms makes it easier to understand all plant life cycles.

Sexual Cycles Can Be Heterosporic or Homosporic

To keep the basics simple, Figure 12.6 omits one important way in which plant life cycles differ: Some plants make just one kind of spore and gametophyte (they are **homosporic**), whereas others are **heterosporic,** making two kinds of spores and gametophytes. The distinction is vital to human life, because our food supply depends on plants that, like the cherry tree, make two kinds of spores and gametophytes.

Why is heterospory important? The answer emerges when you consider the advantage that a heterosporic life cycle gives to a plant. Survival of a young plant depends on how good of a head start the parental plant gives to the offspring. To get a good start in life, two things must be done. First, the offspring must avoid competition with established plants, which are so much larger; and, second, offspring must carry enough stored food to survive until they can make their own food. To avoid competition, it helps greatly for the offspring to differ genetically from the parents, so that some of the offspring may be able to exploit opportunities that the parental plants could not. The greatest diversity comes when eggs are fertilized by sperm from different plants—a process called **outcrossing** (which is distinct from **inbreeding,** where eggs and sperm come from close relatives or even from the same plant). One thing that promotes outcrossing is to produce vast numbers of tiny reproductive units that can travel far. However, tiny offspring cannot

carry much food and are likely to starve before they become established. As you see, then, the need for much stored food conflicts with the need for travel.

Heterospory avoids the conflict in a simple way: The plant makes one kind of spore that is tiny enough to be made in huge numbers and carried far away by wind or animals and another kind of spore that is too heavy to travel but is stuffed with food. Most of the tiny traveling male gametophytes will die of starvation, but when one of them reaches a stationary female gametophyte, the resulting diploid offspring will have a great food supply. Heterospory is effective, and most plant species have it. It is important for humans because the seeds that form our basic food supply come from the food-rich female structures—which could not have evolved without the simultaneous presence of tiny mobile male structures (pollen).

Thousands of plant species—most notably mosses and other seedless plants such as ferns—make only one kind of spore and one kind of gametophyte; they are homosporous. All of their spores are too tiny to travel far, and none are important in our food supply, except as emergency foods. Their continued existence does show, however, that homospory is an effective lifestyle in some ecologic situations. An example is the growth of mosses on tree trunks and rocks, especially in moist cool areas; another example is the aquatic environment where most algae dwell.

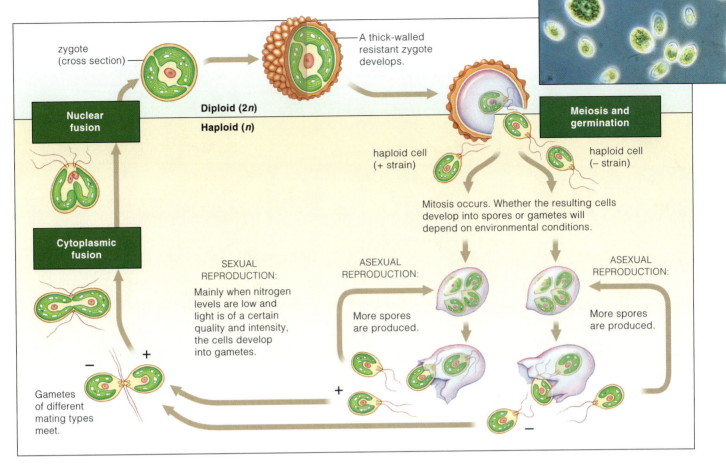

Figure 12.7 Zygotic life cycle of a single-celled species of *Chlamydomonas*, one of the most common green algae of freshwater habitats. *Chlamydomonas* reproduces asexually most of the time. It also reproduces sexually under certain environmental conditions. (Photo by D. J. Patterson/Seaphot Limited: Planet Earth Pictures. Art by Raychel Ciemma.)

Only One Generation Is Multicellular in Zygotic or Gametic Life Cycles

Plant life cycles vary in which kinds of cells (haploid or diploid) undergo many mitotic divisions to build multicellular bodies (sporophytes and gametophytes). A life cycle is said to be **sporic** if it includes alternating sporophyte and gametophyte bodies. All embryophytes, such as flowering plants and mosses, have sporic life cycles. The same is true of some algae, such as sea lettuce (*Ulva*) and the brown kelps seen on beaches.

Many algae lack sporophytes and have what is called a **zygotic life cycle.** This is the case with the green alga *Chlamydomonas* **(Fig. 12.7).** As shown in the bottom right corner of Figure 12.7, *Chlamydomonas* gametophytes are single, motile cells commonly found in freshwater habitats. Each cell has a single haploid nucleus. Cells appear to be similar, but genetically they exist as either plus or

minus mating types. Occasionally, a gametophyte nucleus will undergo mitosis, producing several haploid spores. The parent cell bursts, releasing these spores, and each spore matures into a new gametophyte-generation cell. Thus, in this case, the gametophyte cells have acted as spores in an asexual part of the life cycle. Under other conditions, one plus and one minus gametophyte cell are attracted to each other in pairs (Fig. 12.7, *bottom left*). Plasmogamy and karyogamy occur, resulting in a 2*n* zygote cell. In this case, the gametophyte cells have acted as gametes in a sexual part of the life cycle. The zygote may rest in a dormant stage for some time, but it does not develop into a multicellular sporophyte-generation organism. The zygote eventually will undergo meiosis (Fig. 12.7, *top right*) and release many haploid cells. Each cell will mature into either a plus or minus gametophyte-generation cell. *Chlamydomonas* has a zygotic life cycle because there is no multicellular 2*n* phase.

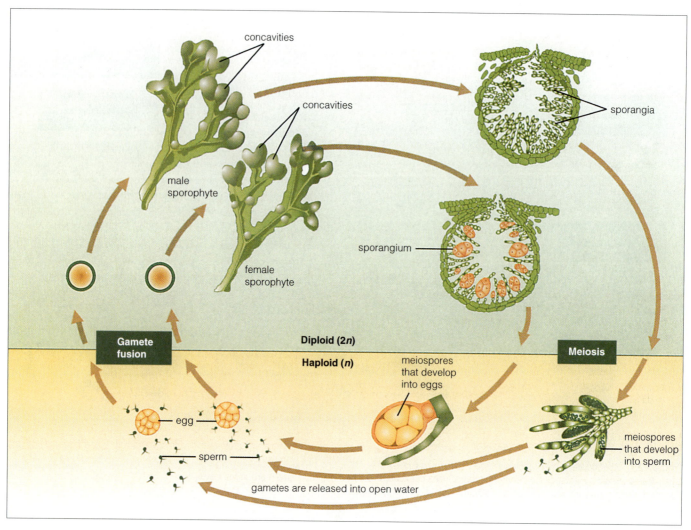

concavities

concavities

sporangia

sporangium

male
sporophyte

female
sporophyte

**Gamete
fusion**

Diploid (2n)

Haploid (n)

Meiosis

meiospores
that develop
into eggs

egg

sperm

meiospores
that develop
into sperm

gametes are released into open water

Figure 12.8 Gametic life cycle of the intertidal rockweed *Fucus*. In this life cycle, only the unicellular gametes
are haploid.

A **gametic life cycle,** such as that of the brown alga rockweed (*Fucus;* Fig. 12.8), begins with a multicellular sporophyte. The sporophyte is relatively large and complex, with rootlike, stemlike, and leaflike regions. Within body concavities, special cells enlarge and become sporangia, and their nuclei undergo meiosis. One type of sporangium produces large meiospores, and another type produces small ones. The meiospores are not released from their sporangia at this time. Each large meiospore first differentiates into a female gamete (an egg), and each small meiospore differentiates into a male gamete (a sperm). Then the gametes are released into the surf in such huge numbers that many eggs and sperms are brought together. Only eggs from one parent and sperm from another can fuse; eggs and sperm produced by the same plant will not be attracted to each other. Plasmogamy and karyogamy occur, and the zygote immediately begins to divide and grow into a sporophyte. As it enlarges, it sinks to the bottom of the intertidal zone, becomes attached to a rock, and then grows into

maturity. *Fucus* has a gametic life cycle because the only haploid phase is a single-celled gamete. There is no multicellular gametophyte generation in a gametic life cycle.

The Diploid Generation Has Become Dominant over Evolutionary Time

The fossil record of algae, fungi, and plants is described in some detail later in this textbook. For the purposes of this chapter, it is sufficient to know that the chronologic appearance of photosynthetic organisms on earth began with algae, followed by mosses and lower vascular plants such as ferns, then by conifers, and last by flowering plants. There are parallel relationships between this evolutionary path and life cycles. Gametic and zygotic life cycles are common among algae but absent from any of the more advanced plants. In contrast, sporic life cycles are the rule among the complex, more recently evolved terrestrial plants. (Sporic life cycles

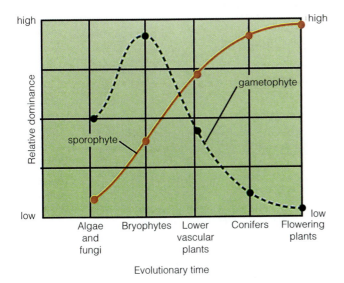

Figure 12.9 Life cycle trends, from the algae and fungi through the flowering plants. The more recently evolved the group and the more adapted it is to dry land, the more dominant is its sporophyte generation. *Dominant* means larger, more complex, and having a longer life span than the gametophyte generation.

do exist among some algae such as kelps and seaweed, but they are the exception.)

The generations of sporic life cycles are not equal among all terrestrial plants. There is a clear trend of increasing dominance by the sporophyte in groups that are more recent in the fossil record **(Fig. 12.9)**. *Dominance* here means that the sporophyte lives longer, is larger, is more structurally complex, and is more independent than the gametophyte. There must be advantages, in our modern environment, to protecting and limiting the haploid phase of the life cycle. What could these advantages be?

The diploid condition permits many recessive genes to be carried along from generation to generation, each one masked by the dominant gene on the other homologous chromosome. Such recessive genes may have no value in the current environment (indeed, they may even be harmful), but they could be valuable in some different, future environment. The recessive genes carried now could contribute to the species' future success, especially because the genes already exist and do not have to be created (with some lag time) by future mutations. There is, however, a potentially unsafe phase in the life cycle for carrying recessive genes, and that is in the gametophyte generation. In the haploid cells of gametophytes, there are no recessive genes because there is only one set of chromosomes; therefore, every gene's expression shows through in this phase. If, however, the gametophyte is small, short-lived, and protected by the sporophyte, the expression of potentially deleterious genes might be tolerable. Perhaps the ability to be genetically diverse in the diploid chromosome condition is the explanation for the modern dominance of sporophytes on land.

KEY TERMS

1*n*	life cycle
2*n*	meiosis
alternation of generations	meiosporangium
anther	meiospore
chromosome set	mitosis
clone	mitosporangium
crossing over	mitospore
cytokinesis	outcrossing
diploid	ovary
egg	ovule
embryo	pistil
embryophytes	plasmogamy
gametangium	pollen grain
gametic life cycle	pollen sac
gamete	pollen tube
gametophytes	pollination
green algae	spore
haploid	sporic life cycle
heterosporic	sporophyte
homologous chromosomes	synapsis
homosporic	zygote
inbreeding	zygotic life cycle
karyogamy	

SUMMARY

1. Life perpetuates itself through reproduction, the transfer of genetic information from one generation to the next, and this transfer is the definition of life cycle used throughout this textbook. Reproduction can be asexual or sexual.

2. Asexual reproduction produces offspring with the same genetic composition as the parent, because the offspring are parts of a single parent organism in which the nuclei were produced by mitosis. Asexual reproduction offers advantages of speed and economy, but the uniformity of asexual offspring is a disadvantage because diversity may be needed to survive changes in the environment.

3. Sexual reproduction involves the union of two haploid cells, called gametes, which are usually produced by two parents. The result is a diploid zygote cell, which may divide mitotically to form a diploid body. Later in the life cycle, a special type of cell division called meiosis converts diploid cells to new haploid cells that carry diverse combinations of genes from the two parents. To generate diversity, meiosis joins homologous chromosomes in pairs (synapsis), in which crossovers exchange parts of chromosomes from the two parents.

4. *Spores,* made of one or a few cells, grow into organisms without the need to fuse with another cell. Meiospores are made to disseminate new combinations of genes from two parents, and are named for the role of meiosis in making the new combinations. Mitospores are made to disseminate a single parent's own combination of genes; and they are named for the role of mitosis in duplicating parental nuclei.

5. The informal term *embryophytes* includes all plants that shelter their offspring as embryos within the parental body. They include all familiar plants of the landscape. But the term *plants* also includes green algae, which do not make embryos.

6. All embryophytes have sporic life cycles, in which both diploid and haploid cells are made into multicellular bodies by mitotic cell divisions. The formation of alternate diploid and haploid bodies is called alternation of generations. Haploid bodies are called gametophytes because they generate gametes and are always small and relatively simple. Diploid bodies, called sporophytes because they make spores, often are large and complex. Trees and grasses are sporophytes.

7. Some algae also have sporic life cycles with complex sporophytes, but most have zygotic or gametic life cycles. A zygotic life cycle has only a single-celled zygote to represent the sporophyte phase; the rest of the life cycle is haploid. The green alga *Chlamydomonas* has a zygotic life cycle. A gametic life cycle has only single-celled gametes to represent the gametophyte phase; the rest of the life cycle is diploid. The seaweed *Fucus* has a gametic life cycle.

8. The fossil record reveals that the most recent groups to evolve—those plants that are most complex and adapted to dry land—have sporic life cycles in which the gametophyte generation is smallest and the sporophyte generation is most dominant. We may imagine, then, that there are advantages to protecting and limiting the haploid phase of the life cycle.

Questions

1. Why do plants reproduce?

2. What are some advantages and disadvantages of asexual and sexual reproduction?

3. What is the key difference between mitosis and the reduction division step of meiosis?

4. What is meant by the expression *alternation of generations?* How do sporophyte and gametophyte generations, in kingdom Plantae, typically compare in terms of size, complexity, dominance, and life span?

5. Summarize the life cycle of a cherry tree. Describe where meiosis occurs, what the gametophyte plants look like, how the two gametes are brought together to form a diploid zygote, and how the zygote develops back into a cherry tree.

6. The cherry life cycle is sporic. Certain algae and fungi have non-sporic life cycles. Fundamentally, how do they differ from plants with a sporic life cycle?

7. Why do you think that the most advanced terrestrial plants have life cycles dominated by the diploid (sporophyte) generation, whereas less advanced plants have life cycles that are dominated by the haploid (gametophyte) generation?

InfoTrac® College Edition

http://infotrac.thomsonlearning.com

Blackwell, W.H. 2003. Two theories of origin of the land-plant sporophyte: Which is left standing? *The Botanical Review* 69:125. (Keywords: "meiosis" and "algae")

Jenkins, C.D., Kirkpatrick, M. 1995. Deleterious mutation and the evolution of genetic life cycles. *Evolution* 49:512. (Keywords: "life cycles" and "mutation")

The Flower and Sexual Reproduction

Visit us on the web at http://biology.brookscole.com/plantbio2 for additional resources, such as flashcards, tutorial quizzes, InfoTrac exercises, further readings, and web links.

1. Sexual reproduction occurs in flowering plants in specialized structures called flowers. They are composed of four whorls of leaflike structures: the calyx (composed of sepals) serves a protective function; the corolla (composed of petals) usually functions to attract insects; the androecium (composed of stamens) is the male part of the flower; and the gynoecium (carpel or pistil) is the female part.

2. Stamens produce pollen. The stamen's female counterpart is the pistil, which is composed of the stigma (collects pollen), the style, and the ovary. The ovary produces eggs.

3. Sexual reproduction in plants involves pollination, which is the transfer of pollen from the anther of one flower to the stigma of the same or a different flower. The pollen lands on the stigma and grows into the style (as the pollen tube) until it penetrates the ovule and releases two sperm.

4. Double fertilization is unique to flowering plants. One sperm fuses to the egg to form the zygote; the other sperm fuses to the polar nuclei to form the primary endosperm nucleus. The zygote forms the embryo, and the primary endosperm nucleus forms the endosperm (a nutritive tissue). The fertilized ovule develops into the seed.

5. There are basically two types of pollination: self-pollination, in which the pollen comes from within the same flower or plant, and cross-pollination, in which the pollen comes from different plants. Pollination strategies involve abiotic factors such as wind and water and interesting relationships between plants and animals, biotic factors.

13.1 THE FLOWER: SITE OF SEXUAL REPRODUCTION

The vivid colors and intricate patterns of flowers have always attracted the notice of humans and captured their imagination. The symmetry and brightness of petals and the interesting scents from flowers are evolutionary designs to attract pollinators (mostly insects and birds) to visit flowers. The result of such visits often is a reward of nectar (sugary water); in exchange, the insect will fly away carrying pollen to a different flower. This fine-tuned relationship between the flower and the animal guarantees that sexual reproduction continues among plants, which are not able to roam about to select a mate. People will always love flowers, but the most important function of the flower is as the site where sexual reproduction occurs, leading to the formation of seeds. Imagine a world without wheat grains to make bread, corn to make products that people use every day, or barley to make beer; we would all live very different lives without seeds.

Of all the characteristics of flowering plants, the **angiosperms,** the flower and fruit are the least affected by changes in the environment such as age, light, water, and nutrition. Consequently, the appearance of flowers and fruits is important to understanding evolutionary relationships among angiosperms.

The development of the flower begins the sexual reproductive cycle in all flowering plants. The function of the flower is to facilitate the important events of gamete formation and fusion. The essential steps of sexual reproduction, meiosis, and fertilization (see discussion in Chapter 12) take place in the flower. The complete sexual cycle involves the following steps: (1) the production of special reproductive cells after meiosis; (2) pollination; (3) fertilization; (4) seed and fruit development; (5) seed and fruit dissemination; and (6) seed germination.

13.2 THE PARTS OF A COMPLETE FLOWER AND THEIR FUNCTIONS

A typical flower is composed of four whorls of modified leaves—sepals, petals, stamens, and carpels—that are all attached to the terminal end of a modified stem, the receptacle (Fig. 13.1).

The **sepals** are generally green and enclose the other flower parts in the bud. All the sepals collectively constitute the **calyx,** and they function to protect the reproductive parts

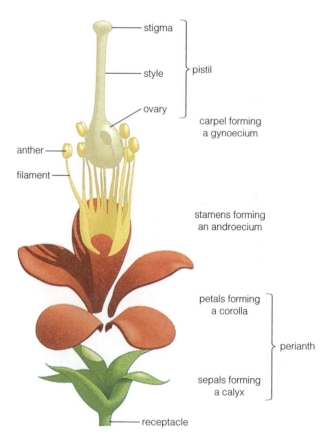

stigma
style — pistil
ovary
carpel forming a gynoecium
anther
filament
stamens forming an androecium
petals forming a corolla
perianth
sepals forming a calyx
receptacle

Figure 13.1 A flower, showing the whorls of parts. The perianth consists of two whorls of sepals and petals (calyx and corolla). There is one whorl of stamens (collectively called the *androecium*). A single carpel (a pistil) forms the central whorl of floral parts, the *gynoecium*.

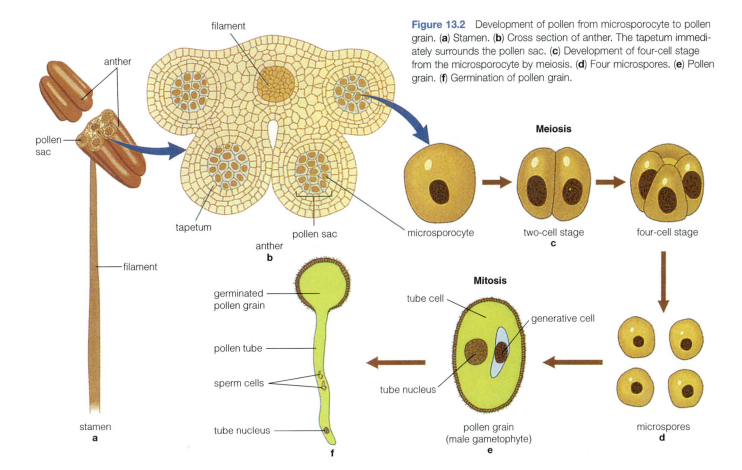

Figure 13.2 Development of pollen from microsporocyte to pollen grain. (**a**) Stamen. (**b**) Cross section of anther. The tapetum immediately surrounds the pollen sac. (**c**) Development of four-cell stage from the microsporocyte by meiosis. (**d**) Four microspores. (**e**) Pollen grain. (**f**) Germination of pollen grain.

Labels in figure: anther; filament; pollen sac; filament; tapetum; anther; pollen sac; stamen; **a**; anther; **b**; germinated pollen grain; pollen tube; sperm cells; tube nucleus; **f**; Meiosis; microsporocyte; two-cell stage; four-cell stage; **c**; Mitosis; tube cell; generative cell; tube nucleus; pollen grain (male gametophyte); **e**; microspores; **d**

inside the flower. The **petals** are usually the conspicuous, colored, attractive flower parts, which together constitute the **corolla.** Their most important function is to catch the attention of pollinators.

The **stamens** form the whorl just inside the corolla. Each stamen has a slender stalk, or **filament,** at the top of which is an **anther,** the pollen-bearing organ. The whorl or grouping of stamens is called the **androecium,** and this constitutes the male part of the flower (Fig. 13.1). The female part of the flower is called the **gynoecium.** It is composed of one or more **carpels,** which are modified leaves folded over and fused to protect the attached ovules. Sometimes an individual carpel or group of fused carpels is called a **pistil.** The term is used simply because its shape is reminiscent of the pistil (or pestle) used to grind objects in a mortar. There may be more than one pistil in the gynoecium. The carpels are usually located at the center of the flower (Fig. 13.1).

Carpels generally have three distinct parts: (1) an expanded basal portion, the **ovary,** which contains the **ovules;** (2) the **style,** a slender supporting stalk; and (3) the **stigma,** at the very tip (Fig. 13.1).

The term **perianth** is applied to the calyx and corolla collectively. The flower organs necessary for sexual reproduction are the stamens (androecium) and carpels (gynoecium). The perianth, composed of calyx and corolla, protects the stamens and pistil(s), and also attracts and guides the movements of some pollinators.

The Male Organs of the Flower Are the Androecium

As mentioned earlier, the androecium is the whorl of stamens, with each stamen consisting of an anther at the end of a filament. The anther usually is made up of four elongated lobes called **pollen sacs (Fig. 13.2a).** Early in the development of the anther, each pollen sac contains a mass of dividing cells called **microsporocytes (Fig. 13.2b).** Each microsporocyte divides by meiosis to form four haploid (*n*) microspores **(Fig. 13.2c,d).** The nucleus of each microspore then divides by mitosis to form a two-celled **pollen grain,** which contains a **tube cell** and a smaller **generative cell (Fig. 13.2e).** The role of this two-celled, haploid, **male gametophyte** (gamete-producing plant) is to produce **sperm cells** for fertilization.

The pollen grain is surrounded by an elaborate cell wall. The pattern on this wall is genetically fixed, and it varies widely among major groups of plants **(Fig. 13.3).** The walls contain *sporopollenin,* a very hard material that resists decay. As a result, pollen grains make good fossils, and botanists have found them useful in studying the evolutionary history of seed plants.

After the pollen grains are mature, the anther wall splits open and the pollen is shed. In various ways discussed later in this chapter, pollen grains are transported to the stigmas of adjacent or distant flowers. This process is called **pollination.** The dry pollen grain then absorbs water from the stigma and becomes hydrated. It also secretes proteins,

T. E. Weier

a

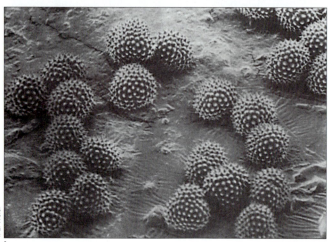

T. E. Weier

b

Figure 13.3 Pollen grains as viewed with the scanning electron microscope. (**a**) Iris. ×174. (**b**) Ragweed (*Ambrosia*). ×700.

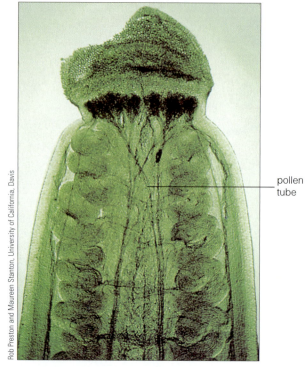

Rob Preston and Maureen Stanton, University of California, Davis

pollen
tube

Figure 13.4 Style of jewel plant (*Strepthanthus*) flower with pollen tubes that have grown toward the ovary. ×62.

including some that are involved in pollen recognition and compatibility reactions with the cells of the stigma. This recognition mechanism is important because it inhibits pollen from germinating and growing on the wrong species. These proteins are the same ones that cause allergies and hay fever.

The pollen grain then germinates to form a **pollen tube,** which grows toward the ovary. The pollen tube grows between cells in the center of the style, a region called the *transmitting tissue* **(Fig. 13.4).** In some plants, such as the lily (*Lilium*), the style is hollow, and the transmitting tissue is reduced to a layer of secretory cells on its inner surface; the cells secrete a special material in which the pollen tube grows. In solid transmitting tissues, the pollen tube secretes digestive enzymes that partially break down the transmitting tissue cells, making it easier for the pollen tube to grow.

The pollen tube grows by elongation very near its tip. It is filled with many small vesicles containing cell wall building materials needed for tube growth. Endoplasmic reticulum and other organelles also are present. Rapid cytoplasmic streaming (see Chapter 3) occurs in pollen tubes.

The Female Organs of the Flower Are the Gynoecium

In its simplest form, the gynoecium consists of a single folded carpel or **simple pistil (Fig. 13.5a)** in which the ovary resides. In more complex flowers, the gynoecium may consist of several separate carpels or several groups of fused carpels. A group of fused carpels is called a **compound pistil (Fig. 13.5b).**

The ovary is a hollow structure having from one to several chambers, or **locules** (Fig. 13.5b). The number of carpels in a compound pistil is generally related to the number of stigmas, the number of locules, and the number of sides on the ovary. The pea (*Pisum*) has a single stigma, and its ovary has one locule (Fig. 13.5a). The tulip (*Tulipa*) pistil has three stigmas, its ovary has three sides, and it contains three locules and three carpels (Fig. 13.5b).

The tissue within the ovary to which an ovule is attached is called the **placenta (Fig. 13.6).** The manner in which placentae are distributed in the ovary is termed **placentation.** When the placentae are on the ovary wall, as in bleeding heart (*Dicentra;* Fig. 13.6a), the placentation is **parietal.** When they arise on the axis of an ovary that has several locules, as in *Fuchsia*, the placentation is **axile** (Fig. 13.6b). Less frequently, the ovules form on a central column, which is **central placentation** (*Primula;* Fig. 13.6c).

Emanating from the ovary is the style, a stalk whose length and shape vary by species, with a stigma at its upper end (Fig. 13.1). It is through stylar tissue that the pollen tube grows (Fig. 13.4). In general, the style withers after pollination. Often, the surface of the stigma is covered with short

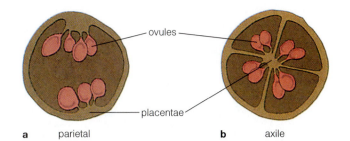

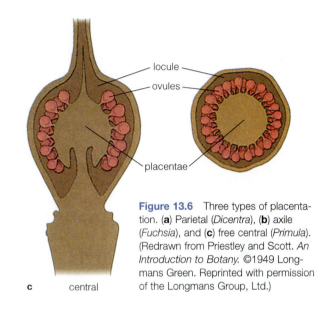

Figure 13.6 Three types of placentation. (**a**) Parietal (*Dicentra*), (**b**) axile (*Fuchsia*), and (**c**) free central (*Primula*). (Redrawn from Priestley and Scott. *An Introduction to Botany.* ©1949 Longmans Green. Reprinted with permission of the Longmans Group, Ltd.)

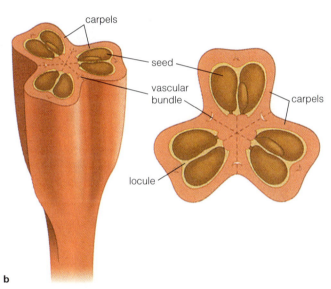

Figure 13.5 Simple and compound pistils compared. (**a**) A section through the simple pistil of pea (*Pisum*), showing an ovary composed of a single carpel. (**b**) A section through the compound pistil of tulip (*Tulipa*), showing an ovary with three fused carpels.

hairs that aid in holding the pollen grains, and sometimes they secrete a sticky fluid that stimulates pollen tube growth. In many wind-pollinated plants, such as the grasses, the stigma are elaborately branched or feather-like.

Within the ovary is the ovule, the structure that (if fertilized) will eventually become the seed. As the ovule matures (Fig. 13.7), it forms one or two outer protective layers, the **integuments.** The integuments do not fuse, leaving a small opening called the **micropyle** (Fig. 13.7b). At the same time, one of the internal dividing cells of the ovule, the **megasporocyte,** is enlarging in preparation for meiosis (Fig. 13.7c). The megasporocyte is embedded in a tissue called the **nucellus.** The ovule thus is composed of one or two outer protect-

ing integuments, together with the micropyle, megasporocyte, and nucellus.

The Embryo Sac Is the Female Gametophyte Plant

As a result of meiosis of the megasporocyte, a row of four cells called **megaspores** is produced in the nucellus (Fig. 13.7d). Each megaspore is haploid *(n)*. As a rule, the three cells nearest the micropyle disintegrate and disappear, whereas the one farthest from the micropyle enlarges greatly. This megaspore develops into the mature **embryo sac** in several stages: (1) a series of three mitotic divisions occurs to form an eight-nucleate embryo sac (Fig. 13.7e–g); (2) the nuclei migrate; and (3) a cell wall forms around the nuclei.

At the end of this process (Fig. 13.7g), an **egg cell** and two **synergid cells** are positioned at the micropylar end of the embryo sac. Because it is frequently difficult to differentiate the egg cell from the other two cells, all three cells are sometimes referred to as the **egg apparatus.** The two nuclei that migrate toward the center are **polar nuclei;** they lie in the center of the large **central cell.** The three nuclei at the end of the embryo sac opposite the micropyle form three **antipodal cells.** As described in Chapter 12 (see Fig. 12.5), the embryo sac is the **female gametophyte** phase of the flowering plant's life cycle; it is a haploid or *n* plant.

Not every species of flowering plant produces an embryo sac with the same number of cells, and they do not all follow

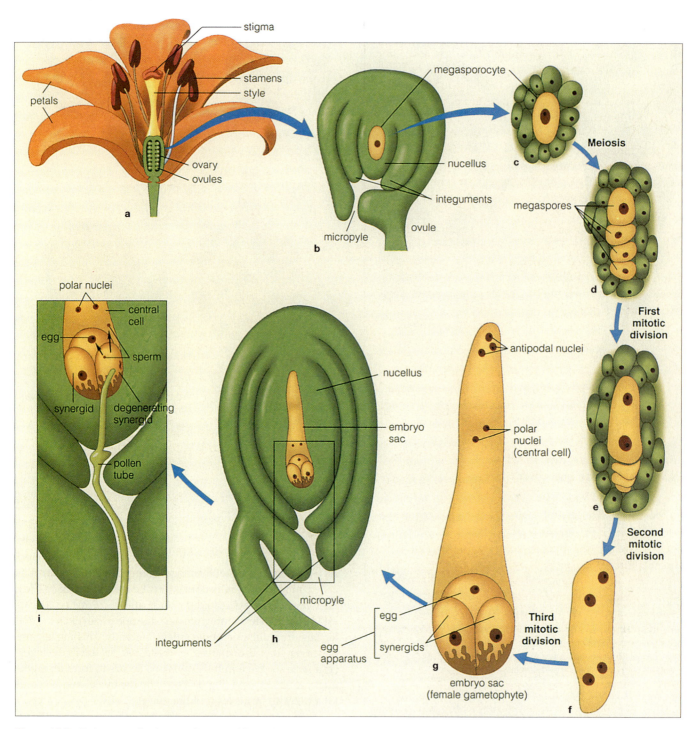

Figure 13.7 Embryo sac development in cotton (*Gossypium hirsutum*).
(**a**) Mature flower. (**b**) Magnified longitudinal section of an ovule.
(**c**) Megasporocyte before meiosis. (**d**) Four *n* megaspores after meiotic division of the megasporocyte; the three megaspores closest to the micropyle disintegrate. (**e**) First mitotic division of the megaspore.
(**f**) Second mitotic division. (**g**) Third mitotic division and the resulting three antipodal nuclei, the egg, two synergids, and the two polar nuclei. All have an *n* chromosome number. (**h**) Mature embryo sac; the antipodals disintegrate at this stage in cotton. (**i**) Pollen tube penetration through the nucellus into a partially degenerated synergid. (b–h, Redrawn from Gore U.R. 1932. Development of the female gametophyte and embryo in cotton. *Am J Bot* 19:795–807; i, modified and redrawn from Jensen, W.A. 1965. The ultrastructure and composition of the egg and central cell of cotton. *Am J Bot* 52:781–797.)

the same developmental sequence. In cotton (*Gossypium hirsutum*), for example, the antipodal cells disintegrate before fertilization (Fig. 13.7h). However, the mature embryo sac must contain, at a minimum, an egg and polar nuclei.

Double Fertilization Produces an Embryo and the Endosperm

As shown in this chapter, germination of a pollen grain produces a pollen tube, which grows down through the stigma and style and enters the ovary (Fig. 13.7i). As the pollen tube is growing, the generative cell within it divides by mitosis to form two sperm cells. In some plants, such as sunflower (*Helianthus annuus*), the sperm cells form even before the pollen is shed from the anther. Many pollen grains may germinate, and their pollen tubes may grow through the pistil, but only one usually enters an ovule and its embryo sac.

How is the pollen tube directed into the embryo sac? The answer has been partially worked out for cotton. In the embryo sac, one or both of the synergid cells begin to shrivel and die before the pollen tube enters. It is conjectured that, in the process, the synergids release chemicals that may influence the direction in which the pollen tube grows. Two observations favor this idea. First, the cell wall at the base of the synergids is highly convoluted (Fig. 13.7g); in other parts of the plant, cell walls like these (transfer cells; see Chapter 4) are associated with active transport or absorption. The second bit of circumstantial evidence is that sometimes the pollen tube actually penetrates one of the synergids and, on rupturing, empties its contents into the synergid. The two sperm then pass through an incomplete upper cell wall of the synergid, one moving to fuse with the egg and the other with the central cell.

The sperm and egg nuclei fuse to form a diploid (2*n*) **zygote,** which will grow into the embryo. The other sperm nucleus fuses with the polar nuclei of the central cell (Fig. 13.7i) to form a triploid (3*n*) **primary endosperm nucleus,** which will divide to become a food reserve tissue called the **endosperm.** This double fusion of egg with sperm and polar nuclei with sperm is called **double fertilization.** The antipodals and synergids usually degenerate. Now the conditions are set for development of the seed and fruit (see Chapter 14 for discussion of this topic).

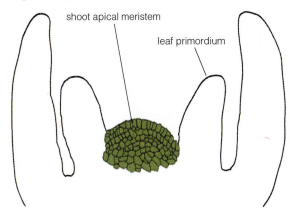

Vegetative apex

shoot apical meristem

leaf primordium

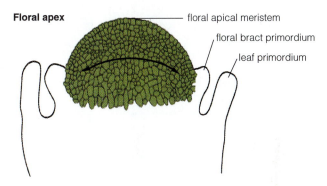

Floral apex

floral apical meristem

floral bract primordium

leaf primordium

Figure 13.8 Comparison of vegetative and floral apices.

13.3 APICAL MERISTEMS: SITES OF FLOWER DEVELOPMENT

During vegetative growth, the shoot apex produces stems and leaves. At some time during the growing season, a signal (for example, day length or temperature) will trigger a change in the metabolism of the shoot apex, thereby starting its transformation into a floral apex. The first step in the transition of the apex is a broadening of the apical dome **(Fig. 13.8),** accompanied by a general increase in RNA and protein synthesis and in the rate of cell division in the apical dome. The first organs to form from the floral apex are bracts. These modified leaves develop at the lower periphery of the floral apex. Floral organs usually form in whorls or in spirals, and the internodes between successive sets of floral organs are usually short. The floral organs are modified leaves; therefore, the flower itself is really a shortened and modified branch.

In pheasant-eye (*Adonis aestivalis*—**Fig. 13.9a**), the sepals and petals form after the bracts, followed by several spirals of stamens and carpels **(Fig. 13.9b–d).** Each new floral organ forms closer to the floral apex. Eventually, carpels develop at the tip, terminating any further growth of the floral apex **(Fig. 13.9d,e).**

The development of an inflorescence, a group of flowers from the same apex, follows a spatial sequence: bracts form first, followed by flowers starting at the periphery of the inflorescence apex and progressing toward the center and the tip. The parts of the flowers (sepals, petals, stamens, and carpels) on the inflorescence develop in the same sequence as they do in individual flowers.

Flowers Vary in Their Architecture

Botanists have been studying flowers and plant reproduction for a long time; consequently, there are many terms to describe flower types. Flowers may be complete or incomplete, perfect or imperfect. A flower that develops all four sets of floral leaves—sepals, petals, stamens, and carpels—is said to be a **complete flower.** An **incomplete flower** lacks one or more of these four sets.

Unisexual flowers are either **staminate** (stamen bearing) or **pistillate** (pistil bearing) and are said to be **imperfect.** Bisexual flowers are **perfect.** When staminate and pistillate flowers occur on the same individual plant, as they do in corn (*Zea mays*), English walnut (*Juglans regia*), and many

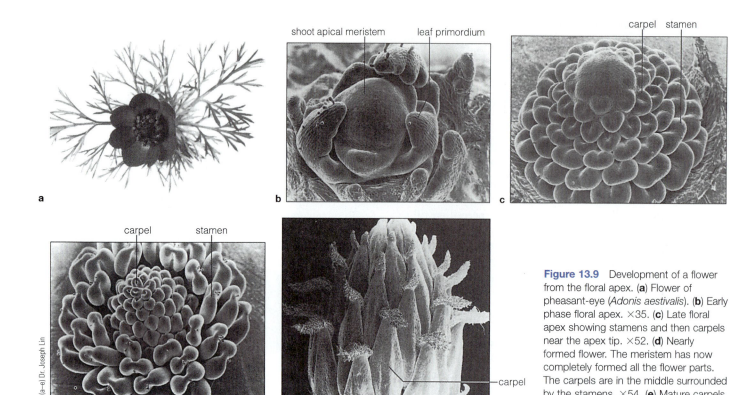

shoot apical meristem   leaf primordium

carpel   stamen

carpel   stamen

carpel

(a–e) Dr. Joseph Lin

a

b

c

d

e

Figure 13.9 Development of a flower from the floral apex. (**a**) Flower of pheasant-eye (*Adonis aestivalis*). (**b**) Early phase floral apex. ×35. (**c**) Late floral apex showing stamens and then carpels near the apex tip. ×52. (**d**) Nearly formed flower. The meristem has now completely formed all the flower parts. The carpels are in the middle surrounded by the stamens. ×54. (**e**) Mature carpels with feathery stigmas. ×17.

pistillate flower

stigma

ovary

pistillate catkin

staminate (male) catkin

one stamen

anther

staminate flowers

a

b

c

Figure 13.10 Flowers of English walnut (*Juglans regia*), a monoecious plant. (**a**) Branch with staminate and pistillate inflorescences (catkins). (**b**) Pistillate flower. (**c**) Staminate flowers.

other species, the plant is called **monoecious (Fig. 13.10).** When staminate and pistillate flowers are borne on separate individual plants, as in *Asparagus* or willow (*Salix*), the plant is said to be **dioecious.**

Flowers may also be classified according to their symmetry, which may be regular or irregular. In many flowers, such as stonecrop (*Sedum* sp.) **(Fig. 13.11a),** the corolla is made up of petals of similar shape that radiate from the center of the flower and are equidistant from each other. Such flowers are said to have **regular symmetry.** In these cases, even though there may be an uneven number of parts in the perianth, any line drawn through the center of the flower will divide the flower into two similar halves. The halves are either exact duplicates or mirror images of each other.

Flowers with **irregular symmetry,** such as garden pea (*Pisum sativum*) **(Fig. 13.11b),** have parts arranged in such a way that only one line can divide the flower into equal halves; the halves usually are mirror images of each other.

The parts of the flower may be free or united. In the flower of stonecrop (*Sedum* sp.; Fig. 13.11a), the parts of the flower are separate and distinct. Each sepal, petal, stamen, and carpel is attached at its base to the receptacle. In many flowers, however, members of one or more whorls are to

Figure 13.11 Floral symmetry. (a) Regular flower of stonecrop (*Sedum* sp.). (b) Irregular flower of garden pea (*Pisum sativum*).

T. E. Weier

a

Thomas L. Rost

b

some degree united with one another. The union of parts of the same whorl is termed **connation.** Union of flower parts from two different whorls is known as **adnation.**

The position of the ovary differs among flowers. In tulip (*Tulipa*) flowers **(Fig. 13.12a),** the receptacle is convex or conical, and the different flower parts are arranged one on top of another. The gynoecium (whorl of pistils) thus is situated on the receptacle above the points of origin of the perianth (whorls of sepals and petals) and androecium (whorl of stamens). An ovary in this position is said to be a **superior ovary.** In the daffodil (*Narcissus pseudonarcissus*; **Fig. 13.12b**), the ovary appears to be below the apparent points of attachment of the perianth parts and the stamens. This is an **inferior ovary.** In a flower with an inferior ovary, the lower portions of the three outer whorls—calyx, corolla, and androecium—have fused to form a tube, the **hypanthium,** which also is fused to the ovary. In some flowers, such as those of cherry, peach, and almond (all in the genus *Prunus*), the hypanthium does not become fused to the ovary **(Fig. 13.12c).**

Often Flowers Occur in Clusters, or Inflorescences

In many flowering plants, flowers are borne in clusters or groups known as inflorescences. An **inflorescence** is a flowering branch; some of the most common ones are illustrated in **Figure 13.13.**

T. E. Weier (a–c)

— petal

— stigma

— anther

— ovary
— filament

— sepal scar

a

— stigma

— anther
— style

— filament
— petal

— sepal

hypanthium —

— ovary

— receptacle

c

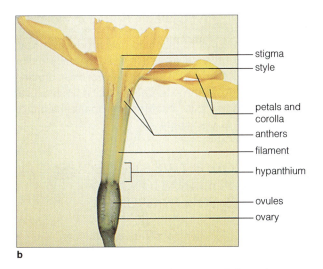

— stigma
— style

— petals and corolla

— anthers

— filament

— hypanthium

— ovules
— ovary

b

Figure 13.12 Variations in the elevation of floral parts relative to the ovary. (a) Superior ovary in the tulip flower (*Tulipa*). (b) Inferior ovary in daffodil flower (*Narcissus pseudonarcissus*). (c) In almond flower (*Prunus* sp.), the hypanthium is not fused to the ovary.

Figure 13.13 Types of inflorescence.

A simple type of inflorescence, called a raceme, is found in such plants as currant (Ribes) and radish (Raphanus). The main axis has short branches, each of which terminates in a flower; the short branches are pedicels. The oldest flowers are at the base of the inflorescence, and the youngest are at the apex. A branched raceme is called a panicle (Fig. 13.13).

In a **spike,** the main axis of the inflorescence is elongated, as in a raceme or panicle, but the flowers have no pedicels. A **catkin** is a spike that usually bears only pistillate or staminate flowers. Examples are walnut (Fig. 13.10) and willow (Fig. 13.20).

An **umbel** inflorescence has a short floral axis, and the flowers arise umbrella-like from approximately the same level (Fig. 13.13); onion (*Allium cepa*) and carrot (*Daucus carrota*) are examples.

A **head** is an inflorescence in which the flowers lack pedicels and are crowded together on a short axis. Members of the sunflower family (Asteraceae) have this type of inflorescence (Fig. 13.13).

In a **cyme,** the apex of the main axis produces a flower that involves the entire apical meristem, so that the axis itself does not elongate (Fig. 13.13). Other flowers arise on lateral branches farther down the axis. The youngest flowers in any cluster occur farthest from the tip of the main stalk; chickweed (*Cerastium*) is an example.

13.4 REPRODUCTIVE STRATEGIES: SELF-POLLINATION AND CROSS-POLLINATION

The importance of pollen to plant reproduction was first demonstrated in the 1760s by Joseph Koelreuter, who created hybrid plants by dusting the stigma on one species with the pollen of another. Three decades later, Christian Sprengel correctly distinguished between self-pollinating and cross-pollinating species and described the role of wind and insects as pollen vectors. Koelreuter and Sprengel are the founders of a field of study called *pollination ecology.*

There are essentially two different kinds of pollination. Transfer of pollen from the anther to the stigma in the same flower and transfer of pollen from one flower to another on the same plant are types of **self-pollination** or **selfing.** Genetic recombination (see Chapter 16) does not result from such crosses because only one plant is involved. The other kind of pollination, **cross-pollination** or **outcrossing,** involves transfer of pollen from one genetically distinct plant to the stigma of another. Cross-pollination results in genetic recombination, leading to genetically diverse offspring.

Various plant traits encourage the success of either selfing or outcrossing. Separation of sexes onto different individual plants ensures outcrossing. Plants that have both sexes on the same individual can prevent selfing by inhibiting pollen tube growth through the style or by inhibiting the formation of a zygote. Selfing also is prevented when the anthers mature or release their pollen some time before (or after) the stigmas on the same plant mature and are receptive to pollen.

Selfing does have advantages, even though genetic diversity is sacrificed. Selfing is a necessary means of reproduction for scattered populations in extreme habitats such as arctic tundra, where pollinating insects are few and the odds of pollen from one plant reaching another are low. Selfing also is common among plants of disturbed habitats, and it permits weedy species (for example, many grasses) to multiply and spread even if only one parent reaches a site. Selfing also saves pollen (and the metabolic energy required to produce it) because selfers generally produce fewer pollen grains per flower or per ovule than outcrossers. Selfing also increases the probability that pollen will reach a stigma because the distance traveled and the time needed for travel are so short. The time element is important because pollen is quite sensitive to low humidity and to ultraviolet rays of sunlight; consequently, pollen ordinarily has a short life span, measured in hours or days for most species.

Extreme examples of self-pollinating species have flowers that never open. They remain small and budlike, and the pollen falls from the anther sacs onto the stigma while the flower parts are still close together.

Production of Some Seeds Does Not Require Fertilization

Apomixis refers to a form of asexual reproduction in which no fusion of sperm and egg occurs. Normally, egg nuclei will not embark on the series of changes that leads to an embryo unless fertilization has occurred. Occasionally, however, an embryo does develop from an unfertilized egg. This type of apomixis is called **parthenogenesis.** Another type occurs when the embryo plant arises from diploid tissue surrounding the embryo sac. These *adventitious* embryos occur in *Citrus* and other plants.

Apomixis also has other variations, but all forms of this phenomenon involve the origin of new individuals without nuclear or cellular fusion of sperm and egg. In some plants, deposition of pollen on the stigma is a prerequisite to apomictic embryo development, even though a pollen tube does not grow down the style and nuclear fusion does not take place. In such instances, there is evidence that hormones formed in the stigma, or furnished by the pollen, move to the unfertilized, but diploid, egg cell (in this case, meiosis is omitted during embryo sac formation), initiating embryo development.

Pollination Is Affected by Vectors

Pollen may be moved by either biotic (animal) or abiotic (wind and water) vectors. The set of unique flower and pollen traits that adapt a plant for pollination by a particular vector is its **pollination syndrome (Table 13.1).**

Animal pollinators are not altruistic. They visit flowers for some reward, and only incidentally do they transfer

Table 13.1 Pollination Syndromes: Traits of Flowers Pollinated by Different Vectors

Trait	Vector							
	Beetle	Fly	Bee	Butterfly	Moth	Bird	Bat	Wind
Color	Dull white or green	Pale and dull to dark brown or purple; sometimes flecked with translucent patches	Bright white, red yellow, blue, or ultraviolet	Bright, including red and purple	Pale and dull red, purple, pink, or white	Scarlet, orange, red, or white	Dull white, green, or purple	Dull green, brown, or colorless; petals may be absent or reduced
Nectar guides	Absent	Absent	Present	Present	Absent	Absent	Absent	Absent
Odor	None to strongly fruity or fetid	Putrid	Fresh, mild, pleasant	Faint but fresh	Strong and sweet; emitted at night	None	Strong and musty; emitted at night	None
Nectar	Sometimes present; not hidden	Usually absent	Usually present; somewhat hidden	Ample; deeply hidden	Abundant; deeply hidden	Abundant; deeply hidden	Abundant; somewhat hidden	None
Pollen	Ample	Modest in amount	Limited; often sticky and scented	Limited	Limited	Modest	Ample	Abundant; small, smooth, and not sticky
Flower shape	Large, regular dishlike; erect	Funnel-like or a complex trap	Regular or irregular; often tubular with a lip; erect	Regular; tubular with a lip; erect	Regular; tubular without a lip; closed by day; pendant or horizontal	Regular or irregular; tubular without a lip; pendant or horizontal	Regular; trumpet-like; closed by day; pendant or borne on truck	Regular; small; anthers and stigmas exserted
Examples	Tulip tree, magnolia, dogwood	Skunk cabbage, philodendron	Larkspur, snapdragon, violet	Phlox	Tobacco, Easter lily, some cacti	Fuchsia, hibiscus	Banana, sausage tree, agave	Walnut, grasses

PLANTS, PEOPLE, AND THE ENVIRONMENT:
Bee-pollinated Flowers

Figure 3 The scotch broom (*Cytisus scoparius*) landing platform corresponds to bee size and shape. A bee's weight forces its petals apart, releasing the stamens. These dust the bee with pollen. A bee grooms itself and packs pollen (the orange mass) inside "baskets" of leg hairs.

Figure 4 Some orchids imitate the odor, shape, and color of insects, and they attract pollinators without offering any rewards. Species of *Ophrys* attract male bees who attempt to copulate with the flower.

Bees are among the most "intelligent" insects. Their sense of color is well developed, in contrast to beetles and flies. They are receptive to blue, yellow, white, and ultraviolet wavelengths. They cannot see red, but they will visit red flowers that also reflect ultraviolet light (Fig. 1). These colors are produced by carotenoid and flavonoid pigments in the chromoplasts (see Chapter 3) or vacuoles of petal cells. Splotches, dots, or lines of contrasting color, called nectar guides, often lead toward the sexual organs (Fig. 2).

Bees are sensitive to sweet odors and to sugary nectar, and both often are present in bee-pollinated flowers, especially during the day (Fig. 2). Odors come from organic molecules in the pollen or petals. Bees can discern the intensity of sweetness, and they favor the sweetest nectars. Unless the sugar concentration is greater than 18%, honeybees operate at a loss—that is, their expenditure of metabolic energy exceeds that of the reward.

Pollen is a major reward, and bees carry the pollen in their crop, on abdominal hairs, or in hairy "baskets" on their hind legs (Fig. 3). One insect can carry as many as a million pollen grains. A bee's proboscis (tubelike snout) is very sensitive, and petal hairs often force the insect to enter the flower by a certain path that leads it past both anthers and stigmatic surfaces.

Only recently has the importance of oil as a reward for pollinating bees been recognized. More than 2,000 species in 13 plant families produce oil instead of nectar. Their flowers are visited by specialized bees with long, densely hairy forelegs; these forelegs slice open *elaiophores* (the oil-containing equivalent of nectaries) and soak up oil like a wick. The bees also wag their abdomens over the petals to mop up oil onto body hairs. The energy content of oil—twice that of sugar—makes it a suitable food for bee larvae.

A few genera of orchids (including *Calypso* and *Ophrys*) mimic the odor, color, and shape of female bees (Fig. 4). Males are attracted, land on the flower, and try to copulate with it. In the process, wads of sticky pollen become attached to their bodies, and these are transferred to another orchid when copulation is again attempted. Some South American species of the orchid *Oncidium* vibrate in the wind and mimic male bees; these are attacked by territorial bees. The diving attack often results in pollen becoming attached to the bee and then transferred to another flower. Other orchids mimic insect prey of female wasps, which land on and sting the petals. They pick up sticky pollen in the process and take it to the next mimic. In these cases, there is no reward for the insect.

a   **b**

Figure 1 A flower that reflects ultraviolet (UV) light. Evening primrose (*Oenothera fruticosa*) is photographed with normal light-sensitive film (**a**) and with UV-sensitive film (**b**). Unlike humans, bees are able to see the reflected UV light.

a   **b**

Figure 2 Nectar guides. (**a**) In *Viola*, splashes of color and lines lead to the throat of the corolla. (**b**) *Nasturtium* flower sectioned to show guide lines and hairs leading to the entrance of the tubular spur where the nectar accumulates.

Websites for further study:

Bee Pollination of Georgia Crop Plants:
http://www.ces.uga.edu/pubcd/b1106-w.html

Insect Pollination of Cultivated Crop Plants:
http://gears.tucson.ars.ag.gov/book/index.html

Pollination Adaptations:
http://koning.ecsu.ctstateu.edu/Plants_Human/pollenadapt.html

T. E. Weier

Figure 13.15 A fly-pollinated plant. Skunk cabbage (*Lysichitum americanum*), a member of the arum family. Many small flowers give off a putrid odor. The funnel-shaped "corolla" is actually a modified leaf called a spathe.

Figure 13.14 A beetle-pollinated plant. Tulip tree (*Liriodendron tulipifera*) has the characteristic syndrome of beetle-pollinated flowers, although it may be visited by other insects as well. Some of the numerous stamens and carpels will be destroyed by the chewing behavior of pollinating beetles.

pollen. The most common rewards are pollen and nectar, but sometimes they are waxes or oils.

Because of the great cost and consequences of wasting rewards, plants have evolved various strategies to prevent rewards from being stolen by nonpollinating animals. For example, the rewards are offered in such a precise way that only pollinators are attracted or able to reach them. In more than a few cases, instead of offering any real reward, plants mimic insects or food, which tricks the pollinator into visiting its flower. In effect, such plants are parasitizing animals for their energy as carriers of pollen. When a reward is actually offered, the interaction is mutualistic—that is, of benefit to both partners.

Pollen is an excellent food for animals. Its tough outer wall resists digestion, but once it is pierced or chewed, the cytoplasm within typically provides protein (15–30% by weight), sugar (about 15%), fat (3–13%), starch (1–7%), in addition to trace amounts of vitamins, essential elements, and secondary substances. Most pollen is yellow to orange and is highly noticeable. Many pollen grains have a distinctive odor. The timing of pollen maturation can be precise, coinciding with the seasonal or daily activity of pollinators. The anthers of corn (*Zea mays*), for example, split open in the morning; those of crocus (*Crocus*) split in midday; and those of apple split in the afternoon. Certain bat-pollinated flowers release pollen only at night.

Nectar is sugary water transported by the phloem into specialized secretory structures called **nectaries.** The sugar solution from nectaries may run down into elongated petal spurs, or it may form pools at the base of the floral tube. Like pollen, it may have a limited, precise time of availability. Nec-

tar usually contains 15% to 75% sugar (glucose, fructose, and sucrose). Amino acids are present only in minor amounts, but these are significant to pollinators such as butterflies that are completely dependent on nectar for all their nutrition for a period of several months. All 13 essential amino acids for insects are present. Lipid is often a minor component as well.

BIOTIC POLLEN VECTORS Beetles, flies, bees, butterflies, moths, birds, and bats are all pollinators. Beetles are among the oldest insect groups, already in existence when flowering plants first evolved. The flowers pollinated by beetles today have many primitive traits: regular symmetry, large single flowers, bowl-shaped architecture, and many floral parts that are not fused. Beetles are clumsy fliers. The bowl-shaped flowers attract beetles by odor (Table 13.1), and they provide a large target. Beetles chew a path through the sexual organs. By possessing many stamens and pistils, flowers are sure to have some that not only survive the attack but are pollinated. Many beetle-pollinated species are tropical. The temperate-zone tulip tree (*Liriodendron tulipifera*) flower **(Fig. 13.14)** is a good example of this syndrome.

Flies are diverse, and some have evolved to mimic bees, bumblebees, wasps, and hawk moths. For this reason, there is no single syndrome of floral traits for fly pollination. However, the skunk cabbage (*Lysichitum americanum*), in the fly-pollinated arum family **(Fig. 13.15),** is unique. Their trumpet-shaped petals or bracts often have a checkered mosaic of opaque patches. In a breeze, this moving pattern, together with the epidermal hairs flicking in the wind, attracts flies. The effect may be a mimicry of many moving flies, which will attract other flies. These flowers emit the odor of decaying protein in carrion or dung. Some fly-pollinated flowers trap the insects for several hours.

Bees and butterflies are active by day, and they must have a landing platform, usually the flower petals. Moths can hover; therefore, they do not require a corolla lip. They are

Figure 13.16 *Phlox diffusa*, a butterfly-pollinated flower.

Figure 13.17 Bird-pollinated flowers. (**a**) Columbine (*Aquilegia formosa*), pollinated by hummingbirds, is a tubular type. (**b**) *Dryandra hewardiana*, pollinated by honeyeaters, is a brush type.

active by night or at dawn and dusk. Bees, butterflies, and moths harvest nectar as their reward, but the flowers they pollinate are strikingly different (Table 13.1). Butterfly flowers are vividly colored, emit faint odors, have a broad blossom rim, are erect, and exhibit prominent **nectar guides (Fig. 13.16).** These are various markings that direct the pollinator to the flower's sexual organs and source of nectar. Moth flowers are white or faintly colored, emit heavy odors that penetrate the night air, have a fringed blossom rim, are pendant or horizontal, and have no nectar guides. Moth flowers often are closed during the day. Both butterfly and moth flowers have long, narrow tubes (impossible for bees and beetles to enter) with pools of nectar at their base. (Bee pollination is discussed in the sidebar "PLANTS, PEOPLE, AND THE ENVIRONMENT: Bee-pollinated Flowers.")

Birds were not recognized by botanists as pollinators until relatively recently. Now we know that thousands of plant species are pollinated by birds in many parts of the world—by hummingbirds in North and South America, sunbirds and sugar birds in Africa and Asia, honeyeaters and honey creepers on Pacific islands, and honey parrots or lorikeets in Australia. Flowers that birds visit for nectar have strikingly similar syndromes. The flowers range from scarlet to red to orange and generally lack nectar guides. They have deep tubes usually without a landing platform, are pendant or horizontal, and have abundant nectar but emit no odor (see Table 13.1). Columbine (*Aquilegia formosa*) flowers have this syndrome **(Fig. 13.17a).** Other bird-pollinated flowers have brush-type flowers, with elongate stamens that stick out in all directions **(Fig. 13.17b).**

Most bats eat insects, but some are vegetarian; these bats have longer snouts and tongues, smaller teeth, larger eyes, and a better sense of smell. Bat flowers open at night, just as moth flowers do. They are positioned below the foliage of the parent tree-hanging pendant on a long pedicel or attached to the trunk or low limbs. Bats are color-blind; therefore, bat flowers are drab white, green, or purple. The flowers exude a strong musty odor at night, reminiscent of fermenting fluid, cabbage, or bats themselves. Bat flowers are large and tough, with lots of pollen (some have more than 1,300 anthers) and nectar (7–15 mL). Bat flowers include *Musa* (banana; **Fig. 13.18**), *Adansonia* (baobab), *Kigelia* (sausage tree), *Ceiba* (kapok), and *Agave*.

Figure 13.18 Banana (*Musa velutina*), a bat-pollinated flower.

Figure 13.19 Wind-pollinated flower. The architecture of jojoba flowers (*Simondsia chinensis*) creates vortices of air currents that bring pollen grains to the stigmatic surfaces. (Redrawn from Niklas, K.J., Buchmann, S.L. 1985. Aerodynamics of wind pollenation in *Simmondsia chinensis* (Link) Schneider. *Am J Bot* 72:530–539, with permission.)

ABIOTIC POLLEN VECTORS Wind and water both carry pollen. Typical wind-pollinated flowers are small, colorless, odorless, and lacking in nectar. Petals often are lacking or are reduced to small scales. The flowers or inflorescences are positioned to dangle or wave in the open. Trees such as walnut (*Juglans*), hazelnut (*Corylus*), and oak (*Quercus*) produce flowers before new leaves emerge in the spring. Grasses and sedges (for example, *Cyperus*) position their flowers well above the leaves so they are exposed to wind currents.

The pollen grains of such plants are generally smaller, smoother, and drier than those of animal-pollinated species. Wind-carried pollen grains are 20 to 60 µm in diameter, whereas insect-carried pollen grains are 13 to 300 µm in diameter. The pollen often changes shape from spherical to Frisbee-like on release to dry air, improving its aerodynamic form. Wind-pollinated flowers also produce more pollen grains per ovule (500–2,500,000) than animal-pollinated flowers (5–500). The volume of pollen released is tremendous. A single rye (*Secale cereale*) or corn (*Zea mays*) plant releases 20 million grains, and one plant of dock (*Rumex acetosa*) can release 400 million.

Stigmatic surfaces are enlarged and elaborate, often extending outside the flower. Architecture of the flower and the inflorescence creates vortices that trap pollen and permit the grains to settle onto stigmas at a rate greater than predicted by chance (Fig. 13.19).

To their advantage, many wind-pollinated species also are visited by insects. For example, arroyo willow (*Salix lasiolepis*; Fig. 13.20) flowers have nectaries, and bees are responsible for more than 90% of seed set. Pollen of corn and of some sedge (*Carex*) species is scented and attractive to insects. Bees are known to collect pollen from grass flowers, even though they lack nectar, color, and odor. Therefore, wind pollination and insect pollination syndromes are not mutually exclusive.

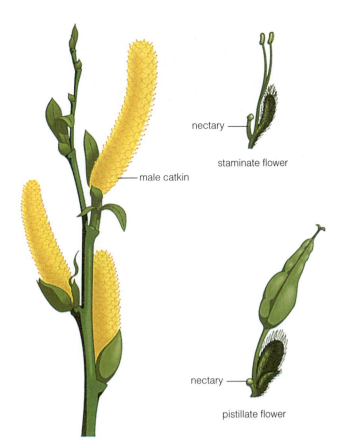

Figure 13.20 Wind- and insect-pollinated flowers. The staminate and pistillate flowers of willow (*Salix*), a dioecious plant, have typical features of a wind-pollinated plant; they lack a corolla and are grouped into catkins that appear before the leaves emerge in spring. Yet, the flowers have nectaries and the pollen grains are rather large and intensely yellow. Some species of willow rely predominantly on bees as pollen vectors, wind being relatively unimportant.

Most aquatic plants produce flowers that project above the water surface. Some of these, such as pondweed (*Potamogeton*), are pollinated by wind; others, including water lily (*Nymphaea*), are pollinated by insects. Other plants produce flowers just at the water surface, and the pollen floats from anther to stigma; ditch-grass (*Ruppia*) is an example. Only a few plants actually transfer pollen in the water, below the surface, and these use a variety of syndromes. The pollen grains of water nymph (*Najas*) are heavy and spherical and sink from the anthers into trumpet-shaped stigmas below them. In eelgrass (*Zostera*), the pollen grains are elongated and threadlike. They actively coil around any narrow object, such as a stigma, that they encounter.

KEY TERMS

adnation	apomixis
androecium	axile
angiosperms	calyx
anther	carpels
antipodal cells	catkin

central cell
complete flower
compound pistil
connation
corolla
cyme
dioecious
double fertilization
egg apparatus
egg cell
embryo sac
endosperm
female gametophyte
filament
generative cell
gynoecium
hypanthium
imperfect
incomplete flower
inferior ovary
inflorescence
integuments
irregular symmetry
locules
male gametophyte
megaspores
megasporocyte
micropyle
microsporocytes
monoecious
nectar guides
nectaries
nucellus
ovary
ovules

panicle
parietal
parthenogenesis
pedicels
perfect
perianth
petals
pistil
pistillate
placenta
placentation
polar nuclei
pollen grain
pollen sacs
pollen tube
pollination
pollination syndrome
primary endosperm nucleus
raceme
regular symmetry
self-pollination (selfing)
sepals
simple pistil
sperm cells
spike
stamens
staminate
stigma
style
superior ovary
synergid cells
tube cell
umbel
zygote

SUMMARY

1. The flower, the distinguishing structure of the flowering plants (angiosperms), is formed of four whorls of parts specialized to carry out sexual reproduction, including pollination, fertilization, and seed production.

2. The four whorls are (a) the calyx, composed of sepals; (b) the corolla, composed of petals; (c) the androecium, composed of stamens; and (d) the gynoecium, composed of carpels.

3. The stamen has two parts: the pollen-producing anther and the filament.

4. The gynoecium consists of one or more pistils. A simple pistil is made of only one carpel; if it consists of two or more fused carpels, it is a compound pistil.

5. A pistil consists of three parts: the stigma, the style, and the ovary. The stigma is receptive to pollen, and the ovary encloses the ovules. At the time of fertilization, the ovules consist of integuments, the nucellus, and the embryo sac.

6. Meiosis takes place in both the anthers and the ovule.

7. Pollen is a two-celled haploid plant or male gametophyte, protected by a cell wall, which is frequently elaborately sculptured. Two sperm cells are produced by the generative cell.

8. The embryo sac is a seven-celled haploid plant, or female gametophyte. There are two synergids, one egg, three antipodals, and a central cell with the two polar nuclei (seven cells and eight nuclei).

9. Pollination is the transfer of pollen from anthers to stigmas. A cellular recognition system regulates the germination of compatible pollen. A pollen tube that grows down the style to the ovule carries the two sperm to the embryo sac.

10. Double fertilization involves the union of one sperm with the egg (to form the zygote) and the union of a second sperm with the polar nuclei (to form the primary endosperm nucleus). This double process is unique to the flowering plants.

11. The embryo plant develops from the zygote. The endosperm develops from the primary endosperm nucleus, and it becomes the nutrient source for the developing embryo.

12. A floral apex has lost its ability to elongate and its potential for vegetative growth.

13. Species that have male and female sexes on different plants are dioecious. Species with staminate and pistillate flowers on the same plant are monoecious.

14. Variation in floral architecture arises from (a) variation in number of parts in a given whorl; (b) symmetry in floral parts; (c) connation of floral parts; (d) adnation of floral parts; (e) superior or inferior ovary; and (f) the presence or absence of certain whorls.

15. Flowers are either solitary on flower stalks or grouped in various inflorescences such as heads, spikes, catkins, umbels, panicles, racemes, and cymes.

16. There are essentially two types of pollination: cross-pollination, which involves more than one plant and results in much variation in progeny, and self-pollination, which involves a single plant and results in progeny having great similarities.

17. Apomixis is a type of asexual reproduction in which an unfertilized egg develops into an embryo.

18. Pollination occurs by means of biotic and abiotic vectors. Biotic vectors are animals that transfer pollen accidentally and secondarily; their primary objective is to obtain rewards for visiting flowers. The rewards commonly include pollen, nectar, oil, wax, and odors. The rewards cost the

plant metabolic energy to produce, but this cost is balanced against the surety of cross-pollination. The plant–animal interaction is mutualistic because there are benefits to both partners. Some flowers attract pollinators without providing any rewards; in such cases, the relationship is parasitic rather than mutualistic.

19. Animal vectors include beetles, flies, bees, butterflies, moths, birds, and bats. Each animal visits flowers with a particular syndrome of traits. The traits—shape, color, odor, timing of opening, nature of rewards offered—attract the animal and permit it to function as a pollinator.

20. Abiotic vectors include wind and water. Wind-dispersed pollen is distinctly different from animal-dispersed pollen; it is smaller, the outer wall is smoother, and the grains are not sticky. The amount of pollen produced also is much greater, and the ratio of pollen grains to ovules is orders of magnitude larger.

Questions

1. Define each of the following terms:

sepal, petal

pistil, carpel, stigma, style, ovary

stamen, filament, anther

calyx, corolla, gynoecium, androecium

2. Make a detailed diagram showing the steps that occur in the anther leading to the formation of the pollen grains (the male gametophyte).

3. Make a detailed diagram showing the steps that occur in the ovule leading to the formation of the embryo sac (the female gametophyte).

4. Make labeled diagrams of a mature pollen grain and a mature embryo sac. Show all of the cells and nuclei found in each of them.

5. Describe the process of double fertilization. What are the products of each sperm fusion?

6. Explain why a flower is actually a shortened shoot and the flower parts are modified leaves.

7. Explain the following terms: regular versus irregular flowers, complete versus incomplete flowers, and free versus united flower parts.

8. Distinguish between a simple flower and an inflorescence.

9. Explain the difference between self-pollination and cross-pollination. Is one of these strategies better than the other? Why?

10. Pollination syndromes involve pollination strategies that have evolved between particular flowers and particular vectors (for example, bees). Describe an example of a pollination syndrome involving a biotic vector and another involving an abiotic vector.

InfoTrac® College Edition

http://infotrac.thomsonlearning.com

Pollination

Janz, N., Nylin, S. 1998. Butterflies and plants: A phylogenetic study. *Evolution* 52:486. (Keywords: "phylogeny" and "plant")

Nabhan, G.P. 2001. On the nectar trail. *Audubon* 103:80. (Keywords: "nectar" and "trail")

Summers, A. 2002. Sex is in the air: Birds do it. Bees do it. Sometimes even gentle breezes do it. *Natural History* 111:70. (Keywords: "pollination" and "reproduction")

Wilhelmi, L.K., Preuss, D. 1996. Self-sterility in *Arabidopsis* due to defective pollen tube guidance. *Science* 274:1535. (Keywords: "pollen tube" and "self-sterility")

Withgott, J. 1999. Pollination migrates to top of conservation agenda (efforts of biologists bring to public attention the vital role played by migratory pollinators). *BioScience* 49:857. (Keywords: "migratory" and "pollinators")

Seeds and Fruits

Visit us on the web at http://biology.brookscole.com/plantbio2
for additional resources, such as flashcards, tutorial quizzes,
InfoTrac exercises, further readings, and web links.

1. In flowering plants, seeds are the structures containing the embryo plant for the next generation. Seeds are surrounded by a seed coat and contain the embryo axis and the cotyledons. They contain either one (monocotyledonous plants) or two cotyledons (dicotyledonous plants). Cotyledons contain stored food.

2. Germination of seeds involves the activation of processes in the embryo, such as mobilization of food reserves and starting cell division and elongation. The embryo radicle becomes the root system of the seedling plant, and the epicotyl becomes the shoot system.

3. There are several different types of seeds. Seeds have differing mechanisms and specialized structures for dispersal.

4. A fruit is a ripened ovary. There are several different types of fruits.

5. The function of the fruit is to aid in dispersal of the seeds. Several different vectors—wind, water, and animals—are involved in fruit and seed dispersal.

14.1 SEEDS

Seeds and fruits are without doubt the most important source of food for people and animals, and they always have been. Seeds and fruits are filled with stored foods intended to help the embryo germinate and grow or to attract an animal to eat the fruit and inadvertently carry the seeds away to spread them elsewhere. Rice (*Oryza sativa*), corn (*Zea mays*), and barley (*Hordeum vulgare*) grains are used for food by the majority of the people in the world. Early people recognized the nutritional value of seeds and fruits; they harvested them from wild plants, and later they figured out how to grow them for food.

Biologically, seeds are mature ovules that contain the embryonic plants of the next generation. The tremendous production of seeds ensures the renewal of plant populations. Each seed is constructed and packaged to ensure its dispersal to a favorable site for successful germination and growth. The fruit is the packaging structure for the seeds of flowering plants. This chapter discusses the structure and development of seeds and fruits and their adaptations for dispersal.

The Seed Is a Mature Ovule

The seed completes the process of reproduction initiated in the flower. After fertilization, the zygote develops into an embryo, the primary endosperm nucleus develops into the endosperm, and the integuments of the ovule develop into the seed coat.

For a short time after fertilization **(Fig. 14.1b,d),** the zygote nucleus divides frequently while the **primary endosperm nucleus** divides rapidly to form the **endosperm,** the nutrient-rich storage tissue that will feed the seed when it germinates. After the endosperm has developed, the zygote nucleus divides to form a filament of several cells **(Fig. 14.1c).** The cell farthest from the micropyle begins a series of divisions that produces the early-stage embryo or proembryo. At about the same time, the cell closest to the micropyle elongates and divides, becoming the **suspensor,** which supports the embryo in the endosperm (Fig. 14.1d). Further divisions result in a globular stage **(Fig. 14.1e)** and, finally, a heart-shaped stage, after the two **cotyledons** have developed **(Fig. 14.1f).** In addition, the embryo develops a radicle (the embryonic root) at one end and a shoot tip at the other **(Fig. 14.1g).**

The specific steps just described apply to cotton (*Gossypium hirsutum*); there are many variations in the details of embryo development for other plants. For example, the major distinction between embryos of **dicotyledonous** and **monocotyledonous** plants is the number of cotyledons (two or one, respectively).

While the embryo is developing, the **nucellus,** endosperm, and **integuments** also are undergoing changes that are characteristic of the group of plants to which the seed belongs. In the great majority of plants, the nucellus and endosperm are required only for the initial stages of embryo development. This is particularly true of the nucellus, which is generally used as a nutritive source in early embryo stages. It persists as a food storage tissue, the **perisperm,** in seeds of sugar beet and many other species. The endosperm persists as a food reserve in seeds of many monocot plants, such as onion (*Allium cepa*) **(Fig. 14.2);** these include grasses of such major economic importance as rice (*Oryza sativa*) and serious weed pests such as yellow foxtail (*Setaria lutescens*) **(Fig. 14.3).** Endosperm persists as a food storage tissue in relatively few dicot seeds, with castor bean (*Ricinus communis*) being an exception **(Fig. 14.4).**

When food storage occurs within the embryo, the normal vascular tissues of the embryo convey the solubilized food to the meristems of the emerging plant, where it is required for growth. Food stored in the endosperm, outside the embryo, is absorbed through epidermal cells of the embryo axis.

The integuments become the **seed coats** in the mature seed. Scanning electron microscopy shows the seed coats to be variously and sometimes beautifully sculptured **(Fig. 14.5).** The seed coat acts as a protective shell around the embryo and sometimes contains chemical substances that inhibit the seed from germinating until the temperature, light, or moisture conditions are exactly right for germination.

Seed Structures Vary

Seed structure varies widely among species. This means that plants have evolved many solutions to propagating themselves successfully. This chapter briefly describes, as examples of variations in seed structure, the seeds of two dicot plants—bean and castor bean—and two monocot plants—a grass and onion.

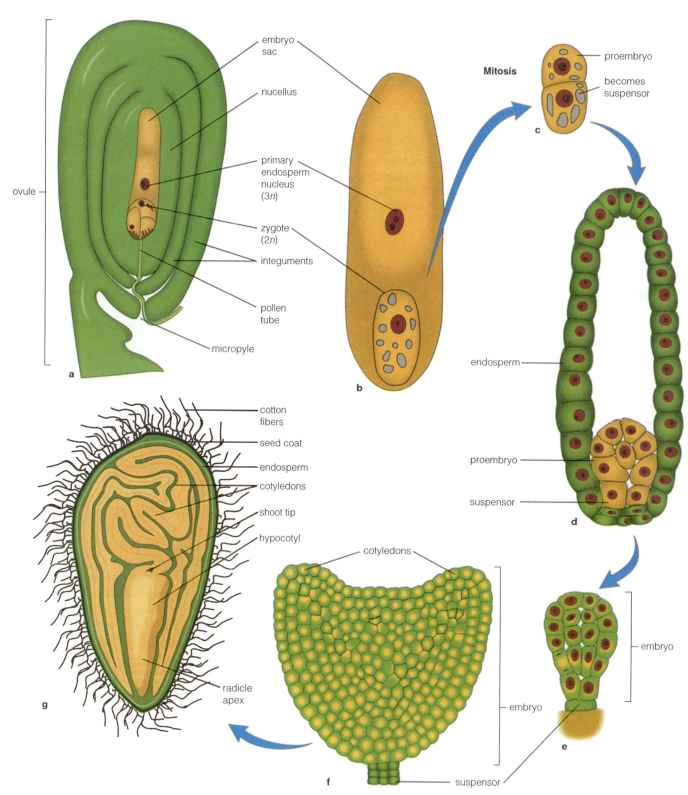

Figure 14.1 Embryo and seed development in cotton (*Gossypium hirsutum*). (**a**) Ovule after double fertilization. (**b**) Embryo sac after fertilization. (**c**) The zygote divides by mitosis; one of the two cells is destined to become the embryo and the other the suspensor. (**d**) Early stage of embryo (proembryo) and endosperm development. (**e**) Early embryo as a small globe of cells. (**f**) Heart-shaped stage of the embryo with newly formed cotyledons. (**g**) The cotton seed is the mature embryo, with highly folded cotyledons, surrounded by a seed coat. Note the seed coat fibers (really epidermal hairs) for which cotton is harvested. (a, c–e, Redrawn from Gore, U.R. 1932. Development of the female gametophyte and embryo in cotton. *Am J Bot* 19:795–807.)

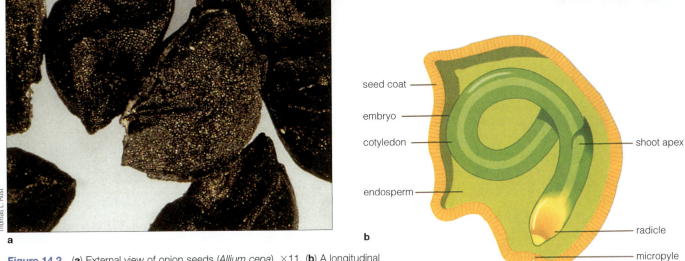

Figure 14.2 (**a**) External view of onion seeds (*Allium cepa*). ×11. (**b**) A longitudinal section through an onion seed showing the embryo coiled within the endosperm.

Figure 14.3 Caryopses (grains) of yellow foxtail grass (*Setaria lutescens*). (**a**) External views. ×17. (**b**) Median section.

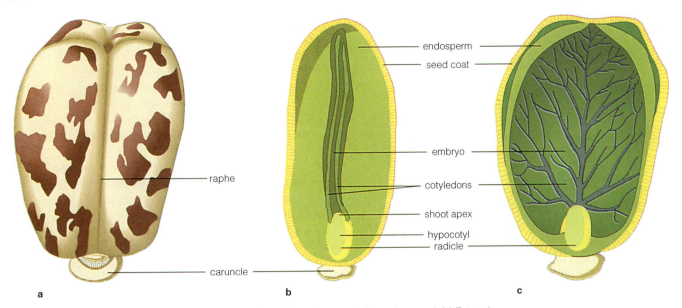

Figure 14.4 Castor bean (*Ricinus communis*) seed, a dicot with endosperm in the mature seed. (**a**) External view. (**b**) Section showing edge view of embryo. (**c**) Section showing flat view of embryo.

PLANTS, PEOPLE, AND THE ENVIRONMENT:

Doctrine of Signatures

Because seeds and fruits are so important, stories, myths, and legends abound. Even the Bible describes "forbidden fruit" in the Garden of Eden. In the 1500s, the Doctrine of Signatures was widely believed. The idea of the doctrine was that the appearance of a plant or plant part would reveal its inner secrets and possible uses to people. Others expanded this idea to include shape, color, and smell as indicators of use. The scales of a pinecone, for example, look like teeth, so medieval people made a concoction of pinecones mixed with vinegar to gargle for teeth and gum problems. (Actually, the vinegar probably did the trick by itself.) Seeds of viper's bugloss (*Echium*) resemble a snake's head, so the belief arose that a mixture made from these seeds could be used as a remedy for snakebite. Seeds of snapdragon (*Antirrhinum majus*) worn in a linen bag around the neck were supposed to prevent one from being bewitched. Some of these stories may have held a little truth simply by happenstance, but their real importance lies in that they arose in the first place because people found plants, and their seeds and fruits, important to their very survival.

 Websites for further study:

Sing Fefur Organic Herbs:
http://www.herbsorganic.co.za/pages/Doctrine%20of%20Signatures.htm

An Introduction to the Doctrine of Signatures:
http://www.holysmoke.org/wb/wb0081.htm

T. E. Weier

a

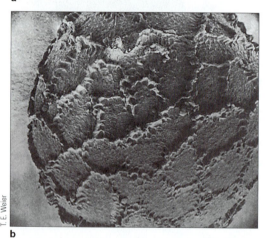

T. E. Weier

b

Figure 14.5 Sculptured seed coats. (**a**) Field bindweed (*Convolvulus arvensis*). ×19. (**b**) California poppy (*Eschscholzia californica*). ×118.

COMMON BEAN The bean (*Phaseolus vulgaris*) seed is kidney-shaped in outside view. External structures on the seed are the hilum, micropyle, and raphe (**Fig. 14.6**). The **hilum** is a large oval scar left when the seed breaks away from its placental connection, the funiculus. The **micropyle** is a small opening in the seed coat at one end of the hilum; it is the opening through which the pollen tube enters the ovule. The **raphe** is a ridge at the end of the hilum opposite the micropyle and is the base of the funiculus.

When the seed coat of a soaked bean is removed, what remains is the embryo; no endosperm is present. The bean embryo consists of two fleshy cotyledons and the *embryo axis*. The embryo axis is composed of the embryonic root or **radicle** at one end and the embryonic shoot or **epicotyl** at the other end. The **hypocotyl** is just below the cotyledons.

CASTOR BEAN The castor bean (*Ricinus communis*) seed has an external structure called the **caruncle,** which is a spongy outgrowth of the outer seed coat. The hilum and micropyle of the castor bean are covered by the caruncle, and the raphe runs the full length of the seed (Fig. 14.4). The caruncle functions in absorbing water, which is needed during germination.

The castor bean embryo is embedded in a massive endosperm. The embryo consists of two thin cotyledons, a short hypocotyl, a small epicotyl, and a small radicle. Castor bean oil was used by the Egyptians as a laxative; a paste made from the seeds also was used as a treatment for toothaches. These seeds should be handled with care, how-

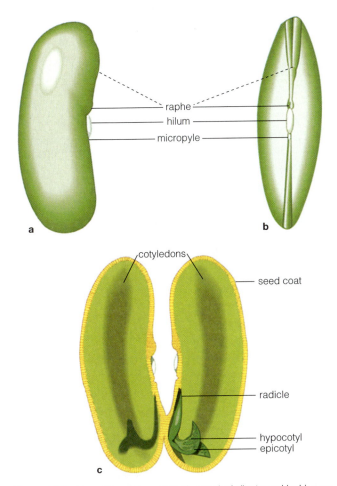

cotyledons

seed coat

radicle

hypocotyl
epicotyl

c

raphe
hilum
micropyle

a

b

Figure 14.6 Bean (*Phaseolus vulgaris*), a typical dicot seed lacking an endosperm when mature. (**a**) External side view. (**b**) External face view. (**c**) Opened to expose embryo.

ever, because they contain a toxic substance called *ricin*. If ingested, ricin causes nausea, muscle spasms, and convulsions; as few as eight seeds can cause death in people who are particularly sensitive to the seed.

GRASSES The so-called seed of grasses is really a fruit, the **caryopsis** or **grain** (Fig. 14.3). It is a one-seeded, dry fruit in which the pericarp or ovary wall is firmly attached to the seed coat and seed. The starchy endosperm constitutes the bulk of the caryopsis and is surrounded by a layer of cells, the **aleurone layer,** that contain proteins and fats but little or no starch.

The grass embryo has an axis with a shoot apex and a root apex. The shoot apex, together with several rudimentary leaves, is ensheathed by a **coleoptile.** The radicle is surrounded by the **coleorhiza.** A relatively large part of the grass embryo is a specialized, shield-shaped cotyledon called the **scutellum.** The outer cells of the scutellum secrete enzymes that digest the adjacent stored foods in the endosperm. These digested foods move from the endosperm through the scutellum to the growing parts of the embryo.

The process of milling to make polished white rice removes the caryopsis fruit coat, the entire protein-rich aleurone layer, and the outer layers of the endosperm. This means that most of the nutritious protein of the grain is removed before the rice is packaged for human consumption. The milled material, called *bran*, is now itself a popular food.

Popcorn also is the fruit of a grass plant. Popcorn is a popular food snack in the United States; Americans eat billions of gallons of it every year. Have you ever wondered why popcorn pops? Have you ever wondered why so many kernels are left unpopped at the bottom of the kettle? Scientists at the University of Illinois conducted experiments to answer these questions. Popcorn pops when heated because steam builds up inside the grain. The resistance of the fruit coat holds the steam back until the pressure becomes so great that the grain bursts. All the starch inside becomes fluffy. The "duds" do not pop because their fruit coat has cracks in it that allow the steam to escape.

ONION Like grasses, the onion (*Allium cepa*) is a monocot; but unlike grasses, its seed coat encloses only a small amount of endosperm. The embryo is simple, the radicle and single cotyledon being quite prominent. The shoot apex is located close to the midpoint of the axis and appears as a notch. The embryo is coiled, with the radicle end usually pointing toward the micropyle (Fig. 14.2).

14.2 GERMINATION

Germination, the first step in the growth of the embryo, begins with the uptake or **imbibition,** of water. This is a critical step because seeds are quite dry, containing only 5% to 10% water. The cells of dry seeds are tightly packed with stored proteins, starch, and lipids **(Fig. 14.7a).** This stored food is packaged into cytoplasmic organelles called *protein bodies*, *lipid bodies*, and *amyloplasts*, which store starch (see Chapter 3). After imbibition, enzymes are activated and rapidly released to digest the stored food into smaller molecules that can then be transported and converted into energy needed for growth. Consequently, the cells of imbibed embryos contain fewer storage organelles and more mitochondria, ribosomes, and endoplasmic reticulum, which are organelles involved in metabolism **(Fig. 14.7b).**

The first indication that germination has begun is generally the swelling of the radicle. It imbibes water rapidly and, bursting the seed coat and other coverings that may be present, starts to grow downward into the soil.

The Germination Process Differs among Plants

Although the succeeding steps of germination are essentially similar in all plants, there are variations. In the germination of beans, peas, castor beans, and onions, a structure with a sharp hook is first forced upward through the soil. The structure forming the hook is different in each case. In beans

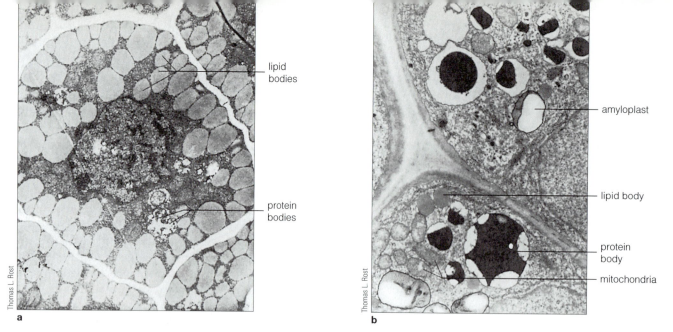

Thomas L. Rost

lipid
bodies

protein
bodies

a

amyloplast

lipid body

protein
body

mitochondria

Thomas L. Rost

b

Figure 14.7 Cellular changes brought about by germination, demonstrated in transmission electron micrographs of yellow foxtail grass (*Setaria lutescens*). (**a**) Scutellum cell in dry caryopsis before germination; note the abundance of storage organelles such as lipid bodies and protein bodies. ×8,600. (**b**) Root cell after 65 hours of germination; the cell now has fewer storage organelles and more organelles involved in metabolism such as mitochondria and leucoplasts. ×6,500.

(*Phaseolus vulgaris*), the hypocotyl elongates (**Fig. 14.8**). In peas (*Pisum sativum*), the epicotyl elongates (**Fig. 14.9**). In both cases, cotyledons and shoot apex remain below ground at first. The hook straightens once it is aboveground and is exposed to light. In the case of beans, the straightening of the hypocotyl raises the cotyledons and shoot apex toward the light. This is called **epigeal** germination. When the pea epi-

cotyl straightens, the cotyledons remain belowground, and only the apex and first leaf are raised upward. This is called **hypogeal** germination (Fig. 14.9).

The castor bean (*Ricinus communis*) has epigeal germination (**Fig. 14.10**). It is different from common beans in that its cotyledons first function as absorbing organs, facilitating the transfer of food from the endosperm to the rest of the seedling.

Figure 14.8 Stages in the epigeal germination of a bean (*Phaseolus vulgaris*) seed.

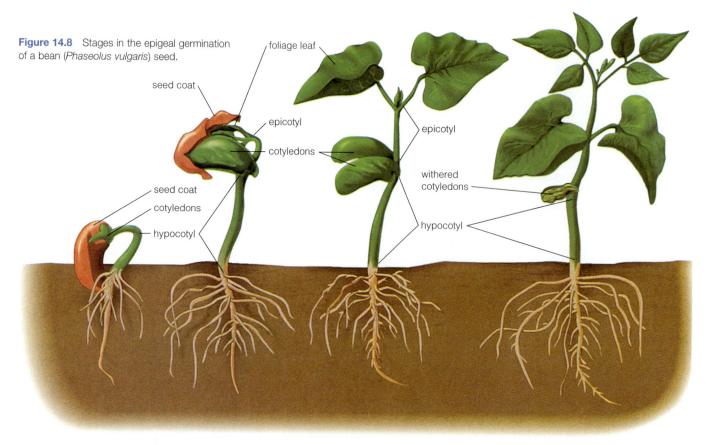

foliage leaf

seed coat

epicotyl

cotyledons

epicotyl

withered
cotyledons

seed coat

cotyledons

hypocotyl

hypocotyl

Figure 14.9 Stages in the hypogeal germination of a pea (*Pisum sativum*) seed.

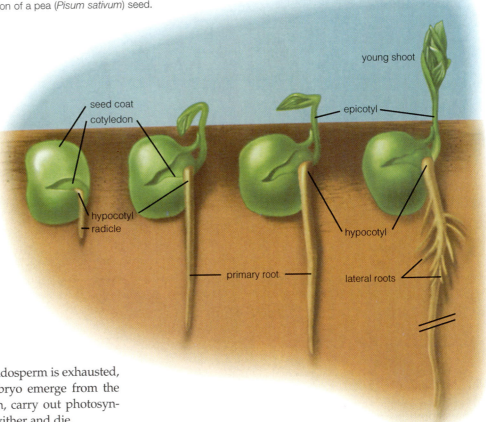

When the reserve food supply in the endosperm is exhausted, the cotyledons of the castor bean embryo emerge from the seed coat. They enlarge, become green, carry out photosynthesis for a time, and then eventually wither and die.

In the onion (*Allium cepa*), a sharply bent cotyledon breaks the soil surface and slowly straightens out. The cotyledon of the onion is tubular, and its base encloses the

Figure 14.10 Stages in the germination of a castor bean (*Ricinus communis*) seed.

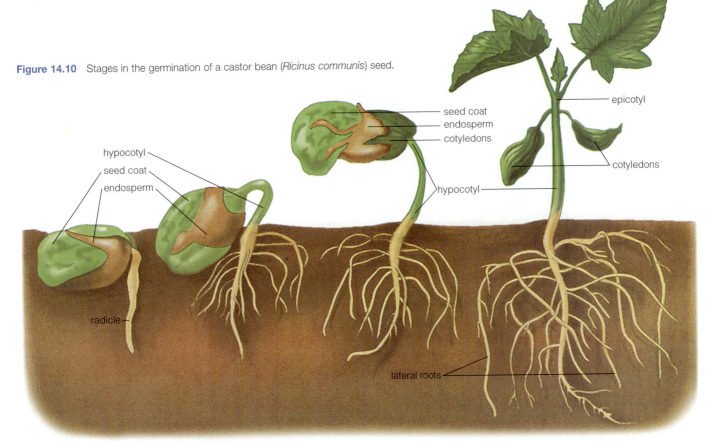

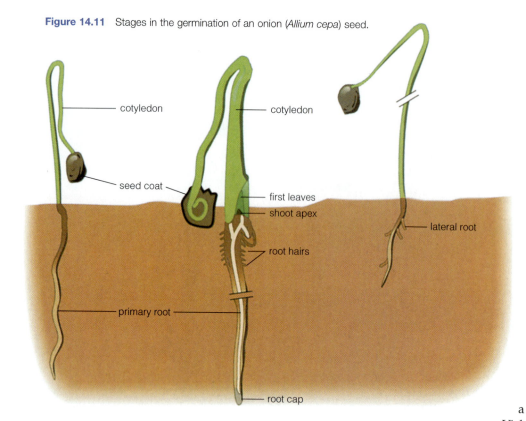

Figure 14.11 Stages in the germination of an onion (*Allium cepa*) seed.

cotyledon

cotyledon

seed coat

first leaves

shoot apex

lateral root

root hairs

primary root

root cap

tious roots then arise from the lower nodes of the stem. The coleoptile elongates and emerges aboveground, becoming 2 to 4 cm in length. At this time, the uppermost leaf pushes its way through the coleoptile and, growing rapidly, becomes part of the photosynthesizing shoot.

Germination May Be Delayed by Dormancy

Seeds can remain viable for remarkably long periods. In one study, jars containing seeds from several different plant species were buried. At 5- and 10-year intervals, the jars were opened, and the seeds were tested for germination. Most species remained viable for at least 10 years, and one species, the moth mullein (*Verbascum blattaria*), germinated after more than 100 years. This is not a record for seed longevity, however. Viable seeds from the Oriental lotus (*Nelumbo nucifera*) have been removed from archaeological sites known to be more than 1,000 years old.

shoot apex **(Fig. 14.11).** The first leaf finally emerges through a small opening at the base of the cotyledon.

In grasses such as corn (*Zea mays*), the situation is more complex. The shoot and root are enveloped by tubular sheaths of the coleoptile **(Fig. 14.12)** and coleorhiza. The primary root rapidly pushes through the coleorhiza. Adventi-

Many viable seeds will not germinate even when supplied with water, oxygen, and a favorable temperature because they are in a state of dormancy (inability to germinate because of reduced physiological activity). Various fac-

Figure 14.12 Stages in the germination of corn (*Zea mays*). After the primary root emerges, it branches to form the root system. Adventitious roots emerge from the lower stem, and prop roots form to hold the stem upright. The emerging young leaves are protected by a sheath-like coleoptile.

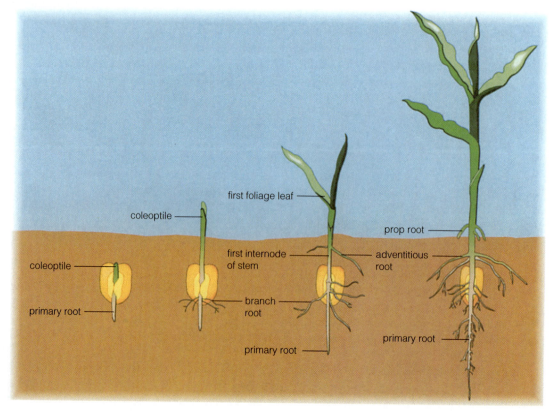

first foliage leaf

coleoptile

prop root

coleoptile

first internode of stem

adventitious root

primary root

branch root

primary root

primary root

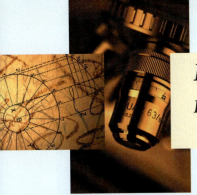

IN DEPTH:

Key to Fruits

A key is a tool to help students identify things when there are several possible choices. This key can be used to identify the different types of fruits. It is a dichotomous key, in which there are two choices at each level. For example, the first level asks if the fruit is formed from a single ovary or from several. Progress through each level in the key to determine the fruit type.

I. Fruit formed from a single ovary of one flower: *Simple fruits*
  A. Pericarp fleshy
    1. The ovary wall fleshy and containing one or more carpels and seeds: *Berry* (tomato, *Lycopersicon* sp.)
      a. Ovary wall with a hard rind: *Pepo* (watermelon, *Cucumis melo*)
      b. Ovary wall with a leathery rind: *Hesperidium* (orange, *Citrus* sp.)
    2. Only a portion of the pericarp fleshy
      a. Exocarp thin; mesocarp fleshy; endocarp stony; single seed and carpel: *Drupe* (cherry, *Prunus* sp.)
      b. Outer portion of pericarp fleshy, inner portion papery, floral tube fleshy; several seeds and carpels: *Pome* (apple, *Malus* sp.)
  B. Pericarp dry
    1. Dehiscent fruits
      a. Composed of one carpel
        i. Splitting along two margins: *Legume* or *Pod* (pea, *Pisum* sp.)
        ii. Splitting along one margin: *Follicle* (individual fruits in *Magnolia* multiple fruit)
      b. Composed of two or more carpels
        i. Dehiscing in one of four different ways: *Capsule* (*poppy*)
        ii. Separating at maturity, leaving a persistent partition wall: *Silique* (mustard)
    2. Indehiscent fruits
      a. Pericarp bearing a winglike growth: *Samara* (maple)
      b. Pericarp not bearing a winglike growth
        i. Two or more carpels, united when immature, splitting apart at maturity: *Schizocarp* (carrot)
        ii. One carpel; if more, not splitting apart at maturity; one-seeded fruits
          a) Seed united to the pericarp all around: *Caryopsis* or *Grain* (rice)
          b) Seed not united to the pericarp all around
            i) Fruit large, with thick, stony wall: *Nut* (walnut)
            ii) Fruit small, with thin wall: *Achene* (sunflower)

II. Fruits formed from several ovaries
  A. Fruits developing from one flower: *Aggregate fruit* (for example, strawberry) (classify the individual fruits in key for simple fruits)
  B. Fruits formed from several flowers: *Multiple fruit* (for example, pineapple) (classify the individual fruits in key for simple fruits)

 Websites for further study:

Identification of Major Fruit Types:
http://waynesword.palomar.edu/fruitid1.htm

Reproductive Characters:
http://www.csdl.tamu.edu/FLORA/tfplab/reproch.htm

---

tors can break dormancy. For instance, light is necessary for the germination of some lettuce (*Lactuca*) species. Scarring or breaking through the seed coat is required before the seeds of some plants (including many kinds of legumes) will germinate; their hard, dense seed coats restrict the movement of water and gases. In orchids, seeds are dispersed while the embryos are immature, and they must develop further before germinating. Seeds of some cool temperate zone plants (for instance, gooseberry [*Ribes speciosum*]) will not germinate unless, while moist, they are subjected for a time to temperatures close to freezing. At the other extreme, the seeds of some pines (*Pinus*) will not germinate unless they have been subjected to the rather high heat of fire. Another type of dormancy is produced by natural chemical inhibitors, which occur in many fruits or in seed coats—for example, yellow foxtail grass (*Setaria lutescens*).

14.3 **FRUITS: RIPENED OVARIES**

A **fruit**, the ripened ovary of a flower, is an important auxiliary structure in the sexual life cycle of angiosperms. Fruits protect seeds, aid in their dispersal, and may be a factor in timing their germination. Because fruits are highly constant in structure, even when grown in different environments, they play an important role in the classification of angiosperms.

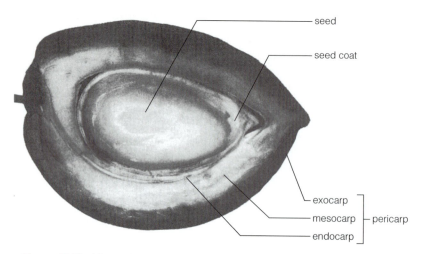

Figure 14.13 Mature almond (*Prunus amygdalus*) fruit showing split outer portion of ovary walls (pericarp)—endocarp, mesocarp, and exocarp. (Photo by T. E. Weier.)

Figure 14.14 Strawberry (*Fragaria*), an aggregate fruit. The tiny individual fruits, called achenes, are simple, dry fruits. They are embedded in an enlarged fleshy receptacle.

The sidebar "IN DEPTH: Key to Fruits" is a key to help you identify the different types of fruits described in this section.

In everyday usage, the term *fruit* usually refers to a juicy and edible structure such as an apple (*Malus* sp.), plum (*Prunus* sp.), peach (*Prunus persica*), or grape (*Vitis vinifera*). Structures that are commonly called *vegetables*, such as string beans (*Phaseolus vulgaris*), eggplant (*Solanum melongena*), okra (*Hibiscus esculentus*), squash (*Cucurbita* sp.), tomato (*Lycopersicon esculentum*), and cucumber (*Cucumis sativus*), are all fruits in a botanical sense, as are grains of corn (*Zea mays*), oats (*Avena sativa*), and other cereals.

The Nature of the Ovary Determines the Structure of the Fruit

Despite considerable variation along family lines, fruits share basic developmental and anatomical characteristics. For instance, all fruit development is initiated by fertilization, which stimulates the ovary wall to undergo development and differentiation into three layers. The fruit wall (which develops from the ovary wall) is called the **pericarp;** its three more or less distinct layers (in order, beginning with the outermost) are **exocarp, mesocarp,** and **endocarp (Fig. 14.13).** When the fruit is mature, floral structures—such as pedicel, calyx, withered stamens, style and stigma, and even remnants of the corolla—may also be present.

Tissues other than the ovary wall that form part of a fruit are referred to as accessory. Much of the fruit of pineapple (*Ammas* sp.), apple (*Malus* sp.), and strawberry (*Fragaria* sp.) can be called accessory. In the strawberry, the edible portion is a thickened, pulpy central receptacle in which achenes (dry, one-seeded fruits) are embedded.

Fruits May Be Simple or Compound

There are three main categories of fruits: simple, aggregate, and multiple fruits. Simple fruits are derived from a single ovary. They may be dry or fleshy, the ovary may be composed of one or more carpels, and the fruit may be **dehiscent** (splits open when mature) or indehiscent (does not split open). Compound fruits are composed of more than one fruit. There are two types of compound fruits: aggregate and multiple fruits. **Aggregate fruits** are derived from many separate ovaries of a single flower, all attached to a single receptacle (for example, strawberry; **Fig. 14.14**). **Multiple fruits,** such as pineapple **(Fig. 14.15),** are the enlarged ovaries of several flowers grown more or less together into a single mass. The receptacle of some multiple fruits, such as fig (*Ficus* sp.), enlarges and is actually the edible part **(Fig. 14.16).**

In classifying the different kinds of fruits, the following criteria are taken into account:

1. the structure of the flower from which the fruit develops
2. the number of ovaries involved in fruit formation
3. the number of carpels in each ovary
4. the nature of the mature pericarp (whether the fruit wall is dry or fleshy)
5. whether the pericarp splits (dehisces) at maturity
6. if the pericarp dehisces, the manner of its splitting
7. the role, if any, that accessory tissues may play in formation of the mature fruit

These criteria were used in creating the fruit key shown in the sidebar.

Simple Fruits Are from Single Ovaries

Simple fruits come in several forms. Their pericarp (fruit wall) may be dry or fleshy. If dry, the pericarp may or may not dehisce.

PERICARP DRY AND DEHISCENT The **legume** or **pod** is the type of fruit found in nearly all members of the pea family (Fabaceae). A pod arises from a single carpel, which at matu-

J. W. Perry

a

J. W. Perry

b

Figure 14.15 Pineapple (*Ananas comosus*), a multiple fruit. The individual fruits, called berries, are embedded in the swollen edible inflorescence axis. Jagged-edged bracts extend out over the fruits. (**a**) Exterior view. (**b**) Section through center showing fruits.

berry

central axis

floral parts from several separate flowers

rity generally dehisces along two sides **(Fig. 14.17).** In the pea (*Pisum sativa*) pod, the shell is the pericarp, and the pea is the seed. Pods may be spirally twisted or curved, as in Scotch broom (*Cytissus scoparius*). However, a number of legumes such as alfalfa (*Medicago sativa*) have pods that do not dehisce.

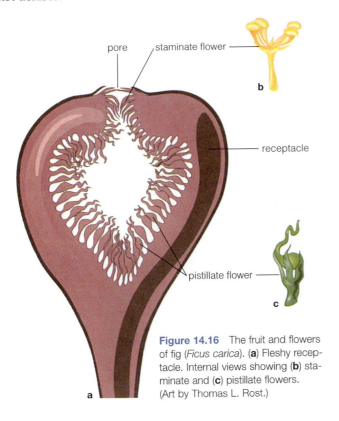

pore
staminate flower
receptacle
pistillate flower

b

c

a

Figure 14.16 The fruit and flowers of fig (*Ficus carica*). (**a**) Fleshy receptacle. Internal views showing (**b**) staminate and (**c**) pistillate flowers. (Art by Thomas L. Rost.)

Thomas L. Rost

seed
funiculus
pericarp
style
stigma

Figure 14.17 Opened pea pod (*Pisum sativum*) showing developing seeds attached to carpel margins.

Michael G. Barbour

Figure 14.18 Dehiscing follicle of a *Magnolia* sp.

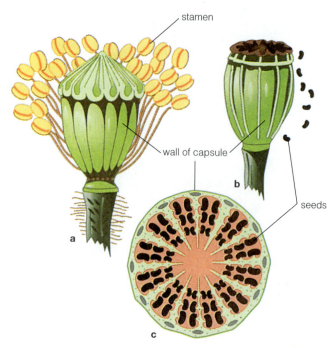

Figure 14.19 Poppy (*Papaver* sp.) capsule. (**a**) Side view before dehiscence. (**b**) Mature poppy capsule dehiscing by pores at the top. (**c**) Cross section of poppy capsule showing the position of seeds. (Redrawn from Krosmo, E. 1930. *Unkraut in Ackerbau der Neuzeit.* Berlin: Springer-Verlag. ©1930 by Springer-Verlag.)

Thomas L. Rost

Figure 14.20
Silique of *Mathiola* sp.

An example of a **follicle** fruit is the magnolia (*Magnolia grandiflora*) **(Fig. 14.18).** The follicle develops from a single carpel and opens along only one side.

Capsules are simple fruits derived from compound ovaries (an ovary composed of more than one carpel). Each carpel produces a few too many seeds. Capsules dehisce in various ways along the top surface. Poppy (*Papaver* sp.) is an example **(Fig. 14.19).**

The **silique** is the characteristic fruit of members of the mustard family (Brassicaceae). The silique **(Fig. 14.20)** is a dry fruit derived from a superior ovary consisting of two locules. At maturity, the dry pericarp separates into three portions; the seeds are attached to the central, persistent portion.

PERICARP DRY AND INDEHISCENT The **achene** is a dry, one-seeded fruit. Sunflower (*Helianthus annuus*) achenes are usually called seeds, but carefully breaking open the pericarp reveals that the seed is inside. The pericarp is easily separated from the seed coat, which is a thin, filmy layer surrounding the sunflower embryo **(Fig. 14.21).**

The **caryopsis** or **grain** is the fruit of the grass family (Poaceae), which includes rice (*Oryza sativa*) and wheat (*Triticum aestivum*). The grain is a dry, one-seeded, indehiscent fruit (Fig. 14.3). It differs from the achene in that pericarp and seed coat are firmly united all the way around the embryo.

The **samara** may be a one-seeded simple fruit, as in elm (*Ulmus* sp.), or a two-seeded one, as in maple (*Acer* sp.)

Thomas L. Rost

Figure 14.21 Achene of sunflower (*Helianthus annuus*), unopened and opened to show attachment of seed.

(Fig. 14.22). These fruits are typified by an outgrowth of the ovary wall, which forms a winglike structure that aids in seed dispersal.

The **schizocarp** is a fruit characteristic of the carrot family (Apiaceae), which includes celery (*Apium graveolens*). The schizocarp consists of two carpels that split, when mature, along the midline into two one-seeded, indehiscent halves (Fig. 14.23).

The term **nut** is popularly applied to a number of hard-shelled fruits and seeds. Botanically speaking, a typical nut is a one-seeded, indehiscent dry fruit with a hard or stony peri-carp (shell). Examples are chestnut (*Castanea* sp.) and walnut (*Juglans* sp.). An acorn, the fruit of the oak (*Quercus* sp.) (Fig. 14.24), is partially enclosed by a hardened cup. The outer husk of the walnut, which is removed during processing, is composed of bracts, perianth, and the outer layer of the pericarp. The hard shell is the remainder of the pericarp. Note that unshelled almonds (*Prunus* sp.) are really not nuts but rather fleshy fruits known as drupes, from which the hulls, exocarp and mesocarp, have been removed. Brazil nuts (*Bertholleda excelsa*) are seeds, not nuts, and the unshelled peanut (*Arachis hypogaea*) is really a pod.

PERICARP FLESHY The fruits in this category are popular for food. They feature a fleshy fruit wall (pericarp). The fleshy part usually is attractive to animals; animals eat the fruit and, in turn, carry away the seeds. The seeds found in these fruits tend to have a hard seed coat that is not broken down as the seed passes through the animal and is deposited in its feces.

Cherry, almond, peach, and apricot (all are *Prunus* sp.), in the rose family (Rosaceae), are examples of **drupes.** The olive (*Oleo* sp.; Oleaceae family) fruit also is a drupe. Derived from a single carpel, the drupe usually is one-seeded. It has a hard endocarp consisting of thick-walled sclereids (see Chapter 4), and a thin exocarp forms the skin. The mesocarp is the edible fleshy portion. The pit of a cherry (*Prunus* sp.) is a seed, with a thin seed coat, and the stony inner layer (endocarp) of the ovary wall. In almond (*Prunus* sp.) fruit (Fig. 14.13), the mesocarp is fleshy like a typical drupe when the fruit is young. As it develops, however, the mesocarp becomes hard and dry and forms the hull. The shell of the almond is endocarp. This is an instance in which the seed, not the outer part, is the edible part of a drupe.

A **berry** is a fleshy type of fruit that is derived from a compound ovary. Usually, many seeds are embedded in the flesh (Fig. 14.25), which is pericarp, although the line of demarcation may be difficult to see. It comes as a surprise to many that tomatoes, lemons (*Citrus limon*), and cucumbers (*Cucurbita* sp.) are berries, but strawberries (*Fragaria* sp.) and blackberries (*Rubus* sp.), despite their common names, are not. These are good examples of botanical names and common names that do not match.

Lemons, oranges, limes, and grapefruits (all *Citrus* sps.) are a type of berry called a **hesperidium.** Its thick, leathery rind (peel) with numerous oil cavities is exocarp and mesocarp; the thick, juicy pulp segments

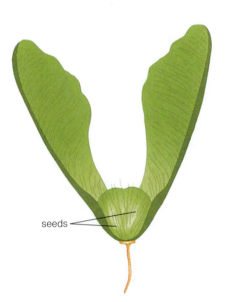

Figure 14.22 Two-seeded samara of maple (*Acer*).

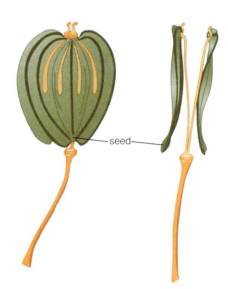

seed

Figure 14.23 Schizocarp of cow-parsnip (*Heracleum hirsutum*); left is face view, and right is side view.

a

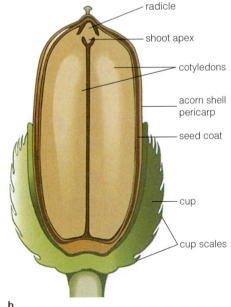

radicle

shoot apex

cotyledons

acorn shell pericarp

seed coat

cup

cup scales

b

Figure 14.24 Nut. (**a**) An acorn of oak (*Quercus*) is actually a nut; the cup is a fused bract. (**b**) Internal view of acorn, showing the embryo.

J. W. Perry

a

b   seeds

Labels on image 1: sepal, style, pericarp, placenta, locule

Figure 14.25   Berry of tomato (*Lycopersicon esculentum*). (**a**) External view. (**b**) Cross section.

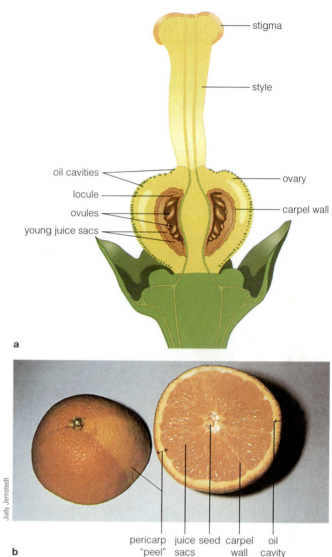

Judy Jernstedt

a

b   pericarp   juice   seed   carpel   oil
      "peel"   sacs   wall   cavity

Labels on image 2: stigma, style, oil cavities, locule, ovules, young juice sacs, ovary, carpel wall

Figure 14.26   A citrus fruit is a hesperidium. (**a**) Flower of orange (*Citrus sinensis*), showing a lengthwise section of maturing ovary. (**b**) Cross section and external view of mature fruit.

(endocarp) are composed of several wedge-shaped locules **(Fig. 14.26).** The juice forms in juice sacs or vesicles that are outgrowths from the endocarp walls. Each mature juice sac is composed of many living cells filled with juice. The fruits of watermelon (*Citrullus vulgaris*), cucumber, and squash—all members of the cucumber family (Cucurbitaceae)—are a kind of berry called a **pepo** (Fig. 14.27). The outer wall (rind) of the fruit consists of receptacle tissue that surrounds it and is fused with the exocarp. The flesh of the fruit is principally mesocarp and endocarp.

Apples (*Malus* sp.) and pears (*Pyrus* sp.) (in the Rosaceae family) are examples of **pomes.** This fruit is derived from a flower with an inferior ovary **(Fig. 14.28).** The flesh is enlarged hypanthium (a fleshy floral tube), and the core is from the ovary.

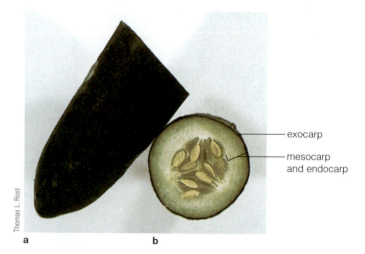

Thomas L. Rost

a   **b**

Labels on image 3: exocarp, mesocarp and endocarp

Figure 14.27   Cucumber (*Cucumis sativus*) fruit is a berry. (**a**) External view. (**b**) Transverse section.

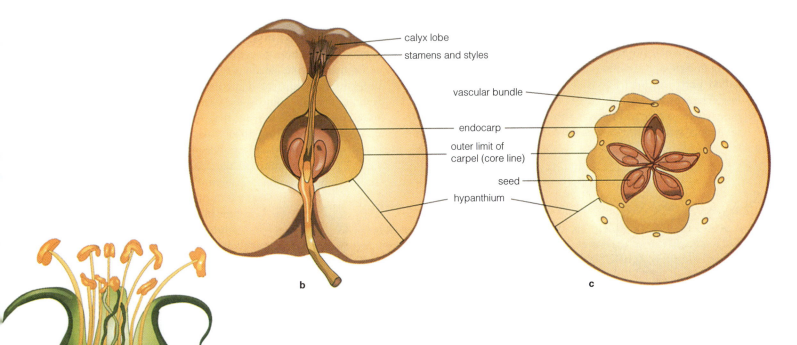

calyx lobe
stamens and styles
vascular bundle
endocarp
outer limit of
carpel (core line)
seed
hypanthium

b

c

styles
hypanthium
ovary (carpel)
ovule

a

Figure 14.28 Pome fruit of apple (*Malus sylvestris*). (**a**) Median longitudinal section of flower showing maturing fruit. (**b**) Median section of fruit. (**c**) Cross section of fruit.

Compound Fruits Develop from Several Ovaries

An **aggregate fruit** is formed from numerous carpels of one individual flower. These fruits are made up of many simple fruits attached to a fleshy receptacle. The strawberry flower has numerous separate carpels on a single receptacle. Each carpel develops into an achene (Fig. 14.14). Flowers of raspberry, blackberry, and other species of *Rubus* have essentially the same structure as strawberry, except that the attached fruits are small drupes (Fig. 14.29).

A **multiple fruit** is formed from individual ovaries of several flowers, all clumped together. The fig (*Ficus* sp.) (Fig. 14.16) and the pineapple (*Ananas comosus*) (Fig. 14.15) are examples of multiple fruits; the individual fruits composing them are drupes in fig and berries in pineapple. The fig fruit that humans eat is an enlarged, fleshy receptacle. Its flowers are small and are attached to the inner wall of the receptacle.

one aggregate fruit

drupes

J. W. Perry

Figure 14.29 Aggregate fruit of the blackberry (*Rubus ursinus*); the individual fruits are drupes.

Not All Fruits Have Seeds

In some plants, normal fruit may develop without seeds being enclosed. Fruits that develop without fertilization are called **parthenocarpic;** consequently, such fruits are seedless. Thompson seedless grapes (*Vitis* sp.) were thought to be parthenocarpic until it was shown that fertilization does take place but that the ovules fail to mature into seeds. Such situations have led to a broader use of the word *parthenocarpy* to mean, simply, seedless fruits.

Parthenocarpic (seedless) fruits are quite regularly produced in such cultivated plants as eggplant, navel orange, banana, pineapple, and some varieties of apple and pear. In certain plants, seedless fruits may be induced by pollen that is incapable of fertilizing the ovules. For example, in some orchids, placing dead pollen or a water extract of pollen on the stigma may start fruit development. Parthenocarpy is commercially induced in some plants by spraying the blossoms with dilute aqueous solutions of growth substances such as auxin.

Figure 14.30 Fruits and seeds showing various devices aiding in dispersal. Wind: (**a**) *Clematis* (×4); (**b**) dandelion (*Taraxacum vulgare*) (×5); (**c**) Coulter's big-cone pine (*Pinus coulteri*) (×1). Attachment: (**d**) cranesbill (*Geranium*) (×3); (**e**) foxtail (*Hordeum hispida*) (×3); (**f**) bur clover (*Medicago denticulata*) (×4); (**g**) fleshy edible fruit of *Cotoneaster* (×2). Violent dehiscence of pericarp: (**h**) vetch (*Vicia sativa*) (×2); (**i**) California poppy (*Eschscholzia californica*) (×2).

14.4 ADAPTATIONS FOR SEED DISPERSAL

The role of ripe fruit is threefold: to aid in the dispersal of the seeds inside, to deter inappropriate seed-dispersing animals from taking the fruit or seed, and to protect the seeds from herbivores who merely consume seeds but do not disperse them. It is important to realize that there is no nutritional relationship between the fruit and the seeds within it. That is, the stored food in the fruit cannot be used by dormant seeds or by germinating seedlings. The only stored food available to seedlings is in the endosperm and cotyledons within the seed itself.

Both seeds and fruits are rich in a variety of chemical resources, including sugar, starch, protein, lipid, amino acids, and a variety of secondary compounds. The average caloric value of this material is about 5,100 kilocalories (kcal) per gram dry weight, a value approaching that of healthy animal tissue (about 6,000 kcal/g). In contrast, leaf, root, stem, and other vegetative tissues average only 4,000 kcal per gram. This means that it costs the plant more to make seeds and fruits than it does to make vegetative organs. This expense is necessary to ensure that the materials are present inside seeds and fruits to guarantee the successful dissemination of the seeds. Several different mechanisms have developed to aid their dispersal. Fruits and seeds may catch the wind and fly (Fig. 14.30a–c), float in seawater, attract the eye and the stomach of a bird (Fig. 14.30g), entice an ant to carry a seed, or permit the fruit to hook onto the hairs of a passing mammal (Fig. 14.30d–f). Just as in pollination, the animal vectors used are sometimes rewarded and sometimes exploited—that is, the relationship is sometimes mutualistic and other times parasitic.

Wind, Water, and Animals May Disperse Seeds

Common **abiotic** vectors for fruit and seed dispersal are wind and water. Winged and plumed fruits (Fig. 14.30a–c) are common adaptations for wind dispersal. In some cases, the seeds are ballistically exploded by a violent dehiscence of the pericarp (Fig. 14.30h,i). Some sedges (*Carex* sp.) have a fruit with a membranous envelope containing air, and these are spread by floating on water. The coconut (*Cocos* sp.) is a tropical group of plants famous for growing on midoceanic islands thousands of miles from other land. The coconut fruit is capable of floating for many days and then germinating when washed onto a sandy beach and leached of salt by rainwater (Fig. 14.31a). Many other tropical beach plants have similar (although much smaller) floating fruits. A number of weed species of irrigated farmland are dispersed by water along irrigation ditches. In deserts, for example, the smoke tree (*Dalea*) and desert willow (*Chilopsis*) growing along arroyos (dry stream beds) have hard seeds that are carried away from the parent by flash floods (Fig. 14.31b). The rushing water also pushes the seeds against rocks, scraping the seed coats and scarifying them. Without that scarification, the seeds would remain dormant. Water in this case is more than a dispersal agent; it pretreats the seed and makes it receptive to germination cues.

Common animal (**biotic**) vectors include ants, birds, bats, rodents, fish, ruminants, and primates. They are attracted to fruit by its color, position, seasonal availability, odor, and taste. Sometimes, the vector eats only the fruit and discards the seeds; this is true of some primates. In other cases, the vector swallows the seeds unchewed; after passing unharmed through the gut, the seeds are excreted some distance from where they were consumed. This often is the case with birds. Birds are attracted to fleshy, colored berries and, after dining

Figure 14.31 Plants with water-dispersed fruits or seeds. (**a**) Coconut germinating on a tropical beach. The coconut fruit, or husk, is fibrous, and the seed within is large and buoyant, capable of floating hundreds or even thousands of miles in seawater. (**b**) The seeds of the desert smoke tree (*Dalea*) are carried away from the parent by flash flood waters in the adjacent arroyo.

Robert S. Boyd, Auburn University (a–b)

a

b

Figure 14.32 Elaiosomes. (**a**) Flannel bush (*Fremontodendron californicum*), a chaparral shrub of California. (**b**) Each seed of the flannel bush has an elaiosome at one end of each seed. The seeds are harvested by ants, taken to their nests, stripped of their elaiosomes, and then discarded, whereon the seeds are free to germinate without competing with the parent plant.

on them, may fly long distances before regurgitating or excreting the hard seeds. Cattle eat legume pods of mesquite (*Prosopis fuliflora*) and *Acacia* in southwestern grasslands and later pass many undamaged seeds out in their excrement. These seeds germinate, and the seedlings grow well in their fertilized microenvironment. In yet other cases, the animals eat many seeds but cache others. Squirrels and jays, for example, may carry walnuts, hickory nuts, acorns, and pine seeds from parent trees to distant hiding places. Apparently, they forget some of the caches and never revisit them; in effect, they have planted these seeds, and clusters of seedlings will emerge later.

Ants are responsible for dispersing many seeds of herbs in temperate zone forests and grasslands. They harvest the hard, small seeds and deposit them in granaries below ground. Some seeds escape consumption and germinate. Other plant species have **elaiosomes,** or food bodies, at one end of their seeds (Fig. 14.32). Ants harvest the seeds only for that reward and then toss them out. Discarded outside the ant nest (several meters from the parent plant), the unharmed seed may then germinate.

The relationships described above are mutualistic because there is some reward for the animal. Plants also can use animals in a more parasitic fashion; in such cases, there is no reward. For instance, seeds of some aquatic and marsh plants stick to the feet of birds in mud and are carried for long distances. Mistletoe (*Phoradendron* sp.), a parasite of other plants, has naturally sticky seeds, and birds can carry them on their feet to new host trees. Seeds with beards, spines, hooks, or barbs catch a ride to a new site by adhering to animal hair and human clothing (Fig. 14.30e,f).

Some Plants Have Evolved Antiherbivore Mechanisms

At the same time that the fruit attracts dispersal vectors, it must repel herbivores. Mechanisms to discourage herbivores include reducing the time of fruit availability, making the fruit or seed coat physically hard, and making the fruit or endosperm chemically repellent.

Many perennial plant species do not reproduce every year, or at least the magnitude of their reproduction varies from year to year. Such species produce fruit and seed abundantly only during what are called mast years. Because food supply is a limiting factor in population size, the relatively low amount of seed produced in off years keeps the number of seed eaters in check. As a result, seed-eating mammal, bird, and insect populations are not large enough to consume all the seeds available during a mast year; therefore, some seeds escape consumption and germinate.

Those species of plants that do reproduce every year often limit the time when ripe fruit is available. Large fruits that require a long time to develop remain green, hard, and relatively small until just before the final maturation stages. Then color, texture, size, and sweetness change suddenly, and the fruit is available to herbivores for only a brief time. This limits the number of fruits (and seeds) they can consume before dispersal.

Our common notion of fruit is a juicy, soft organ, but many plants produce fruits that are partly or completely dry and hard. Examples include drupes, which have a hard endocarp, and such completely hard fruits as nuts. Biting or boring insects are prevented from invading the fruit or seed by the sclerenchyma tissue (usually sclereids; see Chapter 4), which also prevents the seed from being damaged by the grinding action in the crops of birds or the mouths of chewing mammals. Legume seed coats are notoriously hard and often pass through animal guts unharmed.

Chemical protection mechanisms are widespread and diverse. Many fruits are rich in secondary compounds, which are chemicals produced by a plant partly or entirely for the effect they have on other organisms. In the case of herbivore defense, the effect is negative and often toxic. Secondary compounds in fruits and seeds include lectins (which cause red blood cells to clump), enzyme inhibitors,

Table 14.1 Legume Defenses and Seed Weevil Adaptations

Plant Defense	Weevil Adaptation
Gum production by seed pods after penetration of first larva from egg mass; this may push off remaining eggs or drown or otherwise obstruct young larvae.	A period of quiescence in embryonic development in the egg until seed maturation is completed. Eggs laid singly instead of in clusters.
Dehiscence, fragmentation, or explosion of pods, scattering the seeds to escape from larvae coming through the pod walls.	Eggs laid on seeds only after they have been scattered. Attachment of seeds to one another or to the pod valve.
Production of a pod free of surface cracks, which prevents eggs from adhering. Also, indehiscent pods, excluding weevils, which lay eggs only on exposed seeds.	Eggs laid into the soft, fleshy exocarp. Chewing holes in the pod walls, in which eggs are then laid. Gluing the eggs to the pod surface.
A layer of material on the seed surface that swells when the pod opens and detaches the attached eggs.	Attachment of the eggs by anchoring strands to allow for substrate expansion. Entry of the larva through the pod into the seed before the pod opens.
Production of allelochemic substances.	Mechanisms for avoidance and detoxification.
Flaking of the seed pod surface, which may remove eggs laid on that surface.	Eggs laid beneath the flaking exocarp. Rapid embryonic development or feeding on immature seeds so that entry occurs before flaking begins.
Immature seeds remaining small throughout the year and then abruptly growing to maturity just before being dispersed.	Entry and feeding on immature seeds. Delay of embryonic maturation until seeds are mature.
Seeds so small or thin that weevils cannot mature in them.	Use of seed contents more completely. Feeding on several seeds.

From Center, T.D., Johnson, C.D. 1974. Coevolution of some seed beetles (Coleoptera: *Bruchidae*) and their hosts. *Ecology* 55:1096–1103, with permission. Copyright ©1974 by the Ecological Society of America.

cyanogens (which release cyanide, a potent nerve toxin), saponins (a detergent), alkaloids (such as opium), and unusual amino acids. When present in the endosperm, these secondary compounds may later have a primary function as well; that is, they may be metabolized by the seedling into valuable nontoxic resources.

During the course of evolution, it appears that the metabolic quirks of at least one species of animal have successfully defused each chemical defense originated by seeds and fruits. This is referred to as *coevolution*. Thus, there are specialized insects capable of eating plant tissue toxic to nearly every other animal. In some cases (for example, the larvae of the monarch butterfly, which is able to feed on milkweeds rich in metabolic byproducts called cardiac glycosides), the animal not only eats the plant but uses the toxin for its own benefit: to deter a predator.

An excellent example of plant–herbivore coevolution is the series of plant defenses and herbivore responses summarized in Table 14.1 for a group of closely related tropical legumes. The plants exhibit a range of herbivore defenses including chemistry, texture, size, and timing. For each defense, however, some species of weevil have evolved a solution.

Distant Dispersal of Seeds Is Not a Universal Aim

The benefit of fruit and seed dispersal is the spread of a species far from its parent. There is a cost, however. Many fruits and seeds are wasted because they are eaten and deposited in places inappropriate for successful germination and seedling establishment. In some stressful habitats, only a few safe sites exist, and these are scattered within a large, hostile area. Because parent plants ordinarily already reside within one of the safe sites, it is advantageous to prevent or limit dispersal away from the parents.

One method of limiting dispersal is self-planting. The morphology of the fruit lends itself to lodging near the parent and drilling into the ground. Many grasses have long, bent *awns* (slender bristles) that twist as air humidity fluctuates. The awns function as levers that drive the grain into the soil. Stiff hairs at the base of the grain prevent it from pulling backward, out of the soil. A few nongrass herbs, such as cranesbill (Fig. 14.30d), use a similar technique.

The peanut (*Arachis hypogaea*) inclines its fertilized flowers down to the ground, and the fruits become buried as they mature. Seeds never leave the immediate proximity of the parent.

Sea rocket, a common annual beach plant of temperate zone shores, has a bipartite fruit. The top half is easily dislodged when mature, and its corky texture allows the enclosed seed to float and be carried to distant beaches through ocean currents. The bottom half is firmly attached to the parent, and its seed is buried with the dead parent by shifting sand at the end of the growing season (Fig. 14.33). The following year, hundreds of seedlings mark the place where the parent plant grew the year before. This two-pronged dispersal strategy serves both to spread the species and to maintain it in safe sites year after year, even though it is an annual plant.

Plant species of isolated islands, when compared with close relatives that grow on distant continents, often exhibit a loss of dispersal mechanisms. It is possible that they

a
b

Figure 14.33 Two-pronged seed-dispersal strategy. (**a**) Sea rocket (*Cakile maritima*) is a common succulent annual plant along many coastlines. Note the immature green fruits along some stems. (**b**) Diagram of the fruit. Each part has one seed. The top half is carried away by ocean currents, and the bottom half stays with the parent. (Photo by Michael G. Barbour.)

evolved on continents first and then by rare chance were spread to islands, where the process of evolution modified their seeds and fruits so that dispersal became limited, in keeping with the limited size of the islands.

KEY TERMS

abiotic	hypocotyls
achene	hypogeal
aggregate fruit	imbibition
aleurone layer	integuments
berry	legume
biotic	mesocarp
capules	micropyle
caruncle	monocotyledonous
caryopsis	multiple fruit
coleoptile	nucellus
coleorhiza	nut
cotyledons	parthenocarpic
dehiscent	pepo
dicotyledonous	pericarp
drupes	perisperm
elaiosomes	pod
endocarp	pomes
endosperm	radicle
epicotyl	raphe
epigeal	samara
exocarp	schizocarp
follicle	scutellum
fruit	seed coats
grain	silique
hesperidium	suspensor
hilum	

SUMMARY

1. A seed consists of a plant embryo surrounded by a seed coat. Seeds may store food within or outside the embryo. In most dicotyledonous plants such as beans, food is stored in the two cotyledons. Cotyledons may also serve as absorbing and, later, as photosynthesizing organs. Food may be stored in an endosperm rather than in the cotyledons.

2. In seeds of monocotyledonous plants, food usually is stored in an endosperm. In grasses, such as corn, the single cotyledon-like structure (scutellum) is a specialized organ that absorbs the nutrients from the endosperm. In seeds such as those of onion, the cotyledon emerges from the seed coat and becomes green, but its tip continues to absorb food from the endosperm.

3. The first step in germination is the imbibition of water. This water facilitates the activation of enzymes involved in digesting stored food, which is converted to energy for growth.

4. In germination, the cotyledons may be elevated above the ground (epigeal), sometimes becoming photosynthetically active, or they may remain below the ground (hypogeal).

5. Mature seeds may be dormant and, depending on the species and the immediate environment, may remain viable and dormant from a few months to many years.

6. Dormancy usually is broken by providing the seed with moisture, oxygen, and a favorable temperature. Other factors, such as light, the removal of chemical inhibitors, or the destruction of the seed coat, may be required in some instances.

7. A fruit is a ripened ovary and other closely associated floral parts.

8. There are three different kinds of fruits, classified on the basis of the number of ovaries and flowers involved in their formation:

   a. simple fruits—derived from a single ovary

   b. aggregate fruits—derived from a number of ovaries belonging to a single flower and on a single receptacle

   c. multiple fruits—derived from a number of ovaries of several flowers more or less grown together into one mass

9. Simple fruits may have either a dry or a fleshy pericarp. If the pericarp is dry, it may be either dehiscent (splitting at maturity to allow seeds to escape) or indehiscent (not splitting).

10. The role of the fruit is to aid in the dispersal of the seeds within and to deter herbivores from eating the seeds without dispersing them. There is no nutritional link between the fruit and the seed of the germinating seedling.

11. Vectors of fruit and seed dispersal include wind, water, and animals. Common animal vectors include ants, birds, bats, rodents, ruminants, and primates. The animals sometimes are rewarded with food for their dispersal activities, but at other times they simply carry seeds and fruits to other sites.

12. Fruits protect seeds from herbivores by the timing of their ripening, their hardness, and their chemical composition. Fruits may contain secondary compounds that deter feeding by all but the most metabolically specialized herbivores.

13. Some fruits prevent or limit dispersal, thus ensuring that the site occupied by the parent plant will be occupied by its offspring well into future growing seasons.

Questions

1. A seed is actually a mature ovule. Define each of the following terms:

integuments

seed coat

suspensor

embryo

cotyledon

hilum

raphe

micropyle

radicle

epicotyl

2. Describe the processes that occur during germination.

3. What is seed dormancy? Why is it an important process?

4. A fruit is a ripened ovary. What are the functions of fruits?

5. What are the differences among simple, aggregate, and multiple fruits?

6. Buy several different fruits at the grocery store and use the information provided in this chapter (see "IN DEPTH: Key to Fruits") to identify their fruit types.

7. One role of fruits is to aid in seed dispersal. Describe two different ways that fruits aid in dispersal by abiotic factors and two different ways that fruits aid in dispersal by biotic factors.

InfoTrac® College Edition

http://infotrac.thomsonlearning.com

Seed Structure

Brown, K. 2001. Patience yields secrets of seed longevity. *Science* 291:1884. (Keywords: "seed" and "longevity")

Comis, D. 1999. Winterfat seeds take ice stakes through the heart. *Agricultural Research* 47:24. (Keywords: "winterfat" and "seeds")

Seed Germination

Fischer, T. 1992. Seed germination: Theory and practice (book review). *Horticulture, The Magazine of American Gardening* 70:85. (Keywords: "Fischer" and "germination")

Keeley, J.E., Fotheringham, C.J. 1997. Trace gas emissions and smoke-induced seed germination. *Science* 276: 1248–1250 (Keywords: "smoke" and "germination")

Seed Dispersal

Barlow, C. 2001. Ghost stories from the Ice Age (methods of seed dispersal that date back to Ice Age). *Natural History* 110:62. (Keywords: "Barlow" and "dispersal")

Control of Growth and Development

Visit us on the web at http://biology.brookscole.com/plantbio2
for additional resources, such as flashcards, tutorial quizzes,
InfoTrac exercises, further readings, and web links.

1. The development of a plant includes morphogenesis, the production of new organs of defined shapes, and differentiation, the creation of specific characteristics in a cell or tissue. Morphogenesis involves cell division and expansion; differentiation involves the controlled expression of genes at particular times and places.

2. Hormones are chemical signals that coordinate the development of an organ with what is happening in other parts of the plant and with environmental conditions. The well-characterized plant hormones include auxins, gibberellins, cytokinins, ethylene, and abscisic acid.

3. Light influences plant development through different receptors that detect different colors of light. Phytochromes, which detect red and far-red light, influence the timing of seed germination, the greening of seedlings, and internode elongation in adult plants. They also are involved in the plant's mechanism for measuring day length.

4. Because changes in the relative lengths of day and night forecast changes in seasons, plants can control developmental processes in a way that adapts them to coming seasons by measuring the length of day and night.

15.1 PRINCIPLES OF PLANT DEVELOPMENT

A plant is a whole organism, not a collection of independent cells. As it grows and matures, tissues and organs develop in predictable patterns. Some depend only on internal conditions (governed by heredity, their position in the plant, or the age or size of the plant), and some respond to the external environment. The shapes of leaves and their arrangement on the mature stem, for instance, form a fairly constant set of inherited traits, which is similar among plants of a species no matter where they grow. The length of a stem and the direction in which it grows, in contrast, may depend more on the light available in its environment than on heredity. For either hereditary or environmentally influenced traits, it is difficult to imagine that the patterns of growth and development could occur without signals—signals that somehow communicate to the constituent cells what is happening throughout the plant and outside it. This chapter describes some of these signals and their effects, and it outlines some of the recent research on how these signals control growth and developmental processes.

What Are the Processes of Development?

The term *development* includes many events that occur during the life cycle of a plant. It includes *growth*—cell division and enlargement—and *differentiation* in both shoot and root. The formation of storage tissues and the events that lead to the mobilization of storage materials when they are needed—for example, when a seed germinates or when a tree's buds start to grow in the spring—also are considered to be developmental processes. Development also includes the induced changes that allow cells to resist herbivory by insects or infection by pathogenic microorganisms, and it includes the distinctive growth patterns that distinguish juvenile, adult, and flowering shoots (Fig. 15.1).

Some of these events reflect *morphogenesis*—developmental changes that lead to the formation of specific shapes such as the cylindrical shape of a shoot or root, the flat shape of a leaf (perhaps with a sculpted outline), or the specialized shape of flower petals. Because plant cells are attached to their neighbors by common cell walls, they cannot move around within the plant. This means that the shapes of new organs are determined by the directions in which cells divide.

Other events represent differentiation—any process that makes cells functionally specialized and different from one another. Differentiation often occurs through the expression of genes. Gene expression includes all those steps by which genes direct the production of proteins. Many proteins, in one way or another, make a cell distinctive. For instance, the chlorophyll-binding proteins of photosynthetic cells, enzymes that make pigments in the cells of flower petals, and carriers in root cell plasma membranes that take up mineral nutrients give their cells special traits.

Plants Have Indeterminate as Well as Determinate Growth Patterns

The pattern of growth in plants differs in a fundamental way from that in animals. In most animals, the pattern of growth is **determinate,** which means "having defined limits." In the context of developmental biology, it means that the cells formed from the zygote proceed through a predictable series of cell divisions, movements, and differentiation processes. At the end of the series, cell division and differentiation stop, and the result is the final, adult organism.

In plants, the growth of shoots and roots is **indeterminate,** that is, the shoots and roots will continue to grow until stopped by an environmental or internal signal. The ability to divide is maintained among the cells in the meristem (see Chapter 4) and is not lost at any predictable point. The cells of the meristem divide continuously, producing new cells. Half of these stay with the meristem, and half become part of the plant body, dividing a few times, enlarging, and then differentiating into the various tissues. The meristem cells do eventually lose the ability to divide, but they do not have an inherent, limited number of divisions after which they must stop.

Not all plant growth is indeterminate. Dicot leaves and organs that are formed from modified leaves—such as bud scales, bracts, petals, sepals, stamens, and carpels—all show a limited, determinate growth pattern, which varies from species to species.

One important implication of the indeterminate growth pattern of plants is that much of development occurs by the

Figure 15.1 Events in the growth and development of plants. (**a**) An embryo dissected from a seed of mouse-eared cress (*Arabidopsis thaliana*). (**b**) Spring bud break in a London plane-tree (*Platanus acerifolia*). (**c**) Juvenile and mature *Eucalyptus* shoots. The blunt, oval leaves (*top left*) mark a juvenile shoot; the long, sharp leaves (*right*) denote a mature shoot. (**d**) Induction of flower shoots in an orchid (*Masdevallia veitchiana*). (**e**) The brilliant color of a senescent *Liquidambar styraciflua* leaf in the fall. Each picture represents a new or changed phase of growth and development.

repetition of a small number of "programs." A module of the shoot—which includes the leaf, lateral bud, and internode (see Chapter 5)—might result from one of these programs. The instructions for forming leaves, starting branch shoots, and growing secondary tissues might be thought of as subroutines within the module program.

Determination and Competence Reveal Stages in Differentiation

Developmental biologists have demonstrated that cells go through a series of stages on their way to becoming mature, differentiated components of an adult organism. This is especially clear in the embryonic development of an animal. The first cell in the developmental pathway, the zygote, is totipotent. This means that it has the capability of making, through cell division and the other processes of development, all the cells in the future organism. The two cells resulting from the first division may also be totipotent, meaning that each one could make a complete organism if it

were separated from the other and incubated under the appropriate conditions. However, after three or four divisions, the cells formed are no longer totipotent; separated from the others, they might form only a few tissues, or they might die. These cells are said to be **determined,** which means that their potential to differentiate is limited. (The term *determined* should not be confused with *determinate* or *indeterminate,* as previously defined.)

A plant zygote, like an animal zygote, is totipotent. To a degree, plant cells may also become determined. During the formation of the embryo from the zygote, certain cells become shoot cells, and others become root cells. The formation of a shoot or root meristem is a first step in determining the course of further development. It is unusual for cells in a shoot meristem to make rootlike structures, and vice versa; thus, plant cells are at least partially determined. Then, during the primary growth of the shoot, the cells just below the apical meristem become parts of the three primary meristems: the protoderm, ground meristem, and procambium. Cells in these three primary meristems are even

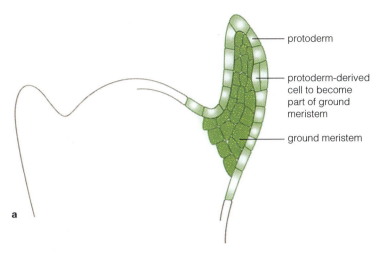

— protoderm

— protoderm-derived cell to become part of ground meristem

— ground meristem

a

b

T. M. Murphy

Figure 15.2 An example of reversible determination in plant cells. The protoderm and epidermal cells of English ivy (*Hedera helix*) lack the ability to make chlorophyll, but the ground meristem and its derivatives in a leaf produce chlorophyll normally. When a protoderm cell divides so that one of its resulting cells becomes part of the ground meristem (**a**), mesophyll in the section of leaf produced from that cell will lack chlorophyll and be light green or white (**b**). The cell that became part of the ground meristem lost its determination as protoderm.

more determined, in that they have started to show characteristics that distinguish each from the other and that suggest they are not interconvertible.

The determination of plant cells often is reversible, however. For instance, adventitious roots form from shoot tissue (see Fig. 7.5) and shoots can form from roots, an indication that the determination of cells as shoot or root cells may be changed. Sometimes, protoderm cells divide so that the new cell wall is parallel to the surface of the apex (forming inner and outer cells). Then, the inner cells become part of the ground meristem and produce leaves (Fig. 15.2), an indication that the protoderm cells lose their determination as presumptive epidermal cells and form mesophyll cells. Also, removing pieces of tissue from shoots or roots and placing them in culture conditions can lead to the formation of whole plants (the conditions needed for this are discussed later in this chapter). This means that at least some cells within the mature shoot or root tissue—cells that have differentiated—can regain their totipotency.

Just as determination in a developmental program channels cells toward certain possible fates and away from others, other processes prepare cells for differentiation. This idea comes from the observation that some differentiation processes occur after the cell has received a stimulus. For example, mesophyll cells produce chlorophyll only after being illuminated, and procambial cells produce secondary cell walls after being stimulated by sucrose. However, not all cells respond to stimuli in the same way. Procambial cells do not turn green when exposed to light, and palisade parenchyma cells do not make secondary cell walls in the presence of sucrose. The cells that do respond appropriately must have gone through preparatory steps that make them competent to respond to the stimulus.

Gene Expression Controls the Development of Traits

The processes of differentiation, by which cells gain properties that allow them to play specialized roles in the life of the organism, depend on the expression of genes. To explain how this works, we must describe how the information in genes is used to direct the synthesis of enzymes and other proteins. Recall that genetic information is encoded in the sequence of bases in DNA (see Chapter 2). Chromosomes containing the DNA are passed to each daughter cell by mitosis every time a cell divides (see Chapter 3).

There are several steps and many components to the process by which information locked in the sequence of bases in DNA produces the enzymes needed for a cell's life and growth. The process is amazing both for the simplicity of the basic principles behind it and the complexity of their implementation. An early step is to *transcribe* the genetic code of DNA onto an RNA molecule (Fig. 15.3). Transcription involves using the base sequence of a section of DNA as a template. The two strands of DNA separate, and an enzyme moves along one of the strands, assembling an RNA molecule with a base sequence that is complementary to that of the DNA strand. *Complementary* means that at each position the RNA base fits with the base on the DNA: A fits to T, C to G, G to C, and U to A. Thus, the base sequence of the DNA template specifies the base sequence of the RNA strand that is produced.

There are several types of RNA synthesized, each with its own base sequence specified by different sections of the template DNA. One type is *ribosomal* RNA (rRNA). Three separate rRNAs, in combination with several proteins, form the basic machinery (ribosomes) for making proteins. A second type is *transfer* RNA (tRNA). Transfer RNA serves as a decoding molecule, translating a base sequence into an

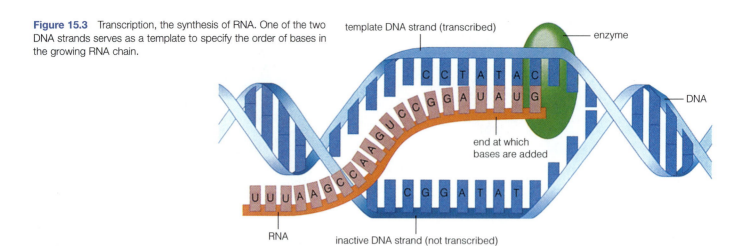

Figure 15.3 Transcription, the synthesis of RNA. One of the two DNA strands serves as a template to specify the order of bases in the growing RNA chain.

template DNA strand (transcribed)

enzyme

DNA

end at which bases are added

RNA

inactive DNA strand (not transcribed)

amino acid sequence. The RNA molecules that specify the amino acid sequences of particular proteins are called *messenger* RNAs (mRNAs). In plant cells, mRNAs carry the code (message) for the protein from the nucleus, where the genetic information is stored, to the cytoplasm, where the protein is synthesized.

The mRNA is translated to make a protein by interacting with ribosomes and tRNAs **(Fig. 15.4).** In translation, the ribosomes bind to the mRNA and then move along the mRNA three bases at a time while binding the appropriate tRNAs. A sequence of three mRNA bases is called a *codon.* There are 64 different codons, each representing a particular

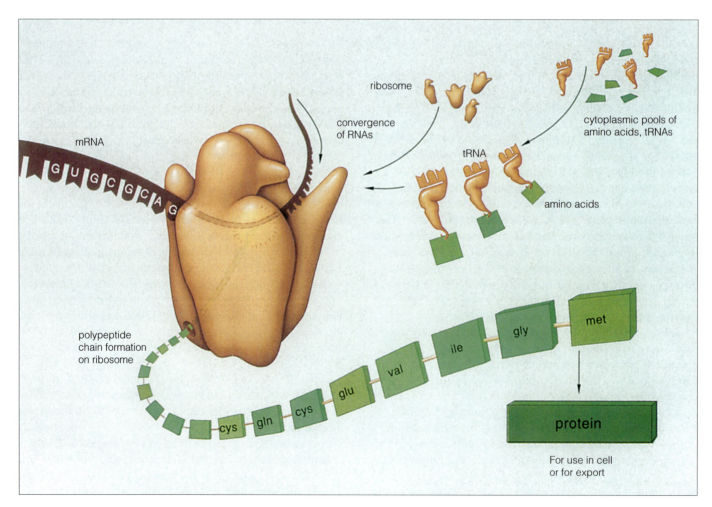

ribosome

convergence of RNAs

cytoplasmic pools of amino acids, tRNAs

mRNA

tRNA

amino acids

polypeptide chain formation on ribosome

cys gln cys glu val ile gly met

protein

For use in cell or for export

Figure 15.4 Translation, the synthesis of a protein. The ribosome serves as the central point where transfer RNA (tRNA) molecules match amino acids to the sequence of codons on the messenger RNA (mRNA). Through enzymatic action by the ribosome, the amino acids are joined to form a polypeptide chain.

amino acid, except for three codons that are "stop" signals. When the ribosome reaches a particular codon for an amino acid, it finds a tRNA with the complementary set of bases, known as the *anticodon*. This tRNA is carrying the amino acid specified by the codon on the mRNA. The ribosome connects the amino acid of the tRNA to the preceding amino acid with a peptide bond. More than one ribosome may work in this fashion on one mRNA, each ribosome forming one polypeptide chain. In some cases, the polypeptide chain automatically coils into the three-dimensional structure specified by its amino acid sequence. However, some polypeptide chains must be modified before they become active. Whatever the modifications, it is the DNA that provides the main information for producing each enzyme and other functional proteins in every cell.

An example of the use of genetic information to make proteins is seen in the production of red and purple pigments in the petals of petunia (*Petunia* sp.) flowers. These pigments, called *anthocyanins*, are made in petal cells by a sequence of chemical reactions, each of which is catalyzed by an enzyme protein. The structure of each enzyme is specified by its own DNA template. The sequence of enzymes modifies the pigment molecules to make them absorb light more effectively. If a gene is mutated and an enzyme late in the sequence is missing, the flower will have a less intense color (Fig. 15.5).

One particularly important enzyme early in the sequence is called *chalcone synthase*. Chalcone synthase is coded by the *Chs* gene, which is a segment of DNA on a chromosome in the nucleus of the cells. The expression of the *Chs* gene involves a long sequence of steps, the first of which is activating the DNA in the portion of the chromosome containing its gene. Inactive parts of chromosomes are wound around proteins to form nucleosomes and more complex structures that fit into the nucleus. Activation may involve the removal of nucleosome proteins and unwinding of the DNA. The exact steps are still somewhat mysterious, but they may occur during the early development of the flower (rather than just before the synthesis of the enzyme). It is possible that the activation step is one of the processes leading to determination of the petal cells.

The next step is transcription of the *Chs* gene. As noted earlier, this involves the enzyme RNA polymerase and other proteins, which form a transcription complex, a structure that starts the process of RNA synthesis. Some of these proteins are always present in the cell; some may be synthesized specifically to start the transcription of a particular gene. For instance, certain genes must be expressed before the *Chs* gene itself can be expressed. It could be that they produce proteins needed for the transcription complex at the *Chs* gene. It is possible that the production of transcription complex proteins is one of the steps in a developmental program that makes a cell competent to respond to a stimulus.

The RNA molecule transcribed from the *Chs* DNA is called a nuclear RNA (nRNA). It is processed into mRNA and transported to the cytoplasm, where it is translated by ribosomes to make the protein. The protein may require further processing—for instance, the removal of a short sequence of amino acids—before being transported to the area in the cell where it functions. The *Chs* gene product, chalcone synthase, catalyzes a biochemical reaction in the cytoplasm. This reaction is the production of chalcone, which serves as a substrate (reactant) for the production of the red or purple pigments in the flower.

Each step in a gene's expression must occur for the gene to be expressed and the enzyme to be functional. The lack or inhibition of any one step would inhibit the synthesis of the enzyme and prevent the production of chalcone. Thus, turning off just one step has a major effect (for example, no color—that is, a white petal); turning it on reverses that effect. To learn how a gene's expression relates to development, it is necessary to learn what steps are turned off during some stages of development and what signals turn them on at the appropriate stages.

The Coordination of Development Requires a Series of Signals

Cells, tissues, and organs form in predictable patterns. The existence of patterns implies that the cells within a tissue or organ "sense" their position and develop accordingly. This in turn means that there must be short-range signals, presumably from adjacent cells, that identify the position within a tissue or organ. Furthermore, a plant generally has a characteristic shoot branching pattern, a pattern of root growth and function, and a pattern of flowering—all of which suggest coordination among the different organs. For such coordination to occur, long-range signals must exist that inform one part of the plant about conditions in another part. By the same reasoning, for plants to respond to the environment in appropriate ways, they must perceive signals from the environment. These signals include light, temperature, day

T. M. Murphy

Figure 15.5 Differential effects of gene expression on the color of petunia petals. The dark purple flower (second from left) expresses all the genes needed to make the most complex anthocyanin pigment. The flowers third, fourth, and fifth from the left, by losing genes in the anthocyanin biosynthetic pathway, have progressively simpler pigments that are less effective in absorbing light. The pink flower on the left has all the genes in the anthocyanin biosynthetic pathway, but it also has a gene that changes the pH of the vacuole, and thus the color of the pigment.

length, water, nutrients such as calcium ions, and mechanical disturbances such as wind and wounding by herbivores.

Some signals may stimulate new patterns of gene expression; other signals may activate preexisting proteins or other cell components. In either case, the signal must be perceived by the target cells. That means that the cells must have a receptor, a protein that is activated when it detects the presence of the signal. Although it is possible that the perception process may end there, recent studies of developmental control suggest that reception of the original signal generally starts a signal *cascade*—a series of events in which one signal leads to another and another **(Fig. 15.6).** Thus, the original signal might be perceived by receptors on the plasma membrane (many are found there), yet, through a cascade system, trigger

events at a distance, such as transcription in the nucleus. In addition, a single signal and receptor can, through a branched cascade, affect several different developmental processes.

The original signals that coordinate the growth and development among cells are generally hormones, chemicals that diffuse or are transported from one part of the plant to another. Other possible signals are transient electrical impulses (that travel like nerve impulses from cell to cell along a stem or root) and hydraulic impulses (changes in the tension of the water column in the xylem). Environmental stimuli, such as light or temperature, also can serve as initial signals.

The signal receptors are proteins. Hormone receptors have binding sites with a three-dimensional shape into which the hormone fits. One common effect of activating a receptor is a transient increase of calcium in the cytoplasm. The normal calcium concentration in plant cell cytoplasm is low, but there is much calcium both outside the cell and inside, in the vacuole and in vesicles. Some stimulated receptors allow calcium to enter the cytoplasm, increasing the concentration by as much as 50 times. Because several enzymes are sensitive to calcium concentration, this change can have several effects. One possible effect is the activation of an enzyme that attaches a phosphate group onto other proteins, thereby either activating or inhibiting them. These proteins may lead to another link in the cascade of events. The cascade finally ends either in the activation (or inactivation) of a transcription factor—a protein that stimulates the reading of a particular gene—or an enzyme that produces the required differentiation of the cell.

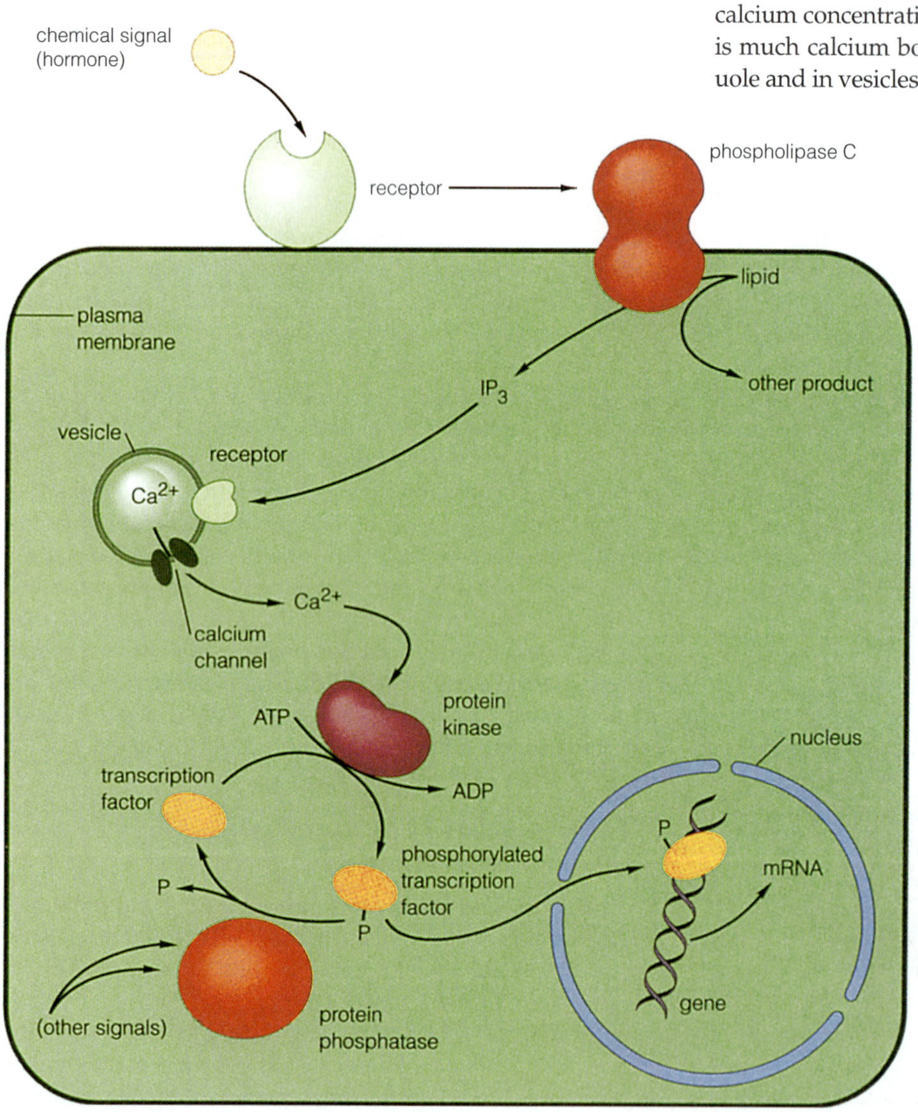

Special Signals Regulate the Cell Cycle

One signal cascade is particularly important for the control of growth. This is the signal that allows cells to pass checkpoints in the cell cycle, and thus to divide, forming new cells. Chapter 3 describes how cells fail to enter the S (DNA synthesis) or M (mitosis) phases of the cell cycle if they are starved. Researchers Lee Hartwell, Paul Nurse, and Timothy Hunt received the 2001 Nobel Prize for their discoveries of genes and proteins that control the cell cycles of yeast and animal cells. Plants have similar genes and proteins, although less is known about them.

One gene, the *cdc2* gene, is known to provide the genetic information for an enzyme protein involved in **phosphorylation** reactions, *cyclin-dependent protein kinase* (C-PK). This enzyme *adds*

Figure 15.6 Proposed steps of a signal cascade in a plant cell. *Arrows* represent a sequence of activations or enzymatic reactions. The receptor is a protein in the plasma membrane that binds the hormone and is stimulated to activate phospholipase C. Phospholipase C is an enzyme in the plasma membrane that breaks down certain membrane lipids. Inositol 1,4,5-triphosphate (IP_3), a product of the reaction, stimulates a receptor on a vesicle, which releases calcium ions (Ca^{2+}) into the cytoplasm. The Ca^{2+} stimulates an enzyme, protein kinase, which adds a phosphate group (from adenosine triphosphate [ATP]) onto various proteins. One of the proteins may be a transcription factor, which stimulates the expression of a particular gene. A separate signal cascade could stimulate a protein phosphatase, which would inactivate the transcription factor. ADP, Adenosine diphosphate; mRNA, messenger RNA.

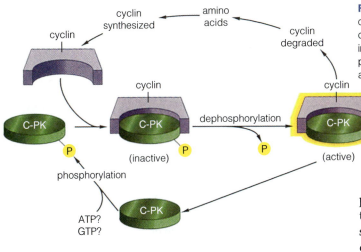

Figure 15.7 Diagram showing how the *cdc2* protein kinase (C-PK) and cyclin trigger cell cycle progression in cycling cells. Starting in the top left, cyclin has joined to C-PK in a phosphorylated form to make a new, but inactive, protein. An enzyme removes the phosphate (P) to activate this protein. This form of the protein acts as the trigger to switch on S phase and M phase. It is recycled by separating C-PK, which is phosphorylated again and recycled. The cyclin is broken down into amino acid subunits and must be resynthesized. ATP, Adenosine triphosphate; GTP, guanosine triphosphate.

cyclin synthesized

amino acids

cyclin

cyclin degraded

cyclin

cyclin

C-PK

C-PK

dephosphorylation

C-PK

(inactive)

(active)

P

P

P

triggers mitosis or DNA synthesis

phosphorylation

C-PK

ATP?
GTP?

a phosphate functional group ($-O-PO_3H_3$) to another protein. Another kind of enzyme catalyzes **dephosphorylation** reactions. This *removes* phosphate from proteins. Proteins that have phosphate added or removed will change their three-dimensional shape, and therefore their biological activity. In some cases, the phosphorylated form will be the active form of the protein, but in other cases, the dephosphorylated form will be the active form. This kind of phosphorylation/dephosphorylation reaction seems to be the key to regulating the cell cycle.

C-PK works in conjunction with other proteins, called *cyclins* (Fig. 15.7). C-PK must itself be phosphorylated before it can combine with a cyclin. The initial combination is inactive. Then a phosphatase removes one phosphate, activating the C-PK-cyclin complex. The active form of the combined

proteins acts as a protein kinase, and it is this enzyme that is the actual trigger to start the S or M phases. (Whether it starts an S or M phase depends on the type of cyclin.) After doing this, the complex is recycled in a three-step process. In the first step, the complex is broken into its two parts. The C-PK can be reused by having new phosphates added to it. The cyclin protein is degraded into component amino acids, and new cyclins must be resynthesized to restart the process. The need to regenerate C-PK and cyclin ensures that the cell must be healthy (not starved) to synthesize more DNA and divide again.

15.2 HORMONES

Low concentrations of certain chemicals produced by a plant can promote or inhibit the growth or differentiation of various plant cells and coordinate development in different parts of the plant (Table 15.1). These chemicals have found many

Table 15.1 Coordination of Plant Development by Hormones

Hormone	Site of Synthesis	Site of Effect	Effect
Auxin	Stem apex, young leaves	Expanding tissues	Promotes cell elongation
		Roots	Initiates lateral roots
		Axillary buds	Inhibits growth (apical dominance)
		Cambium	Promotes xylem differentiation
		Leaves, fruits	Inhibits abscission
	Developing embryos	Ovary	Promotes fruit development
Gibberellins	Stem apex, young leaves	Stem internode	Promotes cell division, cell elongation
	Embryo	Seed	Promotes germination
	Embryo (grass)	Endosperm	Promotes starch hydrolysis
Cytokinin	Root apex	Stem apex, axillary buds	Promotes cell division (release of apical dominance)
		Leaves	Inhibits senescence
Abscisic acid	Leaves	Guard cells	Closes stomata
		Stem apex	Promotes dormant bud formation
	Ovule	Seed coat	Inhibits seed germination
Ethylene	Wounded tissues, aged tissues	Stem	Inhibits cell elongation
		Leaves	Promotes senescence
		Fruits	Promotes ripening

ECONOMIC BOTANY:

Using Plant Hormones

Plant hormones are used extensively to manage plant growth and development in home gardens and in agriculture. The first hormone to be discovered, auxin, also is the one most used by home gardeners. However, it is generally not easy to tell when an auxin is present by inspecting the labels of products in a garden shop. The word *auxin* is seldom mentioned on labels, and the chemicals are generally synthetic analogs of the natural auxin (indoleacetic acid). Concentrations of auxins greater than the optimal inhibit plant growth and kill meristems. Dicots are particularly sensitive, but monocots are much less sensitive. Therefore, auxins such as 2,4-dichlorophenoxyacetic acid (2,4-D) are used as weed killers to remove dicot weeds from monocot grass lawns. Auxins stimulate fruit growth; therefore, a preparation of chlorophenoxyacetic acid is available for spraying on tomato blooms to increase fruit set. Because auxins stimulate the formation of roots, indole-3-butyric acid is sold as a "root stimulator and starter."

Following are just a few of the uses of hormones in commercial horticulture and agriculture:

- Auxins suppress dicot weeds in fields of corn and other monocots; stimulate root growth in cuttings of roses and other nursery plants; and delay fruit and leaf drop.

- Gibberellins stimulate germination of seeds (many species); promote the growth of grape inflorescence internodes, leading to more open berry clusters that are less susceptible to fungal infections; promote berry growth in grapes; increase malt production for brewers; stimulate sugarcane elongation in winter, leading to an increase in yield of 0.7 metric tons per hectare (2 tons per acre); and stimulate early cone formation in conifers, leading to faster breeding programs.

- Cytokinins, in a mixture with gibberellins, promote the growth and

change the shape of apple fruits. Together with auxin, cytokinins are used in tissue culture propagation of, for instance, orchids.

- Ethylene (applied as a solution of 2-chloroethane phosphonic acid, known as *ethephon* or by the trade name Ethrel, which releases ethylene when taken up by plant cells) stimulates and synchronizes flowering of pineapples; thins (causes a partial drop of) the fruit of cotton, cherries, and walnuts; and induces the ripening of apples, tomatoes, and other fruits. Reducing the amount of ethylene or inhibiting ethylene action is an important method of delaying fruit ripening, allowing fruits to be stored longer and shipped to markets farther away. Apples, for example, can be stored for long periods under high CO_2 concentration and low atmospheric pressure, which inhibits ethylene action.

Websites for further study:

Plant Hormones: http://www.plant-hormones.info/

Biology Online—Plant Hormones and Directional Growth: http://www.biology-online.org/3/5_plant_hormones.htm

Botany Online—Plant Hormones: http://www.biologie.uni-hamburg.de/b-online/e31/31a.htm

uses in agriculture (see sidebar "ECONOMIC BOTANY: Using Plant Hormones"). By analogy with chemicals in animals that are secreted by glands into the bloodstream and influence development in different parts of the body, the chemical signals in plants have been called *hormones*. However, some scientists apply a rather strict definition to the term *hormone*: a chemical secreted from one organ in the body that influences another organ in specific ways. Although the chemicals in plants are diffusible, they often influence the same cells that produce them (as well as others); therefore, they are sometimes called growth regulators to distinguish them from the hormones that carry systemic or long-distance signals.

A variety of chemical compounds serve as hormones or growth regulators: auxins, gibberellins, cytokinins, abscisic acid, and ethylene are the five best studied.

Plant Hormones Are Discovered by Studying Plant Developmental Processes

Auxin was discovered during studies of the growth of grass coleoptiles, a model system that has been used for investigations of growth since Charles Darwin and his son Francis worked on "The Power of Movement in Plants" in the 1880s. A coleoptile is a modified tubular leaf that surrounds the first true leaves and the stem of a germinating seedling of a

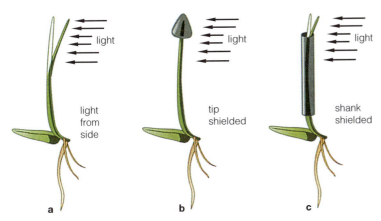

Figure 15.8 The Darwins' experiments on the dependence of grass coleoptile growth on light. The coleoptile is a modified hollow leaf. It grows upward in the dark but bends toward a source of light (**a**). If the tip is shielded from the light, there is no bending (**b**); if the shaft is shielded but the tip is exposed, there is bending (**c**), showing that the perception of the light occurs at the tip.

member of the grass family (Poaceae). Moistened and incubated in the dark, a grass seed germinates, and its coleoptile grows rapidly straight up. By the time it is 1 to 2 cm in length, all cell divisions have ceased, and further elongation reflects only the expansion of the cells.

The Darwins' experiments demonstrated that the coleoptile tip was necessary for the elongation of the shaft (Fig. 15.8). A hypothesis attempting to explain this suggests that a substance produced at the tip is transported down the shaft, where it stimulates growth. The coleoptile also could be induced to bend toward a light, a movement called **phototropism** (blue light was most effective). If the coleoptile

was placed on its side, it would bend upward, a movement called **gravitropism.** A reasonable explanation is that light and gravity cause a movement of the growth-promoting substance to one side of the shaft. The faster growth at that side causes the bending.

Experiments to test the idea of a diffusible growth-promoting substance were devised by N. Cholodny and Frits Went in the 1920s (Fig. 15.9). They cut off a coleoptile tip and placed it on a small block of agar (a gelatin-like substance), leaving it there long enough for diffusible substances to move into the agar. When they placed the agar on a decapitated coleoptile, the shaft resumed growth, showing that the agar had received chemicals from the tip that stimulated the growth of the coleoptile (Fig. 15.9c). If the agar was placed on one side of the coleoptile, the growth was uneven, and the coleoptile bent away from the side that received the agar (Fig. 15.9d,e).

The growth of decapitated coleoptiles demonstrated the presence of a growth-promoting substance, which was named *auxin*. A substance that acted like an auxin was first purified from urine and identified as indoleacetic acid. (In our bodies, proteins are continually being broken down to their constituent amino acids, including one named tryptophan. A portion of the tryptophan is metabolized to indoleacetic acid and excreted in urine; it is simply a waste product.) Later, the same substance was found in plant extracts.

Gibberellins are a group of similar plant hormones that stimulate the elongation of stem internodes. They were discovered in Japan by plant physiologists studying a disease of rice (*bakanae*, or *foolish seedling disease*) caused by a fungus named *Gibberella fujikuroi*. The infected rice seedlings grew

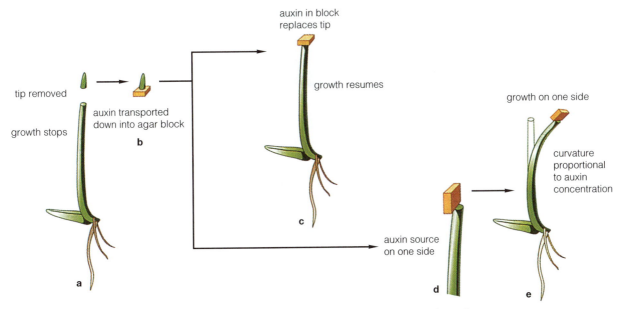

Figure 15.9 The dependence of coleoptile growth on auxin. If the tip of the coleoptile is removed, growth ceases (**a**); thus, the tip produces some factor needed for growth. If the tip is placed on an agar block (**b**), and later the block is placed on the decapitated shaft, growth resumes (**c**). This shows that the growth factor is a diffusable chemical (auxin). If the agar block is placed on one side of the shaft (**d**), the shaft bends as it grows (**e**), suggesting that the light-induced curvature may be caused by a movement of auxin to the side away from the light.

much faster than uninfected ones, but they fell over and died before they could produce grain. The physiologists showed that chemicals produced by the fungus stimulated the early growth. Later, it was found that the same chemicals are normally made in low amounts in young leaves and transported throughout the plant in the phloem.

Cytokinin was discovered during experiments designed to define the conditions needed for culturing plant tissues. In the tissue culture procedure, pieces of stem, leaf, or root are removed from a plant. They are briefly sterilized on their surface to remove any microbes, and then are placed in an enclosed flask containing a nutrient medium. The medium contains a carbon source such as sucrose or glucose, nutrient minerals (see Chapter 11), and certain vitamins. To induce the cells to divide and grow, hormones must also be added. Auxin is important, but most plant cells placed in a medium containing only auxin as a hormone enlarge without dividing. Early studies showed that a second hormone, found in solutions of boiled DNA and in coconut milk (which is a liquid endosperm), also is necessary. The active ingredients turned out to be cytokinins, modified forms of adenine

(a component of DNA and RNA). In a medium containing nutrient components plus auxin and cytokinin, plant cells could divide and grow rapidly.

Further experiments showed that auxin and cytokinin influence the development, as well as the growth, of plant tissue cultures **(Fig. 15.10).** When there is about 10 times more auxin than cytokinin added to the cell culture, growth is undifferentiated. The amorphous mass of cells, often loose and fluffy, is called a *callus.* When the auxin concentration is further increased, or when the nutrient concentration in the medium is reduced, the callus produces roots. But when the cytokinin concentration is increased, the callus becomes green and compact and produces shoots. These are general observations; in practice, the precise response of plant tissues to different auxin and cytokinin concentrations depends on the plant species and other growth conditions.

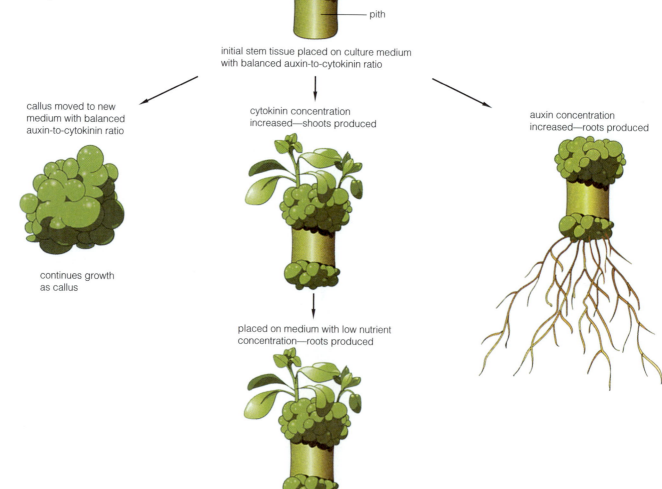

Figure 15.10 The interplay of auxin and cytokinin. Differences in the amounts of auxin and cytokinin can change the development of tobacco (*Nicotiana tabacum*) tissues in culture.

The hormone **abscisic acid** is more associated with the suspension of growth than with its stimulation. It was discovered at nearly the same time by two different groups of researchers. While at the University of California at Los Angeles, F. Addicott discovered a compound that promoted the abscission (breaking off; see Chapter 6) of cotton cotyledon petioles from their stem. In Wales, P.F. Wareing and his coworkers found a compound that was associated with the dormancy of woody shoots in the winter. These compounds turned out to be the same—abscisic acid.

Ethylene is another hormone associated with the inhibition and modification of growth. Ethylene is a simple compound, with the composition C_2H_4. It is an unusual hormone because it is a gas at normal temperatures and pressures. The developmental effects of ethylene were discovered when the gas, produced by oil or kerosene heaters in greenhouses, stimulated the senescence of flowers and caused lemons and oranges to ripen (lose green color). Ethylene is produced by almost any wounded plant tissue and by unwounded tissues whose growth has been restricted, and it can move by diffusion to other nearby organs.

The following sections describe some phases of plant development that are initiated, accelerated, or inhibited by these compounds and by some newly discovered ones.

Signals from the Shoot Apex Promote Growth

In plants, auxin and gibberellins are normally produced by shoot apical meristems, young leaves, and developing fruits and seeds. From shoot apices and young leaves, they are actively transported down the stem toward the roots. They stimulate primary growth of the stem and have other effects.

The mechanism by which auxin stimulates the expansion of cells has inspired active research since the hormone was discovered. Cells enlarge when the internal pressure (turgor pressure) causes an irreversible stretching of the cell wall. It has been shown that the major effect of auxin is to increase the plasticity of the cell wall. This was accomplished by removing turgor pressure and exerting a known force on untreated and auxin-treated coleoptiles. The auxin-treated coleoptile deformed more easily than the untreated coleoptile and failed to snap back as far after the force was removed. We can assume that the same thing happens when the cell walls of auxin-treated cells are stretched normally by internal turgor pressure, and that this increased plasticity represents the reason for the increased growth rate induced by the auxin.

How does auxin stimulate the increase in plasticity? There are two hypotheses, and both may be correct. The first hypothesis, called the *acid-growth hypothesis*, suggests that the main effect of auxin is to cause cells to secrete acid (H^+ ions, protons), and the acid stimulates the changes in plasticity. Experimental evidence to support this hypothesis comes from an increase in the rate of acid secretion by coleoptiles or stem tissue when they are treated with auxin. Auxin stimulates the activity of the plasma membrane proton pump, which results in the active secretion of H^+ ions.

Also, coleoptile or stem sections that have been frozen and thawed several times (a treatment that kills their cells but leaves the components of their cell walls intact) become more plastic when they are soaked in acidic solutions. Notice that this hypothesis does not require that auxin treatment stimulate the expression of any genes.

The second hypothesis suggests that auxin works by inducing the expression of genes that make growth-promoting proteins. There also is evidence to support this *induced gene expression hypothesis*. Translating mRNA is an essential step in gene expression. The application of chemicals that stop ribosomes from translating mRNA also prevents cells from increasing their growth rate in response to auxin. In addition, it has been possible to identify mRNAs that increase in concentration rapidly within 10 to 20 minutes after stem sections have been treated with auxin. Soybean hypocotyls that bend under a gravitropic stimulus (lying on their side) show an asymmetric distribution of auxin-stimulated mRNAs (with more appearing on the lower side of the hypocotyl, before the bending starts). Although we still do not know what proteins all these mRNAs encode, these observations suggest that auxin-stimulated mRNAs, and the proteins for which they code, might be involved in the bending process, which is related to growth.

There is some evidence indicating the existence of special proteins that affect the cell wall's plasticity. Their activities may promote acidic conditions. For instance, one of the proteins induced by auxin stimulates the activity of the plasma membrane proton pump, which acidifies the cell wall. This is one way to explain how both of the hypotheses could be true. Other proteins may depend on acidic conditions. A protein called *expansin* has been extracted from apical regions of cucumber seedlings. When added to stretched, heated stem segments, it increases their plasticity, but only at pH values less than 6.0 (acidic conditions). Another protein has an enzymatic activity (*transglycosylation*) that fits what might be expected for an enzyme that promotes plasticity and growth. It breaks carbohydrate chains and re-forms them in new configurations that can result in a more extended wall **(Fig. 15.11)**. It seems that the exact mechanism of auxin-induced cell expansion will be understood soon.

The elongation of stem internodes also depends on a supply of gibberellins. Evidence for that suggestion comes from experiments in which plants are sprayed with a chemical that inhibits gibberellin synthesis. Such plants develop shorter, stumpier stems with short internodes. This stimulation of internode growth reflects both new cell divisions and increased cell elongation.

Auxin and gibberellins signal the presence and activity of the shoot apical meristem and help control further development of the shoot. Together with gibberellins, auxin stimulates cell division in the vascular cambium. Auxin promotes the formation of secondary xylem, whereas gibberellins stimulate formation of secondary phloem. Auxin also tends to inhibit the activity of axillary meristems near the apical meristem, restricting the formation of shoot branches, a phenomenon called apical dominance. When a

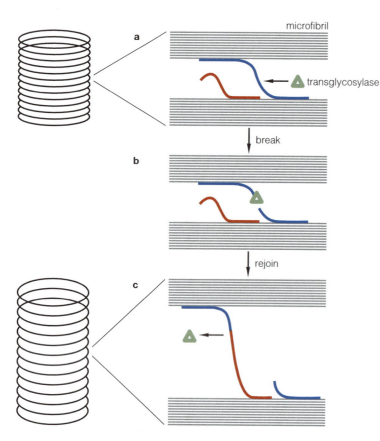

microfibril

a

transglycosylase

↓ break

b

↓ rejoin

c

Figure 15.11 Model for the mechanism of cell wall elongation. (**a**) A cell is surrounded by cellulose microfibrils (*gray*), which are held together by other polysaccharides (*red, blue*). (**b**) A transglycosylation reaction breaks bonds between sugar molecules in cell wall polysaccharides and re-forms longer polysaccharides in a manner that allows the (**c**) microfibrils to separate and the cell to elongate. (Adapted from a model by S.C. Fry, University of Edinburgh, Edinburgh, U.K.)

Cytokinin Coordinates Shoot with Root Growth

Cytokinin is found in embryos and endosperm, where it promotes growth by stimulating the cell cycle, possibly by promoting cyclin synthesis. In mature plants, it has a similar role: It is produced in the roots and transported to the shoots in the xylem sap. A small amount of cytokinin applied to lateral buds of pea (*Pisum sativum*) shoots stimulates them to grow, even though their growth was previously inhibited by the apical dominance effect of auxin.

Cytokinin also delays senescence, the well-controlled process of deterioration that leads to the death of cells. The senescence of leaves, seen as they turn yellow, involves the breakdown of chlorophyll and proteins and the export of the products through the phloem to the meristems or other parts of the plant where they can be recycled. Leaves detached from the stem and placed in water undergo this process, but adding cytokinin delays it by several days **(Fig. 15.13).** Because cytokinin is synthesized by the roots and moves in the xylem stream, the presence of cytokinin signals to the shoots the presence of healthy roots. Without roots, or when roots stop growing, the lack of cytokinin stops shoot growth and allows leaves to senesce.

plant is pruned, apical segments are removed, including the apical meristem and its auxin. With the inhibitory influence of the auxin gone, the axillary buds grow, producing a bushier plant. Finally, auxin stimulates the formation of new root apical meristems, both meristems forming lateral roots and meristems forming adventitious roots from the basal parts of shoots **(Fig. 15.12).** Thus, auxin helps coordinate root growth with shoot growth.

Figure 15.12 Holly (*Ilex opaca*) shoots form roots at their bases faster when the bases are treated with an auxin. The ends of these shoots were dipped for 5 seconds in solutions containing 50% ethanol and (from left to right) 0%, 0.1%, and 0.5% naphthalene acetic acid, a synthetic auxin. They were then rooted in moist vermiculite for 2 weeks.

Figure 15.13 The effect of cytokinin on senescence. Cytokinin applied to the right-hand primary leaf of this bean (*Phaseolus vulgaris*) seedling inhibited its senescence. The left-hand leaf did not get cytokinin.

A. Lang

a   b   c

Figure 15.14 Gibberellic acid substitutes for a cold requirement in the bolting and flowering of a carrot plant (*Daucus carota*). (**a**) The rosette form of the carrot, maintained without a cold or gibberellin treatment. (**b**) A carrot plant that flowered in response to a treatment with gibberellin (no cold treatment). (**c**) A carrot plant that flowered in response to a treatment in the cold (no gibberellin treatment).

Gibberellins, Abscisic Acid, and Ethylene Influence Shoot Growth in Response to Environmental Signals

A spectacular demonstration of the effect of gibberellins on internode growth is found in the *bolting* of plants with a rosette morphology (Fig. 15.14). Rosettes are plants with internodes so short that the leaves look almost as if they spring from a single node. Iceberg lettuce (*Lactuca sativa*), cabbage (*Brassica oleracea*), carrots (*Daucus carota*), and beets (*Beta vulgaris*) are all rosette plants. When these plants flower, the internodes of the stem elongate quickly. Flowers form on the new parts of the stem. Under natural conditions, bolting is triggered by environmental signals, such as a lengthening of the day or a period of cold temperature (Fig. 15.14c). It also can be stimulated by spraying the plant with gibberellin (Fig. 15.14b). Furthermore, it can be shown that the natural signal that stimulates bolting first induces the plant to synthesize gibberellin. Together, these two observations suggest that rosette plants are originally deficient in gibberellin and that they bolt when an environmental signal stimulates them to produce this substance.

In the autumn, the shoot apices of most temperate region perennial plants go dormant. Dormancy is more complex than a simple inhibition of cell activity in the apical meristem. It involves growing protective bud scales (modified leaves that are small, nonphotosynthetic, dry, and tough) around the meristem before cell division stops. As days get shorter or temperatures decrease, abscisic acid accumulates in the shoots of perennial plants and stimulates the formation of a dormant bud at each shoot apical meristem.

Ethylene slows the growth of stems and roots (Fig. 15.15), when it is produced by wounded cells or by organs meeting a physical obstacle. A hypothesis has been advanced to explain this. In very small concentrations the gas disrupts the organization of microtubules close to the cell wall. These microtubules normally lie in parallel rows, and they are thought to control the direction of synthesis of the cellulose microfibrils so that the microfibrils lie parallel to the microtubules (see Chapter 3). Growth then occurs perpendicular to that direction. In the presence of ethylene, the microtubules are absent or disorganized, and the cellulose microfibrils lie in all directions. Growth, too, then occurs in all directions. Cells that develop that way are short and round, rather than long and thin.

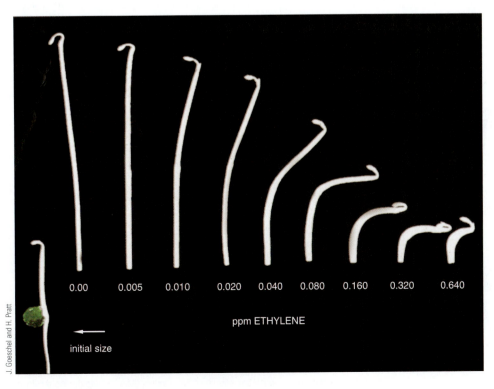

J. Goeschel and H. Pratt

0.00   0.005   0.010   0.020   0.040   0.080   0.160   0.320   0.640

ppm ETHYLENE

initial size

Figure 15.15 The effect of ethylene on growth. An increase in the concentration of ethylene makes dark-grown pea (*Pisum sativum*) seedlings shorter and thicker. Low concentrations of ethylene are effective. In this case, even the most strongly affected seedling has been treated with less than one part of ethylene per million parts of air. ppm, Parts per million.

A stem or root growing under the influence of ethylene is short and stumpy because it is formed from short, round cells.

Abscisic Acid and Gibberellins Control Seed Development and Germination

Abscisic acid plays a role in the formation of viable seeds. The presence of the hormone late in seed development induces the formation of large amounts of certain proteins by stimulating the transcription of their genes. These proteins are thought to store nitrogen, other elements, and energy for use by the embryo when it germinates.

Abscisic acid also is associated with dormancy of some seeds. It is accumulated in the seed coat during the seed's development. In the presence of this hormone, the embryo does not germinate, even if it is hydrated. Some of these seeds (for example, apple [*Malus sylvestris*] or cherry [*Prunus* sp.] seeds) require a long period under cool, wet conditions before they can germinate, conditions that stimulate the breakdown of the abscisic acid.

Initial experiments suggest that the abscisic acid receptor works through a signal cascade that includes phosphory-

lated proteins (Fig. 15.4). A mutant of mouse-eared cress (*Arabidopsis thaliana*) that lacks the ability to respond to abscisic acid lacks a protein phosphatase (an enzyme that removes the phosphate groups from various proteins). This suggests that removing phosphates is one step in the signal chain by which, in various tissues, abscisic acid stimulates storage protein synthesis or maintains dormancy.

Gibberellin promotes the germination of many types of seeds. In these systems, both cell division and cell elongation are stimulated. The mechanism of the stimulation remains obscure, although pea stems treated with gibberellin have an increased concentration of one of the cell wall-loosening enzymes described in the section on auxin-induced cell expansion (p. 245). In addition, gibberellin apparently causes a reorientation of microtubules and, as a result, cellulose microfibrils, so that fewer microfibrils oppose the elongation of cells.

In at least one well-studied monocot system, barley (*Hordeum vulgare*) seeds, gibberellin promotes the metabolic breakdown of storage materials (Fig. 15.16). Barley seeds have been studied for many years because the conversion of their starch to sugar is the key event in the malting process,

Intact seed

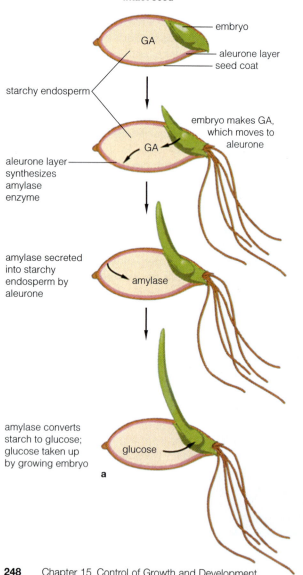

Embryo-less half-seed

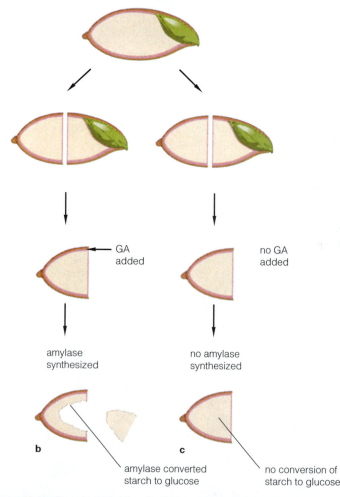

Figure 15.16 The effect of gibberellin (GA) on the germination of a barley (*Hordeum vulgare*) seed. (**a**) In an intact seed, GA from the embryo stimulates the synthesis of amylase, which breaks down starch to form the glucose that nourishes the embryo. Removing the embryo removes the source of GA. Unless GA is added (**b**), no amylase will be synthesized, and the endosperm starch will not break down into glucose (**c**).

PLANTS, PEOPLE, AND THE ENVIRONMENT:

Lorenzo's Oil and Seed Germination

There is plenty of evidence that new chemical controls of growth and development are yet to be discovered, and the field of genetics (see Chapter 16) will grease the discovery process. Researchers at Long Ashton Research Station, Bristol, U.K., have been studying the genetic control of seed germination. They identified a mutant of mouse-ear cress (*Arabidopsis thaliana*) that was defective in germination. They then isolated and sequenced the mutant gene, which they named *Comatose* and abbreviated *cts*. They inferred that the wild-type allele of the gene, *Cts,* is vital for breaking dormancy and allowing germination.

Once the scientists had determined the base sequence of the *Cts* gene, they realized that it was similar in sequence to a gene in humans. A mutation in the human gene leads to the buildup of very long chain saturated fatty acids (VLCSFAs) and a disease called adrenoleukodystrophy (ALD). The symptoms (dementia and loss of sight, hearing, speech, and ambulation) are believed to be caused by solubilization and removal of the myelin sheath around neurons by VLCSFAs in the neurons. Like human patients with ALD, *cts* mutant seeds accumulate VLCSFAs.

Augusto Odone developed "Lorenzo's Oil" for his son, who has ALD disease; this was the subject of the 1992 movie *Lorenzo's Oil.* The oil, made of unsaturated fatty acids, is thought to be effective in delaying and perhaps alleviating symptoms of patients with ALD. The Long Ashton scientists found that it also stimulated germination in the *cts* mutant seeds.

What do the proteins coded by the *Ald* and *Cts* genes do, and how does Lorenzo's Oil work? The human ALD protein transports VLCSFAs across peroxisome membranes of the cell, where they can be degraded. Lorenzo's Oil inhibits the synthesis of the VLCSFAs. But a firm answer to how these act in plants requires more research. The genetic information may be applicable to a close relative of mouse-eared cress, oilseed rape (*Brassica napus*), in which germination frequency is associated with quality and performance. Oilseed rape, incidentally, is a rich source of erucic acid, one of the components of Lorenzo's Oil.

 Websites for further study:

The Myelin Project—Information about Lorenzo's Oil:
http://www.myelin.org/aboutoil.htm

Lorenzo's Oil Works for Plants:
http://www.bbsrc.ac.uk/media/pressreleases/02_12_30_lorenzo.html

Literature, Arts, and Medicine Database—Lorenzo's Oil: http://endeavor.
med.nyu.edu/lit-med/lit-med-db/webdocs/webfilms/lorenzo.s.oil25-film-.html

---

the initial process in the brewing of beer. In malting, barley seeds are moistened with water, which begins the germination process. As the embryos start to grow, the seeds form enzymes (α-amylase, maltase) that convert the starch in the endosperm of the seed to the simple sugar glucose. The glucose would be used to nourish the growing seedling, except at this point the brewmaster bakes the *malted* seeds at high temperature to kill the seedlings and inactivate the enzymes. Caramelization of the sugars in the baking process gives the beer some of its flavor. The dry malt is ground up and then dissolved in water, ready for fermentation.

Without the embryos, moistened barley seeds soften, but they do not produce the enzymes that break down the starch. However, if the softened seeds are treated with low concentrations of gibberellins, a special layer of cells on the outer edge of the endosperm (the *aleurone* layer) synthesizes the enzymes and releases them into the starchy tissue. The gibberellins induce the transcription of the genes for α-amylase in that layer of cells. Newly made mRNA serves as a template for the synthesis of the enzyme. Under normal condi-

tions, the gibberellins are made by the germinating embryo. Thus, they represent a signal from the embryo to the endosperm, announcing the need for nutrients.

Ethylene Stimulates Senescence

By far the most characteristic effect of ethylene is its stimulation of senescence. It takes special enzymes—chlorophyllases and proteases, by which chlorophyll and proteins are broken down—to start this process. Ethylene triggers the expression of genes leading to the synthesis of these enzymes. In many plants, senescence is associated with abscission, which is caused by enzymes that digest the cell walls in a localized region at the base of the petiole (the abscission zone). The petiole breaks from the stem at that point. The activity of the enzymes involved in both senescence and abscission is increased by ethylene. The mechanism by which ethylene induces the synthesis of these enzymes is not well understood, but a specific protein receptor that recognizes and binds to ethylene has been found.

Figure 15.17 The effect of ethylene on the ripening of fruit. The box of tomatoes on the right was kept for 3 days in a room with an atmosphere containing 100 parts of ethylene per million parts of air. The fruit on the left had not yet been treated with ethylene.

Figure 15.18 Systemic acquired resistance. Both mouse-eared cress (*Arabidopsis thaliana*) plants were inoculated with the pathogenic bacterium, *Pseudomonas syringae*. To stimulate its resistance mechanisms, the plant on the left had been treated earlier with a synthetic compound thought to mimic one of the natural signals that induces systemic acquired resistance. The plant on the right did not receive this treatment.

Many other pieces of the puzzle probably will be discovered in the next few years.

The ripening of fruit is a variation on the process of senescence. Depending on the type of plant, it may involve the conversion of starch or organic acids to sugars and the softening of cell walls to form a fleshy fruit, or the rupturing of the cell membrane with the resulting loss of cell fluid to form a dry fruit. In either case, ripening is stimulated by ethylene **(Fig. 15.17).** There is an autocatalytic (self-promoting) aspect to this effect. Like wounded plant tissues, senescent plant tissues (including ripe fruit) form ethylene. An overripe banana or apple is a potent source of ethylene. The ethylene that these fruits emit can stimulate senescence (ripening) in other, adjacent fruits. This is the physiological truth behind the statement, "One bad apple spoils the barrel." By the same token, the ripening of fruit can be controlled by reducing the concentration of ethylene in the atmosphere. Picked apples, for instance, often are stored for months, without ripening, in an atmosphere containing low concentrations of ethylene and high concentrations of CO_2 (which inhibits the effect of ethylene, possibly by binding to its receptor).

A Variety of Compounds Serve as Stress Signals

Abscisic acid has another role, which is not concerned with development, but rather with the control of the photosynthetic system under stress. When water becomes so scarce that leaves wilt, it is important that the stomata close to prevent any further water loss. This is true even though closing stomata will cut off the supply of CO_2 and shut down the photosynthetic system. Abscisic acid is a signal of this emergency situation. Under drought conditions, wilted mesophyll cells of a leaf rapidly synthesize and excrete abscisic acid. This abscisic acid diffuses to the guard cells, where an abscisic acid receptor recognizes the presence of the hormone and acts to

release potassium ion (K^+), chloride ion (Cl^-), and H_2O, closing the stomata (see Chapter 11).

As the attention of plant scientists has shifted to include developmental responses to biotic stress and infection, new signaling compounds have been discovered. One type of stress experienced by plants is attack by fungi and fungus-like parasitic protists. Once they detect a fungal infection or the threat of an infection, plants may develop various resistance mechanisms near the site of infection. They may, for instance, produce hydrogen peroxide (H_2O_2), which is thought to act as an antibiotic, killing the infectious agent. They may produce enzymes that break down fungal cell walls. The plant cells surrounding the infection site may die, producing tannins (organic polymers related to lignin) in the process. This is effective because food and water are cut off from the site— and therefore from the fungus—and the tannins are hard for the fungus to digest. Thus, the fungus has difficulty spreading through the dead area to new, live cells. Plants recognize the presence of a fungus from oligosaccharides, short chains of sugar molecules (see Chapter 2) released from the fungal cell wall or from the plant cell wall under fungal attack. There are many kinds of oligosaccharides, but only the few kinds released from fungal cell walls signal the presence of fungus. It is possible that other oligosaccharides may influence developmental processes of healthy plants. For instance, certain oligosaccharides stimulate the formation of flowers from sections of leaves grown in tissue culture.

The resistance induced by the perception of an infection may spread throughout a plant. That is, a plant that has been infected by a virus, bacterium, or fungus and survived may become less susceptible to a later invasion into other parts of its body by the same pathogen (and some other pathogens, too). This phenomenon is known as systemic acquired resistance **(Fig. 15.18).** The mechanism is not fully understood, but it involves the transmission of a signal from the infected organ to new leaves, probably through the phloem, and the induction of several types of antibiotic proteins in the new leaves. (It does not involve antibodies of the type that appear in an immunized animal.)

Recent research has focused on identifying the signals that induce resistance in leaves distant from the primary site of infection. Several compounds have been tentatively identified as being involved. Salicylic acid, a compound related to aspirin, is formed in infected plants, and it induces systemic resistance when it is applied to a plant. Both jasmonic acid, which is formed by the oxidation of a membrane lipid molecule during infections, and hydrogen peroxide can stimulate resistance responses when applied to leaves. Perhaps the most interesting compound is an 18-amino-acid polypeptide chain named *systemin*, which is produced in response to infections and is required for establishing systemic resistance. If systemin is the signal that moves from the infected leaf to induce resistance in new leaves, it is the first polypeptide hormone to be discovered in plants.

15.3 LIGHT AND PLANT DEVELOPMENT

Plant development is strongly influenced by environmental factors, as well as by internally produced chemicals. Light is the most important of these factors. The form, growth rate, and reproduction of plants often are influenced by the light in the environment. Recognizing and responding to light is a major way in which plants adapt to their surroundings and match their activities to the time of day and season of the year. These responses result from mechanisms in addition to photosynthesis, which is itself a major light-influenced process. Plants "sense" three colors of light, which correspond to three (or more) distinct light receptors. Red light activates receptors known as phytochromes; blue and near-ultraviolet (black) light stimulate different receptors, called *cryptochromes*; intermediate-wavelength ultraviolet light (the ultraviolet B [UVB] radiation from the sun that causes sunburn) is detected by a third, currently unnamed, receptor.

The Red/Far-Red Response Acts Like an On/Off Switch

Some small seeds do not germinate in darkness. But after a brief soaking, a 1-minute exposure to red light will induce them to germinate. Surprisingly, the effect of the red light can be canceled if, immediately after the red light, the seeds are exposed to 5 to 10 minutes of far-red light. Far-red light includes those wavelengths of the spectrum that are longer than visible red light but shorter than the infrared radiation commonly thought of as heat radiation. A second exposure to red light, given after the far-red light, will stimulate germination again, showing that the far-red light reverses the first red signal, but does not inhibit the seeds in any irreversible way. It is as though an on/off switch were controlled by the two wavelengths of light.

The red/far-red response governs many aspects of plant development, including the growth of seedlings. A pea seedling grown in the dark is long and light yellow, with unexpanded leaves and a tight hook just below the apical meristem (Fig. 15.19). This syndrome is called etiolation. Its purpose is to allow the shoot apex of these seeds, which may germinate several centimeters underground, to reach the surface as rapidly as possible. Exposure of the seedlings to red light starts a process of de-etiolation: The hook opens, the leaves expand, and the growth of the stem slows. However, the de-etiolation process is retarded if the red signal is followed immediately by far-red light.

The red/far-red reversibility reflects the structure and function of phytochromes. Phytochromes are a type of protein containing a pigment molecule related to heme (the oxygen-carrying molecule in animal blood cells). When phytochrome is synthesized in the dark, its pigment can initially absorb red light. This form of phytochrome is called P_r (*r* for red-absorbing); it is thought to be inactive. If this phytochrome is irradiated with red light, it changes its form, and its pigment becomes capable of absorbing far-red light. This is the active form of the phytochrome, called P_{fr}, the form that stimulates seed germination and de-etiolation. When P_{fr} is exposed to far-red light, most of the phytochrome returns to its original form. Because P_r is inactive, whereas P_{fr} is active, irradiating phytochrome with red and then far-red light is equivalent to turning a switch on and then off. Scientists have not identified with certainty what a phytochrome actually does, but one possibility is that it attaches phosphates to other proteins.

In at least some plants, there are actually three or more types of phytochromes, each coded by a different gene. All types are present in young, dark-grown seedlings. One type, present in greater concentrations than the others, is

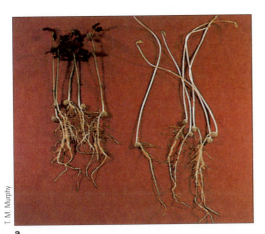

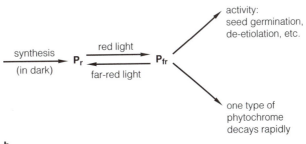

b

Figure 15.19 The red/far-red response. (**a**) Pea (*Pisum sativum*) seedlings grown in the light (*left*) and in the dark (*right*). (**b**) Light changes the form of phytochromes.

T. M. Murphy

a

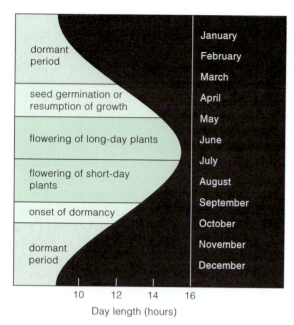

flowers

a

Figure 15.20 Effect of day length on flowering and other plant activities in temperate regions of the northern hemisphere.

degraded by light. That is, in the light its P_{fr} form is rapidly broken down, and the gene that codes the protein is turned off. In adult plants growing in the light, only the more stable types of phytochromes are present. These, however, are adequate to make sure that adult plants show some red/far-red responses. For instance, the internodes of many plants are longer in the shade of a leafy canopy and shorter in bright light, a response that helps shoots grow out of darker areas and into the sun. This response occurs because chlorophyll in the canopy leaves absorbs red light more than far-red light. Thus, the ratio of far-red light to red light is higher under the canopy than in unshaded areas. This means there will be more of the P_r form of phytochrome, which allows (does not inhibit) internode growth. Out from under the canopy, there is a higher ratio of red to far-red light; therefore, more P_{fr} is present, which inhibits internode growth.

Photoperiodic Responses Are Controlled by a Biological Clock

Over most of the earth's surface, there are pronounced seasonal differences in temperature, water availability, and illumination. Many plants show developmental changes that prepare them for the coming of both harsh and mild seasons. Before the winter, dormant buds are formed at the shoot apical meristems. The leaves of deciduous plants become senescent, turning colors as their chlorophyll and proteins are broken down. As spring comes, buds break open and start to grow, and the plant may flower. Of all the earth's seasonal variables, the most reliable and the most useful for anticipating changes in climate is the annual cycling of day and night lengths (Fig. 15.20). Plants have a system that measures the lengths of the days and nights. The control of development by this system is called photoperiodism.

b

Figure 15.21 Flowering in response to day (night) lengths. (a) Spinach (*Spinacia oleracia*), a long-day plant, initiates flowers and elongates its stem when the days are 14 hours or longer (nights are 10 hours or shorter). The plant on the right was kept in a growth chamber in 14-hour days; the plant on the left was kept in 12-hour days. (b) *Chrysanthemum*, a short-day plant, flowers in the autumn, when the nights are long. Both plants were kept in short days, but the plant at the right received 1 hour of light near the middle of each night and did not flower.

Many plants use photoperiodism to time their flowering. These plants fall into two major groups, which traditionally are called long-day plants and short-day plants (Fig. 15.21). As you will see, these names are somewhat misleading, but they are too firmly embedded in the language of physiologists to be changed. (Many other plants time their flowering

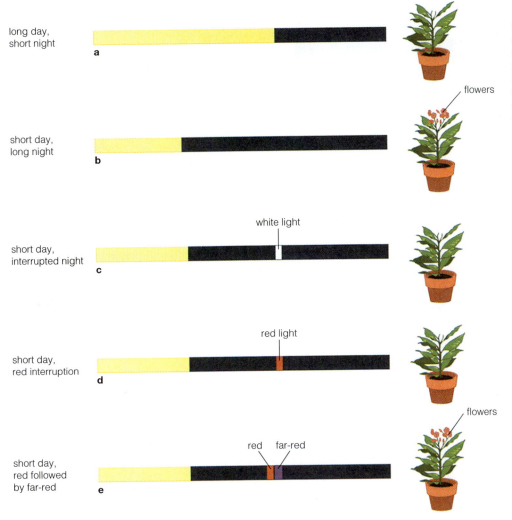

long day,
short night
a

short day,
long night
b

flowers

short day,
interrupted night
c

white light

short day,
red interruption
d

red light

short day,
red followed
by far-red
e

red   far-red

flowers

Figure 15.22 Effect of night length on short-day plants. (**a–c**) Experiments demonstrate that the length of night, not day, is the critical signal that stimulates flowering. (**d,e**) Experiments show that phytochrome is the receptor by which the plant perceives an interruption of night.

without reference to the length of the day, and these are called day-neutral plants.) Long-day plants begin flowering sometime between January and June, when the days are getting longer. Each variety has a characteristic day length and begins flowering when the days are longer than that day length. Similarly, short-day plants begin their flowering between July and December, when the days are getting shorter. Each variety of short-day plant begins flowering when the days become shorter than its characteristic day length.

A simple experiment demonstrates that the plants actually measure the length of the night, rather than the day (Fig. 15.22a–c). Short-day plants are grown to maturity under conditions of long days (light periods) and short nights (dark periods); then they are placed in a growth room in which the days are shorter than the characteristic length and the nights are correspondingly longer. Under these conditions, the plants will flower. However, if a plant receives a pulse of light in the middle of the night, it does not flower. A pulse of darkness during the day has no effect. Because a light pulse at night blocks the signal to start flowering, it can be concluded that it is the length of the uninterrupted night that is the important part of the signal.

This same type of experiment can be used to show that phytochrome is the receptor by which the short-day plant perceives light (Fig. 15.22d,e). If different wavelengths (colors) of light are used to interrupt the long night period, it is red light that inhibits flowering, and only a short pulse of red light is needed. However, if a short pulse of red light is followed by 10 minutes of far-red light, flowering proceeds as though the interruption never happened. The red/far-red reversibility is a characteristic of phytochrome.

Finally, this experiment can be modified to show that a biological clock is involved in measuring the length of the night. Biological clocks are known in all eukaryotic organisms. They are seen, for example, as daily (*diurnal*) rhythms in the internal temperature of animals and in the pattern of spore formation in fungi. The leaflets of some plants, such as *Mimosa*, open during the day and close at night. These **nyctinastic movements**, which are caused by the transport of ions and the resulting changes in the turgor pressures of cells on opposite sides of the petiole, provide clear evidence for the operation of a clock. Biological clocks are generally reset everyday by exposure to light; but because they continue to operate even in continuous darkness, they are called endogenous—that is, coming from within. They probably represent a basic biochemical or physical oscillation within cells, although the nature of this oscillation remains a mystery.

In one experiment, mature short-day plants are given their short day and then placed in the dark for 48 hours. Different

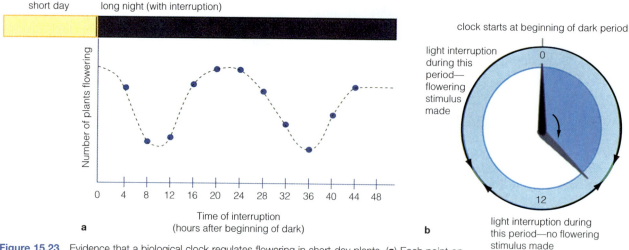

short day long night (with interruption)

clock starts at beginning of dark period

light interruption during this period— flowering stimulus made

light interruption during this period—no flowering stimulus made

Number of plants flowering

Time of interruption
(hours after beginning of dark)

a

b

Figure 15.23 Evidence that a biological clock regulates flowering in short-day plants. (**a**) Each point on the graph represents a set of plants that was placed in the dark but was given a light pulse (interruption of the dark) at the time shown on the *x*-axis. Plants interrupted at different times showed different degrees of flowering, revealing a sensitivity to the light pulse that varied with a 24-hour period. (**b**) An interpretation of the night interruption experiment. An internal biological clock starts running when the lights are turned off. When the lights are turned on, the clock is reset. The effect of light depends on the position of the clock when it is reset.

plants are given pulses of light at different times after the start of the dark period. The effect of the light depends on the time it is given, with a periodicity that cycles about every 24 hours **(Fig. 15.23a)**. Because the effect of light on the flowering responses of the experimental plants changes periodically, the responses are thought to be controlled by interaction between light and the endogenous biological clock. In this experiment, the action of the biological clock is made visible by how the plants respond to light, as if the clock turned on and off some aspect of the light-reception process **(Fig. 15.23b)**.

The signal pathway that initiates flowering may be complex. The perception of day or night length occurs in the leaves, but flower development occurs at the apical or axillary meristem. Stimulating a single leaf by exposing it to a critical night length can be enough to channel development of the meristem toward flowering; in fact, grafting a stimulated leaf onto an unstimulated plant can induce the plant to flower. These observations have led to the hypothesis that a chemical compound, named florigen, is produced in the leaf and transported to the meristems. However, the chemical nature of florigen has not been determined, and even its existence is in doubt. Gibberellins stimulate flowering in some plants, especially long-day plants and plants that require a period of cold temperatures (Fig. 15.14), but gibberellins do not work on many of the plants thought to respond to florigen. Auxin transport is needed for flower formation, but auxins are present in the meristems even when they are not flowering. Neither gibberellins nor auxins appear to be florigen. It is possible that florigen is a compound (such as an oligosaccharide or a protein) that is difficult to preserve once it has been extracted from a plant because enzymes destroy it in the extract; or perhaps it is difficult to detect because it cannot normally be taken up into plant tissues once it has been extracted.

Other responses that are controlled by photoperiodism, probably through the interaction between light and the endogenous clock, include dormancy and senescence in the fall and the resumption of growth in the spring.

Plants Respond to Light in Many Ways

Many responses of plants to light involve the blue part of the spectrum. These include such responses as bending toward light—phototropism (Fig. 15.8). They also include the induction of enzymes that synthesize the red pigments in the skins of such fruits as apples and plums. Actually, the latter example is a good way of demonstrating the complexity of light responses. The red pigments are anthocyanins, water-soluble compounds often seen in the lower epidermis of leaves and the petals of flowers. The induction of one of the enzymes involved, chalcone synthase, was described earlier in this chapter. In some seedlings, these pigments are formed after irradiation with red light, with phytochrome as the receptor. In seedlings of other species, these pigments are formed only after blue light irradiation. Still other plants require both red and blue light for a full response, although either red or blue light will stimulate the formation of some pigment. It seems that induction of the enzymes synthesizing these compounds may be connected to one or both light-signaling pathways.

Chlorophyll synthesis is another process that requires more than one color of light. Although bean seedlings that are grown in the dark will start the process of de-etiolation after a short red-light exposure, their leaves will not turn green if that is the only light they receive. The green color is chlorophyll, and the synthesis of chlorophyll requires a longer exposure to red or blue light. The receptor, protochlorophyll, is a precursor to chlorophyll, and it must be excited by light photons before it can be converted to chlorophyll.

Thus, the response of plants to light is complex, both at the molecular level and at the level of the whole plant. This should not be surprising, given the importance of light to the survival of a photosynthetic organism.

KEY TERMS

abscisic acid	gibberellins
auxin	gravitropism
cytokinin	indeterminate
dephosphorylation	nyctinastic movements
determinate	phosphorylation
determined	phototropism
ethylene	

SUMMARY

1. The development of plants depends on both morphogenesis and differentiation. Morphogenesis is the creation of shape, determined by the frequency and direction of cell division and cell elongation. Differentiation is the acquisition of properties (determined by differential gene expression) that distinguish a cell, tissue, or organ from other cells, tissues, or organs.

2. Shoots and roots have indeterminate growth patterns: They may continue to grow without an obvious stopping point so long as conditions are favorable. An indeterminate growth pattern depends on the existence of a meristem. Some plant organs, such as flowers and dicot leaves, have a determinate growth pattern, which means that their growth stops at a predetermined size or age.

3. A *totipotent* cell can give rise to any cell type in a plant. Once a cell has become *determined,* however, it and its progeny have a limited number of possible differentiated states. A *competent* cell is primed to respond to a signal by differentiating in a predetermined manner.

4. The expression of a gene includes all of the following steps: activation of chromosomal DNA, transcription, processing of the transcript (nRNA) to form mRNA, translation, processing of the translation product to form an enzyme, transport of the enzyme to its site of action, activation of the enzyme, and formation of the product of the enzymatic reaction.

5. Chemical signals (hormones) coordinate development among different parts of a plant. Signals are perceived by receptors in a cell; perception often starts a chain of events (a cascade) that results in the activation of specific genes, resulting in growth or differentiation. The *cdc2* gene in yeast, and other genes in plant cells, code for enzymes that are essential for allowing cells to start DNA synthesis or mitosis, and thus to divide to form new cells.

6. The hormone auxin stimulates growth through cell enlargement in leaves, stems, and roots. It also stimulates cell division and differentiation in the vascular cambium, promotes the formation of root meristems, and inhibits stem branching (through apical dominance).

7. The stimulation of cell enlargement by auxin reflects an increase in cell wall plasticity. Two hypotheses have been advanced to explain this. The acid-growth hypothesis states that auxin works by stimulating the excretion of protons from the cell; and the induced gene expression hypothesis states that auxin works by inducing the synthesis of cell wall proteins.

8. The hormone gibberellin stimulates the growth of stems, the germination of seeds, and the resumption of growth in winter-dormant buds. Gibberellin from the embryo triggers the metabolism of stored starch in germinating monocot (barley) seeds.

9. The hormone cytokinin stimulates cell division and inhibits senescence. Different ratios of cytokinin to auxin concentrations lead to the induction of different types of organs (callus, shoots, or roots) in plant tissue cultures.

10. The hormone abscisic acid is responsible for inducing dormancy in the shoot buds of perennial plants and for maintaining dormancy in seeds. It stimulates abscission of petioles from cotton seedlings and induces the synthesis of certain storage proteins during seed development. It also is a signal of drought stress in leaves, stimulating the closure of stomata.

11. Ethylene, a gaseous plant hormone produced by senescent plant tissues, stimulates senescence of leaves and fruits. It also inhibits stem and root growth by disrupting the organization of cellulose microfibrils in the cell wall.

12. Several chemicals with hormone-like signaling activity are involved in the induction of resistance to disease. These include oligosaccharides (parts of fungal cell walls), salicylic acid, hydrogen peroxide, jasmonic acid, and a polypeptide chain called *systemin.*

13. Plants have several light-receptive systems that influence their development: Phytochromes respond to red and far-red light; cryptochrome(s) respond to blue and near-ultraviolet light; and a separate system responds to intermediate-wavelength ultraviolet (UVB) radiation.

14. Phytochromes are proteins that have two interconvertible forms: P_r, which absorbs red light, and P_{fr}, which absorbs far-red light. P_r is converted to P_{fr} in the presence of red light; P_{fr} is converted to P_r in the presence of far-red light. Seedlings that develop in darkness have only the P_r form. They have long stems, poorly developed leaves, undeveloped chloroplasts, and are said to be *etiolated.* The conversion of P_r to P_{fr} by red light de-etiolates the seedlings by inhibiting stem growth, stimulating leaf growth, and starting the maturation of the chloroplasts.

15. Plants use night length to control developmental processes in a way that adapts them to coming seasons. Plants that flower in response to short nights are called long-day plants, and those that flower in response to long nights are called short-day plants. In perennials, night length also influences the establishment of winter dormancy and the regrowth of buds in the spring.

16. The perception of day length by plants involves phytochrome and an endogenous oscillator (the biological clock). Plants seemingly measure the period of darkness between two light events; therefore, a light pulse (which forms P_{fr}) at a critical time during a dark period will stimulate flowering in a long-day plant and inhibit flowering in a short-day plant.

Questions

1. Which of the following plant organs show indeterminate growth: leaf, vegetative shoot, flowering shoot, carpel, root, stolon?

2. Describe how each of the following concepts applies to plant development: the formation of determined cells; the appearance of competent cells; the preservation of totipotent cells; indeterminate growth and modules.

3. If pieces of stem tissue are placed in the appropriate tissue culture medium, they will regenerate a whole plant. Discuss possible reasons why this can be done with plants and not with humans.

4. The expression of genetic information in the production of a protein involves several sequential steps. Place the following steps in the correct order: translation; activation of chromosomal DNA; transcription; activation of the enzyme; processing of nuclear RNA; processing and transport of the translation product; enzyme activity.

5. Discuss the following statement: Each hormone affects plant development by turning on the expression of a specific gene.

6. Using the concept of a signal cascade, describe how it is possible for both auxin and gibberellin to stimulate the elongation of young internode cells.

7. Name the hormone that is probably involved in the following responses:

a. In early summer, your cabbage plants bolt and flower.

b. In the fall, your ash tree stops growing and develops protective bud scales over its shoot apexes.

c. The ficus tree in your apartment is leaning toward the window.

d. The last of your peaches is getting soft.

e. You spray your ficus tree with a new chemical, which causes lateral buds on the tree to start growing branch shoots.

8. Four oat coleoptiles have been cut from their seed. The first (a) has been slit vertically; the second (b) has its lower section covered by a light-tight shield; the third (c) has its upper section covered by a light-tight shield; and the fourth (d) has had its tip removed. The coleoptiles have been placed in a support with light coming from one side. Which coleoptiles will bend toward the light? Assume that no coleoptile shades another coleoptile.

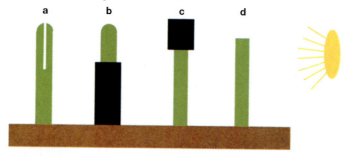

9. Mature plants become leggy—that is, they have longer than normal internodes—when they grow under a thick canopy of leaves. Chlorophyll in the canopy absorbs more red than far-red light from sunlight. It is hypothesized that one form of phytochrome inhibits stem growth more than the other. In this hypothesis, which form of phytochrome, P_r or P_{fr}, inhibits stem growth? Explain your choice.

10. A scientist has suggested that the synthesis of phytochrome is controlled by phytochrome itself because the enzyme is synthesized less rapidly in the light than in the dark. Which of the observations below would most convincingly support this suggestion? Discuss your answer.

The mRNA for phytochrome is present only in the light.

High intensities of light are necessary for the plant to make normal amounts of phytochrome.

The rate of synthesis of phytochrome is greater in plants treated with a pulse of red light followed by a pulse of far-red light than in plants treated only with red light.

Phytochrome is synthesized only in red light.

InfoTrac® College Edition

http://infotrac.thomsonlearning.com

Plant Hormones

Jones, A.M., Im, K-H., Savka, M.A., Wu, M-J., Dewitt, N.G., Shillito, R., Binns, A.N. 1998. Auxin-dependent cell expansion mediated by over-expressed auxin-binding protein 1. *Science* 282:1114. (Keywords: "auxin" and "cell expansion")

Raloff, J. 1987. Plant hormone: Key to ozone toxicity? *Science News* 131:357. (Keywords: "hormone" and "ozone")

Riou-Khamlichi, C., Huntley, R., Jacqmard, A., Murray, J.A.H. 1999. Cytokinin activation of arabidopsis cell division through a D-type cyclin. *Science* 283:1541. (Keywords: "cytokinin" and "cyclin")

Rouse, D., Leyser, O. 1998. Changes in auxin response from mutations in an AUX/IAA gene. *Science* 279:1371. (Keywords: "auxin" and "AUX")

Stelljes, K.B. 2001. Abscisic acid—the plant stress hormone. *Agricultural Research* 49:20. (Keywords: "Stelljes" and "abscisic")

Effects of Light

Raloff, J. 1997. When tomatoes see red: The horticultural tricks colored mulch can play. *Science News* 152:376. (Keywords: "Raloff" and "tomatoes")

Robison, S.A., McCarthy, B.C. 1999. Growth responses of Carya ovata (Juglandaceae) seedlings to experimental sun patches. *The American Midland Naturalist* 141:69. (Keywords: "seedlings" and "sun patches")

Genetics

Visit us on the web at http://biology.brookscole.com/plantbio2
for additional resources, such as flashcards, tutorial quizzes,
InfoTrac exercises, further readings, and web links.

1. The hereditary traits of an organism are determined by units of information called *genes* and are encoded in the base sequence of the organism's DNA. A mutation in a DNA base sequence produces a new allele of a gene, which can sometimes be identified by a detectable change in a trait.

2. In plants, DNA is divided among several chromosomes, with each gene occurring at a specific place along the length of one of the chromosomes.

3. Meiosis and fertilization work together to assure that different combinations of alleles appear in new generations.

4. The transmission of chromosomes, genes, and alleles from parents to progeny follows clear rules, so that the traits of progeny can be predicted from the traits of their parents and more distant ancestors. However, the processes of meiosis and fertilization include chance events; therefore, the predictions describe probabilities rather than certainties.

5. A knowledge of genetics allows plant breeders to develop crops with improved qualities and yields.

16.1 TRAITS, GENES, AND ALLELES

Plants of a particular species can be described by a series of **characters,** such as flower shape and color, stem length, leaf shape and arrangement, fruit type, and seed shape. All these characters are specified to some degree by internal factors called **genes.** Genes provide the instructions to the plant cells on how to grow and develop and how to respond to environmental cues. Genetic information is inherited—that is, passed from parents to progeny, from generation to generation in the process of sexual reproduction. **Genetics** is the study of how genes work and the rules by which they are inherited.

Different Alleles Produce Different Traits

Some basic principles of genetics were discovered by Gregor Mendel, an Austrian monk, botanist, and teacher who used garden peas (*Pisum sativum*) as his experimental organism. His report, published in 1866, was generally unappreciated until his principles were rediscovered around 1900. One of Mendel's principles was the idea that some plant characters are determined by heritable factors. When Mendel looked at his pea plants, he noticed that certain characters occurred in more than one form. For instance, for the character of stem length, pea plants could be short or tall; the character of flower color could be red or white; seed shape could be round or wrinkled. He reasoned that the factors for those characters came in more than one form. Today, each variant is called a **trait;** the factors are called genes; the alternate

forms of a gene are called **alleles.** One allele of a gene for stem length makes a pea plant tall; another makes it short.

Traits Reflect the DNA Code

Genes are sequences of nucleotides in DNA, or **base sequences** (referring to the parts of the nucleotides that give them their identities; see Chapters 2 and 15). What we know about genes has come from the work of many scientists, but no doubt the most famous breakthrough in knowledge was made by James Watson and Francis Crick, the discoverers of the double helical structure of DNA. They recognized how the sequence of bases along a DNA molecule could act as a code to specify the sequence of amino acids in a protein. Proteins have numerous important functions in a cell, including their role as enzymes.

The functions of proteins depend on their three-dimensional structures. An example is glucosyl transferase, an enzyme needed for the synthesis of amylose, a type of starch. This protein works because it has a three-dimensional structure that binds to the reactants (uridine-diphosphoglucose—an active form of glucose—and a growing starch chain) and catalyzes the reaction that adds glucose to the end of the chain. That three-dimensional structure is a consequence of the sequence of the amino acids in the protein molecule. The sequence of the amino acids is determined by the base sequence of its gene.

Sometimes a plant with a different form, a **mutant,** may appear spontaneously in a collection of similar plants from the same parental stock. **Figure 16.1** and **Table 16.1** compare several mutant plants to their "normal," or **wild-type,** relatives. Different forms of a trait arise through **mutation,** changes in the base sequence of DNA. Sometimes, the mutation is as simple as a change in one base. At other times, the mutation is more drastic: a complete disruption of the base sequence by the addition or deletion of a line of bases. When a mutation changes the base sequence of a gene that previously coded for an enzyme, the gene no longer contains the information to code for the correct amino acid sequence. Then the enzyme either is not made, or it does not function properly. For instance, a mutation in the gene encoding glucosyl transferase results in a corn plant not being able to make the functional enzyme. The mutant corn plant lacks the ability to form amylose in its seeds. Without that type of starch, the seeds appear "waxy"; therefore, the gene involved is called "waxy" (Table 16.1).

Mutations are the raw material for evolution through natural selection. Although most mutations are "bad," some may simply make a plant different; others may make a plant more suited to its environment. Once a mutation occurs, the altered DNA will be passed from the mutant organism to its progeny according to the rules that Mendel discovered. If the progeny that expresses the mutant trait reproduce more effectively, the mutation and its trait will spread through the population, and the population will have evolved.

Figure 16.1 Mutations in maize plants. (**a**) An albino mutant seedling among wild-type green seedlings. (**b**) An attached-leaf mutant in which the cuticle has changed, causing the leaves to stick together as they develop. (**c**) A stripe mutant, in which chlorophyll accumulation decreases periodically as the leaf grows from the base. (**d**) Mature ears, in which different kernels are expressing different combinations of alleles of genes controlling the production of anthocyanins (red and purple pigments).

Table 16.1 A Small Selection of Mutations In Three Plant Species

Wild-type Form	Alternate (Mutant) Form	Name of Gene
Pisum sativum (garden peas)		
Yellow cotyledons	Green cotyledons	I
Red flower petals	White flower petals	A_1
Smooth seed surface	Wrinkled seed surface	R
Tall (more than 20 internodes)	Short (10–20 internodes)	T
Green foliage	Yellow-green foliage	O
Axillary flowers	Terminal flowers	Fa
Straight pod	Curved pod	Cp
Tendrils	No tendrils	N
Zea mays (maize, corn)		
Filled endosperm	Shrunken endosperm (lacks sucrose synthase)	sh
Yellow endosperm	White endosperm	y
Colored (red) endosperm	Yellow endosperm	R
Normal endosperm	Waxy endosperm (altered starch-synthesizing enzyme)	wx
Dormant seed	Viviparous (germinates on cob)	Vp
Has isocitrate dehydrogenase	Lacks isocitrate dehydrogenase	idh
Brassica campestris (rapid-cycling Brassicas, fast plants)		
Normal internode length	Short internode (gibberellin-insensitive)	dwf
Normal internode length	Short internode (rosette–gibberellin-sensitive)	ros
Red pigment on hypocotyl	Lacks pigment	anl
Waxy covering on leaves, stem	Lacks waxy covering (glossy)	glo
Green leaves	Yellow-green (chlorophyll deficient)	ygr
Normal internodes	Elongated internodes (lacks phytochrome B)	ein

Chromosomes Carry Genes

In plants, as in all other eukaryotic organisms, most of the DNA is found in the nucleus (see Chapter 3) and is combined with proteins to form **chromatin.** The chromatin is divided into **chromosomes,** each of which contains a single linear strand of DNA. Each gene is a portion of the DNA between 300 and 3,000 bases in length (or longer). Every gene has a particular position—its **locus**—on one of the chromosomes **(Fig. 16.2).** It is separated from the adjacent genes by long stretches of DNA believed to be nonfunctional. Because the DNA in a chromosome may be a hundred million bases in length, there is plenty of room for several hundred genes on one chromosome. The collection of all the genes in an organism is called its **genome.** The smallest number of chromosomes in a plant is four, found in the desert plant *Haplopappus gracilis* (a relative of the sunflower) and some other species. The coast redwood (*Sequoia sempervirens*) has nearly a hundred chromosomes; some ferns have several hundred.

All the cells that make up the root and shoot of a plant are formed from an original single cell by a sequence of mitotic cell divisions. As described in Chapter 3, mitotic cell divisions follow the S phase of the cell cycle, during which the DNA in each chromosome is faithfully replicated. Mitosis assures that each of two progeny cells has one of every chromosome. This means that every allele (mutant or wild-type) found in the original cell also will be present in all the cells of the plant. When plants reproduce by vegetative reproduction, which occurs solely through mitotic cell divisions, the progeny plants have the same alleles as the parent plants.

Each different type of chromosome has a different set of genes. However, as explained in Chapter 12, all diploid organisms, including flowering plants, have two copies of each type of chromosome, one from the sperm and one from the egg. The two copies are said to be homologous because they have the same set of genes. That does not mean, however, that the genes at the same locus on two homologous chromosomes are necessarily identical; they may have different alleles.

Meiosis Segregates Alleles, and Fertilization Combines Them

Diploid organisms—including plants—reproduce sexually, a mechanism that offers greater genetic variety than mutation alone. As described in Chapter 12, the gametes of a plant—sperm and egg—are haploid, each having only one chromosome of each type. Their union forms a new single cell, the zygote, which has two sets of chromosomes. Thus, the zygote, and the plant that develops from it, is diploid. To reproduce, the diploid plant must form new haploid cells by meiosis.

If we assume that a plant (and thus every cell in that plant) has two different alleles (which we designate *A* and *a*) at some locus on a chromosome, we can trace the allocation of those alleles to gametes during the process of meiosis **(Fig. 16.3).** The preparation for meiosis begins in the preceding S phase with the synthesis of new DNA, after which each chromosome has two identical sister **chromatids** and two copies of each gene. Meiosis opens with prophase I, in which the chromosomes coil, shorten, and thicken, becoming more visible. During this period, homologous chromosomes come together to form pairs (*synapsis*). The nuclear

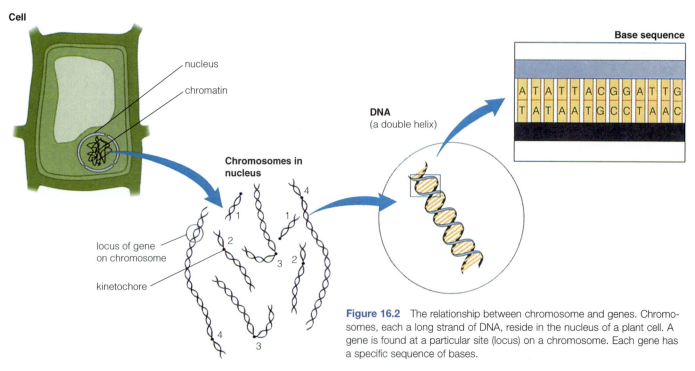

Figure 16.2 The relationship between chromosome and genes. Chromosomes, each a long strand of DNA, reside in the nucleus of a plant cell. A gene is found at a particular site (locus) on a chromosome. Each gene has a specific sequence of bases.

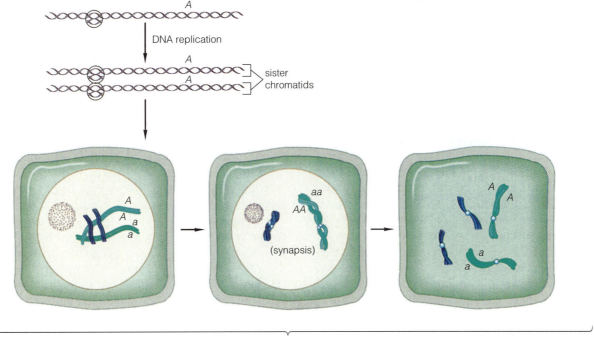

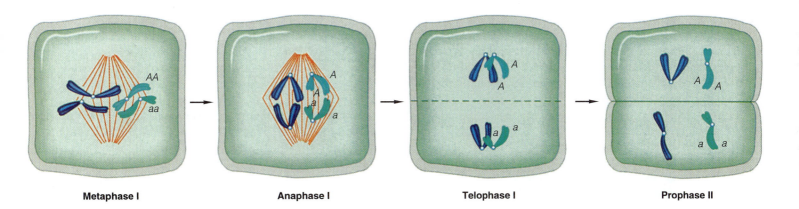

Metaphase I   **Anaphase I**   **Telophase I**   **Prophase II**

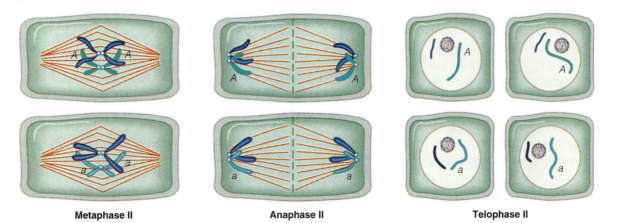

Metaphase II   **Anaphase II**   **Telophase II**

Figure 16.3   The separation of alleles of one gene (one locus on a chromosome) during meiosis. During the preceding S phase, the DNA in each chromosome is replicated. Each sister chromatid has a copy of each gene. During prophase I, the chromosomes condense, and homologous chromosomes come together (synapsis). At this point, there may be an exchange of pieces among the four chromatids in each chromosome pair (see Fig. 16.4). A spindle forms, and during anaphase I, the two homologous chromosomes separate. After intervening stages (telophase I, prophase II), a second division (anaphase II) separates the two sister chromatids of each chromosome.

envelope disappears, a spindle forms, and in the first division, the spindle separates the homologous chromosomes. A second division separates the chromatids, so each daughter cell gets one chromatid (now referred to as a chromosome) of each type and only one copy of each gene (see anaphase II and telophase II in Fig. 16.3). The original diploid cell had different alleles for a particular gene (*A* and *a*); each resulting haploid gamete got only one of the two alleles (*A* or *a*). In Mendel's language, the two alleles segregated (separated) during meiosis. (This is sometimes known as Mendel's Law of Segregation.) Amazingly, even though Mendel did not know about chromosomes, genes, alleles, and meiosis, he deduced from his experiments how traits are passed from parents to progeny.

An unusual process occurs during prophase I **(Fig. 16.4).** During synapsis, the chromatids of homologous chromosomes may exchange some corresponding pieces with each other in a process called **crossing over.** The cross formed by the chromatids during the exchange (and visible with a microscope) is known as a **chiasma** (derived from the Greek word for "cross," plural *chiasmata*). The process results in the formation of rearranged chromatids possessing fragments from both of the homologous chromosomes. This is important when we consider more than one gene. If the homologous chromosomes of the parent cells have different alleles for both genes, then crossing over produces chromatids with new combinations of alleles (Fig. 16.4), an effect known as **recombination.** In contrast to mutation, which actually changes alleles, recombination simply forms new combinations of alleles already present in the genome.

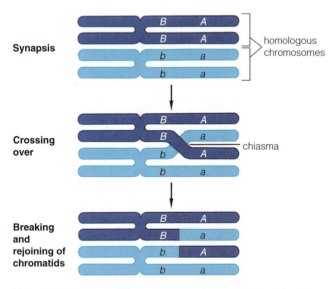

Synapsis

Crossing over

chiasma

Breaking and rejoining of chromatids

homologous chromosomes

Figure 16.4 Steps leading to the recombination of alleles. Crossing over between two homologous chromosomes during prophase I of meiosis can recombine alleles for the genes on that chromosome. *A* and *a* represent different alleles of one gene at a particular locus on the chromosome; *B* and *b* represent different alleles of another gene. In the crossing over shown, nonsister chromatids of a homologous pair of chromosomes exchange *A* and *a* alleles.

Simple Crosses Yield Predictable Results

Mendel was not the first person to study heredity, but he was the first to identify clear rules that predicted and explained the inheritance of traits. His conceptual breakthrough was the identification of several individual characters in pea plants, each of which occurred in one of two forms. Mendel grew plants that were identical except for differences in one, two, or three of these characters, so that he could focus on the inheritance of the alleles of their genes.

SINGLE GENES The most basic type of genetic experiment involves a mating between two plants that are genetically identical except for one character, in which they express different traits. If the character is controlled by one gene, then the two plants have different alleles for that gene.

An example comes from one of Mendel's experiments. He carefully took pollen from the anthers of a dwarf variety of peas and dusted it on the stigma of a tall variety. The resulting seeds were planted the next season. All the plants that grew from these seeds—called the F1 (first filial) generation—were tall. The same was true when pollen from tall plants was dusted on stigmas of dwarf plants. Then in a second mating, flowers of the tall progeny were self-pollinated; that is, pollen from each flower was dusted on the stigma of the same flower. The seeds resulting from these self-pollinated flowers—the F2 (second filial) generation—produced 787 tall plants and 277 dwarf plants. In other words, about three fourths of these progeny were tall, and one fourth of the progeny were dwarf plants.

Why were all the plants tall in the offspring of the first mating if one of the parents was dwarf? And how did some of the plants from the second mating come to be dwarf when none of their parents were? These results can be explained by the model shown in **Figure 16.5.** Because there are two traits for this character, tall and dwarf, we can assume that there are two alleles involved, called *T* (for tall) and *t* (for dwarf). Every adult diploid plant has two copies of each gene because it has two sets of chromosomes. Mendel took pains to make sure that his varieties were genetically homogeneous; therefore, the tall variety had two tall alleles, and the dwarf variety had two dwarf alleles. Geneticists call the visible traits of an organism its **phenotype** and its collection of alleles its **genotype.** In this case, the plant with the tall phenotype had the genotype *TT*, and the plant with the dwarf phenotype had the genotype *tt*. Plants with two copies of the same allele are called **homozygous** for that allele because the zygotes from which all of the plants' cells originated had two copies of the same allele. When plants form gametes, every gamete gets one allele from the parent (in this case, *T* or *t*, depending on which parent). Gametes from a homozygous parent all carry the same allele. The genotype of a zygote is the combination of the alleles from the two gametes that fused to form that zygote.

In the first mating, the progeny all received one *T* allele and one *t* allele, and so they all had the genotype *Tt*. Plants that have different alleles of a gene are said to be

Figure 16.5 Diagram of a cross between tall *(TT)* and dwarf *(tt)* pea plants. All plants in the first filial generation (F1) are heterozygous *(Tt)*, meaning that the zygote— and therefore the cells—have both alleles for the trait, one from each parent. The dwarf allele *(t)* is not outwardly visible in the F1 plants, which all exhibit a tall phenotype. When the F2 plants are self-pollinated, the second filial generation (F2) produces (on average) three plants with the tall phenotype *(TT, Tt, Tt)* to every one that has the dwarf *(tt)* phenotype.

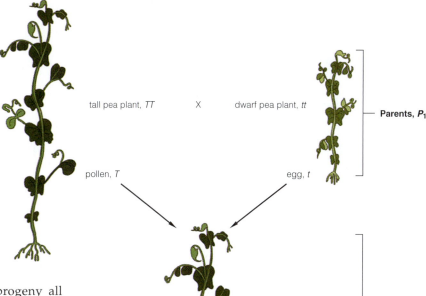

tall pea plant, *TT*   X   dwarf pea plant, *tt*

pollen, *T*   egg, *t*

Parents, *P₁*

tall pea plants (hybrids), *Tt*

First filial generation, *F₁*

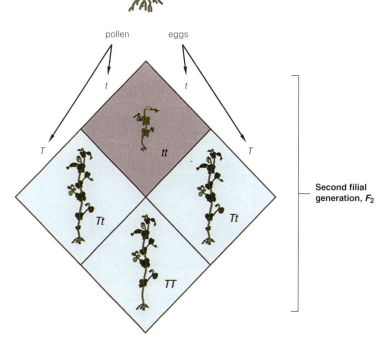

pollen   eggs

Second filial generation, *F₂*

heterozygous. The model suggests that the progeny all had the same genotype; this fits with the observation that they all had the same phenotype.

But why were they tall, if they had copies of both alleles? To explain why the progeny were tall, we have to assume that the expression of the tall allele in some way overshadowed the expression of the dwarf allele. We say that the tall allele was **dominant;** the dwarf allele was **recessive.** This means that the presence of one *T* allele is enough to produce the tall phenotype. Only a *tt* genotype will produce a dwarf plant.

The relationship among alleles is not always dominant–recessive; for some genes, heterozygous plants show traits intermediate between those of the parental homozygotes. In these cases, the alleles are said to be **codominant** or **incompletely dominant.** For instance, in snapdragons (*Antirrhinum majus*), the combination of a red allele and a white allele for the flower color gene produces pink flowers.

To explain the results of the second mating, additional assumptions need to be made. The first assumption is that the two alleles in the *Tt* parents are distributed randomly to their gametes; therefore, half the gametes receive a *T* allele, and half receive a *t* allele. The second assumption is that the gametes fertilize randomly, so that combinations of alleles are formed in proportion to their frequency in the gametes.

According to the model, the *Tt* parent that donated pollen produced half *T* and half *t* pollen. The *Tt* parent that was pollinated produced half *T* eggs and half *t* eggs. At fertilization, there were four combinations, and they occurred in equal proportions. One-fourth had the *TT* genotype, half had the *Tt* genotype, and one-fourth had the *tt* genotype. Because the *TT* and *Tt* genotypes yield tall plants, three fourths of the progeny were tall; the remaining one fourth of the progeny were dwarf plants.

A Punnett square (named for the genetics professor who popularized it) is useful for keeping track of the combinations of alleles formed during fertilization. It is constructed by drawing a square and listing alleles of the male gametes available for the mating along one side of the square and those of the female gametes along the perpendicular side (Fig. 16.5). The combinations of alleles at the intersections of the columns and rows show the expected genotypes of the progeny. The mathematical principles of probability can be used to make quantitative predictions about the phenotypes and genotypes of the progeny (see sidebar "IN DEPTH: Probability and Mendelian Genetics").

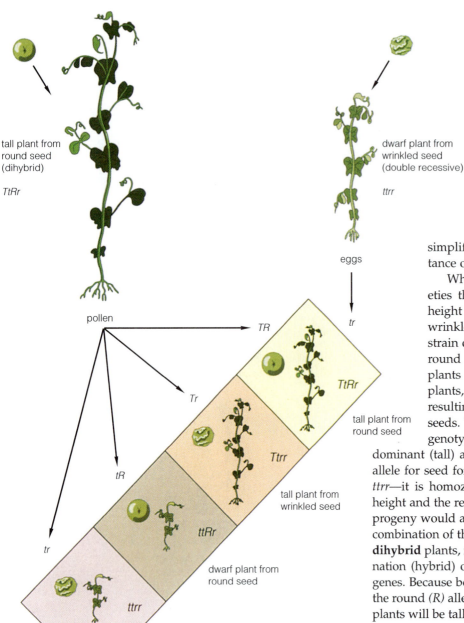

tall plant from
round seed
(dihybrid)

TtRr

dwarf plant from
wrinkled seed
(double recessive)

ttrr

eggs

pollen

TR

tr

Tr

TtRr

tall plant from
round seed

Tr

Ttrr

tall plant from
wrinkled seed

tR

ttRr

dwarf plant from
round seed

tr

ttrr

dwarf plant from
wrinkled seed

Figure 16.6 Inheritance of combinations of alleles of unlinked genes (genes on different chromosomes). *R* is the round allele for seed shape; *r* is the wrinkled allele. *T* is the tall allele for height; *t* is the dwarf allele. A cross between a dihybrid pea plant, heterozygous for two unlinked genes *(RrTt)*, with a plant that is homozygous recessive for both genes *(rrtt)* gives progeny with four different genotypes and phenotypes. The phenotypes of the progeny reflect the genotypes of the gametes produced by the dihybrid parent.

simplified this question by studying the inheritance of pairs of genes.

What will be the result of mating two pea varieties that differ in two characters—for instance, height (tall or dwarf) and form of seeds (round or wrinkled)? Let us assume that we have available a strain of plants that breeds true for tall plants and round seeds and one that breeds true for dwarf plants and wrinkled seeds. When we mate these plants, we will find—as Mendel did—that the resulting progeny are all tall and have round seeds. This fits a model in which one strain has a genotype of *TTRR*—it is homozygous for both the dominant (tall) allele for height and the dominant (round) allele for seed form—and the other strain has the genotype *ttrr*—it is homozygous for the recessive (dwarf) allele for height and the recessive (wrinkled) allele for seed form. The progeny would all have the genotype *TtRr,* representing the combination of the genes from *TR* and *tr* gametes. These are **dihybrid** plants, meaning that these offspring are the combination (hybrid) of plants that differ in their alleles for two genes. Because both the tall *(T)* allele of the height gene and the round *(R)* allele of the seed gene are dominant, all the F1 plants will be tall and will come from round seeds.

What happens when these dihybrid progeny plants are used as parents? When cells undergo meiosis, the alleles from these two genes *(TtRr)* are distributed so that each gamete receives one allele of each gene. There are four possible combinations in the gametes: *TR, Tr, tR,* and *tr.* If the two genes are on different chromosomes (which act independently in meiosis), the alleles will be distributed randomly and the four combinations of alleles will occur in equal numbers. If such a plant is mated with the homozygous recessive parent *(ttrr),* the phenotypes of the progeny will reflect the genotypes of the gametes from the dihybrid **(Fig. 16.6)**—that is, four different phenotypes will occur in equal numbers. Mating an experimental plant, such as the dihybrid described earlier, with a plant known to be homozygous recessive in all the genes of interest is called a **test cross** because the phenotypes of the progeny indicate the genotypes of the gametes from the plant to be tested. It is an easy technique for determining whether combinations of alleles are produced in equal numbers in the gametes of a dihybrid plant.

Performing this experiment with two plants that differ in one trait constitutes a test of the assumptions we have made. If the progeny of the first cross were not all identical, we would have to abandon the assumption that the parents were homozygous. If the progeny of the first cross were identical, but the proportions of the second cross did not come out 3:1, we would need to question the assumption that the trait was controlled by a single gene.

TWO GENES Plants, like other organisms, are guided by thousands of genes acting in concert. The overall phenotype—for instance, the branching pattern of a tree together with the shape of its flowers—is determined by combinations of alleles of many genes, rather than by the alleles of a single gene. How are combinations of alleles inherited? Mendel

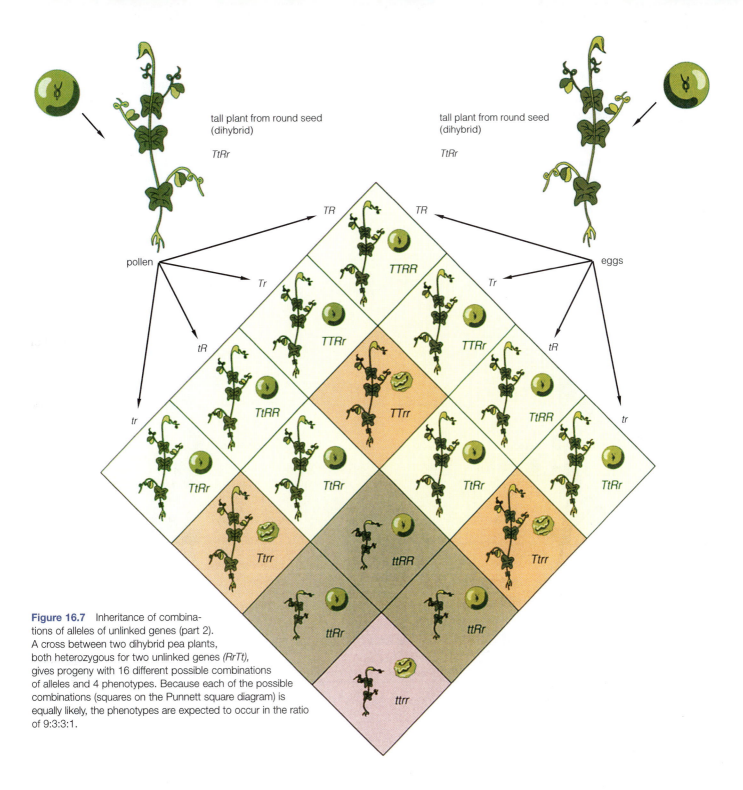

tall plant from round seed
(dihybrid)

TtRr

tall plant from round seed
(dihybrid)

TtRr

pollen

TR TR

Tr Tr

tR tR

tr tr

eggs

TTRR

TTRr TTRr

TtRR TTrr TtRR

TtRr TtRr TtRr TtRr

Ttrr ttRR Ttrr

ttRr ttRr

ttrr

Figure 16.7 Inheritance of combinations of alleles of unlinked genes (part 2). A cross between two dihybrid pea plants, both heterozygous for two unlinked genes *(RrTt)*, gives progeny with 16 different possible combinations of alleles and 4 phenotypes. Because each of the possible combinations (squares on the Punnett square diagram) is equally likely, the phenotypes are expected to occur in the ratio of 9:3:3:1.

A more complicated case occurs if we use dihybrid plants for both parents. In this case, there are four possible genotypes for both the male and the female gametes. Combining these genotypes at fertilization gives 16 possible genotypes in the zygote **(Fig. 16.7).** The Punnett square now becomes a useful tool for arranging the zygote genotypes and grouping the phenotypes for the different genotypes. We find that the mating of two dihybrids gives tall plants with round seeds, tall plants with wrinkled seeds, dwarf plants with round seeds, and dwarf plants with wrinkled seeds in the ratio 9:3:3:1. The 9:3:3:1 ratio is typical for all crosses in which both parents are dihybrid for the same genes.

Mendel, who knew nothing of meiosis or chromosomes, correctly interpreted data from crosses such as these to mean that pairs of heritable factors assort independently in gamete formation. (This is sometimes called Mendel's Law of Independent Assortment.) Today, it is said that two genes that are independently assorted are unlinked, meaning that they are on different chromosomes or that they are so far apart on the same chromosome that they need not be transferred together during meiosis.

When Genes Are Linked, Traits Are Inherited Together

What happens if, in a test cross of a dihybrid plant, equal numbers of the four possible genotypes are not found in the progeny? This result may mean that the two genes are on the same chromosome. In this model **(Fig. 16.8),** one chromosome of the dihybrid parent carries the dominant alleles of two genes—for example, genes for tendrils and seed shape—and the homologous chromosome carries the recessive alleles. We say the genes are linked because during the formation of gametes, these alleles migrate as a unit rather than independently. So, in an uncomplicated course of events, when the chromatids separate in meiosis, half the gametes receive the two dominant alleles, and the other half receives the two recessive alleles (Fig. 16.8a). The plants that come from round seeds have tendrils; the plants that come from wrinkled seeds do not.

But it is not quite that simple. In prophase I, chromatids may exchange alleles through crossing over (Fig. 16.4b). This can result in the formation of chromatids with recombinant

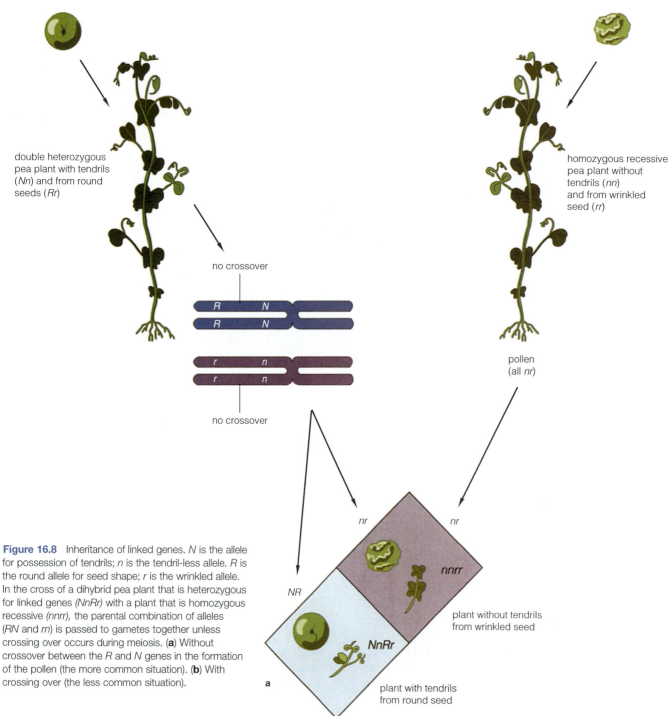

Figure 16.8 Inheritance of linked genes. *N* is the allele for possession of tendrils; *n* is the tendril-less allele. *R* is the round allele for seed shape; *r* is the wrinkled allele. In the cross of a dihybrid pea plant that is heterozygous for linked genes *(NnRr)* with a plant that is homozygous recessive *(nnrr)*, the parental combination of alleles *(RN* and *rn)* is passed to gametes together unless crossing over occurs during meiosis. (**a**) Without crossover between the *R* and *N* genes in the formation of the pollen (the more common situation). (**b**) With crossing over (the less common situation).

double heterozygous pea plant with tendrils (*Nn*) and from round seeds (*Rr*)

homozygous recessive pea plant without tendrils (*nn*) and from wrinkled seed (*rr*)

no crossover

no crossover

pollen (all *nr*)

nr

nr

NR

nnrr

plant without tendrils from wrinkled seed

NnRr

plant with tendrils from round seed

a

combinations of alleles (Fig. 16.8b). In the situation we postulated, in which the two dominant and two recessive alleles are the original (parental) combinations, a recombinant combination of alleles would be the dominant allele of one gene and the recessive allele of the other. Some plants from round seeds will lack tendrils, and some from wrinkled seeds will have them. Notice, however, that the situation could be reversed if different parents had been chosen: The parental combinations could be the dominant allele of one gene and the recessive allele of the other.

The probability of crossover events occurring depends on how close together the genes are on the chromosome. Assuming that crossing over occurs randomly along the chromosome, the farther apart two genes are, the more likely crossing over will occur between them, breaking their linkage and forming recombinant combinations of alleles. However, no matter how far apart the two genes are, the fraction of gametes with recombinant combinations is never greater than the fraction of gametes with the parental combinations of alleles (that is, never greater than 50%).

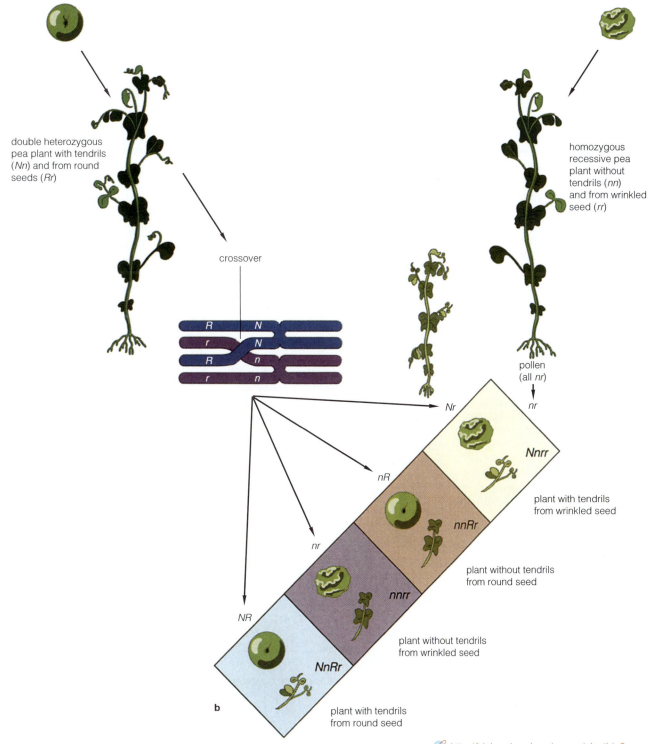

double heterozygous pea plant with tendrils (*Nn*) and from round seeds (*Rr*)

homozygous recessive pea plant without tendrils (*nn*) and from wrinkled seed (*rr*)

crossover

pollen (all *nr*)

Nr   *nr*

nR

nr

NR

Nnrr

plant with tendrils from wrinkled seed

nnRr

plant without tendrils from round seed

nnrr

plant without tendrils from wrinkled seed

NnRr

plant with tendrils from round seed

b

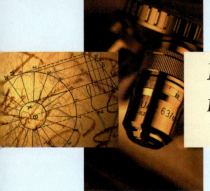

IN DEPTH:

Probability in Mendelian Genetics

One of Mendel's most important innovations was his use of the concepts of mathematical probability and statistics in the study of heredity. These concepts, first developed by Blaise Pascal and Pierre de Fermat to analyze games of chance, can be applied to any event or experiment in which the results are subject to both random fluctuations and well-ordered processes. The purposes of such application are to separate the well-ordered features of the experiment from the random features, to identify the patterns that the experiment follows, and to use these patterns to predict future results.

A critical concept in probability is the idea of the *outcome* of an event or experiment. The outcomes are unit results, only one of which can occur at any one trial of the experiment and all of which together constitute all possible results of the experiment. If we toss a coin to see which side lands up, the outcomes are "heads" and "tails." If we mate two parents and observe the sex of a resulting offspring, the outcomes are male and female.

A basic operation in the analysis of probability is to assign numbers to each of the various outcomes, numbers that indicate our feeling about the likelihood that each outcome will occur. If we are certain that a particular outcome will occur, we assign it a probability of 1.0 (denoted $P = 1.0$); if we are certain that an outcome will not occur, we assign it a probability of 0.0 ($P = 0.0$); if we feel an outcome will occur half the time, we assign it a probability of 0.5 ($P = 0.5$).

Some of the principles useful in assigning and interpreting numerical probabilities are described below, together with examples taken from problems in Mendelian genetics.

1. In an experiment with n equally likely outcomes, the probability of the occurrence of any one outcome is $1/n$.

Example: In the mating of two hybrid plants with genotypes *Gg*, a Punnett square diagram indicates that the progeny may have any of four different genotypes (that is, outcomes): *GG*, *Gg*, *gG*, and *gg*. The probability of any one occurring is $1/4$.

2. If the probability of a particular outcome is $1/n$, then in a *large* number of trials, that outcome will tend to represent $1/n$th of the total.

Example: In the experiment described above, the probability of finding the *gg* genotype in any one seed is $1/4$. If we looked at 1,000 progeny seeds, we would expect to find $1/4(1,000) = 250$ seeds with the *gg* genotype. We also would expect to find close to 250 seeds each with the *GG*, *Gg*, and *gG* genotypes. Notice, however, that if we look only at *small* numbers of seeds, we should not be surprised if the outcomes do not reflect their respective probabilities. If we look at 10 seeds, we do not expect 2.5 seeds with each genotype; if we look at 100 seeds, we do not expect *exactly* 25 seeds with each genotype. But the more seeds we consider, the closer the relative frequency of appearance of each genotype should be to its probability.

3. The probability of an occurrence represented by two or more different (mutually exclusive) outcomes is the sum of the probabilities of the separate outcomes.

Example: Assume that the experiment above refers to seed color. The dominant allele, *G*, gives yellow seeds; the recessive allele, *g*, gives green seeds when present in a homozygote. The occurrence, "A seed is yellow," is represented by three different outcomes (the genotypes *GG*, *Gg*, and *gG*). The three outcomes are mutually exclusive because only one can occur in any one seed. The probability that a seed is yellow, therefore, is the sum of the probabilities for these three genotypes.

$$P(\text{seed is yellow}) = P(GG) + P(Gg) + P(gG)$$
$$= 1/4 + 1/4 + 1/4 = 3/4$$

Likewise, the probability that the seed is heterozygous is:

$$P(\text{seed is heterozygous}) = P(Gg) + P(gG)$$
$$= 1/4 + 1/4 = 1/2$$

4. If the probability of a certain outcome in one experiment is $1/n_1$, and the probability of another outcome in a second (independent) experiment is $1/n_2$, the probability that *both* outcomes will occur is $(1/n_1)(1/n_2)$.

Examples: In the experiment described above, the probability of finding a green seed, genotype *gg*, is $1/4$; the probability of finding a yellow seed, genotype *GG* or *Gg* or *gG*, is $3/4$. Each seed represents an independent mating event, so the probability that the first seed investigated is green and the second seed investigated is yellow is:

$$P(\text{first seed green } and \text{ second seed yellow})$$
$$= P(gg) \cdot P(GG \text{ or } Gg \text{ or } gG) = (1/4)(3/4) = 3/16$$

The probability that the first seed, the second seed, and the third seed are all green is:

$$P(\text{first seed green } and \text{ second seed green } and$$
$$\text{third seed green}) = P(gg) \cdot P(gg) \cdot P(gg)$$
$$= (1/4)(1/4)(1/4) = 1/64$$

 Websites for further study:

Probability of Inheritance:
http://anthro.palomar.edu/mendel/mendel_2.htm

Probability and Genetics:
http://www.facstaff.bucknell.edu/udaepp/090/w3/matthewr.htm

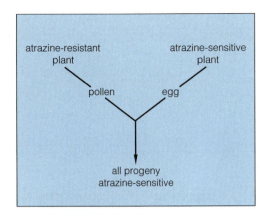

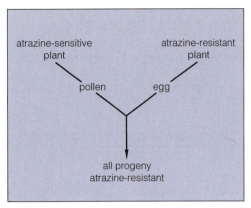

Figure 16.9 Maternal inheritance of a gene. The gene for resistance to the herbicide atrazine is on chloroplast DNA. Chloroplasts and chloroplast DNA in an embryo come only from the egg.

Maternal Inheritance Involves Organellar Chromosomes

Some characters do not follow the rules of inheritance described in the preceding sections. One of these is the response to the herbicide atrazine. Atrazine is a chemical that blocks electron flow in photosynthesis. Most plants are sensitive to it and will die after an atrazine treatment and exposure to light; a few mutant plants, however, show resistance. Atrazine sensitivity (*atr^s^*) and atrazine resistance (*atr^r^*) are inherited traits; they represent alleles of one gene.

If an atrazine-sensitive stigma is pollinated with pollen from an atrazine-resistant plant, all the progeny plants will be sensitive **(Fig. 16.9).** If, in contrast, an atrazine-resistant stigma is pollinated with pollen from an atrazine-sensitive plant, all the progeny plants will be resistant. This is not at all what would be expected under the Mendelian model described earlier. Regardless of whether the sensitive allele or the resistant allele was dominant, or both alleles were codominant, the results of these two matings should have been the same. Instead, it seems that the progeny inherited their sensitivity (or resistance) allele only from the maternal parent.

To explain this unusual type of inheritance, we must recognize that although most of the genes in a plant cell are located on the nuclear chromosomes, not all of them are. There are chromosomes of DNA in the plastids and in the mitochondria as well. These organellar chromosomes are smaller than those in the nucleus, and they are different in other ways; but they do contain genes, and these genes can mutate. During fertilization, only the chloroplasts and mitochondria from the egg are incorporated into the zygote. Chloroplasts and mitochondria from the sperm cells either do not enter the egg or degenerate during the fertilization process. Thus, the chloroplast and mitochondria genes in the zygote all come from the egg, and all the alleles of these genes show maternal inheritance.

16.3 PLANT BREEDING

In the last 40 years, the earth's population has doubled, but the number of acres of land under cultivation has decreased. Despite the increase in population and the decrease in farmed land, the amount of food produced per person is at least as great as it was in the 1950s. The ability of farmers to increase their output results in great part from the breeding of new, more productive plants. New varieties have been formed with characteristics that make them easier to grow or harvest, with resistance to disease or stress, or with edible parts that are more attractive or nutritious. For instance, modern rice (*Oryza sativa*) varieties are shorter and less likely to fall over when the grain is mature. (Plants that fall over are difficult to harvest, and their grain is more likely to mold.) Modern tomato (*Lycopersicon esculentum*) varieties are bred with resistance to *Verticillium* and *Fusarium* fungi.

Mating Plants Combines Useful Traits

The principles of genetics are used extensively to develop plants with new and useful traits **(Fig. 16.10).** However, there are some tricks to plant breeding that go beyond the basic genetic principles.

Pioneer Hi-Bred International, Inc. © 1996

Figure 16.10 Breeding trials for maize varieties at a research farm of Pioneer Hi-Bred International, Inc. Each plot represents a different stage in the development of a new variety or a test of different varieties under different conditions (for example, soil type or amount of fertilization or irrigation).

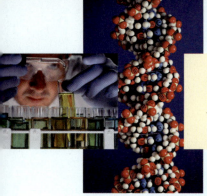

BIOTECHNOLOGY:

Plant Breeding and the Green Revolution

In 1970, the Nobel Peace Prize was awarded to Norman Borlaug, a plant breeder and plant pathologist who, funded by the Rockefeller Foundation, had started a program of breeding high-yield corn and wheat in Mexico. His ideas about breeding, irrigation, and fertilization spread throughout much of the world. Internationally recognized centers such as CIMMYT (the abbreviation of the Spanish name for the International Maize and Wheat Improvement Center) in Mexico and IRRI (International Rice Research Center) in the Philippines produced new varieties that tripled yields between 1950 and 1990. The director of the United States Agency for International Development coined the term *Green Revolution* to point out the spectacular increases in productivity resulting from the application of scientific knowledge to farming.

Much of the advantage of the new varieties of wheat and rice came from breeding for a seemingly simple morphological change: height. The original varieties grew very tall. It is possible that height gave these plants an evolutionary advantage in dispersing pollen, because their flowers are at the top of the stalk and their pollen is carried by the wind. However, there is an agricultural disadvantage to tall grasses. They tend to fall over ("lodge") when disturbed by wind, rain, animals, or when they are too crowded. The Green Revolution breeding programs selected much shorter plants. These resisted lodging, and the extra energy not used in growing upward could be used to produce more seed mass.

Another advantage from breeding programs lies in the development of pathogen resistance. Fungi and bacteria can account for the loss of a large fraction of a wheat or rice harvest. There are, however, strains that develop resistance to infection by a specific pathogen, and the genes that produce the resistance can be transferred to varieties in combination with other desirable traits. This is an ongoing activity, because new fungal or bacterial strains that overcome the resistance continually arise through mutation, recombination, and natural selection. One of the goals of biotechnologists (see Chapter 17) is to analyze the molecular mechanisms of resistance and develop methods that are less easily circumvented by pathogen evolution.

A plant breeding project starts by identifying genetic plants with new or improved traits. These might come from spontaneous mutations in a population of crop plants, from chemical or physical mutagenic treatments, or from wild relatives. To find wild plants with useful traits, it is important to know the center of origin of the crop of interest (**Fig. 1**). Once a plant with a valuable trait is located, it is crossed to the best crop plants, and progeny with the desired trait are selected for subsequent crosses ("backcrosses") to the crop. It is important to perform many cycles of backcrosses to ensure that the valuable traits of the crop plant are not lost. Because the desired trait may be recessive, or even dependent on several alleles, so that its phenotype does not appear in the initial progeny, it may be necessary to use molecular techniques for selecting plants with the appropriate alleles. Chapter 18 discusses some useful methods, such as polymerase chain reaction.

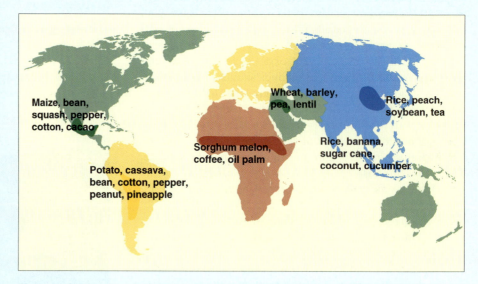

Figure 1 Centers of origin of some important crops. (Information from P. Gepts, University of California, Davis, CA.)

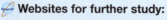 **Websites for further study:**

International Maize and Wheat Improvement Center:
http://www.cimmyt.org/

International Rice Research Center:
http://www.irri.org/

Suppose we want to breed a tomato variety that will resist infection by a fungus that has recently appeared. We may mate a successful but fungus-susceptible commercial variety with a wild variety that shows resistance but has inedible fruit. If the resistance allele is dominant (let us assume that it is), the progeny will be resistant, but they probably will have inedible fruit. We then would mate the progeny with the commercial variety (a back cross) and test the progeny of that mating for resistance. By chance recombination, some of the resistant progeny also will have acquired some of the genes needed for edible fruit. The most resistant progeny would again be mated with the commercial variety, and the most resistant progeny again selected. After several cycles, a strain that has both commercial fruit and resistance to the fungus would result.

Multiple Genes Explain Continuous Variation

The traits discussed in the previous paragraphs occur in one of two (or three or four) forms; they are said to have discrete variation. Many important traits, however, vary *continuously* over a certain range; these are said to have continuous variation. Examples in crop plants are size of the harvested organ (Fig. 16.11), sugar content, or firmness of the fruit.

There are several factors that can lead to a continuous variation. First, the most important factor is the involvement of multiple genes, each of which influences the trait of inter-est. Individually, alleles of the different genes may have rather small effects on the phenotype. Together, they can combine to provide a wide range of variation. If there are many genes, there are many combinations of alleles, and this situation leads to a distribution of phenotypes in which the differences among many individual forms are small, relative to the total range of possibilities. Genetic experiments have shown that the length of the corolla of a *Nicotiana* flower is controlled by at least four genes. Second, sometimes a gene has multiple alleles, each with a different degree of activity. This increases the number of possible phenotypic forms. It may occur because different changes of the base sequence of a gene produce different amino acid substitutions in an enzyme, leading to different amounts of enzyme activity. The *R* gene of peas occurs in several allelic forms, which give different seed colors. Third, environmental effects may alter the form of the phenotype. For instance, crowded conditions produce shading and tend to make bean plants grow longer internodes; hot, dry conditions will give grape berries a greater sugar concentration. The randomness of environmental effects tends to blur the distinction among genotypes.

The inheritance of multiple genes involved in a continuously variable trait is no different from that of other genes. Alleles can be combined by mating plants with different phenotypes, and progeny can be selected on the basis of their phenotypes for further matings with one another. It must be recognized, however, that if the number of genes is unknown, how many genes are heterozygous and how many are homozygous also is unknown. Obtaining a strain in which all the relevant genes are homozygous may be important (such a strain will breed true). But with a continuously variable trait, it may be difficult to tell when this has been achieved.

Heterosis Can Give Vigorous Progeny

It is sometimes found that the progeny from the mating of two inbred (highly homozygous) strains are much larger and healthier than either of the parents. This effect is called **hybrid vigor** or **heterosis**. Most corn (*Zea mays*) planted in the United States comes from hybrid seed. It poses a special problem for breeders because hybrids (heterozygotes) do not breed true—that is, when mated with themselves, they produce both heterozygous and homozygous progeny.

One solution is to produce new hybrid plants solely through vegetative reproduction. Potato (*Solanum tuberosum*) varieties that are reproduced by germinating buds cut from potato tubers all have the same genotype as the plants that produced the tubers.

Another solution is to produce hybrid seed by mating two homozygous strains. To do this, seed companies have developed strains that are male-sterile: They do not produce anthers or viable pollen. When a homozygous male-sterile strain of corn is planted close to another homozygous strain, it gets pollen only from the other strain, and all its seed will be heterozygous.

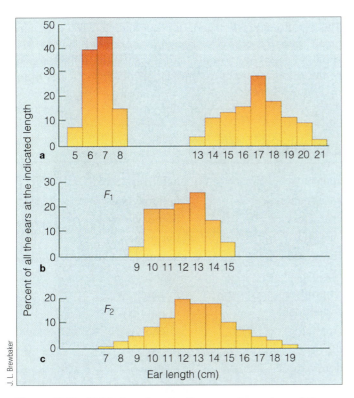

Figure 16.11 Distribution of ear lengths among two maize varieties and their progeny. Both parental varieties (**a**) and the F1 (**b**) and F2 (**c**) progeny show wide and continuous variation in ear lengths, probably because several genes contribute to the trait (1 cm = 0.4 inch).

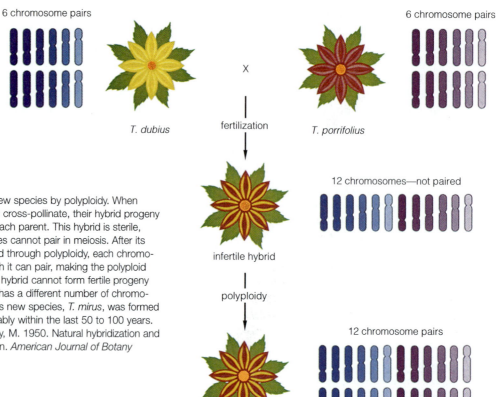

6 chromosome pairs

T. dubius

X

fertilization

T. porrifolius

6 chromosome pairs

infertile hybrid

12 chromosomes—not paired

polyploidy

12 chromosome pairs

T. mirus
(fertile polyploid hybrid)

Figure 16.12 The formation of a new species by polyploidy. When *Tragopogon dubius* and *T. porrifolius* cross-pollinate, their hybrid progeny has one set of chromosomes from each parent. This hybrid is sterile, because its two sets of chromosomes cannot pair in meiosis. After its chromosome complement is doubled through polyploidy, each chromosome has an identical one with which it can pair, making the polyploid hybrid fertile. However, the polyploid hybrid cannot form fertile progeny with either of its parents, because it has a different number of chromosomes; thus, it is a new species. This new species, *T. mirus*, was formed naturally in the American west, probably within the last 50 to 100 years. (Adapted from information in Ownbey, M. 1950. Natural hybridization and amphiploidy in the genus Tragopogon. *American Journal of Botany* 37:487–499.)

Plants Often Are Polyploid

In contrast to animals, plants become polyploid relatively easily. **Polyploidy** means having more than two sets of chromosomes. This may occur spontaneously, when a cell reduplicates its DNA and separates the resulting chromatids, but then fails to complete cell division. This commonly occurs in the last stages of development of tracheary elements and storage tissues, although it is less common in the meristem. If a meristematic cell becomes polyploid, the new nucleus, with twice the original number of chromosomes (tetraploid, if the parent cell was diploid), can undergo new rounds of DNA synthesis and mitosis. It may take over the meristem and produce a shoot whose cells are all polyploid. The cells are still capable of undergoing meiosis; therefore, they can form spores, and the spores can form gametes. If a polyploid plant fertilizes itself, its progeny also will be polyploid.

Plant breeders have found that polyploid plants often are larger and more vigorous than their parental types. They also have found that certain chemicals that interfere with the formation of the spindle in mitosis (colchicine, for example) can increase the probability of forming polyploid cells, and these chemicals have been used to produce new horticultural varieties. In nature, polyploids seem more tolerant of such environmental stresses as short, cool growing seasons, aridity, or high temperatures.

In plants, it is sometimes possible for different species to interbreed—that is, to form viable progeny. Often, the original progeny are sterile, but they can become fertile if their cells become polyploid **(Fig. 16.12).** The most important requirement for fertility is successful meiosis, and this requires that each chromosome be present in two copies.

This is automatically true in a polyploid cell. The common bread wheat (*Triticum aestivum*) is hexaploid, having six copies of each chromosome, and it is thought to have arisen in the Middle East by hybridization of three species from two related genera.

KEY TERMS

alleles	heterosis
base sequences	heterozygous
characters	homozygous
chiasma	hybrid vigor
chromatids	incompletely dominant
chromatin	locus
chromosomes	mutant
codominant	mutation
crossing over	phenotype
dihybrid	polyploidy
dominant	recessive
genes	recombination
genetics	test cross
genome	trait
genotype	wild-type

SUMMARY

1. The characters of plants are controlled by genes. Different alleles of a gene (mutant or wild-type) produce different traits, alternate forms of the character. The alleles, and thus the traits, are inherited.

2. Chemically, genes are made of DNA, and they control the production of proteins, which serve many functions in the cell.

3. Mutations are changes in the base sequence of the DNA. The modified base sequence may then change the amino acid sequence of an enzyme, or it may affect the rate or timing of expression of another gene.

4. Each gene is found at a particular locus on a chromosome. Different plants have different numbers of chromosomes, but each diploid cell in a plant has two copies of each chromosome.

5. Mitosis preserves the number of chromosomes and the alleles that are found on each chromosome. Vegetative reproduction occurs through mitosis and yields progeny that are genetically identical to the parent plant.

6. The sexual life cycle is dependent on meiosis. In meiosis, the two alleles of a gene in the diploid cell *segregate* (separate), so that half the haploid gametes receive one allele from the parent and half receive the other allele.

7. If two genes are on different chromosomes, the segregation of the alleles of one gene does not affect the segregation of the alleles of the other. This means that if there are four possible combinations of alleles from the two genes, each combination is equally likely, and one fourth of the gametes will receive each combination.

8. If two genes are linked (on the same chromosome), they tend to move into a gamete as a unit, rather than separately. The result is that the two parental combinations of alleles are more likely to occur in gametes than the two recombinant combinations of alleles. Recombinant combinations of alleles can be formed by crossing over of chromatids during meiosis.

9. Genetic techniques can be used to combine useful alleles of different genes in agricultural plants, even if the relationship between genes and traits is not simple.

10. Continuously variable traits can be explained by the influence of multiple genes, by the possibility of several different alleles, and by environmental effects. Even if a desired trait is controlled by several genes, a series of matings (with careful selection of progeny) can usually isolate plants with the correct combination of alleles.

11. Hybrids (heterozygotes) often are particularly vigorous. Special techniques are needed to produce hybrid seed for agricultural use.

12. Polyploidy (formation of plants with more than two chromosomes of each type) sometimes produces particularly vigorous plants. Polyploidy can also allow hybrids of distantly related plants to be fertile and form new plant species.

Questions

1. Provide the appropriate terms:

a. *Gene* is to *character* as _____ is to *trait*.

b. *Phenotype* is to *character* as *genotype* is to _____.

c. A *base sequence* is to a *gene* as _____ is to a *protein*.

d. _____ chromosomes have similar sets of genes.

e. _____ organisms have two identical alleles for one gene; _____ organisms have two different alleles for one gene.

f. A new allele can be produced by _____; a new combination of alleles can be produced by _____ during meiosis.

2. Trace the pathway by which information in the genes becomes a visible trait, by listing (in order) the processes that must occur.

3. Some wild radish plants have white flowers, whereas others have yellow flowers. Explain what you would do to determine whether the flower color is controlled by one gene with two alleles.

4. Explain why it takes plant breeders many growing seasons to produce a new crop plant (for instance, a tomato plant) with both desirable fruit and a new useful trait, such as disease resistance.

Genetics Problems

The abstract but regular nature of heredity has led biology teachers to adopt a tradition of assigning word problems in genetics. Following are some problems, all concerning plants, that will test your understanding of the concepts described in this chapter.

1. In peas, the dominant allele of the A_1 gene for flower color gives red flowers; the recessive allele a_1 (when homozygous) gives white flowers. Which is more likely: two plants with red flowers producing progeny plants with white flowers, or two plants with white flowers producing progeny plants with red flowers?

2. The wild-type, dominant allele of the *ein* gene in a mustard plant produces phytochrome, and the phytochrome regulates growth so that the plants have normal-length internodes. The recessive allele is associated with elongated internodes. In a greenhouse full of normal-sized mustard plants, one plant stands higher than the rest. If you assume that the increased height of this plant is because of a mutant allele of *ein*, then do you also assume that the plant is homozygous or heterozygous for this allele?

3. The color of pea cotyledons is controlled by several genes. One gene (probably the one that Mendel studied) has a dominant allele that suppresses the green color and gives yellow cotyledons. The recessive allele gives green cotyledons when it is homozygous. Assume that this is the gene involved in the following problem: A plant that grew from a green seed is pollinated with pollen from a yellow-seeded plant; a yellow seed from this mating is grown, and the resulting plant is pollinated with pollen from a plant that grew from a green seed. When the pods from this second mating mature, what proportion of the seeds will be yellow? What proportion will be green?

4. A geneticist collected seeds from a pea plant whose genetic constitution she did not know. All the seeds had yellow cotyledons. One pea plant, which grew from one of those yellow seeds, self-pollinated; the first pod she opened had one green and nine yellow seeds. She then collected 100 more seeds. Of these 100, how many did she expect would be green? Would she expect the same proportion of green seeds in a second self-pollinated plant grown from another of the original yellow seeds?

5. The flower color of sweet peas is controlled by two genes (*C* and *P*). To have red flowers, a plant must have a dominant allele of each gene (for example, *CCPP*, *CCPp*, *CcPp*, *CcPP*; note that the dominant alleles are capitalized). If a plant has two recessive alleles for either gene (or both) (for example, *ccpp*, *ccPP*, *CCpp*, *ccpP*, *Ccpp*), its flower will be white. Two white plants are mated, and they produce 100% red-flowered progeny. What were the genotypes of the parents?

6. In corn, the genes for red endosperm and shrunken endosperm are unlinked. The dominant allele of the red gene gives red seeds (*RR*, *Rr*); homozygous recessive seeds (*rr*) are yellow. The dominant allele of the shrunken gene gives normal-shaped seeds (*ShSh*, *Shsh*); homozygous recessive seeds (*shsh*) are shrunken. A plant that is heterozygous for the red gene and homozygous recessive for the shrunken gene is used to pollinate the silks (stamens) on a plant that is homozygous recessive for the red gene and heterozygous for the shrunken gene. What will be the genotypes and phenotypes of the seeds in the resulting cob, and in what proportions will they be expected to appear?

7. A corn plant that is heterozygous for three genes (*AaBbCc*) is crossed with a plant that is homozygous recessive for all three genes (*aabbcc*). Four phenotypes of progeny seedlings are found, in equal numbers. The four phenotypes correspond to the following combinations of alleles: *ABC*, *aBC*, *Abc*, *abc*. Which genes are linked, and which are unlinked?

8. In corn, the gene for red endosperm, *R*, is located on the same chromosome as the gene for white seedling, *W2*. A test cross of a plant that was heterozygous for the two genes with a homozygous recessive plant produced progeny with the following genotypes and in these numbers: *RrW2w2*, 20; *Rrw2w2*, 172; *rrW2w2*, 180; *rrw2w2*, 28. (*R* and *W2* are the dominant alleles; *r* and *w2* are recessive.) Draw a diagram of the homologous chromosomes in the heterozygous parent, showing which alleles were on which chromosomes (that is, show the parental combinations of alleles). What fraction of the pollen produced by the heterozygous parent was recombinant?

9. As a breeder of tomatoes, you have discovered a strain of wild tomato that is resistant to a destructive virus. The resistance is controlled by the dominant allele of a single gene (*Rv*). Unfortunately, the wild tomato has bad-tasting fruit and other undesirable traits; therefore, you want to breed a domestic tomato with only the resistant gene from the wild strain. You cross the wild tomato with a domestic tomato (cross 1) and select resistant progeny; you cross the progeny from the first cross with the domestic tomato (cross 2) and select resis-

tant progeny; and so on. What fraction of the progeny of cross 1 will be resistant? Of cross 2? Of cross 10? What fraction of the progeny of cross 1 will have alleles from the wild tomato of genes unlinked to *Rv*? Of cross 2? Of cross 10?

10. Several years ago, corn geneticists isolated a mutant in which anthers did not develop. The mutant was called *male-sterile*. They pollinated this mutant with wild-type pollen; all the resulting plants also were male-sterile. They pollinated these plants with wild-type pollen; all these progeny plants also were male-sterile. At this point, the geneticists decided that the gene for male sterility was maternally inherited. Explain why they drew this conclusion.

InfoTrac® College Edition

http://infotrac.thomsonlearning.com

Plant Breeding

Borlaug, N.E. 1983. Contributions of conventional plant breeding to food production. *Science* 219:689. (Keywords: "plant breeding" and "food")

Kawano, K. 2003. Thirty years of cassava breeding for productivity—biological and social factors for success (crop breeding, genetics & cytology). *Crop Science* 43:1325. (Keywords: "cassava " and "productivity")

Radin, J.W. 1996. Today's plant breeders increase tomorrow's food supply (genetic innovations: shaping corn, cattle, soybeans of the future). *Agricultural Research* 44:2. (Keywords: "plant breeders" and "supply")

Polyploidy

Ben-Ari, E.T. 1998. Polyploids come out of the (evolutionary) closet. *BioScience* 48:986. (Keywords: "polyploidy " and "closet")

Miller, J.S., Venable, D.L. 2000. Polyploidy and the evolution of gender dimorphism in plants. *Science* 289:2335. (Keywords: "polyploidy" and "dimorphism")

Milius, S. 2000. Glitch splits hermaphrodite flowers. *Science News* 158:214. (Keywords: "hermaphrodite" and "flowers")

Biotechnology

Visit us on the web at http://biology.brookscole.com/plantbio2 for additional resources, such as flashcards, tutorial quizzes, InfoTrac exercises, further readings, and web links.

1. DNA can be purified. Purified DNA can be broken into pieces that contain individual genes, and different pieces can be recombined to form new combinations of genes.

2. Recombined DNA can be inserted into plant cells. The plant cells may express the genes contained by the inserted DNA.

3. The genetic manipulation of plants can produce useful agricultural and horticultural varieties with traits that would have been extremely difficult or impossible to select by traditional breeding methods.

17.1 USEFUL MATERIALS FROM LIVING ORGANISMS

Biotechnology, a word that has been in the news only since about 1985, could mean many things. Because *bio-* refers to life, and *technology* is the application of science to the creation of products for human use, processes, and services, biotechnology could refer to almost any branch of agricul-

ture, to the fermentation and food-processing industries, and to many industries and services supporting health providers. However, new and spectacular discoveries about the structure and function of genes and enzymes have tended to overshadow more traditional fields, and the word *biotechnology* now generally implies the genetic manipulation of organisms to give them new capabilities or improved characteristics. This chapter describes some of the techniques that allow genetic engineers to modify the genes of plants and some of the advantages that can be expected from the new organisms.

17.2 STEPS TOWARD GENETIC ENGINEERING

DNA Can Be Purified, Recombined, and Cloned

Years of basic research and many technological breakthroughs were needed before scientists learned to manipulate genes in ways that were truly useful.

PLASMIDS   The technology for the chemical manipulation of DNA began in the 1960s with the discovery of **plasmids** in bacteria. Plasmids are small pieces of DNA—separate, autonomous, circular minichromosomes—generally not required for the survival of the bacterial cell. Some plasmids carry genes that help the cell survive in unusual environments, some make the host bacterium resistant to antibiotics or to certain viruses, some carry genes for enzymes that metabolize rare compounds, and some provide unusual biochemical pathways. Plasmids have the ability to be replicated in a cell, just like the main chromosome. An example is a plasmid called pBR322, which was devised for genetic engineering in the gut bacterium *Escherichia coli.* It contains two genes—one that provides resistance to ampicillin, and one that provides resistance to tetracycline—and a site where DNA replication can start.

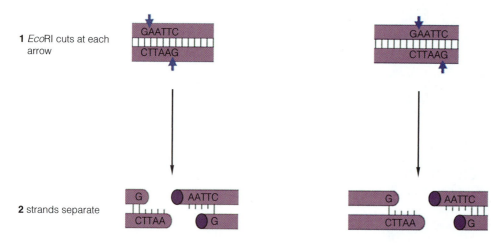

1 *Eco*RI cuts at each arrow

2 strands separate

3 sticky ends reassociate

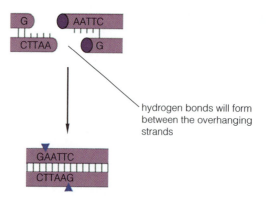

hydrogen bonds will form between the overhanging strands

4 DNA ligase connects strands (at triangles)

Figure 17.1 How restriction enzymes work. A restriction enzyme—here *Eco*R1, one of several restriction enzymes produced by *Escherichia coli*—cuts DNA at particular base sequences. The sticky ends from any two DNAs cut by the same enzyme can form hydrogen bonds and be connected covalently by DNA ligase.

Plasmids are useful because they are easy to purify and work with. The main chromosome of the bacterium *E. coli* has 4.7 million base pairs and thousands of genes, but a plasmid might have only 2,000 base pairs and 2 genes (see Fig. 19.3). This makes the plasmid more stable in a test tube and easier to analyze. Furthermore, bacterial cells can be induced to take up circular plasmids from the surrounding solution, a process known as **transformation.** By this process, a recipient cell obtains new genetic information.

RECOMBINANT DNA   Also in the 1960s, microbiologists discovered that bacteria contain enzymes capable of cutting DNA at specific base sequences. These enzymes are called **restriction endonucleases** or **restriction enzymes** because their function is to protect the cell by restricting invasion by foreign DNA. Cells modify their own DNA so that it is not cleaved by the restriction enzymes. Different restriction enzymes (from different species or even strains of bacteria) recognize different sequences of bases in DNA.

When isolated and purified, restriction enzymes were found to be extremely useful because they allowed scientists to cut purified plasmid DNA in specific, reproducible places **(Fig. 17.1).** Furthermore, the cuts can be reversed. Many restriction enzymes make cuts with *sticky ends*, at which there are overlapping regions of complementary DNA strands. At lower temperatures these ends stick together, and the DNA can be covalently connected (*ligated*) using another enzyme, DNA ligase.

Pieces of DNA from different sources can be combined because the sticky ends formed by a particular restriction enzyme all have the same base sequence, no matter what their source. These pieces will stick together and can be ligated, forming a **recombinant DNA** molecule **(Fig. 17.2).** If the recombination process inserts a new gene into a plasmid, and if the DNA becomes circular (and is not too large), the new gene can be taken up with the plasmid by a receptive bacterium. In this case, the plasmid is called a **vector** for the transfer of the new gene.

When a researcher adds to a test tube a mixture of pieces of DNA (plasmids and isolated copies of a gene, for instance), all cut with the same restriction enzyme, and allows them to recombine, the result is a random mixture of DNA molecules of various sizes. Some pieces of DNA stick their own ends together to form small circles; two plasmids (or two genes) may combine together just as well as one plasmid with one gene; finally, three, four, or more pieces may join end to end to form really large circles. Cutting the DNA with *two* restriction enzymes simplifies the situation. Because each molecule has sticky ends with different base sequences, single pieces cannot re-form circles. Still, much of the art in genetic engineering is in selecting the desired combination of ligated pieces of DNA—through a procedure known as *cloning.*

CLONING   A **clone** is a colony or group of cells or organisms, all of which have the same genes. Cloning is replication of the cells in the colony. In the context of DNA

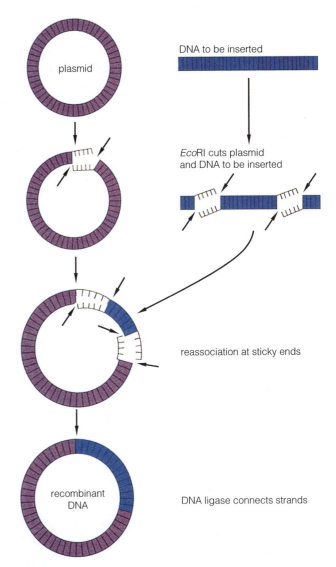

Figure 17.2   The formation of recombinant DNA. Restriction enzymes (such as *Eco*R1) can be used to insert a piece of DNA into a plasmid. If the plasmid has only one site recognized by the restriction enzyme, the foreign DNA can only be inserted there.

biotechnology, cloning is a simple method for separating and eventually characterizing individual molecules of DNA. It works because an individual molecule inserted into a single bacterial cell can be replicated many times as the cell divides, until the colony contains hundreds of thousands of copies of the same molecule.

To see how it works, assume that we prepare a mixture of recombinant DNA molecules formed from a plasmid and a gene of particular interest to us. We use a special plasmid, pUC19, which has the gene for resistance to the antibiotic ampicillin and a gene that makes the enzyme β-galactosidase. We treat the plasmid with a restriction enzyme that makes one cut in the middle of the β-galactosidase gene, then add the new gene cut with the same enzyme, and ligate. We combine the mixture of DNA molecules with a suspension of bacterial cells in such a way that each cell will, on the average, take up only one DNA molecule. Only circular molecules will be taken up. Next, the bacteria are spread out on a Petri dish containing nutrient agar and ampicillin (Fig. 17.3). The bacteria are diluted enough so that each cell is separated from every other cell. Those bacteria that did not take up plasmids will be prevented from growing by the ampicillin. Those bacteria that do contain plasmids will survive and grow into individual colonies (clones). We know that each clone must contain a plasmid. In the nutrient agar is a colorless chemical that turns blue in the presence of β-galactosidase. Those colonies of bacteria that contain intact β-galactosidase genes will synthesize β-galactosidase and turn blue. Colonies that contain plasmids in which the gene of interest has been inserted into the middle of the β-galactosidase gene will be unable to make β-galactosidase and will remain white. We check the bacteria from a white colony to verify that they contain the gene we want. By growing cells from the selected colony in nutrient medium, we can have an inexhaustible supply of that particular gene.

We also can determine the exact base sequence of the DNA we are interested in, although this step is not necessary for us to reproduce the DNA, transfer it to other bacteria, or recombine it with other pieces of DNA. The basic strategy is to synthesize DNA, starting at a fixed point and stopping the synthesis at particular bases, so that the size of the fragments relates to the positions of the bases (see sidebar "IN DEPTH: Determining the Base Sequence of DNA"). It is possible to separate DNA fragments of different sizes by electrophoresis, a technique that moves DNA molecules through an electric field at a rate proportional to their molecular weights (or lengths), resolving fragments that differ by just one base pair. This allows us to see directly the relative positions of different bases—the base sequence.

There Are Many Ways to Isolate a Useful Gene

To use cloning to produce multiple copies of a gene, it is first necessary to identify and isolate an initial sample of that gene. There are several strategies to do this; the choice depends on the information available. One set of methods starts with a purified protein, whereas another set starts with the base sequence of related genes.

REVERSE TRANSCRIPTASE AND CDNAS Imagine that you have discovered a fungus that produces a protein that is toxic to insect pests. You want to clone the gene for this protein. You know that the base sequence of the gene is embodied among the many messenger RNAs (mRNAs) in the fungal cells that are producing the protein.

The first step is to extract the mRNAs and reproduce their base sequences in DNA molecules. An enzyme called *reverse transcriptase* can produce DNA using an RNA template. Starting with the mRNAs, a "primer," which is a small piece of DNA complementary in base sequence to the mRNAs, and substrates (nucleoside triphosphates; see Chapter 2), the enzyme adds nucleotides to the primer to form single strands of DNA with base sequences complementary to the mRNA templates. The primer often is a line of T bases, "oligo(dT)," which is complementary to the poly(A) tail of most eukaryotic mRNAs. The result is a mixture of "complementary" or "copy" DNAs, abbreviated *cDNAs*.

POLYMERASE CHAIN REACTION The next step is to isolate the cDNA for the protein of interest from among the cDNAs made with reverse transcriptase. Again, there are several strategies. One is useful if you know the base sequences at the ends of the cDNA. Even though you do not know the base sequence of the whole gene, you may be able to guess the base sequence of one end. There are chemical ways of determining the amino acid sequence of the toxic protein, starting with one end (the "N-terminal" end). Given the sequence of about seven amino acids, you can infer mRNA codons that produced that amino acid sequence and synthesize a DNA oligonucleotide with the same base sequence as the codons. That oligonucleotide, together with poly(dT), defines the ends of the cDNA you want.

A DNA *polymerase chain reaction* (PCR; Fig. 17.4) will produce multiple copies of the desired gene. The reaction combines the cDNAs with the oligonucleotides (now serving as *primers*), nucleoside triphosphates, and a special DNA polymerase, the enzyme that synthesizes DNA. The special DNA polymerase is not destroyed by high temperature, and, in fact, it needs a fairly high temperature to function. Each reaction cycle includes three steps at different temperatures. (1) The reaction solution is heated, almost to boiling, to separate any complementary strands of DNA, making each

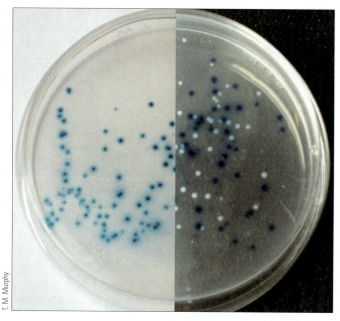

Figure 17.3 Colonies (or clones) of *Escherichia coli* on an agar plate. The cells in the blue colonies (most visible at the left) are making β-galactosidase, which liberates the blue dye from its colorless complex with the sugar galactose. The white colonies (most visible at the right) do not make β-galactosidase, because their gene for that enzyme has been interrupted by the insertion of a foreign piece of DNA.

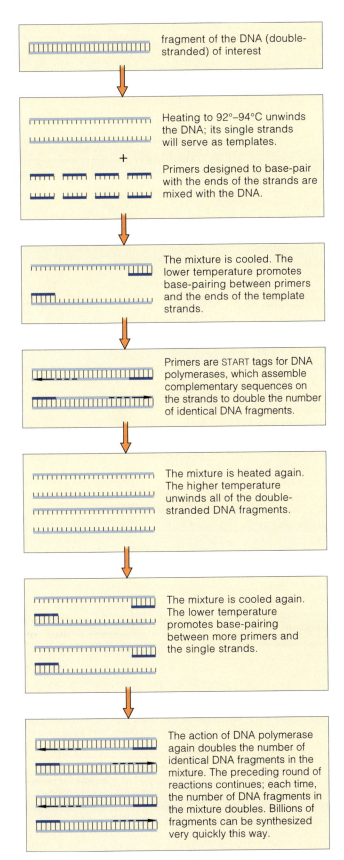

	fragment of the DNA (double-stranded) of interest
	Heating to 92°–94°C unwinds the DNA; its single strands will serve as templates.
+	Primers designed to base-pair with the ends of the strands are mixed with the DNA.
	The mixture is cooled. The lower temperature promotes base-pairing between primers and the ends of the template strands.
	Primers are START tags for DNA polymerases, which assemble complementary sequences on the strands to double the number of identical DNA fragments.
	The mixture is heated again. The higher temperature unwinds all of the double-stranded DNA fragments.
	The mixture is cooled again. The lower temperature promotes base-pairing between more primers and the single strands.
	The action of DNA polymerase again doubles the number of identical DNA fragments in the mixture. The preceding round of reactions continues; each time, the number of DNA fragments in the mixture doubles. Billions of fragments can be synthesized very quickly this way.

Figure 17.4   The polymerase chain reaction (PCR).

strand a potential template. (2) The reaction solution is cooled; this allows primers to bind to the ends of any DNA with complementary base sequences. (3) Finally, the reaction solution is heated to the optimum temperature for the DNA polymerase, allowing the synthesis of new DNA by the addition of nucleotides to the primers. Each cycle produces a new DNA for every template–primer complex. If two primers are specific for the gene of interest, the concentration of that gene doubles with each cycle, and after about 30 cycles, almost all the DNA in the reaction solution consists of that gene.

PCR is a flexible technique. It is possible, for instance, to use it to detect traces of animal or plant genes in criminal investigations. It also is possible to use it to synthesize a gene with added restriction sites at the ends. This allows the gene to be inserted into a plasmid and cloned in bacteria, a useful preparation for transforming plants.

GENOMICS   In the last few years, scientists have deciphered the base sequences of the entire genetic material of many viruses, bacteria, and even human cells, as well as two species of plants, *Arabidopsis thaliana* (mouse-eared cress, a relative of mustard) and *Oryza sativa* (rice). The full assemblage of genetic material of a cell is called the **genome;** therefore, the study of genome structure, function, and evolution is called **genomics.**

The information from genomics is useful in identifying genes of interest. This is because genes with similar functions have similar base sequences. Computer programs can compare base sequences, identifying similarities and differences. Thus, it is possible to find a gene in the rice genome for a particular enzyme by its similarity to a corresponding gene in humans. Once the rice gene base sequence is known, a scientist can design primers and use PCR to produce many copies of the gene or to isolate the same gene from a different plant. Also, by comparing the sequences of a gene in plants, animals, and fungi, it is possible to identify the base sequences of the genes (and thus the amino acid sequences of their proteins) that are essential for the proteins' function and get ideas about how to improve the function. Perhaps most important, the information from genomics is teaching us how networks of genes are regulated. By identifying genes and proteins that control the expression of other genes, we may learn to influence the development of plants in complex ways.

Scientists Use a Special Bacterium to Insert DNA into Plants

The techniques described in the previous sections were developed using DNA from bacteria, fungi, and animals. DNA in plants has the same chemical structure as DNA in other organisms, and it can be extracted from plant cells and manipulated in a test tube using many of the same techniques, including reverse transcription, PCR, and cloning. This means that we can select and purify plant genes. It does

IN DEPTH:

Determining the Base Sequence of DNA

All the information in DNA is encoded in its sequence of bases. Determining the sequence is useful because it allows researchers to tell whether the DNA could code for a protein and, if so, what the amino acid sequence of the protein would be. It allows them to compare the DNA being studied with other known DNAs to detect mutations or potential evolutionary relationships. Sequencing DNA is currently a major activity of biotechnologists, and many millions of DNA sequences are stored in computer databases.

The procedure for determining DNA sequences works by making DNA fragments of sizes that correspond to the positions of particular bases. Originally devised by F. Sanger and his colleagues, it uses DNA synthesis to make the different-sized fragments. Assume that you have cloned a piece of DNA of unknown sequence in a vector. The cloned DNA is double-stranded; you heat the strands to separate them so that you can use one strand as a template for DNA synthesis. Then you add a short piece of DNA with a sequence complementary to the vector DNA that is next to the sequence to be determined. This short piece, the primer, binds to its complementary sequence in the vector DNA and serves as the starting point for DNA synthesis.

To this template–primer complex you add a DNA-synthesis reaction mixture. The mixture contains the DNA-synthesizing enzyme DNA polymerase and activated nucleotides, which are substrates (reactants) for making the new DNA. The mixture also contains a small amount of special substrates that stop DNA synthesis at particular bases. One substrate stops synthesis at A bases, one at G bases, one at C bases, and one at T bases **(Fig. 1a)**. Each substrate is linked to a dye that fluoresces a particular color.

When you run the DNA-synthesis reaction, you will be making four sets of DNA fragments. Each set is itself a mixture of DNAs of different lengths. The mixture with the A-terminating reagent, for instance, has a mixture of DNAs. All of them started at the primer, but some of them stopped at the first A of the synthesized DNA, some at the second, some at the third, and so forth. The relative sizes of these fragments indicate the relative positions of the A bases (and their complementary T bases in the template). All of these fragments glow green **(Fig. 1b)**.

The newly synthesized DNA is separated by electrophoresis. The mixture is placed in an artificial plastic gel and subjected to an electric field. The DNA fragments all have negative ionic charges, so they move away from the negative and toward the positive pole. The holes between the gel molecules are small, so fragments move at different speeds, with short fragments moving through the gel more quickly than long ones. Even fragments that differ by only one base in length can be distinguished. At the end of a period of electrophoresis, the positions of the different fragments are visualized by their fluorescence. The presence of a band of a certain length that glows green means that there was an A at that position. The locations of the bands that are one base longer and shorter show the bases adjacent to that A. The sequence of the DNA can be read directly off the electropherogram (as shown in **Fig. 1c**).

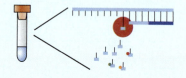

single-stranded DNA fragments of unknown nucleotide sequence

primers

DNA polymerase

nucleotide substrates, including
A T G C

four types of terminator nucleotides, each with a fluorescent dye
A* T* G* C*

a The components are mixed in a test tube and the DNA polymerase is allowed to work

b New DNA strands are assembled. At random times, a terminator nucleotide stops DNA synthesis. The color of fluorescence of the DNA indicates the base at that particular size of strand.

c The mixture is placed in one lane of gel and subject to electrophoresis. The nucleotide sequence can be read from the sequence of bands on the gel. This picture shows six vertical lanes, representing six DNAs, from a real gel.

Figure 1 The sequence of a section of DNA can be determined by synthesizing pieces of DNA. Each band in the electropherogram (*right*) represents a piece of a particular length: the closer the band is to the top, the larger the piece of DNA. All bands start at the same base, determined by the base sequence of the primer; all bands with the same color end at the same type of base (G, A, T, or C), determined by the base on the terminator nucleotide. The relative positions of the bands then give the relative positions (sequence) of the different bases.

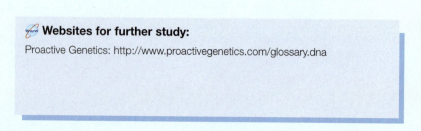

Websites for further study:

Proactive Genetics: http://www.proactivegenetics.com/glossary.dna

not, however, automatically mean that we can put the genes back into plant cells in a way that allows the cells to survive and to use the new genetic information.

In searching for a way to insert genes into plant cells, scientists focused on a condition called crown gall disease. This disease, caused by the soil bacterium *Agrobacterium tumefaciens*, results in tumors (galls) growing in places where a plant has been wounded **(Fig. 17.5).** The disease is called *crown gall* because plants often are wounded at the *crown*, the junction between shoot and root. *A. tumefaciens* cells attach to the walls of plant cells exposed in the wound and cause the plant cells to start dividing. That the plant cells continue to divide even when the bacteria are killed with antibiotics shows that the bacteria *transform* the plant cells, apparently by turning off their normal mechanisms for limiting cell division. The result is rather like an animal cancer, and, in fact, much of the early research on crown gall disease was financed by the National Cancer Institute.

What mechanism do the bacteria use to transform the plant cells? For many years, biologists suspected that a transfer of genes was responsible, and in the mid-1970s, the DNA that contained those genes was identified. Infectious strains of *A. tumefaciens* have a large plasmid, called the *Ti (tumor-inducing) plasmid* **(Fig. 17.6),** part of which the bacteria inject into plant cells **(Fig. 17.7).** The region that is injected (the T-DNA) contains three main genes that cause the cells to divide and grow. Two of the genes code for enzymes that make the plant hormone auxin (indoleacetic acid); one of the genes codes for an enzyme that makes another plant hormone, a cytokinin (isopentenyl adenine). It has been known for years that an excess of auxin and

T. M. Murphy

Figure 17.5 Crown galls are formed when *Agrobacterium tumefaciens* infects wounded plant tissue. The wounds often occur around the crown (area between stem and root), but can also be higher on the stem, like the gall on this walnut (*Juglans* sp.) tree. The gall tissue grows actively in the laboratory.

cytokinin will cause plant cells, especially the parenchyma cells at the surface of a wound, to divide and grow.

Another T-DNA gene is for an enzyme that synthesizes an unusual type of amino acid called an *opine*. Opines generally are combinations of two normal amino acids (there can be different combinations, depending on the enzyme). Once synthesized in the plant cells, opines leak out into the

Figure 17.6 The Ti plasmid from a strain of *Agrobacterium tumefaciens*, showing the original T-DNA that is transferred to plant cells and stimulates growth of galls and a typical replacement T-DNA used by genetic engineers to add a genetic function to a plant. The promoter controls when and in what tissue the gene of interest is expressed. The "reporter" gene allows the selection of transgenic plants: It might provide resistance to an antibiotic or an herbicide, allow the plant to grow on an exotic nutrient source, or make the cells fluorescent or luminescent (Fig. 17.8).

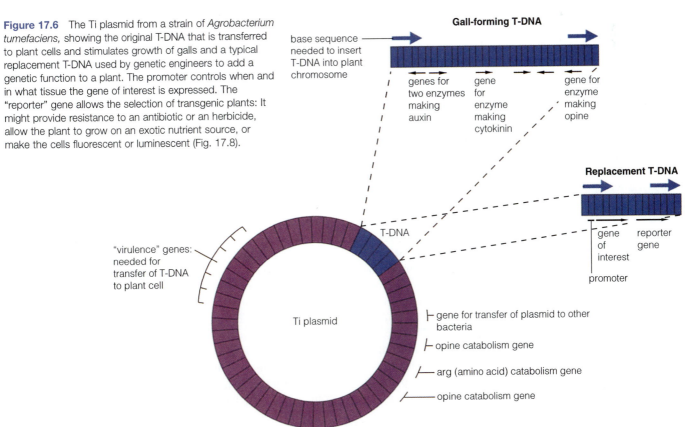

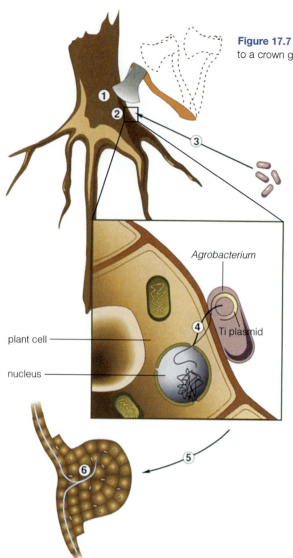

Figure 17.7 The sequence of events that leads to a crown gall tumor.

1 Plant tissue is wounded.

2 Plant secretes acetosyringone, a chemical that attracts *Agrobacterium tumefaciens.*

3 Bacteria swim to wound and attach to cell walls of wounded cells.

4 *Agrobacterium* cell inserts a copy of its T-DNA in plant cell. T-DNA moves to nucleus and is incorporated into plant DNA.

5 Stimulated by auxin and cytokinin produced by the enzymes coded in the T-DNA, the cell repeatedly divides, forming a tumor.

6 The growing tumor serves as a sink for phloem transport. Nutrients delivered by the phloem are in part used to make opines, which are secreted. Bacteria living in the spaces between the plant cells take up the opines and catabolize them (break them into components to use for growth).

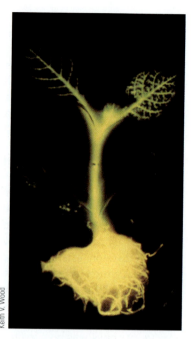

Figure 17.8 Plants can be transformed when infection by a genetically engineered *Agrobacterium tumefaciens* Ti plasmid introduces new genetic information. This tobacco (*Nicotiana tabacum*) plant glows in the dark because the new gene that was inserted (which came from a firefly) produces the enzyme luciferase.

intercellular spaces. The bacteria growing in the intercellular spaces of the tumor also make enzymes that allow them to take up and metabolize opines, using them as a source of carbon, nitrogen, and energy. Thus, the plant cells transformed by the bacteria—that is, injected with new bacterial genes—nourish their bacterial parasites. In short, *A. tumefaciens* has evolved a natural mechanism for conducting genetic engineering.

For human genetic engineers, this system is valuable because the T-DNA of the Ti plasmid can be modified to serve as a vector to carry genes into plant cells (Fig. 17.7). In practice, researchers start with T-DNA that has lost the genes for auxin and cytokinin synthesis, so that it does not form tumors. They insert the gene of interest, controlled by a promoter that regulates when and in what tissues it is turned on, together with a "reporter" gene that allows them to select for cells that have stably incorporated the T-DNA (for instance, a gene coding for the enzyme luciferase; Fig. 17.8). This recombinant T-DNA, generally in the form of a miniplasmid, is transferred to an *A. tumefaciens* cell that has a Ti plasmid lacking its own T-DNA. Spread on the cut surface of a piece of leaf, the bacteria transfer the recombinant T-DNA to the plant cells. The leaf is transferred to a medium containing antibiotics that kill the bacterial cells, and the engineers select for those plant cells that have incorporated the reporter gene in the T-DNA. After these steps, the plant cells can be coaxed to regenerate new plants by using standard techniques of tissue culture (see Chapter 15). The plants, with new genetic information, are called *transgenic*.

There Are Other Methods for Injecting DNA into Plants

Although the *A. tumefaciens* system is the favorite method of transforming plant cells, there are a few other ways to add new genetic material to these cells. One is called *biolistics*, a play on the word ballistics. The apparatus also is called a gene gun. DNA containing the gene of interest is adsorbed onto the surface of small (subcellular-sized) particles of gold or tungsten. These particles are pressed onto the front of a bullet. The bullet is then loaded into a gun, and the gun is fired at the plant tissue. Originally, a common .22 caliber gunpowder-powered bullet was used, but a pellet for a high-pressure air (or helium) pistol also is effective and avoids the toxic gases produced by the gunpowder. A special metal plate with a hole that is smaller than the cross section of the bullet abruptly stops the flight of the bullet before it reaches the tissue, but the gold or tungsten particles keep going with enough energy to penetrate the cells. They are so small that they appear not to leave holes in the cell membranes after they pass through; they also appear not to harm the cells in any other way. After the particles come to rest inside the cells, the adsorbed DNA dissolves into the cytoplasm. A new method, based on the same principle, uses thin silicon carbide fibers coated with DNA. Mixed with a suspension of plant cells and agitated briskly, some fibers penetrate the cells and release their DNA. In at least some cells, the DNA reaches a place where it can be used as a template for RNA synthesis, and its genetic information is expressed.

Another method for getting DNA into a plant cell is *electroporation*. This is based on the discovery that a short, high-voltage charge of electricity can produce temporary holes in a plasma membrane without permanently harming the cell. For this technique, it is necessary to make **protoplasts**—cells without walls—from the recipient plant cells. The cell walls can be removed by treatment with polysaccharide-hydrolyzing enzymes (cellulase and polygalacturonase) while the cells float in an osmotically balanced medium to prevent water from entering and breaking them open. The protoplasts are then placed between two electrodes in an ice-cold solution that contains the DNA. A few pulses of electricity, each less than 0.1 second in duration, produce the membrane holes. These close up in seconds or minutes, but during this time some DNA can enter the cells. Cultured under the proper conditions, protoplasts regenerate their cell walls, begin dividing, and even regenerate whole plants—plants that may express the genes of the DNA that entered the protoplasts.

Viruses also can be used to inject genes into plants. Plant viruses (subcellular particles of nucleic acid and protein) infect plant cells, multiply, and spread throughout the plant (see Chapter 19). The genes of most plant viruses are usually made of RNA rather than DNA, but genetic engineers can produce DNA copies of the viral genes and connect them to genes of interest. In the plant, the introduced genes are expressed in the cytoplasm of cells wherever the virus spreads. This method does not produce a permanently transformed plant because the viral and introduced genes are not incorporated into the plant's nuclear DNA, and they are not passed to seed formed by the infected plant. However, proteins made by the infected plant in response to the introduced genes can be extremely useful.

17.3 APPLICATIONS OF BIOTECHNOLOGY

Because bacteria were the first organisms to be transformed with new genes, they were the first to be used to produce new commercial products. The most valuable have been proteins. For instance, patients with diabetes who require daily injections of insulin currently receive human insulin produced by bacterial cells transformed with the human gene for insulin. Previously, they received pig insulin, purified from the pancreases of slaughtered pigs; this procedure was costly, and the insulin was not as satisfactory. Other proteins having medical uses that are currently produced by genetic engineering include somatotropin, erythropoietin, clotting factors, and interferon. Genetically engineered bacteria (or yeasts) also produce the enzymes added to laundry detergents. And, of course, many of the restriction enzymes and DNA polymerases on which genetic engineering depends are produced in large quantities by genetically engineered bacteria.

Plants Can Serve as a Source of Biochemicals for Medicine and Industry

Although bacteria can grow rapidly, nothing can beat plants for the economical production of large quantities of protein. Through photosynthesis, plants acquire carbon and harness solar energy for protein synthesis, whereas bacteria require a supply of growth medium. It is possible to harness the metabolic capacity of transgenic plants to produce large quantities of a protein valuable for industrial uses.

Plants are being engineered to produce vaccines that are cheaper and more acceptable to people. A vaccine contains an inactivated or weakened bacterial or viral pathogen (or a part of the pathogen). Injected into a person's body, it stimulates the immune system to produce antibodies or immune cells that will kill an active pathogen if it infects the person at a later time. Some vaccines (for instance, the Sabin polio vaccine) can be taken orally. With this in mind, scientists are designing and testing food plants that contain genes for proteins from important pathogens. Two intriguing examples are a banana (*Musa sapientum*) that makes a protein from the hepatitis B virus and an alfalfa (*Medicago sativa*) sprout that contains a part of the cholera toxin. In the future, for the cost of a few seeds and some careful farming, even an impoverished region may be able to protect its population from serious lethal diseases through such vaccination.

In addition to proteins, plants synthesize oils as storage materials for their seeds. By changing the enzymes of the synthetic pathway, it is possible to design plants that produce specialized oils or waxes. These materials could be useful in detergents, cosmetics, pharmaceuticals, and lubricants. Most biological oils are biodegradable. Previously, this has been seen as undesirable in an industrial product such as motor oil because it means that the product has an uncertain (and possibly short) lifetime. The risk of polluting the environment and the cost of disposing of nonbiodegradable wastes, however, have emphasized the advantages of easy degradability.

Gene Manipulations Can Produce Many Types of Useful New Plants

NEW HORTICULTURAL VARIETIES Transforming plants with new or altered genes may yield plants with new traits that are useful or attractive. Much is known about the genes for the enzymes that make anthocyanins—the red, blue, and purple pigments found in plant flowers, as well as in other organs. Plant molecular biologists have produced plants that have additional genes for enzymes in the anthocyanin

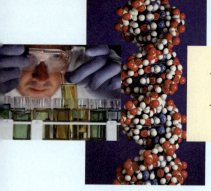

BIOTECHNOLOGY:

RNA Gives Plants the Silent Treatment

We know that RNA is involved in gene transcription and translation, but RNA can also turn off ("silence") gene expression. Antisense technology and ribozymes have been used for gene silencing for a while, but a new mechanism has been discovered and it works better! Actually, Rich Jorgensen from the University of Arizona first stumbled onto this cosuppression mechanism in 1990. He accidentally silenced genes in petunias, when trying to deepen the purple color through genetic engineering. He initially thought it was a bizarre attribute of the species. Hairpin gene silencing was first discovered and described in plants in the mid-1990s. Dr. Peter Waterhouse, a molecular geneticist from CSIRO (Commonwealth Scientific & Industrial Research Organization) Plant Industry in Australia, described the first successful gene silencing in tobacco in 1995, but there was a delay in filing a patent until 1998.

The term *RNAi* (RNA interference or hairpin gene silencing) was coined in 1998 by Andrew Fire (then at the Carnegie Institution of Washington) and Craig Mello (University of Massachusetts Medical School) in *Nature*. RNAi was called the Breakthrough of 2002 in RNAi News (www.RNAinews.com). It has been called the *third revolution in biotechnology*. The first was recombinant DNA technology and the second was the PCR technique.

RNAi is an evolutionarily ancient mechanism. It is believed that RNAi is part of a plant's natural defense mechanism against double-stranded (ds) RNA viruses and transposable elements. RNAi may also be involved in normal genetic control during plant development. Short strands of dsRNA called *siRNA* and microRNA (*miRNA*) exist naturally in the cells of many species, where they inactivate mRNAs with either complete or partial base-sequence similarity, respectively. siRNA mediates mRNA degradation, whereas miRNA is thought to act predominately as a translational regulator.

RNAi is straightforward, although incompletely understood. The dsRNA

What is post-transcriptional gene silencing (PTGS)?

Inducer: dsRNA

DICER

Effectors: siRNAs

RISC: nuclease/helicase

RNA Induced Silencing Complex

Sequence-specific mRNA degradation

Native mRNA

PTGS

mRNA degradation

Figure 1 Mechanism of RNAi (RNA interference): double-stranded RNA (dsRNA), acted on by Dicer enzyme, provides the instructions (siRNA) for the RISC nuclease/helicase to locate and cleave the messenger RNA (mRNA) with the same base sequence of the dsRNA. This blocks the expression of the gene that produced the mRNA.

(complementary to the target mRNA sequence) is cleaved by the **Dicer enzyme** into many siRNA molecules (~21–23 base pairs). These siRNAs bind to the **RISC** (RNA-induced silencing complex) protein complex, which then forms a complex with the target mRNA, which leads to its degradation and post-transcriptional gene silencing **(PTGS)**. The silent state can spread systematically, demonstrating that mobile silencing signals are present **(Fig. 1)**.

The use of RNAi (PTGS) tools is revolutionizing the field of functional genomics. siRNA can be synthesized and used to silence genes quickly and easily. Scientists can see how genes are involved in cellular processes by observing the phenotypic effects. The laboratory of Professor Abhaya Dandekar at the University of California at Davis is using this gene silencing technique in model plant systems to silence the crown gall oncogenes and prevent crown gall tumor formation (induced by *Agrobacterium tumefaciens*) **(Fig. 2)**.

a   b

Figure 2 RNAi-induced silencing of crown gall oncogenes (tumor-inducing genes) in tomato. (**a**) The tomato plant has been given a gene that produces a dsRNA complementary to one of the oncogenes that crown gall bacterium injected into the stem. Because of this, no tumor (gall) is formed. (**b**) In contrast, the unprotected plant forms a gall.

Websites for further study:

Orbigen: http://www.orbigen.com/RNAi_Orbigen.html

RNA Interface and Gene Silencing:
http://www.ambion.com/techlib/hottopics/p19_071102.html

Figure 17.9 Petunia plants that are transformed with additional flower color genes produce unusual patterns, including sectors that have no pigment at all.

pathway and that, as a result, have flowers with unusual colors or patterns (Fig. 17.9). There is hope of producing a blue rose, something not found in searches among wild varieties and random mutants.

MORE PEST RESISTANCE   Predation by insects and damage caused by viral, bacterial, and fungal diseases are among the major factors that limit the productive harvest from food crops. Classical plant breeding traditionally has selected varieties that resist these pests, and with genetic mating techniques it has been possible to combine resistance genes

Figure 17.10 Plants can be transformed with genes that provide resistance to insects and viruses, as well as bacteria and fungi. (**a**) Caterpillars dine on a cotton (*Gossypium hirsutum*) plant, but (**b**) not if it is genetically modified to resist attack. (**c**) A normal potato (*Solanum tuberosum*) plant is susceptible to viral attack, but (**d**) a genetically modified strain is resistant.

with genes for high-quality vegetables, cereal grains, and fruit. But classical genetic techniques are inefficient, requiring many cycles of back crossing and selection. As we identify the genes for resistance, we can transfer them quickly and efficiently using molecular techniques (Fig. 17.10). For instance, seeds of common beans produce a protein that blocks the digestion of starch by two insect pests, cowpea weevil and Azuki-bean weevil. The gene for this protein has now been transferred to the garden pea. In tests, it protected stored pea seeds from infestation by these insects, killing or slowing the development of the larvae.

Furthermore, we can identify methods of resistance that plants never evolved. *Bacillus thuringiensis* is a bacterium that makes a protein toxin that kills insects. Scientists have already inserted the gene for *B. thuringiensis* toxin into important crop plants, including potato, tomato, corn, and cotton. These plants synthesize the toxin and kill many of the insects that graze on them. There are other antibacterial proteins that might also be useful. For instance, an apple sapling containing the insect gene for the attacin E protein is less susceptible to fire blight, a disease caused by the bacterium *Erwinia amylovora*.

Tobacco mosaic virus infects crops in the plant family Solanaceae (tomato, potato, eggplant, green pepper). Inserting the gene for the viral coat protein into the plant genome for unknown reasons makes the plant resistant to infection by the virus. Scientists in Mexico have used the same technique to produce potato varieties immune to potato viruses X and Y, and in Hawaii, a genetically engineered papaya resistant to a ringspot virus saved the Hawaiian papaya industry from impending disaster (Fig. 17.11). Protecting agricultural plants against viral infection increases the yields of fruits or other useful crops that can be harvested from those plants.

Protecting plants against competition from weeds takes a different strategy. Some of the first plants to be engineered were plants resistant to herbicides. In extensive use are varieties of soybeans, cotton, sugar beet, corn, and canola that are resistant to glyphosate (trade name Roundup). Many of the same crop species, as well as others, such as radicchio and rice, have been engineered for tolerance to the herbicides glyfosinate, sulfonylurea, and bromoxynil. A resistant crop allows farmers to use herbicides to kill weeds in the middle of a field of crop plants. Properly applied, the crop allows a more discriminating use of safer herbicides.

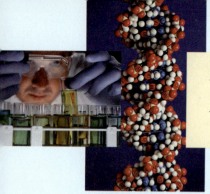

BIOTECHNOLOGY:

Biotechnology in Pest Management: Opportunities and Threats

Diseases caused by viruses, bacteria, and fungi; herbivory by insects, birds, and mammals; and competition by weeds all can take a major toll on crop production. Improved pest management has strongly reduced the loss of production to pests, but more needs to be done. Richard Roush, Ph.D., Director of the University of California Statewide Integrated Pest Management Program (UC IPM) makes these points concerning the role of biotechnology in pest management.

Biotechnology is already contributing enormously to pest management. Following are some of the benefits that can be credited to genetically modified (GM) crops:

- Pesticide use has been significantly reduced by GM crops. In 2001, 46 million less pounds of pesticide were applied to 8 GM crops in the United States alone.

- Farmers benefit from increased yields (4 billion pounds) and profitability ($1.2 billion saved in production costs). GM crops promote the adoption of reduced tillage, which saves energy and soil quality.

- In China, human pesticide poisonings have been reduced by about 75% by insect-resistant ("Bt") transgenic cotton. In China, Bt cotton returns almost all its financial benefits to people farming less than a few acres.

- Insect-resistant corn shows great potential for reducing throat cancer and liver problems in southern Africa. These diseases arise from fungal toxins (fumonisins) produced on insect-damaged corn.

- The papaya industry on the islands of Hawaii was rescued by GM virus-resistant varieties.

Virus, insect, and herbicide resistance can and should be extended to some additional key crops, and there is a critical need for further advancing the management of bacterial and fungal diseases, animal pests, and weeds. It may be efficient to accomplish this through new GM crop plants.

There also are concerns about the extensive use of GM plants, including:

- Public acceptance of GM foods

- Impact on nontarget species and populations: the fear that Bt-containing corn pollen could poison the Monarch butterfly population is one example

- Evolution of resistance to new control techniques among the pests: additional research and strategies can reduce this threat, as illustrated by the considerable progress that has been made in slowing resistance to Bt transgenic crops

- Gene flow of GM traits to nontarget populations: this is important in that it may aggravate the impact on nontarget species and speed selection of resistant pest varieties

GM techniques can be valuable in agriculture, but like all other techniques they must be applied with attendance to ecological and evolutionary principles and concern for the native biological community.

Websites for further study:

Current and Potential Impact for Improving Pest Management in U.S. Agriculture: http://www.ncfap.org/40CaseStudies.htm

Genetically Engineered Crops for Pest Management in U.S. Agriculture: http://www.ers.usda.gov/epubs/pdf/aer786/

Figure 17.11 A stand of papaya trees in Hawaii. The trees on the left were infected with papaya ringspot virus and show serious symptoms of decline. The trees on the right contain the gene for the virus coat protein and are resistant to infection by the virus.

T. M. Murphy

IMPROVED QUALITY OF FRUIT AFTER HARVEST In many countries, a large proportion of harvested crops never reaches consumers because it spoils en route to the market. In America, a great deal of the energy and cost associated with food production results from the complications of getting food to market in edible and attractive form. Controlling the atmosphere (limiting the amounts of oxygen and ethylene) and reducing the temperature are two expensive techniques for preventing certain fruits from spoiling. Many fruits are picked green, stored, and then artificially ripened with an ethylene treatment. How much more effective and efficient it would be if we could insert genes that slowed the rate of senescence (aging), thus slowing spoilage. One of the enzymes needed for the softening of tomatoes as they rot is polygalacturonase. The first bioengineered food crop approved in the United States was the "FlavrSavr" tomato, which contained a gene that blocked the synthesis of polygalacturonase, thus delaying the senescence of the fruit. A more recent strategy to slow the over-ripening of tomatoes, also applied to cantaloupe, has been to insert genes that reduce the synthesis of the ripening hormone ethylene.

IMPROVED NUTRITION Some crops that are dietary staples are not necessarily the most nutritious. A well-known example is corn, which is low in lysine and tryptophan. Both lysine and tryptophan are essential amino acids in the diets of human and nonruminant animals. High-lysine corn varieties have been developed by classical genetic techniques. But vitamin and mineral deficiencies are still responsible for much suffering in underdeveloped countries, suffering that could be alleviated by more nutritious plant products.

A deficiency in vitamin A results in blindness. It is estimated that 124 million children worldwide are vitamin A–deficient; and in southeast Asia, 250,000 children go blind each year because of this deficiency. Scientists in Switzerland have designed a variety of rice that produces seed with endosperm rich in β-carotene. β-carotene is the precursor for vitamin A. The new rice has enough β-carotene to color the seeds, leading to its name, "golden rice." Another project has produced a rice that is rich in ferritin, a protein that binds iron. Iron deficiency, a problem for 1 to 2 billion people worldwide, results in anemia, increased susceptibility to infection, and impaired learning ability in children.

A variety of canola (*Brassica napus,* or oilseed rape) has been given the gene for a fungal enzyme, phytase. Phytic acid is a storage form of phosphate in plants. By releasing the phosphate from phytic acid in pig and chicken feed, phytase improves the nutrition of these animals—they grow faster and stronger. This is partly because of the improved availability of phosphate, and partly because of the destruction of phytic acid, which inhibits the uptake of iron in the gut. Not incidentally, the phytase also reduces the phosphate excreted in manure, thus reducing environmental pollution by farm effluents.

For many decades, margarines made from plant oils have served as butter substitutes. The fatty acids in plant oils are generally unsaturated (have double bonds), in contrast to fatty acids in butter fat, which are saturated. Humans require some unsaturated oils in their diet, and saturated fatty acids are considered a health risk. Yet, polyunsaturated fatty acids (two or more double bonds) are generally liquid at room temperature—inconvenient for a bread spread—and they oxidize to form chemically reactive free radicals, leading to a lower shelf life and possible toxic effects. The ideal fatty acid might be oleic acid, a 12-carbon fatty acid with one double bond. DuPont scientists have produced a soybean with a gene that causes it to accumulate high amounts of oleic acid.

IMPROVED TOLERANCE TO ENVIRONMENTAL STRESS One of the most difficult challenges is to produce plants that grow better under stressful conditions. We know that certain plant varieties have genetic information that allows them to survive saline conditions, toxic metals, droughts, high light intensities, or chilling temperatures. In some cases, we even know how the resistance works, in a general way. But seldom do we understand what genes or proteins are involved.

One strategy used by plants to survive and even thrive in saline soils is to transport salt into their cells' vacuoles, where it does not interfere with cellular processes. A specific protein, called a Na^+/H^+ antiporter, serves as the channel (see Chapter 11) between cytoplasm and vacuole. Researchers have inserted into a tomato plant a gene that synthesizes high levels of this channel **(Fig. 17.12)**. The plant grows, flowers, and bears fruit when irrigated with 200 mM NaCl, a con-

Figure 17.12 Salt tolerance of wild-type tomato (*Lycopersicon esculentum*) plants and transgenic salt-tolerant plants. (**a**) Fruits from wild-type plants grown normally (*left*) and transgenic plants grown in the presence of 200mM NaCl (*right*). (**b**) Wild-type plants. (**c**) Transgenic plants. The transgenic plants synthesize high levels of a sodium carrier, which allows the leaf cells to sequester NaCl in their vacuoles. (Adapted from Zhang, H.X., Blumwald, E. 2001. Transgenic salt-tolerant tomato plants accumulate salt in foliage but not in fruit. *Nature Biotechnology* 19:765–768, with permission.)

centration 50 times the normal limit for tomato. The salt accumulates in the vacuoles of the leaves, but not the fruit; therefore, the fruits are edible, and harvesting the leaves can help clear a field of salt.

This same strategy can be used to produce plants that will take up toxic heavy metals, such as lead, cadmium, and uranium from contaminated soil and water, a process known as *phytoremediation.*

It is thought that resistance to some stresses depends on several genes, which complicates efforts to increase it. Much current research is directed toward identifying the genes that differ between stress-tolerant and stress-sensitive varieties. The objective is to produce varieties of plants for a wider range of climate and soil conditions.

17.4   IS BIOTECHNOLOGY SAFE?

In 2002, a survey conducted by the International Service for the Acquisition of Agri-biotech Applications found that genetically modified crops were planted on 59 million hectares (148 million acres) of farmland. These crops included more than 20% of the world's soybeans, corn, cotton, and canola, and 75% of soybeans in the United States. The use of genetically modified crops, which is expanding rapidly, has not been without controversy.

In 1992, a group of 1,000 chefs signed a pact promising not to use any genetically engineered foods in their cooking. In 1998, the European Union instituted a moratorium on the planting and use of genetically modified crops. What are people worried about? In part, the worry reflects recent experience that new technologies may cause problems. If the technologies are widely and carelessly applied, these problems can become serious. The problem of disposing of radioactive and toxic chemical wastes is one example. In part, the worry reflects general antitechnology feelings. Much of the controversy about genetically engineered plants reflects the public perception that little is known about genes, especially about the genes that are "unnaturally" introduced into organisms.

One counterargument points out that crops have been genetically manipulated for centuries—by breeding and selection. Familiarity and our positive experiences with the resulting new plant varieties have made their use acceptable. Few people realize how little is known about the genetic composition of new crop varieties that have been developed through random mutagenesis (production of mutants) or crossbreeding with wild types. We may be able to say that we know more about changes in the genes of a crop modified by genetic engineering than about changes in the genes of one developed through classical breeding.

In the early days of genetic engineering, when little was known about engineered organisms, scientists themselves placed regulations and limits on different types of experiments involving recombinant DNA. Different types of experiments required different levels of care. For instance, experiments with cancer viruses and pathogens that infect mammals required (and still require) stringent containment procedures, whereas simple experiments cloning genes in a genetically weakened and well-understood strain of bacteria could be performed in a regular laboratory with simple aseptic procedures. The manipulation of plant genomes is considered a low-risk experiment. The marketing of engineered plants has more risks, but these have been examined, and ways for eliminating uncertainties have been identified. The scientific issues to be evaluated in the approval of a genetically engineered food are: (1) Does the product contain any new allergenic material that might affect especially sensitive groups? (2) Are new toxic compounds introduced into the food supply, or are existing toxins increased to unacceptable levels? (3) Are nutrient levels adversely affected? (4) Will the use of genes for antibiotic resistance (used to indicate when a plant has been stably transformed) compromise the use of important therapeutic drugs? If the answers to these questions are all negative, the food is considered to be safe.

A second type of problem involves environmental effects—for instance, the impact of new plants on wildlife or the possibility that new genes from the desired recipient crop species could be transferred to a related wild, weedy species. This is a concern when the new gene confers protection against natural pests or chemical herbicides. A 1999 report that pollen from corn carrying the *B. thuringiensis* toxin gene could kill Monarch butterfly larvae caused consternation among ecologists and biotechnologists alike. Glyphosate resistance has been found in weedy *Brassica* species near canola fields in Canada. Understanding how serious these findings are will require long-term ecological research; this is one reason why some groups urge governments to delay the widespread release of transgenic plants.

The problems involved with plant biotechnology should be balanced against the advantages. Biotechnology provides an excellent example of information-intensive solutions replacing energy-intensive solutions to problems. Planting crops that resist pests takes much less energy (in terms of fuel, chemicals, and labor) than applying insecticides. Designing crops to avoid spoilage reduces the waste of energy (fertilizers, labor, and fuel for both farm operations and distribution) that occurs when spoiled crops are discarded. For this reason, it is not surprising that biotechnologists sometimes display frustration when the products of their work are not quickly accepted. In this growing field, research is the key. It is certain that the more we understand about plant and animal physiology and ecology, the more safely and effectively we can use biotechnology to improve our lives.

KEY TERMS

clone	restriction enzymes
plasmids	transformation
recombinant DNA	vector
restriction endonucleases	

SUMMARY

1. Biotechnology is the application of biology to the creation of products for human use, including particularly the genetic manipulation of organisms to give them new capabilities or improved characteristics.

2. DNA can be manipulated through chemical and biochemical techniques. This allows a scientist to isolate, identify, and reproduce (through cloning) a gene.

3. Genes can be inserted into plant cells by using the Ti plasmid from *Agrobacterium tumefaciens,* a parasitic bacterium that naturally transfers DNA to plant cells, or by using physical techniques, including biolistics and electroporation. In either case, if the inserted genes include one that is selectable (for instance, that provides resistance to a toxic antibiotic), it is possible to find cells that have stably incorporated the genes into their own chromosomes.

4. The techniques of molecular biology and genetic engineering are used to produce bacteria and fungi that synthesize new pharmaceutical drugs and industrial compounds.

5. New varieties of plants produced by genetic engineering can serve as a source of biochemicals—for instance, proteins or lipids useful in industrial processes.

6. Changing the genes of plants can produce new horticultural varieties with attractive and unusual flowers.

7. Genetic engineering may produce pest-resistant crops more efficiently than classical genetic methods. And by creating plants resistant to herbicides, genetic engineering may allow chemical methods of weed control to be used more safely.

8. Certain genes may be altered by genetic engineering to produce fruit that maintains its quality longer after harvest.

9. Genetic engineers are working to identify genes that will give plants improved tolerance to environmental stresses such as salt, heat, cold, or drought.

10. The genetic engineering of plants, and especially the presence of genetically engineered foodstuffs on the market, has engendered concern among members of the public. Assuring the safety of new plant varieties is no more difficult for transgenic plants than for plants produced by classical breeding techniques. Understanding the ecological implications of transgenic plants requires long-term research.

Questions

1. A mutant *E. coli* bacterium without a plasmid cannot grow on a medium containing the antibiotic ampicillin or streptomycin; it fails to grow. When the same strain of *E. coli* is infected with the pBR322 plasmid, it can grow on ampicillin or on tetracycline. Inserting a new gene into the middle of the ampicillin-resistance or the tetracycline-resistance gene (in the pBR322 plasmid) inactivates that gene. Assume that Strain A of *E. coli* contains a pBR322 plasmid that has a piece of DNA inserted into the ampicillin gene and that Strain B contains a different pBR322 plasmid that has a piece of DNA inserted into the tetracycline gene. Complete the table below.

	Grows on ampicillin	Grows on tetracycline
Bacterium without pBR322 plasmid	No	No
Bacterium with normal pBR322 plasmid	Yes	Yes
Strain A	?	?
Strain B	?	?

If you had a suspension that contained a mixture of Strain A and Strain B, how could you select and grow a culture of Strain A cells?

2. Some investigators have found DNA associated with ancient plant fossils. They suggest that this DNA comes from the cells of the original plants. (Alternatively, the DNA might come from ancient bacteria or fungi or from a modern contaminant.) To analyze the DNA, they must have large quantities of individual pieces, which they can obtain by cloning the pieces. Describe the process by which they might obtain these quantities.

3. Match the following terms on the left with their corresponding definition on the right:

polygalacturonase	piece of bacterial DNA independent of the chromosome
anthocyanin	plasmid used to carry a gene into a cell
Agrobacterium tumefaciens	enzyme that digests plant cell walls
cloning	enzyme that cuts DNA at specific sites
restriction enzyme	technique for making plasma membrane permeable to DNA
plasmid	technique for isolating a single strain of bacterium or sequence of DNA
electroporation	colored compound in a plant cell
vector	bacterial strain used to transfer plant cells

4. Outline the arguments for and against the following projects:

a. Producing a bacterium that could live in the human gut and break down cellulose to the sugar glucose. Such a bacterium might allow humans to derive more nutrition from what is currently indigestible plant material.

b. Producing a strain of bacterium that could fix nitrogen symbiotically with any species of plant (as *Rhizobium* does with legumes).

c. Producing a strain of wheat, the kernels of which produce a toxic compound that kills mildew fungi.

d. Producing a banana that makes a protein from the hepatitis B virus.

5. Suggest a trait that could be added to an existing plant to improve it in some way, then outline arguments for and against your proposal.

6. New techniques allow genetic engineers to produce plants that lack genes and traits of their species—rather than gain new ones. Suggest useful plants that might be created using these techniques.

InfoTrac® College Edition

http://infotrac.thomsonlearning.com

Plant Biotechnology

Bradley, D. 1996. Shining, unhappy plants (plant genetic engineering). *American Scientist* 84:25. (Keywords: "shining " and "plants")

Gepts, P. 2002. A comparison between crop domestication, classical plant breeding, and genetic engineering. *Crop Science* 42:1780. (Keywords: "domestication " and "engineering")

Henkel, J. 1995. Genetic engineering: Fast forwarding to future foods (includes related article on genetically engineered tomatoes). *FDA Consumer* 29:6. (Keywords: "Henkel" and "genetic engineering")

Ethical and Ecological Aspects

Deneen, S. 2003. Food fight: Genetic engineering vs. organics: The good, the bad and the ugly. *E* 14:26. (Keywords: "genetic" and "organics")

Marvier, M. 2001. Ecology of transgenic crops. *American Scientist* 89:160. (Keywords: "ecology" and "transgenic")

Thompson, P.B. 1997. Food biotechnology's challenge to cultural integrity and individual consent. *The Hastings Center Report* 27:34. (Keywords: "food biotechnology" and "integrity")

CHAPTER **18**

Evolution and Systematics

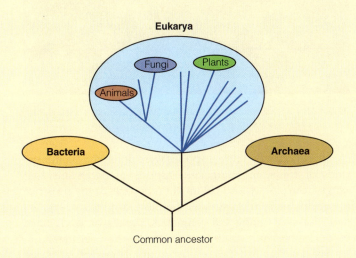

Eukarya

Fungi

Plants

Animals

Bacteria

Archaea

Common ancestor

Visit us on the web at http://biology.brookscole.com/plantbio2
for additional resources, such as flashcards, tutorial quizzes,
InfoTrac exercises, further readings, and web links.

1. Linnaean taxonomy organizes and groups organisms into a hierarchy of taxa (species, genera, families, orders, classes, phyla, kingdoms, and domains) on the basis of comparisons of traits.

2. Life evolves because DNA can be altered at random by mutation and recombination, leading to organisms with varied characteristics. Some of the variants reproduce more than others because they respond better to environmental factors that impose natural selection. A species can split into two species by dividing into populations that evolve separately (speciation).

3. By comparing traits in modern and fossil species, phylogenetic systematists seek the paths of evolution (phylogeny) that led to modern taxa. The paths are expressed in diagrams called phylogenetic trees or cladograms.

4. Cladistics provides quantitative, computerized methods for deducing evolutionary relationships among taxa. The cladistic ideal of limiting formal names to monophyletic groups helps make the taxonomic system more logical and predictive.

18.1 THE DIVERSITY OF LIFE

Ranging from oak trees to bacteria and from whales to mushrooms, life takes on a colossal number of forms. Scientists have named about 1.5 million species, which may be less than 10% of the species currently alive. Fossils suggest that a hundred times that many species arose, flourished, and died out in the past. The multitude of species makes up the diversity of life, or **biodiversity.**

How did so many species of life come into being? How can we sort them for study? Did they descend from common ancestors? If they did, how are modern species related? We have many answers, but much is still unknown. Currently, a young science called cladistics is reshaping old ideas and bringing new insights. This is a great time to be studying the diversity of life.

The most relevant fields of study regarding the diversity of life include taxonomy, evolution, and systematics. **Taxonomy** is the process of sorting and naming the multitude of life forms—a practical task of great importance, for without an orderly system of names, biologists could not tell each other which organisms they study. **Evolution** is the process by which living species change and new species come into being. **Systematics** is the effort to find how modern life forms are related. Today, most systematists look for the evolutionary steps that led from ancient to modern forms of life. Those steps are called **phylogeny,** a term that means "the origin of groups." This chapter introduces all these fields of study, beginning with the question of how we distinguish between species in the first place.

18.2 DEFINING SPECIES

If you spend time in a wild area, you find that organisms are not infinitely variable—there are recognizable kinds, such as the orange poppies in **Figure 18.1.** Each kind is a distinct **species**—a group of organisms that are more closely related to one another than to organisms of any other kind. Members of a given species may also look more like one another and interbreed more freely with one another than with organisms outside the group.

Michael G. Barbour

Figure 18.1 Several kinds of wildflowers are easy to distinguish in this California field. Each kind is a different species.

Most currently known species were defined by combinations of traits, or **characters**—aspects of organisms that range from shapes and colors of body parts to the presence of specific molecules, such as the oils that give mustard its acrid flavor. Among the characters, details of DNA have become increasingly important in the last few decades.

Species that are defined by combinations of traits are called **phenetic species.** *Citrus* trees are an example; based partly on distinctions between their fruits, orange trees are assigned to the species *Citrus sinensis*, lemon trees to the species *Citrus limon*, and grapefruit trees to the species *Citrus paradisi*. Few people would mistake one of the fruits for another; they differ in characteristic ways, as do the trees that bear them.

With *Citrus* species as examples, we can discuss the difficulties in defining species. Let us begin with a practical problem: Given an individual organism, how can we tell whether it belongs to a particular species? One way is to compare the organism with a specimen that a trained taxonomist identified and placed in a museum or botanical garden. The final authority is a **type specimen** that was placed on file when the species was first named. But your specimen will differ somewhat from the museum sample. How much can organisms differ without being separate species? The answer depends on how much natural variation occurs within the species. Differences occur with age, and environmental factors such as climate can affect body form. Individuals of the same species can even differ slightly in DNA, leading to variations such as those that distinguish one human being from another. Thus, to define a species we need many specimens, of various ages and from varied environments, to display the range of variation within the species.

The *Citrus* example illustrates yet another problem in defining species: The boundaries drawn depend on the traits used for comparison. Yet, there is no specific rule about what kinds of traits must be considered and how many traits must differ between two populations to consider them as separate species.

To avoid these difficulties, many biologists prefer to define species by means of a **mating test:** If organisms from two populations mate and produce fertile progeny (offspring) under natural conditions, then the two populations belong to the same species. Species defined by the mating test are called **biological species.** The mating test can have surprising results at times: Two populations may turn out to be different biological species even though they look alike, because they cannot mate and produce fertile offspring. In other cases, two populations belong to the same biological species despite having many differences. This is true of the many breeds of dogs, all of which belong to the same biological species.

But the mating test has its own problems. Most importantly, it does not apply to organisms that lack sexual reproduction. Examples include thousands of species of bacteria and fungi. By the mating test, each individual in an asexual population would be a separate species—a conclusion that would defeat the purpose of taxonomy. Then, too, extinct organisms that are known only from fossils cannot be subjected to a mating test. For life forms that are extinct or entirely asexual, species can only be defined phenetically, by comparing combinations of characters.

A second problem with the mating test is that many plant species can interbreed with closely related species, producing progeny that are weakly fertile. How fertile must they be to pass the mating test? Any dividing line that might be set would be arbitrary and controversial.

Because of such difficulties, the assignment of life forms to species is always open to debate and change. Nevertheless, the great majority of named species are well accepted. They provide a sound basis for exploring life's diversity and how it came about.

18.3 TAXONOMY: NAMING LIFE FORMS

People have always grouped familiar organisms into species that are given informal names, such as "cats." But common names vary so that a certain name may refer to different species in different locations. For example, the bird called the robin in the United States is a completely different species from the robin of England. For scientific communication, we need names that everyone accepts, and a formal system for assigning new names. This is the function of taxonomy.

The Linnaean System Is Hierarchical

The taxonomy in common use today is based on a **hierarchy,** meaning that it has levels, and the groups at one level are nested within groups at the next level. The taxonomic hierarchy began with the work of a Swedish physician and

botanist named Carolus Linnaeus. In 1753, Linnaeus published a book in which he named about 6,000 species of plants and assigned them to 1,000 groups called *genera* (singular, *genus*). A **genus** is a group of species that are similar enough to be obviously related, as in the *Citrus* example.

For each species, Linnaeus wrote a short descriptive phrase in Latin. He regarded this phrase as the formal species name, but for convenience, he also wrote a single word in the margin that could be combined with the genus to give an abbreviated two-word name, the **binomial.** Currently, every species is given a binomial, or **species name,** which is always printed in italics. The first word (always capitalized) is the genus. The second word (never capitalized) is the **specific epithet.**

Linnaeus also saw that two or more genera can resemble one another more than others, making it possible to group genera into larger sets. Genera with similar traits make up a **family.** For instance, all the plants that have roselike flowers belong to the family *Rosaceae.* This family includes garden roses, cherry trees, almond trees, and many others. Modern taxonomists group families into **orders,** orders into **classes,** classes into **phyla** (singular, *phylum;* also called **divisions**), phyla into **kingdoms,** and kingdoms into **domains. Table 18.1** illustrates these categories. There you see that the names at certain levels have standard endings.

Table 18.1 Linnaean Classification Illustrated by the Common Garden Nasturtium (*Tropaeolum majus*)

Linnaean Rank	Name	Ending
Domain	Eukarya	*-a*
Kingdom	Plantae	*
Phylum (division)	Magnoliophyta	*-ophyta*
Class	Magnoliopsida	*-opsida*
Subclass	Rosidae	*-idae*
Order	Brassicales	*-ales*
Family	Brassicaceae	*-aceae*
Genus	*Tropaeolum*	*
Species name	*Tropaeolum majus*	*
Specific epithet	*majus*	*

*Names at this level do not have a consistent ending.

You might think the levels in Table 18.1 would be enough to classify all the forms of life, but taxonomists need extra levels to divide up the multitude of species. For example, a family may be divided into subfamilies, and several families may be grouped into a superfamily. A species can be divided into subspecies, varieties (or races, among animals), and forms. Subdivisions below the species level are important in cultivated plants, where breeding programs have led to variants called cultivars. The term **cultivar** is equivalent to variety, but it is only used to describe products of human selection within a species. Cabbage, cauliflower, Brussels sprouts, and broccoli are cultivars of the same species, *Brassica oleracea.*

In discussing taxonomy and systematics, there often is a need to speak of taxonomic groups in general. Here the term **taxon** (plural, *taxa*) is useful. A taxonomic group at any level is a taxon. For instance, a species is a taxon, and a kingdom is another taxon.

Biologists Define Domains and Kingdoms of Life

The taxonomic system is a work in progress, changing every year as new discoveries challenge old boundaries. Nothing illustrates this point better than the kingdom concept.

Until the nineteenth century, scientists placed all forms of life into two kingdoms: animals, which move actively and consume prey; and plants, which do not. But then biologists found that some microscopic organisms combine animal-like mobility with the plantlike ability to take energy from sunlight. Also, studies of cell structure and metabolism showed that fungi (mushrooms and their relatives) have more in common with animals than plants, even though fungi traditionally had been grouped with plants. Finally, cells of plants resemble those of animals much more than they resemble bacteria, even though bacteria traditionally had been put in the plant kingdom. In response, early twentieth-century biologists divided the old plant kingdom into four new kingdoms. Bacteria and similar organisms were assigned to kingdom **Monera,** fungi went into kingdom **Fungi,** the green plants remained in kingdom **Plantae,** and all other forms of life except animals were placed in a catchall kingdom, **Protista.**

In the mid-twentieth century, studies with the electron microscope confirmed the wisdom of giving bacteria their own kingdom, showing that bacteria have much simpler cell structure than other life forms. Among other complexities, the cells of plants, animals, fungi, and protists are *eukaryotic,* meaning that they enclose most of their DNA in a membranous envelope, forming a true nucleus. Cells of bacteria have no envelope around their DNA and are *prokaryotic.*

In the late twentieth century, biologists toppled kingdom Monera when they began to compare the DNA of various organisms. Led by American bacteriologist Carl Woese, they found that prokaryotes include two distinct groups of organisms, different enough to have evolved separately since the early days of life **(Fig. 18.2).** The main evidence came from studies of the DNA that tells cells how to make ribosomes, the organelles that build proteins (see Chapter 2). Each ribosome contains several molecules of a substance called *RNA,* and the DNA that specifies ribosomal RNA is called **rDNA.** Woese's group compared rDNA from many organisms and found far more variation within each group of prokaryotes than among all plants, animals, and fungi (Fig. 18.2).

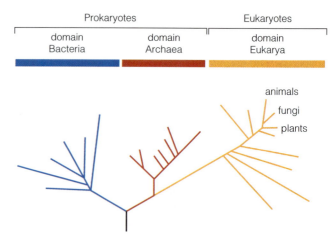

Figure 18.2 Variation in rDNA between prokaryotes and eukaryotes, supporting the domain concept. Animals, plants, and fungi are at the short line segments as labeled. Other terminal lines represent other groups included in the study. Their unfamiliar names are omitted for simplicity. The difference in rDNA between two groups is proportional to the length of line segments between them. Bacteria (*blue*) and Archaea (*red*) are prokaryotic. Eukarya (*orange*) are eukaryotic. (Redrawn with permission of M. L. Wheelis from Fig. 1 in Wheelis, Kandler, and Woese, "On the nature of global classification," *Proc. Natl. Acad. Sci. USA* 89:2930–2934 (April 1992).)

Biologists had long known that prokaryotes vary in metabolic abilities far more than animals do. That fact, and the variation in DNA, suggested that each of the two prokaryotic groups should have the same taxonomic status as all eukaryotes put together. But the kingdom concept was already firmly associated with animals, plants, and fungi—groups within the eukaryotes. Thus, to express the diversity of prokaryotes, biologists had to invent a new, higher category above the level of the kingdom: the domain, a group that contains one or more kingdoms.

Currently, biologists divide life forms into two prokaryotic domains (**Bacteria** and **Archaea**) and one eukaryotic domain (**Eukarya).** Animals, plants, and fungi are well-established kingdoms within the Eukarya, but the old "kingdom" Protista has fallen apart, leaving many smaller groups that do not fit into the three established kingdoms **(Fig. 18.3).** Scientists are currently debating the status of the smaller eukaryotic groups. Two of them, the *Heterokonts* (also called *Stramenopila*) and the *Alveolates* (see Chapter 21), have been proposed for kingdom status, but the issue is not yet settled. Likewise, it remains to be seen how the domains Bacteria and Archaea will be divided into kingdoms.

Modern biologists widely agree that all life probably evolved from a single common ancestor, and the most logical classification would group organisms by their evolutionary relationships, just as we group humans into families on the basis of parent–child relationships. We now have tools (see later in this chapter) to deduce the paths of evolution that led from the original ancestor to the millions of modern living species. Because evolution is the cause of all this diversity, the next section discusses the study of how and why life evolves.

18.4 EVOLUTION: THE CAUSE OF LIFE'S DIVERSITY

Through most of recorded history, scholars viewed life's diversity as a divine plan of creation, in which the Creator established an unchanging ideal body form for each species. One belief was that growing organisms strive to match the ideal patterns, and they differ because the physical world is imperfect. The scholar's goal was to deduce the ideal pattern for each species by examining the common features of individuals.

This way of thinking began to recede in the nineteenth century as scientists explored **fossils** (relics of life such as bones and leaves that are embedded in stone). Observing the way rocks form today, scientists concluded that more deeply buried fossils formed earlier in history. Older fossils differ from more recent ones, challenging the view that each species is unchanging. The difference between dinosaurs and modern reptiles is a familiar example, but equally great changes have taken place among plants. For instance, all the modern marsh plants called horsetails (genus *Equisetum;* see Chapter 23) are herbs. But 300 million years ago, there were tree-sized horsetails with secondary growth and wood; and 600 million years ago, there were no land plants at all. The only photosynthetic organisms were prokaryotes (see Chapter 19) and algae (see Chapter 21).

Darwin Proposed a Model for Evolution

By the 1850s, fossil studies had made a convincing impression that life's history emphasizes change rather than constancy. Consequently, many scientists were ready to believe that the hereditary characteristics of species could change, or evolve, over the course of many generations. All that scientists needed was a clear idea of how such evolution might occur. Charles Darwin and Alfred Wallace, who were English natu-

ralists, came up with the required idea at about the same time. Darwin's ideas about the mechanism of evolution began to take shape during a trip around the world, which began when he was 22 years old. Many sights (such as fossils and an earthquake in South America) influenced Darwin's thinking, but a brief stop at the Galápagos Islands, 900 km off the coast of Ecuador, made the strongest impression. There he found 14 species of birds that resembled finches he had seen in Ecuador, but they also differed in many ways. It occurred to Darwin that finches could have migrated from the mainland to the islands long ago, where they encountered new conditions that somehow altered their heredity. Here was born the important idea that the environment could change the patterns of heredity in a species, shaping life into new forms.

Darwin resolved to keep his thoughts to himself until he could present a convincing argument, and many years passed as he carefully built his case. Meanwhile, Alfred Wallace built up similar ideas in the rain forests of Southeast Asia. When Darwin received a letter from Wallace that revealed the coincidence, Darwin knew the time had come to publish. He presented his ideas and arguments in *The Origin of Species*, published in 1859. In his book, Darwin proposed a mechanism of evolution based on the following four assertions:

1. Changes in heredity occur in the individuals of a population, leading to varied progeny.

2. Populations produce more progeny than the environment can support. This leads to competition among the progeny.

3. The progeny that are best adapted to the environment will reproduce most abundantly. **Natural selection** was Darwin's term for this environmental effect.

4. Repeated over many generations, the preceding three factors could lead to great changes in heredity, and, hence, great changes in the forms of life.

Darwin's arguments were so convincing that most biologists adopted his view within a few decades. His ideas suggested that there is no ideal body form for each species, and the forms can change as the environment changes. Among scientists, the quest for perfect patterns was replaced by a search for paths of evolution that led to modern forms. It was a revolutionary change in the way scientists thought about the diversity of life. And as knowledge grew, scientists refined Darwin's ideas to build the more complete view of evolution that is presented below.

Mutations and Recombination Alter Heredity

To explore current ideas about evolution, this section begins with how and why hereditary variations occur within every species. Darwin sharpened his own ideas by experiments with pigeon-breeding, which clearly showed that organisms of the same species can vary in ways that are passed on to future generations, and that new variations arise from time to time. Little was known about heredity in those days, so Darwin could only guess at how variations might occur. Now

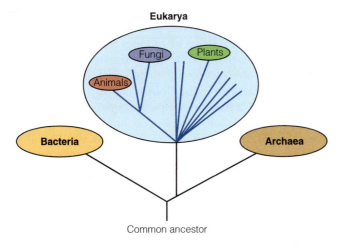

Figure 18.3   The domains of life (Bacteria, Archaea, Eukarya). Domain Eukarya contains three accepted kingdoms (Animals, Plants, and Fungi) and many smaller groups that are not currently given kingdom status (*blue lines*). These smaller groups were previously assigned to an artificial kingdom called Protista and are still called "protists" for convenience, even though kingdom Protista is no longer recognized.

we know that hereditary information is stored in molecules of DNA, and we know how cells copy DNA and pass it on to progeny (see Chapters 2 and 12). We also know how information in DNA can be changed—indeed, scientists change it daily to create new patterns of heredity in the quest for knowledge and more productive plants and farm animals. The two main sources of change in DNA, called mutation and recombination, are discussed in the following sections.

MUTATION Many events can cause changes at random points along a DNA molecule. These random changes in DNA, called **mutations,** are vital to evolution. They are the primary source of new hereditary information. To understand mutation, a brief review of the way DNA stores information is necessary.

DNA is a giant molecule consisting of two paired chains of many subunits called *nucleotides* (see Chapter 2). There are four kinds of nucleotides, which differ in one of their components, called a DNA base. Information is stored in the sequence of bases along DNA, just as a book stores information in the sequence of letters. The information in a book is organized as a series of messages. The same is true of DNA, where each message is a segment of DNA called a *gene*. Hundreds or thousands of genes occur along each DNA molecule. The messages stored in genes are concerned with making protein molecules that build the body.

With this background, we can explore the way mutations occur. Some mutations take place when cells copy their DNA before dividing. The copying process is extremely accurate, but heat energy keeps the molecules in motion, causing collisions that sometimes lead to insertion of the wrong base at some point along the copy. Such changes, called *base substitutions,* are much like errors you would make if you typed a page from a book (Fig. 18.4).

Original DNA **CGGTAGTGACCTGGATGC**
Mutant copy **CGGTAGTGGCCTGGATGC**

↑
base substitution

Figure 18.4 A base substitution, the simplest kind of mutation. For simplicity, only the bases along each segment of DNA are shown, with the letters *A, C, G,* and *T.* In the copy, base G substitutes for base A in the original.

The proteins that copy DNA can correct many base substitutions, but some are missed. As a result, one out of every billion nucleotide pairs is miscopied each time DNA is copied. The errors occur at random locations, so any one spot in DNA is only rarely miscopied. But a human cell contains about 6 billion nucleotide pairs, so there are likely to be several new mutations every time a human cell copies its DNA. When such errors occur in the DNA of reproductive cells, the altered gene may produce new hereditary characteristics in the progeny (Fig. 18.5).

Agents that cause mutations are called **mutagens.** For the base substitutions described previously, body heat was the mutagen. But other mutagens, coming from the environment or the body's metabolism, can greatly increase the frequency of copy errors, or cause more extensive changes. Some muta-

normal stamen   normal fused carpels   carpel replacing stamen

normal flower   mutant flower

Figure 18.5 Flowers from normal and mutant plants of the species *Arabidopsis thaliana,* which is widely used in research on plant gene control. Outer parts of flower buds were removed to show female structures (carpels) and male structures (stamens). In the mutant, carpel-like structures occur where stamens would normally form. The mutant trait resulted from changes in a single gene, which was altered by treating parental plants with a chemical mutagen.

gens are high-energy *radiations,* such as dental X-rays, ultraviolet light from the sun, and high-energy particles that are released when nuclei of radioactive elements decay. Radiations strike DNA at random locations and sometimes break the DNA or cause bases to fuse together. Cells have proteins that repair some of the changes, but sometimes the repair is imperfect and leads to extra or missing segments of DNA, or wrong connections when multiple breaks occur and the ends are reconnected. Other mutagens are *chemicals* that occur in foods, industrial products, and the natural environment. An example is benzopyrene, which occurs in smoke and causes errors when DNA is copied. Even normal metabolism makes mutagens, such as small amounts of highly unstable molecules called reactive oxygen species that damage DNA. In summary, mutagens strike at random locations along the DNA, and they normally strike any one gene only rarely. But mutagens are always present; therefore, a few mutations may occur almost every time DNA is copied.

Occasionally, a mutation helps the organism and spreads through the population, contributing to evolution. Without these rare positive mutations, all organisms would still be nearly identical to the original forms of life on Earth. A familiar example of positive mutation is the appearance of antibiotic resistance in bacteria that cause human disease. When an antibiotic is applied, any resistant bacteria rapidly replace more vulnerable bacteria. This example shows that the value of a mutation may depend on the environment, because bacteria that spend energy on making resistance genes would probably die out in a world that lacks antibiotics.

Although some mutations are beneficial, most mutations have little or no effect on evolution because they cause damage that leads to their elimination. Most genes have been perfected by millions of years of evolution, so random changes in a gene usually detract from the quality of the message, just as random typographical errors detract from the information in a book. Sometimes changing a single DNA base has no effect at all. But if the change occurs at a critical point in the gene for a vital protein, the mutated cell will make bad copies of the protein, leading to cell death. Worse yet, when mutations damage genes for proteins that control cell division, the cells may multiply without limit and produce tumors and, in animals, cancers.

RECOMBINATION Large changes in the forms of life, such as the evolution of flowering plants from mosses, take an immense length of time because they require positive changes in many genes—and positive mutations are rare. But evolution would be far slower without **recombination,** a process that creates new combinations of genes by joining parts of DNA molecules from separate organisms. Recombination can quickly produce valuable combinations of genes that would take thousands of generations to form by mutation alone.

Recombination occurs in several ways. In **transduction,** viruses sometimes carry DNA of one host organism to another. In **transformation,** bacteria take up segments of DNA that are released from decaying organisms, and enzymes insert compatible portions of the foreign DNA into the cell's own DNA. In **conjugation,** some bacteria pass a copy of their own DNA into another bacterium of the same species, and enzymes exchange parts of the host's own DNA for some of the transferred DNA. But in cells of eukaryotes, **sexual reproduction** is the most common source of recombination.

Figure 18.6 outlines the way sexual reproduction recombines genes. (For detailed diagrams, see Chapters 12 and 16.) Briefly, sexual reproduction brings together two haploid reproductive cells (*gametes*), each carrying one complete set of chromosomes from a parental organism. Two gametes fuse to become a diploid cell (the *zygote*) with two sets of chromosomes. The zygote grows and divides to make an adult body with diploid cells. When the diploid organism prepares to reproduce, some of its cells go through a special division process called *meiosis*, during which the chromosomes first double and then line up side by side. With the DNA of both parents side by side, enzymes break the chromosomes at precisely the same point and reattach the broken ends in such a way that the chromosomes exchange segments. This event, called *crossing over*, happens at many random points along most chromosomes. As a result, each chromosome emerges with the proper set of genes, but some segments come from one parent and some from the other parent. Then two successive divisions produce four haploid cells that become gametes with new combinations of genes. Because the multiple crossover points are randomly located, no two gametes are likely to have the same combination of parental chromosome segments.*

---

*Even if there were no such thing as crossing over, meiosis would yield considerable recombination because each pair of chromosomes lines up at the cell's equator before cell division without regard to the orientation of parental chromosomes. As a result, any two haploid cells emerging from meiosis are likely to have different combinations of parental chromosomes. However, crossing over yields far more recombination.

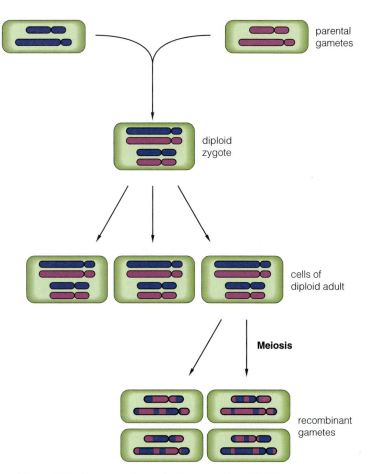

Figure 18.6 How sexual reproduction recombines DNA from two organisms, illustrated with a species that has two kinds of chromosomes. To keep track of parental origins of DNA segments, chromosomes from the two parental gametes are shown with different colors.

HYBRIDIZATION As an agent for bringing together useful genes, sexual reproduction is limited in that many organisms mate only with members of their own species. This is particularly true in animals. But among plants, mating often occurs between members of different species. When two species mate successfully, the process is called **hybridization,** and the progeny are **hybrids.** Usually, only species of the same genus (such as *Citrus*) can hybridize; less closely related species (for example, from different genera) usually differ too much for hybrid progeny to survive.

Hybrid plants often cannot reproduce sexually because the mismatch between chromosomes disrupts meiosis. Nevertheless, hybrids can be vigorous and may multiply by asexual reproduction. Where two related species coexist, many hybrids may occur because hybridization occurs repeatedly.

Hybrid plants sometimes transfer genes between the two parent species, a process called **introgression.** Such transfer requires **back-crossing,** a mating in which a hybrid plant mates with a member of one parental species. Back-crossing is possible if some of the hybrid's meiotic divisions produce viable gametes that are similar enough to those of the parent species. When that happens, the parent species acquires genes the hybrid brought in from the other species.

Biologists agree that hybridization has been important in the evolution of some plant species, but it is uncertain how often it occurs among plants on the whole. One recent study in five parts of the world revealed hybrid populations in only 6% to 16% of wild plant genera. However common it is, hybridization and introgression can be important to humankind because of fears that it may allow genes from genetically engineered plants to escape into wild populations. Also, plant breeders use hybridization and back-crossing to improve crops. For example, commercial varieties of tomatoes produce abundant fruit but have lower resistance to certain diseases than wild species of the same genus from Peru. Tomato breeders have been hybridizing and back-crossing these species to make highly productive tomato plants that also have high disease resistance.

ENDOSYMBIOSIS Strong barriers, such as flight responses or chemical antagonisms, usually prevent very different organisms from merging their DNA. These mechanisms avoid wasting reproductive resources on progeny that have fatal conflicts between parental genes. However, DNA from very different organisms may come together in an association called **endosymbiosis,** in which cells of one species reside inside cells of another species. For example, certain bacteria live inside cells of host organisms. In most cases, both organisms retain their genetic identity. But if endosymbiosis lasts for many generations, DNA can pass from the guest to the host, adding to the host's nuclear DNA and leaving the guest as a dependent organelle. This can give the host cell important new capabilities.

Strong evidence indicates that endosymbiosis led to the energy-processing organelles called *mitochondria* and *chloroplasts*. Mitochondria occur in nearly all eukaryotic organisms, where they release food energy by using respiration to oxidize fuel molecules (see Chapter 9). Chloroplasts, found only in plants and some protists, make sugars by using light energy in *photosynthesis* (see Chapter 10).

Biologists have concluded that mitochondria and chloroplasts evolved from endosymbiotic bacteria that were capable of respiration and photosynthesis. Mitochondria may have evolved as shown in **Figure 18.7.** The process began when free-living bacteria, which had evolved respiration, took up residence inside primitive eukaryotic cells that could draw energy only from the less efficient process of fermentation. The bacteria paid for shelter by giving off energy-rich molecules that the host could use. In time, most of the bacterial DNA was transferred to the nucleus. The transfer left the residents unable to survive outside the host but able to reproduce with help of the nucleus. The bacteria had become mitochondria.

Later, some of the respiring eukaryotic cells engulfed photosynthetic bacteria, giving the host cells an ability to use light energy. These bacteria transferred most of their DNA to the nucleus and became chloroplasts. The origin of

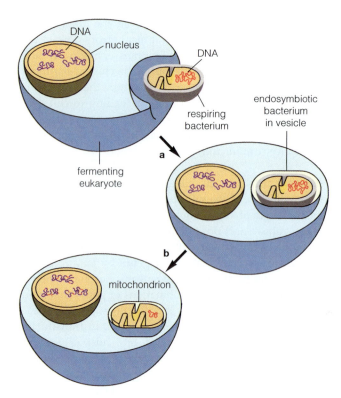

Figure 18.7 Proposed endosymbiotic origin of mitochondria. (**a**) A respiring bacterium took up residence inside a primitive eukaryotic cell, where it divided along with the host cell. (**b**) Descendants of the respiring guest bacterium lost their walls and most of their DNA, to become mitochondria. Some of the lost DNA was transferred to the nucleus. Similar events led to the endosymbiotic origin of chloroplasts.

mitochondria and chloroplasts from bacteria is called *primary endosymbiosis*. Still later, other eukaryotic predators gained chloroplasts through endosymbiotic partnership with eukaryotes that already had chloroplasts. This *secondary endosymbiosis* led to brown algae and certain other protists (see Chapter 21).

By merging DNA from very different organisms, endosymbiotic associations can open the way to major innovations and great increases in the diversity of life. Mitochondria and chloroplasts demonstrate this point. Mitochondria gave eukaryotes the respiratory energy needed to grow large and complex, and chloroplasts allowed some eukaryotes to evolve into algae and land plants.

Natural Selection Guides Evolution

For large changes in the forms of life (such as the evolution of trees from mosslike ancestors), evolution needs a guiding agent that is much less random than mutation and recombination. Darwin and Wallace proposed that selective agents in the environment guide evolution by influencing the reproductive success of variant organisms within each population.

To prove that selection can lead to evolution, Darwin cited the artificial selection that farmers practice to improve livestock and crop plants. For thousands of years, farmers have used the most productive animals and plants for breeding. The farmer acts as a **selective agent,** and the favored trait is high productivity. Darwin proposed that the environment has its own selective agents, which impose natural selection on wild species. Natural selective agents include nonliving (*abiotic*) factors such as climate, water supply, and light, as well as living (*biotic*) factors such as competing organisms, predators, and prey. Consider a field where tall and short plants grow in the presence of grazing animals. Tall herbs attract attention and often are bitten, whereas shorter plants often escape notice. In this way, the selective agent (the grazing animal) gives small size a selective advantage.

Multiple selective agents may act at the same time, sometimes affecting the same trait in opposite ways. For example, tall plants capture more sunlight and spread seeds farther. Here the selective factors include light and agents that disperse seeds and affect seedling survival. Their influence favors tall size and offsets the negative effect of attracting predators. We can expect most cases of natural selection to involve such compromises among effects of selective agents. It is therefore unwise to settle on any simple idea of why a given trait evolved without conducting careful experiments to explore the effect of many selective agents.

DIRECTIONAL SELECTION Because of natural selection, prolonged exposure to a stable environment can cause a species to accumulate hereditary traits that enhance success in that environment. Such favorable traits are called **adaptations.** But once the population is fully adapted, any further changes—such as new mutations—are likely to have a negative impact, lessening the degree of adaptation. Now the same agents that led to progressive evolution tend to keep the species stable. To acknowledge these facts, biologists distinguish between **directional selection,** which leads to new adaptations, and **stabilizing selection,** which maintains existing adaptations.

Directional selection can be illustrated by events that may have happened as cacti were adapting to life in deserts. Modern cacti have spines in positions where other plants have leaves, suggesting that spines evolved by modifying genes that originally called for broad leaves **(Fig. 18.8).**

To illustrate the components of directional selection, Figure 18.8 shows 1 year of an adaptive process that would have taken many generations to be complete. Figure 18.8a shows how leaf width varied in the parental population. In Figure 18.8b, mutation and recombination have led to young plants with more variation. But in the desert, a selective agent (very dry air) acted against progeny with broad leaves, which lose much water. Progeny that reached reproductive age under these conditions

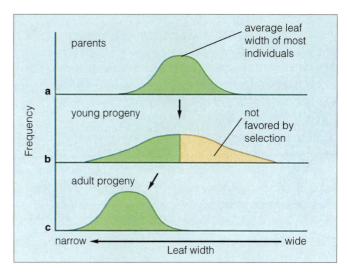

Figure 18.8 Directional selection, illustrated by adaptive narrowing of cactus leaves (ultimately converting leaves to spines) in response to desert conditions. (**a**) Variation in leaf width among parental plants. (**b**) Variation in leaf width among young progeny. Mutations and recombination expanded the range of variation, but directional selection in the desert acts against progeny with broad leaves (*orange*). (**c**) Variation in leaf widths among progeny that reach reproductive age. The average leaf width is less than in the preceding generation.

had narrower average leaf widths than their parents (Fig. 18.8c). Repeating these events over many generations, this directional selection could have narrowed the leaves to the width of spines.

You may wonder why adaptation to the desert did not eliminate leaves entirely. Part of the answer is that spines deter animal attack. Recent studies also suggest that spines help a cactus to collect rain water by channeling the flow of water. These ideas suggest that Figure 18.8 shows just one of several simultaneous selective effects that led to the conversion of cactus leaves to spines. Also, since spines lack photosynthesis (they are dead at maturity), all these changes would have been fatal without the previous or simultaneous transfer of photosynthesis from green leaves to the stem.

STABILIZING SELECTION When directional selection has brought a trait to the best condition for the prevailing environment, the same forces that previously caused change will now exert stabilizing selection, which holds the trait in its present condition. For example, suppose a cactus population in the desert already has highly adapted spines. Narrower spines are too weak to protect the plant, and broader spines waste matter and energy on excessive strength. In every generation, mutation and recombination produce some offspring with spines that are too narrow or too broad, but natural selection acts against them. Since any change from the current mean is harmful, selective forces now act equally against variants on both sides of the mean,

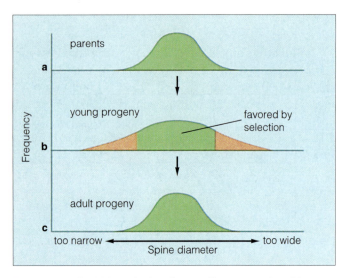

Figure 18.9 Stabilizing selection, illustrated by preservation of the most adaptive spine diameter in cactus plants. (**a**) Variation among parental plants. (**b**) Variation among young progeny. Mutation and recombination expanded the variation, but natural selection favors progeny with spine diameters most like the parents (*green*). (**c**) Variation among progeny that reach reproductive age. The final variation in spine diameter is the same as in the preceding generation.

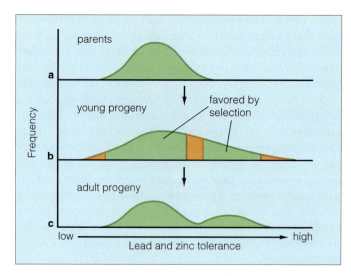

Figure 18.10 Diversifying selection, illustrated by the evolution of metal tolerance where lead-mine tailings cover some normal soil. (**a**) Variation among parental plants before mine tailings were present. (**b**) Variation among young progeny. Mutation and recombination broadened the range of variation, but growth on mine tailings selects for plants with high metal tolerance, whereas growth on normal soil selects for plants with low tolerance. (**c**) Variation among progeny that reach reproductive age. The progeny vary more widely than the parents.

leaving the next generation of adults with the same average spine diameter as the generation before (Fig. 18.9).

DIVERSIFYING SELECTION Both directional and stabilizing selection tend to reduce the variety within a population. But too much loss of variety can be costly. This is well known in agriculture, where some crops are so genetically uniform that local diseases can easily become major epidemics. A fine example is the black Sigatoka disease of banana. Caused by a fungus that infects both leaves and fruits, the disease was first seen Fiji in 1963, but it spread worldwide by the early 1970s because most commercial banana trees are of a single uniform cultivar with low resistance to the disease.

In nature, the very fact that genetic uniformity results in epidemics gives a selective advantage to variants that are not susceptible. The result is a form of natural selection that *increases* genetic diversity in a population—the opposite of the effect that comes from directional and stabilizing selection. This **diversifying selection** is defined as natural selection that increases genetic variation. It can be caused not only by disease agents, but also by other factors that favor two or more distinct types in a population.

Diversifying selection has been well studied in a grass (*Agrostis tenuis*) that grows on lead-mine tailings in Wales. The mine tailings are piles of rock that are rich in lead and zinc. *A. tenuis* plants occur on mine tailings and also on surrounding normal soil. Plants on mine tailings and plants on normal soil belong to the same biological species; they can, and do, exchange pollen and produce fertile progeny. But the progeny vary widely in tolerance for lead and zinc.

Plants that thrive on normal soil have low tolerance; they fail when transplanted to mine tailings. Likewise, plants that thrive on mine tailings grow poorly on normal soil. Only the presence of mine tailings beside normal soil permits tolerant and intolerant plants to persist simultaneously in the population.

The mine tailings have existed for less than 200 years, so we infer that before then, all the plants grew only on normal soil where stabilizing selection kept their metal tolerance low. From that starting point, **Figure 18.10** shows how diversifying selection may have increased metal tolerance within the population.

For simplicity, Figure 18.10 compresses adaptive evolution that probably spanned several decades into one generation. Before adaptation is complete, diversifying selection is a special case of directional selection, differing only in that selective agents favor more than one type. There are two results: The adult progeny are more diverse than the parents, and their mean metal tolerance is greater than in the parents. Figure 18.10 does not show what happens when the population is fully adapted, but it can be easily inferred. If mine tailings persist together with normal soil, diversifying selection becomes a special case of stabilizing selection. Mine tailings and normal soil will continue to favor two tolerance levels, trimming extremes and intermediate types. The adapted population's broad variation in metal tolerance will be maintained from one generation to the next.

DIVERGENT EVOLUTION The concept of directional selection helps to explain the great diversity of life on Earth. If a population moves to a new environment, directional selection will gradually change the population's hereditary traits in ways that better fit the new environment. The population will come to differ from related populations that remained in the old environment. This increase in genetic differences among groups is called **divergent evolution.** The cacti and their close relatives provide an example. Where climatic changes converted moist habitats to deserts, broad-leafed plants evolved water-saving adaptations such as fleshy green stems and leaves that are reduced to spines **(Fig. 18.11a).** These desert-adapted cactus plants diverged strongly from broad-leafed relatives that still thrive in moist tropical forests.

With directional selection in differing local environments, countless cases of divergent evolution have occurred in the billions of years that life has existed on Earth. The net result is the great diversity of life forms on this planet.

CONVERGENT EVOLUTION Directional selection can also cause quite different species to evolve similar traits. Such an increase in similarity between two taxa is called **convergent evolution.** It occurs when differing populations are exposed to similar environments over many generations. For example, some plants in African deserts resemble cacti of American deserts **(Fig. 18.11b).** The plants in both areas have spines, fleshy green stems, and reduced leaves. This might tempt us to put the African and American species in the same family. But other traits, including differences in flowers and DNA, show that the African and American plants belong to different families, the Euphorbiaceae and the Cactaceae. Desert members of the families evolved similar traits because they experienced similar directional selection for a long time.

COEVOLUTION Many species interact as competitors, predator and prey, or symbiotic partners. In such interacting pairs, each species may exert selective pressure on the other species. The result is simultaneous evolution of both species, a process called **coevolution.** In many cases, coevolution leads to a close match between the two species. For example, plants that are pollinated by moths often produce nectar at the base of long, slender tubes or spurs, beyond the reach of other pollinators but ideal for the long tongues of moths. An exclusive pollinating relationship helps both partners: Pollen transfer is more efficient if the pollinator visits just one plant species, and the pollinators get a private food supply. The mutual benefit suggests that moth pollination favored the evolution of long spurs in the flowers, as well as long tongues in the moths.

An extension of the preceding example shows how coevolution can increase the diversity of life. Competition for pollinators reduces a plant's reproductive success, leading to directional selection for new traits that reduce competition—new flowering dates and new flower colors and shapes. As new kinds of plants appear, they become new resources for pollinators that can exploit them. The

a   b

Figure 18.11 Desert plants of the southwest United States (**a**) and Africa (**b**). The similarities are attributed to convergent evolution. On the basis of floral and anatomical features, confirmed by DNA comparisons, these plants are placed in separate families: the *Cactaceae* (**a**) and the *Euphorbiaceae* (**b**). The two families also include plants that do not have desert adaptations and are much less similar than those shown here.

result is directional selection for pollinators that have matching changes. Over time, this coevolution can result in many new species of flowering plants and pollinators.

Population Genetics Reveals Evolution and Its Causes

Darwin's theory of natural selection was a vital first step in building the modern theory of evolution, but his ideas were handicapped by lack of knowledge about how hereditary information is stored, altered, and passed to progeny. Not until the twentieth century was genetics advanced enough to show the molecular basis of evolution. By then, geneticists had adopted Mendel's idea that hereditary information is carried on particles (now called genes), but they wondered why different versions of the same gene (called alleles) persist in a population, even though one allele is more abundant or is expressed more strongly than the other. It is easy to see the relevance of these questions to evolution, for evolution in its most basic form consists of changes in the relative abundance of alternative alleles. Such questions attracted scientists who love to create mathematical models of real events. In 1908, the English mathematician G. H. Hardy and the German physician G. Weinberg simultaneously published the same model to answer such questions. Their analysis postulated an ideal population in which the following five conditions apply:

1. Mutations do not occur.
2. Organisms do not migrate between populations.
3. Reproduction is limited to random sexual mating.
4. There is no natural selection.
5. The population is very large.

The analysis by Hardy and Weinberg showed that under such conditions, two alleles for the same gene will remain indefinitely in the population at a fixed ratio, even if one allele is dominant over the other. This conclusion, called the **Hardy-Weinberg equilibrium,** became the basis of a new discipline known as **population genetics**—a field of study that integrates genetics and evolution. Because the mathematical details are not necessary to understand what this text says about evolution, this chapter provides none of the mathematics. You can find details in any genetics textbook. However, a brief look at the implications of the Hardy-Weinberg equilibrium will show how it reveals that evolution is taking place and how it helps to identify the agents that guide evolution.

To determine whether a wild population is at Hardy-Weinberg equilibrium with respect to a pair of alleles, the researcher takes some organisms from the population and determines the ratio of alleles they collectively contain, and then measures the same ratio in the next generation and compares the two results. If the ratios are similar enough that any differences can be explained by random errors, the pair of alleles is at Hardy-Weinberg equilibrium. Many genes and alleles in wild populations have been studied in this way. In many cases, some alleles are at equilibrium, whereas others are not. Because evolution consists of changes in allele ratios, those that are not at equilibrium are evolving. In this way, the Hardy-Weinberg equilibrium yields clear evidence that much evolution is taking place on a micro-scale today, and that one gene can be stable while another gene in the same species is evolving.

Studies related to the Hardy-Weinberg equilibrium can identify factors that are causing evolution. This is true because all the conditions specified for Hardy-Weinberg equilibrium relate to causal factors. To show this, the following list considers the five conditions in turn:

1. Mutations convert one allele to another, and therefore alter the ratio of alleles, unless forward and reverse mutations exactly balance.

2. If many individuals enter or leave the population, the allele ratio will change unless the migrating individuals have alleles in exactly the same ratio as the overall population.

3. If mating is not random, some allele combinations may be reproduced disproportionately often.

4. Natural selection favors the reproduction of individuals with a certain allele combination over others.

5. If the population is very small, chance can determine which individuals reproduce (see next section).

Thus, failure to meet any one of the Hardy-Weinberg conditions can result in evolution (disequilibrium). All five causes of disequilibrium have been measured many times, revealing, for example, that natural selection typically has much larger effects than mutation, migration, and nonrandom mating. With such information, many evolutionists use population genetics as a tool to predict changes and to explore the causes of evolution.

EFFECTS OF CHANCE ON SMALL POPULATIONS Darwin attributed all evolution to natural selection, but population genetics predicts that *chance* can affect evolution in small populations. The prediction has been verified in field studies. In small populations, the best-adapted individuals do not always leave the most offspring. If a population is so small that only a few individuals have a certain valuable trait, a random accident such as a fire or epidemic may accidentally eliminate all the individuals that have the best allele, whereas some of those with other alleles are spared. Such a random change in the allele ratio is called **genetic drift.**

Another effect of chance on small populations, the **founder effect,** occurs when a few individuals from a large population establish a small, isolated population. Chance may determine which of the main population's alleles are present in the founders. As a result, the founders may have a combination of traits that is uncommon in the old population. This may start the new population on a new path of evolution. The founder effect often is seen in studies of oceanic islands, where wind, water, and birds occasionally bring seeds from mainland plants. The island plants are related to mainland species, but their traits often differ in many ways.

Speciation Multiplies Species

Natural selection by itself does not increase the number of species; it changes species that already exist. Nevertheless, fossil studies show many cases where the number of species increased with time. Evolutionists explain such multiplication by a process called **speciation,** which splits one species into two. Speciation involves the processes discussed below.

REPRODUCTIVE ISOLATION AND DIRECTIONAL SELECTION For one species to become two, an original population must divide into two populations that encounter different directional selection, resulting in divergent evolution. But even with different patterns of selection, the populations will remain a single species if they exchange genes through sexual reproduction. Thus, in sexual life forms, speciation can occur only if some factor prevents the populations from exchanging genes. Such a block to gene exchange is called **reproductive isolation.** Common causes of reproductive isolation are described in the next sections.

GEOGRAPHIC ISOLATION In many cases, speciation begins when geographical barriers prevent populations from meeting to exchange genes. Such a separation is called **geographic isolation.** For example, a storm may blow individuals from a mainland population to a distant land with new environmental conditions and selective agents. In time, the resulting divergence can become so great that individuals from the two populations no longer crossbreed even if they are brought back together. This may have happened in the case of Darwin's finches on the Galápagos Islands. Geographic isolation also can result from slow geological events, such as the creation of a mountain range—events that leave parts of a population on opposite sides of an impassable barrier.

POLYPLOIDY   The possession of more than two chromosome sets per cell—a condition called **polyploidy**—is an important source of new species in plants. Comparisons of chromosomes suggest that up to 70% of flowering plant species have a polyploid origin. Wheat, potatoes, and cotton are examples. Until the time comes for making gametes, reproductive cells in seed plants remain diploid by dividing every time they duplicate their chromosomes. But occasionally, a cell with duplicated chromosomes fails to divide. This cell and its descendants have four sets of chromosomes instead of the normal two sets. If polyploid cells give rise to a stem that forms roots, an independent polyploid plant is formed.

A new polyploid plant with four sets of chromosomes per cell is instantly unable to exchange genes with its diploid relatives (it is reproductively isolated), because mating with its diploid ancestors will not produce fertile offspring (Fig. 18.12). Gametes from the polyploid plant have two sets of chromosomes, whereas gametes from a diploid plant have one set of chromosomes. If the two kinds of gametes fuse, the resulting triploid plant may be vigorous, but its gametes are defective because complete pairing in meiosis requires an even number of chromosome sets.

HYBRIDIZATION   Hybridization is another source of reproductive isolation that can lead to speciation. As discussed earlier, a hybrid arises by fusing the gametes of two species. If the species are close relatives, the hybrid may be vigorous and even fertile. An example is the widely planted London Plane tree (*Platanus X acerifolia*), which arose naturally when European and American species hybridized. Likewise, some grasses and some composites (relatives of the sunflower) generated many species by hybridization.

New hybrids often are sterile, usually because the two parent species have different numbers or kinds of chromosomes (Fig. 18.13). If the chromosomes differ too much, meiosis fails because the chromosomes do not pair properly. But fertility can be restored if a cell at the tip of a hybrid plant becomes polyploid and initiates a polyploid shoot that forms gametes. When a polyploid cell undergoes meiosis, each chromosome has a partner, so pairing can occur and meiosis can be completed, thereby making gametes. The polyploid plant is reproductively isolated from both parent species because its gametes have a different number of chromosomes than gametes of either parent species. An example is *Triticosecale*, a human-made hybrid between wheat (*Triticum*) and rye (*Secale*) that combines the high productivity of wheat with the disease resistance of rye. The sterile hybrids became fertile when plant breeders doubled the chromosome number. Some crop plants, such as wheat, arose by hybridization followed by polyploidy.

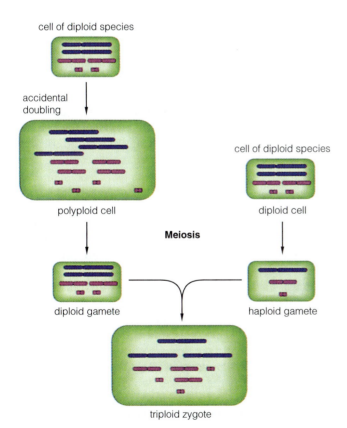

Figure 18.12 Polyploidy can lead to reproductive isolation, and thereby to a new species. The diploid species has two sets of chromosomes. Failure to divide after doubling chromosomes results in a polyploid cell with four sets of chromosomes. When a polyploid cell goes through meiosis, it makes gametes with two sets of chromosomes, whereas gametes of the originating diploid plants have one set. A joining of diploid and haploid gametes creates a triploid plant, which is sterile because meiotic pairing requires an even number of chromosome sets.

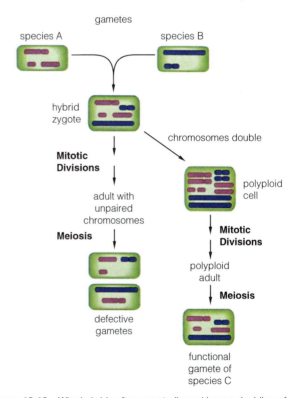

Figure 18.13 Why hybrids often are sterile and how a doubling of chromosomes (polyploidy) restores fertility. (*Top*) A hybrid forms by fusing gametes of two species with differing chromosome sets. (*Left branch*) The hybrid's chromosomes do not form matching pairs; hence, they are distributed abnormally in meiosis, yielding nonfunctional gametes. (*Right branch*) Chromosome doubling permits normal pairs in meiosis; gametes receive one chromosome of each kind and are fully functional.

Macroevolution Generates New Forms of Life

All evolution is based on changes in the information stored in DNA—information that guides development and leads to distinctive body forms, metabolic products, and behavior patterns. Based on the scale of change, biologists recognize two levels of evolution. **Microevolution** consists of changes too small to alter the fundamental nature of the species, such as alterations in flower color. **Macroevolution** consists of changes large enough to represent the emergence of a new life form, such as the evolution of flowering plants from mosslike ancestors. The difference is important when considering the kinds of research that are needed to explore the two types of evolution.

Microevolution is rapid, easy to observe, and easy to produce artificially in the laboratory. We are surrounded by everyday examples, such as our success in breeding turkeys with increased white meat, breeding corn (maize) for greater yields, and the increase in antibiotic resistance that has occurred in bacteria because of our use of antibiotics. Population genetics is concerned with microevolution.

Macroevolution can be the sum of many microevolutionary changes over long periods, or it may involve larger abrupt changes, such as chromosome rearrangements. Sudden radical changes usually are fatal, but occasionally they lead to a form that has great success.

Macroevolution is more difficult to observe than microevolution. Major changes in body form often require changes in many genes. Constructive changes are rare, so the time needed to accumulate many such changes is far longer than a human life. As a result, most ideas about macroevolution are based on indirect evidence. For instance, fossils show that ancient life forms differed greatly from modern forms. We also find clues about macroevolution by comparing cellular and molecular traits of modern organisms. It was a surprise when biologists found that plants and animals have similar cell structures. The discovery suggested that all life evolved from a common ancestor. The idea gained strength when twentieth-century biologists found that all organisms use the same language to encode information in DNA.

To the great majority of modern biologists, these facts are evidence beyond a reasonable doubt that macroevolution generated all modern forms of life from microscopic organisms that first populated Earth some 3.8 billion years ago. With that point established, the current task is to explore the paths of evolution that produced today's millions of diverse species. These studies make up the discipline called **phylogenetic systematics.**

18.5 PHYLOGENETIC SYSTEMATICS

If all life evolved from the same original ancestor, then a diagram that puts all the paths of evolution together would resemble a tree that grew from a single seed, starting with one stem and adding millions of branches. Such a diagram

is called a **phylogenetic tree.** The task of phylogenetic systematics is to find the most accurate tree.

Figure 18.14 shows the main features of a phylogenetic tree. Tips of branches are the most recent products of evolution along each branch, such as modern liverworts and flowering plants (*B* and *G* in Fig. 18.14). Each branch point is an act of speciation, where one species divided into two.

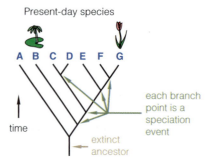

Figure 18.14 Basic components of a phylogenetic tree. Each branch point is an act of speciation, launching new taxa. The time axis is upward, and *A* through *G* represent the most recent products of evolution along each branch.

Today, phylogenetic systematics is one of the most active fields in science. Three developments late in the twentieth century made it so. The first was a new set of methods and concepts called cladistics, which provided a more orderly way to explore evolutionary relations among life forms. Equally important was the invention of fast, inexpensive computers that made it practical to analyze large amounts of data. The third big push came when molecular biologists invented quick ways to read information stored in DNA.

Phylogenetic Systematics Has Practical Value

Before exploring systematics, let us ask why the study is worth doing. The answer is that great practical rewards can come from knowing how evolution led to present-day species. Consider the search for new medicines: Most pharmaceutical products, such as antibiotics, were first found in living organisms. Suppose a researcher finds that trees of a rare, slow-growing species make a compound that easily cures colon cancer. The compound is too complex to make artificially. How could we find a better source? One way is to look for it in fast-growing relatives of the trees. But how would we recognize the relatives, which may barely resemble the trees? With the use of phylogenetic systematics, we would just type the name of the tree into a computer, ask for close relatives, and out would come a list of species and their descriptions.

Similarly, we may find ways to stop parasites that attack food plants by experimenting with relatives of the parasites that can be grown without a host. Phylogenetic systematics can identify such relatives.

Cladistics Explores Clades by Means of Cladograms

Loosely defined, **cladistics** is a set of quantitative methods and concepts for exploring the evolutionary relations among taxa (see the Glossary for a more complete definition). A cladistic analysis compares many modern species to deduce the most probable point in evolution where each species branched off from another evolving group. The name *cladistics* (derived from the Greek *klados*, meaning "tree branch") reflects this focus on branch points. A **clade** is a branch in the tree of life, consisting of an originating taxon and all of its descendant taxa. Before cladistics was invented, scientists had no quantitative basis for deciding which groups of species make up true clades. Cladistics offered a more orderly method, and the phylogenetic trees that it produces are called **cladograms** to reflect the method used in their development.

To illustrate the basic features of a cladogram, **Figure 18.15** shows the relationships among three taxa, each represented by a different kind of fruit tree. The tip of each branch represents the most recent product of evolution along the branch, in this case, three kinds of fruit trees. New branches arise by speciation at the nearest branch point, which cladists call a **node.** At each node, an ancestral species split to produce two new species, while the ancestor itself ceased to exist. The oldest node is called the **root** of the cladogram.

Cladograms rarely include more than a small sampling of species that evolved from the ancestor. Only species that contributed data to the study are listed. The cladogram would have many more nodes if all branch points and terminal species were included. For example, Figure 18.15 shows the "apple" branch evolving from the root node without branching. But many other fruit trees (peach, pear, cherry, and so on) arose by separate speciations along that branch. Thus, the word *apple* in the cladogram merely names a representative group that occurs on the branch. The same may be true of any terminal branch, such as the branch that ends with orange.

With this information, we are ready to identify clades in Figure 18.15. As stated earlier, a clade is an ancestral taxon and all of its descendants. Five groups in Figure 18.15 meet this definition; each is surrounded by a differently colored enclosure. The overall clade includes the ancestor at the root, all three named taxa, and many more taxa that are not included in the study. Four more clades are nested within the overall clade. The largest of these includes the ancestor of all *Citrus* species and its descendants, represented by orange and lemon. Each terminal branch (orange, lemon, apple) also is a clade, which may include many species that are not shown. An enclosure delimiting a terminal clade does not include the preceding node, for that node represents the ancestor of *both* branches. The ancestor of a terminal branch is one of the two species that *emerged* from the node.

There are several ways to draw cladograms for the same set of species **(Fig. 18.16).** The first three cladograms in Figure 18.16 are rooted, whereas Figure 18.16d is unrooted. The

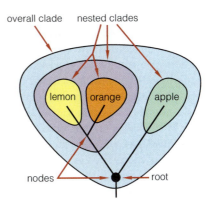

Figure 18.15 Basic features of a cladogram, illustrated with three kinds of fruit trees. Each of the five colored enclosures surrounds a clade (an ancestor and all of its descendants). Each branch point is called a node and is a speciation event. The oldest node is the root of the cladogram.

difference is important. **Rooted cladograms** identify the node in the cladogram that occurred first, thereby showing the direction of evolution throughout the clade. **Unrooted cladograms** do not show which node is closest to the root. As a result, they leave the direction of evolution between each pair of nodes unspecified.

In rooted cladograms, there are different ways to show branching. In one method, descendants diverge from a branch point like arms of the letter *Y* (Fig. 18.16a,b). The other method keeps the arms parallel by showing all the divergence from a branch point at once, then making a right-angle bend in each arm, like the letter *U* (Fig. 18.16c). This method makes it easier to write names of taxa at the ends of the arms, especially if the cladogram is on its side (for example, in Fig. 18.16c). To indicate that a branch includes many species, the tip may be expanded to a triangle.

Alternative cladograms are equally valid as long as they agree on the number of nodes that separate any two taxa. Differences in orientation of branches are not important. Figure 18.16 illustrates this point by using arrows to show that

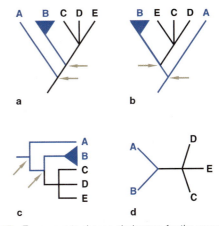

Figure 18.16 Four ways to draw a cladogram for the same taxa. **(a–c)** Equivalent rooted cladograms. All have the same number of nodes between any two taxa, illustrated for taxa *A* and *B* with *arrows* at nodes. **(d)** An unrooted cladogram, which does not show a root node.

each rooted cladogram has two nodes between taxa *A* and *B*. There is similar agreement for all other taxa; thus, the three rooted cladograms are equivalent. Because they differ in branch orientation, you cannot judge which taxa are closest relatives by looking at their positions along the ends of the branches. To determine evolutionary relationships, you must count the nodes that separate the taxa.

A postulate of cladistics is that each act of speciation replaces one parent species with two new species. If a cladogram shows three or more species arising from the same point, it usually means that further studies are needed to determine which species branched off first. However, multiple species may occasionally branch off so closely in time that no study could reveal the branching sequence.

ALTERNATIVE CLADOGRAMS Any set of species might be related in a variety of ways, with a different cladogram for each—and the more species we include in the study, the more cladograms are possible. The number of possible unrooted cladograms depends only on the number of species. To illustrate, for any set of 5 species, there are 15 possible unrooted cladograms, 3 of which are shown in **Figure 18.17**. The cladograms differ as to which pairs of species are the closest relatives. In Figure 18.17a, only one node separates *A* and *B*, three nodes separate them in Figure 18.17b, and two nodes separate them in Figure 18.17c. Thus, alternative cladograms differ in how many steps of evolution stand between each pair of species. To cement this point, draw some of the 12 other unrooted cladograms that are possible. As you work, keep in mind that two cladograms are identical if one can be converted to the other by flipping it over, changing its branch angles, or rotating the diagram or its branches.

Cladists Find the Best Cladogram by Comparing Character States

Every cladogram is a hypothesis about evolution, and there are many alternative cladograms. How do cladists identify the cladogram that most likely reflects the real paths of evolution? The cladistic method compares species with respect to various characters. **Morphological characters** are related to body form, such as the number of flower petals, flower color, and growth habit (herb, tree, and so on). **Molecular characters** are chemical traits such as the detailed structure in a certain segment of DNA, or the ability to make a particular kind of molecule. To be useful in selecting the best cladogram, a character must occur in all the species being considered, and its details—called the **character states**—must differ among some of the species. For example, flower petals might be red in one species and white in another. Here the character is flower color, and the character states are red and white. Cladistics views a change in character state as the basic event in evolution.

To trace evolutionary relationships among species, systematists must compare traits that have a common evolutionary origin. Traits that arose from the same ancestral trait are said to be **homologous.** The wings of a bird and the

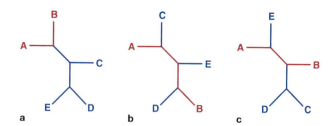

Figure 18.17 Three alternative unrooted cladograms that show ways in which any five species (*A, B, C, D, E*) might be related. Twelve other cladograms are possible. *Red* is used to emphasize differences in number of nodes separating species *A* and *B*.

forelegs of a horse are homologous, because both evolved from the bony forelimbs of a common ancestor. In cladistic terms, homologous traits are alternative states of the same character. Bird wings and horse forelegs are alternate states of the character "forelimb." By contrast, the wings of insects and the wings of birds are not homologous. Despite their similar function, they evolved from entirely different ancestral structures. They are merely **analogous,** meaning that they have a similar form or function, but evolved from different structures. Analogous structures are not alternate states of the same character. They are states of different characters.

Major errors can occur if a classification is based on comparisons of analogous structures. The character "has wings" is a bad choice for defining groups among animals, because it would promote the false conclusion that birds are related more closely to insects than they are to horses.

To compare characters among species, cladists list the character states in a table called a **character matrix,** such as **Figure 18.18.** In the matrix, taxa are listed along the left mar-

Characters

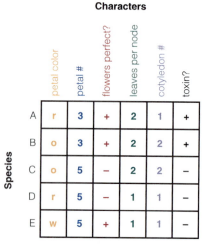

Species	petal color	petal #	flowers perfect?	leaves per node	cotyledon #	toxin?
A	r	3	+	2	1	+
B	o	3	+	2	2	+
C	o	5	−	2	2	−
D	r	5	−	1	1	−
E	w	5	+	1	1	−

Figure 18.18 A character matrix. There are five species (*A, B, C, D, E*). Characters are listed across the top. Each entry in the matrix is a state for the character listed above it. Character states: Petal colors are red (*r*), orange (*o*), or white (*w*). Number of petals in each flower is 3 or 5. Flowers can be perfect (+) or imperfect (−). Plants can have 1 or 2 leaves at each node, and 1 or 2 cotyledons in the embryo. Plants can have (+) or lack (−) a certain toxin.

gin. Characters are listed across the top, and the boxes show the state of each character for each species.

THE IMPORTANCE OF MOLECULAR CHARACTERS In the early days of systematics, biologists relied almost entirely on morphological characters to deduce paths of evolution. As biochemistry brought new information, molecular characters entered the picture—traits such as the ability to make a particular protein. But the choice of characters had the greatest increase when biologists learned how to read DNA. Systematists eagerly turned to DNA base sequences as a source of characters for comparing species. After all, evolution is based on changes in the information stored in DNA. Even morphological characters result from an organism following the instructions in DNA. By comparing DNA among species, we may hope to get the most detailed look at paths of evolution.

DNA offers a multitude of characters for comparison. Each nucleotide position along the DNA can be a character for cladistic analysis, and DNA contains millions of nucleotides. There are four possible character states at each position: The base can be adenine (A), cytosine (C), guanine (G), or thymine (T). Currently, countless cladistic studies include DNA data. Often, they confirm relationships that were traditionally accepted from morphological data. In other cases, they reveal new relationships.

To understand how cladistics works, there is no substitute for solving cladistic problems. The sidebar "IN DEPTH: Do a Cladistic Analysis of DNA" leads you from constructing a DNA character matrix through finding and rooting the best cladogram. Although you will not need a computer, you will appreciate why computers are important in systematics. To find the best cladogram, you must analyze all the possibilities—and as the number of species and characters increases, the number of alternatives rapidly becomes far too great to handle without a computer.

THE PRINCIPLE OF PARSIMONY With many possible cladograms, how do cladists choose the one that shows the real evolutionary relations among taxa? This is the key problem in cladistics. We were not present when the species evolved, so we can never know definitively which cladogram is correct. But several statistical methods help to choose the cladogram that is *most likely* to be correct. One popular method is based on the **principle of parsimony,** which postulates that the cladogram requiring the fewest evolutionary events is most likely to be correct. That cladogram is said to be the most parsimonious.

Although a cladogram may be most parsimonious, we can never be sure it is correct. It is simply a good hypothesis about the paths of evolution, and later studies with new evidence may prove it wrong. With the help of computer programs, cladists use parsimony and several additional methods to find candidates for the best cladogram. The methods often pick slightly different cladograms and a comparison of the choices reveals points on which all methods agree. Those points are said to be *strongly supported.* A compromise diagram, called a **consensus tree** or *consensus cladogram,* is

drawn that includes all the points of agreement while leaving points of disagreement unresolved as nodes from which more than two branches depart. Points of disagreement need further study with additional species and characters.

Rooted Cladograms Show the Sequence of Evolutionary Change

Finding the root of a cladogram is one of the most important tasks of cladistics, because it reveals the direction of evolution. To show its value, **Figure 18.19** compares unrooted and rooted cladograms for the same five species (numbered *1* through *5*). Each numbered species represents the most recent product along its branch. This means recent evolution always runs toward each present species, as shown by the arrowheads in the unrooted cladogram (Fig. 18.19a). But the unrooted cladogram says nothing about the direction of evolution in the internal segments that lie between the three nodes. We do not know which node is the oldest and closest to the point where the ancestor started the evolution of the five species. In contrast, a rooted cladogram (Fig. 18.19b) shows that the common ancestor first split to form the ancestor of species 1 and the ancestor of all the remaining species. With that information, we suddenly know the direction of evolution between all nodes (orange arrows). With a root in place, the cladogram can be redrawn with the root at the bottom, current species at the top, and time flowing upward (Fig. 18.19c).

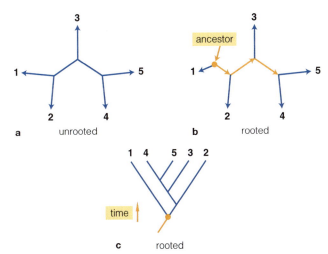

Figure 18.19 Rooting a cladogram reveals the direction of evolution. (**a**) An unrooted cladogram. Recent evolution (*arrows*) is toward all present species, but the direction between nodes is unspecified. (**b**) A rooted cladogram. The *dot* marks the root. Now we know the direction of evolution between all nodes (*orange arrows*). (**c**) A rooted cladogram redrawn with root at bottom and time flowing upward.

How do cladists locate the root? The answer is to include data on additional taxa, called **outgroups,** together with character data on the **ingroup** (the set of taxa that is the target of the study). Good outgroups are taxa that share many characters with the ingroup but differ too much to be part of the ingroup clade. Researchers often include several outgroups, in case some of them are poor choices. For example, green algae are good outgroups for a study of evolution

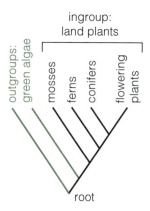

Figure 18.20 Use of green algae as outgroups to root a cladogram in which four kinds of land plants make up the ingroup. The rooted cladogram shows the sequence in which land plant groups evolved.

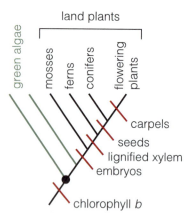

Figure 18.21 Locating innovations on a rooted cladogram. Labeled tick marks show where each innovation arose; they are located by finding the points where their origin accounts for the array of taxa with the fewest evolutionary changes. For example, all land plants have embryos, whereas green algae never do; the most parsimonious postulate is that embryo formation evolved in the land plant clade before mosses branched off.

among land plants. The researcher simply includes data on green algae in a character matrix together with data on land plants, and then runs a computer program to find the best cladogram. **Figure 18.20** shows a typical result. At the root node, an unknown ancestor split to launch one of the outgroups (left) and the ancestor of all the other species (right). The right-hand branch split again, launching the second outgroup (left) and the land plant clade (right). Among other things, the rooted cladogram shows that mosses arose before ferns, and ferns arose before conifers and flowering plants.

The results of one cladistic study can suggest good outgroups for more detailed studies. For example, there are thousands of fern species. How did they evolve? To answer, we could make a character matrix of ferns and include mosses as outgroups. For such a study, mosses are better outgroups than green algae because they share more characters with ferns than do algae. This makes it possible to include many more characters in the matrix.

Because a rooted cladogram can show the sequence in which taxa arose, it also reveals the sequence in which important character states evolved. Cladograms often use labeled tick marks to show where such innovations arose **(Fig. 18.21).** For example, all green algae and all land plants have the photosynthetic pigment chlorophyll *b;* therefore, green algae and land plants probably inherited this trait from their common ancestor. In contrast, all land plants form embryos as part of their life cycle, whereas no green algae have embryos. Thus, the ability to make embryos probably evolved just once in the lines that led to modern green organisms in the land plant clade before mosses branched off.

ANCESTRAL AND DERIVED CHARACTER STATES By showing where a character state first arose, a rooted cladogram allows us to distinguish between **ancestral character states,** which a clade inherited from its ancestor, and **derived character states,** which evolved later. Among seed plants, the ability to make seeds and carpels illustrates the distinction **(Fig. 18.22).** The ability to make seeds evolved in the ancestor of all seed plants, and is therefore ancestral for the clade

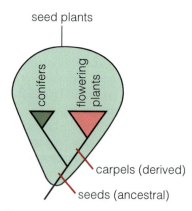

Figure 18.22 Ancestral and derived character states. For seed plants as a whole, represented in this cladogram by conifers and flowering plants, the ability to make seeds is an ancestral character state, inherited from the ancestor of the entire clade. The ability to make carpels is a derived character state, evolved after the first node in the clade.

as a whole. In contrast, the ability to make carpels (the female parts of flowers) evolved later, in the line that led to flowering plants. Thus, for the seed plant clade as a whole, the ability to make carpels is a derived character state.

Use of the terms *ancestral* and *derived* requires care, because the judgment depends on the point of view (which must always be specified). For example, in discussing the seed plant clade as a whole, we concluded that the ability to make carpels is a derived character state (Fig. 18.22), but if attention is limited to flowering plants, the ability to make carpels is ancestral, because all flowering plants inherited it from their ancestor.

Cladistics Reveals Convergent Evolution

The greatest problem in classifying organisms comes from assuming that a shared character state implies common ancestry. The assumption often is sound, but sometimes it leads to wrong conclusions. Errors occur because similar character states sometimes arise independently in two groups of organisms. This is convergent evolution (see earlier). It occurs when populations with different origins evolve under similar selection pressures and develop similar character states. To minimize errors caused by convergent evolution, systematists include many characters in the analysis. Then, even though a few similarities result from convergent evolution, they affect the analysis much less than the differences that come from different ancestry.

With enough characters, a cladogram can reveal which characters arose through convergent evolution. Such

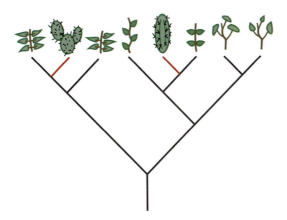

Figure 18.23 A cladogram can reveal convergent evolution. The two taxa on *red lines* have desert adaptations (fleshy green stems, leaves reduced to spines). Other taxa do not. The most parsimonious explanation is that the two desert-adapted taxa evolved their adaptations independently.

insights occur when a well-supported cladogram, based on many characters, places two taxa in separate clades even though they share a character state that is not found in other members of the clades. For example, in **Figure 18.23,** either the ancestor of all the taxa had desert adaptations and most taxa lost the adaptations, or the ancestor lacked the adaptations, and the two desert-adapted taxa evolved their adaptations independently through convergent evolution. The latter explanation is the most parsimonious, considering how many taxa would have to change otherwise.

All Formally Named Taxa Should Be Monophyletic

Ever since the concept of evolution led to phylogenetic systematics, a major goal in taxonomy has been to group organisms by common ancestry. In cladistic terms, the goal will be accomplished when every formally named taxon is a true clade—an ancestor, all of its descendants, and nothing else. Such a taxon is said to be **monophyletic (Fig. 18.24).**

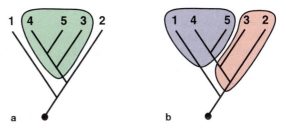

Figure 18.24 A monophyletic group and alternatives. (**a**) The outlined group is monophyletic. It includes an ancestor, all of the ancestor's known descendants, and nothing else. (**b**) Neither of the circled groups is monophyletic. The 1, 4, 5 group omits the group's ancestor and some of its ancestor's descendants, and the 2, 3 group omits some of its ancestor's descendants.

Each currently accepted domain and kingdom of life is believed to be monophyletic. But many traditional taxa at lower levels are still not monophyletic, despite more than 250 years of careful study. It takes time to correct taxonomic problems, for each systematic study is just one hypothesis, and further studies can lead to different views. The work is well worth the effort, though. As discussed earlier, a truly phylogenetic system of taxonomy has much more predictive value than the alternatives, and it can speed such goals as improving crops and discovering new cures for human disease.

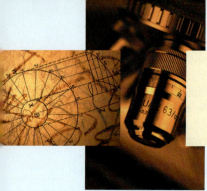

IN DEPTH:

Do a Cladistic Analysis of DNA

Do a Cladistic Analysis of DNA

This sidebar presents a practice problem in cladistics, using concepts from Section 18.5, "Phylogenetic Systematics." Your task is to determine how five given species most likely evolved from their common ancestor. To do it by the cladistic method, you must compare a set of characters that vary among the species. This exercise will take you from the selection of characters through the full cladistic analysis.

Select a Gene and Correct Any Displacements

In a cladistic analysis based on DNA data, the first step is to choose a gene that occurs in all the species you wish to study. Real genes contain thousands of bases along their length, and real studies compare many taxa. But you can learn how to generate a DNA character matrix more easily with a simpler case. We will examine how 5 species differ in the first 32 bases of an imaginary gene **(Fig. 1)**.

Position on DNA strand

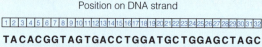

Figure 1 The first 32 bases for the same gene in five species. Numbers at the top are nucleotide positions along the gene. *A, C, G,* and *T* indicate the DNA base at each position in each species.

The array in Figure 1 already looks like a valid character matrix, in which each numbered position is a character, and the base at that position is the character state for a given species. This view would be correct if genes only evolved by replacing one base with another. But sometimes genes lose or add nucleotides at various positions. If that happens, all nucleotides to the right of the alteration are displaced from the position they had in the ancestor of all the species. Unless we discover such displacements and correct them, comparisons of the DNA will lead to false conclusions about evolutionary relationships among the species. The error is the same as comparing analogous rather than homologous structures. If there is no displacement at a given position, the bases at that position in all species are homologous because they descended from the same position in the ancestor's DNA. If a nucleotide is deleted, its place is filled with a nucleotide that began elsewhere in the ancestor's DNA.

To find and correct displacements, look for lengthy regions in Figure 1 where a series of bases is the same in all the organisms under study, and color them for visibility. If a lengthy base sequence is the same in all the species, it is called a **conserved**

sequence, because it probably descended from the ancestor with little or no change. The longer the conserved sequence, the more sure we are that it is truly a shared heritage rather than an accidental similarity among molecules. Our result is shown in **Figure 2.**

Position on DNA strand

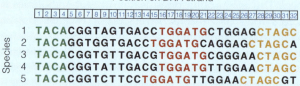

Figure 2 Three conserved sequences, with different colors for visibility. Loss or gain of nucleotides displaced the second and third consensus sequences in some species.

TACA, the first conserved sequence in Figure 2, is already lined up in all five species, occupying positions 1 through 4. This must be where the sequence occurred in the common ancestor. The other two sequences are not lined up, so nucleotides must have been removed or added in some species.

The next step is to line up the second and third conserved sequences by breaking the DNA and moving the broken parts to the right, leaving gaps. But how do you choose which species to break, and where to make the breaks? Inspection tells us the simplest choice is to break DNA in species 2 and 5. The first break must be in the black region between the first two conserved sequences—but where, exactly? With a trial-and-error approach, find a point where a shift to the right will reveal new conserved sequences. The more conservation your move reveals, the more likely it is that you found the real positions where DNA changed during the evolution of the species. Our result is shown in **Figure 3.**

Position on DNA strand

Figure 3 To align the conserved sequences, the sequences in species 2 and 5 have been broken, and the right-hand parts have been moved to the right.

In species 2, we placed the break between positions 4 and 5, then shifted the right-hand part of the DNA one position to the right, leaving position 5 empty. In species 5, we placed the break between positions 11 and 12, and shifted the resulting DNA fragment two positions to the right. As shown by the blue regions in **Figure 4,** these choices reveal conserved sequences that were overlooked before.

Position on DNA strand

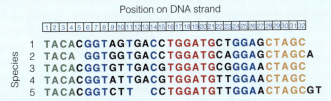

Figure 4 Aligned DNA, with newly revealed conserved regions marked in *blue.*

Choose Characters and Build a Matrix

With the DNA aligned, Figure 4 shows that most DNA positions in all five species are conserved. Changes (shown in *black*) occurred at just 9 of the 32 positions along the gene (forget the bases to the right of position 32; they arose outside the first 32 positions.)

Now build a character matrix from the positions that were *not* conserved. These are all the positions, such as position 9, where the five species do not all have the same base. Evolution occurred at these locations, changing DNA after the species parted from the ancestor. Every nonconserved position is a character that may be added to the matrix. But to keep the workload down, only include the six positions where all five species have a base. Your character matrix should list the species along the left side, list the characters (DNA positions) along the top, and fill in the matrix with bases (character states) from the aligned DNA.

Our matrix is shown in **Figure 5.** For later use, we colored the six DNA positions that contribute to the matrix, and the resulting matrix is shown at the bottom.

Position on DNA strand

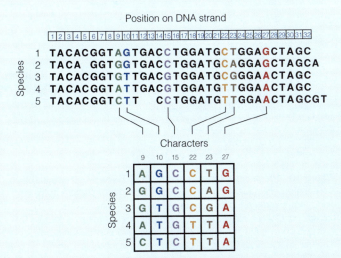

Figure 5 Building the character matrix. Colors indicate nonconserved positions in DNA where every species has a base, and their character states are entered into the matrix (*bottom*). The characters are positions 9, 10, 15, 22, 23, and 27 along the DNA, and the character states are A, C, G, and T.

Draw All the Possible Cladograms

Your next task is to draw all the possible cladograms. For any 5 species, there are 15 possibilities. **Figure 6** shows two of them.

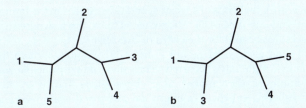

Figure 6 Two of the 15 unrooted cladograms that are possible for any 5 species.

As you draw the other 13 possible cladograms, remember that two cladograms are equivalent unless they differ in how many branch points separate any two species. To save time, you may want to share the task with a partner. You need all the possible cladograms to decide which one most likely reflects the evolution of the species.

Determine Which Cladogram Is Most Parsimonious

Cladistics gives a logical procedure to decide which cladogram most likely shows the real evolutionary relationships among taxa. To make the decision, a variety of methods exist. We will use a method based on the principle of parsimony. It judges that the most probable cladogram is the one that needs the fewest evolutionary events to account for the differences among species.

To apply the principle, work with one cladogram and one character at a time. **Figure 7** illustrates the method by showing how many events are needed to explain the evolution of the first character in the matrix (position 9) if the five species are related as in Figure 6a. A description of the steps follows the figure.

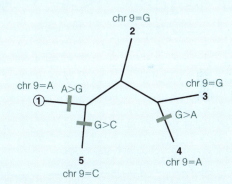

Figure 7 Possible evolution of character 9, assuming the ancestor was like species 1, and the five species are related as in Figure 6a. Current state of character 9 in each species is shown. *Labeled tick marks* show evolutionary events (for example, A>G means A was replaced by G).

Our first step in the analysis was to choose a point in the cladogram where the ancestor of the five species might have entered to initiate the group. We chose the ancestor to be exactly like species 1, entering at the position of species 1. This is noted in the diagram by circling species 1. How do we know this is the real starting point? The answer is that we do not know—but it does not matter, because the outcome of the parsimony analysis does not depend on the starting point you choose—as long as you make the same choice for all cladograms.

With species 1 as the starting point, we guessed that A changed to G before species 1 split, as shown by the tick mark labeled A>G. If that happened, then species 2 and 3 need no further change, because they got G from their ancestor. But species 4 and 5 differ, so two more changes are needed: G became A in the branch that ends at species 4, and G became C in the branch that ends at species 5. Altogether, it took three events (tick marks) to account for the states of character 9.

In the analysis of Figure 7, how did we guess where the changes might have occurred? The only guiding principle was to minimize the number of changes (steps shown by tick marks). The real changes might have occurred elsewhere, and we tried many alternative changes. Some of them took the same number of steps as Figure 7, others took more steps, but none took fewer steps. The point is that we kept trying until it was clear that no choice would take fewer steps than ours.

With this background, your task is to determine how many changes (tick marks) are needed to account for the other five characters in the matrix. Work with one character at a time, following the same approach as in Figure 7. For each character, start the analysis from species 1 so you can compare your result with ours, which is given in **Figure 8.** You will gain more from this exercise if you try it before checking our answer.

To keep track of the changes, Figure 8 uses ticks and labels of a different color for each character, and shows all the character states of each species. Our answer needed 12 events to account for the changes in all 6 characters. In your own analysis, did your tick marks add up to 12? If you got a lower number, recheck your work.

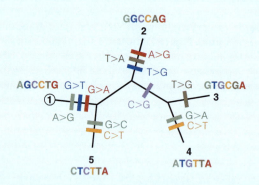

Figure 8 Possible evolution of all six characters in the matrix of Figure 5, if the species are related as in Figure 6a and the ancestor was like species 1. States of all six characters are listed beside each species. *Colored tick marks* show points where each character might have changed. Alternative points of change are possible.

By now you know that the cladogram in Figure 6a needed 12 character changes to explain the data matrix, if the ancestor was like species 1. But this is only 1 of 15 possible cladograms. To find which cladogram takes the fewest steps, you must conduct a similar analysis for all 14 of the remaining cladograms. To start, analyze the cladogram in Figure 6b. Remember that you must use the same starting point and ancestral character states that you used for Figure 6a.

Our results **(Fig. 9)** show that the cladogram in Figure 6b needs 10 steps to generate the present character states, if the

evolution began with an ancestor identical to species 1. This makes Figure 6b more parsimonious than Figure 6a. Do you agree?

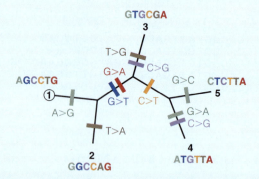

Figure 9 Possible evolution of all six characters in the matrix of Figure 5, starting from an ancestor identical to species 1, if the species are related as in Figure 6b. Other symbolism is described in the legend of Figure 8.

Now complete the search for the most parsimonious cladogram by analyzing the other 13 possible cladograms. To save time, you may wish to divide the work among partners.

Our own analysis showed that 2 of the 15 possible cladograms **(Fig. 10a,b)** take 10 steps, which is fewer than any of the other cladograms. Those two cladograms are equally parsimonious. When two or more cladograms are equally good, you must either do more studies with more data, or settle for a compromise, called a *consensus tree* (Fig. 10c).

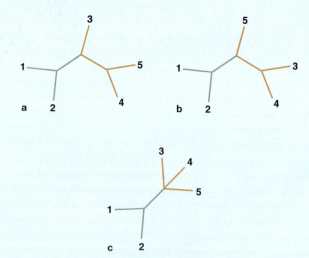

Figure 10 Preparing a consensus tree. **(a,b)** These two cladograms are equally parsimonious; each takes 10 steps. Nodes where both cladograms agree are shown in *green*; disagreements are shown in *orange*. **(c)** Consensus cladogram, with branches leading to species 3, 4, and 5 coming from a single point because **a** and **b** disagree on whether species 3 or 5 is more closely related to species 1 and 2.

A consensus tree (or consensus cladogram) is a diagram that has all the features shared by the equally parsimonious cladograms and leaves the conflicts unresolved (Fig. 10c). Like the two rival cladograms, the consensus tree shows species 1 and 2 arising from a single branch point. Cladists say species 1 and 2 are *sisters*, because they arose from the same node. But the two cladograms disagree about which of the three remaining species is most closely related to species 1 and 2. To reflect the disagreement, the consensus tree has species 3, 4, and 5 all branching from the same point. We normally expect just three lines to meet at each node. Therefore, the meeting of four lines is a signal that we have not resolved the sequence in which the three species evolved. Many such cases occur in published cladograms.

Root the Cladogram

The final task is to find the root of the chosen cladogram (or consensus tree). This is done by including *outgroups* in the character matrix **(Fig. 11).** Here we confess to a secret: Among the species we have been studying, species 5 was included as an outgroup. Species 1, 2, 3, and 4 make up the *ingroup*, the target of the study (Fig. 11a).

Now rearrange the consensus tree (Fig. 11a) so it shows more clearly how the four ingroup species are related. To rearrange the tree properly, remember that Species 5 is *not* the ancestor of the ingroup clade—it is a modern species, as are the four ingroup species. To represent that fact, the first step in rooting the cladogram is to put all five species at the top (Fig. 11b,c).

Next, locate the point where the ancestor of all five species enters the cladogram. Find that point by inspecting Figure 11a again. Species 5 joins the ingroup at the node where species 3 and 4 arise. Between that node and species 5, the ancestor must have split to give species 5 and the ancestor of the ingroup. Thus, Figure 11b inserts a node between species 5 and the ingroup. That node is the root, the oldest node in the cladogram, represented by a dot. It is the point where the ancestor split to launch the evolution of all five species. The rest of the branching in Figure 11b is the same as in Figure 11a, with adjustments in the orientation of the lines. Figure 11c shows the same thing with all lines straightened out.

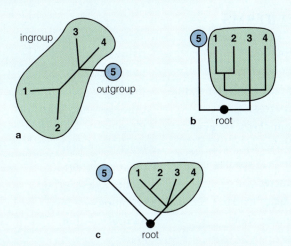

Figure 11 Rooting the consensus tree. **(a)** Species 5 was included in the study as an outgroup; species 1 through 4 are the ingroup clade (*green*) that is the focus of study. **(b,c)** Two ways to rearrange **a,** placing the oldest node (the root) at the bottom.

With the root in place and the outgroup at one side, the cladogram shows that species 1 and 2 are probably sisters, related to each other more closely than to any other species in the clade. But this consensus cladogram does not reveal the sequence in which species 3 and 4 separated from the ancestor of species 1 and 2. To resolve that point, we would need to repeat the study with more data so as to find which branch split off first.

In addition to providing you with first-hand experience in cladistics, we hope this lesson showed why cheap, powerful desktop computers had an explosive impact on phylogenetic systematics. With the small character matrix of Figure 5, a computer could come up with the rooted consensus cladogram in less than a second, saving hours of manual labor. For studies with a few more species than this, the number of possible cladograms is far too many to analyze by hand, and a computer is absolutely essential.

Websites for further study:

Introduction to Cladistics: http://www.ucmp.berkeley.edu/clad/clad1.html

KEY TERMS

adaptation	hybrid
analogous traits	hybridization
ancestral character state	ingroup
back-crossing	introgression
binomial species name	kingdom
biodiversity	macroevolution
biological species	microevolution
character	monophyletic group
character matrix	mutation
character state	natural selection
clade	node
cladistics	outgroup
cladogram	phenetic species
coevolution	phylogenetic systematics
consensus tree	phylogenetic tree
conserved sequence	phylogeny
convergent evolution	polyploidy
derived character state	population genetics
directional selection	principle of parsimony
divergent evolution	recombination
diversifying selection	reproductive isolation
domain	root
endosymbiosis	rooted cladogram
evolution	speciation
founder effect	species
genetic drift	stabilizing selection
genus	taxon
geographic isolation	taxonomic hierarchy
Hardy-Weinberg	taxonomy
equilibrium	unrooted cladogram
homologous traits	

SUMMARY

1. Linnaean taxonomy groups organisms into a hierarchy of taxa, including species, genera, families, orders, classes, phyla (divisions), kingdoms, and domains. Organisms are assigned to a phenetic species by comparing combinations of characters and to a biological species by applying a mating test.

2. Biologists divide life forms between two prokaryotic domains (Bacteria and Archaea) and one eukaryotic domain (Eukarya). Domain Eukarya includes three monophyletic kingdoms (Plantae, Animalia, and Fungi) and many smaller groups that often are informally called protists, some of which are candidates for kingdom status. Domains Bacteria and Archaea are not yet subdivided into widely accepted kingdoms.

3. Studies of fossils, DNA, metabolism, cell structure, and reproduction suggest that all organisms evolved from a common ancestor through processes that alter the DNA in organisms and that permit some variants to reproduce more effectively than others (natural selection).

4. DNA carried by organisms in a population can be altered by mutations, which produce alternative versions (alleles) of genes; by recombination, which brings together DNA that arose in different organisms; and by migration, which can transfer organisms between populations of the species. For recombination, DNA from different organisms can be brought together by endosymbiosis, bacterial transformation and conjugation, viral transduction, and sexual reproduction within or between species.

5. Natural selection results from environmental interactions that allow some variants in a population to reproduce more effectively than others. Selective agents in the environment include food supply, climate, predation, and other factors. Natural selection can improve the adaptation of a species to the environment (directional selection) or maintain the current adaptation (stabilizing selection). Diversifying selection can maintain or increase the genetic diversity within a species.

6. Population genetics integrates the fields of genetics and evolution and provides mathematical tools of great predictive value that help determine whether evolution is taking place and how it is caused.

7. In small populations, chance events can affect the direction of evolution independently of natural selection. Such changes make up genetic drift and lead to a founder effect when small colonies are established.

8. Speciation splits one species into two when part of a population becomes reproductively isolated from the rest, followed by divergent evolution in response to directional selection. Reproductive isolation may result from migration, formation of geological barriers, polyploidy, or hybridization.

9. Phylogenetic systematists seek the paths of evolution that led from a common ancestor to modern species. Their findings often are expressed in phylogenetic trees, which include cladograms. Given a set of species, many different cladograms can be drawn that reflect possible evolutionary relationships among the species.

10. The set of methods and concepts called cladistics provides an orderly, computerized way to select the cladogram that most probably reflects the real evolutionary relationships among taxa. To determine the direction of evolution, cladistic studies include outgroup taxa together with ingroup taxa. This reveals a root node in the cladogram, from which all evolution proceeded.

11. Cladistic analysis can determine the sequence in which character states and taxa evolved, distinguish between ancestral and derived character states, and detect instances of convergent and divergent evolution.

12. A tenet of phylogenetic systematics is that all named taxa should be monophyletic. By revealing cases where named taxa are not monophyletic, cladistic analysis points the way to changes that increase the predictive value of the taxonomic system.

Questions

1. How do biologists determine whether two organisms belong to different species?

2. What are the main taxonomic levels between the domain and the species, and why do the levels constitute a hierarchy?

3. Describe the roles in evolution of mutagens, sexual reproduction, the founder effect, hybridization, and interactions between organism and environment.

4. Under what circumstance is natural selection likely to be directional rather than stabilizing?

5. How does population genetics show whether a pair of alleles is undergoing evolution in a wild population?

6. How does polyploidy lead to a new species?

7. What can be determined from a rooted cladogram but not from an unrooted cladogram?

8. Would insects be useful outgroups for rooting a cladogram of plant species? Why or why not?

9. How do outgroups increase the value of a cladistic analysis?

10. How does cladistics reveal cases of convergent evolution?

11. Copy the cladogram of Figure 18.14, and then draw a circle around a group of species to illustrate the concept of a monophyletic group. Why is it important for named taxa to be monophyletic?

12. To practice the methods used in cladistics, do the exercise in the sidebar "IN DEPTH: Do a Cladistic Analysis of DNA."

InfoTrac® College Edition

http://infotrac.thomsonlearning.com

Cladistic Analysis

Doyle, J.J. 1993. DNA, phylogeny, and the flowering of plant systematics (difficulties in the analysis of data on molecular technology in the field of plant systematics). *BioScience* 43:380. (Keywords: "Doyle " and "plant systematics")

Pennisi, E. 2001. Linnaeus's last stand? (introducing PhyloCode). *Science* 29:2304. (Keywords: "Pennisi " and "PhyloCode")

Watanabe, M. 2002. Describing the "Tree of Life": Attainable goal or stuff of dreams? *BioScience* 52:875. (Keywords: "tree of life" and "dreams")

CHAPTER

19

Archaea, Bacteria, and Viruses

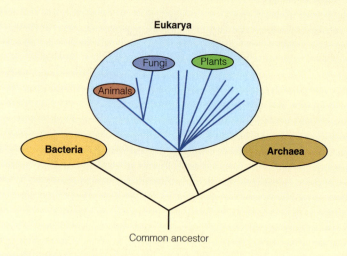

PROKARYOTES, VIRUSES, AND THE STUDY OF PLANTS

PROKARYOTIC CELL STRUCTURE

Many Prokaryotic Cells Have Simple Structures

Some Prokaryotic Cells Have Modified Extracellular and Intracellular Structures

Some Bacterial Cells Form Endospores

LIFESTYLES OF SELECTED GROUPS OF PROKARYOTES

Archaea Inhabit Harsh Environments

Bacteria Include Many Diverse Species

◼ *BIOTECHNOLOGY:*
Bacteria and Biomining

◼ *PLANTS, PEOPLE, AND THE ENVIRONMENT:*
Microbes and Global Warming

PROKARYOTES THAT FORM SYMBIOTIC RELATIONSHIPS WITH PLANTS

Rhizobium Forms a Mutualistic Association with Legumes

Bacteria Can Be Plant Parasites

VIRUSES

◼ *BIOTECHNOLOGY:*
Bacteria as Biocontrol Agents

Viruses Are Infectious Genes

Viral Infections Stunt Plant Growth

SUMMARY

Visit us on the web at http://biology.brookscole.com/plantbio2
for additional resources, such as flashcards, tutorial quizzes,
InfoTrac exercises, further readings, and web links.

1. All prokaryotic organisms can be divided into two domains: the Archaea and the Bacteria. Members of the Archaea dominate harsh environments such as hot springs, salt flats, and anaerobic mud flats, and they also are found in more equable habitats. Members of the Bacteria domain show great variation, to some degree in structure, but especially in metabolic capabilities, habitat, and life histories.

2. The cellular organization of the prokaryotes (Archaea and Bacteria) seems much less complex than that of eukaryotic cells. However, some prokaryotic cells develop complex, specialized structures such as flagella, thylakoid membranes, and spores.

3. Prokaryotes obtain energy for movement and metabolism and carbon for growth and reproduction from various sources. Chemotrophs get energy by oxidizing organic or inorganic compounds; phototrophs capture the energy of light (photosynthesis). Autotrophs obtain carbon from an inorganic source, typically by reducing CO_2; heterotrophs obtain carbon by metabolizing organic compounds.

4. Many bacteria have close, symbiotic associations with other organisms. Some associations are mutualistic: They benefit the bacterium and its host. The relationship between the nitrogen-fixing *Rhizobium* and legumes is one example. Some associations are parasitic: The bacterium harms the host. *Erwinia amylovora,* which causes fire blight in apples, is an example of a plant parasite.

5. Viruses are genes, wrapped in a coat of protein, that infect cells. They appropriate the biochemical machinery of the cells, using it to reproduce themselves and sometimes to kill the cells. Viral diseases of plants cause serious reductions in crop yield and quality.

19.1 PROKARYOTES, VIRUSES, AND THE STUDY OF PLANTS

Prokaryotes is one term to describe all the organisms with cells that lack a nucleus. These organisms generally have a simpler cell structure than do plants, animals, or other **eukaryotes.** The terms *prokaryote* and *eukaryote* were introduced in the 1920s by Edouard Chatton, based on microscope observations. Eukaryotic cells had clear nuclei and other inclusions that were lacking in prokaryotic cells. (The words *eukaryote* and *nucleus* are related: *Karyon* is Greek, and *nucleus* is derived from the Latin *nuculeus,* meaning "kernel.")

In the late 1970s, a microbiologist named Carl Woese proposed using the small subunit ribosomal RNA (rRNA) gene to create a universal tree of life. This gene is critical for a major cell function, protein synthesis, and it is therefore ubiquitous and constrained in its rate of evolution. The slow rate at which the gene evolves allows a comparison of even the most evolutionarily distant organisms. Woese's tree had immediate and profound effects on our views of evolutionary history. The most important of Woese's findings was that

the prokaryotes are not a coherent group, as previously thought; rather, they are composed of two separate groups, which he named the Eubacteria (*eu-* is derived from the Greek, meaning "true") and the Archaebacteria (*archae-* is derived from the Greek, meaning "ancient"). Later, in recognition that this division was as basic as the division between prokaryotes and eukaryotes, biologists coined the grouping *domain,* and named the three domains **Archaea** (the Archaebacteria), **Bacteria** (the Eubacteria), and **Eukarya** (the eukaryotes). Although Archaea and Bacteria differ in fundamental ways, in this textbook it is convenient to consider them together. The word *bacteria* with a lowercase *b* often is used as a common noun for all prokaryotes.

Most often, the prokaryotes are single-celled organisms, although many form colonies, and some form structures with a degree of morphological differentiation. As small and apparently simple as they are, they pervade the world and represent a large fraction of the earth's biomass. Although their structures are simple, their biochemical abilities often are complex and sometimes unique—that is, they possess enzymes and metabolic pathways not found in any eukaryote. Some of these metabolic capabilities are absolutely essential for maintaining the physical and chemical characteristics of the earth in a state suitable for life. The group of prokaryotes is much too large a subject to be covered in any detail by a textbook on plants. Yet, there are three important reasons why an individual who studies plants should also be familiar with prokaryotes:

1. Many of the biochemical compounds, enzymes, and metabolic pathways of plants also are found in prokaryotes. Photosynthesis, the uptake of mineral nutrients, and responses to environmental stresses (such as drought) occur in both prokaryotes and plants. The discoveries made in studying prokaryotes can be used to guide plant research.

2. The evolutionary ancestors of plants were prokaryotes. Not only did the first eukaryotic cell probably evolve from one or more unknown prokaryotes, including an Archaea, but the large organelles in plant cells—the mitochondria and plastids—are probably related to two different types of Bacteria. Studying prokaryotes is necessary for understanding the origin of plants.

3. Plants form ecological associations with prokaryotes. Some of these associations are mutualistic symbioses (for instance, the nitrogen-fixing association between the bacterium *Rhizobium* and legumes); in some associations, the prokaryote is a parasite on the plant. In either case, it is necessary to study the prokaryotes to understand how the plant responds to its environment.

Viruses are particles constructed of a nucleic acid genome (either RNA or DNA) and a protein coat. They are not prokaryotes. They are not even cells and cannot live independently. To reproduce, they must infect the cells of an organism, and they are parasites on the cells they infect. Viruses should be studied by plant biologists for many of the same reasons that prokaryotes should be studied. Many of the basic properties of genes and proteins can be investigated using viruses. The discoveries obtained by studying viruses can be used to guide plant research. Some viruses are plant parasites, and thus are important in the lives of many plants. The diseases that viruses cause, the ways in which they are transmitted from plant to plant, and the methods plants use to combat viral diseases are all part of the life histories of plants.

PROKARYOTIC CELL STRUCTURE

Many Prokaryotic Cells Have Simple Structures

A prokaryotic cell appears in an electron micrograph to be simpler than a eukaryotic cell, because it lacks the internal membrane-enclosed organelles (compare Figs. 19.1 and 3.3). The prokaryotic cell is surrounded by a plasma membrane. This plasma membrane fulfills all the same roles that it does in plants (and sometimes more). It accumulates cell compo-

nents (such as enzymes) in the cytoplasm; it excludes toxic compounds; it allows charge separation, which is necessary for energy generation; and it is sensitive to aspects of the environment, such as water potential or the presence of nutrients. Archaea and Bacteria differ in the chemical compositions of their plasma membranes. Bacterial plasma membrane lipids are similar to those of eukaryotes. Archaeal plasma membrane lipids are unlike any others and are held together by stronger bonds, perhaps essential, given the severe environments in which many of them live.

Outside the plasma membrane, Archaea and Bacteria differ markedly. Bacterial cell surfaces fit into one of two categories, termed Gram-positive or Gram-negative, on the basis of a differential staining technique devised by the Dutch physician Gram (Fig. 19.2). **Gram-negative cells** have a thin cell wall that, like plant cell walls, functions to prevent cells from bursting in a hypotonic solution (solution more

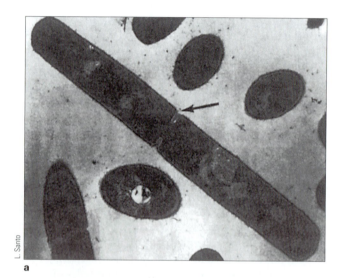

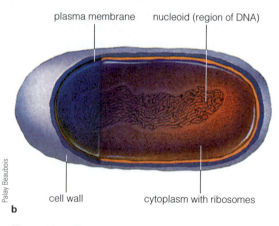

L. Santo

a

Palay Beaubois

b

plasma membrane   nucleoid (region of DNA)

cell wall   cytoplasm with ribosomes

Figure 19.1 The prokaryotic cell. (**a**) A transmission electron micrograph of *Bacillus cereus* showing the light-colored nucleoid and darker cytoplasm. At the time the cell was prepared for microscopy, it had recently divided and a new cross-wall (*arrow*) was forming. (**b**) Organization of the major components of a prokaryotic cell.

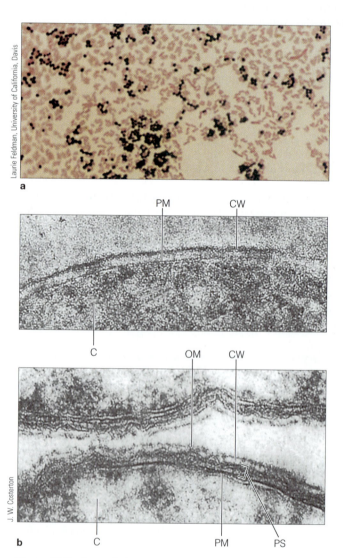

Laurie Feldman, University of California, Davis

J. W. Costerton

a

PM   CW

C

C   OM   CW

b   C   PM   PS

Figure 19.2 Some distinguishing features of Gram-positive and Gram-negative Bacteria. (**a**) Light micrograph of stained Gram-positive cocci (purple spheres) and Gram-negative bacilli (pink rods). (**b**) An electron microscopic comparison of the plasma membrane and cell wall regions of Gram-positive (*top*) and Gram-negative (*bottom*) Bacteria. C, cytoplasm; OM, outer membrane (lipopolysaccharide layer); PM, plasma membrane; PS, periplasmic space; W, cell wall.

dilute than cytoplasm) and determines the shape of the cells: **cocci** (small, round cells), **bacilli** (rods), **vibrios** (bent or hooked rods), **spirilla** (helical forms), and stalked forms. Unlike plant cell walls, the Bacterial cell wall is composed of **peptidoglycan,** a combination of amino acids and sugars that surrounds the cell like a net. Outside of the cell wall is a second membrane composed of phospholipids, polysaccharide, and protein, termed the outer membrane or **lipopolysaccharide** layer. The area between the two membranes is called the **periplasmic space,** a compartment for certain cell functions that contains specific enzymes and other proteins, including those that sense the osmotic potential of the environment.

Gram-positive cells have a thick peptidoglycan cell wall. Although no Gram-positive cells have a second membrane, some have a waxy polysaccharide capsule that is analogous to an outer membrane. The capsule protects some Gram-positive human pathogens from being recognized by microbe-eating cells of the immune system, such as macrophages. One of the best-known families of antibiotics, which includes penicillin, acts by inhibiting the formation of the Bacterial cell wall and is particularly effective against Gram-positive cells. Bacteria treated with these antibiotics become osmotically sensitive and lyse in hypotonic solutions, including human body fluids.

Most Archaea have a paracrystalline surface layer (called an S layer), composed of protein or glycoprotein, that performs the same general functions as a Bacterial cell wall. Because the outer layer is protein, Archaeal cells often are sensitive to proteases (enzymes that degrade proteins) and surfactants. Alternatively, some Archaea have pseudo-peptidoglycan, which is similar to Bacterial peptidoglycan, or thick walls of polysaccharide. Archaea typically lack an outer membrane.

In the cytoplasm is a full set of genes and a complete apparatus for expressing them to make proteins. The genes are encoded in DNA, just as in eukaryotic cells. Typically, the "housekeeping" genes (genes needed for the basic functions of life) are on a single circular chromosome. This chromosome may be very long (in circumference). In the best-studied bacterium, *Escherichia coli*, the chromosome is 1.4 mm in length and contains 4.6 million nucleotide pairs **(Fig. 19.3),** which is a large amount of DNA, considering that the *E. coli* cell is only about 1 μm in length and 0.5 μm in width. The chromosome is wound in the cell and is localized in an area called the **nucleoid.** Unlike eukaryotes, the prokaryotic chromosome is not surrounded by a nuclear envelope, so there is no defined nucleus. It used to be thought that the Bacterial chromosome had no structure and was packed randomly into the cell; however, recent investigations have found that it is complexed with specific structural proteins that organize it into loops. The Archaeal chromosome is complexed with histone proteins, similar to the chromosomes of eukaryotes.

Many prokaryotes have sets of accessory genes called **plasmids,** which are relatively small circles of DNA, about 2,000 to 200,000 nucleotide pairs in length. They often contain functionally related sets of genes. Some plasmids, for

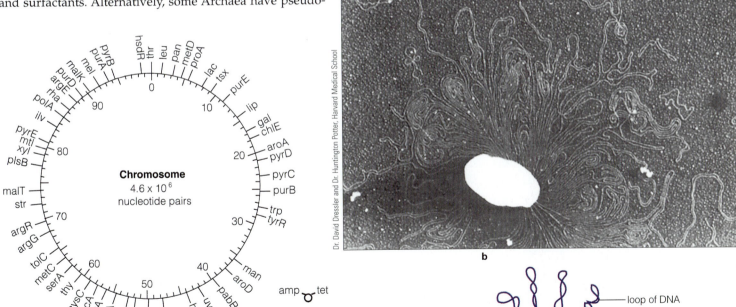

Figure 19.3 A prokaryotic chromosome. (**a**) This diagram of an *Escherichia coli* chromosome shows a fraction of the identified genes. The order of the genes was determined by gene-transfer experiments using F plasmids. The sequence of all 4.6×10^6 nucleotide pairs is known. There are 4,288 genes identified from the sequence. Next to the chromosome is a diagram of a genetically engineered plasmid, showing the positions of two genes that confer resistance to two antibiotics. Note that the plasmid is actually one-thousandth the size of the chromosome. (**b**) An *E. coli* nucleoid spread out for electron microscopy. The loops are held in place by a central matrix of proteins.

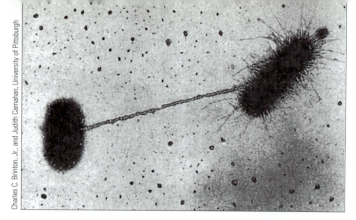

Charles C. Brinton, Jr. and Judith Carnahan, University of Pittsburgh

Figure 19.4 Conjugation between two *Escherichia coli* cells. The donor (F⁺) cell is on the right; the recipient (F⁻) cell is on the left. The long connecting arm is an F pilus. Note that the donor cell also has many smaller pili.

instance, have genes that confer a resistance to antibiotics. Some plasmids in parasitic Bacteria have genes that make them pathogenic (disease-causing) in their hosts. The plasmids can be replicated (reproduced) independently of the chromosome. Sometimes they are replicated faster than the chromosome and the cell, so there are many copies in a cell. Sometimes, such as at high temperatures, the plasmids are replicated more slowly; therefore, daughter cells are formed that do not have a plasmid.

Some plasmids have genes for enzymes and structural proteins that transfer copies of the plasmid to Bacteria that do not have any. This is one method by which Bacteria acquire new genetic capabilities. Although most plasmids are transferred only among cells of one species, some can be transferred to many species of Bacteria. This is how it has been possible for resistance to antibiotics to spread so quickly to many pathogenic Bacteria in the past few years, a situation that threatens to neutralize many of our modern defenses against disease.

A special type of plasmid, the F plasmid, has the ability to incorporate itself into the main chromosome. The F plas-

mid contains genes for making a tube, called an F pilus, that connects its cell with one that lacks an F plasmid. An F pilus is a conduit for the transfer of plasmid or chromosomal DNA from a donor (the cell with the pilus) to a receiver cell **(Fig. 19.4).** When the F plasmid is part of the main chromosome and the chromosomal DNA is replicated, part of one copy of the chromosome can be transferred to the recipient cell through the F pilus. Once the new DNA is in the recipient cell, there is a possibility of an exchange of pieces of the new DNA with the recipient's original chromosome. The transfer, called **conjugation,** allows for a genetic recombination of chromosomal genes. Although it is entirely unlike meiosis, conjugation may play a similar role (promoting diversity) in the population genetics of prokaryotes.

The prokaryotic cell contains ribosomes with a general composition and structure similar to those of eukaryotes. The ribosomes have two subunits, each made of RNA and protein, although prokaryotic ribosomes are smaller than eukaryotic ribosomes. In nucleotide sequence, the rRNAs of Archaea are more similar to the rRNAs of eukaryotes than those of Bacteria, evidence that an Archaea was an ancestor of the original eukaryote.

Prokaryotic cells reproduce by **binary fission,** meaning that the cell splits in two. Preceding this step, the DNA in the main chromosome replicates, so that there are two full copies of the chromosome. It is thought that each copy of the chromosome may be attached to the plasma membrane and that the separation of the points of attachment (as the cell grows and the membrane enlarges) may pull the chromosome copies apart **(Fig. 19.5).** This process is analogous to the separation of chromosomes in mitosis of eukaryotic cells (see Chapter 3), but it differs from mitosis because prokaryotes do not have microtubules, and thus they do not have a spindle apparatus for separating the chromosome copies. A new cross-wall (Fig. 19.1a) forms between the two chromo-

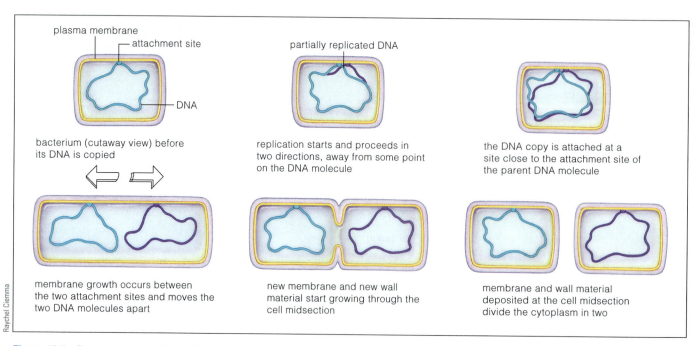

Raychel Ciemma

plasma membrane
attachment site
DNA

partially replicated DNA

bacterium (cutaway view) before its DNA is copied

replication starts and proceeds in two directions, away from some point on the DNA molecule

the DNA copy is attached at a site close to the attachment site of the parent DNA molecule

membrane growth occurs between the two attachment sites and moves the two DNA molecules apart

new membrane and new wall material start growing through the cell midsection

membrane and wall material deposited at the cell midsection divide the cytoplasm in two

Figure 19.5 Bacterial reproduction by binary fission.

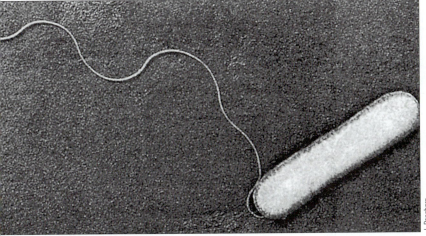

J. Pangborn

a

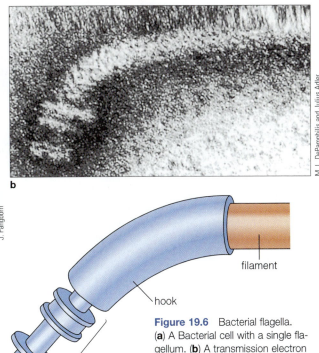

M. L. DePamphilis and Julius Adler

b

somes, so that each progeny cell receives a copy. The other organelles, ribosomes, enzymes, and plasmids are probably divided randomly as the cross-wall forms. Although cell division in prokaryotes seems much simpler than it is in eukaryotes, this may be because of the small size of the cells. Only recently have researchers begun visualizing the intricate details of the division process.

Notably lacking in the above description is any component of the eukaryotic endomembrane system: endoplasmic reticulum, Golgi apparatus, vesicles, or vacuole. Prokaryotes have no mitochondria or plastids and no known cytoskeleton (although genes related to the eukaryotic tubulin gene recently have been identified in both Archaea and Bacteria). It would be wrong, however, to think that there are no complex adaptations among the prokaryotes. The following sections describe some of the more common modifications to the basic prokaryotic cell structure.

Some Prokaryotic Cells Have Modified Extracellular and Intracellular Structures

Many Bacterial and some Archaeal cells swim using **flagella,** projections that propel the cell **(Fig. 19.6).** The prokaryotic flagellum is not related in any way to the eukaryotic flagellum. It is formed from many subunits of the protein flagellin,

Figure 19.6 Bacterial flagella. (**a**) A Bacterial cell with a single flagellum. (**b**) A transmission electron micrograph of an isolated Bacterial flagellum (*top*) with an interpretive diagram (*bottom*).

which line up to form a helical (corkscrew-shaped) filament. The filament rotates, propelled by a motor apparatus—the basal body—where it penetrates the plasma membrane. The rotation of the basal body is powered by a proton and electrical gradient produced across the plasma membrane. The rotating flagellum pulls the cell through the liquid medium. In certain Bacteria covered with many flagella, the motors can reverse direction. They generally do this when the cell senses an unsatisfactory change in the environment. Because the helical flagella are not as rigid under the reverse stress, they bend (flop), and the cell tumbles. After a short period, the motors reverse again, the flagella regain their original conformation, and the cell starts swimming in a straight line. If once again the environment is unsatisfactory, the procedure is repeated, and this continues until the cell finds an improvement in its conditions **(Fig. 19.7).**

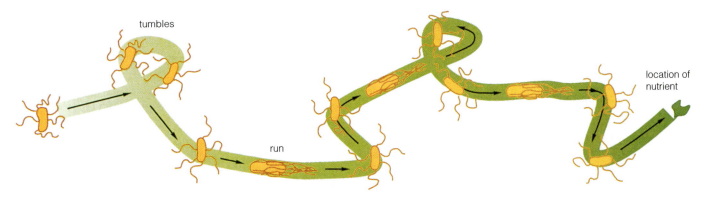

Figure 19.7 Bacteria control their flagella to seek nutrients and avoid stress or toxic compounds. When the flagella turn counterclockwise, the cell moves in a straight line (run); when the flagella turn clockwise, the cell tumbles. When conditions are getting better, the runs are longer.

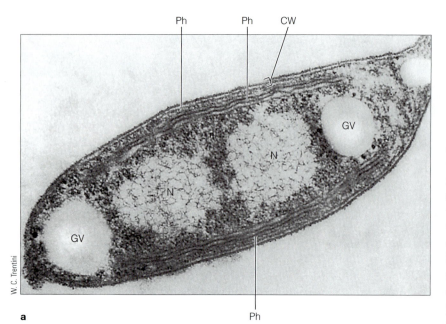

Ph   Ph   CW

GV

N

N

GV

Ph

W. C. Trentini

a

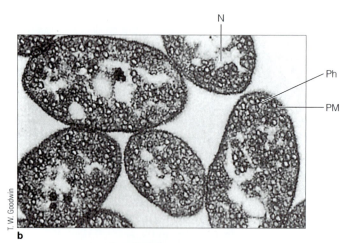

N

Ph

PM

T. W. Goodwin

b

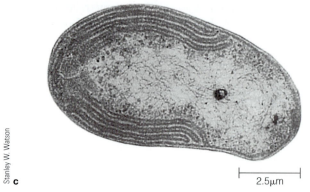

2.5µm

Stanley W. Watson

c

Figure 19.8 Prokaryotes with internal membranes. **(a,b)** Photosynthetic bacteria, showing two different arrangements of thylakoid membranes. **(c)** A chemotroph with internal cisternae. CW, cell wall; GV, gas vesicle (for flotation); N, nucleoid; Ph, photosynthetic thylakoids; PM, plasma membrane.

(Fig. 19.8). These structures are related to the cell's energy metabolism. As in chloroplasts, the thylakoid membranes of prokaryotes function in the light reactions of photosynthesis. They separate two compartments with different pH values and electrical potentials, thereby enabling the cell to store the free energy generated by electron transport reactions across the membranes (see Chapter 10). Prokaryote cells called **chemotrophs** derive energy from inorganic oxidation–reduction reactions, rather than from photosynthesis, and some of these also have cisternae for energy storage. The oxidation–reduction reactions presumably transport electrical charge and protons across the cisternal membranes; as in photosynthesis, the charge and proton gradients can be used to synthesize the high-energy compound adenosine triphosphate (ATP). But not all photosynthetic and chemotrophic prokaryotes have internal membranes. Some accomplish the same process by establishing their pH and electrical gradients across the plasma membrane.

Some Bacterial Cells Form Endospores

Under harsh environmental conditions, prokaryotic cells will die, but some species have developed ways of surviving by forming tough **endospores.** Typically, bacterial endospores are small, desiccated cells in a condition of suspended animation. Covered with a specialized, hardened cell wall, they contain a complete genome and sufficient enzymes and metabolites to germinate and re-establish growth when conditions improve. The combination of the hardened wall and the dry, inanimate state makes these spores resistant to many environmental insults, including boiling and the action of oxidizing agents (such as sodium hypochlorite, the active ingredient of household bleach) and antibiotic compounds.

The formation of endospores is a complex developmental process, involving the activation of specialized genes. In *Clostridium tetani,* an endospore forms within the vegetative (growing) cell **(Fig. 19.9).** The nucleoid and some ribosomes are surrounded by the spore wall, which, in turn, is surrounded by the endospore coat. Because the nucleoid is isolated from it, the rest of the cell degenerates. In other Bacteria, endospores develop on specialized structures. In Actinobacteria, which are abundant in soil, a group of spores forms on a vertical stalk that raises the endospores above the substrate so they can be blown to new sites by air currents. Myxobacteria, another group of soil Bacteria, form sacs full of endospores that are released when the sac is hydrated **(Fig. 19.10).**

Another extracellular organelle is the **pilus** (Latin, meaning "hair"), a thin, hollow, nonmotile projection from the cell. At the end are proteins that attach the cell to solid surfaces or to receptors on other cells. A pathogenic bacterium may recognize and bind to a host cell through a pilus.

Although prokaryotes usually are described as not having internal membranes, this is not true in all cases. Some cells have cisternae or thylakoid membranes, flattened bladders that enclose separate compartments within the cytoplasm

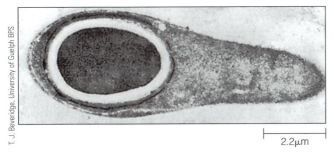

Figure 19.9 An endospore (oval region) inside *Clostridium*, an anaerobic bacterium.

2.2μm

T. J. Beveridge, University of Guelph BPS

Figure 19.10 A "fruiting" form of the myxobacterium *Chondromyces crocatus.* The organism is a large (0.2-mm) colony with some specialized cells. The raised appendages are filled with spores.

Hans Reichenbach

Every component of a prokaryotic cell plays a role in the cell's survival, growth, or dispersal. The next section describes the processes by which various prokaryotes find the energy to live, grow, and reproduce.

19.3 LIFESTYLES OF SELECTED GROUPS OF PROKARYOTES

A useful classification of prokaryotes is based on their nutritional requirements, particularly the ways they obtain carbon and energy. **Autotrophs** (*auto-*, "self"; *-troph*, "related to feeding") incorporate carbon into organic molecules from inorganic sources, typically by reducing CO_2 or HCO_3^-. **Heterotrophs**

Figure 19.11 Evaporation ponds near Blenheim on the north coast of the south island of New Zealand. The red color is caused by *Halobacteria*, among other microorganisms, which thrive in salt-saturated water. Similar colored ponds can be seen at the south end of San Francisco Bay, where bay water is evaporated and the salt crystals are collected for commercial use.

Richard Cowen, University of California, Davis

(*hetero-*, "other") derive carbon from the breakdown of organic compounds. **Chemotrophs** (*chemo-*, "chemical") derive energy from catalyzing inorganic reactions. **Phototrophs** (*photo-*, "light") derive energy by absorbing light photons. All four combinations—chemoautotrophs, chemoheterotrophs, photoautotrophs, and photoheterotrophs—are found in nature. These differences reflect basic genetic capabilities, and they have ecological implications because they may determine where a species lives and how it affects its environment.

Table 19.1 lists selected prokaryotes and some of their characteristics. Much of the chemistry by which prokaryotes obtain carbon and energy involves oxidation–reduction reactions. (Review Chapter 2 if these concepts are unfamiliar to you.)

Archaea Inhabit Harsh Environments

Many Archaea live in places that we would consider stressful—even deadly. For example, **methanogens** (methane-generating Archaea) derive energy from the following reaction:

$$CO_2 + 4H_2 \rightarrow CH_4 + 2H_2O$$

CH_4 is methane, a component of natural gas. In reducing (oxygen-poor) environments, such as swamp mud and the rumens of cows, this reaction is a downhill (thermodynamically favorable) reaction; therefore, it can produce usable free energy. In our normal, oxygen-rich atmosphere, methane tends to be oxidized to CO_2; but without oxygen, and in the presence of H_2, the formation of methane is favored. Therefore, methanogenic Archaea actually *require* anoxic environments to obtain the free energy they need. All methane on Earth, including natural gas, comes from this reaction, catalyzed by methanogenic Archaea. Methanogenic Archaea also incorporate CO_2 into bio-organic molecules; therefore, they are examples of chemoautotrophs.

Another group of Archaea, the **halophiles**, live in saturated salt solutions, such as soda lakes and drying salt flats **(Fig. 19.11).** Some halophiles have adapted so well to these conditions that they have (and need) little or no cell wall. If

Table 19.1 Selected Types of Prokaryotes

Domain and Genus	Phylum	Comment	Physiological Type
Archaea			
Methanococcus, Methanospirillum	Euryarchaeota	Methanogenic $CO_2 + 4H_2 \rightarrow CH_4 + 2H_2O$	Chemoautotroph
Halobacterium, Natronococcus	Euryarchaeota	Extreme halophiles, grow in saturated salt solutions	Photoheterotroph
Thermoplasma	Euryarchaeota	Thermoacidophile, grows at pH 1–4 and 33–67°C	Chemoautotroph
Pyrolobus	Crenarchaeota	Extreme thermophile; grows up to 113°C	Chemoautotroph
Bacteria			
Rhizobium	Proteobacteria	N_2-fixing plant symbionts	Chemoheterotroph
Frankia	Actinobacteria	N_2-fixing plant symbionts	Chemoheterotroph
Erwinia, Agrobacterium, Pseudomonas syringae	Proteobacteria	Plant pathogens	Chemoheterotroph
Rhodopseudomonas, Chromatium	Proteobacteria	Anaerobic phototrophs; purple nonsulfur and purple sulfur groups	Photoautotroph or chemoheterotroph
Chlorobium	Chlorobi	Anaerobic phototrophs; green sulfur group	Photoautotroph
Anabaena, Nostoc, Prochloron	Cyanobacteria	Oxygen producers	Photoautotroph
Desulfovibrio, Desulfomonas	Proteobacteria	Sulfate-reducing bacteria	Chemoheterotroph
Stigmatella, Chondromyces	Proteobacteria	Myxobacteria; colonial spore formers	Chemoheterotroph
Streptomyces	Actinobacteria	Antibiotic producers	Chemoheterotroph
Nitrosomonas, Nitrobacter, Methylomonas	Proteobacteria	Nitrogen, methane oxidizers	Chemoautotroph
Spirochaeta, Treponema	Spirochaetes	Long, thin, spiral-shaped; some pathogenic	Chemoheterotroph
Bacillus	Firmicutes	Aerobic endospore formers	Chemoheterotroph
Escherichia	Proteobacteria	Enteric; model organism	Chemoheterotroph

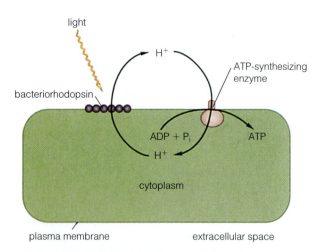

Figure 19.12 The mechanism of photosynthetic energy metabolism in *Halobacterium halobium*. ADP, adenosine diphosphate; ATP, adenosine triphosphate; Pi, inorganic phospate.

such a halophile is moved from its normal environment to distilled water, it will burst. A representative halophile, *Halobacterium halobium,* has a unique type of photosynthesis. A photoreceptor, bacteriorhodopsin, which is more similar to the mammalian eye photoreceptor than to any type of chlorophyll, is embedded in its plasma membrane. There is no associated electron transport system. When this photoreceptor absorbs light, it forces a proton from the cytoplasm to the outside of the cell, forming a pH and electrical gradient **(Fig. 19.12).** This gradient can be used to synthesize ATP by the same mechanism that occurs in mitochondria. The Archaeal cell derives useful chemical energy from light, but because there is no photosynthetic electron transport chain, it cannot make carbohydrates by reducing CO_2. Thus, it is a photoheterotroph.

A third group of Archaea, the **thermoacidophiles,** live in hot, acid environments such as volcanic hot springs. For them the optimum temperature is 70 to 75°C, with a maxi-

Table 19.2 Oxidation–reduction Reactions Used by Chemotrophs to Obtain Energy and by Autotrophs to Incorporate Carbon into Organic Molecules

Bacteria/Genus Example or Type	Electron Donors	Electron Acceptors	Products
Chemoheterotrophs			
Escherichia	Organics	O_2	CO_2, H_2O
Escherichia	Organics	NO_3^-	CO_2, NO_2^-
Desulfovibrio	Organics	SO_4^{3-}	CO_2, S°, H_2S
Lactobacillus	Carbohydrates	None	Lactic, acetic, formic, carbonic acids
Chemoautotrophs			
Nitrosomonas	NH_3	O_2, CO_2	NO_2^-, carbohydrate
Nitrobacter	NO_2^-	O_2, CO_2	NO_3^-, carbohydrate
Thiobacillus	S^{2-}	O_2, CO_2	SO_4^{2-}, carbohydrate
Thiobacillus denitrificans	S	NO_3^-, CO_2	SO_4^{2-}, N_2, carbohydrate
Methylomonas	CH_4	O_2	CO_2, H_2O
Photoautotrophs			
Green sulfur Bacteria	H_2	CO_2	Carbohydrate, H_2O
Green, purple sulfur Bacteria	H_2S	CO_2	Carbohydrate, H_2O, S
Purple nonsulfur Bacteria	Reduced organics	CO_2	Carbohydrate, H_2O, oxidized organics
Cyanobacteria, chloroxybacteria	H_2O	CO_2	Carbohydrate, H_2O, O_2

mum of 88°C; the optimum pH is 2 to 3 (minimum pH 0.9). One of the reasons that Archaea, and especially thermoacidophiles, are thought to be ancient is that these environments are expected to have prevailed on primitive Earth several billion years ago.

Recently, researchers have found Archaea in equable environments, such as forest soil; therefore, they may be more widely distributed than is currently appreciated. The habitats and lifestyles of Archaea are a fascinating area of active research.

Bacteria Include Many Diverse Species

Bacteria as a group seem much more diverse than Archaea, especially in the various ways they derive energy from their environment (although this situation may change as more is learned about Archaea). Three lifestyles of Bacteria are described in the following sections.

CHEMOHETEROTROPHS Bacterial chemoheterotrophs live on the organic compounds of living or dead tissue or on the excretions of other organisms. In living organisms, these chemoheterotrophs may be harmful parasites, but many that live on skin or in the gut are beneficial because they compete for niches with potential pathogens. Gut chemoheterotrophs may be essential for nutrition—for example, they provide humans with vitamin K. In dead tissue and on excretions, chemoheterotrophs play the valuable ecological role of recycling carbon, nitrogen, and other elements that are locked in unused and otherwise unusable materials.

Usually, energy is released for the use of the heterotrophs through oxidation of the organic compounds. Various species of Bacteria use different electron acceptors in these reactions, including oxygen, nitrate, nitrite, ferrous iron, and sulfate (Table 19.2).

In the absence of oxygen or another inorganic electron acceptor, some Bacteria derive energy from complex organic molecules by converting them to more stable states, a process called **fermentation.** For instance, carbohydrates—starch or sugars—can be broken up into smaller molecules of lactic, acetic, formic, or carbonic acid. Bacteria of the genus *Lactobacillus,* using these reactions, put the "sour" in sourdough bread.

The most intensively studied bacterium, *E. coli,* is a chemoheterotroph (Figs. 19.2 and 19.3). It belongs to the group Proteobacteria and the family Enterobacteriaceae, whose members live in the soil and in the intestines of animals. (They often are called enteric or coliform Bacteria, referring to the colon, the large intestine.) In intestines, they obtain their energy by metabolizing undigested foods; in the soil, they get it by metabolizing litter, excretions, dead microbes, and other sources of organic material. The presence of coliform Bacteria in water supplies is an indicator that the water is contaminated with sewage, which is a problem because human sewage carries pathogenic Bacteria and viruses. The particular coliform Bacteria that are detected in a water test may be perfectly safe—some are, after all, natural inhabitants of healthy animals—but some strains of *E. coli* produce toxins that cause severe infections.

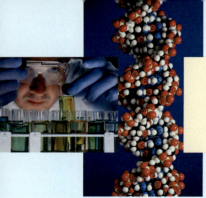

BIOTECHNOLOGY:

Bacteria and Biomining

Prokaryotes are impressive, both in the variety of chemical reactions they catalyze to obtain energy and in the scale at which they operate. Much of the earth's surface has been molded by the chemical activity of bacteria. Scientists believe that before the appearance of bacteria, virtually all the iron on Earth was present in a chemically reduced form; currently, most beds of iron ore are oxidized. Iron- and manganese-oxidizing bacteria have been isolated and studied.

The abilities of some species of bacteria to chemically modify various elements are now being used in the smelting of ores to extract copper, gold, and phosphate. Copper (Cu) often is found in rocks as an insoluble sulfide, CuS. Traditional energy-intensive smelting methods require that the rock be crushed and heated to extract the metal. A much easier and cheaper method is to soak the ore in an acidic solution containing *Thiobacillus ferroxidans*, a bacterium that obtains energy by oxidizing sulfide to sulfate. This bacterium functions happily under acidic conditions, which also promote the oxidation of reduced copper. The resulting copper sulfate is soluble and can be leached from the remaining insoluble rocks. The metal then is prepared from the copper sulfate solution. Currently, 25% of all copper produced in the world comes from this *biomining* technique.

The same bacterium can help extract gold from sulfide-containing ores. In low-quality ores, small particles of gold are trapped in an insoluble matrix of copper and iron sulfides. Oxidation of the sulfur and leaching of the remaining sulfates release the gold, which can be recovered by extraction with cyanide. Treating the ore with suspensions of *T. ferroxidans* is cheaper than the traditional high-pressure, high-temperature method of oxidation, and it extracts a greater percentage of the gold.

Phosphate, needed as fertilizer in agriculture and in some industrial processes, is traditionally extracted from ore by using high temperatures or sulfuric acid. A new process dissolves the phosphates in organic acids, metabolic byproducts of the bacteria *Pseudomonas cepacia E-37* and *Erwinia herbicola*. This method is more environmentally friendly because the organic acids are themselves biodegradable through the actions of other bacteria.

A current research project is aimed at using films of the bacterium *Pseudomonas aeruginosa* to precipitate metallic flakes from solutions containing ions of gold, silver, or platinum.

🌐 **Websites for further study:**

Access Excellence—Biomining:
http://www.accessexcellence.org/AB/BA/Biomining.html

Biomining and Bioprocessing: http://energy.inel.gov/mining/biomining.shtml

Figure 19.13 A steam vent in Lassen Volcanic National Park in northern California. Some thermophilic prokaryotes derive energy from reduced sulfur compounds, easily detected by their smell, which are dissolved in the water.

Lawrence Migdale © 1985

CHEMOAUTOTROPHS The accumulation of any compound, organic or inorganic, in a relatively reduced (electron-saturated) state provides an organism in an oxygen atmosphere with the opportunity to obtain energy by catalyzing the transfer of electrons from the reduced compound to the oxygen (or a relatively oxidized substitute). Some of the electrons can be transferred to CO_2, reducing the carbon to the oxidation state of carbohydrate. Molecules of carbohydrate then can be used to synthesize the other compounds needed for growth and reproduction.

A group of chemoautotrophs, sometimes called **lithotrophs** (*litho-*, "rock"), specialize in the oxidation of inorganic compounds (Table 19.2). These Bacteria are especially important in recycling elements such as nitrogen and sulfur, so that these elements can be used by other organisms **(Fig. 19.13).**

The oxidation of sulfur by Bacteria also releases the elements that were chemically bound to it, which can help in the smelting of metals (see sidebar "BIOTECHNOLOGY: Bacte-

Methane formation plays an important role in human affairs, because it is a major component of natural gas. But methane is important for another reason. It is a greenhouse gas, 25 times more effective than CO_2 in trapping infrared radiation. An increase in the amount of methane in the atmosphere could result in seriously increased temperatures on the earth's surface.

Geochemists estimate that the mud of the ocean floor contains 10 trillion tons of methane, and methanogenic Archaea below the surface of the ocean produce 300 million more tons of methane per year. But little or none of this methane reaches the atmosphere. From studies of the mud in the Black Sea and Eel River Canyon off California, scientists have found that Archaea, Bacteria, and symbiotic combinations of both use the methane as a carbon and energy source. However, the mud on the ocean bottom is anaerobic; therefore, the oxidation of the methane, needed to harvest its energy, requires another electron acceptor. The electron acceptor for this system is sulfate. The reduction of sulfate produces hydrogen sulfide, which may be useful as an energy source for other prokaryotes.

This is one example of prokaryotic symbiosis, and one that may have played a role in regulating the climate of Earth. Scientists suggest that in the ancient Earth, when the sun was not as bright, the presence of methane in the atmosphere warmed Earth to a temperature comfortable for life. As the sun's radiance increased, the evolution of methane-eating prokaryotes reduced this greenhouse gas and kept Earth's temperature equable.

Websites for further study:

Methanogenesis, Mesospheric Clouds, and Global Habitability:
http://geo.arc.nasa.gov/sgp/aero/aerocloud3.html

Biomethanogenesis: Principles:
http://www.agen.ufl.edu/~chyn/age4660/lect/lect_08x/lect_08.htm

ria and Biomining"). The oxidation of methane, produced by methanogenic Archaea, may be an important determinant of the earth's climate (see sidebar "PLANTS, PEOPLE, AND THE ENVIRONMENT: Microbes and Global Warming").

PHOTOAUTOTROPHS Like plants, Bacterial photoautotrophs (Figs. 19.8a,b and 19.14) derive their energy from sunlight.

They have light-absorbing pigments (bacteriochlorophyll or chlorophyll) that collect light energy, which excites electrons and promotes their transfer from one molecule to another. In other words, these organisms store light energy by transferring electrons from relatively stable to relatively unstable chemical compounds. There are several types of photoautotrophic Bacteria. They all reduce carbon in CO_2, but they

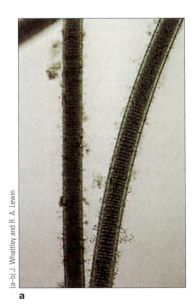

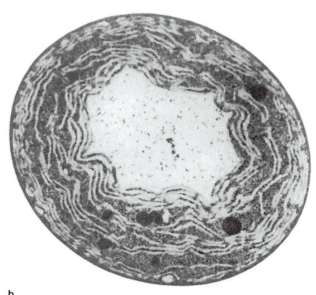

Figure 19.14 Phototrophic Bacteria. (**a**) A light micrograph of a species of *Lyngbya*, a filamentous cyanobacterium. (**b**) An electron micrograph of *Prochloron*, a member of the chloroxybacteria (phylum Cyanobacteria), which has been found only as a symbiont in the gut of a tunicate (a marine invertebrate).

(a–b) J. Whattley and R. A. Lewin

a

b

differ in their cell structures, ecological habitats, and the substrates from which they derive electrons.

Photosynthetic Bacteria include the green sulfur Bacteria (phylum Chlorobi), purple sulfur Bacteria and purple nonsulfur Bacteria (both within phylum Proteobacteria), and the cyanobacteria (phylum Cyanobacteria) (Table 19.2). In each case, carbon is reduced from an oxidized state, CO_2, to the more reduced state of carbohydrate. Also, an electron-rich compound loses electrons (and protons) and appears in its oxidized state.

The electron-rich compound may be H_2, H_2S, an organic compound such as H_3CCH_2OH (ethanol), or H_2O. The corresponding oxidized products would be nothing (from H_2), S, acetaldehyde, and O_2. Only cyanobacteria are able to oxidize water and produce O_2 like the plastids of plants. The green and purple sulfur Bacteria and the purple nonsulfur Bacteria are poisoned by oxygen and must grow under anaerobic conditions. The evolution of the capability to produce O_2 was a major innovation in photosynthetic metabolism. It is thought to be an important stage in the development of the earth as we now know it, because all the O_2 in the atmosphere was first formed by the ancestors of cyanobacteria.

There also are major differences among the Bacterial phototrophs in their photosynthetic pigments. The anaerobic phototrophs have bacteriochlorophyll, a pigment related to but different from chlorophyll. Cyanobacteria have chlorophyll *a* in their photosystems, but their major light-harvesting complexes contain proteins called *phycobilins* in a package (called a *phycobilosome*) attached to the surface of the thylakoid membranes. The phycobilin in cyanobacteria, phycocyanin, absorbs yellow-green light. This is why the cyanobacteria look relatively blue-green. A group within the cyanobacteria, sometimes called chloroxybacteria, have chlorophylls *a* and *b* in their light-harvesting complexes and no phycobilins.

PHOTOAUTOTROPHS AND ENDOSYMBIOSIS On the basis of the photosynthetic pigments of their present-day members, some primitive cyanobacteria and chloroxybacteria are thought to be evolutionary precursors of the plastids of photosynthetic eukaryotes. Researchers propose that Bacteria originally formed an **endosymbiotic relationship** with a eukaryotic precursor cell (*endo-*, "within"; *symbiotic*, "living together"). Over time, a substantial portion of the genetic material of the Bacteria was somehow transferred to the nucleus of the eukaryote, and this endosymbiont became an organelle, an actual part of the eukaryotic cell. Cyanobacteria and the eukaryotic Rhodophyta (red algae) have chlorophyll *a* and phycobilosomes; both also have phycocyanin, although members of the Rhodophyta also have a second phycobilin, phycoerythrin. Chloroxybacteria and Chlorophyta (green algae) have chlorophylls *a* and *b,* and both lack phycobilins.

Biologists consider that the similarities between the light-harvesting complexes of cyanobacteria and red algae and of chloroxybacteria and green algae provide strong evidence for the evolution of plastids through an endosymbiotic mechanism. The base sequences of genes for photosynthetic enzymes in Bacteria and photosynthetic eukaryotes also are very similar, and plastids have small subunit rRNA genes similar to those of cyanobacteria, providing further evidence for this mechanism. What is unknown—and a subject of controversy among biologists—is whether the loss of phycobilins and appearance of chlorophyll *b,* traits that distinguish chloroxybacteria from other cyanobacteria, occurred before or after the endosymbiotic event leading to green algae.

It is likely that the endosymbiotic mechanism applies to the formation of other organelles. Mitochondria probably arose from a Bacterium with the capability to use oxygen as an electron acceptor. The nucleus may have arisen from an endosymbiotic Archaea. Plastids of some algae (e.g., dinoflagellates and euglenids; see Chapter 21) have structures suggesting that they arose from an already-formed plastid, a process known as secondary endosymbiosis, when a photosynthetic eukaryotic cell became an endosymbiote in another, nonphotosynthetic cell.

This section has described the variety of ways by which various prokaryotes obtain their carbon and energy for growth and reproduction. Although obtaining nutrients is not the only significant activity of a cell, it is one of the more important ones. The chemical formulas just described especially demonstrate that prokaryotes play a major role in the interconversion of inorganic elements and compounds, many of which are nutrients on which plant metabolism depends. Thus, plants depend on prokaryotes for their existence on Earth. Plants sometimes form more intimate relationships with prokaryotes, as described in the following section.

19.4 PROKARYOTES THAT FORM SYMBIOTIC RELATIONSHIPS WITH PLANTS

Bacteria and plants can form **symbiotic relationships,** in which the Bacteria live on or within the plant. One kind of symbiosis, called *mutualism,* benefits both the Bacteria and the plant, but other kinds of symbiosis are damaging to the plant.

Rhizobium Forms a Mutualistic Association with Legumes

As described in Chapter 11, nitrogen represents a special nutritional problem for a plant. The world's major store of nitrogen is N_2 in the atmosphere. Plants cannot use this N_2, but some Bacteria can. The infection of plants with such Bacteria can be beneficial and even sometimes essential for the plant.

Rhizobium is a chemoheterotrophic bacterium that lives in soil. Like several other prokaryotes, it synthesizes an enzyme called *nitrogenase,* which gives it the ability to *fix* nitrogen—that is, to convert N_2 to ammonium (NH_4^+). Ammonium can be incorporated by the Bacteria and plants into the structures of amino acids and other nitrogen-containing organic compounds. Nitrogen fixation is an energy-using reaction; the formation of 2 NH_4^+ ions from 1 N_2 molecule requires the addition of 8 high-energy electrons and the hydrolysis of 16 ATP. The Bacteria get this energy by metabolizing other organic compounds, especially carbohydrates. Nitrogen fixation works best in a low-oxygen atmosphere because oxygen inactivates the nitrogenase enzyme.

Rhizobium has developed an ability to form a close mutualistic association with legumes, plants of the family Fabaceae. Both partners in the association contribute, and both partners derive benefits. The plant contributes high-energy carbohydrate and a protected environment; the bacterium contributes nitrogenase and other enzymes. The benefit to both partners is the supply of fixed nitrogen. The association occurs in special organs called **root nodules** (see Fig. 7.20).

The establishment of the symbiotic association is a cooperative process in which each partner triggers steps in the other, including the induction of genes to make proteins that are used only in the active nodules. The sequence of events, as far as we now know it, is as follows:

1. The root secretes an attractive chemical.

2. The chemical induces *Rhizobium* Bacteria in the vicinity to swim toward the root. It also begins the induction of nitrogen fixation genes in the Bacteria.

3. The Bacteria enter at a root hair and move inward through an infection thread (a tube of plasma membrane), losing their cell wall and synthesizing nitrogen-fixing enzymes as they do so.

4. When the Bacteria reach the root cortex, they are released from the infection thread into several cells. The Bacteria without their cell walls are called *bacteroids;* they become surrounded by a special membrane, the peribacteroid membrane.

5. Chemicals secreted by the Bacteria (or bacteroids) during the formation of the infection thread later induce cell division in the root cortex and pericycle, forming the nodule. These chemicals also induce the synthesis of specialized nodule proteins, including a type of hemoglobin (leghemoglobin) that buffers the oxygen concentration in the central part of the nodule, where the nitrogen is fixed.

A typical nodule in soybean has a core of bacteroid-infected cells surrounded by a cortex of parenchymal cells that contain extensive vascular bundles. The photosynthetically produced carbohydrate sucrose is transported to the nodule through the phloem **(Fig. 19.15).** The sucrose then diffuses or is actively transported to *Rhizobium* cells in the core of the nodule. There, oxidative respiration of the carbohydrate provides the high-energy electrons and ATP necessary to reduce N_2 to NH_4^+. The oxidative reactions not only provide energy but also reduce the concentration of O_2, a necessary step because O_2 inactivates the nitrogenase enzyme. Leghemoglobin, present in the plant cells that host

the bacteroids, also helps regulate or buffer the concentration of O_2 (Fig. 19.15). The NH_4^+ produced by the bacteroids is used in the synthesis of a key amino acid, glutamine (gln), which can donate nitrogen to other compounds. A substantial fraction of this amino acid is exported to the plant cells in the cortex of the nodule, which convert it into an easily transportable nitrogenous compound. This is secreted to the apoplast and transported up the xylem to the growing parts of the shoot.

Although best known, the *Rhizobium*–legume symbiosis is not the only Bacteria–plant symbiosis that fixes nitrogen. *Frankia*, a member of the Actinobacteria, can live within the cells of root nodules of alder trees and other plants. *Anabaena*, a cyanobacterium, forms a symbiotic association with the water fern, *Azolla. Nostoc,* another cyanobacterium, invades cavities in the gametophytes of hornworts (a type of bryophyte; see Chapter 22) and specialized roots of cycads (a type of gymnosperm; see Chapter 24). In each case, the Bacterial partner fixes nitrogen, and the reduced nitrogen compounds nourish both the Bacterium and the plant. Growing *Anabaena-Azolla* in rice paddies is a traditional method of fertilizing rice in Southeast Asia.

Bacteria Can Be Plant Parasites

Symbiotic associations are not always mutualistic. Those in which one organism benefits at the expense of the other are known as *parasitism. Agrobacterium tumefaciens* (see Chapter 17) is a Bacterial plant parasite, and there are many others. Two important examples are *Pseudomonas syringae* and *Erwinia amylovora.* Bacterial plant pathogens are divided into subgroups called *pathovars,* according to the plants they infect. Pathovars of *P. syringae* cause the wildfire disease of tobacco; blights of beans, peas, and soybeans; and diseases of several other crops. Pathovars of *E. amylovora* cause fire blight of apple and pear. The names "wildfire" and "fire blight" suggest the speed at which a Bacterial infection can spread through a field of tobacco or a large apple tree.

Bacteria are carried to uninfected plants by water, insects, humans, or other animals. *E. amylovora*, for example, is carried from flower to flower in apple or pear orchards by bees, as they make their pollination rounds. The Bacteria can enter the plant tissues through natural openings: stomata, lenticels, hydathodes (the openings at the tips of leaves through which guttation water is extruded), and nectarthodes (the openings in flowers from which nectar comes). Once inside the plant, these heterotrophs multiply quickly, absorbing nutrients from the plant cells. Some Bacteria secrete enzymes that break down the plant cell walls, some produce chemical toxins, and some apparently cause their damage simply by absorbing nutrients and multiplying. The infected plant tissues often turn brown and die. A large portion of the plant may be involved. Some Bacterial pathogens can overwinter in the dead tissue, returning to infect new plant tissue at the next growing season.

Plants have some defenses against the infecting Bacteria. Sometimes, depending on the plant and the infecting agent,

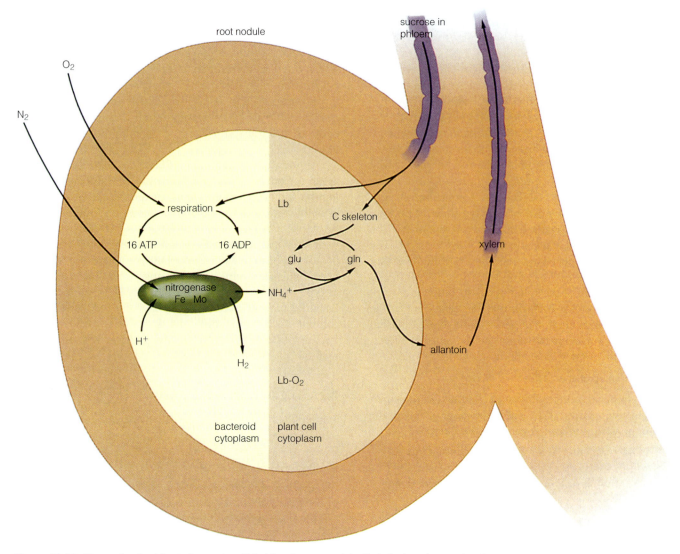

Figure 19.15 The synthesis of fixed nitrogen in a *Rhizobium*–legume nodule. Carbohydrate (sucrose) arrives through the phloem. Some of the carbohydrate is donated to the bacteroids for use in respiration; some is used to make carbon skeletons to which NH_4^+ can be attached. N_2 and O_2 arrive at the nodule through air spaces in the soil; they diffuse into the core. O_2 is used quickly in Bacterial respiration; this keeps the oxygen concentration low and protects the nitrogenase enzyme with its iron (Fe) and molybdenum (Mo) cofactors from inactivation by oxygen. Changes in the oxygen concentration are buffered by leghemoglobin (Lb), an oxygen-binding protein made by the plant. Glutamic acid (glu) and glutamine (gln) are nitrogen-rich amino acids; allantoin is another nitrogen-rich compound. These carry nitrogen to the plant shoot in the xylem stream. ADP, adenosine diphosphate; ATP, adenosine triphosphate.

a plant will react by producing antibiotic compounds, including phytoalexins and hydrogen peroxide. This is called the **hypersensitivity response.** The antibiotics may directly kill some pathogenic cells, but hydrogen peroxide may also restrict the spread of infection by triggering the death of adjacent plant cells. In such a case, the infection may produce a spot of necrosis in a leaf, but it will not kill the whole leaf or the plant.

These descriptions of symbiotic associations between a plant and a Bacterium demonstrate the complexity of the relationships between members of the two kingdoms. They also show how important it is to consider the prokaryotes in the environment when describing the biology and ecology of plants.

Viruses, although they are not free-living and do not influence the physical environment, are parasites of plants and also form complex relationships with them. The next section describes some of these relationships.

19.5 VIRUSES

Viruses are subcellular parasites. They are said to be subcellular because they have neither the internal structures found in prokaryotic or eukaryotic cells nor the capacity to reproduce on their own. They are considered parasites because they invade cells and use their host's metabolism to produce more of their kind. Many biologists have debated the ques-

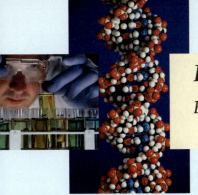

BIOTECHNOLOGY:

Bacteria as Biocontrol Agents

Biological control (or biocontrol) involves the use of "natural enemies" to restrict the growth of pests, such as insects, molds, and weeds. It is a major component of the agricultural practice known as Integrated Pest Management, which seeks to limit the loss of crop value to pests through ecologically sound methods. Bacteria can play important roles in biocontrol strategies.

A good example of a bacterium used for biocontrol of insect pests is *Bacillus thuringiensis* (sometimes abbreviated Bt). Bt cells synthesize a protein that kills insect larvae. Once a larva has ingested the Bt, the protein is released and quickly binds to receptors in the gut. The larva stops eating, and the gut dissolves, releasing bacteria into the bloodstream. The larva dies from starvation and infection. Different strains of Bt produce different types of protein, so it is possible to control specific types of insects. For instance, there is a strain that kills wax moth larvae in beehives without harming the bees. Bt is used to protect cotton plants and forest trees, and it can be sprayed on fruits and vegetables. It has no harmful effect on humans, even in huge amounts. (In early tests, some subjects even ate Bt as a paste on crackers.)

Biocontrol of invasive plants—weeds—usually is accomplished with insects or fungi, not bacteria. However, there is at least one commercial product that uses bacteria for pest control. An isolate of *Xanthomonas campestris*, a wilt-inducing bacterium, was isolated in Japan from *Poa annua* (annual bluegrass or wintergrass). Suspensions of the bacterial cells are registered in Japan as the bioherbicide CAMPERICO® to control annual bluegrass in golf courses.

Pseudomonas syringae, the bacterium that causes wildfire disease in plants, forms the basis of an unusual biocontrol agent. Strains of *P. syringae* that lack the ability to produce disease in plants have been isolated. These form the basis of commercial preparations. When suspensions of these strains are sprayed on fruits, such as apples, pears, bananas, and oranges, they prevent mold fungi from infecting and destroying the fruits.

As global trade and travel increase, so does the incidence of biological invasions—pests from other parts of the world reproducing and spreading without limitation by the herbivores, predators, or pathogens of their native habitat. The strategy of biocontrol is to find organisms that restore the ecological balance. The same strategy can be a useful and economic management technique in agriculture.

Websites for further study:

Biological Control: A Guide to Natural Enemies in North America:
http://www.nysaes.cornell.edu/ent/biocontrol

---

tion of whether viruses are alive. Their inability to reproduce independently suggests that they cannot be considered truly alive; yet, what functions they do possess are all based on physical and chemical principles basic to normal living organisms.

Viruses Are Infectious Genes

All types of living organisms have viruses that infect them. You are undoubtedly acquainted with the effects of various human viruses: Chicken pox, herpes, polio, and influenza are all caused by specific viruses, as is acquired immune deficiency syndrome (AIDS). Insects and other animals, fungi, and bacteria all have their own types of viruses, and so do plants.

Viruses have a simple structure, compared with cells. They contain a nucleic acid genome surrounded by a coat of protein. (The coat sometimes includes lipids.) Surprisingly, the genome may be either RNA or DNA, depending on the type of virus, and it may be single-stranded or double-stranded. Whatever its structure in the virus, the genome contains the information for making virus-specific proteins by using the same genetic code that operates in all cells. The coat may take different forms. It can be a rod, a polyhedron, or a more complex shape. Whatever its shape and composition, it protects the nucleic acid as it is transferred from one organism to the next and is sometimes instrumental in the infection process. Some viruses also contain one or more enzymes that play a role in the infection process, but outside a cell no virus has any metabolic activity (respiration or protein synthesis, for example). One might consider a virus to be an infectious gene.

Tobacco mosaic virus is a well-studied plant virus. It has a single-stranded RNA genome complexed with coat protein

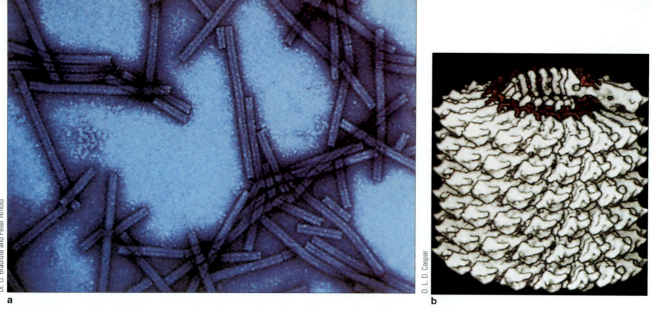

Figure 19.16 The tobacco mosaic virus. (a) An electron micrograph. (b) An interpretive diagram of the virus's arrangement of nucleic acid (in red) and protein.

Dr. O. Bradfute and Peter Arnold

D. L. D. Caspar

a

b

subunits, all wound into a helical rod (Fig. 19.16). When RNA and protein are separated chemically and applied separately to test plants, a plant that receives the RNA will develop a viral infection, whereas a plant that receives the protein will not. The experiment demonstrating this was among the early pieces of evidence that nucleic acids are hereditary material, carrying the instructions for reproduction of the entity from which they came.

Viral Infections Stunt Plant Growth

When the tobacco mosaic virus infects a cell of a tobacco, tomato, or other related plant, the coat protein is removed, and the viral RNA serves as a messenger RNA, forming complexes with the cell's ribosomes and using the cell's amino acids, transfer RNAs, and energy source to make viral proteins. The first protein formed is an enzyme (called an RNA-dependent RNA polymerase) that forms a complementary strand of RNA (that is, a single-stranded RNA with a base sequence complementary to that of the viral genome), using the viral genome as a template. This complementary RNA then can serve as a template for the formation of new viral genomes. Later, another protein, the coat protein, will be synthesized. This protein will bind to and coat the new viral genomes, forming complete viral particles. The viral genome codes at least one other protein, which promotes the transmission of the virus among plant cells.

Plant viruses, including tobacco mosaic virus, cannot infect plant cells without help. They are too large to pass through a normal primary cell wall, much less the barriers, such as epidermis or cork, that protect a plant from invasions. In nature, plant viral infections are almost always spread by insects, which pick up the viral particles as they chew or suck on infected plants and then transmit them to uninfected plants. Mites (arachnids, related to spiders), nematodes (roundworms), and fungi also can infect plants when they penetrate the cells.

In the laboratory, viral infections are induced by rubbing abrasive-treated leaves with a viral suspension. The abrasive (for example, Carborundum™) makes holes in the cell walls, through which the virus enters. Once in the cell and forming new viral particles, the virus may become systemic—that is, it may spread throughout the plant. Recent evidence indicates that viruses move through the plasmodesmata and phloem and that a viral protein is responsible for opening the plasmodesmata so that particles as large as a virus can pass through. A mutant virus has been found that cannot move systemically. A study of this mutant was instrumental in demonstrating that the sizes of the openings of plasmodesmata can be regulated. There also are mutant tobacco plants that do not allow tobacco mosaic virus to move systemically, but instead confine an infection to a local lesion (a small area of dying or dead cells). These plants have been used to study the hypersensitivity response, a reaction sometimes seen in plants with bacterial infections.

Generally, viral infections do not kill plants. Because further transmission of a virus that killed its host and made the host unappetizing to insects would be unlikely, natural selection may eliminate plant viruses that are excessively virulent. However, infected plants are usually stunted relative to uninfected plants, and the infection often (but not always) causes changes in the color or shape of the foliage (Fig. 19.17a). The mosaic of tobacco mosaic virus, the symptom of a systemic infection, is a pattern of dark and light green regions on a leaf. The necrotic local lesions of tobacco mosaic virus are small yellow or brown spots about 1 mm in diameter (Fig. 19.17b). Other viruses cause chlorosis (yel-

Figure 19.17 Symptoms of viral diseases in plants. (**a**) A mosaic from a systemic infection in tobacco (*Nicotiana tabacum*). (**b**) Necrotic local lesions on a tobacco (*N. tabacum*) leaf. (**c**) Variegation in tulips (*Tulipa gesneriana*), also called color breaking.

a

T. M. Murphy

b

Kenneth Corbett

c

Brecks, Peoria, IL.

lows), concentric white rings on a leaf (ringspot), transparent areas next to the leaf veins (vein clearing), or tumors (galls). Viral infections in flowers sometimes cause beautiful, multicolored patterns (color breaking). Variegated tulips are a famous example (Fig. 19.17c).

KEY TERMS

Archaea	heterotrophs
autotrophs	hypersensitivity response
bacilli	lipopolysaccharide layer
Bacteria	lithotrophs
binary fission	methanogens
chemotrophs	nucleoid
cocci	peptidoglycan
conjugation	periplasmic space
endospores	phototrophs
endosymbiotic relationship	pilus
Eukarya	plasmids
eukaryote	prokaryote
fermentation	root nodules
flagella	spirilla
Gram-negative cells	symbiotic relationship
Gram-positive cells	thermoacidophiles
halophiles	vibrios

SUMMARY

1. The basic structure of a prokaryotic cell is simple, relative to a eukaryotic cell. A cell wall and plasma membrane surround the cytoplasm and an undifferentiated nucleoid. Many prokaryotic cells have no other organelles recognizable by electron microscopy. Prokaryotes never have a complex endomembrane system, as eukaryotes do, but some prokaryotes contain internal membranes that provide sites for electron transport reactions and store electrochemical energy.

2. The basic genetic structure of a prokaryote also is simple. Most genes are lined up on a single circular chromosome. Some specialized genes may be found on smaller DNA circles called plasmids. Plasmids can be transferred from cell to cell. The F plasmid, when incorporated into the chromosome, mediates the transfer of chromosomal genes from a donor cell to a recipient.

3. Some prokaryotes have complex modifications to their extracellular structures, such as an outer lipopolysaccharide membrane (in Gram-negative cells), one or more flagella for motility, and a pilus for attachment to other cells. Some prokaryotes form specialized spores for reproduction, dispersal, and resistance to harsh environmental conditions.

4. Prokaryotes can be divided into two major groups. Archaea live in hot, O_2-poor, or salt-rich environments. Bacteria are more diverse and specialized and often have more complex structures and metabolic capabilities.

5. Prokaryotes can be classified according to the ways in which they obtain carbon and energy. Chemoheterotrophs live on the organic compounds of living or dead tissue or on the excretions of other organisms. Chemoautotrophs get energy from the oxidation and reduction of inorganic compounds. Phototrophs derive their energy from sunlight. Chemoheterotrophs and chemoautotrophs play major ecological roles by recycling the elements in organic and inorganic waste materials. Photoautotrophs contribute to this role; like plants, they also contribute to the input of carbon and energy into the organic components of an ecosystem.

6. Nitrogen represents a special nutritional problem for plants. Some plants form mutualistic associations with prokaryotes, cooperating in the fixation of N_2 to form NH_4^+. The plants contribute carbohydrate as a source of energy and provide an oxygen-protected environment; the prokaryotes contribute the enzyme nitrogenase. The development of a root nodule in the *Rhizobium*–legume symbiotic association is a complex process that involves the expression of special genes in both partners.

7. Some prokaryotes are parasitic on plants, invading leaves and other organs through wounds, stomata, and hydathodes. Prokaryotic cells multiply quickly in the intercellular spaces of the plant, secreting enzymes to break down cell walls and metabolizing and absorbing the contents of the cells. The blighted parts of the plant are destroyed.

8. Plant viruses are subcellular parasites, formed from a nucleic acid genome and a protein coat; they appropriate the plant cell's protein-synthesizing machinery, using it to reproduce huge numbers of themselves. Plant viruses are normally transmitted among plants by insects. Viral infection weakens a plant and causes various visible symptoms, but it does not generally kill the plant.

Questions

1. From the list below, indicate the organelles (a) that are found in every prokaryotic cell, (b) that are found only in some prokaryotic cells, and (c) that are found only in plant (or other eukaryotic) cells.

cell wall	plasmid
peptidoglycan	endoplasmic reticulum
nuclear envelope	nucleoid
spore	lipopolysaccharide
ribosome	flagellum
thylakoid	plasma membrane

2. Distinguish between prokaryotes and viruses. Which do you think appeared first in the evolution of life?

3. Provide a critical analysis for each of the following statements.

a. The only way a bacterium can acquire new genetic information is by mutation.

b. Conjugation in Bacteria is just like meiosis in plants and other eukaryotic cells.

4. Archaea are thought to represent an ancient line of organisms that may have existed on Earth for billions of years. Give one reason why they have been able to resist competition from more highly evolved organisms such as Bacteria and Eukarya. (Hint: Where do Archaea live?)

5. For each of the reactions below, indicate whether the reaction would be catalyzed by a chemoheterotroph, chemoautotroph, or photoautotroph. Which of these reactions also would be catalyzed by a plant cell? (Note: CH_2O represents carbohydrate.)

a. $2H_2O + CO_2 + light \rightarrow H_2O + CH_2O + O_2$

b. $organics + NO_3^- \rightarrow CO_2 + N_2$

c. $2NH_3 + 3O_2 \rightarrow 2H^+ + 2NO_2^- + 2H_2O$

d. $carbohydrates \rightarrow lactic, acetic, formic, and carbonic acids$

e. $2H_2S + CO_2 + light \rightarrow H_2O + CH_2O + 2S$

6. Describe the advantages to each partner of the mutualistic symbiotic association between *Rhizobium* bacteria and legume plants.

7. Legumes often are more successful than other plants in sandy soils that have little ability to retain anions, such as nitrate. Suggest one reason why this is true.

8. Distinguish between the methods by which the two plant parasites *Erwinia amylovora* and *Agrobacterium tumefaciens* (see Chapter 17) derive nutrition by infecting plant cells.

9. A sample of tobacco mosaic virus is separated by chemical procedures into RNA and protein fractions. The leaf of one plant is rubbed with the protein solution. The leaf of another plant is rubbed with the solution of RNA. Which, if either, of these plants will be infected? Will the infected plant produce viral protein, viral RNA, or whole viral particles (RNA plus protein)?

10. Explain the following statement: Agriculturalists have found that the most effective way to stop the spread of a plant viral disease is by the use of insecticides.

11. Find an example of biocontrol that is (or would be) useful in your region of the country.

 InfoTrac® College Edition

http://infotrac.thomsonlearning.com

Archaea

Jarrell, K.F., Bayley, D.P., Correia, J.D., Thomas, N.A. 1999. Recent excitement about the Archaea (Archaebacteria). *BioScience* 49:530. (Keywords: "excitement" and "Archaea")

Lyons, S. 2002. Thomas Kuhn is alive and well: The evolutionary relationships of simple life forms—a paradigm under siege? *Perspectives in Biology and Medicine* 45:359. (Keywords: "Kuhn" and "life forms")

Smith, D.C. 2001. Expansion of the marine Archaea. *Science* 293:56. (Keywords: "marine" and "Archaea")

Bacterial Ecology

Fenchel, T. 2001. Marine bugs and carbon flow. *Science* 292:2444. (Keywords: "marine" and "bugs")

Souza, V., Castillo, A., Eguiarte, L.E. 2002. The evolutionary ecology of *Escherichia coli*: Abundantly studied and much feared, *E. coli* has more genomic plasticity than once believed and may have followed various routes to become a pathogen. *American Scientist* 90:332. (Keywords: "*Escherichia*" and "ecology")

Wuethrich, B. 1995. Bacterial virulence genes lead double life (cause infection in both plants and higher animals). *Science* 268:1850. (Keywords: "Bacterial," "virulence," and "plants")

Plant Viruses

DeNoon, D.J. 1996. Plant-produced vaccine antigens: Safe, cheap, and stable. *AIDS Weekly Plus* April 1, p. 4. (Keywords: "plant-produced" and "vaccine")

Moffat, A.S. 1999. Geminiviruses emerge as serious crop threat. *Science* 286:1835. (Keywords: "Geminiviruses" and "crop")

Kingdom Fungi

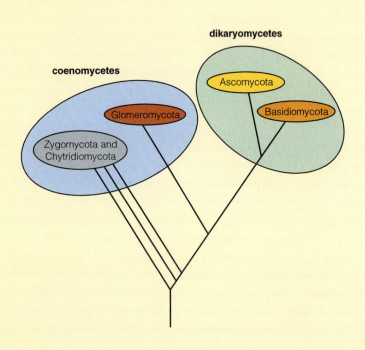

coenomycetes

dikaryomycetes

Ascomycota

Glomeromycota

Basidiomycota

Zygomycota and
Chytridiomycota

Visit us on the web at http://biology.brookscole.com/plantbio2
for additional resources, such as flashcards, tutorial quizzes,
InfoTrac exercises, further readings, and web links.

1. Kingdom Fungi (the true fungi) is a monophyletic group of eukaryotic heterotrophs that reproduce with spores and have chitinous cell walls. The most familiar fungi are kitchen molds and mushrooms. The kingdom may include 1.5 million species, of which about 80,000 species have been named and described.

2. Some fungi destroy crops and stored food. Others are valuable decomposers or symbionts that cohabit with algae and cyanobacteria or assist plant growth. Baker's yeast is a fungus, and penicillin is a fungal product.

3. Most fungi develop a mycelium, composed of branching threads (hyphae) that collect nutrients and produce reproductive structures. Some fungi have a simpler thallus or live as microscopic unicells (yeasts). Dimorphic fungi make both mycelia and yeasts.

4. Many fungi make asexual spores to multiply and sexual spores for diversity. Exceptions include mushroom fungi (which use sexual spores to multiply) and mitosporic fungi (which have not been observed to reproduce sexually). However, nearly all tested fungi show signs of recent genetic recombination.

5. Two large phyla (Ascomycota and Basidiomycota) contain 95% of named species in kingdom Fungi and are informally called dikaryomycetes because their sexual life cycle has a unique dikaryotic stage. The remaining 5% of named species are divided between three phyla (Glomeromycota, Zygomycota, and Chytridiomycota) and are informally called coenomycetes because their hyphae lack the regular septation found in dikaryomycetes.

6. Kingdom Fungi excludes some organisms that traditionally are called fungi, and adds other organisms that were previously left out. New studies are changing classification within the kingdom.

20.1 FUNGI: FRIENDS AND FOES

Everyone has met the fungi, for better or worse (Fig. 20.1). We all know about mushrooms and moldy food, but fungi are much more important than that. Many fungi reduce human hunger by aiding the growth of plants. Others attack crop plants, farm animals, and humans with costly results. Many fungi are valuable decomposers that break down wastes and release the elements for reuse. But this function takes a dark side when fungi attack stored food, lumber, and clothing.

We make bread, beer, and wine with the help of yeast fungi. We also use common baker's yeast, *Saccharomyces cerevisiae,* as a model in the quest for cancer cures, and its genome was the first among eukaryotes to be sequenced. Another fungus yielded the first evidence of how genes act. In medicine, countless lives have been saved by antibiotics that were first discovered in fungi, and organ transplants were made possible by fungal molecules that suppress the immune system. All in all, fungi have a remarkably diverse set of relationships with humankind.

Figure 20.1 A mushroom, the reproductive body of a fungus. Made of filaments tightly packed together, the mushroom is only part of the body. The rest consists of branching, underground filaments. This fungus may be living in symbiosis with a forest tree.

20.2 TRAITS OF TRUE FUNGI

Members of kingdom Fungi vary immensely in size, body form, and life patterns. Judging from details of their DNA, they share a common ancestry and form a monophyletic clade. Not all organisms commonly called fungi belong to the kingdom, and we reserve the term **true fungi** for those that do belong.

Kingdom Fungi may have branched off from the rest of the eukaryotes a billion years ago. With all that time to evolve, it makes sense that they would be highly diverse today. Still, they retain several traits that must have come from their common ancestor.

True Fungi Are Eukaryotic, Spore-producing Heterotrophs with Chitinous Walls

Like plants and animals, fungi are **eukaryotes;** their cells have true nuclei. In fact, their cells have all the organelles that occur in animal cells. But fungal cells lack the one organelle that is most characteristic of plants: They have no chloroplasts and cannot perform photosynthesis. Lacking photosynthesis, fungi are **chemoheterotrophs** (see Chapter 19); they get energy and carbon as animals and most bacteria do, by taking organic molecules from the environment.

Fungal cells are surrounded by a **chitinous cell wall,** a protective coating that contains the substance **chitin** and other molecules. The cell walls led early biologists to view fungi as plants, for in both fungi and plants, cell walls precluded the evolution of muscles and nervous systems by limiting mobility.

Cell walls have advantages and disadvantages. Like a rubber tire, the wall is flexible but resists expansion, so it gives shape to the pressurized cell. Fungi depend on cell shape to hold reproductive structures aboveground. The strength of cell walls limits invasion by other organisms, and it prevents the cell from bursting under pressure. Offsetting those benefits, walls prevent the fungus from engulfing solid foods as animals do; nutrient molecules must pass one by one through the wall to enter the cell, giving all fungi **absorptive nutrition.** The wall also limits growth and must be softened locally for a cell to expand. To explore new territory without the help of muscles, fungi rely on growth and abundant reproduction.

To reproduce, fungi release **spores**—units consisting of only one or a few cells **(Fig. 20.2).** Spores are not unique to fungi; plants and some bacteria and protists also make them. A fungal spore can resist dehydration and is light enough to drift in a gentle breeze, remaining dormant until it reaches a moist environment that contains food. The most productive fungi release trillions of spores per year, and in the height of allergy season, a cubic meter of air may contain thousands of spores. The ever-present spores make it likely that a sandwich or a pile of manure will soon become moldy.

To summarize, members of kingdom Fungi are spore-making, chitin-walled, eukaryotic chemoheterotrophs. Keep this in mind if you see a mold and wonder if it is a true fungus. Everyone recognizes molds as furry, colored patches on decaying organic matter—but not all molds are true fungi. For example, slime molds (see Chapter 21) do not have chitin in their walls. The absence of chitin is one of many small clues that slime molds came from different ancestors. Hence, slime molds are not included in kingdom Fungi.

Most Fungi Have Bodies Called Mycelia

Fungi are simpler than animals and plants. A body as simple as that of a fungus often is called a **thallus** (plural, *thalli*)—a multicellular body without specialized conducting tissue.

To explore, feed, and make reproductive structures, most fungi grow a unique type of thallus known as a **mycelium (Fig. 20.3),** composed of slender, branching tubes called **hyphae** (singular, *hypha*). Individual hyphae are extremely slender and almost colorless, making them hard to see. But at the surface of a food mass such as bread, countless exploratory hyphae grow into the air and make a visible fuzz. When reproduction starts, colored spores may cover the surface.

Hyphal growth begins when a spore absorbs water, swells, and germinates. The enclosed cell softens the spore wall, allowing it to expand. As the hypha grows, soft wall material is added to the tip, while older material stiffens along the sides of the hypha. The result is a tube that expands only at the tip. Sensing chemicals that diffuse from a food mass, hyphae grow toward the source. This orientation response to chemicals is called **chemotropism.** As the hypha lengthens, at intervals the tip lays down the beginnings of branch hyphae, small bumps that are left behind as the tip grows forward. Later, some of the bumps become branch hyphae, which can form more branches. In this way, a single hypha quickly branches into a mycelium. The mycelium may also form cross-bridges by the fusion of hyphae, creating a web.

A mycelium is well equipped to exploit a food mass such as a rotting peach. Pushed through the food by internal pressure that causes growth, each hyphal tip secretes digestive enzymes into the food mass, breaking proteins and other large food molecules into small molecules such as amino

Figure 20.2 Spore production by *Aspergillus*, a common kitchen mold. Tiny spores, called conidia, form in chains at the tips of stalks called conidiophores. They often form black patches on moldy bread. ×300.

acids and sugars **(Fig. 20.4).** The small molecules pass through the cell wall and meet the cell membrane. In the membrane, proteins spend energy to pull food molecules into the cell (an example of *active transport;* see Chapter 11). Accumulation of food causes water to enter by osmosis. The added water exerts pressure that stretches the cell wall, elongating the hypha and driving it deeper into the food mass.

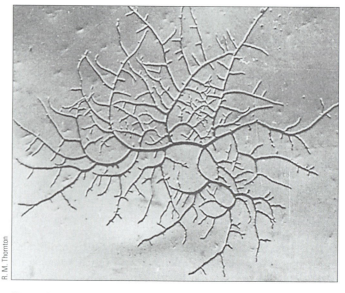

Figure 20.3 Hyphae making up a young mycelium of *Phycomyces blakesleeanus*, a zygomycete fungus. ×400.

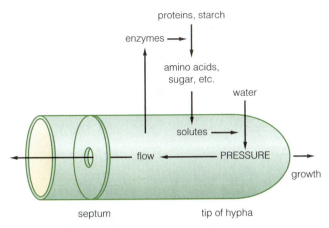

proteins, starch

enzymes →

amino acids,
sugar, etc.

water

solutes →

flow ← PRESSURE →

growth

septum

tip of hypha

Figure 20.4 The act of feeding pushes a hypha into a food mass and drives food through the mycelium. The hypha secretes enzymes that break up large molecules. Proteins in the cell membrane carry small solute molecules into the hypha, causing water to enter. The added volume creates pressure that expands the soft tip, pushing forward. Pressure and cytoplasmic streaming create a flow that carries food to older parts of the mycelium. Cross walls (septa) may occur, but they have perforations that allow a flow of material.

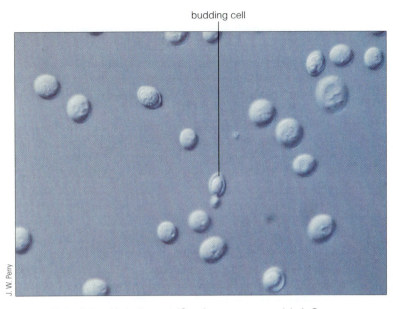

budding cell

J. W. Perry

Figure 20.5 Cells of baker's yeast (*Saccharomyces cerevisiae*). One cell is reproducing by budding. ×1,700.

Because these events require water, mycelia become inactive when the substrate dries.

Injured mycelia have remarkable powers of regeneration. If you cut a mycelium into fragments, the cut hyphae seal their broken ends and continue growing as separate fungi. Thus, injury can multiply the number of growing mycelia. But hyphae can distinguish self from nonself, and if hyphae from two genetically identical mycelia touch, they may fuse to make the two mycelia into a single unit.

The level of coordination between parts of a mycelium is much less than between parts of the animal body, but hyphae do share materials. A pressure-driven mass transport of material, like the flow of water in a pipe, runs from feeding hyphae to sites of reproduction. In addition, high magnification shows a slower, more controlled movement of particles along cytoskeletal elements in the hyphae. This controlled flow, called **cytoplasmic streaming,** can either oppose or augment the mass transport. Such movements carry food and organelles to hyphae that can make the best use of them.

The simplest fungi, **yeasts,** lack mycelia and live as microscopic, rounded cells **(Fig. 20.5).** Each time the growing cell doubles in size, it divides into two independent cells. In baker's yeast (*Saccharomyces cerevisiae*), the parent cell forms a bud that is cut off and released when it equals the parent cell's size. Some other yeasts simply double their size and split crosswise. Some fungi exhibit **dimorphism,** the occurrence of two growth forms. Dimorphic fungi can switch between mycelial and yeast growth, depending on the growing conditions.

Most fungi can make nearly all the molecules they need from a few environmental raw materials. They need water, a few minerals, and an organic compound for carbon and energy. Many fungi also need a few vitamins, including the same vitamin B_1 that humans need. Almost any organic material can be food for one or more fungi, including wood, paper, glue, leather, manure, and so on.

Mycelia Compete Well with Bacteria

As walled heterotrophs, fungi compete with bacteria for food—and bacteria evolved superb scavenging abilities a billion years before the first fungi. How could fungi succeed against the fierce bacterial competition? The answer relates partly to mycelial growth. To reach new food sources, a fungus or bacterium must feed quickly, reproduce abundantly, and disperse offspring widely. Bacteria excel in fast feeding and reproduction; their tiny cells can double in number every half hour, and each cell is a reproductive unit. But if bacteria live deep within a food source, they can be slow to reach the surface for dispersal. Even at the food surface, bacteria are slow to reach new regions after they exhaust the food in their immediate vicinity. A single bacterium quickly forms a small colony, but most colonies remain small after that. By contrast, fungal hyphae rapidly grow into new territory; a mycelium can expand without limit (its size is indeterminate). Furthermore, the mycelium is a system of connected pipes in which pressure can rapidly move nutrients from buried feeding hyphae to aerial hyphae that make and disperse spores.

Rapid transport through a mycelium is possible because hyphae are not divided into distinct cells. Most fungi make cross walls known as **septa** (singular, *septum*) at regular intervals along the hyphae (Fig. 20.4). But the middle of each septum has a hole or **septal pore** (also called a perforation) that allows material to move between compartments. Thus, most of the mycelium is a single giant cell.

Given that septa are perforated, you might wonder what value they have. Don't septa slow down transport, reducing the mycelium's advantage over bacteria? Why do hyphae make septa at all? Part of the answer may be that septa aid in damage control. Some of the fastest-growing fungi lack septa and are said to be *aseptate*. Their fast growth may derive partly from having wide-open pipelines for transport. But if an aseptate hypha is cut through, much material leaks out before repair mechanisms seal the break. Septate hyphae leak less when broken.

In addition to reaching and transporting food quickly, fungi exploit several chemical tools to compete with bacteria. All fungi secrete acid (H^+) as part of their feeding system. A side benefit is that acid slows bacterial growth. Acidic soils, such as those beneath conifers or heaths, are rich in fungal decomposers and poor in soil bacteria. Some fungi also secrete specific antibiotics that poison bacteria. Penicillin was the first fungal antibiotic to be recognized as such in science. It causes bacteria to burst by inhibiting the enzymes that build the bacterial cell wall—an effect that leaves fungi (and humans) unharmed because eukaryotes do not have the enzyme that penicillin blocks. The alcohol made by baker's and brewer's yeasts also is an antibiotic that slows bacterial growth. Alcohol is a fine antiseptic that was used in medicine long before the discovery of penicillin, and yeast cells tolerate it more readily than do bacteria.

Finally, some fungi outcompete bacteria with the help of superior food-digesting systems. Fungi that attack wood are an important example. Wood contains cellulose, a valuable nutrient, which is protected by a coating of the hardening agent *lignin*. Lignin resists attack by most organisms, but some fungi have enzymes that degrade lignin to expose the cellulose.

20.3 ECOLOGICAL STRATEGIES OF FUNGI

Fungi have several lifestyles that make them vital agents in the world ecosystem. To get food from living host organisms, many fungi engage in **symbiotic relationships** comparable to those of bacteria (see Chapter 19): There is **mutualism,** which benefits both the fungus and its host; and there is **parasitism,** in which the fungus benefits while its host is harmed. Some fungi take a step further, killing the host organism as they feed. This is true of fungi that kill insects and nematode worms.

Many fungi are adept at mutualism, where they partner with organisms from every kingdom of life. The most important fungal mutualisms involve photosynthetic organisms—plants, algae, and bacteria. Every plant has fungi living in its leaves or stems. These opportunists may harm the plant, or they may aid the plant by secreting chemicals that deter predators or parasites. More organized and important associations are **mycorrhizae,** in which fungi colonize roots. Mycorrhizal fungi take organic nutrients from the plant, in exchange for minerals and water that hyphae get from soil. In **lichens (Fig. 20.6),** fungi partner with green algae or cyanobacteria. The fungus shelters the algae or bacteria and provides minerals and water, taking energy-rich carbon compounds in return.

Figure 20.6 An Alaskan forest lichen occupies most of the photograph. In this lichen symbiosis, an ascomycete fungus (genus *Peltigera*) shelters photosynthetic green algae or cyanobacteria. More than 13,500 fungal species take part in lichens.

Fungi also form mutualistic symbioses with animals. Cows depend on tiny fungi to help them digest grass. The fungi dwell in one of the cow's several stomachs (the rumen), where they break down cellulose and other substances that the cow's own enzymes cannot attack. A more complex symbiosis involves leaf-cutter ants, tropical fungi, and bacteria. The ants cut leaves into small pieces and carry them underground where they infect the leaves with small bits of mycelium. The ants carefully tend their fungal gardens, and feed on bits of the mycelium. When the ants move to a new location and start a new nest, they carry tiny bits of the fungus with them, to establish new fungal colonies. The ants may also carry bacteria that inhibit the growth of competing fungi—a form of biological pest control.

Among the many species of fungi that live as parasites, some attack other fungi or animals, but the great majority attack plants. Rust and smut fungi are especially notable because some of them attack crop plants such as corn and wheat.

Not all fungi get food from living hosts; many species live all or part of the time as **saprobes,** organisms that feed on dead organic matter. In ecological terms, saprobic fungi are *decomposers*—organisms that recycle chemical elements by breaking down organic molecules. Without decomposers, the nitrogen, carbon, and phosphorus needed for life would be increasingly tied up in wastes and dead bodies.

Most saprobic fungi are *opportunists:* They survive by exploiting temporary opportunities, such as a fallen fruit. Many opportunistic fungi are small organisms because they have little time for growth before the food is gone. But saprobic mycelia may grow large in a forest or grassland, where fallen leaves and dead branches regularly replenish the food supply. Other saprobes are less opportunistic and grow large and old by attacking dead trees that contain enough food to support years of growth.

20.4 REPRODUCTIVE STRATEGIES

Fungi rely on spore production to reach distant food sources. More than 1,000 species have swimming spores that actively seek out food. But the great majority of fungi lack swimming cells; the spores drift passively in air or water. Spores that travel through air have walls that limit water loss by means of waxy or oily molecules. Some spores have extensions that catch air currents; others attach to animals by means of hooks or glue.

Only a tiny percentage of spores reach food sources; therefore, a fungus must build huge numbers of spores as fast as possible. Most fungi meet the challenge with *asexual reproduction* (see Chapter 12), releasing tiny parts of the parental body as *mitospores* **(Fig. 20.7)**. Mitospores are made in a variety of ways, some of which will be shown later in the chapter. Those in Figures 20.2 and 20.7 show variations of the most common method, in which spores form one by one at the tips of hyphae. Such spores are called **conidia** (singular, *conidium*), and hyphae that make them are called **conidiophores.** Masses of conidia often give moldy food vivid colors.

Asexual reproduction is the fastest, most economical way to multiply because it does not require a partner. However, all the mitospores are genetically identical to the parent. This uniformity limits the capacity of asexual reproduction to exploit new opportunities.

To make genetically diverse spores, most fungi also have *sexual reproduction.* When compatible cells or hyphae meet, they fuse and bring their nuclei together. Then the nuclei fuse and

go through *meiosis,* a division process in which chromosomes trade parts (see Chapter 12). Nuclei emerge from meiosis with varied combinations of genes from the two parents and are made into *meiospores.* The meiospores grow into mycelia with new traits, some of which may be useful in new habitats.

Thousands of fungal species have never been seen to reproduce sexually and are informally termed **mitosporic fungi.** But in wild populations of mitosporic fungi, studies of DNA nearly always show evidence of recent recombination. Either these fungi do reproduce sexually on occasion, or they recombine genes in other ways. One such method, which has been seen only in the laboratory, is referred to as the **parasexual cycle** because it lacks the meiosis that typifies sexual reproduction. The cycle begins when hyphae of two haploid mycelia fuse together, permitting nuclei of both to mingle. Such a mycelium, with nuclei of more than one genotype in uncontrolled proportions, is said to be **heterokaryotic.** Nuclei of the two kinds sometimes fuse, trade parts of chromosomes, and return to the haploid state by losing chromosomes. The result is a recombinant nucleus that has a mixture of genes from the original nuclei.

20.5 FUNGAL ORIGIN, CLADES, AND GRADES

Before biologists invented molecular phylogenetics (see Chapter 18), organisms were classified mainly by visible characters such as details of body form. This classification obscured the real diversity among the simpler organisms such as fungi, which have many fewer visible characters than animals and plants. Currently, **mycologists**—scientists who study fungi—are updating fungal classification with molecular phylogenetic methods that provide many more characters for comparison. These methods reveal that there are many more species of fungi than previously thought. Some estimates suggest that the kingdom may include as many as 1.5 million species, of which the current 80,000 named species is just a small sample. Other findings help to revise the way fungi are subdivided (see discussion below).

True Fungi Are a Monophyletic Kingdom of Life

After centuries of classifying fungi with plants, systematists in the early twentieth century decided that fungi should be viewed as a separate kingdom. Their view emphasized differences in cell structure and nutrition. Both plants and fungi have cell walls, but the walls differ radically in composition. And whereas green plants are photosynthetic autotrophs (photoautotrophs), fungi never perform photosynthesis and are chemoheterotrophs.

The current emphasis on molecular characters has shown that the original kingdom Fungi included some organisms that did not belong there, and excluded some members that did belong. In so doing, the original kingdom violated a demand of modern systematics: All named groups must be *monophyletic*—that is, they must include one originating ancestor, all of its descendants, and nothing else.

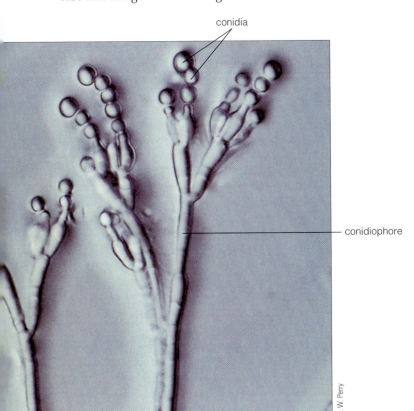

conidia

conidiophore

J. W. Perry

Figure 20.7 Mitospore production by *Penicillium,* the kind of fungus in which penicillin was discovered. Masses of these spores often make green patches on moldy oranges and cheese. Called conidia, the spores are produced one by one at the tips of specialized hyphae (conidiophores). ×550.

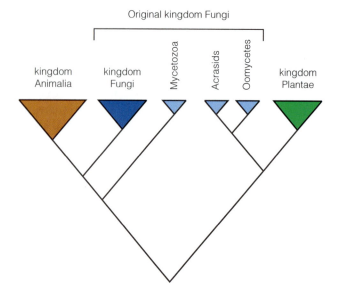

Original kingdom Fungi

kingdom Animalia | kingdom Fungi | Mycetozoa | Acrasids | Oomycetes | kingdom Plantae

Figure 20.8 Evolutionary relationships between kingdom Fungi and other eukaryotic groups, based on comparisons of the DNA that specifies ribosomal RNA. As originally defined, kingdom Fungi was not monophyletic. Since then, systematists made the kingdom monophyletic by excluding the groups shown in *light blue*. Kingdom Fungi is a sister group to animals.

To see the problem with the original kingdom Fungi, it must be viewed in the context of other kingdoms. **Figure 20.8** uses shades of blue for all the groups that were in the original kingdom Fungi. They come from varied origins, so a monophyletic clade that includes all of them would also include the animal and plant kingdoms. With this information, mycologists saw that the only way to make kingdom Fungi monophyletic was to remove the three small groups in the middle of the cladogram. That has now been done, leaving the modern kingdom Fungi monophyletic like the plant and animal kingdoms.

The three omitted groups are still studied by mycologists, for they have many similarities to organisms of kingdom Fungi. In fact, most of them are still informally called "fungi." To avoid confusion, mycologists often refer to members of the monophyletic kingdom as *true fungi.*

Like true fungi, all three omitted groups are spore-producing, eukaryotic chemoheterotrophs that have cell walls at some stage of life. The group that most resembles modern fungi still bears a fungal name—*Oomycota,* meaning "egg fungi." Its members have mycelia and other structures that closely resemble true fungi, but differences in detail suggested a separate origin even before DNA comparisons were possible. Figure 20.8 suggests that the similarities to true fungi came about by convergent evolution, a conclusion that is currently well accepted. The other two groups that traditionally were included in kingdom Fungi are *slime molds*—organisms that spend most of their lives without walls, moving about and engulfing solid foods as amoebae do. They were originally classified with fungi because they make spores like true fungi. Again, the molecular differences suggest that slime molds evolved spores independently. In Figure 20.8, the slime molds

are labeled Acrasids and Mycetozoa. Chapter 21 discusses Oomycota and slime molds.

The DNA comparisons of Figure 20.8 suggest that the closest relatives of kingdom Fungi are animals. These studies confirm earlier suspicions on the basis of other traits: The chitinous cell walls of fungi differ too much from the cellulose/lignin walls of plants to have come from a common ancestor; and elsewhere in the living world, only animals make chitin. Moreover, some fungi have swimming cells with a single trailing flagellum like animal sperm, unlike the paired lateral flagella of plant sperm. Other similarities include storage products and various enzymes that occur in animals and fungi but not in plants. These facts strongly suggest that kingdom Fungi is a sister clade to the animal kingdom and that its relation to plants is much more distant.

Sexual Reproduction and Flagellation Define Traditional Phyla

Mycologists group the members of kingdom Fungi into phyla (also called divisions), with names ending in -*mycota* (derived from the Greek word *mykes,* meaning "fungus"). Within a phylum, class names end in -*mycetes;* other ranks have the same endings as in plants (see Table 18.1).

To define phyla of terrestrial fungi, early mycologists emphasized the details of sexual reproduction. This defined three phyla that differ in the way they make sexual spores **(Fig. 20.9).** Members of phylum **Basidiomycota** extrude sexual spores from the surface of a cell called a **basidium** (Fig. 20.9a), much as children blow bubbles from a pipe. The walls of the resulting spores come from stretching and thickening a local part of the basidial wall. Phylum Basidiomycota contains more than 26,000 named species. In contrast, members of phylum **Ascomycota** make sexual spores *inside* a cell called an **ascus** (Fig. 20.9b). The spores get their wall by dividing the contents of the ascus into cells and making a completely new wall around each cell. Phylum Ascomycota includes more than 32,000 named species. Members of phylum **Zygomycota,** numbering only 1,100 named species, have a third way of making sexual spores: when parental hyphae meet in sexual reproduction, they fuse at the tips to form a large cell called a **zygosporangium,** which hardens its wall and goes to rest as a **zygospore** (Fig. 20.9c).

Details of sexual reproduction are not adequate for classifying all terrestrial fungi, because thousands of fungal species have never been seen to reproduce sexually. Early mycologists classified these fungi by details of mitospore formation and other nonsexual traits and grouped them into an artificial phylum called **Deuteromycota.**

When mycologists studied aquatic forms of life, they found organisms that resemble terrestrial fungi enough to suggest descent from common ancestors. Many of these aquatic fungi have never been seen to reproduce sexually, but one feature sets them apart from all other true fungi: They make reproductive cells that swim by means of a flagellum. They were assigned to a fourth phylum, the **Chytridiomycota.**

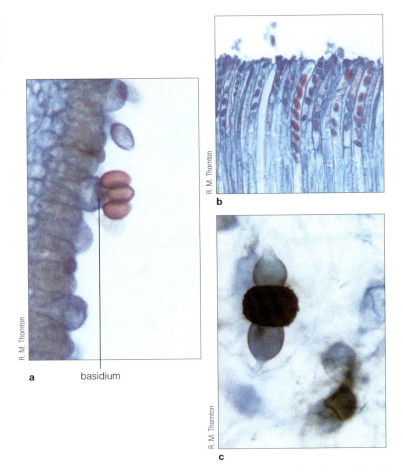

Figure 20.9 Sexual structures that were used to define three traditional phyla of fungi. (**a**) A basidium of *Coprinus* with extruded spores, phylum Basidiomycota. (**b**) Asci of *Peziza*, phylum Ascomycota. Each elongated bag is an ascus with eight spores. (**c**) A zygospore of *Rhizopus*, phylum Zygomycota. The dark object is the resting zygospore, between the two parental hyphae that fused to produce it.

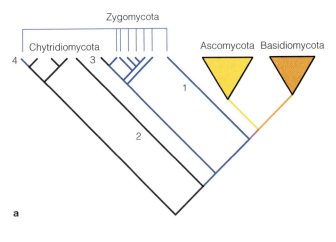

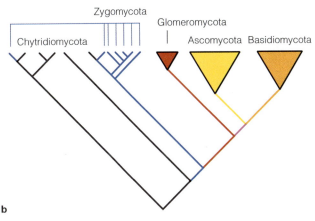

Figure 20.10 Cladograms based on rDNA, showing probable evolutionary relationships among phyla of Fungi. (**a**) If the inferred relationships are correct, the traditional Ascomycota and Basidiomycota are monophyletic, and the other two phyla are not. Zygomycota and Chytridiomycota are shown in *blue* and *black*, respectively. Numerals *1, 2, 3,* and *4* are explained in the text. (**b**) The same cladogram, relabeled to establish lineage 1 as a fifth phylum, the Glomeromycota.

Fungal Systematics Is a Work in Progress

Molecular phylogenetics is having a major impact on the way true fungi are grouped within the kingdom. One result was to eliminate the artificial phylum Deuteromycota. On the basis of similarities in mitospore formation, scientists had always suspected that most asexual fungi were ascomycetes that had lost or severely reduced their sexual function. DNA comparisons confirm the suspicion, and the great majority of mitosporic fungi have now been placed in phylum Ascomycota. Others have been placed in phylum Basidiomycota.

Numerous molecular phylogenetic studies agree that the traditional Ascomycota and Basidiomycota are monophyletic (**Fig. 20.10**). But the same studies agree that Zygomycota and Chytridiomycota are probably *not* monophyletic. If Figure 20.10a is correct, the only way to make all fungal phyla monophyletic is to put groups 1 and 2 into new phyla and reverse the assignments of groups 3 and 4.

A step toward making Zygomycota monophyletic occurred in 2001 when the German mycologist Arthur Schüßler and colleagues proposed moving the group desig-

nated 1 in Figure 20.10a from Zygomycota to a new monophyletic phylum, the **Glomeromycota** (Fig. 20.10b). This text adopts that proposal, as do many mycologists. Phylum Glomeromycota has few named species but is immensely important—its members assist the growth of about 80% of green plants, through symbiotic associations with roots.

As you consider the problems posed by Figure 20.10, keep in mind that the cladogram is just one hypothesis, based on one set of genes (**rDNA,** the genes that specify RNA molecules found in ribosomes). Other studies, using different genes, reach broadly similar conclusions but differ in detail. Thus, like all systematics, the classification of fungi is still a work in progress. Currently, we must simply accept that the evolutionary relationships among zygomycete and chytrid fungi are still uncertain. This is true partly because these fungi branched off from common ancestors in ancient times—perhaps 1.6 billion years ago, by some estimates. It is difficult to find the true sequence of branching when it happened so long ago. Perhaps further studies with additional characters will resolve the uncertainties.

Dikaryomycetes and Coenomycetes Are Grades of Fungal Evolution

If some members of a clade are far more successful than the rest because of special traits they share, the successful members are said to be at a higher **grade of evolution.** So it is with some of the true fungi (**Fig. 20.11**). The Ascomycota and Basidiomycota are sister clades that include more than 95% of named fungal species. All the large and complex fungi are found here. They include mushrooms, many common molds, and most fungi that cause human disease. Their success suggests that they inherited something useful from their immediate ancestor—new traits that placed them at a higher grade of evolution. Reflecting that idea, mycologists once called Ascomycota and Basidiomycota *higher fungi,* leaving the rest of the kingdom as *lower fungi.*

What made higher fungi so successful? Two relevant traits are named in Figure 20.11, with a red tick mark to show when they probably arose. One trait is a unique **dikaryotic stage** in the sexual life cycle—a stage in which hyphae have paired haploid nuclei of two genotypes. To emphasize the dikaryotic stage, the informal term **dikaryomycetes** was coined for fungi that have it. The second relevant trait is the presence of **septate hyphae**—hyphae that have septa at regular intervals.

In contrast, lower fungi never have a dikaryotic stage, and their hyphae are usually **aseptate,** with few or no septa. An aseptate mycelium is said to be **coenocytic,** meaning that the many nuclei are not walled off into separate compartments. To emphasize their coenocytic hyphae, the informal term **coenomycetes** was coined for all fungi that are not dikaryomycetes. This textbook uses the terms dikaryomycetes and coenomycetes rather than higher and lower fungi.

Figure 20.12 illustrates a sexual life cycle that includes a dikaryotic stage. As in other sexual cycles, plasmogamy brings 1*n* nuclei of two parents into a single cell. But instead of fusing, the two kinds of 1*n* nuclei form pairs that multiply by mitosis and guide the growth of new hyphae. The result is a **dikaryotic mycelium,** or **dikaryon,** so named because each

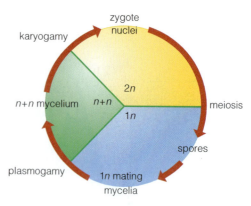

Figure 20.12 A sexual life cycle with a dikaryotic stage. The *green sector* is the dikaryotic (n+n) stage, found only in dikaryomycetes. After plasmogamy, the fusion cell grows a new septate mycelium in which mitotic divisions give each compartment a pair of 1*n* nuclei from the two parents. In coenomycetes, there is no dikaryotic stage; karyogamy directly follows plasmogamy.

of its septate compartments contains paired nuclei of two kinds (*di-,* "two"; *karyon* is derived from Greek meaning "nucleus"). The dikaryon's genetic constitution is called *n+n* because both kinds of nuclei are 1*n*. The dikaryon is the dikaryotic stage in the life cycle (the green sector in Fig. 20.12).

A dikaryon can produce spores with immense genetic diversity. Such diversity can greatly increase the formation of new species, by improving the chance that some offspring will thrive in new environments. **Figure 20.13** shows how a dikaryon increases spore diversity. Focus on meiosis, the source of diversity. From one 2*n* nucleus, meiosis makes four 1*n* nuclei with new combinations of genes. The gene combinations differ in every meiosis, because crossovers at ran-

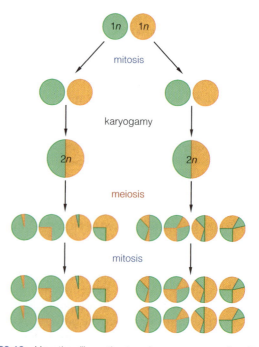

Figure 20.13 How the dikaryotic stage increases spore diversity. Each disk is a nucleus. *Green* and *orange* represent contributions from different parents. In the dikaryon, mitosis multiplies the number of pairs of 1*n* nuclei before karyogamy. Meiosis converts each 2*n* nucleus to four 1*n* nuclei with unique combinations of parental genes. Mitosis after meiosis does not produce new combinations.

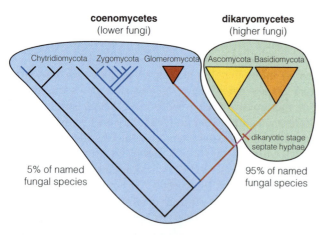

Figure 20.11 Two evolutionary grades of true fungi (dikaryomycetes and coenomycetes). Mycelial dikaryomycetes have a dikaryotic stage and regularly septate hyphae. Coenomycetes usually have aseptate hyphae and never have a dikaryotic stage.

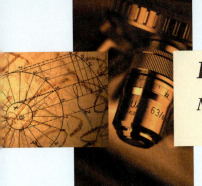

IN DEPTH:

Microsporidia: Fungi in Disguise?

The widespread human and animal parasites called microsporidia may turn out to be highly modified fungi. Although long known to attack animals, microsporidia were ignored as human pests until 1985, when they were found to cause chronic diarrhea in patients with AIDS. Since then, many microsporidia have been found to infect humans.

Microsporidia are unicellular eukaryotes that attack as shown to the right. Between hosts, they are spores that contain one cell with a long membranous tube coiled around it. When the spore senses a host cell, the tube is rapidly thrust out. If it penetrates the host cell, the microsporidian cell moves through the tube into the host. There, the microsporidian cell multiplies and makes new spores.

Microsporidia are eukaryotes, but they lack mitochondria. In cladograms made from rDNA data, microsporidia branch off at the base of the cladogram, far from the plant, animal, and fungal kingdoms. These facts once suggested that microsporidia are relics of a time before eukaryotes acquired mitochondria. But studies based on other genes suggest a more recent origin of microsporidia, either from the base of the fungal clade or from a group of parasitic zygomycete fungi. If so, they once had mitochondria but lost them. Evidence for recent loss is that two microsporidians have a gene for a protein called HSP70, which is widespread among eukaryotes and serves only to carry other proteins into mitochondria. The basal position of microsporidia in rDNA cladograms occurs because rDNA evolves more rapidly in microsporidia than other organisms. Finding that microsporidia are related to fungi may be a help in the quest for effective medications against them.

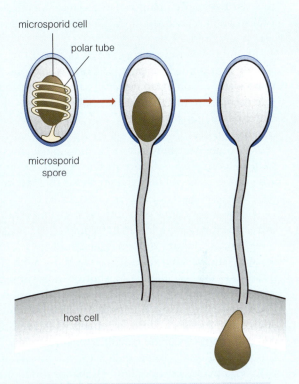

microsporid cell

polar tube

microsporid spore

host cell

Websites for further study:

Para-Site Online—Microsporidia: http://www.med-chem.com/Para/New/mi-intro.htm

Microsporidia: A Handbook of Biology and Research Techniques:
http://pearl.agcomm.okstate.edu/scsb387/content.htm

Protozoa and Microsporidia:
http://www.nysaes.cornell.edu/ent/biocontrol/pathogens/protozoa.html

dom points trade parts of paired homologous chromosomes (see Chapter 12). Therefore, the number of genetically different spores that come from a mating depends on how many $2n$ nuclei go through meiosis.

One way to get more $2n$ nuclei is to divide them by mitosis, as the human body does. But few fungi do that. *Allomyces* (a coenomycete) and *Armillaria* (a basidiomycete) are among the small minority of fungi that build a $2n$ mycelium. The vast majority of fungi put $2n$ nuclei directly through meiosis. Thus, to get more $2n$ nuclei, most fungi depend on events that occur *before* parental $1n$ nuclei fuse.

Some coenomycetes bring many $1n$ nuclei together when tips of mating hyphae fuse. But this yields very limited spore diversity, for the tiny fusion cell has room for only a few hundred nuclei. Karyogamy converts them to hundreds of $2n$ nuclei; hence, no more than a few hundred genetically different spores are produced.

A dikaryotic mycelium is immensely more effective. Its growing hyphae can multiply a single pair of $1n$ nuclei into *trillions* of pairs that fuse to make $2n$ nuclei. In this way, a large dikaryon such as a mushroom fungus can produce trillions of spores, each with a different combination of parental genes.

A dikaryotic stage may also promote the evolution of complex bodies by reducing the impact of harmful mutations. The more complex a body is, the more genes are needed to build it, and the more points exist where failure of a gene can be fatal. In a body with nuclei of two genotypes, even if one kind of nucleus has a bad copy of a gene, the other kind will probably have a good copy. This may explain why the largest, most complex structures in fungi are dikaryotic. Coenomycetes, with haploid mycelia, have only one version of each gene, and they have much simpler bodies.

Regular septation may also be important in making complex bodies. To construct such differentiated parts as a mushroom's cap and stalk, nuclei in different parts of the body must follow different genetic programs. This may be hard to achieve if nuclei wander freely, as they can do in a coenomycete mycelium.

20.6 PHYLUM CHYTRIDIOMYCOTA

With about 1,000 named species, phylum Chytridiomycota contains all the fungi that have swimming cells at some point in their life cycle. None are familiar in everyday life—they are too small to study without a microscope, none cause human disease, and few are significant pests. Nevertheless, they merit attention because they are part of the natural world, and because they give clues about the origin of kingdom Fungi.

If you have a microscope, chytrids are not hard to find. Simply add a pinch of pine pollen to a container of either pond water or soil in tap water. In a day or two, you may see microscopic chytrid thalli on the pollen grains. Chytrids are common in ponds and wet soil, where they decompose organic matter or attack algae, pollen grains, and plant roots. Sometimes they cause "swimmer's itch" in people who bathe in lakes. One parasitic chytrid became a pest when it attacked potatoes in the ground, but currently farmers grow potato strains that resist chytrids.

A few chytrids make highly branched aseptate mycelia, but most have a simpler thallus composed of just one or a few tiny cells. Figure 20.14 diagrams traits that are common to most simple chytrids. The cycle starts with a haploid meiospore that is called a **zoospore** because it swims. The zoospore has no cell wall and it swims with a single trailing flagellum, much like a human sperm cell. Reaching a food source, the zoospore drops or retracts its flagellum, forms a cell wall, and grows into a thallus. Often, slender extensions called **rhizoids** anchor the thallus to a food mass and absorb food. Eventually, part or all of the thallus differentiates into one or more **mitosporangia,** enclosures in which mitospores form. A mitosporangium contains cytoplasm and nuclei that divide by mitosis, then differentiate into many swimming mitospores. Most of the mitospores function in asexual reproduction, making new thalli that are genetically identical to the original meiospore.

Sexual reproduction is unknown in most chytrids. Where it has been seen, rhizoids sometimes fuse to transfer nuclei between thalli. In other cases, some of the swimming mitospores serve as gametes. Two gametes fuse (plasmogamy) to become a single cell with two nuclei and two flagella. The fusion cell expands, forming a thick-walled **meiosporangium**—an enclosure in which meiospores are made. The meiosporangium may lie inert for months. Then, its nuclei fuse (karyogamy) and divide by meiosis to make genetically diverse meiospores, completing the life cycle.

Every chytrid species has its own unique traits, and most have more complex bodies than shown in Figure 20.14. Some chytrids have branching aseptate hyphae. The largest is the much-studied genus *Allomyces,* which can be collected from soil and may grow into a fuzzy mycelium around decaying seeds. *Allomyces* makes alternating haploid and diploid mycelia—a rare trait among fungi.

In tracing the origins of fungi, chytrids are of interest because they have flagellated swimming cells. Eukaryotic flagella are complex; therefore, it is unlikely that more than one group evolved them independently. Similar flagella occur in

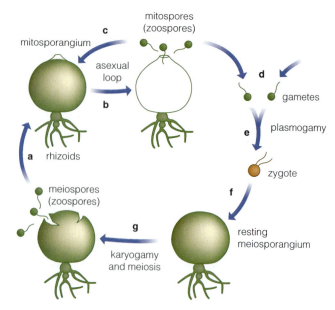

Figure 20.14 Life cycle of a simple chytrid. (**a**) A $1n$ meiospore swims to a food source, loses the flagellum, and grows into a thallus with a mitosporangium. (**b**) Mitotic divisions in the mitosporangium create many cells that become swimming mitospores. (**c**) Some mitospores make new $1n$ thalli, achieving asexual reproduction. (**d**) Other mitospores serve as gametes. (**e**) Plasmogamy results in a swimming zygote. (**f**) A zygote grows into a $2n$ thallus with a meiosporangium. (**g**) Karyogamy and meiosis occur to make cells that become new $1n$ meiospores, completing the cycle.

plants and animals; therefore, flagella probably evolved in the common ancestor of plants, animals, and fungi. This means the first true fungi had flagella. Chytrids retained them. The lack of flagella in other true fungi means that the ability to make them was lost after chytrids branched off. Perhaps the loss happened when fungi moved from water to land, where flagella are easily damaged and less useful.

20.7 PHYLUM ZYGOMYCOTA

Of all land fungi, zygomycetes have the most in common with aquatic chytrids. The similarities make these fungi important in studies of the way fungi moved from water to land more than 400 million years ago.

Only about 1,100 zygomycete species have been named and described, but the members are extremely diverse in lifestyle, and some are quite important as pests in food storage facilities. You may have seen their coarse, cottony mycelia on aging strawberries. Not many zygomycetes attack humans. However, a few cause dangerous diseases called *mucormycoses,* which can be contracted when farm workers inhale spores from dusty fields. Many zygomycetes are saprobes that recycle waste products such as dung and fallen fruits. One of them, *Rhizopus stolonifer,* is used here to illustrate the life history of these important decomposers (Fig. 20.15).

The life history begins when a passively drifting meiospore settles on food and grows into a large, aseptate mycelium (Fig. 20.15a). Burrowing deep into the food mass, hyphae collect food and transport it to numerous reproduc-

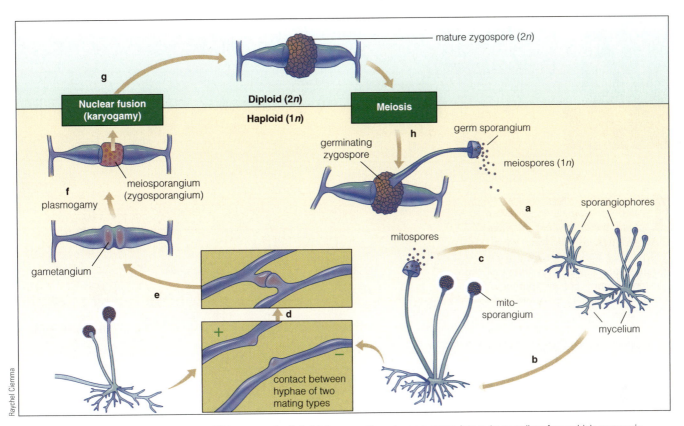

Figure 20.15 Life cycle of a zygomycete (*Rhizopus stolonifer*). (**a**) A nonmotile meiospore grows into a 1*n* mycelium from which sporangiophores grow into the air. (**b**) Tips of sporangiophores swell to become mitosporangia with mitospores. (**c**) The sporangium wall breaks open. The released mitospores grow into new mycelia for asexual reproduction. (**d**) Compatible plus and minus mycelia grow sexual hyphae that meet at the tips. (**e**) Septa form behind the tips of touching plus and minus hyphae, making gametangia. (**f**) Plasmogamy occurs when the walls between gametangia dissolve, making a meiosporangium (the zygosporangium). (**g**) Inside the zygosporangium, karyogamy creates 2*n* zygote nuclei, and the zygosporangium wall thickens and hardens to form a zygospore. (**h**) Meiosis converts the 2*n* nuclei into 1*n* nuclei. A hypha grows out and makes a germ sporangium where recombinant nuclei are packaged into meiospores, completing the life cycle.

tive hyphae called **sporangiophores** that grow into the air. Each sporangiophore swells at the tip to form a mitosporangium, where the contents divide into many mitospores inside the sheltering parental wall (Fig. 20.15b). The mature sporangium breaks open, releasing the spores to drift in the air, settle on food, and make mycelia—an act of asexual reproduction (Fig. 20.15c).

If two compatible *Rhizopus* mycelia meet, they engage in sexual reproduction. To be compatible, mycelia must be of plus and minus *mating types* (see Chapter 12). First, they grow special hyphae that meet at the tips (Fig. 20.15d). Then, cross walls form behind the tips, walling off two cells called *gametangia,* which play the part of gametes (Fig. 20.15e). Next, plasmogamy fuses the gametangia as the walls between them dissolve. The fusion cell, called a **zygosporangium,** is a meiosporangium (Fig. 20.15f). These events bring nuclei from two parents into the same cell without risking dehydration.

Inside the zygosporangium, the two kinds of nuclei fuse in pairs and produce 2*n* zygote nuclei. The wall of the zygosporangium grows thick and hard. With these events the zygosporangium becomes a **zygospore,** which rests for months (Fig. 20.15g). Eventually, meiosis converts the 2*n* nuclei into recombinant 1*n* nuclei, and a hypha grows out of the zygospore (Fig. 20.15h). Its tip swells to form a **germ**

sporangium where the recombinant nuclei are incorporated into meiospores, completing the life cycle.

The life of *Rhizopus*—which is typical of saprobic zygomycetes—shares many features with chytrids, suggesting that these features were present in the common ancestor of all fungi. The shared ancestral traits include making both meiospores and mitospores inside sporangia, putting meiosporangia to rest, and making aseptate hyphae. The chief differences are adaptations to life on land. Swimming spores are replaced by passive spores with water-retaining walls, and swimming gametes are replaced by hyphal tips that deliver gamete nuclei directly to the mating partner.

Rhizopus belongs to a small but important order (the Mucorales) with about 130 named species. Other zygomycetes vary immensely. Many attack insects such as flies, termites, and aphids. If you see a fly walking slowly as if nearly paralyzed, a zygomycete mycelium may be consuming the fly from within. These fungi make tiny mycelia or even simpler thalli inside the body of an insect. Efforts have been made to control insect pests with such fungi, but thus far the results are too variable for commercial use. There also are zygomycetes that parasitize other fungi and small animals such as microscopic nematode worms in the soil.

Large zygomycetes such as *Rhizopus* make extensive aseptate mycelia. But small zygomycetes may have septa, and

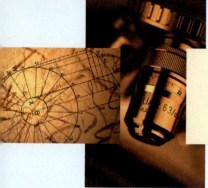

Many zygomycetes let wind, water, or animals carry away spores, but a few are far more active. An example is *Pilobolus*, whose name means "hat thrower." With a hand lens, you may see it growing on dung in a horse pasture. Its tiny golden sporangiophores, a few millimeters in length, look like jewels. They mature in the morning and adjust their growth to point at the rising sun. A swelling under the sporangium focuses light on a sensitive region, causing fine adjustments in growth. Pressure mounts until the swelling bursts, throwing the sporangium several meters through the air. With luck, it sticks to a blade of grass and is swallowed by a grazing animal. The spores survive passage through the animal's intestine and are excreted in manure. Most sporangia are shot away when the sun is about 45 degrees above the horizon—the angle at which projectiles travel the farthest. You can see this effect if you enclose *Pilobolus* in a foil-covered jar with a small hole in the foil to let in light. A day later, look for sporangia on the glass near the hole.

This is just one of many ways in which even the simplest fungi sense and adjust to the environment.

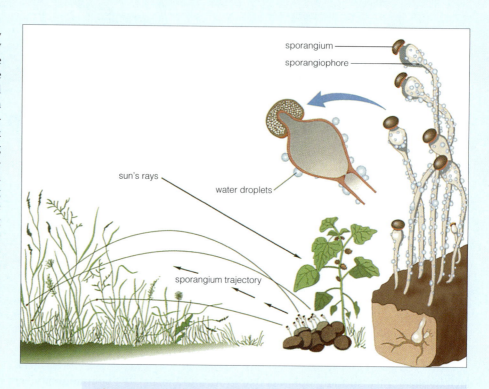

Websites for further study:

Pilobolus—Fungal Shotgun: http://herbarium.usu.edu/fungi/funfacts/Pilobfct.htm

Pilobolus: The Shotgun Fungus:
http://www.biology.usu.edu/biomathlab/pilobolus.htm

some have a tiny thallus without hyphae. Some small zygomycetes have never been seen to reproduce sexually. And although most make asexual sporangia, some make conidia. The great diversity among zygomycetes may result from multiple origins, making this phylum artificial (not monophyletic). However, the early branching that gave rise to zygomycetes occurred too long ago for the relationships to be clear.

Despite their simplicity, some zygomycetes have been studied in great detail because their reproductive hyphae are large and easy to observe, and they are very sensitive to environmental stimuli. Some of them use light and gravity to guide their growth (see sidebar "IN DEPTH: The Hat Thrower"). This helps them release spores where wind and animals are likely to occur. Zygomycetes can be as sensitive to light as the human eye. Scientists who study sensory systems have used these responses as models to guide their thinking. Other experiments showed that zygomycetes find mating partners by exchanging chemical signals called **pheromones,** which differ between species and mating types. We know now that chytrids, ascomycetes, and basidiomycetes also find mating partners by exchanging pheromones.

20.8 PHYLUM GLOMEROMYCOTA

In some ways, the most important fungi may not be mushrooms or molds, but fungi that partner with green plants in the symbiotic associations known as **mycorrhizae.** More than 80% of wild plants are estimated to have such fungal partners, and the relationship may have begun even before primitive plants emerged from water to begin the conquest of dry land. Some of the oldest plant fossils, in rocks that are 400 million years old, contain mycorrhizal fungi.

Until recently, these mycorrhizal fungi were classified in phylum Zygomycota, as the order Glomales. But DNA comparisons have shown them to be a distinct clade, a sister group to the ancestor of dikaryomycetes (Fig. 20.11). They are currently being considered for phylum status with the title **Glomeromycota,** and are treated as such in this chapter.

The Glomeromycota live in soil, where their coarse, aseptate hyphae enter roots or thalli of rootless plants such as liverworts. In roots, the symbiotic associations are called **endomycorrhizae** because hyphae penetrate inside root cells (*endo-* means "within"). In a root, hyphae grow along intercellular spaces between plant cell walls **(Fig. 20.16).** At intervals, a hyphal tip punches through a cell wall of the host root, to branch repeatedly in the space shared with the plant cell. The result is a characteristic structure called an **arbuscule** ("little tree").

The fungus does not penetrate through the plant cell membrane; therefore, the arbuscule is not actually inside the living plant cell. Instead, the plant cell surrounds and coats every hypha in the arbuscule, much as a glove fits each finger of a hand. With this close contact, nutrients can move rapidly between the fungus and the plant. The fungus takes carbon-containing nutrients from the plant cell. In some species, hyphae store excess nutrients in swollen hyphal tips called *vesicles,* which form either in spaces between plant cells or in the space enclosed by the wall of a single plant cell. The fungus pays the plant for nutrients by bringing in water, phosphate, and sometimes nitrogen compounds from the soil.

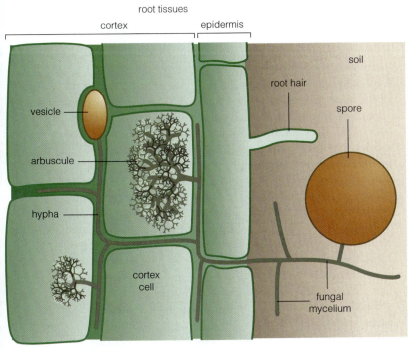

Figure 20.16 Diagram of an arbuscular endomycorrhiza, showing a section through the epidermis and two cortical cell layers of an infected root. Spaces between cells are exaggerated for easier viewing. A mycelium in the soil has grown branching hyphae into the root, where they grow between cells. Hyphae have penetrated through the walls of two cortical cells. They branch repeatedly to form arbuscules. One hyphal tip between root cells has swollen into a food-storing vesicle. A hyphal tip in the soil has swollen to become a spore.

To the plant, the value of having a fungal partner depends on how well the root system could collect water and phosphate on its own. Experiments show that plants benefit most if they have sparse root systems or grow in soils where phosphate is scarce. Thus, the symbiosis can be either mutualistic or parasitic, depending on circumstances. But for many plants, the partnership is vital—the plant grows poorly without its mycorrhizal partner. Perhaps the fungus also helps the plant by repelling parasites and predators.

Sexual reproduction is unknown in the Glomeromycota. They make asexual mitospores in the part of the mycelium that extends into the soil, and some species make mitospores inside the root, either between cells or in the space enclosed by the wall of one plant cell. Unlike chytrids and most zygomycetes, the spores are not made in sporangia. Instead, each spore is simply the swollen tip of a hypha, full of cytoplasm and stored food. The spore may reach a millimeter in diameter—a colossal size compared with spores of other fungi. Sometimes the spores are found in the gut of soil insects, suggesting a possible dispersal mechanism.

Because some of the oldest plant fossils contain similar fungal structures, botanists have postulated that plants and glomeromycetes moved from water to land as partners. If so, these fungi could have been important in helping plants make the transition to land. Early plants would have lacked roots, as algae and nonvascular plants do today. Fungal hyphae, reaching into soil, would have been a great help in collecting water and minerals. The partnerships continued as plants diversified on land. Perhaps this is why arbuscular mycorrhizae are so widespread among modern plants.

Despite their importance, phylum Glomeromycota includes only about 157 named species. But the low total may result from the simplicity of glomeromycetes; they have few structural characters that can be used to distinguish species. Many more species may emerge as systematists analyze the DNA of these important fungi.

20.9 PHYLUM ASCOMYCOTA

Every time you eat bread, drink wine or beer, take penicillin, or throw out a moldy orange, you cross paths with an ascomycete. With more than 32,000 named species, ascomycetes impact our lives in more ways than any other group of fungi.

At least 58 genera of ascomycetes have members that cause human disease. That is a tiny fraction of the 3,409 ascomycete genera, but they threaten the health of many people. Athlete's foot, jock itch, and ringworm are ascomycete infections that the immune system cannot defeat because the fungi grow in dead layers of skin beyond the bloodstream. The immune system blocks most deeper fungal infections, but a few ascomycetes occasionally get into the bloodstream and do great damage, and can even cause death. *Coccidioides immitis* is such a species. Dwelling in soil in the western United States, this species can infect farm workers who breathe contaminated dust. The resulting lung infections may spread to the joints, central nervous system, adrenal glands, and skin. About 95% of patients eventually throw off the fungus; however, in others, the infection can be fatal.

Pathogenic ascomycetes are especially dangerous in individuals with a compromised immune system, such as patients with acquired immune deficiency syndrome (AIDS) and patients whose immune system is suppressed to permit an organ transplant. *Pneumocystis carinii* (also called *Pneumocystis jiroveci*) and *Candida albicans* are particular problems. *Candida* is a common yeast that dwells in the human mouth and vagina, where bacteria and the immune system normally hold it in check. But when the immune system is weak or antibiotics kill the bacteria, *Candida* can cause a painful condition called *thrush*. In patients with human immunodeficiency virus (HIV), thrush often signals the transition to full-blown AIDS. *Pneumocystis* lung infections are more serious, causing fatal pneumonia in many patients with AIDS. At one time about half of all patients with AIDS died of fungal disease. New medicines have decreased the proportion, but fungal diseases are still lethal to many patients with a compromised immune system.

Ascomycetes also cause most fungal diseases in animals and plants. In plants, the most familiar example is the dusty white deposit on leaves called **powdery mildew**, which often is seen on rose bushes, grape vines, barley, and other plants. On grape vines, the infection can seriously reduce the harvest.

On the positive side, ascomycetes have been valuable partners of humankind in the kitchen, laboratory, and hospital. Four kinds of ascomycetes helped scientists win Nobel Prizes: Studies of the mold *Neurospora crassa* first revealed that genes control protein synthesis; *Penicillium notatum* gave us penicillin, launching a revolution in medicine; and the yeasts *Saccharomyces cerevisiae* and *Schizosaccharomyces pombe* vastly improved our understanding of cell division. In the kitchen, morels (*Morchella*) and truffles (*Tuber*) are prized by gourmets, and several ascomycete yeasts are used to bake bread; ferment beer, wine, and soy sauce; and flavor cheeses.

The most valuable fungi of all—*Saccharomyces cerevisiae* and related species, the yeasts of baking and brewing—are ascomycetes. When grown without oxygen, *Saccharomyces* gets energy by alcoholic fermentation, releasing carbon dioxide and alcohol as byproducts. When this happens in bread dough, the carbon dioxide makes the bread rise. In beer and wine, both alcohol and gas bubbles are valued products. The alcohol made by yeast is also an important disinfectant and a fuel.

Ascomycetes Make Septate Hyphae and Conidia

Why did the Ascomycota evolve into such a huge and important phylum, with more than 10 times as many species as all coenomycetes put together? Although some ideas were presented in discussing dikaryomycetes, the full answer is uncertain. For additional ideas, the whole life cycle of a typical ascomycete must be considered **(Fig. 20.17).**

Mycelial ascomycetes begin life as a $1n$ spore, which drifts in space until it lands on a food mass. The spore initiates a branching, septate mycelium (Fig. 20.17a). Ascomycetes spread mainly by making asexual conidia on hyphae at the surface of a food mass (Fig. 20.17b). The green color of moldy oranges and cheese comes from conidia of *Penicillium* (Fig. 20.7). If infected food is exposed to open air in the kitchen, the spores quickly spread to other rooms (a good reason to dispose of infected food while conidial patches are still tiny).

Conidia are made in several ways—some of them so different that they must have evolved independently. For example, some fungi make conidia like blowing bubbles from a pipe, whereas others make them by splitting the conidiophore into short segments.

Ascomycetes Have a Short but Powerful Dikaryotic Stage

If ascomycete mycelia meet and are compatible (usually of plus and minus mating types), they engage in sexual reproduction (Fig. 20.17c). A meeting usually causes *both* mycelia to make female structures called **ascogonia** and smaller male structures called **antheridia.** In plasmogamy, antheridia of each mycelium can fuse with ascogonia of the other mycelium. Nuclei move from the antheridium to the ascogonium (Fig. 20.17d).

The fertilized ascogonium grows $n+n$ hyphae (Fig. 20.17e). The $n+n$ hyphae usually weave together with $1n$ hyphae from the ascogonial parent, to build a fruiting structure called an **ascoma** (plural, *ascomata*; Fig. 20.17f). Ascomata are discussed more fully later in this chapter.

On one side of the ascoma (the upper surface here) many $n+n$ hyphae produce meiosporangia called **asci.** Such an ascus-making hypha is said to be *ascogenous*. One such hypha is enlarged in Figure 20.17g. To make an ascus, the tip of the hypha bends as it grows, making a **crozier** (hook) that contains a nucleus from each parent (Fig. 20.17h). Next, the two nuclei divide mitotically, and walls form to divide the crozier into three cells (Fig. 20.17i,j). The second or *penultimate* cell has two nuclei of opposite types. This cell is the young ascus. Within it, karyogamy occurs to make a $2n$ zygote nucleus (Fig. 20.17k). Also, the tipmost cell of the crozier fuses with the hyphal stalk, regaining the $n+n$ state. This sets the stage for making more asci from the same hypha (an event that is illustrated later in this chapter). Next, the zygote nucleus goes through meiosis to make four recombinant $1n$ nuclei (Fig. 20.17l). A mitotic division usually follows, and the resulting nuclei are packaged with cytoplasm into eight meiospores, which are called **ascospores** (Fig. 20.17m,n).

Ascomycetes vary widely in how they release ascospores. Some species release ascospores passively. In other species, each ascus shoots off its spores, either one spore at a time or all at once in a sticky mass. The shooting force comes from pressure that builds up inside the ascus until it ruptures the tip.

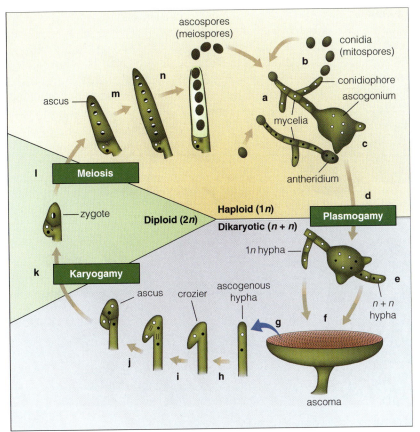

ascospores
(meiospores)

conidia
(mitospores)

b

conidiophore
ascogonium

ascus

m

n

a

mycelia

c

antheridium

d

l

Meiosis

Haploid (1*n*)

Plasmogamy

zygote

Diploid (2*n*)

Dikaryotic (*n* + *n*)

1*n* hypha

k

Karyogamy

ascus   crozier

ascogenous
hypha

n + *n*
hypha

e

g

f

j

i

h

ascoma

Figure 20.17 Life cycle of a mycelial ascomycete.
(**a**) Septate 1*n* mycelia grow from 1*n* spores. (**b**) Conidia
(mitospores) are released for asexual reproduction, making
new 1*n* mycelia. (**c**) If 1*n* mycelia are compatible, they
grow female ascogonia and male antheridia (one of each is
shown). (**d**) In plasmogamy, an antheridium delivers nuclei
to an ascogonium. (**e**) *n*+*n* hyphae grow from the fertilized
ascogonium. (**f**) *n*+*n* hyphae and 1*n* hyphae from the
ascogonial parent make an ascoma. (**g**) One of many *n*+*n*
ascogenous (ascus-making) hyphae in the ascoma is
enlarged here to show ascus formation. (**h**) The tip of the
hypha grows laterally to make a crozier. (**i**) Nuclei divide by
mitosis. (**j**) Two septa divide off cells at the end of the
crozier. The penultimate cell is the ascus, a meiosporangium. (**k**) Nuclei fuse in the ascus (karyogamy) to make
a 2*n* zygote. The cell at the tip of the crozier fuses with the
stalk cell to restore the *n*+*n* condition there. (**l**) Meiosis
makes four recombinant 1*n* nuclei in the ascus. (**m**) 1*n*
nuclei divide mitotically. (**n**) 1*n* nuclei are incorporated into
meiospores (ascospores) that are released, often forcefully.

We hinted in Figure 20.17 that each *n*+*n* hypha can make
many asci, but the full picture requires another diagram
(Fig. 20.18). The key fact is that when a crozier makes an
ascus, it also makes another crozier. Each crozier repeats the
same steps as the one before. In Figure 20.18, the numbers 1,
2, and 3 show where the first three croziers arise. Development of the first ascus, colored blue for visibility, is shown in
full. The other asci will develop the same way. Following the
lettered steps in Figure 20.18: (a) An ascogenous hypha
forms a crozier; (b) the 1*n* nuclei in the crozier divide by
mitosis; (c) septa divide the crozier into three cells; (d) the
crozier's tip cell fuses with the hyphal stalk, restoring the
n+*n* condition in the stalk cell; (e) karyogamy occurs in
the penultimate cell (the ascus) to make a 2*n* zygote nucleus;
(f) meiosis follows, producing four recombinant 1*n* nuclei.
The cell below the first ascus grows an *n*+*n* hypha that
forms a new crozier. (g) Mitosis doubles the number of 1*n*
nuclei in the first ascus, while the second crozier repeats
steps a through e to make a second ascus with a 2*n* nucleus
and restore the *n*+*n* condition in the underlying cell.
(h) Spores mature in the first ascus, while the second ascus
goes through meiosis and a third crozier forms at its base.
The cycle of events is open-ended and can repeat many
times to make many asci.

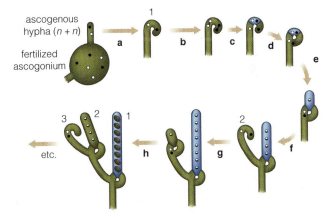

ascogenous
hypha (*n* + *n*)

1

fertilized
ascogonium

a

b

c

d

e

3

2

1

2

etc.

h

g

f

Figure 20.18 One *n*+*n* hypha can make many asci. The first ascus
(*blue*) is followed to maturity; early stages of two more asci also are
shown (*green*). They will follow the same steps as the first.

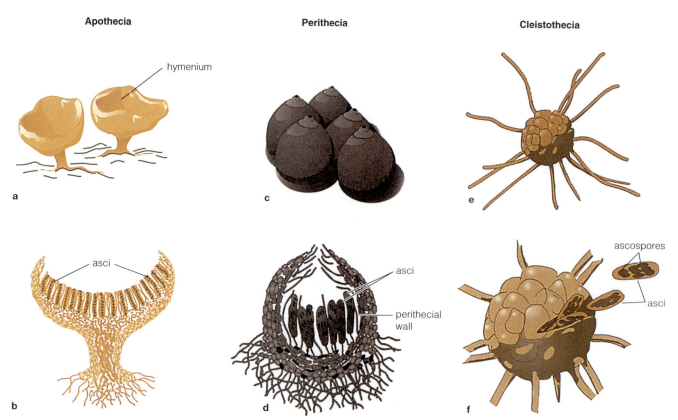

Apothecia · Perithecia · Cleistothecia

Figure 20.19 Ascomata. (**a**) Two apothecia. (**b**) A cross section through an apothecium. ×10. (**c**) Five perithecia. (**d**) Diagram of a perithecium in cross section. ×150. (**e**) A cleistothecium that is still closed. (**f**) A cleistothecium of a mildew that has opened to release asci. Each ascus contains eight spores. ×500. Most cleistothecia do not actively discharge asci or ascospores.

There Are Three Main Types of Ascomata

Most ascomycetes make complex fruiting bodies (*ascomata*) in which both 1*n* hyphae and *n*+*n* hyphae take part. An ascoma makes asci and spores, shelters the immature spores, and assists in dispersing mature spores. This is a step beyond anything that occurs in coenomycetes, where sexual spores are made individually with little supporting structure from other hyphae.

There are three main types of ascomata, which are variations on a theme: a protective layer of nonreproductive hyphae provides a shelter or platform, and multiple asci form in a layer called the **hymenium** on one side of the supporting layer (Fig. 20.19). An ascoma is an **apothecium** if it resembles an open cup (Figs. 20.9b, 20.17, and 20.19a,b). Apothecia range from a millimeter to several centimeters in diameter. Asci line the upper surface of the ascoma, pointing upward or outward. The alignment aids in dispersal when spores are shot away. Shooting the spores upward helps disperse spores because air is stagnant near the face of the ascoma (a region called the *boundary layer*). The ascus shoots spores through the boundary layer, to a height where spores may encounter a breeze.

An ascoma is a **perithecium** if it is nearly closed, like a tiny flask. A pore at the small end opens to release the spores (Fig. 20.19c,d). Here the spores may be either shot off or not shot off, depending on the fungal species. Dutch elm disease is caused by a perithecial fungus, *Ophiostoma ulmi*. The spores spread with the help of beetles that burrow into tree bark. When a spore-carrying beetle makes a tunnel in the bark, spores rub off and grow mycelia into the bark tissue. Ascospores are not shot off. Instead, perithecia open into burrows and extrude a jelly that contains spores. Beetles accidentally pick up the jelly when they pass, and carry spores to other trees. In this way, Dutch elm disease has all but eliminated the elm forests that once existed all across the eastern United States.

Still other ascocarps are tiny closed spheres, or **cleistothecia,** which dry up and shelter the enclosed asci until the cleistothecium is crushed (Fig. 20.19e,f). Then the spores may be shot away or released passively. Powdery mildews make cleistothecia that remain through winter on dead leaves.

Many Ascomycetes Engage in Symbiosis

Many ascomycetes form symbiotic associations with other organisms. Some form mycorrhizae with tree roots. Truffles, the delight of French chefs, are underground ascomata of the genus *Tuber* that live in mycorrhizal association with tree roots. The hyphae in these and other ascomycete mycorrhizae do not enter plant cells, but rather coat the root tip and grow between root cells. Such external associations are called **ectomycorrhizae.** They assist the plant by bringing in nitrogen compounds and water.

A great many ascomycetes engage in symbiotic associations called **lichens,** where they partner with algae, cyanobacteria, or both. More than 42% of named ascomycetes—more than 13,500 species—take part in lichens. Only a few dozen other fungal species (all of them basidiomycetes) form lichens.

Figure 20.6 illustrated a leaflike or **foliose** lichen. But there also are branching or **fruticose** lichens, and most familiar of all are crusty or **crustose** lichens that form green, yellow, red, or orange patches on bark, posts, and rocks (Fig. 20.20).

Lichens are important pioneers in rocky areas without soil, such as newly exposed volcanic islands and alpine cliffs. Secreting acid, lichens break down the rock to release minerals. As their bodies collect dust and finally decay, they build up soil that other organisms can exploit. Lichens with cyanobacteria are important in the nitrogen economy of some conifer forests, because the cyanobacteria convert atmospheric nitrogen to organic forms (the process of *nitrogen fixation*). When the lichen dies and drops from a tree branch to the forest floor, its decomposition releases nitrogen compounds to the soil.

Fruticose lichens of the genus *Usnea* are particularly beautiful, hanging from tree branches near the west coast of the United States as long, gray-green streamers. They often are mistaken for the flowering plant of the south called Spanish moss, and can reach 10 m in length.

Lichens are named for the fungal component, but both partners influence the body form (when each partner grows alone, it bears little resemblance to the lichen). The fungus encloses the photosynthetic partners, holding them in place by peglike hyphae that indent the algal cell wall but usually do not penetrate healthy algal cells. In some lichens, the fungus secretes water-repellant proteins called **hydrophobins** on its exposed surfaces, sealing the contact between hypha and alga. Water flows through the underlying hyphal walls to the alga. Thus, the fungus regulates the alga's water supply.

Scientists debate whether lichen symbiosis is mutualistic or parasitic. When the partners are artificially separated, the algal partner usually grows better than the fungus, and many fungi that form lichens are never found growing alone in nature. These facts have led some researchers to view a lichen fungus as parasitic on the photosynthetic partners. However, the fungus also provides a moist, mineral-rich shelter that permits algae and cyanobacteria to grow in habitats where they could not live alone. Thus, whether the symbiosis is mutualistic or parasitic may depend on the environment.

Most lichens reproduce asexually, often by releasing tiny units called **soredia** (singular, *soredium*) in which a few hyphae tightly wrap around one or more photosynthetic cells. Soredia can drift far in air or water. Other reproductive structures are simple fragments of a lichen, or outgrowths that pinch off. The most common of these are called **isidia** (singular, *isidium*). The fungus may produce conidia as well. For sexual reproduction, the fungal partner makes typical asci, and the algal partners also undergo genetic recombination.

For a fungal spore to initiate a new lichen, it must find a partner. Some spores land on existing lichens, grow mycelia,

R. M. Thornton

Figure 20.20 Fruticose and crustose lichens growing on rocks above the ocean at Pt. Lobos, CA. The branching gray-green lichen in the center is fruticose; the orange crust on the rocks consists of crustose lichens.

kill the lichen's fungal partner, and take over the algae for themselves. But the lichen symbiosis requires a close match between partners. Thus, a fungus growing among free algae will form only loose associations until it meets an alga with compatible genetic traits.

How did so many ascomycetes gain the ability to form lichens? One view is that lichen symbiosis is easy to evolve and has arisen many times in separate fungal groups. If this is true, many groups gained their lichen-forming ability recently. But a recent statistical study supports the opposite view: The ability to make lichens may be very hard to gain, and it may have evolved no more than two or three times—or perhaps just once, early in the evolution of ascomycetes. If the study's views are accepted, then many modern free-living ascomycetes had lichen-forming ancestors. Those fungi include the common kitchen molds *Penicillium* and *Aspergillus*.

Most Mitosporic and Dimorphic Fungi Are Ascomycetes

The great majority of mitosporic fungi are now recognized as ascomycetes. They include the important genera, *Penicillium* and *Aspergillus*, which are among the most useful and most dangerous mitosporic fungi. *Penicillium* has been discussed several times in this chapter. The black-spored kitchen mold *Aspergillus* (Fig. 20.2) also causes food decay and may cause a respiratory disease called aspergillosis. *Aspergillus flavus* makes cancer-producing compounds called **aflatoxins** when it attacks nuts and grains. But some *Aspergillus* species are valuable, contributing to the production of soy sauce and industrial chemicals, as well as the citric acid used in soft drinks. These and other mitotic fungi are extremely widespread, traveling by conidia and growing as molds on organic products as diverse as book bindings and bread.

Hundreds of ascomycete species are dimorphic, switching between mycelial and yeast forms in response to environmental conditions. Even baker's yeast sometimes makes hypha-like filaments. Some dimorphic ascomycetes are dangerous human pathogens, particularly in people with a weakened immune system. *Histoplasma capsulatum* is an example; it grows a mycelium at 25°C, but is a yeast at 37°C, the human body temperature. In the yeast form it can invade the lungs and enter the bloodstream, where its presence can be fatal. *Candida*, by contrast, is a yeast in the environment and a mycelium in the human body.

20.10 PHYLUM BASIDIOMYCOTA

At last we arrive at fungi that make the largest and most complex structures, for example, mushrooms. They occur in phylum Basidiomycota, a group of fungi that is second only to ascomycetes in number of named species (the current total is more than 26,000).

Hymenomycetes Make Long-lived Dikaryons

There are three classes within phylum Basidiomycota. The largest class, the **Hymenomycetes,** contains more than 14,000 named species, including all the basidiomycetes that make complex structures. The other two classes are much less conspicuous but include costly crop parasites (see discussion later in this chapter).

Mushrooms, which make us all familiar with hymenomycetes, are sexual fruiting bodies, or **basidiomata** (singular, *basidioma*). The largest mushrooms are about 1 m in diameter. By contrast, ascomata never exceed a few centimeters in size. Yet, ascomycetes and basidiomycetes are sister clades; both are dikaryomycetes. Why, then, do basidiomata get so much larger? A look at the life cycle of a mushroom fungus provides some clues (Fig. 20.21).

Like any fungus, a hymenomycete usually begins life as a spore (Fig. 20.21a). But the spores are nearly always of sexual origin, because mushroom fungi defy the usual rules of fungal life and make few asexual spores. They use meiospores to multiply, as well as diversify. The haploid meiospore grows into a small $1n$ mycelium. With luck, the $1n$ mycelium soon meets a compatible partner (usually of a different mating type), and mating occurs.

The first events in mating are simple: Compatible hyphae fuse at their tips (plasmogamy), making a fusion cell with one nucleus from each parent (Fig. 20.21b). In many cases, nuclei from each parent slip through septa to other compartments in the partner's mycelium, dividing mitotically as they go, so the joined haploid mycelia become a single $n+n$ mycelium. In other cases, the fusion cell simply begins to make branching $n+n$ hyphae. Either way, the result is an $n+n$ mycelium (Fig. 20.21c).

Eventually, basidiomata—fruiting bodies—form at various points on the $n+n$ mycelium (Fig. 20.21d). To make a basidioma, $n+n$ hyphae branch repeatedly and grow together. Signals must be exchanged between the hyphae, because they develop differently depending on their location in the basidioma. In a gilled mushroom, some hyphae make up a stalk, or **stipe.** Others make up a protective **cap,** within which still other hyphae make **gills,** the spore-producing part of the basidioma (Fig. 20.21e). All these hyphae are dikaryotic.

Spores are made all over the surface of each gill, from tips of $n+n$ hyphae that face the air between the gills. The tipmost cell in each of these hyphae is a **basidium** (one of which is enlarged in Fig. 20.21 for visibility). As the basidium swells, the paired nuclei fuse (karyogamy) to make a $2n$ zygote nucleus (Fig. 20.21f). Meiosis divides the $2n$ nucleus into four $1n$ nuclei, each with a different combination of chromosomes from the two parents (Fig. 20.21g). Next, pressure in the basidium creates spores: The wall of the basidium near each $1n$ nucleus grows a slender stalk or **sterigma;** then the tip of the sterigma balloons into a spore, similar to a person blowing a bubble from a pipe. As pressurized cytoplasm floods through the sterigma, a nucleus is carried with it. This creates four meiospores called **basidiospores** (Fig. 20.21h), each with a different $1n$ genotype. The spore walls harden, and the spores are released (Fig. 20.21i).

Hymenomycetes lengthen the dikaryotic stage of life far longer than ascomycetes, and this may help explain how basidiomata can grow much larger than ascomata. In ascomycetes, the dikaryon is short-lived and serves only to make a fruiting body; the feeding mycelium is haploid. By contrast, hymenomycetes make dikaryons that may feed and live for many years. During its long life, the dikaryon may create basidiomata many times—typically in the warm, wet spring and autumn.

The diameter of a dikaryon can show how old it is. As time passes, the mycelium grows in diameter, until it may cover hectares of meadowland or forest floor. Mushrooms form near the growing edge of the mycelium. Because the mycelium grows in all directions from its point of origin, its perimeter often is more or less circular. Thus, in fruiting season, the mycelium will produce mushrooms in a ring-shaped array, which is called a **fairy ring** from an ancient notion that fairies dance on the forest floor and use mushrooms as stools. Each year's fairy ring is somewhat larger than the last one, permitting estimates of how fast the mycelium expands. On the basis of that and the current diameter, some $n+n$ mycelia appear to be hundreds of years old.

For the $n+n$ mycelium to last so long, it needs a way to give each compartment a dikaryotic pair of nuclei. Both nuclei must divide at the same time, and the four resulting nuclei must be sorted into proper pairs (Fig. 20.22). Some fungi solve the sorting problem by lining up the two divid-

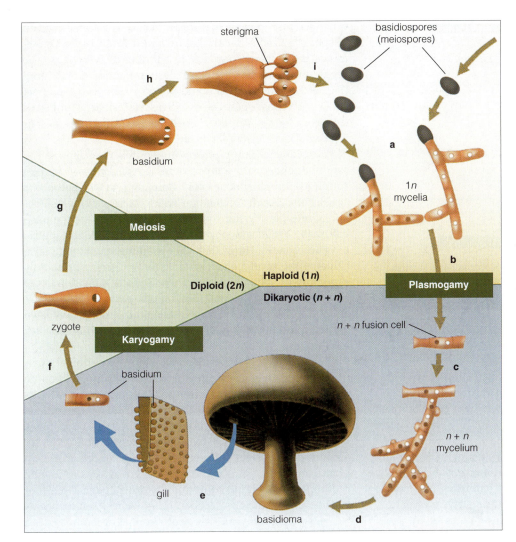

Figure 20.21 Life cycle of a mushroom fungus (class Hymenomycetes). Reproduction is entirely sexual. (a) Basidiospores (1n) from two compatible fungi germinate and form 1n mycelia. (b) The tips of two hyphae fuse (plasmogamy) to make an n+n fusion cell. (c) The n+n cell grows into an n+n mycelium that can live for many years. (d) Many n+n hyphae join to build a basidioma, with spore-producing gills under the cap. (e) Enlargement of a small part of a gill. Young basidia (tips of n+n hyphae) jut out all over the gill surface; one of them is enlarged for visibility. (f) The basidium swells, and its two nuclei fuse (karyogamy) to make a 2n zygote nucleus. (g) Meiosis creates four recombinant 1n nuclei in the basidium. (h) The 1n nuclei and cytoplasm are forced through sterigmata to make four basidiospores (meiospores). (i) With hardened walls, the basidiospores are pushed away to drift down and out of the gill.

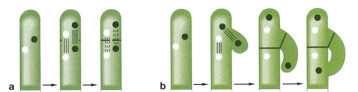

Figure 20.22 Two ways in which basidiomycetes maintain the n+n nuclear condition during the growth of long-lived dikaryotic mycelia. (a) Overlapping mitotic spindles. (b) The formation of a clamp connection.

ing nuclei side by side and making the septum in the plane of the spindle equators (Fig. 20.22a).

Other species maintain the dikaryotic condition in a more complex way (Fig. 20.22b). As the nuclei begin to divide, a backward-pointing branch hypha forms beside the leading nucleus, and mitosis places one product nucleus into the branch. Then septa form at the equators of both spindles, so the leading compartment gets a proper pair of nuclei. The branch tip fuses with the trailing cell, delivering

its nucleus so that the second compartment also has a proper pair. The fused lateral branch remains as a lump on the side of the hypha, positioned at a septum. The fused branch resembles a carpenter's clamp, so mycologists call it a **clamp connection.** The occurrence of clamp connections in a feeding hypha is a sure sign that the hypha was made by an n+n basidiomycete, because no other feeding mycelia have them. However, ascomycetes go through similar steps to make many asci from one n+n hypha (see earlier). Perhaps both phyla inherited genes for maintaining the dikaryotic condition from their common ancestor.

Was it the long dikaryotic stage that enabled hymenomycetes to evolve their impressive reproductive structures? Perhaps. More mitotic divisions result in more pairs of nuclei that can fuse and make meiospores with unique genotypes. By maintaining the dikaryon for many years, a single hymenomycete can produce immensely more recombinant spores than any other fungus. Such diversity is a source of raw material on which natural selection can operate to produce new body forms.

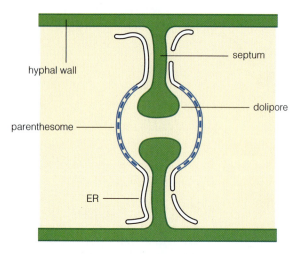

Figure 20.23 A diagram of a dolipore septum, as it appears in a thin section of a hypha. The septum has a ring-shaped thickening (the dolipore) around the central opening. Parenthesomes on each side (*blue*) are cup-shaped and are made from fused layers of endoplasmic reticulum (ER) membranes held together by proteins. Holes through parenthesomes allow passage of molecules and particles but not nuclei.

Hymenomycetes have yet another feature that helps keep the mycelium dikaryotic, and may also promote the evolution of complex body forms: Their $n+n$ hyphae have unique septa of a kind called the **dolipore septum** (Fig. 20.23). Like the simpler septa of the $1n$ mycelium and ascomycetes, the dolipore septum is perforated, with an opening in the center for transport. But the dolipore septum thickens around the opening, and a cup-shaped cap called a **parenthesome** forms on each side of the septum. Made of fused layers of endoplasmic reticulum (ER) membranes held together by proteins, parenthesomes have openings that let small particles and molecules pass, but not nuclei. Thus, dolipore septa assure that each compartment will keep both of its paired nuclei. This may promote the formation of complex basidiomata. To make complex structures, cells must follow different paths of development in different parts of the mycelium. Such local differentiation can be accomplished more easily if nuclei are kept in defined locations.

Hymenomycetes Make Varied Fruiting Bodies

In addition to mushrooms with gills, the Hymenomycetes make several other kinds of basidiomata (Fig. 20.24). Many mushrooms replace the gills with thousands of vertical pores under the cap. Basidia line the pores. **Bracket** or **shelf fungi** (Fig. 20.24a), found on tree trunks and logs, have caps like mushrooms, but they omit the stipe and connect the cap directly to the food source. Brackets depend on a host tree trunk or log to hold the cap aboveground. Some brackets have pores and grow for many years, becoming woody and adding to the ends of the pores every year. **Puffballs** (Fig. 20.24b), common on meadows and the forest floor, make spores inside a leathery bag and rely on raindrops or animals to disperse spores. While internal cells make basidiospores, the outer covering becomes leathery with a hole on top, or it may flake off.

When a raindrop hits a mature puffball or an animal steps on it, spores puff into the air. A **bird's-nest fungus** (Fig. 20.24c) makes basidiomata that look like tiny nests with eggs. The "eggs" are spore-bearing, football-shaped units. **Coral fungi** (Fig. 20.24d) make basidiomata that resemble certain reef corals and release spores all over the surface. In jelly fungi (not shown in Fig. 20.24), the basidioma has supporting tissues with a delicate gelatinous texture.

Basidiomata may produce huge numbers of spores. The common grocery store mushroom, *Agaricus bisporus,* makes about 40 million spores every hour for 2 days. *Ganoderma applanatum,* a shelf fungus that attacks trees, can make 3 billion spores per day for 6 months of each year. A large puffball, made by *Calvatia gigantea,* may produce 7 trillion spores.

The order of fungi that makes edible mushrooms also contains the most poisonous kinds. The genus *Amanita* is especially dangerous. Although poisonous mushrooms are less common than benign ones, mycologists advise enthusiasts to consult an expert before eating mushrooms they have gathered in nature. Deadly species are easily mistaken for edible ones.

Some basidiomycetes are important mycorrhizal partners of forest trees. Many mushrooms on the forest floor are fruiting bodies of an underground mycelium that draws its main nourishment from tree roots. In return, the fungus brings water, nitrogen compounds, and perhaps phosphates from the soil. These associations, like those made by ascomycetes, are ectomycorrhizae. Unlike those made by glomeromycetes, the basidiomycete hyphae do not invade cells of the host plant's roots, but lay down a dense mesh of hyphae around the tips of roots and grow between root cells.

Many basidiomycetes are saprobes that decompose dead wood and leaves. Even mycorrhizal fungi may do that. In decomposing wood, no other organisms can match these basidiomycetes, because they have superior enzymes to degrade the lignin that hardens wood. Lignin—the second most abundant organic compound on Earth—resists decay because few enzymes can attack it, and its breakdown products are somewhat toxic. Some basidiomycete decomposers make the required enzymes and resist toxicity. Without these fungi, most of the world's carbon could be locked up in dead wood.

Unfortunately, the same ability to decompose wood can be expensive for homeowners. To a basidiomycete, exposed wood in a house is just more food, and fungal attack on it causes damage called **dry rot.** The name is misleading, because the fungus only grows and feeds when wood is wet. In dry conditions, the fungus lies dormant. You can protect against dry rot by keeping wood dry.

Ballistospore Release Aids in Dispersing Basidiospores

Mushrooms and brackets drop spores downward to reach open air for dispersal, but gills are closely spaced and pores are narrow. How can a fungus prevent spores from landing on surfaces inside the cap? Basidiomata avoid collisions

a   b   pore   c   rotting wood   "eggs" (spore-bearing structures)

d

Figure 20.24 Some basidiomata other than mushrooms. (**a**) A bracket fungus growing on a tree. (**b**) Puffballs, which passively release spores through the pore when struck. (**c**) A bird's-nest fungus. (**d**) A coral fungus, which produces spores all over its surface.

between spores and walls partly by sensing gravity and orienting growth so that the pores or gills stay nearly vertical. But even with vertical alignment, how can spores be pushed off just hard enough to clear nearby surfaces? The ascomycete way of shooting spores would be far too forceful.

Basidiomycetes evolved a gentler way to push off spores, called **ballistospore release (Fig. 20.25).** Each spore has a small projection where its base connects with a sterigma. When the spore matures, the projection secretes sugars that attract water from the air. A water droplet forms on the projection and grows larger with time. Water also condenses to make a separate thin film elsewhere on the spore. When the droplet grows large enough to touch the film, surface tension pulls the droplet into the film. The laws of physics dictate that any new motion is always accompanied by a countermovement. Here, the movement of the droplet pulls the spore toward the sterigma. As a result, the sterigma is compressed. Being elastic, the sterigma immediately rebounds to its original length, pushing the spore away. The spore moves about halfway to the opposite pore wall or gill before air resistance stops its horizontal motion, leaving the spore to drift slowly downward.

Ballistospore release occurs in members of all the major groups of Basidiomycota; therefore, it probably evolved in the common ancestor of all basidiomycetes.

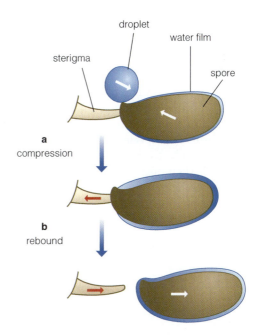

droplet
water film
sterigma
spore

a
compression

b
rebound

Figure 20.25 Ballistospore release. (**a**) Water from the air condenses, making a film on the spore and a separate droplet at the base of the spore. When the growing droplet contacts the film, surface tension pulls the droplet onto the spore. The move pulls the spore to the left, compressing the sterigma (*arrows* on first two diagrams). (**b**) The elastic sterigma rebounds to regain its original length, pushing the spore away (*arrows* on third diagram).

Many Ustilaginomycetes Cause Smut Diseases

In addition to Hymenomycetes, phylum Basidiomycota includes two quite different classes without large fruiting bodies. One class, the **Ustilaginomycetes,** includes 1,064 named species, most of which attack green plants. Most of the resulting diseases are called **smuts** because black, greasy masses of spores erupt from infected plants, as in corn smut **(Fig. 20.26).**

Some smut fungi infect only seedlings, others infect flowers, and still others form localized infections wherever they land on the plant. In flowers, smut spores may be spread by pollinating insects that come for nectar. Fortunately, each smut fungus attacks only one host species and can be controlled by breeding varieties of the host that resist the fungus.

Ustilago maydis (Fig. 20.26) causes a costly smut disease in corn (*Zea mays*). In an average year, *U. maydis* may attack up to 5% of the plants in a cornfield. If conditions are especially good for the fungus, most of the plants may be infected. When the mycelium invades a corn kernel, it takes over control of the kernel's development and stimulates great expansion, converting the kernel to a large gall that combines fungal and corn tissues. The galls drain energy from the rest of the plant and reduce the crop. However, the galls are considered a delicacy in parts of Mexico.

Most smut fungi live as yeasts in the haploid stage and parasitic mycelia in the dikaryotic stage. When compatible haploids meet on a plant, they fuse to make a parasitic $n+n$ mycelium. Later, karyogamy and meiosis generate new $1n$ basidiospores for the next generation.

Urediniomycetes Cause Rust Diseases

The **Urediniomycetes** are the third class of basidiomycetes. There are more than 8,000 named species. Here are found more complex life cycles and more costly pests. The diseases they cause are called **rusts** because their spores often give host leaves a rusty color. You may see rust on lawn grass when spring weather is especially damp.

Some rusts that attack cereal grains belong to the species *Puccinia graminis*. Their host preference is very specific, and taxonomists recognize seven forms within the same species on the basis of host preferences. For example, *P. graminis* form *avenae* attacks only oats, whereas *P. graminis* form *tritici* attacks only wheat.

It is difficult to eradicate rust fungi because many of them have two hosts. In the case of wheat rust, one host is wheat and the second host is a wild shrub called barberry. The two hosts have different roles in the life of the fungus. On wheat, the rust is entirely dikaryotic. Its mycelia make $n+n$ conidia (mitospores) that attack more wheat plants. Being asexual, this reproduction on wheat spreads the parental genotypes through a wheat field. With several generations in a growing season, a single rust spore can start a widespread epidemic. Draining nutrients from the wheat plants without killing them, the rust infection reduces the

Figure 20.26 *Ustilago maydis*, the common corn smut, infecting a cob of sweet corn (*Zea mays*). Each gray lump is an infected kernel, converted to a gall by the mycelium inside. The black masses at the top are mature galls that have opened to expose the smut spores.

harvest and may cost billions of dollars during a bad year. Winter ends the epidemic by killing mycelia and summer conidia. But as the growing season ends, dikaryotic mycelia make special $n+n$ conidia called **teliospores** that survive the winter. In spring, teliospores germinate, go through karyogamy and meiosis, and release $1n$ basidiospores with new genotypes. Basidiospores are pushed away by the same ballistic mechanism as in a mushroom. The basidiospores attack barberry plants. But on barberry, the rust is gentler than on wheat and does not harm the host. The resulting haploid mycelia mate sexually in barberry leaves. Flies assist in the mating by carrying gamete cells between haploid mycelia. The mating results in dikaryotic mycelia with new genotypes. Conidia made by these dikaryotic mycelia attack wheat, launching new epidemics.

Biologists use two approaches to control wheat rust. One approach is to breed new varieties of wheat that resist the common strains of rust. However, new strains of rust fungi arise every year because of sexual recombination on the barberry hosts. Among them are strains that can attack the new varieties of wheat. Within a decade, existing varieties of wheat cannot be planted without risking a rust epidemic. Thus, breeders must continually introduce new wheat varieties to keep ahead of the fungus.

The second approach to controlling wheat rust is to eliminate barberry plants. If successful, it would stop the formation of new rust strains. Unfortunately, barberry still survives our efforts to eradicate it.

In the battle for the wheat fields, rust fungi prove the value of sexual reproduction. Without sex to introduce new rust genotypes, we could eliminate wheat rust in a single year by planting only the latest resistant wheat varieties.

KEY TERMS

absorptive nutrition	hydrophobin
antheridium	hymenium
arbuscule	hypha
ascogonium	isidium
ascoma	lichen
Ascomycota	meiosporangium
ascus	mildew
aseptate hypha	mitosporangium
ballistospore release	mitosporic fungus
basidioma	mycelium
Basidiomycota	mycorrhiza
basidium	parasexual cycle
chemoheterotroph	pheromone
chemotropism	rhizoid
chitinous cell wall	rust fungus
Chytridiomycota	saprobe
clamp connection	septal pore
coenocytic	septate hypha
coenomycetes	septum
conidiophore	smut fungus
conidium	soredium
crozier	sporangiophore
cytoplasmic streaming	sporangium
dikaryomycetes	spore
dikaryon	sterigma
dikaryotic stage	symbiotic relationship
dimorphism	teliospore
dolipore septum	thallus
dry rot	true fungi
ectomycorrhiza	yeast
endomycorrhiza	zoospore
Fungi	Zygomycota
Glomeromycota	zygosporangium
grade of evolution	zygospore
heterokaryotic	

SUMMARY

1. Kingdom Fungi is a monophyletic group of eukaryotic chemoheterotrophs that reproduce with spores and have chitinous cell walls. About 80,000 species have been named, but 1.5 million species may exist. Kingdom Fungi excludes some organisms that are commonly called fungi (mycetozoa, acrasids, and oomycota) because they evolved from different ancestors.

2. Saprobic fungi compete with bacteria as decomposers. Other fungi live symbiotically with algae and cyanobacteria (in lichens) or green plants (in mycorrhizae). Others are parasites on various hosts.

3. Some fungi cause human diseases, particularly in immunocompromised patients. These fungi include *Candida, Aspergillus, Histoplasma, Coccidioides,* and others. Microsporidia parasitize animals and humans, and may be highly specialized fungi. Other fungi produce such useful chemicals as alcohol and penicillin, and some (most notably baker's yeast, *Saccharomyces cerevisiae*) are used in medical research.

4. The fungal body can be a microscopic yeast cell or a thallus, usually a mycelium made of branching hyphae. Growing hyphae secrete enzymes to digest foods. Food absorption leads to water uptake, driving growth and transport within the mycelium. Hyphae provide fast collection of food and fast internal transport of nutrients, and they secrete acids and antibiotics. These features, in addition to a superior ability to decompose lignin in wood, enable fungi to compete with bacteria.

5. Most fungi reproduce both sexually and asexually, releasing mitospores to multiply the parental genotype and meiospores to produce new genotypes. However, many fungi have little asexual reproduction, and others, called mitosporic fungi (previously deuteromycetes), have little or no sexual reproduction. Mitosporic fungi achieve genetic recombination by occasional sexual reproduction, a parasexual process, or other means.

6. Traditional taxonomy defined four phyla in kingdom Fungi, on the basis of details of sexual reproduction and the presence or absence of swimming cells. The two largest traditional phyla (Basidiomycota and Ascomycota) are monophyletic sister groups, whereas the two other traditional phyla (Zygomycota and Chytridiomycota) are not monophyletic. A fifth monophyletic phylum, the Glomeromycota, recently has been recognized.

7. Two grades of evolution—dikaryomycetes and coenomycetes—are informally recognized in kingdom Fungi. A dikaryotic stage and regular septation distinguish dikaryomycetes from coenomycetes. Dikaryomycetes belong to phyla Ascomycota and Basidiomycota and include all fungi that make complex fruiting structures.

8. Members of phylum Chytridiomycota (about 1,000 named species) have swimming reproductive cells. Primarily aquatic, the chytrids sometimes make small mycelia but more often have simpler thalli.

9. Phylum Zygomycota (about 1,100 named species) is named for producing zygospores during sexual reproduction, but some have not been seen to reproduce sexually. This phylum includes important decomposers, parasites of insects and small soil animals, and a few mycorrhizal symbionts. Some have large aseptate mycelia; others have simpler thalli. Mitospores usually are made in sporangia.

10. Members of phylum Glomeromycota (157 named species) nearly always partner with plants to make arbuscular endomycorrhizae. These mutualistic associations may involve more than 80% of plant species. The Glomeromycota have not been seen to reproduce sexually.

11. Members of phylum Ascomycota (more than 32,000 named species) make sexual spores in asci. Most are mycelial, but some are yeasts and others are dimorphic. Mycelia are septate and multiply by releasing mitospores (conidia). Sexual reproduction includes a short-lived dikaryon, usually with complex fruiting bodies (ascomata). Many have not been seen to reproduce sexually. About 42% of named ascomycetes pair with green algae or cyanobacteria in lichens. Others form ectomycorrhizae. Some are parasitic on plants, animals, or humans, and some are saprobes that include food storage pests. Most medically important fungal infections involve ascomycetes. Several ascomycetes are sources of antibiotics and other chemicals or are used in biological research.

12. Members of phylum Basidiomycota (more than 26,000 named species) extrude sexual spores from basidia. A few are yeasts, but most species make septate mycelia. Those in class Hymenomycetes make dikaryons that may last for centuries and also make complex basidiomata such as mushrooms. Some Hymenomycetes live as saprobes, including the causal agent of dry rot in houses. Some partner with trees in ectomycorrhizae. A few cause human disease. Basidiomycetes in the phylum's other two classes make short-lived dikaryons, produce no basidiomata, and cause costly rust and smut diseases in plants.

Questions

1. French chefs flavor foods with truffles, underground fruiting bodies of fungi that are mycorrhizal partners of trees. Each fruiting body contains many unicellular sacs, each containing four or eight meiospores. The hyphae are septate. To what phylum of fungi do truffles belong? Which listed items are clues?

2. Most fungi compete with bacteria as decomposers. What traits allow fungi to compete effectively, and why do the traits help?

3. Offer a hypothesis as to the costs and benefits of regular septation in hyphae. How does your idea explain why dikaryomycetes make complex fruiting bodies and coenomycetes do not?

4. In your job as a farm advisor, a farmer shows you a handful of grain that is infected with a fungus. What is the easiest method to determine whether the fungus is an ascomycete, basidiomycete, zygomycete, chytridiomycete, or glomeromycete?

5. Why would a strict phylogenetic systematist object to calling Zygomycota a phylum?

6. Why did early biologists view fungi as plants, and why do modern biologists view fungi as being more closely related to animals?

7. Two phyla of fungi contribute to a slice of pizza. Name the phyla, and explain how they contribute.

8. The feeding process in fungi automatically leads to growth. Why?

9. What is a dikaryon? Which organisms have it? How long does it last? What is its value?

10. How do mitospores and meiospores contribute to fungal life?

11. Fill in the missing information in the table at the bottom of this page.

 InfoTrac® College Edition

http://infotrac.thomsonlearning.com

Fungi

Adler, T. 1996. Discovering the sexy side of valued fungi (scientists find strain of *Trichoderma* that reproduces sexually rather than asexually). *Science News* 149:118. (Keywords: "sexy" and "fungi")

Anonymous. 1991. Muck and magic (the unique growth of fungi makes cultivating wild mushrooms a problem). *The Economist (US)* 319:82. (Keywords: "muck" and "fungi")

Childs, G., O'Brien, J. 2003. Journey through the kingdom of fungi (tree maintenance). *Arbor Age* 23:10. (Keywords: "journey" and "fungi")

Fungal Symbioses

Bonfante, P. 2003. Plants, mycorrhizal fungi and endobacteria: A dialog among cells and genomes. *The Biological Bulletin* 204:215. (Keywords: "mycorrhizal" and "endobacteria")

Redman, R.S., Sheehan, K.B., Stout, R.G., Rodriguez, R.J., Henson, J.M. 2002. Thermotolerance generated by plant/fungal symbiosis. *Science* 298:1581. (Keywords: "thermotolerance" and "symbiosis")

Sanders, W.B. 2001. Lichens: The interface between mycology and plant morphology. *BioScience* 51:1025. (Keywords: "lichens" and "interface")

Common Features of the Five Phyla in Kingdom Fungi					
Trait	Chytridiomycota	Zygomycota	Glomeromycota	Ascomycota	Basidiomycota
Swimming cells?		No	No		
Grade of evolution	coenomycete			dikaryomycete	
Dikaryotic stage?		No	Unknown		Yes
Septation	Aseptate		Aseptate	Simple perforate	
Characteristic asexual structure		Sporangium	Simple spore		Conidium
Characteristic sexual structure	Resting sporangium			Ascus	
Fruiting body	None				Basidioma
Main habitat		Terrestrial		Terrestrial	

CHAPTER

21

The Protists

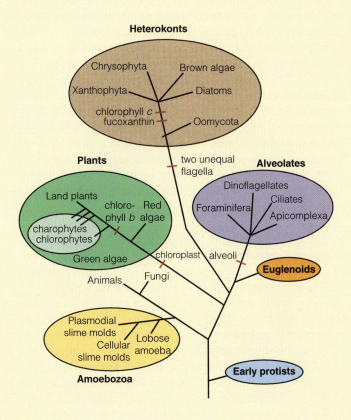

Heterokonts

Chrysophyta
Xanthophyta
Brown algae
Diatoms
chlorophyll *c*
fucoxanthin
Oomycota

Plants
two unequal
flagella
Alveolates

Land plants
chloro-
phyll *b*
Red
algae
Dinoflagellates
Foraminifera
Ciliates
Apicomplexa

charophytes
chlorophytes

Green algae
chloroplast
alveoli
Euglenoids

Animals
Fungi

Plasmodial
slime molds
Cellular
slime molds
Lobose
amoeba

Early protists

Amoebozoa

⟲ Visit us on the web at http://biology.brookscole.com/plantbio2
for additional resources, such as flashcards, tutorial quizzes,
InfoTrac exercises, further readings, and web links.

THE PROTISTS: THE FIRST EUKARYOTES

Relationships among the Eukaryotes Are Incompletely
Understood

■ *IN DEPTH:*
The Mysterious Origin of Chloroplasts

Photosynthetic Protists Are Commonly Called Algae

THE PROTISTS MOST CLOSELY RELATED TO ANIMALS AND FUNGI

ALVEOLATES

Dinoflagellates Cause Red Tides

EUGLENOIDS

HETEROKONTS

Oomycota Have a Great Impact on Humans

Diatoms Are Encased in Glass

■ *PLANTS, PEOPLE, AND THE ENVIRONMENT:*
The Kelp Forest Ecosystem

The Brown Algae Include the Kelps

THE PLANTS

Red Algae Are Adapted to Live at Great Depths

Green Algae Gave Rise to the Land Plants

THE ECOLOGICAL AND ECONOMIC IMPORTANCE OF ALGAE

Planktonic Algae Are at the Base of Aquatic Food Chains

■ *PLANTS, PEOPLE, AND THE ENVIRONMENT:*
Algal Blooms

Algae Help Build Tropical Reefs

Algae Serve as Medicine, Food, and Fertilizer

Algal Cell Walls Have Industrial Uses

ALGAL REPRODUCTION

Ulothrix *Typifies the Zygotic Life Cycle in Algae*

Diatoms Have a Gametic Life Cycle

Ectocarpus *Has a Sporic Life Cycle with Isomorphic*
Generations

Laminaria *Has a Sporic Life Cycle with Heteromorphic*
Generations

SUMMARY

1. The Linnaean kingdom Protista is not a monophyletic group but was created to contain all those eukaryotic organisms that were not plants, animals, or fungi. Modern evolutionary studies have discovered that these organisms represent the earliest diverging lineages of eukaryotes. We collectively call this artificial assemblage protists.

2. The protists are diverse. This chapter focuses on the photosynthetic protists, informally called algae. Although they are not closely related in their evolution, algae often have similar morphology, life cycles, and ecology. Collectively, they represent 20,000 to 30,000 species that are largely aquatic. They range from microscopic single cells to a visible tangle of filaments to large seaweeds, such as kelp, that are differentiated into stemlike and leaflike regions.

3. The protists consist of a number of well-supported lineages. The earliest lineage of protists to appear lacks mitochondria, and it includes a number of pathogens. Another early lineage includes the amoebas, slime molds, animals, and fungi. The remaining lineages all have at least some photosynthetic members.

4. The euglenoids typically are unicellular and can be photosynthetic. They may also have a unique organelle, called an eyespot, that orients them to light. They have no known sexual reproduction and move by means of flagella or by a unique inching motion. They lack cell walls, but they have strips of proteins and microtubules under the cell membrane.

5. The alveolates all have small sacs beneath their cell membrane. They include a lineage of ciliated protists, such as *Paramecium;* a lineage of shelled protists, the foraminifera; a lineage of parasitic or pathogenic organisms called the apicomplexa; and an algal group called the dinoflagellates. Dinoflagellates have unusual chloroplasts that have multiple membranes around them. These may have resulted from the incorporation of another eukaryote that already had chloroplasts. The dinoflagellates cause red tides.

6. The heterokonts are single-celled or multicellular organisms with two unequally sized flagella. Many members tend to have brown or gold photosynthetic pigments. Heterokonts include the water molds, egg fungi, and several lineages of algae. Two major groups are the diatoms, typically single-celled algae with silica cell walls that create vast deposits over time, and brown algae, which comprise the kelps and rockweeds, among other seaweeds, and are important sources of commercial products.

7. A protist clade that this chapter refers to as the plants consists of red and green algae together with the land plants, which are derived within this clade. The red algae are seaweeds that are similar morphologically to the brown algae but smaller in stature and possessing photosynthetic pigments that allow them to live at great depths. They are the sister group to the green plants.

8. Microscopic, floating forms of algae are ecologically important because they are at the base of aquatic food chains. Larger seaweeds and kelps are economically important because of cell wall materials such as agar, carrageenan, and alginates.

9. Algal life cycles include gametic, zygotic, and sporic types. In many cases, however, asexual reproduction (by cell division, fragmentation of filaments, or mitospores) is more common than sexual reproduction.

21.1 THE PROTISTS: THE FIRST EUKARYOTES

The kingdom Protista was created when biologists began to investigate the diversity of microscopic life. They realized that not all eukaryotic organisms could be easily categorized as plant, animal, or fungus. Protista became a catchall category for all eukaryotes that did not fit anywhere else. Modern biologists have realized from extensive evolutionary studies (initially based on morphology, cell structure, and biochemistry, but more recently on genetics) that these diverse organisms do not form a natural group as named. This is because the animal, plant, and fungus kingdoms are lineages derived from within the kingdom Protista (Fig. 21.1). A natural group must include an ancestor and all of its descendants.

As a result, biologists have largely abandoned the Protista, focusing instead on the various well-defined lineages that exist within it. Sometimes, however, it is useful to have a name for these organisms, even though they are not a clade. This chapter uses the common name **protist**. Protists are unicellular, colonial, or relatively simple multicellular organisms that only rarely exhibit specialized cell types, or organs such as leaves or stems. Protist cells are eukaryotic with a double-membrane–enclosed nucleus, double-stranded DNA, and specialized organelles, such as chloroplasts and mitochondria, in the cytoplasm.

Some protists, such as *Paramecium* (Fig. 21.2a), swim actively with beating hairs called cilia and hunt for food in their environment. Other protists, such as various groups of amoebas, crawl along slowly and engulf whatever food they come across. Still other protists, such as the foraminiferans, are covered by a shell and often remain stationary. Some create sticky nets to trap their food, and others eat detritus. These diverse protists (from many separate lineages) that have animal-like feeding strategies often are called *protozoa*.

Some protists are unlike any familiar organisms. Some euglenoids, for example, combine characteristics of animals and plants. Many euglenoids have chloroplasts and produce their own food, but they can also swim actively and consume food like an animal. Slime molds (Fig. 21.2b) can move through the forest like a small slug, and then produce sporangia like a fungus.

Many protists are photosynthetic. These all share the same photosynthetic pigment, chlorophyll *a*, located in the inner membranes of chloroplasts. However, the various lineages have different chloroplast structures, other forms of chlorophyll in addition to type *a,* and a wide variety of unique accessory photosynthetic pigments. Photosynthetic protists often are called algae, and their morphological and habitat diversity is large **(Fig. 21.2c).**

Relationships among the Eukaryotes Are Incompletely Understood

Determining the relationships between the various eukaryotic groups—the protists, plants, animals, and fungi—has been a difficult problem. Some biologists have even predicted that the phylogeny will never be known. Part of the difficulty is that the splits between many of the lineages of eukaryotes are ancient. Another difficulty is that apparently there have been events in which organisms (or parts of organisms) that are not closely related have joined to form new organisms. In addition, not many characters are shared by all members of this group, and often the shared characters yield different results when analyzed cladistically. Recent (and ongoing) phylogenetic work using the genes coding for tubulin, actin, and a few other proteins has given biologists new hope for solving the problem. There are now well-supported phylogenies for the larger lineages of protists, and we are beginning to fill in the missing pieces at finer scales.

Fossils of multicellular coiled filaments of green algae date back to 2.1 billion years ago **(Fig. 21.3).** Biochemical markers that are distinctive to eukaryotes have been found in oil as old as 2.7 billion years. Undoubtedly, the eukaryotes are even older, but fossil evidence is lacking. The currently accepted scenario for the origin of eukaryotes relies on the endosymbiotic hypothesis (see Chapter 18; see also "IN DEPTH: The Mysterious Origin of Chloroplasts" sidebar).

Figure 21.1 A cladogram showing phylogenetic relationships among some groups of protists.

Figure 21.2 Diversity among protists. (**a**) *Paramecium,* a protozoan. (**b**) *Physarum,* a slime mold. (**c**) The feathery red alga *Polysiphonia.*

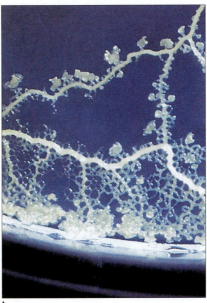

a

b

c

IN DEPTH:

The Mysterious Origin of Chloroplasts

Chloroplasts are so fundamental to plants and the various algal groups that the question of their origin has been the subject of intense research and debate. The subclade that includes the land plants also includes two major algal groups: green and red algae. Biologists agree that the land plants derived their chloroplasts directly from a green alga; however, the question of whether green and red algae both derived their chloroplasts from a single common ancestor is still debated (see Chapter 18).

Cladistic evidence based on several genes, including those coding for proteins involved in photosynthesis, suggests that chloroplasts appeared only once and were then passed down to the other plant groups. However, evidence from recent analyses of chloroplast genomes also is consistent with chloroplasts evolving multiple times. One complicating factor is that lateral gene transfer (movement of genes among distantly related groups) is known to have occurred and may confound efforts to reveal the evolutionary history of chloroplasts.

Another related question concerns how many times secondary endosymbiotic events occurred to produce the great diversity of algal groups present today (see Chapter 18). Solid evidence from comparisons of DNA indicate that all the "nonplant" algae, such as the brown algae, dinoflagellates, euglenoids, and diatoms, acquired their chloroplasts from plants, either green or red algae. Extra sets of membranes surrounding the chloroplasts of the nonplant algae, and in some cases vestigial nuclei, support the idea that secondary endosymbioses have occurred.

Some researchers argue that secondary endosymbioses are rare, and that a mere two events (one involving a green alga and one involving a red alga) are enough to explain the appearance of chloroplasts in all nonplant algal groups. Other researchers maintain that secondary endosymbioses are more common and have occurred as many as seven times.

In some cases, all involving dinoflagellates, there is good evidence for tertiary endosymbiotic events—an organism with a secondarily acquired chloroplast comes to live inside a different organism and functions as a chloroplast. These events are revealed by extra sets of membranes surrounding the chloroplasts.

How many secondary endosymbiotic events occurred to create the great diversity of photosynthetic protists and whether a single primary endosymbiotic event can ultimately account for all chloroplasts remain open questions and the subjects of active research.

Websites for further study:

Evolution Education Wiki:
http://wiki.cotch.net/wiki.phtml?title=Endosymbiotic_theory

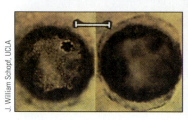

Figure 21.3 Microfossils of green algae in chert approximately 1 billion years old.

J. William Schopf, UCLA

Since their ancient origin, eukaryotes have diversified into a series of lineages. At the base of the eukaryotic tree is a group of protists, some of which lack mitochondria and live as parasites on other organisms (nonmitochondrial protists in Fig. 21.1). The pathogen that causes giardia is in this group.

The rest of the eukaryotes are divided into two major clades. One clade includes the animals, fungi, slime molds, and a small group of amoeboid organisms. There are no photosynthetic members of this clade, and most are motile (the major exception being fungi). The other clade includes a variety of protist groups, many of which are photosynthetic. One of these lineages gave rise to the land plants. Both photosynthetic and nonphotosynthetic protists in this clade may swim about actively, hunting for food or moving toward bright light.

Figure 21.4 Algal diversity. (**a**) A unicellular *Micrasteria.* (**b**) A colonial *Gonium.* (**c**) A colonial *Volvox.* (**d**) Filaments of *Spirogyra.* (**e**) *Caulerpa* with differentiation into rootlike, stemlike, and leaflike regions. All of these are green algae. Seaweeds are (**f**) the intertidal red alga *Porphyra* and (**g**) the brown alga *Fucus.*

a

255 μm

chloroplast

b

c

d

e

nuclei

f

g

holdfast

Photosynthetic Protists Are Commonly Called Algae

The photosynthetic protists in traditional taxonomy often were classified as about a dozen divisions representing 20,000 to 30,000 species. Although we now know these are not all closely related to each other, they are all still commonly called **algae** (singular, *alga;* Fig. 21.4). Algae have a

great variety of life histories, body forms, and ecological roles. The various groups of algae often are named for their distinctive colors (red, green, brown, golden, yellow-green), a result of the photosynthetic pigments they possess. Algae range in shape from unicellular to colonial (clusters of cells) to filamentous or sheetlike. A few are complex enough to exhibit marked differentiation into specialized tissues or even organs. Some species drift or swim in open water; others attach to the bottoms of streams or shallow seas, to surface soil particles, to tree trunks, to other algae, or to rocky cliffs battered by surf. Still others form symbiotic associations with fungi, higher plants, or animals. We commonly see algae growing on the sides of fish tanks, around leaking faucets, in garden pools, and as scum on the surface of ponds in summertime. Table 21.1 summarizes the habitats, morphology, pigments, and storage products of major algal groups.

Multicellular algae occur in wet habitats, where their simple bodies are supported by buoyancy in water. They have no water-conducting or stiffening cells strengthened by a secondary cell wall. Every cell in most algal bodies can carry on photosynthesis and obtain water and nutrients directly from its surroundings by diffusion. Such a simple body is called a **thallus.** Some unicellular algae occur in the most severe habitats on Earth: on snow with perpetually freezing temperatures; in hot springs with temperatures of 70°C; in extremely saline bodies of water, such as the Great Salt Lake in Utah; beneath 274 m of seawater; even surviving—within 1 km of ground zero—a 20-kiloton atomic bomb explosion in Nevada. Many algal species have a semiterrestrial existence by going dormant between wet seasons. The following sections consider five major lineages of protists.

21.2 THE PROTISTS MOST CLOSELY RELATED TO ANIMALS AND FUNGI

The subclade of the eukaryote tree that includes animals and fungi also includes a clade of protists called the Amoebozoa (Fig. 21.1). This clade consists of amoebas and slime molds. These are a diverse group with a unique set of traits that seem to combine aspects of fungi and animals. There are two groups of slime molds: the Myxomycota, or plasmodial slime molds, which contain thousands of nuclei with no membranes separating them; and the Acrasiomycota, or cellular slime molds, which are smaller, have fewer nuclei, and do have membranes between the nuclei.

Slime molds resemble animals in that they lack cell walls, engulf food, and have motile cells at some phase of the life cycle. In contrast, they resemble fungi and plants in that they form sporangia and nonmotile spores with cell walls.

21.3 ALVEOLATES

Modern molecular systematics has recognized several big groups that were not identified in earlier taxonomic schemes. In retrospect, there often were good nonmolecular indications that these clades existed. The **alveolates** are one example. Members of this group share a system of alveoli, or

Table 21.1 The Algae Organized by Evolutionary Groups and Traditional Taxonomic Names

Evolutionary Group	Taxonomic Name	No. Species	Key Characteristics
Euglenoids	Euglenophyta	500	Two equal, anterior flagella; no cell wall; chlorophylls a and b; unicellular; mainly in polluted freshwater
Alveolates			
Dinoflagellates	Pyrrophyta	1,000	Two unequal lateral flagella; no cell wall or wall of cellulose plates; chlorophylls a and b; unicellular; mainly in marine phytoplankton
Heterokonts			
Brown algae	Phaeophyta	1,500	Two unequal lateral flagella; cell wall of algin + cellulose; chlorophylls a and c; filamentous to complex large kelps; mainly in shallow, cool, marine water
Diatoms	Bacillariophyta	8,000	Usually no flagella; cell wall of silica + pectin; chlorophylls a and c; carbohydrates stored as oil; mainly unicellular and free-floating; prominent as freshwater or saltwater phytoplankton
Xanthophyta	Xanthophyta	400	Two flagella (various); cellulose cell wall; chlorophyll a (+ c in some), oil stored; mainly unicellular; mainly in freshwater
Chrysophyta	Chrysophyta	300	Two unequal anterior flagella; cellulose cell wall (+ silica in some); chlorophylls a and c; mainly unicellular and in freshwater
Plants			
Red algae	Rhodophyta	4,000	No flagella; cell wall cellulose + agar or carrageenan; chlorophylls a and d; phycobilin accessory pigments; mostly multicellular seaweeds in deep, warm, marine water
None	Green algae (Chlorophyta and Charophytes)	7,000	Two equal anterior flagella; cellulose cell wall; starch as a storage product; chlorophylls a and b; carotene accessory pigment; diverse morphology; mainly freshwater

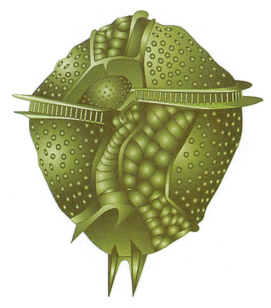

Figure 21.5 The dinoflagellate *Ceratocorys aultii.* Flagella, which ordinarily lie within the grooves or trail behind, are not shown.

tiny membrane-enclosed sacs lying beneath the plasma membrane. These provide structural support and can give rise to distinctive coverings, such as the plates on the surface of dinoflagellates (Fig. 21.5).

The alveolate clade is composed of four ecologically and economically significant lineages (Fig. 21.1). The **ciliates** include many of the actively swimming protists seen in freshwater bodies. Some, such as *Paramecium,* are commonly used in biology laboratories. These protists typically possess cilia and feed by engulfing their food in much the same manner as animals. For this reason, they often are called protozoa. The **foraminifera** are common and diverse protists with a hard shell. Their feeding strategies range from active predation to scavenging to creating sticky webs for trapping food. Their shells are common fossils and are important in dating geological strata and in oil exploration. The **apicomplexa** consist almost entirely of parasitic or pathogenic protists. This group includes *Plasmodium,* the organism that causes malaria, and *Toxoplasma,* the cause of toxoplasmosis. The apicomplexa recently have been found to contain vestigial plastids, and thus may be derived from a photosynthetic ancestor. The fourth lineage of alveolates is the photosynthetic dinoflagellates.

Dinoflagellates Cause Red Tides

The **dinoflagellates** are important members of the phytoplankton. Most species are unicellular, motile, and marine. Some species have plates on their exterior made of cellulose (Fig. 21.5), whereas others lack a cell wall entirely and are enclosed only by a thickened cell membrane. A few are colonial or filamentous. Usually, two flagella are present; both emerge from the same pore, but otherwise they are dif-

ferent. One is flat and ribbonlike and encircles the cell in a groove around the middle; it provides rotational movement. The second flagellum trails behind and provides forward movement.

Dinoflagellates contain chlorophylls *a* and *c* and a brown pigment called fucoxanthin, which gives the cells a greenbrown or orange-brown color. They have unusual chloroplasts that are surrounded by three or four membranes and may contain a remnant nucleus—evidence that the dinoflagellates likely acquired their chloroplasts from other eukaryotic algae.

Some forms produce a bioluminescence and contribute to the glow of water when it is disturbed at night by surf or in the wake of a ship. At certain times of year, along some coasts, the density of dinoflagellates in the phytoplankton multiplies, turning the water a reddish color. This effect is known as a *red tide* (see "PLANTS, PEOPLE, AND THE ENVIRONMENT: Algal Blooms" sidebar). Some dinoflagellates release toxins into the water, which can kill fish, marine mammals, and even humans if they are consumed in great enough quantity.

21.4 EUGLENOIDS

When a stagnant swimming pool or pond turns into a peagreen soup, the most likely cause is a bloom of euglenoids. These organisms have puzzled biologists because they combine characteristics of different lineages, and their relationship to other eukaryotes has been uncertain until the recent use of molecular characteristics.

Euglenoids are typically single-celled organisms (one small genus is colonial) that live primarily in freshwater, but they also can be found in salt or brackish water, and even in soil (Fig. 21.6). A few are parasitic. Euglenoids lack a cell

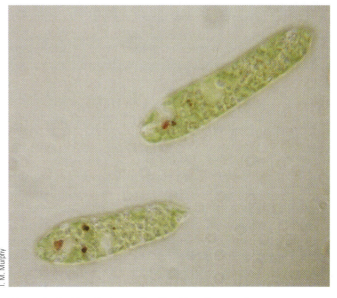

T. M. Murphy

Figure 21.6 Two *Euglena* cells with chloroplasts and eyespots.

wall, but they have flexible strips of proteins and microtubules under their cell membrane. They have two flagella, although in many forms only one emerges from the cell. They also can move by a unique inching motion.

About one third of the 1,000 species of euglenoids have chloroplasts. These contain chlorophylls *a* and *b* and carotenoids, like green plants. Most likely, they obtained this set of pigments through secondary endosymbiosis with a green alga. Three membranes surround euglenoid chloroplasts, supporting the secondary endosymbiosis hypothesis. Many euglenoids also contain a unique red or orange light-sensitive organelle called an **eyespot** that enables the organism to orient itself toward the light. The pigment in the eyespot (astaxanthin) is a carotenoid that has strong antioxidant properties. It is extracted and sold as a health supplement. The remaining two-thirds of euglenoids must find food in their environment. They ingest their food through a pocket-like cavity at one end of the cell.

Euglenoids have never been observed to reproduce sexually. How so many species formed among these asexual organisms remains a mystery, but it may be that euglenoids had some form of genetic recombination in the past. Euglenoids are important in the food chains of freshwater ecosystems, and they can be useful ecological indicators of water rich in organic matter.

21.5 HETEROKONTS

Another large clade recently identified in molecular studies is the **Heterokonta** (also sometimes called Stramenopiles or Chromista). The heterokonts include the **Oomycota** (the water molds and downy mildews), and several algal groups, all of which share a variant of chlorophyll called chlorophyll *c* and a brown accessory pigment called fucoxanthin. Recall that the dinoflagellates also have fucoxanthin and chlorophyll *c*. The heterokonts all have two unequally sized flagella. The heterokonts and alveolates share a common ancestor (Fig. 21.1). In addition to the Oomycota, heterokonts include two lines of golden algae, **Xanthophyta** and **Chrysophyta;** the diatoms; and the brown algae.

Oomycota Have a Great Impact on Humans

Oomycota include egg fungi, downy mildews, and water molds. They often have been classified as fungi in the past, which they resemble in having hyphae, producing spores, and lacking chlorophyll. However, they share cellulose cell walls, swimming spores, certain cellular details, and unique metabolic pathways with some algal groups.

Most Oomycota are benign decomposers that live in soil or freshwater habitats, but a few are pathogens of important crops. Downy mildews, for example, infect beans, grasses, and melons, among other plants. They are easily identified by the dense web of sporangia-bearing hyphae (sporangiophores) that makes the infected leaf appear to be covered with soft down **(Fig. 21.7).**

Figure 21.7 Grape leaves infected with downy mildew, *Plasmopora viticola.*

Downy mildew of grape, *Plasmopora viticola,* nearly destroyed French vineyards in the nineteenth century. Before this time, the disease had been restricted to North America, where it infected wild grapes. These wild species were seen to be valuable rootstocks for European grapes, so they were imported to France. Some carried spores of downy mildew, which spread to European plants and caused disease symptoms that were first noticed in 1878. The disease then spread rapidly. Investors began planting vineyards in countries outside France, believing the French wine industry was doomed. But the mycologist Alexis Millardet discovered that a mix of lime and copper sulfate, applied as a dust on leaves and stems, killed downy mildew. The French wine industry recovered, and opportunistic owners who had overplanted vineyards in other countries, such as Italy, went bankrupt. Millardet called his mixture Bordeaux mix, and it is still currently used.

Another Oomycota changed the history of Ireland. The potato blight of the 1840s was caused by *Phytophthora infestans* (the genus name literally means "plant destroyer"). The potato is a New World plant, originally restricted to the cold, rocky soils of Andean farms. Its value as a food crop on marginal farmland elsewhere was obvious, and it soon became the main crop in Ireland. One farm family could subsist on an acre of potatoes and a cow. An unusual series of warm, humid summers allowed the downy mildew to become epidemic, rotting the potatoes and killing the plants. A quarter of a million people died of starvation, and a million more immigrated to the United States, adding an important new ethnic component to U.S. culture. Decades later, mycologists

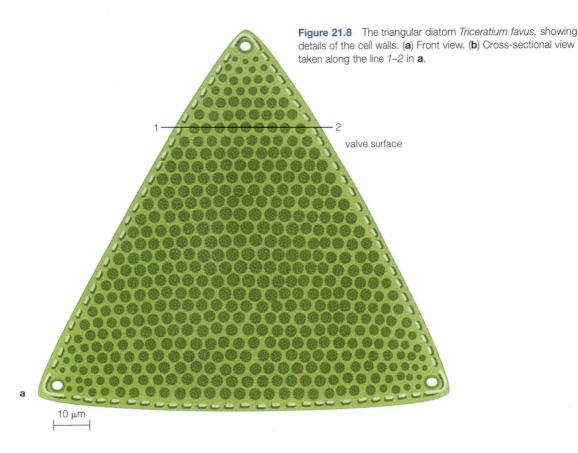

Figure 21.8 The triangular diatom *Triceratium favus*, showing details of the cell walls. (**a**) Front view. (**b**) Cross-sectional view taken along the line *1–2* in **a**.

1 ——————————— 2

valve surface

a

10 μm

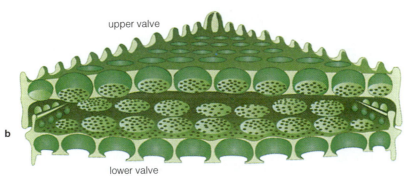

upper valve

b

lower valve

learned that potato blight could be controlled by spraying infected fields with poisons that kill the blight and by carefully disposing of infected potato tubers.

Diatoms Are Encased in Glass

Diatoms are important members of the **phytoplankton,** which are floating, photosynthetic, microscopic algae. Diatoms are unique in that they produce cell walls out of silica, the main component of glass. Viewed from above, their silica walls are exquisitely ornamented with perforations **(Fig. 21.8).** The silica is embedded in a pectin matrix. The shape of the cell can be circular, triangular, oval, diamond shaped, or even more elaborate (Fig. 21.14). In cross section, the cell wall has two parts (valves) that fit over each other like the halves of a Petri dish. Like other heterokonts, diatoms contain chlorophylls *a* and *c* and the accessory pig-

ment fucoxanthin. Fucoxanthin gives these cells their characteristic color, ranging from olive brown to golden brown. Diatoms store food reserves as oils. Some diatoms move along surfaces, even out of water, with a unique gliding motion, though cilia and flagella are absent.

Diatoms also exist as attached, stalked single cells or as filaments. Many stalked diatoms grow as **epiphytes** ("on other plants") on seaweeds and kelps. They do not parasitize the host plant but use it as a base to gain access to light near the water's surface. It is estimated that the productivity of the extra layer of epiphytes can be as great as that of their larger hosts. Thus, epiphytes add greatly to the first tier of the food chain (the transfer of energy in an ecosystem) in shallow water.

Diatoms may also become attached to nonliving surfaces. For example, the undersurface of icebergs is coated with diatoms (particularly in the genera *Navicula, Nitzschia,*

PLANTS, PEOPLE, AND THE ENVIRONMENT:
The Kelp Forest Ecosystem

The California coast is one of the richest areas in the world for kelp. There, giant kelp ("sequoia of the sea," *Macrocystis pyrifera*) dominates nearshore waters on the continental shelf. This is the largest kelp species in the world, and it forms stands so magnificent in size and so dense that they have been called kelp forests or kelp beds **(Fig. 1).**

The range of this species in the northern hemisphere extends from near San Francisco south to the midpoint of the Baja California peninsula. In the southern hemisphere, it is found along the Chilean coast, off scattered islands in the South Atlantic and Indian Oceans, and into the western Pacific all the way to Tasmania and New Zealand.

The species is well known because of its economic importance as a source of algin. Kelp boats cruise offshore, harvesting the blades and stipes that float on, or just below, the surface. The material is then processed onshore to extract and purify the alginates. Giant kelp grows so rapidly that the same bed may be reharvested every 6 weeks. Kelco Company, the largest California-based firm to harvest giant kelp commercially, has monitored kelp bed acreage for many decades, and the data show dynamic fluctuations over time. What has caused these changes, and why haven't kelp beds been stable? The answer seems to be that the environment has not been stable: There have been fluctuations in temperature, turbidity, sewage, sea urchins, and sea otters.

Giant kelp grows in water between 5 and 20 m deep; only the uppermost blades lie partly exposed to air when they float on the surface. The water in which giant kelp grows is relatively cold because local upwelling currents bring deep, cold water to the surface. This deep water also is rich in nutrients. Water temperatures off the California coast normally change little from month to month or year to year, staying in the range of 12 to 20°C. Experiments show that young kelp grow poorly when the water temperature is greater than 20°C and when dissolved nitrate becomes low.

Giant kelp's sensitivity to high temperatures and low nutrient levels means that it experiences declines during El Niño events. Every 3 to 8 years, changes in the pattern of wind and water circulation in the Pacific Ocean cause surface water temperatures to increase several degrees, which produces phytoplankton blooms that use up nitrate in the water. Phytoplankton blooms also reduce the amount of solar radiation that penetrates into water. Mature kelp requires light intensity of about half that of full sun to achieve light saturation and maximum growth. Normally, a depth of 20 m in clear water has light at this level, but even moderate turbidity in the water from phytoplankton or sewage means that optimal light penetrates to only 10 m.

Another factor that affects the vigor of kelp beds is sewage pollution. Sewage creates turbidity and muddy ooze on the nearshore ocean bottom. Young *Macrocystis* must attach to exposed rock. Ooze prevents the organisms from anchoring themselves and may also cover young kelps and kill them. Sewage also increases sea urchin populations; these bottom-dwelling invertebrates then graze more intensively on kelp. Their grazing may sever the upper body of the kelp from the holdfast, setting it free to be washed up as beach drift. Sea urchin populations used to be controlled by a natural predator, the sea otter. By the early twentieth century, this animal had been hunted nearly to extinction for its fur. There are now laws that protect sea otters, and their population numbers are increasing. Where sea otter populations are greatest, kelp forests are most vigorous; where sea otter populations are smallest, kelp forests have declined during this century. The resurgence of sea otters, therefore, pleases kelp harvesters. At the same time, shellfish harvesters are displeased because another favorite prey item of sea otters is abalone, a commercially valuable marine animal highly desired as seafood.

Figure 1 A kelp forest, viewed from underwater off the California coast.

J. W. Perry

Websites for further study:

Monterey Bay Aquarium—Kelp Forest Exhibit:
http://www.mbayaq.org/efc/efc_hp/hp_kelp_exhibit.asp

Oceanlight.com—Kelp Forest: http://www.oceanlight.com/html/kelp.html

Kelp Forest and Rocky Subtidal Habitats:
http://bonita.mbnms.nos.noaa.gov/sitechar/kelp.html

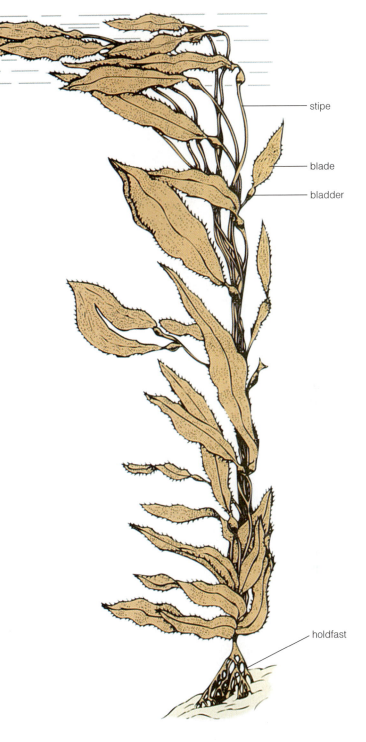

and *Podosira*). Diatoms are abundant enough to stain ice a yellow-brown color, and the base of the arctic food chain is enriched because of them. Diatoms also create *algal turfs*, which coat shallow rocks in quiet freshwater or marine habitats. These turfs have as high a daily productivity per square meter of surface as a tropical rain forest; therefore, they play an important role in the food chain of these habitats. The surfaces of salt marsh mudflats also are dominated by diatoms, which are grazed on by insects.

Diatoms have an extensive fossil record and are important indicator fossils for paleontologists and petroleum exploration. Their accumulated silica shells have created distinctive deposits (see later).

The Brown Algae Include the Kelps

The **brown algae** are almost exclusively marine; but in contrast to red algae, they are most diverse and abundant in cool, shallow waters. The simplest brown algae are filamentous and sheetlike; the most complex are kelps **(Fig. 21.9).** **Kelps** are large seaweeds with well-differentiated regions that resemble stems and leaves and with internally distinctive tissues and regions (see "PLANTS, PEOPLE, AND THE ENVIRONMENT: The Kelp Forest Ecosystem" sidebar). Some brown algae such as *Fucus* (rockweed) are the most common seaweeds on rocky shores. Brown algae provide important food supplements, medicines, and industrial chemicals.

Like other heterokonts, kelps have chlorophylls *a* and *c* and the pigment fucoxanthin. Carbohydrates, stored as mannitol or laminaran, accumulate as granules in the cytoplasm. Chloroplasts do not contain grana. Cell walls are made from cellulose and tough flexible polymers called alginates. Motile cells have two unequal flagella attached along the side of the cell.

Macrocystis is the largest kelp known, and it has one of the fastest growth rates of any multicellular alga. In the course of a single growing season, it grows from a single-celled zygote into a mature, giant, 60-m-long kelp, attached to a rocky bottom with much of the upper part floating on the surface. A mature *Macrocystis* (Fig. 21.9) consists of an anchoring **holdfast,** stemlike **stipes,** and numerous leaflike **blades** that arise all along the stipes. The base of each blade is inflated into a gas-filled bladder, which increases buoyancy.

Kelps are complex anatomically and morphologically. If the stipe is sectioned and examined under the microscope, several regions are apparent **(Fig. 21.10).** Cells in the outermost layer are protective; they also are meristematic and contain chloroplasts. To distinguish this unique tissue from the much simpler epidermis of land plants, it is called **meristoderm.** A broad region of cortex beneath the meristoderm is composed of parenchyma-like cells. Mucilage-secreting cells line canals through the cortex. Loosely packed filaments of cells fill the innermost part of the stipe, a region called the

medulla. Some cells in the transition zone between cortex and medulla function as sieve elements. They have sieve plates, form callose, and adjoin to one another to make continuous tubes, and mannitol moves through them at a rate resembling sugar movement in vascular land plants. The value of a photosynthate-conducting system in these large plants is easy to understand: The mass of floating fronds on the surface shades the lower part of the stipes and the holdfast so much that the shaded parts cannot produce enough carbohydrate to maintain themselves and must have additional amounts translocated to them. Because they live immersed in water, kelps contain no tissue that resembles xylem.

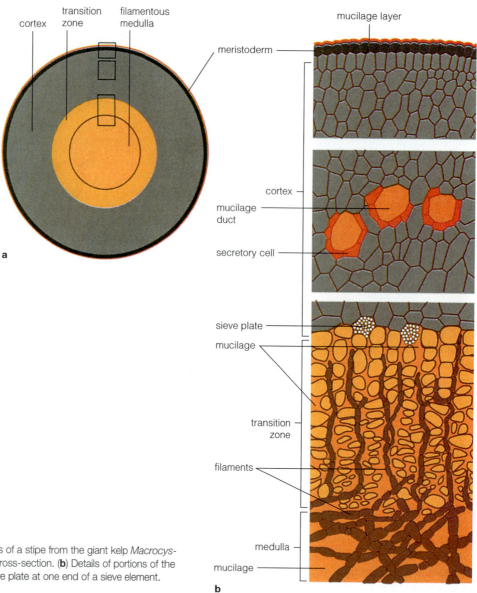

Figure 21.10 Anatomical details of a stipe from the giant kelp *Macrocystis pyrifera*. (**a**) An entire stipe in cross-section. (**b**) Details of portions of the cross section. (**c**) Details of a sieve plate at one end of a sieve element.

21.6 THE PLANTS

A great wealth of molecular data supports the idea that red algae, green algae, and land plants belong in the same clade. Biologists have long suspected—on the basis of cellular details, biochemistry, life cycles, and morphology—that green algae gave rise to land plants. Analyses of proteins and nucleic acids have provided overwhelming support for this suspicion and also showed that red algae belong in this clade. Now that we know these organisms form a clade, it seems likely that their common ancestor was the group originally associated with a photosynthetic prokaryote to form the chloroplast. This original endosymbiosis was passed on to all the descendants (see "IN DEPTH: The Mysterious Origin of Chloroplasts" sidebar).

Green algae are not a natural (monophyletic) group because they gave rise to land plants. To solve this problem,

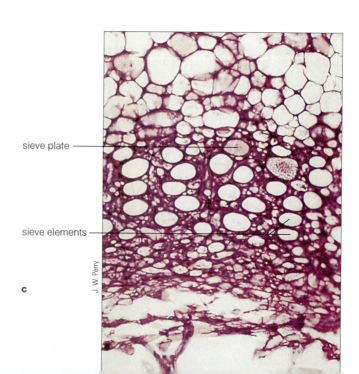

phylogenetic taxonomists simply include at least some green algae in the plant kingdom. Now that we know red algae are sister to this clade and share many characteristics with it (particularly chloroplasts derived from a primary endosymbiotic event), it makes sense to call them plants as well. In this system, the clade that includes plants and green algae can be called simply **green plants** (Fig. 21.1).

Red Algae Are Adapted to Live at Great Depths

Red algae are almost exclusively marine and are most abundant in warm water. They can grow to considerable depth. Most are multicellular and large enough to be called seaweeds. The simplest red algae are small, branched, delicate filaments (Figs. 21.2c and 21.4f). More complex forms (**seaweeds**) have a parenchyma-like tissue, forming a body with a holdfast that anchors the plant to a substrate, a stemlike stipe, and a leaflike blade. Unlike the brown algae, the stipes of red algae are never very long and the blades never have gas bladders. Some forms are encrusted with lime (calcium carbonate) and become parts of tropical reefs. Cell walls contain cellulose, sometimes augmented with agar or carrageenan.

The chloroplasts of red algae retain some characteristics of the prokaryotic cyanobacteria from which they were probably derived. Both typically contain only chlorophyll *a* and have similar accessory pigments called phycobilins. The red algae have a unique food storage molecule called floridean starch.

The accessory pigments of red algae (and cyanobacteria) allow them to grow at greater depths than other photosynthetic organisms. Both the quality and the quantity of light change with passage through water. Red light is completely absorbed in the upper layers, leaving a blue-green twilight to prevail farther down. Experiments have revealed that aquatic algae have adjusted their metabolism to the light at different depths. The action spectrum of photosynthesis for the green alga *Ulva taeniata*, which grows at shallow depths along rocky coasts, and the red alga *Myriogramme spectabilis*, which grows permanently submerged in deeper water, are different **(Fig. 21.11).** You can see that the most important wavelength of light for photosynthesis shifts with depth to center around the blue–green region, 440 to 580 nm. The phycobilins in red algae are able to trap light in this region of the spectrum, transferring the energy to chlorophyll. Eventually, at about 170 m in water of average clarity, the amount of light becomes so depleted that no alga—regardless of pigments—can scavenge enough light to support growth. The record depth for a red alga is 268 m in an unusually clear part of the Caribbean Sea. Algae in such low light grow slowly and are probably hundreds of years old.

Green Algae Gave Rise to the Land Plants

Green algae occur predominantly in freshwater habitats. They also exist in saltwater, on snow, in hot springs, on soil, and on the leaves and branches of terrestrial plants. All green plants (green algae and land plants) share a great number of characteristics that distinguish them as a clade. They have

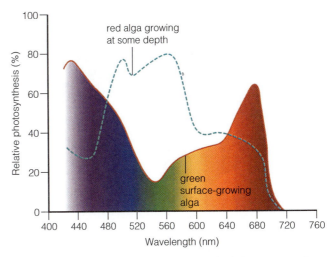

Figure 21.11 An action spectrum of photosynthesis in a green, surface-growing alga (*solid red line*) and a red alga growing at some depth (*dashed blue line*). The red alga can use light in the blue–green wavelengths (440–580 nm), which pass through the upper layers of water.

chlorophylls *a* and *b*, similar carotenoid accessory pigments, starch as a food storage molecule, and cell walls made primarily of cellulose. They also have a unique type of cell division during mitosis (the formation of a phragmoplast) and asymmetrically attached flagella, among other traits.

Before the appearance of land plants, ancient green algae apparently split into two lineages. One lineage, the chlorophytes, persists as a clade that is sister to the land plants in addition to the rest of the green algae. The **chlorophytes** include several monophyletic groups, including the Chlorophyceae and the Ulvophyceae. The other lineage, the **charophytes,** is the group from which land plants are derived.

CHLOROPHYCEAE The Chlorophyceae are distinguished by few morphological characters but many different molecules. This group includes *Chlamydomonas* **(Fig. 21.12).** Each cell has two flagella at the anterior end and a single, large, cup-shaped chloroplast. Most species have a red-colored carotene eyespot, which is capable of sensing light. The two flagella enter the cell at an angle, and they end in basal bodies, which are connected by a bridge of many microtubules. Forward movement results from flagellar motion resembling a swimmer's breaststroke—the flagella extend straight ahead, sweep backward without bending, and then fold and come back to the forward position.

Chlamydomonas-like cells may link together into permanent multicellular colonies; *Gonium* and *Volvox* are examples (Fig. 21.4b,c). *Volvox* is a hollow sphere of cells connected to each other by plasmodesmatal strands of cytoplasm. One colony may contain as many as 20,000 cells. The cells of a colony are clearly integrated because their flagella are synchronized and able to move the colony in a given direction.

ULVOPHYCEAE The Ulvophyceae, or sea lettuces, typically are small, green seaweeds composed of thalli growing attached to a substrate in shallow water. Species in the genus

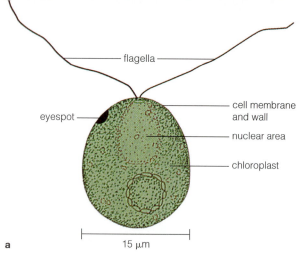

a

15 μm

flagellum (only one of
two flagella is shown)   nucleolus nucleus   chloroplast

H. Hoops

chloroplast   cell wall   pyrenoid   starch   mitochondrion
b

Figure 21.12 The unicellular green alga *Chlamydomonas*. (**a**) Diagram.
(**b**) Transmission electron micrograph. ×19,000.

Ulva, called sea lettuces, consist of an irregularly shaped, bright green, ruffled sheet or thallus two cell layers in thickness attached to a rocky substrate by a rudimentary holdfast. The holdfast often is persistent, whereas the thallus is

lost and regrown on an annual basis. Sea lettuces are consumed as a food in many places.

Some filamentous forms of Ulvophyceae have complex morphologies that resemble land plants. *Caulerpa*, for example, forms extensive mats on the bottom of shallow bays beneath the Mediterranean Sea (Fig. 21.4e). Each individual consists of a horizontal, creeping portion (which sends branching filaments down into the substrate) and feather-like or sheetlike blades that reach upward into the water. Thus, there are rootlike, stemlike, and leaflike regions. This plant was accidentally imported to the Mediterranean Sea from tropical oceans where it grew singly or in small colonies. Apparently, in its new environment, it is free from natural limits to growth, and therefore it has multiplied explosively, pushing out native algae. Even worse, *Caulerpa* produces toxins that are lethal for urchins and some fish, which are important elements of ecosystem food chains. In summary, *Caulerpa* has wreaked ecological havoc on the Mediterranean region. In 2000, *Caulerpa* was discovered off the coast of California.

CHAROPHYTES   Charophytes are a group of ancient green algae composed of a series of lineages that in Linnaean taxonomy were in the Charophyta. However, the full clade consists of a grade of many other lineages, one of which gave rise to land plants. Among living charophytes, *Coleochaete* has long been cited as the closest living relative of the land plants. It shares many characters with land plants. For example, like land plants, *Coleochaete* retains its egg and zygote on the parent plant, although there is no multicellular diploid phase. A recent study, however, based on molecular characters, showed that another alga, *Chara*, is actually more closely related to land plants than *Coleochaete*. *Chara*, or stonewort, is a complex filamentous charophyte with an anchoring region and a stemlike region differentiated into nodes and internodes (**Fig. 21.13**). *Chara* also possesses multicellular gametangia structures, in which the gametes are protected by a shell of sterile cells.

21.7 THE ECOLOGICAL AND ECONOMIC IMPORTANCE OF ALGAE

Algae are important in two basic but quite different ways. They are important to the entire biosphere because they are photosynthetic—that is, they create carbohydrates and release oxygen (and in so doing, form the base of aquatic food chains and build tropical reefs). They also are economically important to humans because they serve as food, fodder, and fertilizer and have many industrial and pharmaceutical uses.

Planktonic Algae Are at the Base of Aquatic Food Chains

Microscopic floating algae called **phytoplankton (Fig. 21.14)** occur in such great numbers over such a large surface of water that they are said to be "the grasses of the sea"; that is,

sex organs

branch

a

female gametangium

male gametangium

b

Figure 21.13 The stonewort *Chara* in the charophytes. (**a**) Part of a cluster of filaments, showing differentiation into nodes and internodes. (**b**) The female gametangium. Note the shell of sterile, helical cells surrounding the gametangium, which protects the gametes within.

J. W. Perry

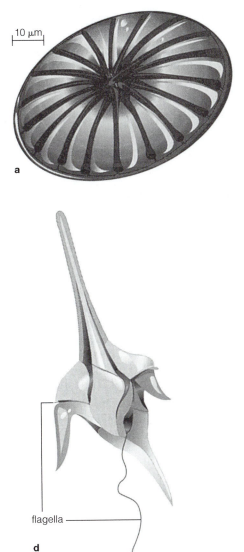

10 μm

a

flagella

d

Figure 21.14 Examples of phytoplankton. Diatoms include (**a**) *Asteromphalos elegans,* (**b**) *Asterionella formosa,* and (**c**) *Biddulphia biddulphia.* (**d**) The dinoflagellate *Ceratium*.

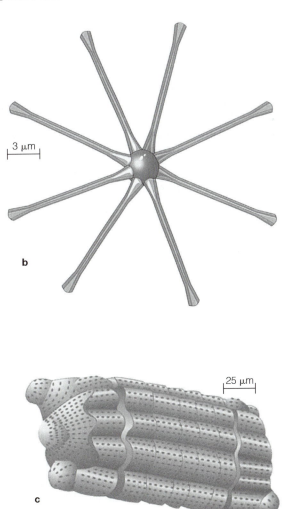

3 μm

b

25 μm

c

Phytoplankton can reproduce extremely quickly, and sometimes their cells become so dense they color the water. Which alga is causing one of these algal blooms often can indicate which environmental factor is responsible. Freshwater often turns a bright green when euglenoids bloom. The Red Sea was named for the frequent recurrence of dinoflagellate blooms. On U.S. coasts, marine dinoflagellate populations periodically bloom, causing red tides.

Algal blooms typically are beneficial because they increase the food supply at the base of the food chain. However, some blooms can adversely affect aquatic life. Sometimes algal blooms can deplete oxygen from the water, causing noxious odors and death of some aquatic life. Certain algae possess sharp spines that lodge in the gills of fish, causing death or disease. Other algae, such as some of the red tide dinoflagellates (*Gymnodinium, Dinophysis, Pseudonitzschia,* and *Alexandrium*), can produce potent, water-soluble toxins that affect vertebrate nervous systems. The toxins are similar to curare poison and are 10 times more toxic than cyanide. Massive fish kills can result from these red tides. Invertebrates such as shellfish can accumulate the toxin without being affected, but humans or other vertebrates that eat tainted shellfish are poisoned.

An illness in humans called *ciguatera poisoning* comes from eating poisoned fish, and *paralytic shellfish poisoning* comes from eating shellfish. The effect of the toxin can be acute, causing respiratory failure and cardiac arrest within 12 hours of consumption. Typically, however, the symptoms are less severe and include nausea, diarrhea, itching, headache, muscle pain, and unusual sensory phenomena such as numbness and the reversal of the sensations of hot and cold. These sensory phenomena can continue for many months after poisoning.

Red tides often are reported along the warm coastal waters of southwest India, southern China, southwest Africa, the eastern United States, the Gulf of Mexico, southern California, Peru, and Japan.

Dinoflagellates are not the only phytoplankton elements to produce dangerous blooms. Blooms caused by the golden alga *Prymnesium parvum* can cause massive fish kills in fresh and brackish water, negatively affecting salmon and trout fisheries. Cyanobacteria, such as *Anabaena, Gloeotrichia,* and *Nostoc,* may also undergo blooms in freshwater. Some of these have poisoned large numbers of wild mammals and birds, as well as domestic animals that drink contaminated water.

Websites for further study:

The Harmful Algae Page: http://www.whoi.edu/redtide/

they produce the vegetation that feeds grazing animals, which in turn feed predators and, ultimately, humans. Phytoplankton are at the base of aquatic, and especially oceanic, food chains. Phytoplankton are unicellular. They must stay close to the surface and avoid sinking below the photosynthetic zone by swimming, storing buoyant oil droplets, altering their ionic balance, and developing fine projections that extend out from the cell wall. The average life span of a given cell is probably measured in hours or days; death is caused by the organism eventually sinking below the photosynthetic zone and ultimately coming to lie on the ocean bottom.

The density of phytoplankton is normally low in the open ocean, perhaps a few thousand cells per liter, but the oceanic expanse is enormous. As a result, on a global scale, phytoplankton produce 3.26 quintillion kcal (3.26×10^{18}) photosynthate each year, which is four times that produced by the earth's croplands. This production of food energy would be even greater if such limiting factors as temperature and mineral nutrients were ameliorated. Deep oceans and freshwater lakes tend to be cool and low in nutrients. Continental edges and shallow lakes are warmer, have more mixing of water through all depths, and have more nutrients; there, productivity by plankton is greater. If nitrogen or phosphorus-rich wastewater is added to such areas—or if water temperature increases for some natural reason—a tremendous increase in the density of phytoplankton called a **bloom** occurs (see "PLANTS, PEOPLE, AND THE ENVIRONMENT: Algal Blooms" sidebar).

Algae Help Build Tropical Reefs

Reefs, which form in shallow tropical seas (most often near islands or on continental shelves), provide an important habitat for an incredible variety of marine life. Partly composed of the stony remains of coralline animals, reefs also are constructed by algae. Certain red and green **coralline algae** create a carbonate exoskeleton resembling coral around their filaments. This exoskeleton becomes a physical part of the reef after the algae die. Coralline algae also help build the reef by connecting the corals' exoskeletons. More than half of the bulk of many reefs is from algae.

In addition, a few species of algae grow symbiotically within the tissues of coralline animals. The tissue of these invertebrate reef-forming animals is photosynthetic because it contains these symbiotic algal cells. The alga, usually the dinoflagellate *Symbiodinium microadriaticum*, has a photosynthetic rate 10 times greater than phytoplankton, probably because it is protected and nourished by the animal cytoplasm around it. The relationship between corals and algae is mutualistic—that is, both organisms benefit. The alga produces sugar and oxygen for the animal, exporting more than 90% of the sugar into the animal's tissue and retaining only a small balance for its own growth and maintenance. In exchange, the cells of the coral contribute carbon dioxide, nitrogen, and mineral nutrients to the alga. Together, both prosper.

Algae Serve as Medicine, Food, and Fertilizer

Seaweeds are marine forms of red, brown, and green algae of moderate size. They usually grow in the rocky intertidal zone, alternately exposed by low tides and covered by high tides. Seaweeds are an important part of the human diet and medicine chest in several parts of the world. In East Asia, seaweed harvesting has been known for 5,000 years. Shen Nong, the legendary Chinese "father of medicine" prescribed seaweed for a variety of ailments in texts dating back to 3,600 years ago. Much later, Confucius also praised their curative value. A million metric tons a year of the brown seaweed *Laminaria* is harvested off the China coast as a source of iodine, which is added in trace amounts to diet for the prevention of goiter (an enlargement of the thyroid gland).

For centuries, the Japanese have used algae as a tasty supplement to their rice diet. The demand for nori, the red alga *Porphyra*, has grown to such an extent that it is cultivated **(Fig. 21.15)**. The Polynesians in Hawaii used and named at least 75 species of *limu* (seaweed) as food sources. Some rare species were cultivated only in marine fishponds belonging to nobility. Dulce, the red seaweed *Palmaria palmate*, has been used as a food for 1,200 years in the British Isles. The Irish discovered that when boiled with milk, small quantities of Irish moss—another red alga, *Chondrus crispus*— would produce a jelly dessert that the French later called *blancmange*.

Nevertheless, with some exceptions, algae do not have much nutritive value; in fact, their major constituents are largely indigestible. Today, algae are used more as condiments, garnishes, or desserts than as staples—much as we use lettuce, watercress, celery, or herbs.

Seaweeds do contain large amounts of potassium, nitrogen, phosphorus, and other minerals characteristic of good

Figure 21.15 Harvesting nori, red alga *Porphyra tenera*, in Sendai Prefecture, Honshu Island, Japan. Hibi nets sit about 30 cm above mean low tide in September, at the beginning of the growing season for nori. Harvest time is January.

fertilizer or cattle feed supplements. In historic times, Native Americans and the Scotch-Irish used Irish moss as a fertilizer to build up poor soils for such crops as corn and potatoes. As a fertilizer, seaweed compares favorably with barnyard manure: It enhances germination, increases the uptake of nutrients in plants, and seems to impart a degree of resistance to frost, pathogens, and insects.

Algal Cell Walls Have Industrial Uses

Physical and chemical characteristics of the cell walls of some algae lead to products with many industrial, pharmaceutical, and dietary applications. It is through these cell wall components—primarily diatomite, agar, carrageenan, and algin—that algae have their greatest direct economic value.

Diatomite (or diatomaceous earth) is a sedimentary rock composed of fossilized diatom cell walls. Recall that diatoms are important members of the phytoplankton. As they flourish and die, their empty walls sink and accumulate as bottom sediments. One of the richest deposits of diatomite is a 300-m-thick layer near Lompoc, California. It formed this way about 15 million years ago beneath a warm, shallow sea. In more recent geologic time, the Lompoc area was uplifted above sea level, and the diatomite deposit was revealed by erosion. Several companies mine the diatomite for industrial and pharmaceutical use. Diatomite makes a superior filter or clarifying material, both because the microscopic wall pores create a large surface area (230 g contain the area of a football field) and because the rigid walls are incompressible. It is used in laboratory and swimming pool filters. Diatomite is inert, and it can be added to many materials to provide bulk, improve flow, and increase stability. In these ways diatomite is used in cement, stucco, plaster, grouting, dental impressions, paper, asphalt, paint, and pesticides. Diatomite also is used as an abrasive.

Agar (or agar-agar) is a polysaccharide analogous to starch or cellulose but chemically different. With cellulose, it is a component of the walls of certain red algae, mainly *Gelidium* and *Gracilaria.* Currently, agar is a common medium on which bacteria are grown. This use was discovered a century ago by a physician's wife, Frau Hesse, who used agar to thicken her jam. The noted microbiologist Robert Koch presented Hesse's idea to the world in his scientific writings during the late nineteenth century. Agarose, purified from agar, is used for gel electrophoresis, a valuable technique of modern molecular biology.

Seaweeds containing agar are commercially harvested by divers who descend 3 to 12 m into warm nearshore waters off Australia, California, China, Japan, Mexico, South America, and the southeastern United States. Some red algae are cultivated for agar production using rocks dropped into shallow bays or ropes suspended from floats. Weeds are removed, and at times solid fertilizer is added to the water to increase production. Agar-producing algae have been cultivated for more than 300 years in Japan.

In addition to its bacteriological use, agar is important in the bakery trade. When added to icing, it retards drying in open air or melting in cellophane packages. Agar is made into a gelatin for consumption in parts of Asia. Because agar is virtually indigestible, it also is used as a bulk laxative.

Carrageenan is another polysaccharide that accompanies cellulose in the walls of red algae, mainly Irish moss. The substance takes its name from the town of Carragheen, County Cork, along the south shore of Ireland, where its properties were first documented. Carrageenan reacts with the proteins in milk to make a stable, creamy, thick solution or gel. Consequently, it is used commercially in ice cream, whipped cream, fruit syrups, chocolate milk, custard, evaporated milk, bread, and even macaroni. It is added to dietetic, low-calorie foods to provide the appropriate "mouth feel." Carrageenan also is used in toothpaste, pharmaceutical jellies, and lotions of many sorts. Irish moss is commercially harvested in the United States off the shore of Maine.

Brown algae have a commercially valuable compound called **algin,** a long-chain polymer made up of repeating organic acid units that is the principal cell wall component. Algin constitutes up to 40% of the cell wall by weight. Water is strongly adsorbed by algin, creating a thick solution. When added to 1 L water, one tablespoon of powdered algin increases the viscosity to that of honey. In nature, algin may be valuable to intertidal algae during low tide because of its ability to retain water. Commercially, it is used as an additive to beer, water-based paints, textile sizing, ceramic glaze, syrup, toothpaste, and hand lotion. Hundreds of algal species contain algin, but only a few are commercially harvested: *Macrocystis pyrifera* along the California coast; species of *Ascophyllum, Fucus,* and *Laminaria* off Maritime Canada, the northeastern United States, England, and the China coast; and *Durvillea* in Australian waters.

21.8   ALGAL REPRODUCTION

The life cycles of many algae are still unknown, but the three basic life cycles (zygotic, gametic, and sporic) have all been documented to exist among algal species. Each life cycle has sexual and asexual portions. Asexual reproduction occurs more often than sexual reproduction; therefore, it is more commonly seen by researchers. Typical methods of asexual reproduction are cell division (for single-celled algae) and fragmentation (the splitting apart of a filament). Asexual reproduction also can be achieved by the formation, liberation, and germination of motile or nonmotile spores produced in sporangia. In many algae, sporangia show little, if any, difference in appearance from ordinary vegetative cells, except that they may be larger. The parent nucleus divides by mitosis several times, and each resulting nucleus accumulates cytoplasm about itself and secretes a surrounding wall, forming a spore.

Ulothrix Typifies the Zygotic Life Cycle in Algae

The filamentous green alga *Ulothrix* has a **zygotic life cycle,** meaning that the only diploid phase of the life cycle is as the single-celled zygote **(Fig. 21.16).** The nucleus of each *Ulothrix* cell is haploid. When sexual reproduction begins, some of the nuclei will divide by repeated mitotic divisions, producing many motile gametes inside the wall of the parent cell. In this case, the parent cell is a gametangium. The gametes of *Ulothrix* look very much like individual *Chlamydomonas* cells: They have two anterior flagella, and they swim about in water. If a *Ulothrix* gamete approaches another suitable gamete, the two will pair and fuse, producing a diploid zygote cell with four flagella. Although all *Ulothrix* gametes are isogametes, meaning they look alike under the microscope, there must be genetic differences recognizable by *Ulothrix* that make fusion possible among some, but not all, gametes—that is, there appear to be two mating types of gametes, known as plus (+) and minus (−). Because a given *Ulothrix* individual produces only plus or minus gametes, gametes from the same individual will not mate, whereas gametes from different individuals will be able to fuse.

The zygote resulting from gametic fusion becomes spherical, loses its flagella, and enters a resting stage during which it is resistant to such environmental extremes as exposure to dry air. Once growing conditions return, the zygote becomes metabolically active and divides by meiosis to produce plus and minus meiospores, each of which is capable of producing a new individual by itself. The meiospores are dispersed, and each can germinate, divide by mitosis, and produce a plus or minus haploid plant.

When *Ulothrix* begins to reproduce asexually, a vegetative cell becomes a sporangium. Eventually, 16 to 64 pearshaped mitospores (spores produced by mitosis) are released; each has 4 flagella. After a period of activity, these motile mitospores settle to the bottom of a pond, lose their flagella, and begin to produce a new plant by mitosis. Some mitospores may enter a dormant phase and become resistant to environmental stress during that time.

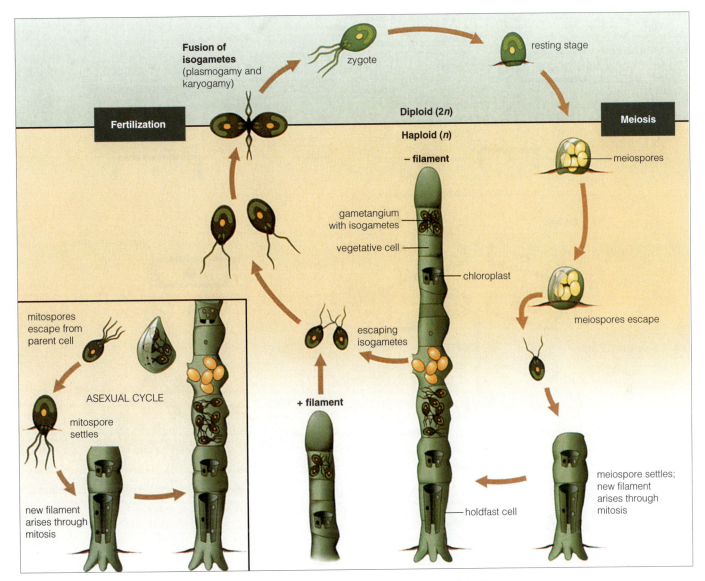

Figure 21.16 The zygotic life cycle of the filamentous green alga *Ulothrix*. Algal filaments are haploid. The nucleus of some cells can divide and produce mitospores (*inset*). Each mitospore settles to the bottom of the pond and produces a new haploid filament. Other cells, however, produce plus (+) or minus (−) gametes, which fuse to produce a resting zygote. The zygote eventually undergoes meiosis and produces meiospores, which germinate and grow to become haploid algal filaments.

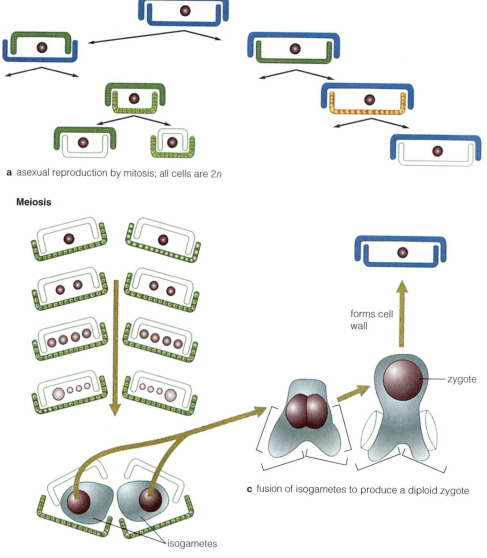

a asexual reproduction by mitosis; all cells are 2*n*

Meiosis

forms cell wall

zygote

c fusion of isogametes to produce a diploid zygote

isogametes

b sexual reproduction by meiosis, leading to isogametes

Figure 21.17 The gametic life cycle of a diatom. **(a)** Repeated asexual cell division results in smaller diploid cells because one of the progeny cells is formed within the wall of the smaller half of the parent cell wall. **(b)** When cells become critically small, meiosis occurs. The products of meiosis are four haploid nuclei within the old cell's membrane and wall. Three of the haploid nuclei degenerate and the single remaining one becomes the nucleus of a gamete. **(c)** The fusion of two gametes forms a diploid zygote.

Diatoms Have a Gametic Life Cycle

Diatom cells are diploid. The nucleus undergoes meiosis, producing four haploid nuclei **(Fig. 21.17).** Only one or two of the nuclei may survive and become gametes. The parental cell has served as a gametangium. If two diatom gametangia are near each other, they open, and the two gametes emerge and fuse, forming a diploid zygote. The zygote increases greatly in size and may rest for a short time. It then secretes a normal, two-part silicate wall around itself and becomes a vegetative diploid diatom.

Diatoms reproduce asexually by cell division. Recall that each cell has two wall segments that fit together like the top and bottom of a Petri dish. After division, a new wall forms within the old one. This means that the progeny cell that inherits the smaller wall segment will make a new wall that is smaller yet. As this pattern continues from generation to generation, one line of cells becomes ever smaller. Eventually, a critically small size is reached and division stops. The cell must then reproduce sexually (see earlier).

The common intertidal rockweed *Fucus* and many other brown algae also have a **gametic life cycle,** meaning that the only haploid phase of the life cycle is as the single-celled gametes.

Figure 21.18 The sporic life cycle of the filamentous brown alga *Ectocarpus*, which has isomerous generations. (**a**) The gametophyte generation releases plus (+) and minus (−) gametes. These fuse in pairs to form diploid zygotes. Each zygote produces a sporophyte plant. A sporophyte can generate asexual mitospores capable of (**b**) producing another sporophyte plant or (**c,d**) meiospores that will germinate to form the gametophyte generation once again.

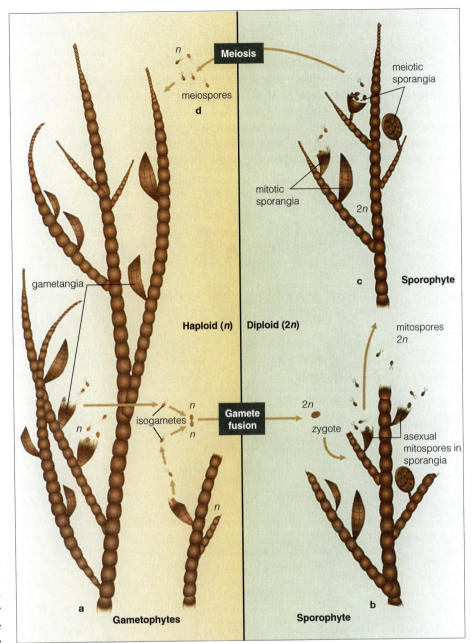

Ectocarpus Has a Sporic Life Cycle with Isomorphic Generations

Ectocarpus is a filamentous brown alga commonly found in cool coastal saltwater. This alga has a **sporic life cycle**, meaning that there are multicellular gametophytes and sporophytes. The life cycle begins with the haploid phase **(Fig. 21.18a)**, which has curved clusters of gametangia on side branches. Mitosis within each gametangium produces many gametes, which swim when released using two unequal flagella that arise from the middle of the cells. The gametes all look alike, even though they represent plus and minus mating types (they are isogametes). Fusion of a plus isogamete and a minus isogamete in open water yields a diploid zygote.

The zygote settles to the bottom, germinates, and divides by mitosis to produce a diploid organism **(Fig. 21.18b,c).** The diploid sporophyte looks identical to the haploid gametophyte. The only difference is the chromosome number within every nucleus. Seaweeds such as *Ectocarpus* that have identical-looking gametophytes and sporophytes are said to have **isomorphic generations.** The sporophyte produces two kinds of reproductive cells.

In one reproductive process, clusters of cells on side branches (looking identical to the gametangia on gametophytes but called asexual sporangia because of their function) liberate motile cells. Although the cells look like gametes, they do not pair and produce zygotes. Instead, each is capable of producing a new individual by itself—that is, each is a diploid mitospore. The individual that each mitospore produces is another sporophyte, identical to the parent sporophyte. A second reproductive process produces a kind of sporangium that is spherical. Meiosis occurs within this sporangium, producing many meiospores **(Fig. 21.18d).** The meiospores, when released, look just like gametes and mitospores. Each one can settle down to the bottom, germinate, and produce a gametophyte, completing the life cycle.

Laminaria Has a Sporic Life Cycle with Heteromorphic Generations

Plants in which the gametophyte and sporophyte are not identical have **heteromorphic generations.** Many groups of algae exhibit this kind of life cycle, but it is most highly developed in the brown algae, particularly in the kelps. The kelp *Laminaria* has a sporophyte generation with a well-developed holdfast and a long unbranched stipe with a cluster of narrow blades extending from the end **(Fig. 21.19a).** Sporangia usually occur in groups just below the meristo-

derm on a blade **(Fig. 21.19b).** Each sporangium is a single cell within which meiosis occurs (followed by one to several rounds of mitosis), producing 8 to 64 meiospores. The meiospores are released, swim about, settle to the bottom, and produce gametophytes.

The gametophytes are small, branched filaments, quite unlike the sporophytes. Some of the gametophytes will produce male gametes **(Fig. 21.19c);** others will produce female gametes **(Fig. 21.19d).** The cells at the tips of some filaments on male gametophytes enlarge and function as **antheridia**

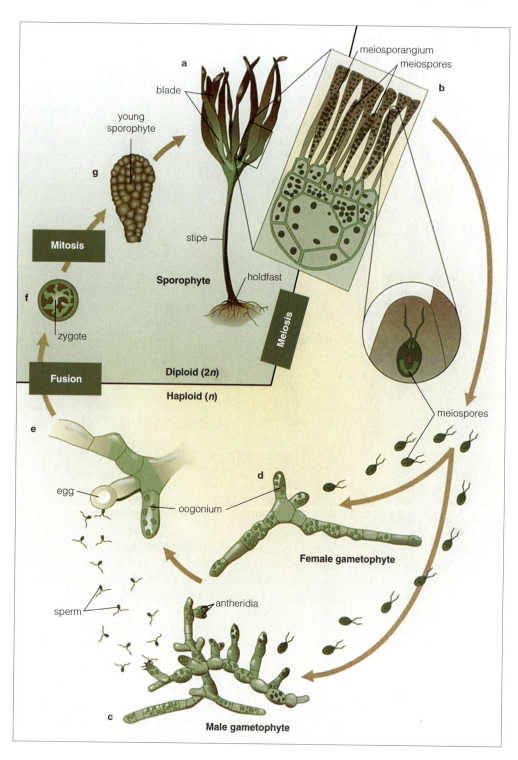

Figure 21.19 The sporic life cycle of the kelp *Laminaria*, which has heteromorphic generations. (**a**) The sporophyte plant. (**b**) Meiosis occurs in sporangia in the blade, producing male and female meiospores. Each meiospore will produce either (**c**) a male or (**d**) female gametophyte individual. The male gametophytes produce antheridia and sperm; the females produce oogonia and eggs. (**e**) The fusion of gametes creates (**f**) a diploid zygote that then forms (**g**) a young sporophyte.

(male gametangia). Their nuclei undergo mitosis, producing many motile sperm. Other cells on filaments of female gametophytes enlarge and function as **oogonia** (female gametangia); their nuclei undergo mitosis, producing one to several large **eggs.** The eggs are extruded nearly out of the oogonium, but they remain attached. A **sperm cell** will fuse with an egg cell, producing first a zygote and then a young sporophyte **(Fig. 21.19e–g).** At this stage, the sporophyte separates from the gametophyte, is carried passively by currents to the bottom, and begins to develop into a mature *Laminaria* sporophyte. The life cycle is completed.

KEY TERMS

agar	green algae
algae	green plants
algin	Heterokonta
alveolates	heteromorphic generations
antheridium	holdfast
apicomplexa	isomorphic generations
blade	kelp
brown algae	medulla
carrageenan	meristoderm
charophytes	oogonium
Chlorophyceae	Oomycota
chlorophytes	phytoplankton
Chrysophyta	Protista
ciliates	protists
diatomite	red algae
diatoms	seaweed
dinoflagellates	sperm cell
egg	sporic life cycle
epiphyte	stipe
euglenoids	thallus
eyespot	Ulvophyceae
foraminifera	Xanthophyta
gametic life cycle	zygotic life cycle

SUMMARY

1. The traditional (Linnaean) kingdom Protista includes nonphotosynthetic organisms and photosynthetic organisms. All organisms are eukaryotic, most are microscopic, and most have simple bodies. Ancestors of the animals, fungi, and land plants are all protists; consequently, the group is not monophyletic.

2. Molecular studies have shown the relationships among the various protist groups and the animals, fungi, and land plants. The photosynthetic protists are the most important for understanding land plant evolution. These are commonly called algae.

3. "Algae" is a common name for a diverse group of about 20,000 to 30,000 species of photosynthetic organisms. Algae can be generally characterized as follows: Their gametangia are single cells, lacking any protective layer of sterile cells around the gametes; they lack an embryo stage; they do not have an epidermis with cuticle and stomata; they are more often aquatic than terrestrial; and their bodies usually are simple and relatively undifferentiated. Body shape includes single-celled forms, colonies, filaments, sheets, and three-dimensional packages of cells. The various groups of algae differ in cell wall construction, number and placement of flagella, types of chlorophylls and pigments, morphology, and habitat ranges.

4. The protists contain a number of well-defined lineages. One clade includes the slime molds, amoebas, animals, and fungi. The four other major clades all include at least some photosynthetic members.

5. Euglenoids are a common algal group that also includes many nonphotosynthetic species. Even the photosynthetic euglenoids can engulf food in the manner of an animal. They are motile (by flagella), may have a unique organ called an eyespot that orients them toward light, and have never been observed to reproduce sexually.

6. Alveolates are a large clade characterized by minute sacs beneath their cell membranes. There are four major subclades: ciliates; foraminifera; apicomplexa; and an algal group, the dinoflagellates. Dinoflagellates have chloroplasts surrounded by multiple membranes that may have been derived from multiple endosymbiotic events. Dinoflagellates cause red tides.

7. Heterokonts are a large protist clade with many photosynthetic members, some of which are multicellular. All heterokonts are characterized by two unequally sized flagella. One important nonphotosynthetic group of heterokonts is the Oomycota, the water molds, which have had a great impact on people as plant pathogens. Diatoms are single-celled, have the brown photosynthetic pigment fucoxanthin in addition to chlorophyll, and have silica (glass) cell walls. The brown algae are a group of marine seaweeds that also contain fucoxanthin. Their cell walls are impregnated with algin, an elastic polymer.

8. Red and green algae comprise a group of algal protists that are particularly important because they gave rise to land plants. Red algae are seaweeds that are capable of living at great depths in the ocean because of their red phycobilin photosynthetic pigments. Green algae primarily occupy freshwater. They comprise two groups: the diverse chlorophytes and the charophytes, which are most closely related to land plants. All green algae contain chlorophylls *a* and *b*, carotenoids, starch as a storage product, and cellulose cell walls. The fossil record of green algae extends back to nearly 1 billion years.

9. Algae are ecologically important in aquatic ecosystems by being at the base of the food chain. Most of their productivity comes from phytoplankton. In polluted or warm, shal-

low water, phytoplankton can become dense enough to color the water, poison vertebrates, and create noxious odors. Algae also are important as reef builders, both as coralline green and red algae and as symbionts with coral animals.

10. Seaweeds have some economic importance to humans as medicine, food, and fertilizer. However, their major value comes in the physical and chemical properties of the cell walls of diatoms (diatomite), red algae (agar, carrageenan), and brown algae (algin).

11. Algae can reproduce sexually by zygotic, gametic, and sporic life cycles. *Ulothrix* is an example of an alga with a zygotic life cycle. Diatoms have a gametic life cycle. *Ectocarpus* and *Laminaria* have sporic life cycles, with multicellular sporophyte and gametophyte generations.

Questions

1. Why is the traditional kingdom Protista no longer considered a valid evolutionary group?

2. Characterize the group of organisms defined as protists in this chapter.

3. What evidence do biologists have to support their hypothesis that land plants evolved from green algae?

4. Make up your own short dichotomous key that separates members of the following algal groups: euglenoids, dinoflagellates, brown algae, red algae, and green algae.

5. How is it possible for microscopic algae to be important enough to fuel oceanic food chains, which culminate in large carnivorous fish and aquatic mammals such as whales? Why is it possible to say that reefs are built by algae as much as they are built by coral animals?

6. Some brown algae are complex marine organisms called kelps. Describe their specialized body regions, their unusual anatomy, and their sporic life cycle.

7. What impact have Oomycota had on humans?

InfoTrac® College Edition

http://infotrac.thomsonlearning.com

Algae

Karol, K.G., McCourt, R.M., Cimino, M.T., Delwiche, C.F. 2001. The closest living relatives of land plants. *Science* 294:2351. (Keywords: "Karol" and "relatives")

Russell, D. 1998. Underwater epidemic (spread of fish diseases and toxins). *The Amicus Journal* 20:28. (Keywords: "underwater" and "epidemic")

Wetherbee, R. 2002. The diatom glasshouse (Perspective: biomineralization). *Science* 298:547. (Keywords: "diatom" and "glasshouse")

Algae: Ecology

Morel, A., Antoine, D. 2002. Small critters—big effects (perspectives on phytoplankton). *Science* 296:1980. (Keywords: "critters" and "phytoplankton")

Raloff, J. 1998. Rogue algae: the Mediterranean floor is being carpeted with a shaggy, aggressive invader. *Science News* 154:8. (Keywords: "rogue" and "algae")

Withgott, J. 2002. California tries to rub out the monster of the lagoon (invasive species of seaweed). *Science* 295:2201. (Keywords: "monster" and "lagoon")

Bryophytes

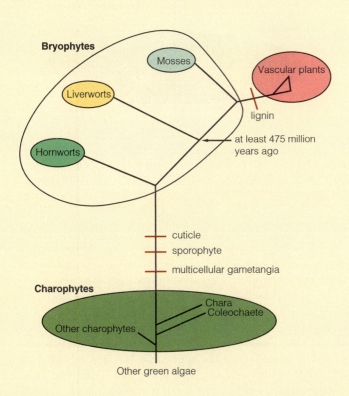

Visit us on the web at http://biology.brookscole.com/plantbio2
for additional resources, such as flashcards, tutorial quizzes,
InfoTrac exercises, further readings, and web links.

1. Bryophytes are land plants that differ from all other plants lacking lignified vascular tissue by having the gametophyte generation dominant and having unbranched sporophytes that produce a single sporangium.

2. Modern bryophytes almost certainly evolved from a single common ancestor, and they likely represent several lineages along the evolutionary path to vascular plants. Recent fossil discoveries push back the earliest appearance of bryophytes to 475 million years ago. Existing bryophytes preserve a suite of ancestral characteristics that give us insight into the origin of land plants. Bryophyte relationships remain uncertain; however, existing bryophytes fall into three lineages: liverworts (Marchantiophyta), hornworts (Anthocerotophyta), and mosses (Bryophyta). Traditionally, liverworts have been considered the earliest evolving lineage of bryophytes, but recent evidence suggests that hornworts may be the earliest. Mosses are likely the closest sister group to vascular plants.

3. Key innovations of the bryophyte radiation, not present in their algal ancestors, include multicellular gametangia (antheridia and archegonia) that protect and insulate gametes from the environment; a multicellular sporophyte that develops from an embryo embedded within and nutritionally dependent on the gametophyte; and the presence of a waxy coating on the shoots (cuticle) and the spores (sporopollenin). The most complex bryophyte sporophytes also contain novel structures such as stomata and water- and sugar-conducting tissue (unlignified vascular tissue). Bryophyte gametophytes are not able to control their water balance, and they dry out rapidly in the absence of free water. The desiccated plants are still alive and can become active within minutes of being rewetted.

4. Mosses are important in many ecosystems. They provide most of the biomass in boreal vegetation such as tundra, they dominate the understory of cool-temperate forests, and they are common in damp microenvironments. Some species are aquatic and most require humid conditions, but some can colonize dry, exposed habitats such as rock outcrops and desert soil surfaces.

22.1   THE LEAP ONTO LAND

Despite being common almost everywhere, **bryophytes** often are ignored because of their small stature, lack of familiar features, and the fact that in many environments they are dormant for much of the year. Yet, bryophytes are exceptionally diverse, with nearly 25,000 named species (among land plants, only flowering plants and ferns have more species). They also are extremely widespread, being present on all continents, including Antarctica. They have a long evolutionary history, and some bryophytes are a sister group to vascular plants **(Fig. 22.1).**

The great diversity of bryophyte species reflects a stunning diversity of habitats, from barren arctic and alpine

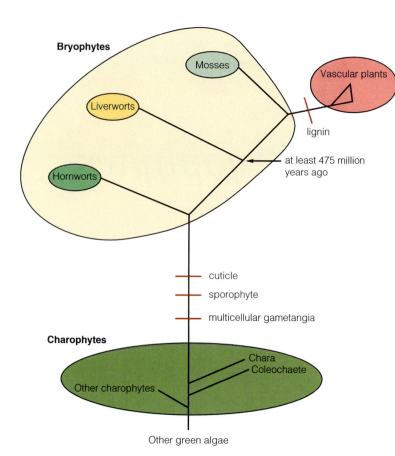

Figure 22.1   A cladogram of relationships between the bryophytes and vascular plants. Hornworts are the basal-most lineage of land plants, although other hypotheses place liverworts in that position. Mosses are almost certainly the sister group to vascular plants.

ground to hot deserts, from the bottom of lakes to the canopy of tropical rain forests. Peat mosses grow submerged in the acidic waters of bogs, whereas various liverworts grow as epiphytes on the leaves of trees. Some mosses grow on rocks exposed to ocean salt spray or intense heat and sunlight. Others tolerate very dim light and grow in the understory of dense forests or inside caves and burrows **(Fig. 22.2).** These adaptable plants can be found nearly everywhere that plant life is possible. But despite being so adaptable, all bryophytes have one major limitation: They require free water (not soil moisture) in their environment. Without it, they cannot reproduce sexually.

As you might expect from the diverse environments inhabited by bryophytes, they have a corresponding diversity of body forms, from the giant moss *Dawsonia superba*, which can reach a height of 70 cm and resembles a pine tree seedling, to *Ptychomnion aciculare*, in which "dwarf" male plants grow attached to the leaves of the female plants.

These relatively inconspicuous and overlooked plants warrant more careful study for a number of reasons. Bryophytes can be important ecologically, by altering pH, absorbing carbon, regulating nutrient cycling, colonizing barren surfaces, creating soils, and reducing erosion. They often are important elements in the local water cycle, absorbing and holding moisture so that other plants benefit.

Figure 22.2 Morphological diversity and habitats of bryophytes. (a) Epiphytic mosses on trees in Olympic National Park, WA. (b) Moss. (c) Granite moss *Grimmea* growing on bare rock outcrop in the Appalachian Mountains. (d) Liverwort. (e) Hornwort.

They also are useful to environmental scientists because the majority of bryophytes, despite their amazing resilience, are intolerant of pollution and often disappear from contaminated areas. This sensitivity makes them good indicators of air and water quality. Bryophytes also possess many physiological adaptations that interest scientists. For example, some bryophytes can survive extended periods (more than 20 years) of desiccation, and then when rewetted revive in a matter of minutes and resume normal growth.

Perhaps the most compelling reason for a closer look at bryophytes is that they are the living representatives of the most ancient lineages of land plants. They preserve early characteristics, and, in effect, give us a glimpse into what was happening at the dawn of the first great plant-adaptive radiation, when plants emerged from the water and took hold of the land.

Bryophytes Faced Many Problems When They Moved onto the Land

Bryophytes retain many of the characteristics of their algal ancestors, including a nutritionally independent (photosynthetic) and complex gametophyte; the photosynthetic pigments chlorophyll *a* and *b*, carotenoids, and xanthophylls; sperm that swim by means of two asymmetrically attached flagella; chloroplasts with conspicuous grana, which store food as starch; and cell walls composed primarily of cellulose and pectin. Bryophytes also engage in a particular type of cell division that is present in charophytes but absent in other green algae: The nuclear envelope breaks down, and microtubules oriented perpendicular to the plane of division form a cell plate, which grows from the center to the outer portion of the cell.

Faced with the great difficulties of a terrestrial life, bryophytes also evolved many new and highly successful adaptations. The primary problems of life on land were preventing death by drying out, dispersing spores through the air, and avoiding damage from weather and intense solar radiation. The responses to these problems, which bryophytes pioneered, were subsequently passed on to all their descendants and have become the fundamental innovations that define land plants.

Key Innovations in Land Plants First Appear in the Bryophytes

Bryophytes evolved important advances in both phases of the land plant life cycle. These key innovations allowed plants to colonize the land, setting off a series of spectacular adaptive radiations, first among bryophytes and later in vascular plants.

SPOROPHYTE EVOLUTION One of the most important innovations of early bryophytes was a multicellular sporophyte (Fig. 22.3). The evolution of this new structure can be accounted for by two simple steps. First, meiosis must be delayed in the zygote, which, as you recall, is retained on the parent gametophyte in land plant ancestors. Second, the zygote must undergo mitotic cell divisions to create a multicellular body.

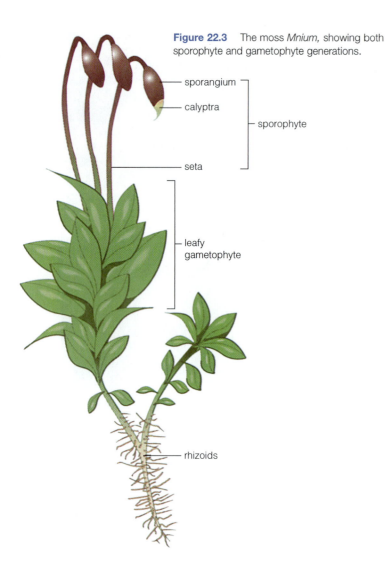

Figure 22.3 The moss *Mnium*, showing both sporophyte and gametophyte generations.

sporangium

calyptra

sporophyte

seta

leafy gametophyte

rhizoids

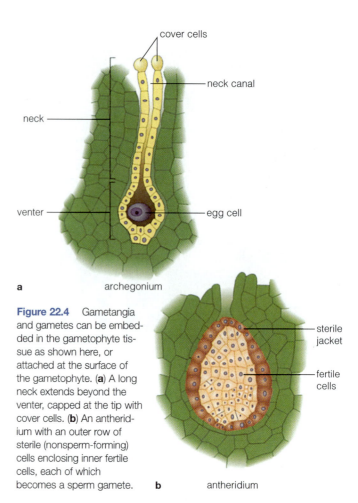

cover cells

neck canal

neck

venter

egg cell

a archegonium

Figure 22.4 Gametangia and gametes can be embedded in the gametophyte tissue as shown here, or attached at the surface of the gametophyte. (**a**) A long neck extends beyond the venter, capped at the tip with cover cells. (**b**) An antheridium with an outer row of sterile (nonsperm-forming) cells enclosing inner fertile cells, each of which becomes a sperm gamete.

sterile jacket

fertile cells

b antheridium

Bryophytes nurture their sporophytes by initially embedding the young sporophyte in the tissues of the gametophyte and providing nutrients, water, and hormones for development. This **embryo** phase of the sporophyte has been retained in all extant land plants, and it is so characteristic of them that **embryophyte** is used as a name for the whole land plant group. In all bryophyte groups, the sporophyte remains dependent on the gametophyte for its entire life. The sporophytes of the earliest land plants were probably nothing more than a "jacket" of cells surrounding a mass of **sporocytes** (the cells that undergo meiosis to produce spores).

The value of these first sporophytes must have been enormous because they would have protected the developing spores from desiccation. They also allowed bryophytes to produce many more spores than their algal ancestors. By using mitosis to multiply the number of diploid cells that can divide by meiosis, bryophytes greatly increased their spore production potential. In the hostile terrestrial environment, where most spores would probably land in unsuitable locations, there would have been great selective pressure to increase the number of spores released. The spores themselves also adapted to the new environment. They lost their

flagella, which would have been useless on land, and became coated with a weather-resistant wall.

Bryophyte sporophytes remain today the leafless, unbranched structures they have been since their first appearance nearly 500 million years ago. However, each of the lineages of bryophytes and the vascular plants have evolved different sporophyte structures and spore dispersal mechanisms, which suggests that these lineages were separate from each other (no longer exchanging genes) before the sporophyte had undergone much evolution. The great differences among them seem to represent four independent ways of elaborating the simple sporophytes of the first land plants.

GAMETOPHYTE EVOLUTION Bryophytes have the largest, most elaborate gametophytes of any living land plant group, some possessing conducting tissue similar to the xylem and phloem in stems and leaves of vascular plants. Bryophyte gametophytes are not only complex structurally, but they often produce a vast array of secondary chemicals to protect against weather, bright light, and herbivory.

Another important innovation of the bryophytes was the multicellular structures surrounding the gametes, called **gametangia.** In these structures, sperm cells are made in globular or club-shaped sacs called **antheridia;** eggs are produced singly in vase-shaped structures called **archegonia (Fig. 22.4).** Gametangia protect gametes during development.

STOMATA AND CUTICLE Stomata and cuticle first appear in the bryophytes. These new features became crucial adaptations to life on land, and virtually all vascular plants possess them. Several lineages of bryophytes have stomata on their sporophytes. Whether bryophytes have stomata on their gametophytes currently is controversial; some produce pores that initially develop in a manner similar to stomata on the sporophyte, and may be homologous, or may represent convergent evolution.

Cuticle may function as it does in most plants today—to reduce the risk for desiccation. However, in many bryophytes, only part of the plant is covered, so the role of the cuticle may be to protect against ultraviolet radiation or fungal infection, as well as to reduce dehydration. Many bryophytes could not afford to completely cover themselves with cuticle because they absorb much of their water and minerals directly through their dermal layer.

22.2 BRYOPHYTES ARE NOT A NATURAL GROUP

Bryophyte is a common name applied to three distinct lineages of plants that lack lignified vascular tissue. Bryophytes do not form a monophyletic group because vascular plants are descended from them. This chapter considers them as three lineages (Table 22.1).

Mosses are familiar to everyone as a green mat in shady, moist places or lining the cracks in rocks and walls (Fig. 22.2b). Although less familiar, the liverworts also are quite common and can be found in many places if you know what to look for (Fig. 22.2d). The hornworts are the least familiar, because they often are rare and inconspicuous (Fig. 22.2e). Because they lack lignified stiffening and vascular tissues, bryophytes remain small. Mosses and liverworts, however, seldom grow alone. Colonies can cover large areas of ground and represent substantial biomass in certain communities.

Some bryophytes have flat, ribbon-like bodies called **thalli** (singular, *thallus*) that often bifurcate as they grow, whereas others have a more familiar upright form with tiny leaves born on short stems. Bryophytes frequently also possess minute projections composed of single or multiple cells that anchor the plants to the soil, and thus resemble roots. In some species, they may even serve to conduct water and minerals from the substrate. However, because these structures are quite different from true roots in development, form, and function, they are called **rhizoids** (Fig. 22.3).

All of the traits discussed in this section are ancestral in the land plant lineage—that is, they were inherited from a common ancestor. Because only derived traits can reveal the exact path of evolution, we must examine differences among bryophyte lineages to determine how their body forms and life histories changed over time in their great leap onto the land.

Table 22.1 Bryophyte Taxonomy

Common Name	Traditional Taxonomic Name
Hornworts	Anthocerotophyta (=Anthocerophyta)
Liverworts	Marchantiophyta (=Hepaticophyta or Hepatophyta)
	Jungermanniopsida
Leafy liverworts	Jungermanniidae
	Porellales
	Porella
	Radulales
	Pleuroziales
	Lepicoleales
	Jungermanniales
Simple thalloids	Metzgeriidae
	Metzgeriales
	Haplomitriales
	Blasiales
	Treubiales
	Fossombroniales
Complex thalloids	Marchantiopsida
	Marchantiales
	Marchantia
	Sphaerocarpales
	Monocleales
	Ricciales
	Riccia
Mosses	Bryophyta
Granite mosses	Andreaeopsida
Peat mosses	Sphagnopsida
	Sphagnum
True mosses	Bryopsida
	Dawsonia, Ptychomnion, Bryum
	Takak Jungermanniidae iopsida

Taxonomic names ending in *-ophyta* are divisions, *-opsida* are classes, *-idae* are subclasses, and *-ales* are orders.
Taxonomic names from Crandall-Stotler, B., and Stotler, R.E. 2000. Morphology and classification of the Marchantiophyta, in Shaw, A.J., Goffinet, B., eds. *Bryophyte Biology.* Cambridge: Cambridge University Press, pp. 21–70, with permission.

Bryophyte Relationships Remain Uncertain

It is currently unclear how the three lineages of bryophytes are related to each other and to the vascular plants. A great number of distinct lines of evidence support the idea that they share a common ancestor among the ancient charophyte algae. Systematists have long suspected that the genus *Coleochaete* is the closest living algal relative of the land plants, but a recent analysis of relationships based on DNA comparisons found that another green alga, *Chara* (see Chapter 21), is one branch point (node) closer to land plants on a cladogram.

One area of great interest is the identity of the first lineage of land plants. Traditionally, on the basis of their relative simplicity of organization and lack of features present in other groups, liverworts have been considered the earliest diverging lineage. However, some recent molecular reappraisals using nucleic acid sequences from each of the three genome compartments (the nucleus, the chloroplast, and the mitochondrion) have revealed that hornworts may represent the earliest lineage. Reassessments of morphological characteristics using more complete data sets support this conclusion. A variety of evidence from morphology, development, and DNA suggests, moreover, that mosses and liverworts are closely related.

Another important question in bryophyte evolution concerns the timing of their first appearance and subsequent radiations. **Megafossils,** fossils that can be seen without a microscope, of bryophytes are rare **(Fig. 22.5).** Until recently, the earliest known megafossils of a bryophyte are liverwort fossils from 425 million years ago (late Silurian period). Curiously, vascular plants, which descended from bryophytes, appear in the fossil record 5 to 45 million years earlier. However, spores of land plants that could be bryophyte in origin can be found as early as 475 million years ago (Ordovician period). Some recent finds from Oman confirm these are, in fact, bryophyte spores and explain the 50-million-year gap between the spores and the plants that produced them. Sifting through organic remains dissolved out

of Ordovician period rocks, paleontologists found masses of early spores preserved inside tiny sporangia. The sporangia were associated with fragmentary fossils of bryophyte plants. These plants had previously escaped discovery because they are extremely small and lack certain features usually associated with bryophytes. Future research undoubtedly will yield even more spectacular evidence of the first plants to colonize the land. The fossil record now agrees with the results of phylogenetic analyses: The first land plants were bryophytes.

Studies of bryophytes have provided a great deal of support for the following hypotheses: (1) Land plants evolved only once from a single algal ancestor, specifically a charophyte; (2) the first land plants were bryophytes and appeared no later than the Ordovician period, about 475 million years ago; (3) the bryophytes form a nonmonophyletic group (that is, the bryophytes gave rise to vascular plants); and (4) each of the three living bryophyte lineages—the hornworts, liverworts, and mosses—is monophyletic.

22.3 HORNWORTS

The **hornworts** have relatively simple gametophytes, consisting of a flat thallus, roughly circular in outline or, especially in epiphytic varieties, long and ribbon-shaped with a prominent midrib **(Fig. 22.6).** The gametophytes of horn-

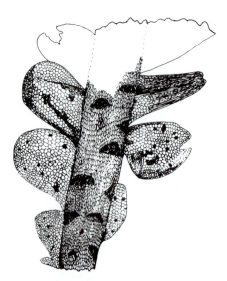

a

b

Figure 22.5 Fossil bryophytes. (**a**) A fragment of a Devonian hepatic. (Reproduced from Scagel, R.F., et al. 1984. *Plants: An Evolutionary Survey.* Belmont, CA: Wadsworth Publishing Co., p. 314 fig. 11–12c, with permission.) (**b**) *Aulocomnium heterostichoides,* an Eocene moss. (Reproduced from Janssens, J.A.P., Horton, D.G., and Basinger, J.F. 1979. Aulacomnium heterostichoides sp. nov.: An Eocene moss from south-central British Columbia. *Canadian Journal of Botany* 57:2150–2161, with permission.)

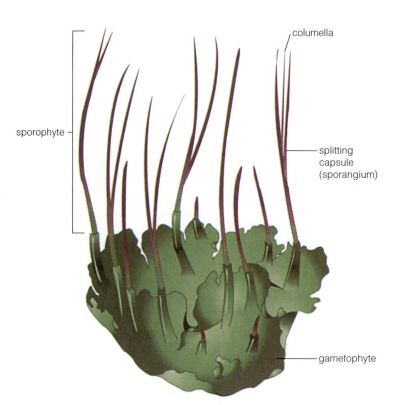

Figure 22.6 The hornwort *Phaeoceros,* showing gametophyte and sporophyte stages. Mature hornwort sporangia split into longitudinal strips, beginning at the tip and opening downward. Rhizoids, which anchor gametophytes to the ground, are not shown.

worts produce copious mucilage inside their thalli, and the ventral portion of the thallus has pores. These may be homologous with the stomata present on the sporophytes. Cyanobacteria (genus *Nostoc*) enter the thallus through these clefts and form symbiotic colonies, which can be seen as blue–green dots when the thallus is held up to the light. *Nostoc* fixes atmospheric nitrogen to ammonia, which the hornworts require; in return, *Nostoc* lives protected in the hornwort gametophyte.

The photosynthetic cells of hornworts are unique among land plants but strikingly similar to algae in that they each typically contain one giant chloroplast with distinct pyrenoids visible within it. Archegonia and antheridia develop embedded in the thallus: The antheridia are clustered in groups of up to 25 in roofed chambers in the upper portion of the thallus, and the archegonia are sunken into the thallus, with only the neck protruding. Sperm are released from the antheridia. They swim toward an archegonium, where they are caught in the mucilage that covers it, and are drawn down into the neck canal. Once inside the archegonium, a single sperm fertilizes the egg, creating a zygote.

The sporophyte of hornworts is unique among plants in that it grows continually from a meristem at its base. It is long and pointed, creating the appearance of horns protruding from the thallus (hence, the name of the division). It consists of a foot (embedded in the gametophyte thallus) and an upright sporangium or **capsule (Fig. 22.7)**. The epidermis of some hornwort sporophytes contains stomata but generally lacks chloroplasts; beneath the epidermis there is a layer of chlorenchyma tissue, and beneath this, there is a mass of sporocytes, which undergo meiosis to produce haploid spores. The spores are intermingled with pseudoelaters, which help separate and disperse the spores. In the center of the sporangium is a central cylinder of sterile tissue, the **columella.**

When mature, the tip of the capsule splits into two **valves** (sections), and spores are released (Fig. 22.7). A meristematic region just above the foot adds new cells to the base of the sporangium so that more sporocytes are continually created. The sporangium thus grows upward from the base much like a blade of grass and continues to release spores over a long period. Its total height may reach several centimeters. If the spores land in a suitable environment, they undergo mitosis and produce new gametophytes.

22.4 LIVERWORTS

Approximately 9,000 species of plants comprise the group known as **liverworts,** or *hepatics* (derived from the Greek meaning "liver"). The name is old, having been recorded in medieval manuscripts as early as the ninth century A.D. It was probably applied to these plants because of their fancied resemblance to the liver and the belief that plants could cure diseases of the organs that they resembled. A prescription for a liver complaint in the sixteenth century called for "liverworts soaked in wine."

As is the case with all bryophytes, the liverwort gametophyte is the more prominent phase of the life cycle. Most grow in moist, shady habitats. Those in temperate regions usually grow as a green ribbon or heart-shaped band of tissue **(Fig. 22.8a).** These thalli can resemble hornworts. The thallus is held to the surface of the damp soil by single-celled rhizoids. Individual plants are small, ranging in size from one to several centimeters across. However, colonies of liverworts can occupy large areas.

Some liverworts are more elaborate, with distinct leaves and stems. The leaves are blunt-tipped or lobed and are attached to the stems in two or three overlapping rows **(Fig. 22.8b).** The gametophytes of liverworts produce an enormous variety of volatile oils, which are stored in a unique single membrane-enclosed organelle called an *oil body.* These oils give many liverworts a distinctive aroma,

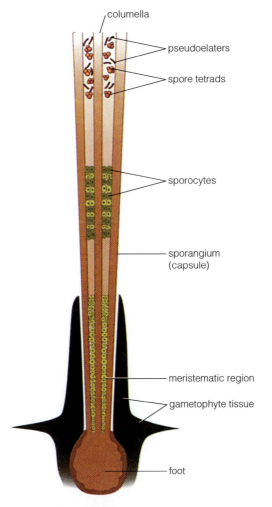

Figure 22.7 A longitudinal section through a hornwort (*Anthoceros*) sporophyte. Only the mid-portion of the sporophyte is shown, omitting the tip of the sporangium and rhizoids at the base. Blackened tissue represents gametophyte tissue in which the sporophyte is embedded. The tubular sporophyte is mainly an elongated sporangium. Sporophyte cells within the sporangium undergo meiosis to produce haploid spores. Other cells produce twisted pseudoelaters, which assist in the ejection of spores from the sporangium.

The figure labels (top to bottom): columella, pseudoelaters, spore tetrads, sporocytes, sporangium (capsule), meristematic region, gametophyte tissue, foot

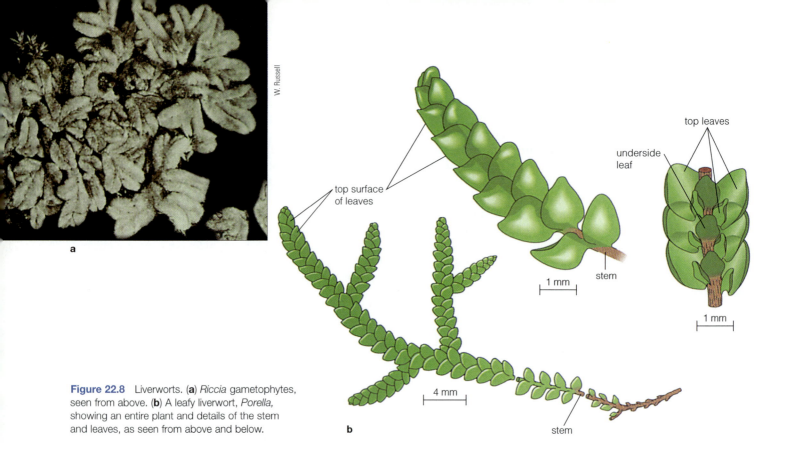

Figure 22.8 Liverworts. (a) *Riccia* gametophytes, seen from above. (b) A leafy liverwort, *Porella*, showing an entire plant and details of the stem and leaves, as seen from above and below.

top surface of leaves

top leaves

underside leaf

stem

1 mm

1 mm

4 mm

stem

a

b

W. Russell

and they may serve to prevent herbivory. Some of these compounds are promising as antibiotics and antitumor agents.

Liverworts produce relatively simple sporophytes that are hidden in folds in the gametophyte tissue. As spores develop, special cells near the base of the sporophyte elongate rapidly to form a stalk, called a **seta,** which pushes the sporangium, or capsule, above the surface of the thallus. The capsule then splits into four valves and releases all of its spores at one time. Special thickened cells inside the capsules, called **elaters,** separate the spores and aid in spore dispersal. If the spores land in a suitable environment, they undergo mitosis and produce one new gametophyte plant per spore.

Our discussion of this group focuses on the genera *Riccia* and *Marchantia,* two types of thalloid liverworts, and the genus *Porella,* a leafy liverwort. These three examples represent the morphological variation of hepatic plants.

Some Liverworts Have Thalloid Bodies

Gametophytes of the **thalloid** (ribbon-shaped) liverworts usually grow Y-shaped branches by a simple forking at the growing tip. In some species, a rosette of branches is formed. About 15% of all liverwort species are thalloid, and of these, about two-thirds have simple undifferentiated thalli. The remaining third have more complex thalli with distinct layers visible.

The gametophyte of *Marchantia* is a good example of a thalloid liverwort. It has a prominent midrib, and the tips of the branches are notched. A typical thallus is 1 to 2 cm

across. The size and degree of branching depend on growing conditions. On the upper epidermis are polygonal areas, each with a conspicuous pore in the center **(Fig. 22.9).** These areas demarcate air chambers below the pore, which bathe chlorenchyma cells in air containing carbon dioxide. The pores perform the same function as stomata. The lower cells of the thallus are colorless parenchyma, modified for carbohydrate storage. Rhizoids and sheets of cells called *scales* project from the lower surface, increasing the surface area in contact with the substrate and anchoring the thallus.

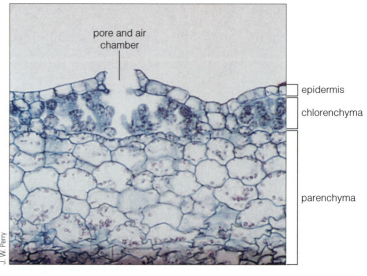

pore and air chamber

epidermis

chlorenchyma

parenchyma

J. W. Perry

Figure 22.9 Cross section of a *Marchantia* thallus, showing the pore, air chamber, epidermis, chlorenchyma, and parenchyma storage tissue.

Marchantia reproduces asexually in two ways: by fragmentation when older parts of the thallus die, separating younger portions that then develop as separate individuals; and by small clumps of tissue called **gemmae** (singular, gamma). These clumps are produced in small **gemmae cups** that form on the upper surface of the thallus **(Fig. 22.10).** When the gemmae are mature, raindrops can break them free of the cup and scatter them away from the thallus. If gemmae land on a suitable substrate, each is capable of developing into a gametophyte plant by mitosis.

Marchantia has stalked structures for sexual reproduction on which the gametangia are produced. The stalks of some gametophytes resemble tiny beach umbrellas and produce only antheridia. These are called **antheridiophores (Fig. 22.11a).** The stalks of other gametophytes resemble tiny palm trees and are called **archegoniophores,** which produce archegonia underneath their "fronds" **(Fig. 22.11b).** Both antheridiophores and archegoniophores serve to hold the gametangia well above the main body of the thallus. After fertilization, the sporophytes develop under the archegoniophores. Eventually, the sporangium capsule becomes visible hanging upside down from the archegoniophore on a short, thick seta. When mature, the sporangia split open and release the spores.

Most Liverworts Have Leafy Bodies

The largest group of liverworts, about 85% of all known species, have gametophytes that are leafy. Leafy liverworts are found from the arctic to the tropics and are especially common in humid climates, where they may be the most abundant bryophytes. Like mosses, which often are found growing with them, leafy liverworts thrive in shady moist areas, but they may also be found in full sunlight. Some

a

antheridiophore   thallus

b   thallus   archegoniophore   gemma cup

Figure 22.11 Gametangium-bearing structures of *Marchantia*. (**a**) Antheridiophores. (**b**) Archegoniophores.

gametophyte   gemmae cups
thallus

Figure 22.10 Gemmae cups, involved with asexual reproduction, as they appear on the upper surface of a *Marchantia* gametophyte thallus.

leafy liverworts can tolerate extreme desiccation, whereas others grow underwater. Some are epiphytes on woody plants, and a few are even epiphytes on other bryophytes.

These plants can be distinguished from mosses by the leaf arrangement. Mosses have spirally arranged leaves, whereas leafy liverworts have two or three distinct rows. In addition, the leaves of leafy liverworts usually are lobed, unlike mosses.

Porella, a common, widespread genus (Fig. 22.8b), forms dense mats on rocks and trees. Young portions of stems are densely clothed with leaves arranged in three ranks. The leaves in the upper two ranks are large and lobed; the lower rank has much smaller, unlobed leaves. The leaves are one cell in thickness and lack a cuticle. Shaded, older portions of stems lack leaves, and new rhizoids arise along those naked segments. Some liverwort species, including *Porella*, have very few rhizoids.

Figure 22.12 Early gametophyte development in the moss *Funaria*. (**a**) The germination of haploid spores. (**b**) A protonema. (**c**) An older protonema with one bud (many others are not shown here).

Leafy liverworts reproduce asexually by gemmae, which often are found attached to leaves or stems. When sexual reproduction occurs, antheridia are produced in the axils of leaves and archegonia grow on specialized short, leafy branches. The production of archegonia is the last series of cell divisions undertaken by the branches that produce archegonia; therefore, the archegonia are always terminal on the stems where they occur. The sporophytes, as with other liverworts, consist of a foot, a seta, and a sporangium. The sporangium splits into four valves when mature, and, in some cases, elaters actively flick out the spores.

22.5 MOSSES

Mosses are the most conspicuous bryophytes because they are larger, have wider distributions and many more species than the liverworts and hornworts, and often cluster to form easily visible tufts or carpets of vegetation on the surface of rocks, soil, or bark. Taxonomists divide them into a number of groups. The most prominent are granite mosses, small, dark-colored plants that grow on rocks in cool climates; peat mosses, which are much larger and are confined mainly to acidic bogs; and the "true" or typical mosses, with the widest habitat range and the most species.

The life cycle of a typical moss, such as *Mnium*, is explored in the following section.

Gametophytes Have Protonemal, Bud, and Leafy Phases

Moss gametophytes have three growth phases, starting when a spore germinates into a branching, filamentous structure called the **protonema** (Fig. 22.12a,b). The second phase

begins when buds form on the protonema (Fig. 22.12c). In the third phase, the buds grow into upright, branching axes, bearing small, spirally arranged leaves and rhizoids. Each protonema may produce a dense population of many genetically identical leafy gametophytes; a single spore thus can produce many separate gametophytes.

As the leafy gametophyte grows, cells in the stems differentiate and mature into specialized tissues. Mosses typically have an epidermal layer of small, thick-walled cells surrounding a homogeneous cortex of parenchyma tissue. Some species have a thin cuticle over the parts of the epidermis, but others lack a cuticle. Stomata are absent. The epidermis of *Sphagnum* is unique in containing large, empty, clear cells (Fig. 22.13). These cells can fill with water through a pore when moistened and serve as a reservoir of moisture for the moss plant.

Many mosses have a more complex stem anatomy, with a central strand of conducting tissue. One kind of conducting tissue is made up of **hydroids**—elongated, thin-walled,

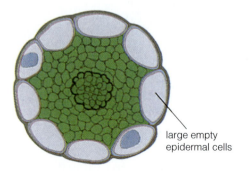

large empty
epidermal cells

Figure 22.13 Cross section of the stem of the moss *Sphagnum*, showing specialized epidermal cells that store water.

PLANTS, PEOPLE, AND THE ENVIRONMENT:

The Ultimate Shade Plants

In the dim light of shallow caves, under 30 cm of snow, or on the floor of a dense forest, it is common to find mosses growing. The lush carpets of green they form may be the only vegetation present in some places. This affinity for low light can even be life-saving: The famous tendency for mosses to grow thickest on the shadier north side of tree trunks enables wood-wise hikers to orient themselves. How can mosses not only survive but thrive in these places where other plants cannot?

Some mosses are sensitive to light. The light compensation point for photosynthesis (where photosynthesis just equals respiration) is reached at 1% of full sun, and the saturation point (where photosynthesis is maximum) at only 4% to 20% of full sun. Some mosses have specialized cells capable of focusing any available light onto chloroplasts. These adaptations allow shade-tolerant mosses to inhabit many places where all other plants are completely excluded. In these extreme habitats, mosses are the dominant vegetation and the basis of food chains—a role they lost everywhere else more than 400 million years ago, when vascular plants appeared.

🌐 **Websites for further study:**

Air Pollution, Lichens, and Mosses:
http://www.mpm.edu/collect/botany/lichens.html

dead, empty cells that conduct water **(Fig. 22.14).** Their end walls are oblique, sometimes very thin, perforated with pores, or partly dissolved. Experiments with dyes show that the translocation of water can occur in this tissue. Some mosses have continuous hydroid tissue from the stems to the midribs of the leaves. Hydroids resemble vessels but lack their specialized pitting and lignified walls.

Some mosses also contain a layer of cells called **leptoids** that resemble the sieve cells of vascular plants (Fig. 22.14). Leptoids surround the hydroids and are living, but their nuclei degenerate. Nearby parenchyma cells may assist the leptoids, like companion cells. Tracer studies show that sugars may move through these cells at rates of up to 30 cm per hour.

Mosses Have Several Forms of Asexual Reproduction

Asexual reproduction is accomplished in several ways. First, the protonema may continue to produce new buds, so a miniature forest of moss plants may spread outward in all directions, matlike, at the edges of the clone. In this respect, the protonema is analogous to a network of runners or rhizomes. Second, leaf tissue placed in wet soil may produce protonemal strands from which buds and new moss individuals develop. Third, rhizoids sometimes can produce buds. Fourth, lens-shaped gemmae may form on rhizoids, leaves, at the ends of special stalks, and even in specialized gemmae cups. Gemmae have the same function in mosses as they do in liverworts: If detached from the parent plant and dispersed to a suitable habitat, each can begin mitotic cell division and will differentiate into a new gametophyte.

Sexual Reproduction Typically Occurs at the Ends of Stems

When undergoing sexual reproduction, most mosses produce gametangia at the gametophyte stem tips **(Fig. 22.15a).** The gametangia often are separated and held upright by sterile filaments called **paraphyses (Fig. 22.15b).** Some moss species produce antheridia and archegonia on separate

leptoids

Figure 22.14 A longitudinal section showing details of conducting cells in the stem of the moss *Polytrichum.*

— hydroid

outside ← → center

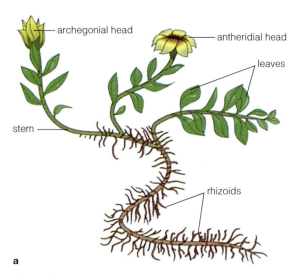

archegonial head

antheridial head

leaves

stem

rhizoids

a

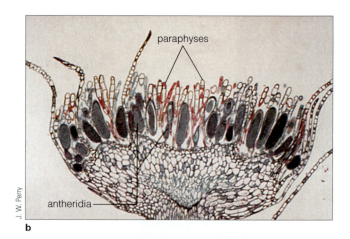

paraphyses

antheridia

b

J. W. Perry

Figure 22.15 Gametangia of *Mnium*, a moss that produces antheridia and archegonia on different heads but on the same plant. (**a**) The general appearance of antheridial and archegonial heads. (**b**) An antheridial head, showing dark-stained antheridia and elongate filaments of cells called paraphyses. (**c**) An archegonial head, showing elongate archegonia enclosing dark-stained eggs, surrounded by many paraphyses.

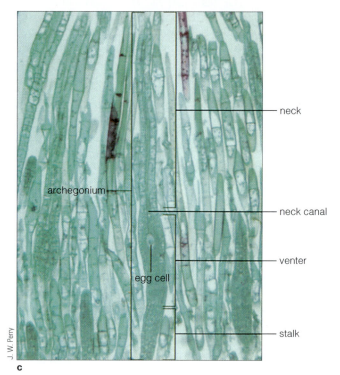

neck

archegonium

neck canal

venter

egg cell

stalk

J. W. Perry

c

plants. Other species have both on a single plant, either together or on separate parts of the plant.

Antheridia release mature sperm when free water is present. Each sperm cell is propelled by two flagella. In some mosses, the leaves that surround antheridial heads spread out like the petals of a flower and function as a splash cup, using the force of raindrops to eject sperm some distance away. Sperm can remain mobile for as long as 6 hours and swim up to 50 cm. The sperm cells of some mosses, such as *Bryum capillare,* on rare occasions become attached to insects that have been attracted to the antheridial heads, either by the red and yellow colors or by secretions from the paraphyses.

Each archegonium has a long neck, a thickened **venter** region that surrounds a single egg, and a long stalk **(Fig. 22.15c).** When the egg is mature, the neck opens, creating a canal. Attracted by a gradient of chemical attractant emitted by the egg, sperm swim down the canal toward the egg. Water is necessary to carry sperm from an antheridium to an egg, and the leaves arranged around archegonial and antheridial heads help retain a film of water over the gametangia. When a sperm cell reaches an egg, it fertilizes the egg, creating a diploid zygote cell.

Moss Sporophytes Have Complex Capsules

Soon after fertilization, the zygote develops into an embryo that differentiates into a foot, seta, and sporangium. The foot penetrates through the venter and into the gametophyte stem, where transfer cells move water and nutrients from the gametophyte to the dependent sporophyte. The seta elongates rapidly, raising the yet-to-be-formed sporangium

above the top of the leafy gametophyte **(Fig. 22.16).** The archegonium increases in size as the sporophyte enlarges, and it is now called the **calyptra** (see Fig. 22.3). It remains for a time as a protective covering for the sporangium. Interestingly, the presence of the calyptra is necessary for normal growth and differentiation of the sporophyte. The mature sporangium (capsule) and seta can have a complex anatomy, with a thick-walled epidermis, a cuticle layer, and stomata; a cortex region; and a central strand of conducting tissue.

Most moss sporophytes contain chlorenchyma and stomata, allowing photosynthesis. The sporophyte can fix 10% to 50% of the carbohydrate needed for growth and maintenance, the rest coming from the gametophyte. Mature sporophytes of the granite mosses, however, lack chlorophyll.

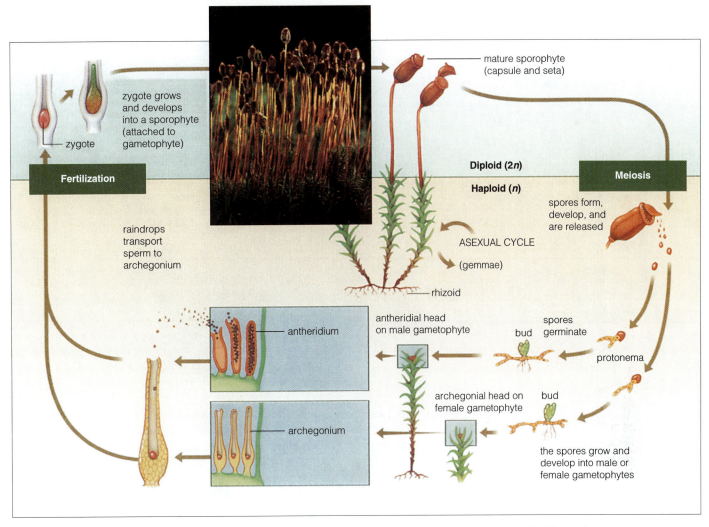

Figure 22.16 Life cycle of *Polytrichum*, a typical moss. (Photograph by Jan Burton/Bruce Coleman Ltd.; art by Rachel Ciemma)

When mature, the seta may elevate the sporophyte capsule more than 10 mm above the gametophyte, allowing for better spore dispersal. Eventually cells inside the capsule undergo meiosis to form thousands of spores. Granite moss capsules open by slits to disperse the spores. Peat moss capsules have lids that blow off when the capsule dries, violently ejecting all the spores at once. In true mosses, the capsule has a central column of sterile tissue, the columella **(Fig. 22.17a)**. When the capsule is mature, it dries and forms a lid, or **operculum (Fig. 22.17b)**. Cells immediately below the operculum form a double row of triangular **peristome teeth.** Eventually, the operculum falls away, exposing the peristome teeth and the spores **(Fig. 22.17c,d)**.

Peristome teeth of many mosses are sensitive to atmospheric humidity. When the air is humid, they bend into the capsule's cavity. When the air is dry, they straighten and lift out some of the spores, which are then disseminated by wind. Spores may travel thousands of kilometers, but typically they fall just a few meters from the parent plant. One family of mosses that grows on dung or carrion (the Splachnaceae) has spores dispersed by insects. These cling together

in a sticky mass on the columella, which projects beyond the operculum at maturity. The capsule takes on purple, red, yellow, and white colors and resembles a small flower from a distance. The colors and mushroom-like odors attract flies, which carry the sticky spores considerable distances away. Moss spores have a waxy covering and are resistant to aridity. They are capable of remaining dormant for decades.

Mosses Have Significant Economic and Ecological Value

Sphagnum, a large genus of peat mosses, is by far the most economically important bryophyte. Because of their special epidermal cells, *Sphagnum* mosses are frequently added to potting soil to increase its water-holding capacity. Tons of *Sphagnum* are harvested throughout the world and then sold in the nursery industry in many countries. During World War I, *Sphagnum* was used on a large scale as a wound dressing because its acidic, sterile tissue acts as both an antiseptic and an absorbent. In some cold-temperate areas, *Sphagnum* and other mosses accumulate as a thick, compacted, semi-

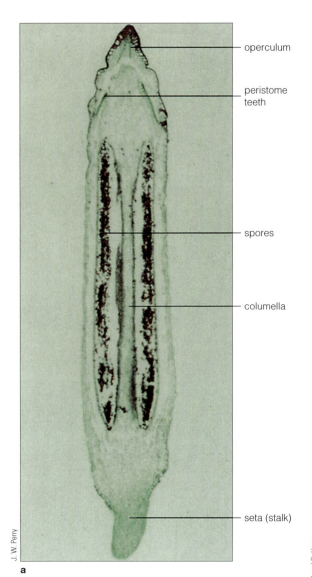

J. W. Perry

a

operculum

peristome
teeth

spores

columella

seta (stalk)

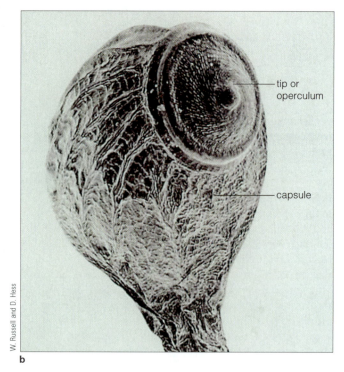

W. Russell and D. Hess

b

tip or
operculum

capsule

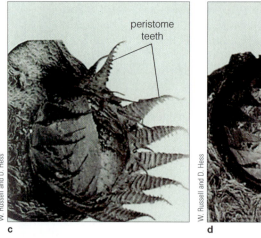

W. Russell and D. Hess

c

peristome
teeth

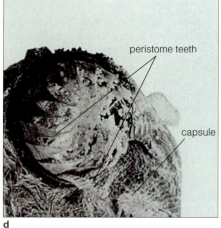

W. Russell and D. Hess

d

peristome teeth

capsule

Figure 22.17 Moss capsules. (**a**) Longitudinal section through a mature *Mnium* capsule, showing the columella and dark-colored spores filling the open space around the columella. (**b**) A mature capsule of *Funaria*. The operculum (lid) is still in place. (**c,d**) Peristome teeth on the top of the capsule that bend in when air is humid and bend out when air is dry, scattering spores that stick to the teeth.

decomposed layer atop the mineral soil. This deposit, called *peat*, can be cut out in blocks, dried, and burned as fuel for cooking and heating (see "ECONOMIC BOTANY: Mining Moss: Peat for Profit" sidebar). On a much larger scale, the Rhode generating station in Ireland burns 2,000 metric tons of peat a day to produce electricity.

Ecologically, bryophytes are important as colonizers of bare rock and sand. Their small bodies trap dust and wind-blown silt, building soil. Mosses create stabilizing soil crusts in such inhospitable places as coastal dunes and inland deserts. Some bryophytes harbor symbiotic nitrogen-fixing cyanobacteria and significantly increase the amount of nitrogen available to the biosphere. In tundra vegetation, bryophytes constitute as

much as 50% of the aboveground biomass, and they are an important component of the food chain that supports animals within the ecosystem. Temperate and cold-temperate forests also have significant amounts of moss biomass.

Bryophytes also are important as tools to improve our basic understanding of fundamental biological processes. Scientists have come to recognize mosses in particular as excellent experimental plants. They are easy to propagate, to grow in small spaces, to clone into sexually identical replicates, and to observe for growth and developmental changes. Many discoveries about moss genetics, tissue development, and ecology in the past several decades have proved widely applicable to land plants in general.

ECONOMIC BOTANY:

Mining Moss: Peat for Profit

Peat is any kind of plant matter that accumulates under water-soaked conditions and that does not completely decompose because the amount of oxygen is limited. Peat is a mixture of freshwater marsh plants such as reeds, sedges, and grasses, and it often includes peat mosses belonging to the genus *Sphagnum*. If conditions are right—cold temperatures, abundant moisture, and water that is acidic and low in nutrients—peat can accumulate in layers at rates as fast as 25 cm per century to a slower 25 cm per millennium. The oldest deposits are tens of meters in thickness. The thousands of square kilometers of peat lands in the cold temperate regions of the world, multiplied by depth, total more than 200 billion metric tons of dry biomass.

Peat has been put to a variety of human uses. It is, first of all, a fuel. During the Stone Age, Europeans found that it could be dug out in blocks, dried, and burned for heat. Currently, Russia, Ireland, Finland, Sweden, Germany, and Poland extensively harvest peat for fuel. An Irish home may use 15 metric tons of dry peat each year, an amount that can be harvested by one person in a month's time. Two-bladed shovels called *slanes* are used by individuals, but commercial operations use specialized tractors, millers, harrows, and harvesters. Such costly investment is needed to harvest the large volumes of peat burned in power plants in Russia and Ireland. The use of peat as a fuel in the United States is minimal, but the energy content is estimated to exceed all current, combined reserves of coal, petroleum, and natural gas.

Peat also is used in horticulture. When added to potting mixes, peat increases the soil's water-holding capacity and lightens the soil, allowing air and water to move more freely. Peat is used as a mulch for acid-loving ornamentals such as rhododendrons and heaths. Compressed peat can be formed into planting containers for seeds, seedlings, cuttings, and root balls. These containers can take up and retain moisture and also decompose over time, releasing the roots to the surrounding soil.

Peat lands also can be cultivated, if drained. Once the peat becomes drier and aerated, it decomposes, releasing nitrogen and other nutrients. Such cool-climate crops as carrots, beets, potatoes, onions, lettuce, cabbage, broccoli, mint, blueberries, and strawberries do well on peat. However, cultivation over the course of many years causes the soil surface to drop because the peat literally oxidizes and blows away. Within a century, several meters of depth can be lost. Agricultural peat lands in California's delta region have to be protected by levees because their surface elevations have dropped as much as 6 m below sea level.

The conservation of peat lands has become an ecological issue in several countries, including the United Kingdom and Ireland, where nearly all of the once-extensive bogs have been stripped. Peat bogs are unique ecosystems, with plant and animal components that have evolved over geologic time. The environmental processes that produce bogs proceed so slowly that it is not possible to restore them in a human lifetime once they have been modified. We certainly have the technology and will to convert entire landscapes and ecosystems, but do we have the right to do so?

 Websites for further study:

Institute of Botany and Botanical Garden:
http://www.botanik.univie.ac.at/pershome/temsch/basics.html

Ireland's Peat Bogs:
http://www.wesleyjohnston.com/users/ireland/geography/bogs.html

KEY TERMS

antheridia	elater	megafossil	rhizoid
antheridiophore	embryo	moss	seta
archegonia	embryophytes	operculum	sporocyte
archegoniophore	gametangium	paraphyses	sporophyte
bryophytes	gemma	pectin	thallus
calyptra	hornworts	peristome teeth	valve
capsule	hydroid	protonema	venter
columella	leptoid		

SUMMARY

1. Bryophytes are small, herbaceous plants. They have in common an unusual life history in which the gametophytes live longer and are more prominent than the sporophytes (a gametophyte-dominant life cycle), and they all possess unbranched sporophytes, each bearing a single sporangium. Their habitats range from aquatic, to humid or wet terrestrial, to epiphytic, to arid.

2. Bryophytes successfully moved into terrestrial habitats by evolving a series of traits that enabled them to resist desiccation, including a multicellular sporophyte phase (the beginning of the sporic life cycle) with an initial embryo stage, specialized gametangia (antheridia and archegonia), and weather-resistant, sporopollenin-coated spores. Some members of the division have leaflike and stemlike organs. They also typically have rhizoids that anchor the plants to the substrate. At the same time, they resemble algae in that free water is essential for the movement of sperm to the egg, there is no lignified supportive tissue, and the gametophyte generation is most prominent.

3. Most plant biologists believe that bryophytes evolved from a single algal ancestor in the charophyta. In this textbook, the 25,000 living species of bryophytes are classified as three separate lineages: hornworts; liverworts, or hepatics, which include thalloid liverworts and leafy liverworts; and mosses, represented by granite mosses, peat mosses, and true mosses.

4. Bryophytes are almost certainly not a monophyletic group because they gave rise to vascular plants. It is unclear which bryophyte lineage represents the first to diverge, but recent molecular and morphological analyses suggest it may have been the hornworts.

5. Hornworts have thalloid gametophytes. They form symbiotic relationships with nitrogen-fixing bacteria and produce pores that may be homologous to stomata. Their sporophytes have an elongated sporangium with indeterminate growth, and sometimes stomata are present.

6. Liverworts include organisms with thalloid gametophytes (*Marchantia*) and others with leafy gametophytes (*Porella*). All liverworts have rhizoids that anchor the plants to their substrate.

7. Sexual reproduction within the liverworts involves archegonia, antheridia, embryos, and small sporophytes; but their morphology and placement depend on the species. *Marchantia* gametangia, for example, are on elevated organs called antheridiophores and archegoniophores.

8. The major points of the moss life cycle are as follows: The life cycle is sporic, the gametophyte generation is dominant, sperm are produced in multicellular antheridia and eggs in multicellular archegonia, water is essential as a medium to allow the sperm to swim from an antheridium to an archegonium, and an embryo stage is present.

9. Moss gametophytes begin growth as a filamentous, alga-like protonemal phase that produces buds, which then develop by mitosis into plants with leaflike appendages on stemlike axes.

10. Specialized tissue in some moss gametophytes includes chlorenchyma, parenchyma, epidermis, water-conducting hydroids, and sugar-conducting leptoids. Some moss gametophytes are capable of tolerating extreme dehydration and then resuming normal activity within minutes of being rewetted. However, despite the presence of conducting tissue in some genera, these organisms are nonvascular plants.

Questions

1. The bryophytes have added several innovations to the algal life cycle, including multicellular gametangia and an embryo stage. How do these innovations contribute to the survival of bryophyte plants?

2. Draw a diagram of the basic life cycle of a bryophyte. Label all the steps and structures involved. Describe each of the structures for a typical liverwort and a typical moss.

3. In what ways is it possible to conclude that the gametophyte generation is dominant over the sporophyte generation in the bryophytes?

4. Describe the difference in morphological and anatomical complexity between the gametophyte of the liverwort *Marchantia* and that of a typical moss, and also between the sporophyte of *Marchantia* and that of a moss.

5. Describe the conducting tissue of some of the most advanced mosses. Why is it not the same as the vascular tissue of higher plants?

6. Mosses cannot control their water balance, and yet they are important elements of the biotic crust of desert soils. How can mosses persist in a desert environment? When would they reproduce?

 InfoTrac® College Edition

http://infotrac.thomsonlearning.com

Cook, M.E., Graham, L.E. 1998. Structural similarities between surface layers of selected charophycean algae and bryophytes and the cuticles of vascular plants. *International Journal of Plant Sciences* 159:780. (Keywords: "charophycean" and "algae")

Gehrke, C. 1999. Impacts of enhanced ultraviolet-B radiation on mosses in a subarctic heath ecosystem. *Ecology* 80:1844. (Keywords: "ultraviolet-B" and "mosses")

The Early Tracheophytes

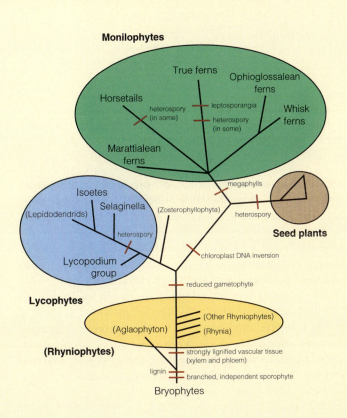

www Visit us on the web at http://biology.brookscole.com/plantbio2 for additional resources, such as flashcards, tutorial quizzes, InfoTrac exercises, further readings, and web links.

1. Tracheophytes, also called vascular plants, possess lignified water-conducting tissue (xylem). Approximately 14,000 species of tracheophytes reproduce by releasing spores and do not make seeds. These are sometimes called seedless vascular plants. Tracheophytes differ from bryophytes in possessing branched sporophytes that are dominant in the life cycle. These sporophytes are more tolerant of life on dry land than those of bryophytes because water movement is controlled by strongly lignified vascular tissue, stomata, and an extensive cuticle. The gametophytes, however, still require a seasonally wet habitat, and water outside the plant is essential for the movement of sperm from antheridia to archegonia.

2. The rhyniophytes were the first tracheophytes. They consisted of dichotomously branching axes, lacking roots and leaves. They are all extinct. They gave rise to two lineages that currently are still present: the lycophytes and the rest of the tracheophytes. This latter group is again divided into two lineages: seed plants and monilophytes.

3. The lycophytes are a group of plants that diverged very early and currently are represented by three lineages, all of which occur in North America. These plants all have leaves, roots, and laterally attached sporangia. One lineage consists of the genus *Lycopodium* (in addition to a number of new genera segregated from it), which is strictly homosporous and has subterranean gametophytes. Each of the other two lineages consists of a single genus. *Selaginella* is heterosporous. *Isoetes*, also heterosporous and having secondary growth, is likely in the lineage that included the extinct lepidodendrids, a major component of the Coal Age swamp forests.

4. The monilophytes consist of five lineages, four of which are commonly called ferns. The whisk ferns or psilophytes are represented by two living genera: *Psilotum* and *Tmesipteris*. Sporophytes of *Psilotum* are reduced, consisting of a green, branching stem, lacking roots and possessing only highly reduced leaves (enations). *Tmesipteris* is an epiphyte with larger leaves. The psilophytes are most closely related to the ophioglossalean ferns. The marattialean ferns are large tropical ferns with upright stems and eusporangia. The horsetails or sphenophytes consist of a single living genus, *Equisetum*. Sporophytes are jointed and ribbed, and their stems have a complex anatomy.

5. The lineage known as monilophytes contains most of the species in the true ferns group. They are diverse in form, and they range from warm tropical regions to the arctic. True ferns are characterized by a distinctive sporangium type, the leptosporangium, which has a long stalk, a single cell layer making up the wall, and an annulus, which aids in dehiscence. Their leaves are complex and varied. A few aquatic ferns are the only heterosporous species. True ferns have food value and other applications for human use.

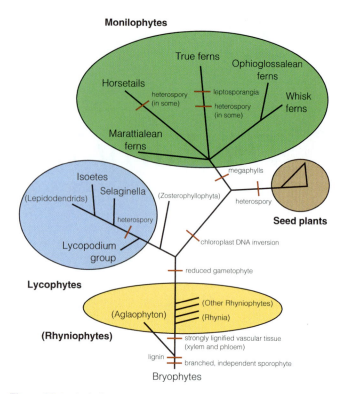

Figure 23.1 A cladogram showing relationships among the early tracheophytes. Relationships among the five lineages of Monilophytes are largely uncertain. Megaphylls and even roots have been lost in some lineages. The names of extinct groups are in parentheses.

with sterile jackets (gametangia), producing multicellular sporophytes, and synthesizing cuticle and other substances. However, these early bryophytes could only survive in sites with abundant water. Soon after achieving a foothold in the terrestrial environment, the bryophytes gave rise to another lineage that became spectacularly specialized for life on land and soon predominated: **tracheophytes (Fig. 23.1)**.

The story of how bryophytes gave rise to tracheophytes and how these plants became increasingly better adapted to harsh terrestrial environments involves profound changes in morphology, life cycle, physiology, and biochemistry. These adaptations reduced the dependence of plants on free water in their environment and allowed plants to occupy a wider range of sites.

If humans had been on Earth about 450 million years ago, they would have witnessed the early stages of this tremendous evolutionary explosion that turned our planet's barren rocks green, stabilized our atmosphere, and became the basis for all the intricate webs of terrestrial life. These new ecosystems became cradles of evolution, giving rise not only to a vast array of plant species, but also to countless other kinds of life, from dinosaurs to butterflies, and, recently, to our own species. However, the initial terrestrial pioneers of 450 million years ago were only a tangled mat of unimpressive, vertical green stems, patchily distributed across the landscape and no taller than this textbook. Within a mere 40 million years, their descendants had produced a rich diversity of forms, including such novel plant structures as roots, wood, leaves, and seeds.

23.1 THE OCCUPATION OF THE LAND

When plants first moved onto the land, they responded to the problem of living and reproducing in a dry environment with key innovations, such as surrounding their gametes

This chapter considers the evolution and diversification of early tracheophytes from their first appearance and initial diversification to the origin of two of the early lineages: lycophytes and monilophytes. Somewhat later, a third lineage of tracheophytes arose with an additional reproductive innovation, the seed. These plants are discussed in Chapters 24 and 25. The tracheophytes discussed in this chapter all disperse by spores. Thus, the plants in this group also are called **seedless vascular plants.**

The First Tracheophytes Were Rhyniophytes

Unmistakable evidence of true xylem appears in the fossil record around the middle of the Silurian period, approximately 430 million years ago, but microfossils suggest that tracheophytes were already established more than 40 million years earlier, in the mid-Ordovician period. We have a clear picture of what these early tracheophytes were like because of extremely well-preserved and abundant fossils found in chert near the village of Rhynie, Scotland. One of the most spectacular fossil beds ever found, it has yielded a vast amount of information on internal and external features, as well as the ecology of the early tracheophyte flora. A variety of plants have been found in these fossil beds. All were small (no taller than 20 cm) and lacked leaves and roots. They consisted of **dichotomously branching** rhizomes with rhizoids attached to them, and vertical aerial stems with sporangia at their tips. The existence of sporangia reveals that these plants were the sporophyte phase of

the life cycle **(Fig. 23.2a).** Unfortunately, few fossil gametophytes have been found.

Rhyniophytes had a simple stem anatomy **(Fig. 23.2b).** There is evidence, as with many early plants, of endosymbiotic fungi living in the stems. These symbiotic relationships may have been crucial to the success of plants on land. The rhyniophyte group gave rise to all other land plants; thus, it is not monophyletic.

Tracheophytes Became Increasingly Better Adapted to the Terrestrial Environment

Examination of fossils and cladistic studies of the earliest land plants have yielded some surprises and have enabled us to reconstruct the most likely path of evolution from bryophyte to tracheophyte. Tracheophyte innovations include: (1) a dichotomously branching sporophyte with multiple terminal sporangia; (2) a free-living, nutritionally independent sporophyte that is prominent in the life cycle; (3) a reduced gametophyte; and (4) lignified vascular tissue (xylem) in the sporophyte.

All of these adaptations were likely important in the colonization of the land. Several dramatic fossils recently have been discovered that help shed light on how and why these changes might have occurred. Although bryophytes succeeded in moving onto the land, they were severely limited by an inability to control their water balance. Therefore, they must have occupied only discrete islands of habitat where conditions were favorable for growth.

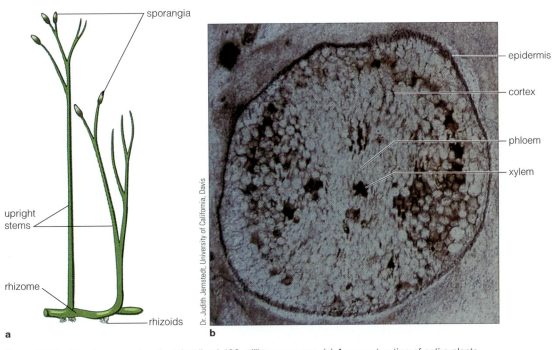

Figure 23.2 *Rhynia,* a vascular plant that lived 400 million years ago. (**a**) A reconstruction of entire plants, about 30 cm tall. (**b**) A cross-section of the stem approximately 3 mm in diameter. A small cylindrical core of xylem is surrounded by a larger cylinder of phloem. An even larger cylinder of cortex, composed of thin-walled parenchyma cells, is packed around the vascular tissue, and the entire stem is covered with an epidermal layer bearing cuticle and stomata.

PLANTS, PEOPLE, AND THE ENVIRONMENT:

Sporophyte Prominence and Survival on Land

Two conspicuous trends in early land plant evolution are the elaboration of the sporophyte and the reduction of the gametophyte. Natural selection for more spores might explain the elaboration of the sporophyte but not its prominence in the life cycle, nor the corresponding reduction in the gametophyte. Why was the sporophyte phase favored over the gametophyte phase in the early stages of land plant evolution? Several hypotheses have been put forward to explain this trend. Some scientists propose that the sporophyte is protected against many deleterious mutations in its genes, because it is diploid and can thus carry harmful mutations as recessive genes. The haploid gametophyte, with only one set of genes, is unable to compensate for a mutated gene by balancing it with a normal version. Evolution, therefore, favored a reduced role in the life cycle for the vulnerable gametophyte and increased prominence of the sporophyte. In addition, the greater genetic diversity of the sporophyte (with two sets of genes) over the gametophyte would have given it a greater capacity to respond to the environment.

Another hypothesis suggests that plants elaborated their sporophytes not because of a genetic superiority of the sporophyte but because gametophytes were limited by the sperm's need to swim to the archegonium. In the competition for resources such as light, water, and nutrients, the gametophytes could grow no larger than the effective dispersal range of the sperm and were limited to locations where water for the sperm to swim through was present. The sporophyte, on the other hand, was not subject to these constraints and was able to increase in size to compete with nearby plants. In an aquatic environment, gametes and spores are equally good at dispersing, and in many algal groups, both are flagellated. On land, spores lost their flagella, became surrounded by weather-resistant walls, and were aided

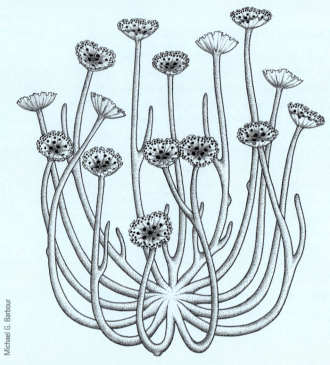

Michael G. Barbour

Figure 1 A reconstruction of the Devonian gametophyte *Sciadophyton*. Gametangia are borne on the terminal disk-shaped gametangiophores. (Kenrick and Crane 1997, p. 304, Fig. 7.29.)

in dispersal by complex sporangial mechanisms, such as the peristome teeth in moss capsules. They became very effective at getting around. Gametes, however, need to fuse with one another to accomplish sexual reproduction. For this, they need water. Gametes are poorly adapted to a terrestrial environment. Thus, evolution may have favored an increase in the prominence of the sporophyte, and a decrease in the prominence of the gametophyte, in the life cycle of land plants.

The shift from gametophyte prominence in the bryophytes to sporophyte prominence in the tracheophytes is dramatic, and no living intermediate plant exists. However, it has been recently discovered that some early fossil plants are actually gametophytes of protracheophytes such as sciadophyton. These gametophytes were as large as the sporophytes, grew upright, branched profusely, and bore sunken gametangia at the tips of their stems **(Fig. 1)**.

One of the strongest selective pressures at this stage must have been exerted by the high mortality of spores; the overwhelming majority of spores released would have fallen on inhospitable ground and failed to germinate. Early plants could have increased spore production by two possible mechanisms: increasing the number of sporophytes produced on each gametophyte, or increasing the number of sporangia per sporophyte. Plants evolved both solutions. Bryophytes, particularly the mosses, produce many sporophytes on each gametophyte. Tracheophytes, in contrast, branched their sporophyte, thereby increasing the number of sporangia the plant could make. These elaborated sporophytes would have soon required stiffening tissue to hold them up, and would eventually have required effective means of transporting water. At some point, to continue this trend of enlarging the sporopyte, a whole new plant architecture would have had to evolve, in which a main stem supports smaller lateral branches.

If the elaboration of the sporophyte was driven by selection for more spores, it would be expected that plants evolved a branched sporophyte first and only later evolved specialized vascular tissues and lateral branching to solve the secondary problems created by branching the sporophyte. Studies of extremely well-preserved rhyniophyte fossils have shown exactly that. Some of the earliest taxa lacked strongly lignified water-conducting cells (tracheids), although they branched profusely (Fig 23.3). These so-called protracheophytes are intermediate between the bryophytes, with unlignified (or no) vascular systems, and tracheophytes, with highly specialized, lignin-stiffened vascular systems. The earliest plants with lignin, such as *Rhynia* (Fig. 23.2b), branched only dichotomously but were eventually superseded by taller plants with a main stem and smaller laterally emerging branches.

Relationships among Early Tracheophytes

Traditionally, tracheophytes that reproduce by releasing spores rather than seeds have been divided into four divisions (Table 23.1). However, botanists have come to realize

Figure 23.3 A reconstruction of the extinct Devonian plant *Horneophyton lignieri*. (Kenrick and Crane, 1997, p. 50 Fig. 2.36a)

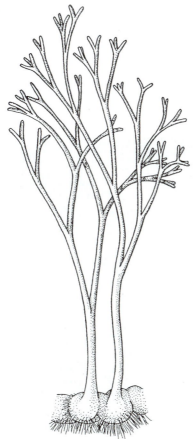

Table 23.1 A Cross-Walk Comparison of Traditional Taxonomic Classification with the Evolutionary Groups of Living Seedless Vascular Plants

Evolutionary Groups	Taxonomic Ranks
Tracheophytes	No corresponding taxonomic rank
Lycophytes	Lycophyta
Club mosses	Lycopodiales
	Lycopodiaceae
	Lycopodium, Phlegmariurus, Huperzia, Diphasiastrum, Palhinhaea, Pseudolycopodiella, Lycopodiella
Spike mosses	Selaginellales
	Selaginellaceae
	Selaginella
Quillworts	Isoetales
	Isoetaceae
	Isoetes
Monilophytes	No corresponding taxonomic rank
Whisk ferns	Psilophyta
	Psilotaceae
	Psilotum, Tmesipteris
Horsetails or Sphenophytes	Sphenophyta (=Equisetophyta)
	Equisetaceae
	Equisetum
	Calamitaceae
	Calamites
Not a monophyletic group	Filicophyta (=Pterophyta, Pteridophyta) [Includes five orders listed below.]
Ophioglossalean ferns	Ophioglossales
	Ophioglossaceae
	Ophioglossum, Botrychium
Marattialean ferns	Marattiales
True ferns: Water ferns	Salviniales
	Azollaceae
	Salvinia
	Azolla
True ferns: Water ferns	Marsileales
	Marsileaceae
	Marsilea, Pilularia, Regnellidium
True ferns	Filicales (5Polypodiales) many families

All living genera present in North America are listed except those in the marratialean and true fern groups, which are too numerous to include.

that this taxonomic scheme does not accurately reflect the evolutionary history of these plants. Studies using cladistic methods have unequivocally demonstrated that living tracheophytes fall into two major clades, which separated more than 400 million years ago (Fig. 23.1). One lineage currently is represented by lycophytes, whereas the second lineage includes all other tracheophytes. The evidence for this ancient split includes a series of morphological features, but also one distinctive molecular feature. There is a region of about 30,000 base pairs in the chloroplast DNA of all plants. This sequence on the chloroplast chromosome "reads" in one direction in the bryophytes and lycophytes, but the opposite direction for all other plants. This inversion must have taken place after the lycophyte line split from the main vascular plant lineage.

Recent studies of the nonlycophyte vascular plants involving morphology and DNA sequences from the chloroplast and the nucleus have yielded an unexpected new result. This group of tracheophytes is itself divided into only two major lineages: the seed plants and a diverse group called the monilophytes. This clade is primarily composed of a variety of plants commonly called ferns, and includes two small, unusual groups that are traditionally treated as divisions, horsetails and whisk ferns.

Currently, we are unsure of the exact pattern of relationships among monilophytes. There appear to be five lineages, and these are divided into four major clades: whisk ferns plus ophioglossalean ferns, horsetails, marattialean ferns, and true ferns. The exact relationships among these four clades have not yet been worked out.

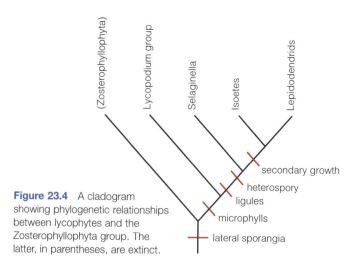

Figure 23.4 A cladogram showing phylogenetic relationships between lycophytes and the Zosterophyllophyta group. The latter, in parentheses, are extinct.

23.2 THE LYCOPHYTES

The lineage that gave rise to **lycophytes** separated from other vascular plants early in the history of tracheophytes **(Fig. 23.4).** Living lycophytes thus preserve a set of different solutions to the problems of a terrestrial life than other vascular plants. The lycophyte line originated in the Devonian or Silurian period. The earliest known members of the lineage are a group called Zosterophyllophyta, which is now extinct. Plants in this group lacked leaves and roots, and they branched dichotomously like the rhyniophytes. However, one unique morphological feature distinguished this lineage from rhyniophytes: the sporangia were attached to stems in a lateral rather than terminal position. The Zosterophyllophyta are important because they either gave rise to or are closely related to the lycophytes. In either case, the lateral sporangial position was present in the common ancestor of both Zosterophyllophyta and lycophytes and is universal among lycophytes.

The lycophyte line reached its peak of diversity and ecological importance in the Coal Age **(Fig. 23.5).** During this period, Earth's climate was warm, topographical relief was low, and lush swamps covered vast coastal lowlands. These swamps supported forests composed largely of trees in a

Field Museum of Natural History

Figure 23.5 A diorama reconstruction of a Coal Age forest.

now extinct lycophyte group called the **lepidodendrids.** One example is *Lepidodendron,* which grew to more than 35 m in height and had secondary growth, although with a different pattern than that of most modern plants **(Fig. 23.6).**

These trees branched dichotomously in both their aerial and underground portions. The fossilized trunks are covered with distinctive diamond-shaped pads where leaves were attached. The underground stems were similarly covered with roots. Lepidodendrids continue to exert a strong ecological and economic influence today because their abundant bodies, compressed under millions of years of accumulated sediment, have been transformed into the extensive coal deposits currently being mined (see "PLANTS, PEOPLE, AND THE ENVIRONMENT: Coal, Smog, and Forest Decline" sidebar).

Lycophytes produce a particular kind of leaf called a **microphyll.** Although the term suggests small size, some of these leaves were quite large. Microphylls are defined by the presence of a single vascular bundle. This bundle (or trace)

does not branch inside the leaf and causes no interruption in the vascular strand of the stem when it peels off to enter the leaf. The most likely candidate for the original structure is a lateral sporangium that was converted to a light-gathering structure to enhance photosynthesis. As the sporangium evolved to more efficiently gather light, it flattened, increased in size, became sterile (lost its spore-producing function), and ultimately became vascularized, forming the microphylls found in all lycophytes, living and extinct.

Today, lycophytes represent a mere remnant of their Coal Age diversity. The group consists of three very different lineages: *Lycopodium* (and related genera), *Selaginella,* and *Isoetes.* Each of these lineages has a worldwide distribution, predominantly in tropical and subtropical forests, but a number occur in temperate and arctic regions. Some have adapted to aquatic environments, completing their entire life cycle fully submerged, whereas others grow in deserts and survive drought much like mosses, by shriveling and becoming dormant.

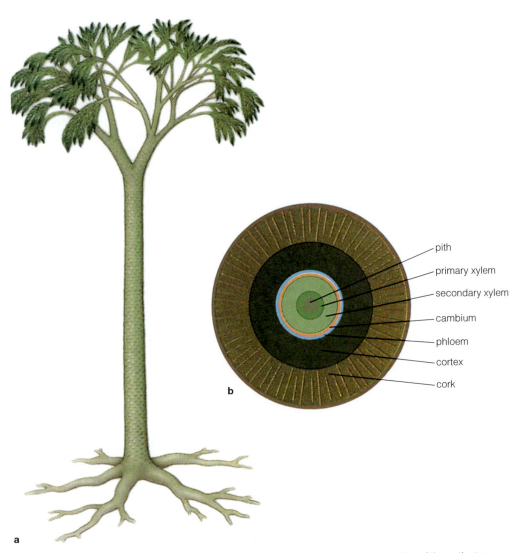

pith
primary xylem
secondary xylem
cambium
phloem
cortex
cork

Figure 23.6 *Lepidodendron,* a dominant tree of the Coal Age forest. (**a**) A reconstruction of the entire tree. (**b**) A fossilized cross-section of the trunk, about 1 m in diameter. (Art by Raychel Ciemma.)

PLANTS, PEOPLE, AND THE ENVIRONMENT:

Coal, Smog, and Forest Decline

All human societies burn fuel, thereby releasing byproducts into the environment. The particular material that is used as fuel and the conditions under which it is burned produce very different types and volumes of byproducts. Technological societies today use fuel for heating, cooking, industrial processes such as smelting, producing mechanical energy, and creating electricity. Although water, wind, sunlight, and nuclear fission are also used for some of the same objectives, most of our demand for energy is met by the carbon-based fuels of wood, coal, natural gas, and crude oil.

It is ironic that we depend on carbon-based fuels, because carbon is a rather uncommon element. It makes up only 0.04% of the earth's mass. Recall that photosynthesis does not create carbon; it merely rearranges it from part of a gas to part of a sugar. Carbon is finite on Earth; therefore, the rate at which it cycles through the biosphere is very critical to life.

Coal is not just a form of carbon, like graphite or diamonds. It is a mixture of organic substances, moisture, and minerals. The major elements that make up the chemical compounds are carbon, hydrogen, oxygen, nitrogen, and sulfur. Coal begins as plant debris deposited in such a way that oxygen is limited. Consequently, the litter is only partially decomposed. Over time, the deposit of litter thickens. The next process in coalification is burial of the organic material in sediment. Over time, if the sediment is deep enough, pressure and heat compress the material further, turning it into a soft rock we call coal.

The heat released by coal when it burns depends upon the pressure and heat it was subjected to over geologic time. Standard names are given to coal of different heat value: lignite (which is the lowest grade), subbituminous coal, bituminous coal, and anthracite. Heat is typically measured in BTUs (British thermal units). One BTU = 252 calories, and 100,000 BTUs = 1 therm = 25,200 kcal. On average, 1 metric ton (1.1 tons) of coal contains 278 therms. Coal satisfies about 30% of the world's annual energy demand.

Coal reserves are not uniformly distributed throughout the world. Of the earth's estimated recoverable coal reserves of 10 trillion metric tons (11 trillion tons), the United States has nearly 50%. The availability of coal has been an economic boon but an ecological disaster for those countries with large coal reserves. Byproducts of coal burning include soot, volatile hydrocarbons, and oxides of nitrogen and sulfur. When the English had exhausted their wood reserves for fuel in the seventeenth century and began to burn coal, London became a notoriously dirty and noxious place to live. Fogs became thicker, because fog droplets could condense around soot particles. Fog became **smog** (smoke + fog). Another health hazard came from the nitrogen and sulfur oxides, which become strong acids in water and can cause respiratory diseases. London became infamous for "killer smogs" as its population and its air pollution climbed into the twentieth century—smogs so unhealthy that people actually died from their effects, particular young people and the elderly. Tall smokestacks alleviated some of the problem locally, but winds then carry the pollutants elsewhere; in the case of England the smog blew into the Baltic countries and into Central Europe. The burning of soft coal, high in sulfur, is no longer allowed in London.

Central and Eastern Europe created their own problems. After World War II, Poland, Czechoslovakia, and Germany became heavily industrialized, burning sulfur-rich coal without smokestack scrubbers, which remove sulfur oxides before they reach the atmosphere. As a result, millions of hectares of conifer forest in Central and Eastern Europe suffered a decline, especially noticeable after 1980. Symptoms include premature shedding of needles and cones, discoloration of needles, abnormal branching patterns, loss of mycorrhizal associations, slowed growth, and death. The combination of symptoms and the wide diversity of affected species do not suggest any known biologically caused disease. Forest ecologists believe that the decline is caused primarily by acid rain from coal burning and automobile exhaust. In North America, major coalfields exist in the Northwest Territories of

Lycopodium Has a Homosporous Life Cycle

The lineage that includes the genus *Lycopodium* is familiar to many as the evergreen trailing plants used in making wreaths **(Fig. 23.7b).** The abundant spores are highly flammable and were once used by magicians and photographers to cause dramatic flashes of light. The spores also were used to coat such latex items as gloves and condoms, but it was eventually discovered that they are extremely irritating to the skin, and their use was discontinued. *Lycopodium clavatum* has been shown experimentally to have hypoglycemic effects.

The large genus *Lycopodium* once contained about 400 species, but it has been broken into 10 to 15 smaller genera. There are seven genera in North America (see Table 23.1). Most are trailing plants, with short upright branches that resemble thick mosses or pine seedlings (hence, the common names club moss and ground pine). The stems branch dichotomously and often are prostrate. They are sheathed

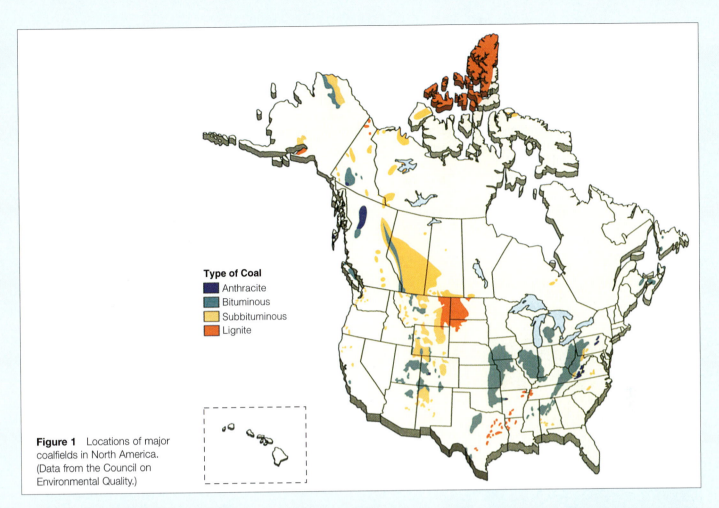

Figure 1 Locations of major coalfields in North America. (Data from the Council on Environmental Quality.)

Type of Coal
- Anthracite
- Bituminous
- Subbituminous
- Lignite

Canada, in the Appalachian and Midwest regions of the United States, in North and South Dakota, and in the Four Corners area of Utah, Colorado, New Mexico, and Arizona **(Fig. 1).** Much of the eastern coal has been mined in underground shafts, whereas western coalfields are open pits. In both cases, noncoal refuse is left aboveground as polluting mine spoil that is very difficult for plants to invade.

Despite the ecological problems that result from the mining and burning of coal, it is likely to remain a prime source of energy for the world for decades to come. Alternative energy sources, such as nuclear fission and hydroelectric dams, have equally damaging or potentially more damaging consequences. We can expect continued technological advances that will clean up coal prior to burning, trap and recycle by-products after combustion, and restore the landscape after mining has ceased.

Websites for further study:

EPA—Acid Rain: http://www.epa.gov/airmarkets/acidrain/

Forest Health—Our Forests Look Healthy, but Are They Really?: http://www.daviesand.com/Perspectives/Forest_Health/

Forest Damage in the Czech Republic: http://www.si.edu/harcourt/h_si/ceps/research/CzechText.html

Traffic, Transportation, and Environment—Emission Legislation and Test System Requirements: http://www.adb.org/Documents/Events/2001/RETA5937/Chongqing/documents/cq_06_preschern.pdf

Figure 23.7 Representatives of the groups discussed in this chapter. (**a**) A whisk fern, *Psilotum nudum.* (**b**) A club moss, *Lycopodium obscurum.* (**c**) A horsetail, *Equisetum telmateia.* (**d**) Chain fern, *Woodwardia fimbriata.*

a — J. W. Perry

b — J. W. Perry

c — Doug Sokell/Visuals Unlimited

d — Michael G. Barbour

with numerous spirally arranged microphylls **(Fig. 23.8).** The stem has a netlike system of interconnected strands of xylem with phloem between them (Fig. 23.8c,d). The xylem contains tracheids, and the phloem contains sieve cells and parenchyma cells. There is no true endodermis, but typically a layer of sclerenchyma encircles the vascular cylinder. Roots arise at the shoot apical meristem and may grow through the cortex for some distance and emerge on the underside of the horizontal stems.

Lycopodium has a homosporous life cycle, meaning that only one type of spore is made and that the gametophytes are bisexual (that is, they produce both archegonia and antheridia) **(Fig. 23.9).** Sporangia are produced on the top surface of **sporophylls** (leaves bearing sporangia). In many members of this group, the sporophylls are aggre-

gated into **strobili** (singular, *strobilus*), soft, conelike structures (Fig. 23.9a).

Meiosis occurs inside sporangia, producing numerous haploid spores (Fig. 23.9c). The spores are shed, germinate on the ground, and develop into gametophytes. These typically are long-living, subterranean, and require an endosymbiotic fungus to survive. Gametophytes of some species grow aboveground and are photosynthetic, but they apparently still require the fungus to develop normally (Fig. 23.9d).

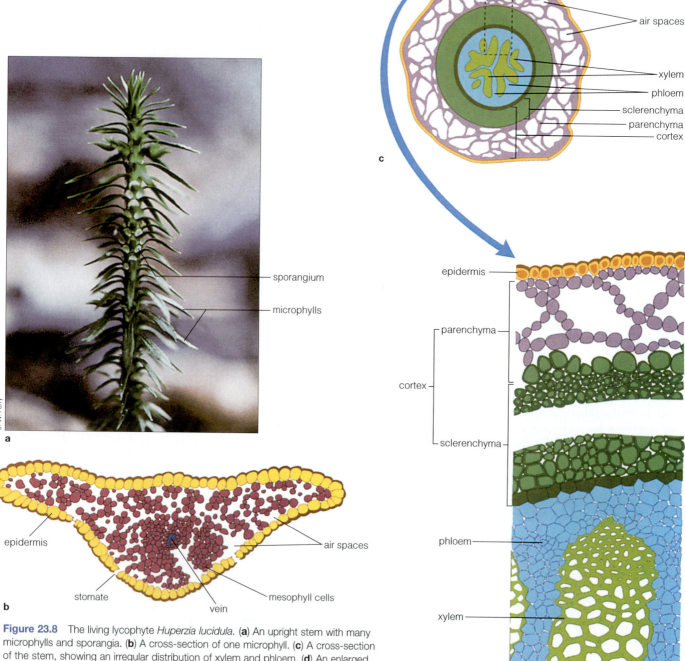

Figure 23.8 The living lycophyte *Huperzia lucidula.* (**a**) An upright stem with many microphylls and sporangia. (**b**) A cross-section of one microphyll. (**c**) A cross-section of the stem, showing an irregular distribution of xylem and phloem. (**d**) An enlarged section of the stem, showing the outer and inner regions.

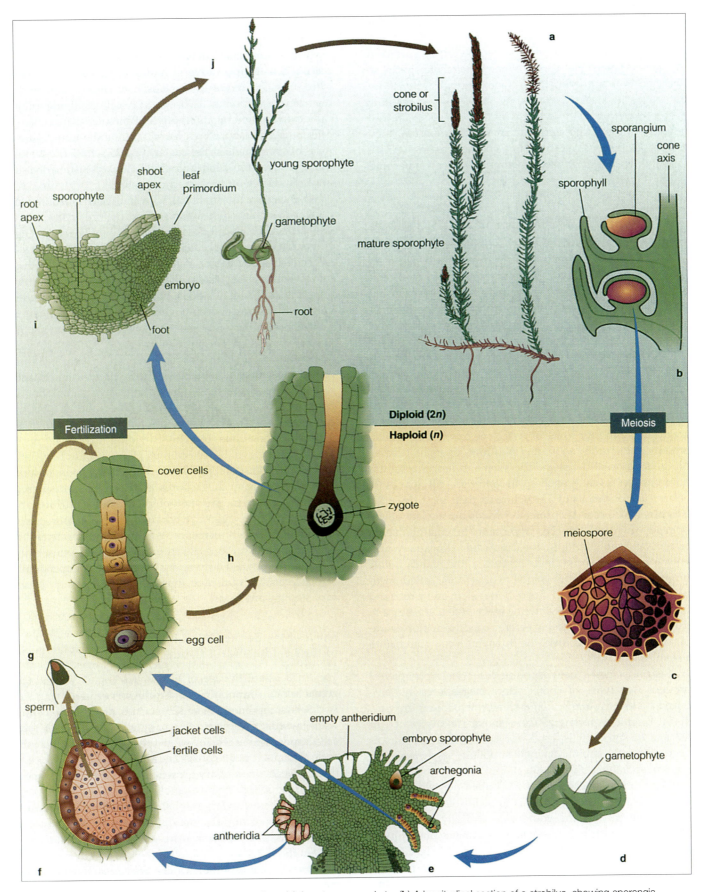

Figure 23.9 The life cycle of a species of a *Lycopodium*. (**a**) A mature sporophyte. (**b**) A longitudinal section of a strobilus, showing sporangia attached to microphylls (sporophylls). Meiosis within the sporangium produces (**c**) haploid meiospores, each of which can germinate, multiply by mitosis, and grow into (**d**) a gametophyte. (**e**) A longitudinal section of a portion of a gametophyte, showing the location of archegonia and antheridia. (**f,g**) Biflagellate sperm are released from antheridia and swim in a film of water to archegonia, where one sperm fuses with one egg to create (**h**) a diploid cell called the zygote. The zygote remains embedded in, and dependent upon, gametophyte tissue. (**i,j**) The zygote divides by mitosis and grows into an embryo and then into a mature sporophyte.

Archegonia and antheridia are borne on the surface of the gametophyte (Fig. 23.9e). When biflagellate sperm are liberated from antheridia, they swim through water to fertilize eggs inside archegonia (Fig. 23.9f–h). The resulting zygote develops into an embryo that possesses a well-developed foot, a short primary root, leaf primordia, and a shoot apex (Fig. 23.9i). The young sporophyte (Fig. 23.9j) is initially dependent on the gametophyte, but soon becomes self-sustaining.

Selaginella Has a Heterosporous Life Cycle

Another lineage of lycophytes has a single living genus, *Selaginella* (spike moss), with approximately 700 species. Most are tropical, but a few tolerate relatively dry or cold habitats. Several species are commercially grown as ornamental plants, including *Selaginella lepidophylla* (resurrection plant, rose of Jericho), *Selaginella willdenovii* (peacock fern), and *Selaginella braunii* (treelet spike moss). In Chinese medicine, *Selaginella doederleinii* is added to a stew of lean pork to treat "immobile masses."

The sporophyte consists of a dichotomously branched stem, which is prostrate in some species (Fig. 23.10a) and upright and slender in others. The microphylls frequently are arranged in four rows, or *ranks,* one row of large leaves on either side of the stem, and two rows of smaller leaves on the topside of the stem. The stem and leaves resemble miniature cypress branches. In addition, all leaves of *Selaginella* possess a small structure on their top side called a **ligule,** which secretes protective fluids during leaf development. Ligules can be important in distinguishing *Selaginella* from plants in the *Lycopodium* lineage, which always lack ligules. Special meristems at the branch points produce an organ that is unique to *Selaginella,* called a **rhizophore.** It has characteristics of both a stem and a root and typically grows downward to the soil where it gives rise to true roots. However, under certain conditions (for example, if the shoot tip is lost), the rhizophore can give rise to a stem.

Selaginella has a heterosporous life cycle. Sporophytes produce two types of spores: large **megaspores** are produced by **megasporangia,** and small **microspores** are produced by **microsporangia.** These sporangia are located in the axil of sporophylls, as in *Lycopodium,* but are always aggregated into strobili. A single strobilus usually has both types of sporangia (Fig. 23.10b,c).

Megasporangia are filled with diploid megasporocytes. One of them divides by meiosis to yield four large megaspores; the rest of the sporocytes typically degenerate. Megaspores divide mitotically to form a **megagametophyte,** but all cells remain inside the thick spore wall (Fig. 23.10d). Immature megagametophytes can be passively released from the sporangium or actively ejected by a snapping motion of the peeled-back sporangium wall. When megagametophytes of *Selaginella* reach maturity, the spore wall cracks open and a cushion of gametophyte tissue protrudes. Archegonia develop in this cushion of tissue, and frequently rhizoids protrude (Fig. 23.10e).

Microsporangia undergo a similar process. They initially are filled with up to several hundred diploid microsporocytes, which divide by meiosis to make four small microspores each (Fig. 23.10f,g). Microspores divide mitotically to form a **microgametophyte.** This consists of a layer of cells inside the spore wall forming an antheridium, and a mass of sperm cells in the center (Fig. 23.10h,i). The mature microgametophyte thus is a minute plant consisting of a single antheridium within a microspore wall.

The sperm cells are liberated when the microspore wall becomes wet, and they swim toward mature archegonia in which a single egg cell resides (Fig. 23.10j). The haploid egg and haploid sperm cell combine to produce a diploid zygote cell, which immediately begins to divide and differentiate into an embryo (Fig. 23.10k,l). The embryo does not become dormant, as it does in a seed, but instead continues to grow into a fully mature sporophyte (returning to Fig. 23.10a).

Heterospory Allows for Greater Parental Investment

Heterospory undoubtedly evolved in *Selaginella* to enhance reproduction. It is a system based on a division of labor. In homosporous species, dispersal of spores is effective, but the spores have no stored food. Consequently, many are viable for only a short time, thereby providing inadequate nutrition to the embryo. Heterosporous plants such as *Selaginella* produce a megaspore, provisioned with a rich store of food. These large, heavy megaspores are not widely dispersed, but the resulting megagametophyte provides nutrition and protection for the zygote, embryo, and young sporophyte. The smaller, lighter microspore has little food resources, but it can disperse over long distances, to bring the sperm to the megagametophyte. Heterospory has evolved independently in many plant groups and represents an important, necessary step toward seeds.

Isoetes May Be the Only Living Member of the Lepidodendrid Group

The third extant lineage of lycophytes contains the single genus *Isoetes,* commonly called quillwort or Merlin's grass. *Isoetes* contains more than 125 species distributed worldwide in aquatic habitats. All are small plants, which typically grow submerged in water for part or all of their life cycle. The plant body consists of a lobed cormlike structure that undergoes secondary growth and produces roots and a tuft of microphylls that resemble grass leaves. The microphylls are filled with large air chambers and have prominent ligules. Most, if not all, microphylls are sporophylls. The entire plant has been likened to a strobilus attached to a bit of stem (Fig. 23.11).

Isoetes is heterosporous. Unlike *Selaginella,* however, each megasporangium produces several hundred megaspores, and each microsporangium produces more than a million microspores. Sperm cells are multiflagellate, whereas those of other living lycophytes are biflagellate. This strange plant shares a number of features with the extinct tree lycophytes of the Carboniferous, including heterospory, ligules, secondary

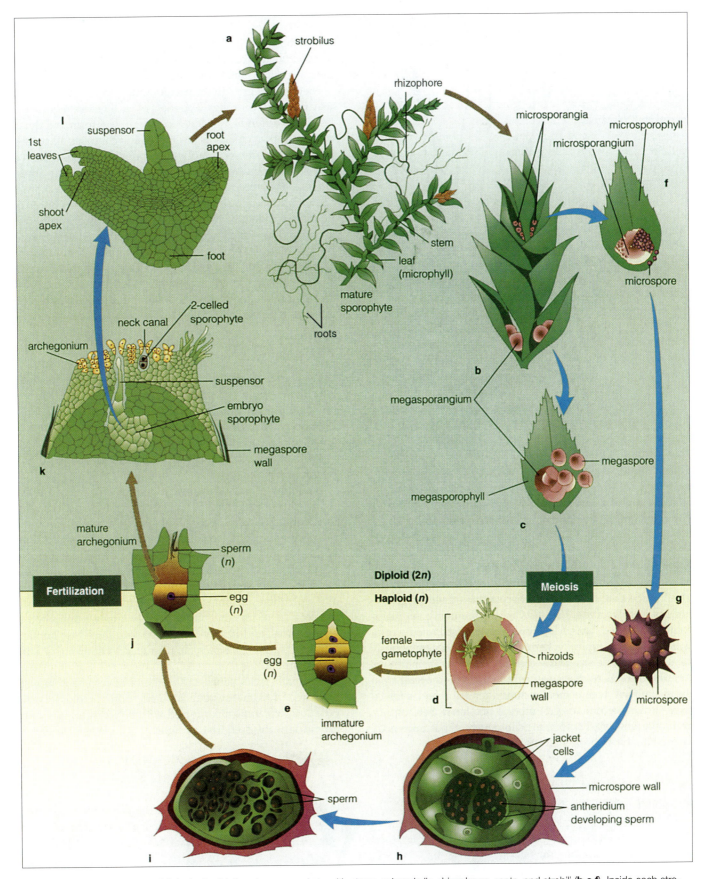

Figure 23.10 The life cylce of *Selaginella*. (**a**) A mature sporophyte with stems, microphylls, rhizophores, roots, and strobili (**b,c,f**). Inside each strobilus are two kinds of sporangia: megasporangia and microsporangia. Meiosis within the larger megasporangium yields four megaspores, each of which divides by mitosis to produce (**d**) megagametophytes. Meiosis within the smaller microsporangium produces many microspores (**g**), each of which develops by mitosis into (**h,i**) microgametophytes, which produce sperm. The motile sperm swim from the antheridium to a nearby archegonium (**d,e,j**), and each can fertilize one egg. A fertilized egg becomes a zygote, which divides by mitosis and grows into an embryo (**k,l**) and then into a mature sporophyte (**a**). The zygote and embryo are retained within, and are dependent upon, gametophyte tissue.

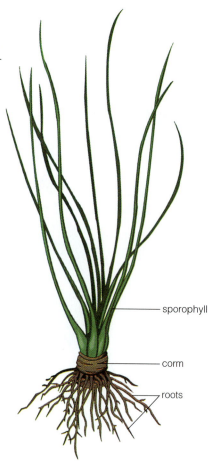

Figure 23.11 *Isoetes*, a living member of the lyco-phyte lineage. The plants are small (3–6 cm tall) and typically grow sub-merged in shallow water for part of each year.

— sporophyll

— corm

— roots

growth, and a distinctive root morphology. Indeed, recent cladistic analyses and the fossil record suggest that *Isoetes* is a living member of the ancient lepidodendrid group.

23.3 THE MONILOPHYTES

The lineage that includes all other seedless tracheophytes—besides the lycophytes—consists predominantly of plants commonly called ferns. **Monilophytes** are distributed world-wide. They are typically herbaceous today, but previously some reached tree size and were important members of Coal Age swamp forests. Because the monilophyte line split from the lycophytes before their common ancestor had evolved roots, leaves, secondary growth, the tree habit, heterospory, and multiflagellated sperm, the subsequent appearance of all

these features in both lineages represents a remarkable series of evolutionary convergences every bit as spectacular as the independent evolution of wings in birds, bats, and pterosaurs.

Secondary growth in monilophytes occurs in a different way than in lycophytes. The roots of each group also are dis-tinctively different, but the variation in leaf form and devel-opment is particularly revealing. The lycophytes all possess microphylls. The monilophytes produce a different kind of leaf called a **megaphyll.** More than one vascular strand enters a megaphyll, and the strands branch extensively in the leaf. In addition, the vascular strands cause an interrup-tion in the xylem of the stem (called a leaf gap) where they branch off to enter the leaf. The difference between micro-phylls and megaphylls reflects their different evolutionary origins. Microphylls are thought to have originated from lat-eral appendages, such as sporangia. Megaphylls, in contrast, are thought to have resulted from the modification of a branch system **(Fig. 23.12).**

Whisk Ferns

Whisk ferns, or **psilophytes,** are unusual plants, and until recently, there has been debate about where they fit in the phylogeny of plants. Some of them look superficially like rhyniophytes, and some botanists have proposed that they are direct descendants of this group. DNA analysis has shown definitively, however, that psilophytes are in the monilophyte clade.

Psilophytes have no known fossil record. There are two living genera: *Psilotum* (Fig. 23.7a) and *Tmesipteris*. These are relatively uncommon herbaceous plants that grow in tropi-cal and subtropical regions, often as epiphytes. *Psilotum* occurs widely through the Pacific Islands (including Hawaii), Africa, and Asia. It also has been reported in a vari-ety of locations across the southern part of the United States, although there is some question as to whether it is truly native in all of its range. It commonly appears uninvited in greenhouses, and it may have been introduced by humans in many places. *Tmesipteris* is restricted to the South Pacific and Australia.

Psilotum lacks roots and consists of a dichotomously branched rhizome system covered with rhizoids. Cortex cells of the rhizome are infected with mycorrhizal fungi that extend into the soil. The upright, aerial stems also branch

Figure 23.12 Hypothetical stages in the evolution of megaphylls. (**a**) A dichotomously branched, leafless plant begins to branch unequally, so that at each fork there is a longer main stem axis and shorter side branches. (**b**) The side branches become splayed into a plane, like a miniature espalier. (**c**) The splayed branch system becomes webbed. (**d**) The webbing becomes the leaf blade and the branches become veins.

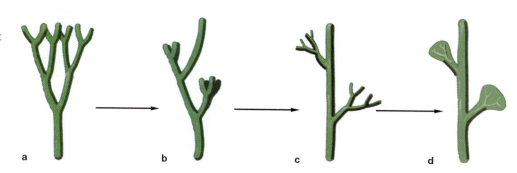

a   b   c   d

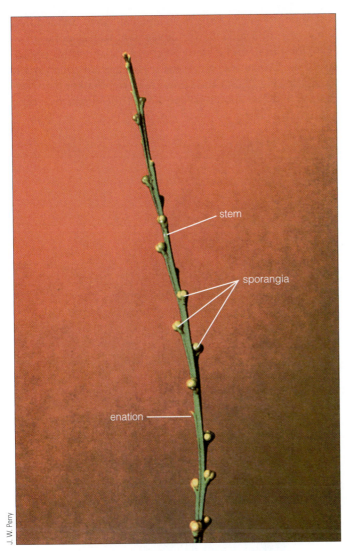

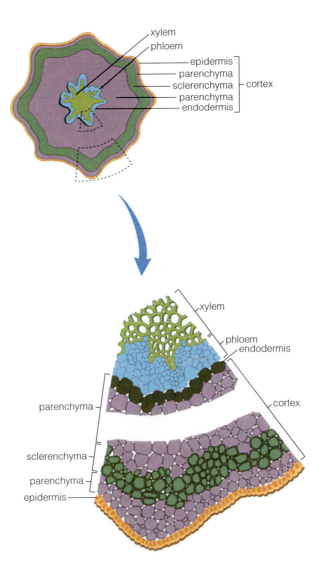

Figure 23.13 A branch of the whisk fern *Psilotum nudum*. Sporangia in all species of this genus occur in aggregations of three (or occasionally more) that are fused together. Small epidermal outgrowths are enations, which may have evolved from sterilized sporangia, or they may be reduced microphylls.

Figure 23.14 A cross-section of a *Psilotum* stem.

dichotomously, and the sporophyte typically is less than 60 cm in height. The aerial stems bear **enations,** small stem outgrowths that are leaflike but lack vascular tissue **(Fig. 23.13).** *Psilotum* is usually described as lacking leaves, but these enations most likely originated from true leaves.

This simple body, lacking roots and possessing only highly reduced leaflike structures, is unique among living tracheophytes but was once common among extinct rhyniophytes and Zosterophyllophyta. In psilophytes, these simple traits are likely adaptations to an epiphyte lifestyle. They have strong associations to mycorrhizal fungi.

The internal stem anatomy is unlike that of the rhyniophytes **(Fig. 23.14).** Rhizomes have a central strand of solid xylem, sometimes ridged or lobed; stems have a pith with fibers, surrounded by a cylinder of xylem. To the outside of the xylem is an endodermis with a Casparian strip, then a cortex with several zones of different cell types, and finally

an epidermis with a thick cuticle and many stomata. Cortex cells often contain chloroplasts, making the entire stem appear green.

Psilophytes are homosporous. Sporangia are borne on short stalks in the axils of enations and typically are fused into groups of two (*Tmesipteris*) or three (*Psilotum*). Meiosis occurs inside sporangia, forming tetrads of spores. The spores are released and, if they land in a suitable habitat, germinate and form a bisexual gametophyte.

The gametophytes lack chlorophyll and associate with endomycorrhizal fungi. Because they are nonphotosynthetic and are fed by fungi, the gametophytes may grow below the soil surface or in crevices in rocks or bark. When mature, they are quite elaborate. They bear rhizoids, grow from an apical meristem, and branch dichotomously. Some even have vascular tissue. Gametangia are scattered over the surface of the gametophyte; flask-shaped archegonia contain one egg each, and spherical antheridia contain spirally coiled, multiflagellate sperm.

The other genus, *Tmesipteris,* is an epiphyte with dangling branches. Although it lacks roots, like *Psilotum,* it has leaves. They have a single, unbranched vascular strand, but they have evolved independently of the microphylls of lycophytes.

Ophioglossalean Ferns

The closest relative of the psilophytes (as determined by DNA comparisons) is a group of about 75 species of unusual plants called **ophioglossalean ferns.** These consist of several genera, including *Botrychium,* the grape fern **(Fig. 23.15a),** and *Ophioglossum,* the adder's-tongue fern **(Fig. 23.15b).** These small plants are members of the spring flora in temperate regions and also are found in disturbed or open tropical sites. A few are epiphytes. *Ophioglossum* has the greatest number of chromosomes of any plant, $2n$ being as high as 1,260 in some species.

Ophioglossalean ferns have unusual leaves divided into two segments: a spikelike fertile segment with a sporangium embedded in it, and a sterile segment expanded for photosynthesis. Unlike true ferns, the leaves are not coiled when young (see Fig. 23.21), the stems are upright rather than horizontal, and the sporangia are not of the distinctive type found in true ferns.

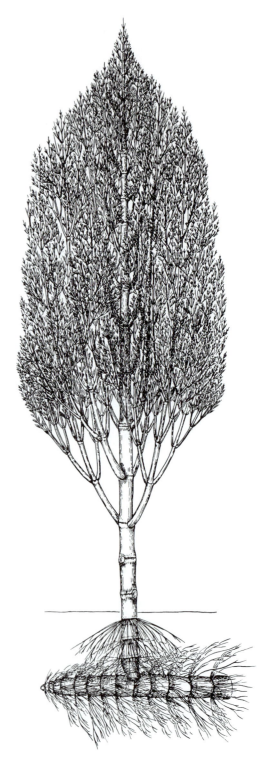

Figure 23.15 Two ophioglossalean ferns. (**a**) A grape fern (*Botrychium multifidum*). (**b**) An adder's-tongue fern (*Ophioglossum*).

Figure 23.16 A reconstruction of *Calamites,* a Coal Age tree, showing the aboveground system of branching stems, the horizontal rhizome, and roots.

The roots of many ophioglossalean ferns also are distinctive. They run horizontally through the soil and produce shoot buds at intervals, allowing the plant to spread vegetatively. In addition, the roots have strong mycorrhizal relationships and lack root hairs.

Horsetails

The **horsetails,** or **sphenophytes,** consist of only one living genus, *Equisetum,* with about 25 species (Fig. 23.7c). *Equisetum* has a worldwide distribution, except for Australia and New Zealand. Many of these species live in moist habitats, but a few, such as *Equisetum arvense,* can grow in seasonally dry places. *Equisetum* contains silica in its stem epidermis, making it abrasive. In pioneer days, *Equisetum* was used to scour pots and pans—hence, it was commonly referred to as scouring rush. The bushy, branching structure of some species gives the genus another common name: horsetail.

Horsetails can be weedy, and some are toxic to livestock or humans because they contain enzymes that break down thiamine. If ingested regularly, symptoms of deficiency can occur. Horsetails have been used medicinally throughout the world for urinary and kidney problems, as well as to reduce bleeding. Native Americans ate the young shoots raw or boiled as a spring green, before silica has been deposited in the stems. The strobili were especially prized. Consuming horsetail was considered cleansing and good for the blood.

Sphenophytes originated in the Devonian period, and, like the lycophytes, they were important members of the Coal Age swamp forests. They were quite diverse, ranging from vines to trees. One example, *Calamites,* looked very much like a giant version of *Equisetum,* with creeping rhizomes and secondary growth, and it grew in dense stands **(Fig. 23.16).** The sporophytes of all sphenophytes, living and extinct, are easily recognized by their jointed and ribbed stems and their whorled appendages.

Living horsetails are typically less than 2 m tall, but one tropical species reaches 5 m in height. All horsetails are perennial and have a branched rhizome from which upright stems arise. Depending on the species, the upright stems may branch profusely, sparingly, or not at all. They are marked by vertical riblike ridges and distinctly jointed nodes. A whorl of branches or leaves, or both, occurs at each node. The leaves are relatively small and lack chlorophyll. The leaves of living *Equisetum* are microphylls, but they were derived by reduction from megaphylls. (Many extinct sphenophytes had megaphylls.) Roots emerge from rhizome nodes but may also arise from stem nodes if the stem is in contact with moist soil.

The stem anatomy of *Equisetum* is quite distinctive, with a large central cavity surrounded by a ring of vascular bundles and smaller cavities, called **vallecular canals (Fig. 23.17).** The stems are hollow except at the nodes, where a diaphragm of tissue forms a solid joint (not shown in Fig. 23.17). The biophysicist Karl Niklas has shown that these hollow stems are

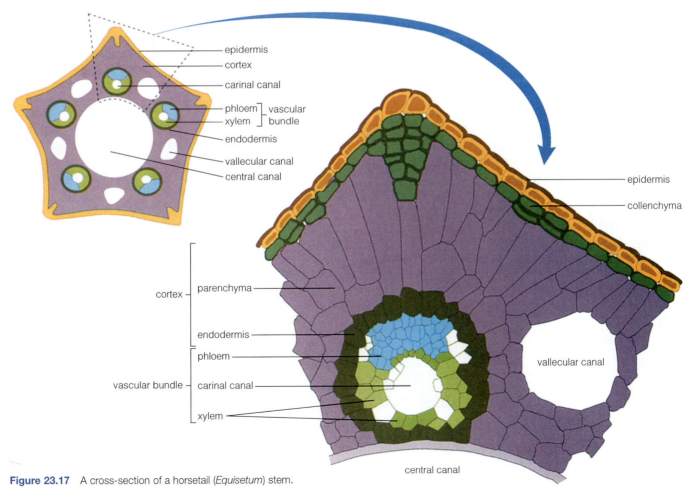

Figure 23.17 A cross-section of a horsetail (*Equisetum*) stem.

PLANTS, PEOPLE, AND THE ENVIRONMENT:

Diversity among the Ferns

True ferns have an amazing variety of growth forms, besides the familiar lacy-leaved, moisture- and shade-loving herbs familiar to most of us. Some of them are so different from typical ferns that many people would fail to recognize them as ferns.

One unusual group, called the climbing ferns, has leaves that grow continuously, eventually reaching lengths of over 30 m. Several of these, such as the Old World climbing fern, *Lygodium microphyllum,* have become serious pests in the southeastern United States, smothering trees and creating dangerous fire conditions.

One highly unusual group of ferns completes its life cycle floating or submersed in water. They are heterosporous and make hard, bean-shaped reproductive structures, which, like seeds, are extremely decay-resistant and may last years in the environment, enabling the plants to survive prolonged drought. The lineage of water ferns is divided into two clades. In one clade, consisting of two genera, the plants are free-floating and never become attached to any substrate. One genus, *Azolla* (mosquito fern), lives in symbiotic relationship with nitrogen-fixing bacteria and is used as a green manure to fertilize rice paddies in Asia. Three species are native in North America. Another genus is not native to North America but has been imported from the tropics and now grows wild in the warmer parts of the United States. Several species of *Salvinia* **(Fig. 1a)** can become pernicious weeds, clogging waterways and pushing out native vegetation. *Salvinia* has round leaves and lacks roots but has a modified leaf that is highly branched and looks superficially root-like.

The other clade of aquatic ferns has three genera: *Marsilea, Pilularia,* and *Reg-nellidium.* These ferns are rooted in mud and may be submersed, floating, or emergent. *Marsilea* **(Fig. 1b)** resembles a four-leaf clover (and is sometimes sold as such), while *Pilularia* lacks a leaf blade altogether. The petioles, however, coil like typical fern leaves. Both of these genera are native to North America.

Another clade of about 1,000 species of true ferns has achieved tree size although no secondary growth occurs (Fig. 1c). Tree ferns are largely tropical, but they are commonly grown in warmer parts of the world or indoors as ornamental plants. They may form monocultures on disturbed sites and are abundant in some areas. Some species are endangered by land clearing for agriculture. The fibrous parts of the stems are used in the nursery trade as an epiphyte-growing substrate.

One strange group of ferns, the filmy ferns (family Hymenophyllaceae) has

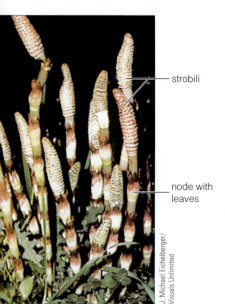

strobili

node with
leaves

J. Michael Eichelberger/
Visuals Unlimited

a

sporangiophores

James W. Richardson/Visuals Unlimited

b

Figure 23.18 Strobili of *Equisetum tolmateia.* (**a**) A cluster of strobili, each at the tip of a stem. (**b**) A longitudinal section through two strobili, showing an outer layer of sporangiophores.

remarkably strong and rigid per unit of weight—stronger, in fact, than solid stems. Smaller canals, called **carinal canals,** are in the center of each vascular bundle. Each bundle consists of two arms of xylem extending from the carinal canal, and phloem lies between the arms. Some species contain vessels with the xylem, but most contain only tracheids. An endodermis may surround each bundle or it may encircle the stem and include all vascular bundles. The cortex consists of parenchyma and chlorenchyma; thus, the stems appear green.

Sporangia are borne in strobili on special structures called **sporangiophores** (sporangium-bearing stalks). These are umbrella-shaped structures borne at right angles to the long axis of the strobilus **(Fig. 23.18).** In most species, strobili occur at the apex of ordinary shoots, but in a few species, they occur only on special shoots that may lack chlorophyll.

Equisetum is considered to be homosporous because it produces one kind of spore. Surprisingly, however, some gametophytes are unisexual. Some extinct sphenophytes were heterosporous. Horsetail spores are green and thin-walled, with long, ribbon-like elaters attached to the spore

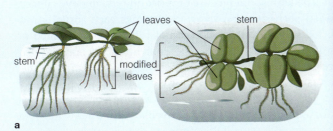

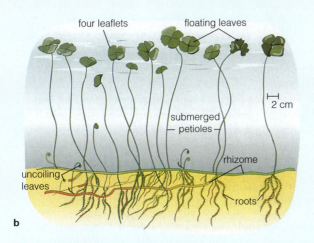

Figure 1 (a) *Salvinia,* a free-floating, heterosporous, aquatic fern. Each node has two round floating leaves and a third modified leaf, which is highly dissected and appears root-like. (b) *Marsilea,* a heterosporous, aquatic fern, which is rooted in mud. These ferns produce hard bean-shaped reproductive structures that can withstand prolonged periods of drought. (c) The tree fern *Alsophila* in a Puerto Rican forest.

reversed the trend toward sporophyte dominance in vascular plants. Filmy ferns sometimes can persist indefinitely in the gametophyte stage of the life cycle, reproducing asexually. In some species, a sporophyte has never been seen.

Websites for further study:

Plant Diversity—Ferns in the Wet Tropics:
http://www.wettropics.gov.au/pa/pa_ferns.html

Royal Botanic Gardens, Kew: Ferns and Fern Allies:
http://www.rbgkew.org.uk/scihort/ferns.html

wall. The **elaters** coil and uncoil in response to humidity, and they help disperse the spores when the sporangium splits open at maturity.

Marattialean Ferns

In appearance, **marattialean ferns** are similar to true ferns, possessing compound leaves that are coiled when young (Fig. 23.19). A synonym for the large, compound leaves that characterize marattialian ferns and true ferns is **frond.** However, like ophioglossalean ferns, they have upright stems, and they possess a distinctive sporangium, which is different from those of true ferns. These plants are largely tropical and possess gigantic leaves that arch above short squat stems. Having a long and extensive fossil record, they were an important element in the Coal Age swamp forest flora.

True Ferns

True ferns make up the majority of living monilophytes, with at least 12,000 species known (Fig. 23.7d). Traditional taxonomic treatments of the ferns divide them into five

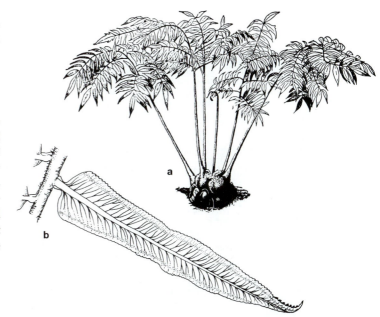

Figure 23.19 A marattialean fern, *Angiopteris evecta.* (a) The above-ground stem and leaves. Note the large stipules surrounding the coiled fiddleheads (not visible). (b) One pinna.

orders to encompass the vast diversity of form. Systematic studies have revealed that two of those taxonomic orders, Marattiales and Ophioglossales (Table 23.1), are not in the true fern lineage, and that two other orders, Salviniales and Marsileales, are really an unusual subclade of the true fern lineage (see "PLANTS, PEOPLE, AND THE ENVIRONMENT: Diversity among the Ferns" sidebar).

The true fern lineage contains the ferns familiar to us as ornamentals and wild plants of shady woods and streams **(Fig. 23.20).** They are readily distinguished from other monilophytes by a single feature: a unique type of sporangium called a **leptosporangium** (Fig. 23.20a). Leptosporangia are distinctive in that they each originate from a single cell in the leaf, have a long thin stalk, have a wall one cell layer in thickness, and have an active form of splitting open (a strip of thick-walled cells called an **annulus** flicks the spores out of the sporangium). The sporangia of all other tracheophytes are called **eusporangia;** these originate from several initial cells, have very short or absent stalks, have thick walls, and never have an annulus. Because this single character reliably identifies true ferns, they sometimes are called **leptosporangiate** ferns.

The leptosporangia are grouped into distinctive clusters called **sori** (singular, *sorus*). Sometimes each sorus is protected by an overarching umbrella-like structure called an **indusium** (Fig. 23.20b,c), and sometimes the sori are protected by the edge of the leaf curling over them as a false indusium (Fig. 23.20d).

The true fern group contains a wide variety of forms, ranging from minute herbs to treelike giants. Some ferns are even vinelike and climb vigorously through trees. Ferns also inhabit a great variety of habitats, from fully aquatic to severely arid. Some grow as epiphytes in tropical forests, and others form dense stands in disturbed areas. The bracken fern, *Pteridium aquilinum,* is the most widespread plant on Earth.

The rest of this chapter focuses on the life cycle of true ferns. True ferns have a rich fossil record extending back to the Devonian period. They were important members of the coal swamp flora during the Carboniferous period and have been important members of many ecosystems ever since.

True Fern Sporophytes Typically Have Underground Stems

True fern sporophytes typically extend themselves by an underground perennial rhizome. The rhizome grows from its tip. Roots and leaves arise from nodes **(Fig. 23.21).** The rhizome can remain alive for centuries in some fern species, spreading over a considerable area. Some bracken fern (*Pteridium aquilinum*) clones are thought to be 500 years old. Most ferns have no stems aboveground; only leaves are visible. Young fern leaves are at first tightly coiled **fiddleheads.** Coiling protects their delicate meristematic tissue as it is pushed up through the soil by the elongating petiole. Above the soil, the leaf uncoils from the base upward, continuing to grow by cell division at its tip.

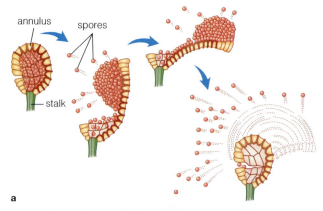

a

b

c

d

Figure 23.20 Components of true ferns. (**a**) A leptosporangium and the role of the annulus in spore dispersal. (**b**) The underside of a holly fern (*Cyrtomium falcatum*) leaf, showing many scattered sori. (**c**) A section of one sorus (of the fern *Crotonium*), showing the umbrella-like indusium that shelters the sporangia. (**d**) The false indusium (the rolled leaf edge) of cliff brake (*Pellaea rotundifolia*).

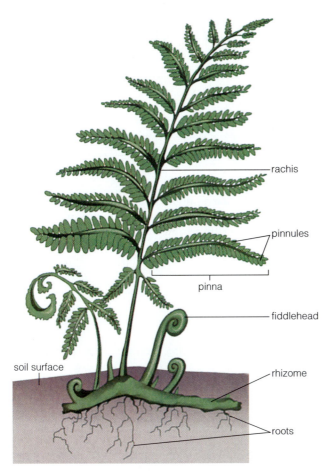

Figure 23.21 The lady fern, *Athyrium felix-femina.* The stem is a horizontal, underground rhizome. Only the leaves (fronds) project above ground. Young leaves emerge through the soil as tightly coiled fiddleheads. Roots anchor the rhizome to the soil and take up nutrients and water.

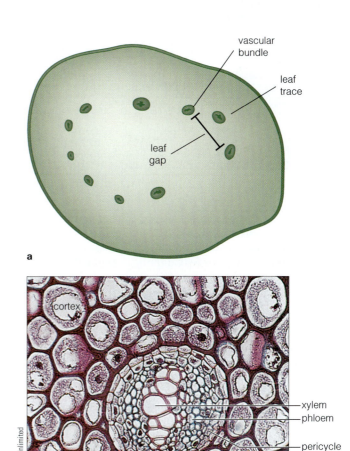

Figure 23.22 A *Polypodium* stem cross section. (**a**) The entire rhizome, showing the arrangement of the vascular system. The center of the stem (the pith) is composed of parenchyma. In this species, the vascular cylinder surrounding the pith is broken into discrete bundles by gaps where vascular tissue peels off to enter a megaphyll. In three dimensions, however, each bundle would merge together, and gaps would form in other places. The vascular system would appear to be a net of interconnecting strands. The gaps between stem vascular bundles are called **leaf gaps;** each gap has one (or sometimes more) leaf traces. (**b**) The details of one vascular bundle.

A typical leaf has a well-developed epidermis with stomata, particularly on the lower surface. The leaf mesophyll may be differentiated into palisade and spongy layers. The petiole is prolonged into a **rachis** on which secondary and tertiary leaflets, called **pinnae** and **pinnules,** respectively, are attached.

The vascular tissue of fern stems is occasionally a solid strand, but more typically it has a well-defined pith of parenchyma cells surrounded by vascular bundles. The xylem of a few genera, such as the bracken fern and the aquatic fern *Marsilea,* contains vessels. Sieve cells are present in the phloem (**Fig. 23.22**).

Sexual Reproduction Usually Is Homosporous

The life cycle of a true fern is shown in **Figure 23.23.** Fern sporophytes become sexually mature in 1 to 10 years. When the fern is mature, sporangia develop. Not all leaves are fertile (capable of producing spores), and, in some species, fertile leaves have a different shape than nonfertile leaves. Clusters of leptosporangia occur most commonly on the lower surface or margins of pinnae along veins. The indu-

sium that covers them (Fig. 23.23b) protects immature sporangia and may shrivel and fall away before spores are shed.

Spores develop within the stalked sporangia (Fig. 23.23c). The annulus in the sporangium wall plays a role in spore dispersal. As it dries, it bends backward, ripping the sporangium open, and then snaps forward, flinging the ripe spores as far as 1 to 2 m. Gravity, wind, water, and even electrostatic factors may carry the spores even farther. In temperate zones, spores tend to be released in the fall, at the end of the growing season; in the tropics, they may be released any month of the year. Charles Darwin, sailing hundreds of kilometers from shore on the *Beagle,* recorded in his journal that some of the air samples he collected contained fern spores (see "PLANTS, PEOPLE, AND THE ENVIRONMENT: Fern Spores" sidebar).

PLANTS, PEOPLE, AND THE ENVIRONMENT:

Fern Spores

Fern spores are incredibly hardy. They are resistant to ultraviolet radiation, low humidity, and very low temperatures. *Dryopteris* (wood fern) spores, for example, have been experimentally held at −254°C for 11 hours without their viability being affected. Fern spores can remain alive from several months to several years in the environment, but most of them land in inhospitable terrain and perish. Consequently, fern spores are produced in prodigious numbers. Individual leaves typically release up to 750 million spores per year, and most ferns have 10–20 leaves per plant. *Dryopteris pseudomas* probably holds the spore production record, releasing 1 billion spores per day per plant! The large number of spores produced by ferns, and their resistance to physical stresses, explain why ferns are the first plants to repopulate devastated areas.

🖱 Websites for further study:

K/T Boundary Impact Hypothesis: http://taggart.glg.msu.edu/isb200/kt.htm

A Brief Introduction to Ferns: http://www.amerfernsoc.org/lernfrnl.html

Fern spores contain phytochrome and other unidentified pigments. Some contain chlorophyll, and the spores of many species require light to germinate. Some species can be inhibited from germinating by exposure to far-red light and stimulated to germinate by red light. Their requirement for red light inhibits spores from germinating if there is not enough sunlight present for adequate growth.

When a spore germinates, a short filament of cells is produced, resembling a moss protonema. Soon, the apical cell of the filament begins to divide in many directions, producing a green thallus. It is usually heart-shaped and 1 to 2 cm across (Fig. 23.23d); however, it can be mitten-shaped, ribbon-shaped, or filamentous and highly branched. Rhizoids attached to the lower surface anchor the plant to the soil. Most fern gametophytes require 2 to 3 months to mature, but extremes exist. Some species colonizing disturbed habitats can mature in only 4 to 6 weeks, whereas the chain fern (*Woodwardia radicans*), an inhabitant of the understory of cold-temperate conifer forests, requires 2 years.

We would expect the homosporous true ferns to have bisexual gametophytes. Yet, some species produce functionally unisexual gametophytes, because the first gametophyte in a particular location to develop archegonia releases a substance that stimulates younger gametophytes to produce only antheridia. This mechanism prevents self-fertilization, which does occur in a few species with bisexual gametophytes; but in many cases, the gametangia mature at different times. Sperm are helical, with as many as 100 flagella. When free water is present, the antheridium bursts (Fig. 23.23e), releasing the sperm. Archegonia produce an attractant, which guides the sperm toward them. Archegonial necks spread open in the presence of free water, making a path for sperm to reach the egg (Fig. 23.23f). As soon as one sperm has penetrated an egg, the plasma membrane of the egg changes, and no other sperm cells can penetrate.

The diploid zygote cell develops into an embryo, with foot, shoot, and root regions (Fig. 23.23g,h). Although one gametophyte may contain many zygotes, only one or two will mature into embryos. In a short time, an embryo develops into a young sporophyte that is larger than the parent gametophyte (Fig. 23.23i)—and becomes nutritionally independent of the gametophyte.

Ferns Have a Variety of Alternative Means of Reproduction

Ferns have a number of ways to propagate vegetatively. In some species, miniature plantlets form on mature leaves and break off, growing into new plants. Other species, called walking ferns, produce propagules at the tip of a frond where it touches the soil. The fern can thus "walk" forward by sequentially forming new plants at the tip of each new frond. Gametophytes also can reproduce vegetatively.

There also are several variants to the normal alternation of generations of a life cycle. A surprising number of fern species can, under certain conditions, produce diploid gametophytes directly out of sporophyte tissue (usually leaf tissue). These gametophytes function normally, but the gametes they produce also are diploid. The resulting sporophyte will have double the chromosome number of the pre-

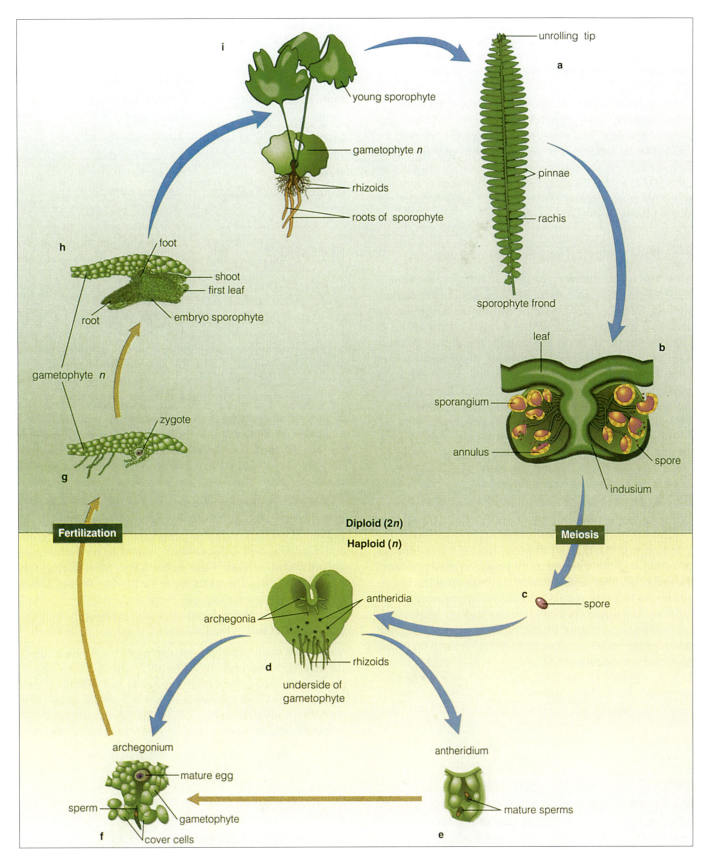

Figure 23.23 The life cycle of a true fern. (**a,b**) On the underside of a sporophyte leaf are many clusters (sori) of leptosporangia. (**c**) Meiosis occurs in each leptosporangium, resulting in many haploid spores. (**d,e,f**) A spore can germinate and grow by mitosis into a heart-shaped gametophyte. The sperm produced by the antheridia are motile and swim through a film of water to reach archegonia and fuse with an egg. (**g,h,i**) The resulting zygote cell is retained within and nourished by gametophyte tissue. It develops into an embryo and then into a young sporophyte that is rooted in the soil and independent of the gametophyte.

vious generation. This form of reproduction is called **apospory** (without spores). In effect, a new species has been created by a single fertilization event. Many ferns have large numbers of chromosomes, the result of numerous past apospory events.

Some ferns engage in **apogamy** (without gametes), in which gametophytes produce sporophytes without any fusion of gametes. This variation occurs when free water for fertilization is lacking. The resulting sporophytes will be haploid, but eventually can undergo normal meiosis by compensating for the haploid chromosome number. Apospory is a relatively rare event, but some ferns reproduce routinely or solely by apogamy.

Ferns Have Ecological and Economic Importance

Ferns are most common in the understory of humid-temperate and tropical forests. They are widely distributed, however, and also grow in arctic and alpine tundra, saline mangrove swamps, semiarid deserts, and on coastal rocks swept by salt spray. Ferns can provide the bulk of biomass in some tropical forests and dominate the understories of some temperate conifer forests. Some ferns—such as *Lygodium* (climbing fern), *Salvinia* (water fern), and *Pteridium aquilinum* (bracken)—are weeds; they smother other vegetation, clog waterways, and poison livestock.

Throughout their life cycle, ferns are remarkably independent of animals. Animals are not essential vectors for spores, sperm, or embryos. Although ferns rarely have mechanical defenses such as thorns, stinging hairs, or hard leaves, animals generally avoid feeding on ferns because of poisonous or unpalatable chemicals present in many species.

Humans have found uses for certain ferns at certain life cycle stages. In the fiddlehead stage, leaves of some species are edible, and these are eaten by people in many cultures. The leaves of other ferns are used as basket-making material. Florists mix fern fronds with flower arrangements, and nurseries propagate ferns as popular indoor houseplants and outdoor landscaping plants. Biologists have found fern gametophytes to be excellent subjects for research on physiology and plant development.

KEY TERMS

annulus	indusium
apogamy	lepidodendrids
apospory	leptosporangium
carinal canal	ligule
dichotomous	lycophytes
elater	marrattialean ferns
enation	megagametophyte
eusporangium	megaphyll
fiddlehead	megasporangium
heterospory	megaspore
horsetails	microgametophyte
microphyll	smog
micropore	sorus
microsporangium	sphenophytes
monilophytes	sporangiophore
ophioglossalean ferns	sporophyll
pinna	strobilus
pinnule	tracheophytes
rachis	vallecular canal
rhizophore	whisk ferns
seedless vascular plants	zosterophyllophyta

SUMMARY

1. Early tracheophytes, the seedless vascular plants, number about 14,000 species. Their shared traits include a well-developed, lignified vascular system (containing tracheids and rarely vessels), a branched independent sporophyte, and a free-living but reduced gametophyte. The life cycle is sporophyte-dominant, and the sporophytes have a well-developed epidermis with cuticle and stomata. Sporophyte dominance was key to success on land.

2. Early tracheophytes were leafless and rootless stems, but evolution was rapid during the 40-million-year Devonian period. Microphylls, megaphylls, true roots, secondary growth, wood, and seeds all appeared for the first time in the Devonian.

3. Early tracheophytes predominated in the world's vegetation during much of the Paleozoic era. Diversity of form peaked in the Carboniferous period, 360 to 286 million years ago. *Calamites* and *Lepidodendron,* related to modern horsetails and club mosses, were major forest trees during this period. Their fossil remains are mined today as coal. Other Carboniferous plants included ferns and early seed plants, the seed ferns. Many forms of early tracheophytes were extinct by the end of the Paleozoic.

4. The first tracheophytes were rhyniophytes. These bridged the gap between bryophytes (with relatively small sporophytes and unlignified vascular systems) and vascular plants with branched sporophytes and lignified xylem tissue. Tracheophytes divided into two major lineages: lycophytes and monilophytes. Monilophytes later gave rise to the seed plants.

5. The lycophytes consist of three living lineages, all of which occur in North America: *Lycopodium* (and a number of related genera), *Selaginella*, and *Isoetes*. The lepidodendrids are an extinct group related to *Isoetes*. All share microphylls and laterally attached sporangia.

6. Sporophytes of *Lycopodium* and related genera (club mosses) have true roots, stems, and microphylls. Sporangia are borne on special microphylls called sporophylls, which may or may not be clustered into cones or strobili near the ends of upright stems. Club mosses are homosporous, and the spores produce small gametophytes that are usually subterranean, surviving by a symbiotic relationship with fungi.

7. *Selaginella* (spike moss) is heterosporous, having megasporangia that produce megaspores and microsporangia that generate microspores. Megaspores develop within the spore wall into megagametophytes; microspores grow within the spore wall into microgametophytes. Gametophytes are reduced relative to homosporous plants. The developing embryo is fed by the megagametophyte.

8. Sporophytes of *Isoetes* (quillwort) have a small tuft of grasslike microphylls and a basal corm. They grow in wet areas, often beneath standing water. They are heterosporous and have secondary growth.

9. The monilophytes include four distinct, closely related lineages. Currently we are uncertain exactly how these are related to each other. Most of the monilophytes have traditionally been thought of as ferns. The monilophytes all have megaphylls or leaves reduced from megaphylls. Most have roots, and they all share a distinct order of genes on the chloroplast chromosome, which differs from the order found in the bryophytes and the lycophytes.

10. The psilophytes, or whisk ferns, were traditionally treated as a division but are now known to be part of the monilophyte clade. They are represented by only two living genera, *Psilotum* and *Tmesipteris*. Sporophytes of *Psilotum* consist of a branched, underground rhizome with rhizoids and an upright, green, branched stem with highly reduced leaves called enations. Psilophytes are homosporous, and the spores produce small, non-photosynthetic gametophytes, which are nourished by symbiotic fungi. Each gametophyte produces both male and female gametes. The embryo sporophyte is at first dependent on the gametophyte but in time becomes independent. The ophioglossalean ferns are closely related to the psilophytes. These are small temperate and tropical plants with unusual leaves consisting of a sterile, photosynthetic segment and a fertile, sporangium-bearing segment.

11. The sphenophytes, or horsetails, have traditionally been treated as a division but are now known to be members of the monilophyte clade. The lineage today consists of a single genus, *Equisetum* (horsetail, scouring rush). Sporophytes have uniquely jointed, ribbed, hollow stems. Whorls of reduced leaves and/or branches are present at the nodes. True roots are present, and vessels are found in the xylem of some species. Sporangia are borne on sporangiophores grouped into complex strobili. Horsetails are homosporous, and spores produce small, photosynthetic gametophytes. Extinct sphenophytes were heterosporous trees or vines and were important members of the Coal Age swamp flora.

12. The marattialean ferns are a distinct lineage that resemble the true ferns, but lack leptosporangia and rhizomatous stems. They are large tropical plants.

13. The true ferns consist of about 12,000 species. This large clade contains a number of subclades, including two traditional orders of aquatic ferns—the Marsileales and the Salviniales—and the tree ferns. The group is homosporous (except for the aquatic ferns) and is distinguished by leptosporangia. These form from a single initial cell, have a thin stalk and wall around the spores, and usually have a row of thick cells (the annulus) that aids in splitting open the sporangium and dispersing the spores.

14. All true ferns are herbs (no secondary growth), and in most cases the stem is an underground rhizome (except in the tree ferns); only fern leaves (fronds) appear above ground. Leaves are megaphylls. Vessels are present in some species. Sporangia, located on the underside of leaves, are usually clustered into sori and often protected by an indusium. Spores germinate to produce small, photosynthetic, thalloid gametophytes.

15. Fern gametophytes and sporophytes sometimes develop by alternative pathways in the life cycle. Gametophytes can produce haploid sporophytes by the process of *apogamy*, and sporophytes can produce diploid gametophytes by the process of *apospory*.

Questions

1. The early tracheophytes include several lineages that today are relics of what they once were. Explain in what ways these groups are relics.

2. Why did plant invasion of land not occur until the Silurian (or possibly Ordovician) period?

3. During which periods and eras were early (seedless) tracheophytes the dominant form of vegetation on land? Describe the composition of a Carboniferous (Coal Age) forest as an example of vegetation dominated by early tracheophytes.

4. In what ways do seedless tracheophytes exercise more control over their internal water balance, as compared to bryophytes?

5. Why can one say that *Psilotum* is not a relict of the rhyniophytes?

6. Describe the difference between a homosporous and a heterosporous life cycle.

7. In mosses, the sporophyte is attached to and dependent on the gametophyte. In ferns, however, the two generations live independent lives. How might this separation be important for the development of the larger, more complex fern sporophyte?

InfoTrac® College Edition

http://infotrac.thomsonlearning.com

Invasive Ferns

Dougherty, R. 2003. Invasive fern smothers plants: Old World climbing fern spreads through Florida parklands. *National Parks*, May–June, p. 12. (Keywords: "Invasive" and "fern")

Wood, M. 2000. Global search for climbing fern's foes. *Agricultural Research* 48:16. (Keywords: "fern's" and "foes")

Systematics and Evolution

Wikstrom, N., and Kenrick, P. 1997. Phylogeny of lycopodiaceae (Lycopsida) and the relationships of *Phylloglossum drummondii* Kunze based on rbcL sequences. *International Journal of Plant Sciences* 158:862. (Keywords: "Phylogeny" and "Lycopsida")

Anon. 1999. Fossils and fern phylogeny. *The Botanical Review* 65:189. (Keywords: "fern" and "phylogeny")

Gymnosperms

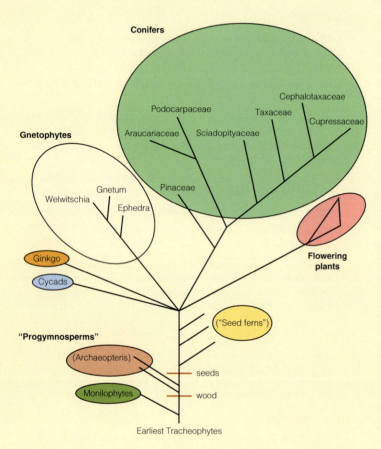

Visit us on the web at http://biology.brookscole.com/plantbio2 for additional resources, such as flashcards, tutorial quizzes, InfoTrac exercises, further readings, and web links.

1. The evolution of seeds, pollen, and wood freed plants from the need for water during reproduction, allowed for more effective dispersal of sperm, increased parental investment in the next generation, and allowed for greater size and strength.

2. Seed plants originated in the Devonian period from a group called the progymnosperms, which possessed wood and heterospory, but reproduced by releasing spores. Currently, five lineages of seed plants survive—the flowering plants plus four groups of gymnosperms: cycads, Ginkgo, conifers, and gnetophytes. Conifers are the best known and most economically important group; it includes pines, firs, spruces, hemlocks, redwoods, cedars, cypress, yews, and several Southern Hemisphere genera.

3. The pine life cycle is heterosporous. Pollen strobili are small and seasonal. Each sporophyll has two microsporangia, in which microspores are formed and divide into immature male gametophytes while still retained in the microsporangia. The gametophytes are released to the wind as pollen grains. Ovulate cones are large and require 2 years to mature. Each cone scale has two megasporangia, in which megaspores are

formed. The megaspores divide into mature female gametophytes while still retained in the megasporangia.

4. Pollen grains sift down into ovulate cones and germinate, producing a pollen tube, which grows through the megasporangium toward a megagametophyte. Both male and female gametophytes are small, simple structures and depend on the sporophyte for nutrition.

5. Fertilization of one egg is sufficient to trigger the development of a seed. Pine seeds contain an embryo sporophyte, stored food (in the form of megagametophyte tissue), and a seed coat. The function of the seed is to protect and disperse the next generation away from the parent plant, in both space and time.

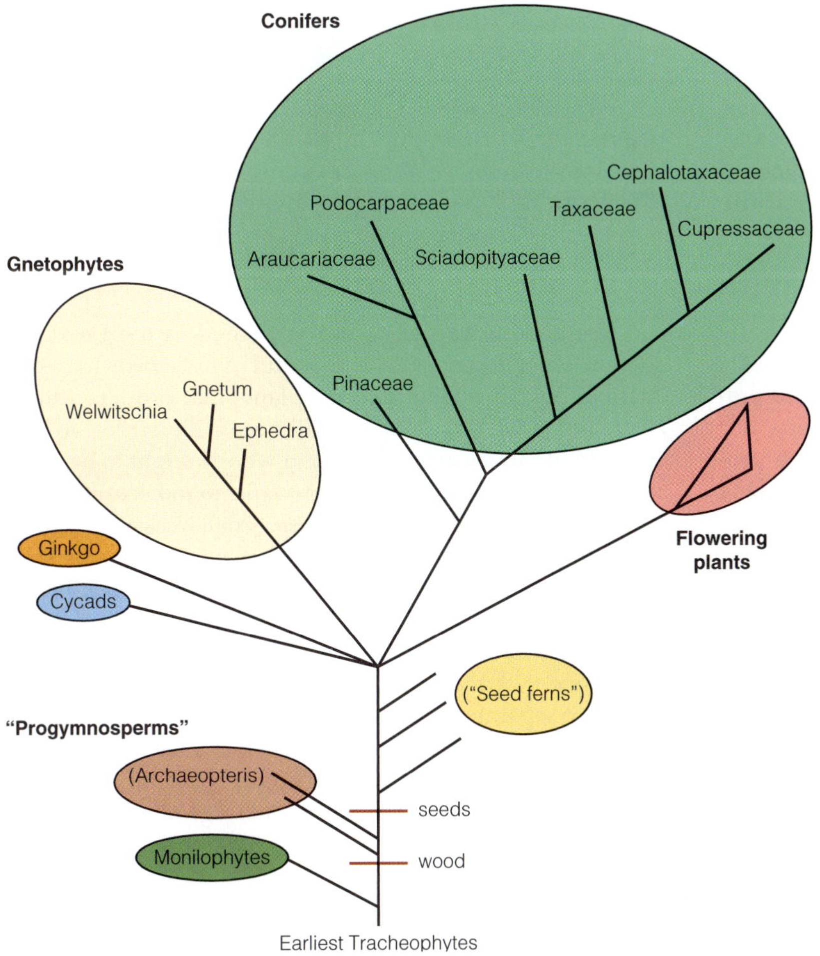

24.1 GYMNOSPERMS: SEEDS, POLLEN, AND WOOD

In the long evolutionary history of plants, few developments have had more profound consequences than the evolution of seeds and pollen. Seed-bearing plants have been prominent in nearly all terrestrial ecosystems from the Paleozoic era to today. Their success is undoubtedly due, in large part, to the evolution of seeds and pollen, which freed them from the films of water needed by other tracheophytes for sperm to swim to the archegonia. The advent of seeds and pollen also provided other benefits to plants. Megagametophytes are retained in the megasporangium, where they receive nutrition and protection. Microgametophytes (pollen) have a brief period of independence that ends when they land on a megasporangium. If they succeed in entering the megasporangium, they receive nutrition and protection and complete their development inside of it. Seeds and pollen thus allow significant parental investment in the next generation.

A prominent characteristic of the seed-bearing plants is extensive secondary growth and the production of wood. Other groups, such as the lepidodendrids and sphenophytes, independently evolved lateral cambia and secondary growth. However, the ancestors of seed plants evolved a different sort of secondary growth that led to the ability to produce exceptionally large and strong plants, something the secondary growth mechanisms of seedless plants never achieved.

The earliest **seed plants** produced seeds that sat unprotected on the surface of leaves or modified stems. Although many of the descendants of these first seed plants hid their seeds beneath cone scales or other specialized structures, they are collectively called **gymnosperms** (derived from the Greek *gymnos* meaning "naked" and *sperma* meaning "seed"). This characteristic distinguishes them from another lineage, the flowering plants, which embed their seeds inside a fruit **(Fig. 24.1).** The nonflowering seed plants

Figure 24.1 Cladogram of the relationships among seed plants. All five lineages of seed plants are shown as emerging from a single point, because there is uncertainty about the actual order of branching. Some studies suggest that gnetophytes are in the conifer subclade or a sister to it. Many analyses show cycads as the basal-most gymnosperm group.

Figure 24.2 Gymnosperms include the pines, which have needlelike leaves borne in clusters and seeds borne naked in woody cones. This species is ponderosa pine (*Pinus ponderosa*).

Figure 24.3 A reconstruction of *Archaeopteris*, a Devonian tree about 25 m in height.

persist today as a diverse group of about 900 species of trees, shrubs, or vines. They are present throughout the world, in most ecosystems, and are particularly prominent in cool-temperate forests.

Common examples of living gymnosperms include cycads (*Cycas, Dioon, Zamia*), maidenhair tree (*Ginkgo*), cedar (*Cedrus*), fir (*Abies*), juniper (*Juniperus*), pine (*Pinus*; **Fig. 24.2**), redwood (*Sequoia, Metasequoia, Sequoiadendron*), spruce (*Picea*), and Mormon tea or joint fir (*Ephedra*). Some of these plants have great economic and ecological importance. This chapter considers the origin and evolution of the key innovations of seed plants and surveys the diversity of living gymnosperms.

The Origin of Seeds, Pollen, and Wood

A variety of intriguing fossils from the Devonian and early Carboniferous periods give us a picture of how seed plants originated. Fossils belonging to an extinct group called **progymnosperms** are particularly important in shaping our understanding of this phase of plant evolution. The progymnosperms represent an intermediate form, or "missing link," revealing the transition from a spore-releasing vascular plant to a seed plant. In the middle to late Devonian period, progymnosperms existed that were small (shrub-sized at most) and homosporous and added wood to their stems from a solid central core of vascular tissue. Beginning in the late Devonian period, another group of progymnosperms appeared that were quite different. One spectacular example was *Archaeopteris*, a tree that may have reached 25 m in height **(Fig. 24.3)**. Fossils show that this plant had a vascular system with a ring of separate bundles around a pith of parenchyma cells and that it produced abundant secondary xylem. The wood and the growth form of these trees were similar to familiar modern trees such as pines and redwoods. However, these gymnosperm-like trunks bore small leaves arranged in flattened sprays reminiscent of fern fronds, and, like ferns, they reproduced by releasing spores into the air.

Plants with true seeds also appeared in the Devonian period. Unattached seeds were found in fossil beds from this period, but for a long time the plants that produced them were unknown **(Fig. 24.4a)**. Small trees with fernlike foliage were found in these beds, too, and were thought to be a sort of tree fern. Paleobotanists eventually found fossils of the seeds attached to the fronds of these small trees and realized they were not ferns but early seed plants. The name seed fern was created to describe them. Several lineages of Paleozoic seed ferns are now known.

The oldest known seed plant, *Elkinsia polymorpha*, was a small tree with the typical fernlike foliage of seed ferns and a ring of separate vascular bundles that gave rise to cambia and secondary growth. Fossils of this plant have been found in West Virginia and Belgium. Another example of an early seed fern is *Medullosa*, a common Coal Age tree **(Fig. 24.4b)**.

Seeds and Pollen Are Key Reproductive Innovations for Life on Land

Seeds are distinctive structures consisting of a protective outer covering (the seed coat) an embryo, and stored food. A seed develops from a megasporangium. Before the mega-

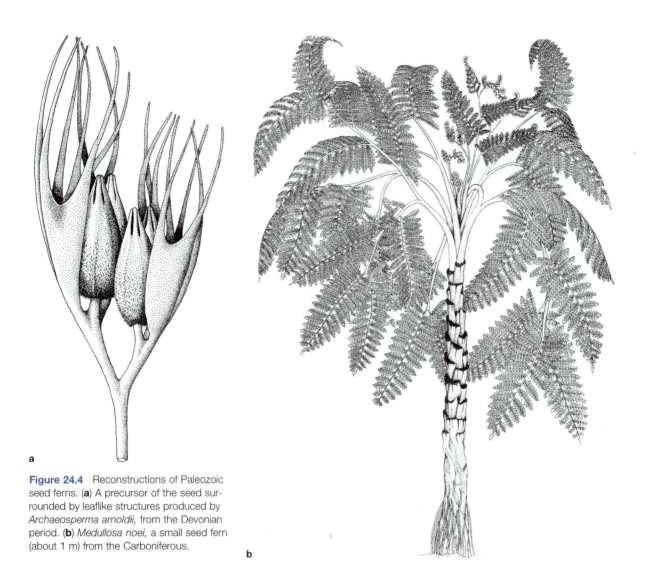

Figure 24.4 Reconstructions of Paleozoic seed ferns. (**a**) A precursor of the seed surrounded by leaflike structures produced by *Archaeosperma arnoldii*, from the Devonian period. (**b**) *Medullosa noei*, a small seed fern (about 1 m) from the Carboniferous.

sporangium matures into a seed, it is called an **ovule.** An ovule does not yet have an embryo, and it is composed of megasporangium tissue inside a protective covering called the **integument** that is equipped with a tiny hole at one end, the **micropyle,** which allows for the entrance of pollen. The megasporangium tissue, called the **nucellus** in seed plants, functions both to give rise to the megaspore by meiosis and to provide nutrition to the developing megagametophyte.

As the ovule develops, the retained megaspore gives rise to a megagametophyte, which ultimately produces egg cells. The ovule becomes a seed when the egg cell is fertilized and an embryo develops, the integument hardens into a seed coat, and the micropyle closes. In the mature seed, the megagametophyte tissue provides nutrition for the embryo and seedling.

How did such a complex structure evolve? A series of small steps can account for the evolution of seeds. The first step must have been the appearance of heterospory. The earliest tracheophytes were homosporous, but heterospory arose independently in many later lineages, including progymnosperms, the ancestors of seed plants. Heterospory is a crucial first step because it creates a division of labor,

enhancing the sedentary nature of one type of gametophyte (*mega-*) and the mobility of the other (*micro-*).

Once heterospory was established, a key innovation was the retention of the megaspore inside the megasporangium. However, this innovation creates a potential problem. When the retained megaspores germinate, the resulting megagametophytes must complete their development within the sporangium. In seedless vascular plants, such as *Selaginella,* four megaspores typically are produced per megasporangium. (This is because spores are the product of meiosis.) In seed plants, the number of functional megaspores must be reduced to one, to avoid competition for space and food; therefore, in seed plants, three of the four megaspores resulting from meiosis typically disintegrate.

A single functional megaspore, giving rise to a megagametophyte while still inside the sporangium, is nearly an ovule. The only missing piece is the integument. How the integument originated has caused much debate. Paleobotanists have shown that some of the early Paleozoic seed ferns from the late Devonian and Carboniferous periods had a series of separate lobes of tissue around their megasporangia. These progressively fused to form more complete coverings. In later plants, they covered the entire megasporangium, except for a region near the tip.

If the megagametophyte remains inside the megasporangium, surrounded by an integument and attached to the parent sporophyte, then another problem arises. How does the sperm get from the microgametophyte to the egg? The sperm cannot swim to the megagametophyte, but rather must be delivered directly. The solution to this problem is **pollen.**

A pollen grain is an immature microgametophyte. In seedless vascular plants with heterospory, the microspore leaves the microsporangium, germinates, and undergoes a series of mitotic divisions to create a microgametophyte within the spore wall. This microgametophyte consists of a layer of antheridial jacket cells and a mass of sperm-producing cells. Pollen begins its development inside the microsporangium, goes through fewer mitotic divisions, and does not produce the antheridial jacket layer. When a pollen grain is ready to leave the microsporangium, it consists of between two and five haploid nuclei. If it reaches an ovule, it goes through several more rounds of mitosis and produces two sperm cells and an elongated structure called a pollen tube, which in most seed plants conveys the sperm to the egg.

Seed Plants Have Distinctive Vegetative Features

In addition to seeds and pollen, gymnosperms are characterized by a number of vegetative features, including megaphylls, originally fernlike, but later typically reduced to simple leaves, needles, or scales; a primary stem vascular system composed of a ring of separate bundles with phloem toward the outside and xylem toward the center; secondary growth from lateral cambia, producing abundant secondary xylem to the inside, and secondary phloem to the outside; and a main stem that has lateral branching (with nodes made up of leaves and axillary buds).

All but a few gymnosperms are woody. Most are trees, some of which have great bulk, such as the coast and Sierran redwoods.

Relationships among Gymnosperms

Based on cladistic studies using morphology in fossil groups and molecules and morphology in living groups, botanists believe that seeds arose only once. The appearance of seeds coincided with a great adaptive radiation of new plant groups. All these early seed plants (the seed ferns) became extinct, but they gave rise to a number of lineages that persisted. Currently, five major lines of seed plants remain, which form a clade (Fig. 24.1). One lineage, the flowering plants, has been particularly successful and dominates the world's flora. The four other surviving lineages are the gymnosperms.

Large-scale molecular analyses have indicated, to the surprise of many systematists, that flowering plants did not evolve from one of the living gymnosperm groups, as had long been assumed. Instead, the living gymnosperms form a clade that is a sister group to flowering plants (Fig. 24.1).

These four gymnosperm groups traditionally have been classified as divisions. Studies indicate that each of the four lineages is monophyletic (with the possible exception of the conifers), but the exact relationships among them have proven to be a stubborn problem. Because of uncertainty concerning the phylogeny of gymnosperms, Figure 24.1 shows them all as originating from the same point.

Despite the uncertainty about relationships, most analyses place cycads at the base of the gymnosperm tree because they possess a series of ancestral features. Ginkgo also retains a number of ancestral traits, but it shares many similarities with conifers. Conifers make up the bulk of gymnosperm species. A distinctive and puzzling group, the gnetophytes, has created a great deal of debate among botanists. They possess certain vegetative and reproductive traits that are amazingly similar to flowering plants, but recent molecular analyses have classified them in the gymnosperm lineage.

24.2 THE MESOZOIC: ERA OF GYMNOSPERM DOMINANCE

The Mesozoic era (245 million to 65 million years ago) was the age of gymnosperms, dinosaurs, and moving continents. During the Permian period, which marked the end of the Paleozoic era, climates became cooler and drier. Dominance of the world's vegetation by seedless vascular plants came to an end. Gymnosperm groups evolved increasingly effective adaptations to arid conditions and spread into many habitats. Our best geologic reconstruction of the world at the start of the Mesozoic era is that all the continents were joined into a single landmass near the equator. Beginning in the Triassic period, this primeval supercontinent, called **Pangaea,** began to break apart. North and South America split from Europe and Africa, and the Atlantic Ocean formed. Prominent Mesozoic gymnosperms included a great variety of cycads, *Ginkgo* relatives, and gnetophytes (Fig. 24.5). Conifers similar to those in our modern flora first appeared in the early Triassic period, including forms of cypress, juniper, monkey puzzle tree, pine, and redwood. Several other groups of gymnosperms that are completely extinct today were also important elements of many Mesozoic floras. These include a diverse assemblage (a nonmonophyletic group) of Mesozoic seed ferns and the curious Cycadeoids, which had large compound leaves and short, unbranched stems like cycads. Despite their similarity to cycads, cycadeoids had a different life history, including flower-like structures that contained both megasporangia and microsporangia (Fig. 24.5b). Because of this characteristic, they have been proposed as possible relatives or ancestors of the flowering plants. Currently, it has not been determined where they fit into the phylogeny of plants, but they were probably not closely related to cycads.

Figure 24.5 Reconstructions of two extinct gymnosperms. (a) *Cordaites*, 9 to 30 m in height, from the Carboniferous period. (b) *Williamsonia*, 5 m in height, a cycadeoid, an extinct group superficially similar to cycads.

a

b

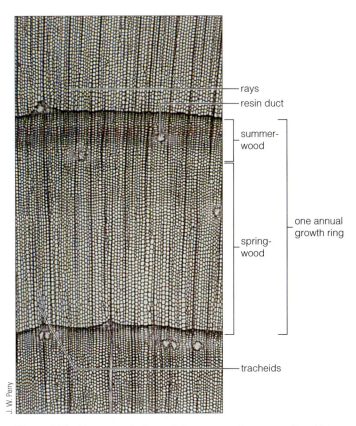

J. W. Perry

rays
resin duct
summer-
wood
one annual
growth ring
spring-
wood
tracheids

Figure 24.6 Three growth rings of pine, as seen in cross section. Note the narrow rays and the absence of vessels. Latewood (summerwood) appears denser than earlywood (springwood) because the tracheids are smaller. Several resin ducts are visible here, but not all conifers have resin ducts in their xylem.

As the Mesozoic era drew to a close, climates and continents continued to change. A southern continent called Gondwanaland split apart, and India became an island. Mountain building created cool weather at high elevations and arid regions in their lee. The climatic gradient from tropics to poles steepened. Many gymnosperms, including all seed ferns, became extinct, and the first flowering plants came to dominance. Dinosaur populations declined, whereas social insects and mammals multiplied and diversified. Today, gymnosperms have fewer species and growth forms, and they occupy fewer habitats than during the Mesozoic era.

24.3 THE VASCULAR SYSTEM OF GYMNOSPERMS

Gymnosperms are anatomically and morphologically more complex and longer lived than any group discussed so far in our survey of plants. They all share a primary stem vascular system composed of a ring of bundles defining a distinct pith and cortex region. Each vascular bundle contains primary phloem to the outside and primary xylem toward the center. As gymnosperms grow, they form a vascular cambium, which produces the secondary tissues. The process is the same for all gymnosperms, but there are some differences in their secondary growth patterns. For example, cycads produce a light wood, containing abundant living parenchyma cells, whereas *Ginkgo* and conifers produce dense wood, composed primarily of cells that are dead at maturity.

All gymnosperm trunks have secondary xylem and phloem, rays, and bark. The anatomy of most gymnosperm wood is extremely regular (Fig. 24.6). Part of the reason for this regularity is the absence of vessels. In many gymnosperms, including a majority of conifers and *Ginkgo,* tracheids produced in the spring are largest in diameter; as the season progresses, new tracheids are smaller in diameter and have thicker walls. Some tracheids are so thick that they resemble fibers and are called fiber-tracheids. The seasonal variation in tracheid size results in annual rings in the wood. Xylem rays are thin and are composed of brick-shaped parenchyma cells and, in some cases, special ray tracheids. Other parenchyma cells are stacked in vertical columns; these are called axial parenchyma (see Chapter 6). Gnetophytes differ from all other conifers in that they may have vessels in their wood.

The phloem of gymnosperms contains **sieve cells** (conducting cells similar to sieve-tube members but not associated with companion cells), fibers, ray parenchyma, and axial parenchyma.

Many gymnosperms, particularly conifers, produce **resin,** a mix of complex organic compounds that are the byproducts of the tree's metabolism. Typically, resin accumulates and flows in long **resin ducts,** chambers enclosed by parenchyma cells (Fig. 24.6). Resin inhibits wood-boring

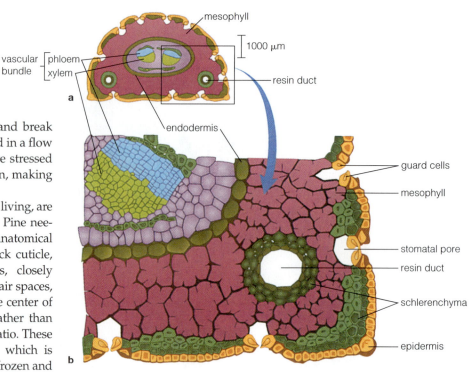

Figure 24.7 Pine needles in cross section. (**a**) The entire needle, with major tissues and regions outlined. (**b**) Cellular details of a part of the needle, from one of the two vascular bundles to the epidermis. Some pines have only one vascular bundle.

Labels in figure:
- mesophyll
- 1000 μm
- vascular bundle [phloem, xylem]
- resin duct
- endodermis
- guard cells
- mesophyll
- stomatal pore
- resin duct
- schlerenchyma
- epidermis
- **a**
- **b**

insects. If insects penetrate the trunk and break through any resin duct, they are encased in a flow of resin and immobilized. Trees that are stressed by drought or disease produce less resin, making them susceptible to insect infestation.

Many gymnosperms, both fossil and living, are supremely adapted to survive drought. Pine needles (leaves), for example, show many anatomical adaptations to aridity. They have a thick cuticle, sunken stomata, a fibrous epidermis, closely packed mesophyll without intercellular air spaces, and veins (vascular bundles) only in the center of the leaf **(Fig. 24.7).** Leaves are thick, rather than thin, yielding a low surface-to-volume ratio. These traits reduce the loss of water vapor, which is important in winter (when soil water is frozen and evergreen leaves are subjected to drying winds of low relative humidity). Needles have a life span that ranges from 3 to 30 years, depending on the species. Although the leaves do not continue growing during that life span, their vascular bundles do undergo secondary growth, producing a limited amount of secondary phloem but no secondary xylem.

24.4 **CYCADS**

Some 200 million years ago, during the early Mesozoic era, **cycads** made up a large part of the earth's vegetation and were probably food for herbivorous dinosaurs. Today, this once-diverse group contains only 11 genera and about 125 species, which grow in widely separated areas (largely in the tropics). Only one species, *Zamia integrifolia*, occurs naturally in the United States, in Florida and southern Georgia. Cycads contain potent toxins, but they can be eaten if prepared correctly. There is archeological evidence for extensive consumption of cycad seeds in Australia up to 13,000 years ago. The stem pith also can yield edible starch, and a commercial starch extraction industry existed in south Florida between 1845 and 1925. The genera *Cycas, Dioon,* and *Zamia* are widely propagated as ornamental plants, either outdoors or in greenhouses **(Fig. 24.8).**

a   b   seed cones

Figure 24.8 Cycads. (**a**) *Cycas revoluta* (sago palm). (**b**) A species of *Zamia*, with ovulate cones.

Many cycads are palmlike in appearance, and some of their common names reflect this nature (*Cycas revoluta* is sago palm, for example). Others, such as *Zamia*, have largely subterranean stems. Cycads are slow-growing: A specimen 2 m in height might be as old as 1,000 years.

Cycads have a number of distinctive features. They are the only gymnosperms to produce large compound leaves. They branch rarely, if at all, and although they have secondary growth, the wood is different from the wood of conifers because it has a lot of parenchyma in the xylem.

All cycads are dioecious. The seed-producing cone (ovulate strobilus or ovulate cone) is large and often protected by sharp prickles or woody plates. The pollen stobili also are large and upright; pollen is transported to the ovules by beetles in some species, or by wind. The microgametophytes grow a pollen tube that infiltrates the tissues of the ovule. Eventually, the microgametophyte produces two large multiflagellated sperm that swim to an enormous egg cell. The seeds of many cycads are covered with a fleshy, brightly colored seed coat to attract animal dispersers.

24.5 GINKGO

A single living representative remains of the ancient ginkgophytes, the maidenhair tree (*Ginkgo biloba*; **Fig. 24.9**). It grows wild today only in warm-temperate forests of China, but it has been grown for centuries on Chinese and Japanese temple grounds as a traditional decorative tree. It also is widely planted throughout the world as an urban street tree, because it is tolerant of pollution.

Chinese herbalists prescribe infusions of the leaves as a medicine for many ailments, and the tree is a symbol of longevity. *Ginkgo* leaf is currently one of the most widely used herbal supplements, taken for brain dysfunction and cardiovascular fitness. The seeds have been an important food in Asia for more than 2,000 years, despite being mildly toxic. They were consumed in Japan in great quantities during World War II because of food shortages, and cases of poisoning were common. Although today poisoning is rare, it still occurs, especially in children.

Like conifers, *Ginkgo* is a large, diffusely branching tree; some trees have trunks more than 3 m in diameter. The dense wood of *Ginkgo* is similar to conifer wood and lacks the abundant parenchyma of cycads. The leaves of *Ginkgo* are unique. They are fan-shaped and often are divided into two lobes. In fall, they turn a brilliant golden color and drop.

Like cycads, *Ginkgo* is strictly dioecious. The pollen is produced in small strobili and transported by wind to ovulate trees. The microgametophytes are similar to those of cycad, and they produce large, flagellated sperm. The ovules are borne in pairs on stalks rather than in a strobilus. Pollen-producing trees are preferred as ornamentals because the seeds are covered with a foul-smelling, fleshy seed coat.

24.6 CONIFERS

The most widely known and economically important gymnosperms are the **conifers,** with approximately 650 species. Woody seed cones are unique and so conspicuous a feature that this group is named for them: *conifer* means "cone-bearer." Conifers that lack woody cones include junipers, podocarps, yews, and plum yews; in these plants, a berry-like tissue surrounds the seeds, making them resemble the fruits of flowering plants. Botanists, however, interpret the fleshy coverings as either modified cone scales or as an elaboration of the integument.

Traditionally, the conifers have been divided into eight taxonomic families. Cladistic analyses using ribosomal RNA genes have revealed that, with one exception, the families are all monophyletic **(Fig. 24.10).** To solve this problem, one family (Taxodiaceae) has been nullified, and all of its members placed in another family (Cupressaceae). This section uses the traditional family names for the groups because they are familiar. However, these names are used in a phylogenetic

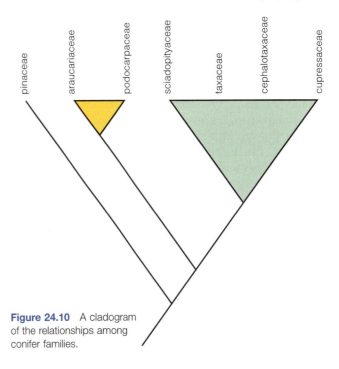

Figure 24.10 A cladogram of the relationships among conifer families.

Figure 24.9 A branch and leaves of *Ginkgo biloba* (maidenhair tree). The leaves have turned yellow and are about to fall off as winter begins.

Michael G. Barbour

Figure 24.11 Fascicles of needles borne on short shoots.

Figure 24.12 Mature ovulate cones of Shasta red fir, *Abies magnifica* var. *shastensis.* Cones in this genus are borne upright. Note the long bracts projecting beyond the cone scales.

context, to designate clades and not taxonomic ranks. Pinaceae is the basal lineage. The remaining conifers fall into two big clades: one is composed of the largely Southern Hemisphere groups Araucariaceae and Podocarpaceae; the other clade includes the Taxaceae, Cupressaceae, and two groups unfamiliar in the United States, Sciadopityaceae and Cephalotaxaceae.

Pinaceae Include the Pines, Firs, and Spruces

Pinaceae are important economically for wood, pulp, turpentine, and resin, as well as for ornamentals. Found primarily in the Northern Hemisphere, this group constitutes the bulk of conifer forests. Leaves are needlelike and are borne singly or in clusters called **fascicles** on special short shoots **(Fig. 24.11).** Most species are monoecious. The seed cone scales have two seeds each, and the seeds usually are provided with a wing made of a thin layer of cone scale tissue. The bract beneath each cone scale is free of (not fused to) the cone scale, at least initially. There are 10 genera in the family.

The pines (*Pinus*) are the largest genus in the family, with 93 species. Pines usually are large, long-lived trees with an asymmetrical shape. Bristlecone pines (*Pinus longaeva*) are the oldest living organisms, some individuals reaching more than 5,000 years of age. Pine needles are clustered, two to five per fascicle (Fig. 24.11b), except for the single-leaf pinyon (*Pinus monophylla*). They are oval to triangular in cross section. Cones are pendant (hanging) and vary greatly in size. Usually, cones are shed once the seeds have matured and spilled out. Some **closed-cone pines,** however, keep their scales closed until heated by fire (which may not occur for many years). The ecological advantage of this behavior is that a rain of seeds falls onto the mineral-rich ashes immediately after a fire has removed all competing vegetation. Although the parent generation of pines is killed by the fire, the species retains its dominance because the site is reoccupied by the next generation.

Firs (*Abies*) have a symmetrical, cylindrical, or pyramidal shape. Firs are neither as long-lived nor as large as pines.

Each year's growth is marked by a symmetrical whorl of branches, and it is possible to determine the age of young trees by counting the number of branch whorls from tip to base. Needles are borne singly. Seed cones are carried erect on the branches **(Fig. 24.12),** and they shatter at maturity rather than falling as a unit. There are about 40 species of firs, all restricted to cooler parts of the Northern Hemisphere.

Spruces (*Picea*) closely resemble the firs, but the needles are angular in cross section and often are sharply pointed **(Fig. 24.13),** rather than flat and blunt as are fir needles. Tree crowns often are narrowly columnar. Seed cones are pendant and fall as a unit when mature, except for those of black spruce (*Picea mariana*), a closed-cone conifer and an important species all across Canada. Large-scale wildfires periodically sweep through the Canadian forest, and the closed cones of black spruce allow the species to quickly reestablish itself.

There are about 40 species of spruce, all in the Northern Hemisphere. They tend to require wetter sites than firs, which explains why spruces are absent from places like the Sierra Nevada where firs are common. Spruces were formerly important for another reason. The Native Americans and later the European immigrants collected the resins from spruce and chewed it. In 1848, John B. Curtis and his brother began to produce what they called State of Maine Pure Spruce Gum. The new gum was slow to catch on, but eventually it became quite popular. Two pieces of spruce gum cost one cent. Gum manufacturers eventually shifted to paraffin and then chicle latex from an angiosperm tree.

Hemlocks (*Tsuga*) are pyramidal with slender, horizontal branches and drooping tops. The needles are somewhat flat,

Figure 24.13 *Picea* (spruce) cones. Although light, they are pendant (hand down).

Figure 24.14 A *Pseudotsuga menziesii* (Douglas fir) branch with needles and cone. The cone is about 8 cm long, and bracts stick out beyond the cone scales. The foliage is soft, in contrast to that of spruce and fir.

with a short petiole. They resemble the leaves of firs but are much shorter. Cones are small and pendant. There are 10 species of hemlock in North America and Asia.

Douglas firs (*Pseudotsuga*) comprise a genus of only five species, two in North America and three in Asia. One of these species (*Pseudotsuga menziesii*) is the most heavily cut timber tree in the United States. It dominates much of the Pacific Northwest, Cascade, and Rocky Mountain regions. It may grow to a height of 60 m and have a trunk diameter of 3 m. Its smaller branches hang down, and the singly borne needles resemble those of spruce but are much softer. Douglas fir seed cones are pendant and easily recognized by the three-lobed bracts, which are longer than the cone scales and stick out from between them (Fig. 24.14).

Larches and tamaracks (*Larix*) are unusual among conifers in being deciduous (some members of the Cupressaceae, such as bald cypress, are the only others). They lose all their needles in the fall. Most gymnosperms lose only some of their leaves each year because the life span of a leaf is several years or longer (the record holder, bristlecone pine, retains its needles for 30 years). Older larch branches have needles grouped into crowded clusters on short shoots. One-year-old shoots, in contrast, have needles arranged singly and spirally. The American larch, or tamarack (*Larix laricina*), frequently is found at the edge of bogs.

Cedars (*Cedrus* sp.) are native to North Africa and Asia. They are widely planted as ornamentals in North America. They vary from asymmetrical to pyramidal in shape, and the needles are borne singly or in fascicles of 10 or more. The cones are thin-scaled and erect on the branches. They usually disintegrate when mature. The cedars have been important timber trees since biblical times.

Cupressaceae Include the Junipers, Cypresses, and Redwoods

Cupressaceae are shrubs or trees, often with small, scalelike or awl-shaped leaves and open branching patterns. Plants can be monoecious or dioecious. Cone scales may be woody, as in cypress (*Cupressus*), or fleshy, as in juniper (*Juniperus*; Fig. 24.15). In either case, the bracts are fused to the cone

Figure 24.15 *Juniperus occidentalis* (mountain juniper), a species in the family Cupressaceae. (a) Massive trees grow on exposed granite at high elevations in the Sierra Nevada. (b) A branch with fleshy female cones commonly called berries. Cones are 1 cm in diameter.

PLANTS, PEOPLE, AND THE ENVIRONMENT:

The California Coast Redwood Forest

Coast redwood, California redwood, and sequoia are names that bring to mind images of fog, looming trees, silent forests, and a sense of prehistoric time. Like other members of the Taxodiaceae family, *Sequoia sempervirens* grows in a humid climate with moderate winters. Over geologic time, its once wide range throughout the north temperate zone shrank to a 700-km-long strip of land hugging the California coast. The redwood belt is less than 55 km wide, and it occupies coast-facing slopes below an elevation of 800 m. The total area of its modern range amounts to less than 800,000 hectares—less than 1% of California's area and less than 0.003% of the earth's surface.

Yet, the coast redwood is well known. It dominates a forest that has a remarkable growth rate, and its biomass (weight of organic matter) is greater than any forest in the world **(Fig. 1)**. Overstory redwood trees commonly exceed 100 m in height and 3 m across at the base of their trunks, and attain ages of 1,000 years. The aboveground biomass is estimated to reach 3,600 metric tons per hectare. The tropical rain forests have only one-seventh as much biomass.

The redwood forest exists where it does because the nearby ocean keeps air temperatures buffered throughout the year. Winters do not have hard frosts, and summers are mild. In addition, cold, upwelling, offshore water creates fog banks that extend inland far enough to bathe the redwoods during most summer days. Fog reduces temperature and moisture stresses around the needles. Also, the tree canopies comb the fog as it passes through them, forming larger droplets that drip to the ground and add the equivalent of 20 cm of rain each year.

Redwood trees are uniquely tolerant of fires, which occur naturally during California's dry fall days when lightning strikes the ground. Redwood bark is thick and usually provides an effective insulation for the vascular cambium deep beneath. If the fire is hot enough to kill the tree's crown, however, the very base of the tree and all the underground roots remain alive (because soil is an even better insulator than bark). At the junction of root and stem lies a band of dormant buds. If the tree trunk is killed, the normal hormone balance is modified, and the buds break dormancy. Many fast-growing shoots emerge through the bark and above the soil. Over time, some of these will mature, producing a circle of trees, all genetically identical to the parent that previously occupied the space in their center. Few other conifers can vegetatively reproduce in this manner.

Fires often lead to floods during the following wet season because ground cover has been removed, leaving nothing to slow the runoff of rainwater. Redwoods also are tolerant of floods, whereas most other species are killed by the layer of silt left behind by floodwaters, which suffocates the roots. Redwood has the capacity to generate adventitious roots

scales. As in Pinaceae, there are two seeds per cone scale. The family contains more than 130 species, with a worldwide distribution. Some cypresses are closed-cone conifers. The fleshy cones of junipers are eaten and used to flavor gin.

A number of the members of this group were previously placed in another traditional taxonomic family, Taxodiaceae, but research has shown that these two families are a single lineage. Species formerly in Taxodiaceae are the dawn redwood of China (*Metasequoia glyptostroboides*), the California coast redwood (*Sequoia sempervirens*), the Sierra redwood (*Sequoiadendron giganteum*), the bald cypress of the eastern United States (*Taxodium distichum*), and several other genera from eastern Asia. The dawn redwood and the bald cypress are deciduous. The coast redwood is possibly the tallest tree in the world; individual trees more than 60 m in height are common, and the greatest recorded height is 112 m (see "PLANTS, PEOPLE, AND THE ENVIRONMENT: The California Coast Redwood Forest" sidebar). Redwood has outstanding lumber qualities, not the least of which is its resistance to decay, conferred by natural byproducts that accumulate in the wood. Sierra redwood is the most massive tree in the world; trunks at breast height reach nearly 10 m in diameter. The trunk alone of the largest living tree, the General Sherman tree in Sequoia-Kings Canyon National Park, is estimated to weigh 625 metric tons.

Redwoods currently are restricted in their range, but they were once more numerous and widely distributed. Early in the Cenozoic era, 20 to 60 million years ago, they were part of a rich forest that covered the cool-temperate zone of the Northern Hemisphere. This forest had a unique mix of gymnosperm and flowering trees—a mix not found anywhere today. Fossil deposits of this forest have been recovered from Asia, North America, Greenland, and Europe. Climatic change and continental drift fragmented the forest and forced redwoods into a continually shrinking habitat. Dawn redwood, in fact, is so rare and narrowly distributed that it was known only from the fossil record until living trees were discovered in China's Szechwan Province

from the buried trunk. New feeder roots then take on the function of the dying deeper roots. Redwood seedlings also survive best on bare mineral soil—either fresh silt or soil burned of its litter. Litter harbors damping-off fungi that otherwise infect and kill virtually every seedling.

In summary, the redwood maintains its hold *because* of episodic catastrophes, not despite them. Other species that compete with redwood for dominance are killed by fire and flood. Thus, park managers who try to protect the redwood forest by suppressing fires may instead be enabling the transition over time from a redwood forest to a fir-hardwood forest. Park managers in many other parts of the world are learning a similar lesson: disturbance is a natural part of the environment, and we cannot preserve some ecosystems without allowing that disturbance to continue.

There are few areas of mature redwood forest left to manage. Because of redwood's timber value, it has been intensively harvested for the last 150 years. Only 3% of the original acreage remains, almost all of it in state and federal parks. Although many young redwoods remain in the cutover areas, species composition and forest architecture there are different. It takes more than 200 years for a redwood tree to attain mature size, and only mature trees are capable of supporting certain animals. Marbled murrelets, spotted owls, flying

squirrels, and tree voles require large trees for nesting or roosting habitat. Certain insects, fungi, and wood-decaying microbes require a thick layer of litter on the forest floor, which is achieved only in old-growth forests. Shade-loving herbs, mycorrhizal fungi, and small vertebrates do not grow and reproduce successfully until the overstory is deep and dense enough to create the required low-light, humid environment.

A wise management scheme for this small remnant of a once-great forest must be a high priority if future generations of humans are to enjoy it. But how exactly does one manage enormous, long-lived organisms that maintain their dominance only with the help of episodic floods and fires that may recur only once a century? Redwood park management needs a long-term plan, extending 100 or more years into the future. The human species has a difficult time planning even a few years ahead. Can we create the will, the public agencies, and the policy to achieve century-long plans?

Figure 1 An old-growth coast redwood forest, showing an overstory dominated by tall, old trees and an understory covered with many species of perennial herbs.

Michael G. Barbour

Websites for further study:

Information Center for the Environment—California Redwoods: http://ice.ucdavis.edu/wits/redwood.html

California Redwood Forest Photographs: http://www.goldenstateimages.com/redwoods.htm

in the 1940s. Seeds have since been distributed all over the world. Dawn redwoods now grow in many botanical gardens and are used as ornamental plants.

Taxaceae Include the Yews, But Plum Yews Belong to Cephalotaxaceae

Taxaceae are shrubs or trees with dark-colored, broadly linear, sharp-pointed leaves. Plants are dioecious. The seeds are borne singly and are covered with a fleshy aril (Fig. 24.16). These are the only conifers to lack cones. One of the family's more than 20 species is the English yew (*Taxus baccata*), famous for excellent bows made of its wood and its connection to English medieval history and folklore. The quality of the wood results from extra spiral thickenings of secondary wall material in tracheids. More recently, the Pacific yew of the western United States (*Taxus brevifolia*) was found to contain an important anticancer compound, TAXOL (paclitaxel; Bristol-Meyers Squibb, New York, NY). TAXOL is present in such small

T. E. Weier

Figure 24.16 A branch of *Taxus baccata* (English yew), a species in the family Taxaceae. The seed is enclosed in an edible, fleshy red aril, attractive to birds, which disperse the seeds.

Figure 24.17 *Phyllocladus*, a genus of New Zealand shrubs in the podocarp family. Most podocarps are Southern Hemisphere shrubs or trees. They lack woody cones, and instead have fleshy cones often mistaken for berries, like those of junipers.

amounts in the bark of Pacific yew (a 100-year-old tree contains only 300 mg) that environmentalists were concerned deforestation would be justified in the name of medicine. However, chemists synthesized the TAXOL molecule in the laboratory in 1994, thereby making it unnecessary to fell trees.

Cephalotaxaceae consist of less than 10 species of Chinese shrubs and trees. Leaves are yew-like. Seeds are borne in pairs and are covered with an aril; they are 2 to 3 cm in length and resemble oval plums. Minute cone scales are associated with the seeds. Plants are dioecious.

Podocarpaceae and Araucariaceae Are Largely Southern Hemisphere Conifers

The Podocarpaceae and Araucariaceae lineages, of predominantly Southern Hemisphere conifers, form a clade, distinguished by a reduction to a single ovule (and seed) per cone scale. Podocarps are shrubs or trees (**Fig. 24.17**). Leaves vary, depending on the species, from short needles to long and broadly oblong blades. Plants may be monoecious, but these are mostly dioecious. The seeds are borne singly and enclosed in a fleshy cone scale. The bracts also are sometimes fleshy. As is also true for juniper berries, the cones of podocarps are attractive to birds that digest the fleshy part but defecate the seed. Many excellent lumber trees of Australasia, Africa, and South America are in this group. *Podocarpus dacrydioides* and *Podocarpus totara* of New Zealand, for example, reach 60 m in height. This is a large lineage, with about 140 species. Most are restricted to the Southern Hemisphere, but some occur in Japan, Central America, and the West Indies. A number of species are widely planted as ornamentals.

Araucariaceae are relatively large trees in the genera *Araucaria* (monkey puzzle, southern pine), *Agathis* (kauri), and the recently discovered *Wollemia* (wollemi pine). The tallest trees in the tropics, reaching up to 89 m, are species of *Araucaria*; and *Agathis australis* of New Zealand has trunks rivaling those of redwood in girth. Leaves vary from needle-like to flat and broad. Cones tend to be large and globose,

Figure 24.18 The genus *Araucaria* (southern pines) includes trees among the tallest and most massive in the world. The trees shown here are *Araucaria heterophylla* (Norfolk Island pine); these trees are widely planted as an ornamental because of their extremely symmetric habit.

and—like fir cones—they disintegrate when ripe. The more than 30 species are exclusively native to the Southern Hemisphere, but some, such as the strikingly symmetrical Norfolk Island pine, *Araucaria heterophylla* (**Fig. 24.18**), are widely planted as ornamentals throughout the world.

24.7   THE LIFE CYCLE OF *PINUS*, A REPRESENTATIVE GYMNOSPERM

This chapter uses pine (*Pinus*) to represent the typical life cycle of gymnosperms (**Fig. 24.19**). Keep in mind that other species of gymnosperms differ from pine in many ways. For instance, only the conifers produce cones (all except Taxaceae), and even among conifers, not all of the cones are large and woody. There are other differences as well. Pine is a convenient plant to use as a model, though, for several reasons: Each tree has both megasporangia and microsporangia, seeds do not have fleshy coverings, and a great deal is known about the life cycle.

Pollen and Ovules Are Produced in Different Kinds of Structures

All pine sporophytes are trees (with the exception of *Pinus mugo* of Europe, which is a shrub). Like all seed plants, they are heterosporous. Pollen is produced in small papery stro-

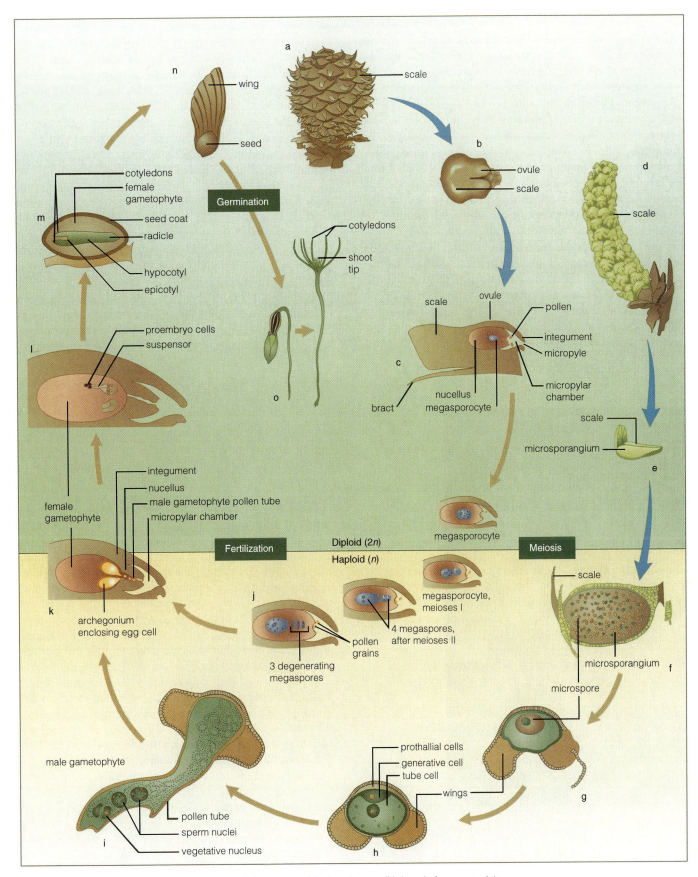

Figure 24.19 Stages in the life cycle of a pine. (**a**) A young ovulate (seed) cone. (**b**) A scale from an ovulate cone with two ovules on its upper surface. (**c**) Longitudinal section of an ovulate cone at the time of pollination. (**d**) A pollen strobilus. (**e**) One scale from a pollen strobilus. (**f**) A section of a microsporangium with mature pollen grains inside. (**g–i**) A developing male gametophyte. (**c,j,k**) A developing female gametophyte. (**l–n**) A developing seed. (**o**) Germination of the seed.

bili; a **strobilus** is a series of densely aggregated sporophylls that are spirally attached to a stem axis. Female gametophytes in conifers typically are produced in larger, woody structures that are called ovulate or seed cones. In conifers, pollen strobili and ovulate cones differ in size, architecture, longevity, and location on the tree.

The pollen strobili of pine, and other conifers, are similar to the pollen strobili of cycads and *Ginkgo*. The ovulate cones of conifers, however, are unique; they are composed of a branch system—a series of modified branches attached to a central axis. The cone scales of conifers, therefore, are not derived from leaves, but from stems. The evidence for this includes a good fossil record of the stages in the evolution of the cone, as well as a morphological feature: a bract (modified leaf) subtends each cone scale, just as a leaf subtends each branch in a typical plant. Cycads produce their seeds on sporophylls and *Ginkgo* produces reduced ovulate structures.

Pollen strobili average 1 cm in length and 5 mm in diameter **(Fig. 24.20a).** They are borne in groups, usually on the lower branches of trees. Each strobilus is composed of a large number of small microsporophylls attached spirally to

an axis. Two microsporangia (more in some conifers) develop on the underside of each sporophyll **(Fig. 24.20b).** The microsporangium is lined with a layer of nutritive cells called the **tapetum.** Inside are microsporocytes that undergo meiosis and produce haploid microspores. The nucleus within each microspore divides several times by mitosis within the spore wall. The resulting pollen grain contains two nuclei that will undergo further divisions and several squashed vegetative cells (Fig. 24.19g). A pollen grain is an immature male gametophyte **(Fig. 24.20c).**

Enormous numbers of pollen grains eventually are shed from the microsporangia of a single tree. The pollen grains are yellow and light in weight (for dispersal by wind). Windrows of pollen are visible on the ground during the period of release, which is generally in spring. Pine pollen grains have two inflated wings that are outgrowths of the wall; these help orient the pollen grain on the pollination droplet of the ovule. The pollen grains of many other gymnosperm species lack such wings or have a different number.

The mature ovulate cone is the familiar pine cone that is commonly associated with pines and other conifers. It is

pollen

strobili

leaf needle

a

J. W. Perry (a–c)

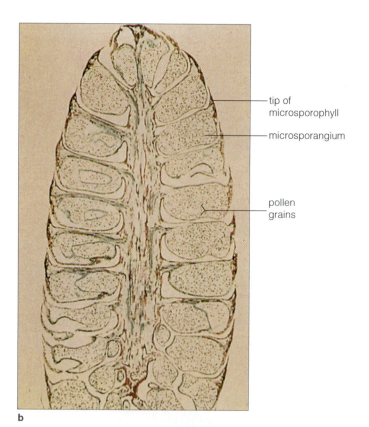

tip of microsporophyll

microsporangium

pollen grains

b

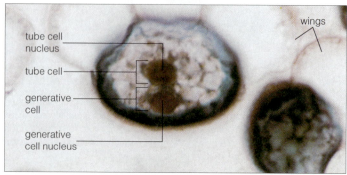

tube cell nucleus

tube cell

generative cell

generative cell nucleus

wings

c

Figure 24.20 Pollen strobili. (**a**) A cluster of strobili near the end of a branch at the time of pollen release. (**b**) A longitudinal section of one strobilus, showing microsporangia attached to the underside of each scale. (**c**) Each microsporangium is filled with pollen grains; the pollen grains consist of only two cells, each with its own nucleus.

composed of many woody scales attached to an axis in a spiral arrangement. Each scale has a bract attached to the axis beneath it. This bract is relatively small and obscure in pines but is quite large and even longer than the cone scale in some genera (such as Douglas fir, *Pseudotsuga menziesii*, and Shasta red fir, *Abies magnifica* var. *shastensis*, see Figure 24.12).

When the ovulate cone is young, it often is reddish and is softer and smaller than the male strobilus. Cones develop singly in early spring at the tips of young branches in the upper part of the tree **(Fig. 24.21a).** Two ovules develop on the upper surface of each scale **(Fig. 24.21b).** Remember, a megasporangium and its integument layer form the ovule of pine. The ovules first appear as small protuberances close to the axis of the cone. The integument, which forms the outer-

most protective layer of the ovule, possesses an open pore facing the axis of the cone. This pore is the micropyle, through which pollen grains later enter. In pine, only one of the many cells filling the young ovule is a megasporocyte **(Fig. 24.21c).** The rest of the megasporangium forms the nucellus, a nutritive tissue. When the megasporocyte divides by meiosis, the result is four megaspores arranged in single file. Only one of the megaspores develops into a megagametophyte; the other three degenerate.

The megaspore grows slowly into a female gametophyte. Most conifers require several months, and pine takes just over a year. The development of the gametophyte takes place entirely within the ovule. Two or more archegonia differentiate at the micropylar end of the growing gameto-

J. W. Perry (a–c)

two-month-old female cones

leaves (needles)

fourteen-month-old female cones

scale

ovule

bract

a

b

resin duct

ovule

integument

megasporocyte

micropyle

sterile bract

scale

c

Figure 24.21 Ovulate cones at the time of pollination. (**a**) Two-month-old cones at the tip of a young branch. Below are larger cones just older than 1 year. (**b**) Longitudinal section through a 2-month-old cone, showing ovules on top of each scale and bracts attached to the bottom of each scale. (**c**) A closer view of one ovule (*dashed area*). Note the large megasporocyte inside the nucellus and the micropyle and micropylar chamber. Meiosis has not yet occurred.

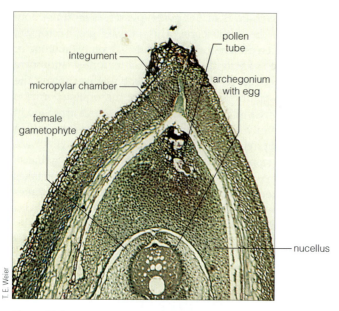

Figure 24.22 An older ovulate cone at the time of fertilization. Only half of the ovule is shown, but the closed micropyle, a pollen tube growing through the nucellus, and a differentiated archegonium with a large egg cell are visible.

phyte, using energy from digestion of the nearby nucellus. Thus, the gametophyte is completely dependent on sporophyte tissue.

At maturity, the ovule consists of an integument, a thin layer of remaining nucellus, and an ovoid female gametophyte that is undifferentiated except for several archegonia at one end, each with an enclosed egg (Fig. 24.22). Notice that there is space between the micropyle and the nucellus; this is the **micropylar chamber,** where pollen grains begin to grow pollen tubes.

Pollination Replaces the Need for Free Water

In flowers, **pollination** is the transfer of pollen from anther to stigma. In conifers, it is the transfer of pollen from male strobilus to ovulate cone. Wind is the vector that carries conifer pollen. In pine, the placement of ovulate cones above pollen strobili makes it more likely that pollen will be carried to another tree rather than settling on a cone of the same tree. Thus, cross-pollination is usual. For those conifer groups that are dioecious—the Cephalotaxaceae, Cupressaceae, Podocarpaceae, and Taxaceae—self-pollination is impossible; cross-pollination always occurs. Cross-pollination is valuable because it creates more genetic variation in the next generation, increasing the capacity of a species to take advantage of diversity in the environment.

Studies of the aerodynamics of pollen near cones have demonstrated that the shape of the young ovulate cone can create unique air currents and eddies that bring pollen grains of the appropriate species close to the open scales, increasing the probability that pollen will land in the cone. Thus, wind pollination is not entirely random.

Pollination occurs when the ovulate cone is about nine months old **(Table 24.1).** Female cone buds are initiated in the late summer preceding the spring of pollination. The cone is fully formed in the bud and merely enlarges when the bud breaks in spring. Cone scales at this stage turn slightly away from the cone axis, providing space for pollen grains to drift down to the ovules. A sticky **pollination drop** exudes from the micropyle. The drop is chemically similar to the nectar of flowers, containing about 8% sugar (glucose and fructose) and traces of amino acids. It does not seem to attract animals, but it passively traps pollen grains that touch it. A chemical signal diffuses from the trapped pollen to the ovule, triggering rapid absorption of the liquid. This draws pollen grains through the micropyle and into the micropylar chamber, where they come to lie on the surface of the nucellus.

The pollen grain germinates, slowly developing an elongating pollen tube that grows through the nucellus toward an egg (Fig. 24.22). Several nuclear divisions occur in the tube, but no cell walls are formed. The final division produces two sperm nuclei. The pollen tube, containing two sperm nuclei and several vegetative nuclei, is the mature microgametophyte.

Fertilization Leads to Seed Formation

At the time of pollination, the megasporocyte is undergoing meiosis (Table 24.1). As the pollen tube grows, the female gametophyte forms. The small reddish cone grows larger and turns green, and the scales become tightly closed. Timing of male and female gamete formation is coordinated so that the egg is ready for fertilization by the time the pollen tube with its sperm nuclei has reached the archegonium. In pine, this development takes about 12 months.

The sperm nuclei, together with the cytoplasmic contents of the pollen tube, are discharged directly around the egg cell. Sperm nuclei do not possess flagella; therefore, they are not motile. One sperm nucleus comes in contact with the egg and enters it. This is not the end of its journey; it must pass through the egg cytoplasm to reach the egg nucleus and fuse with it. Pine egg cells have hundreds of times the volume of a sperm nucleus. How a sperm nucleus finds its way through that huge space to an egg nucleus is unknown.

The fertilized egg becomes a diploid zygote, and it begins to divide immediately into a relatively elaborate **proembryo,** the apical cells of which develop into an embryo (Fig. 24.19m).

Sometimes more than one embryo will form. The multiple embryos can originate in two ways. Each female gametophyte has more than one archegonium, and all can be fertilized if there are enough pollen grains. Furthermore, each proembryo is capable of forming as many as four embryos. In the first case, fraternal siblings would be formed; in the second, identical siblings would be formed. Generally, however, only a single embryo survives to maturity.

Table 24.1 The Timing of Development in the Life History of Pine

Event	Structure		Season, Year
Cone development	Pine tree (sporophyte)		
	Male cone	Female cone	
	↓	↓	
Meiosis	Microsporocyte	Megasporocyte	Summer, 0
	↓	↓	
	Microspore	Megaspore	Spring, 1
	↓		
Pollination and pollen germination	Pollen grain (immature male gametophyte)		Early summer, 1
	↓		
	Pollen tube (mature male gametophyte)	Female gametophyte	
	↓		
Fertilization	Sperm nuclei Egg		Late spring, 2
	Zygote		
	↓		
	Proembryo		
	↓		
Dispersal	Mature embryo within a modified ovule (a seed)		Summer, 2
	↓		
	Cone opens		Fall, 2
Stratification	↓		
	Seed outside the cone		Winter, 3
	↓		
Germination	Seedling		Spring, 4

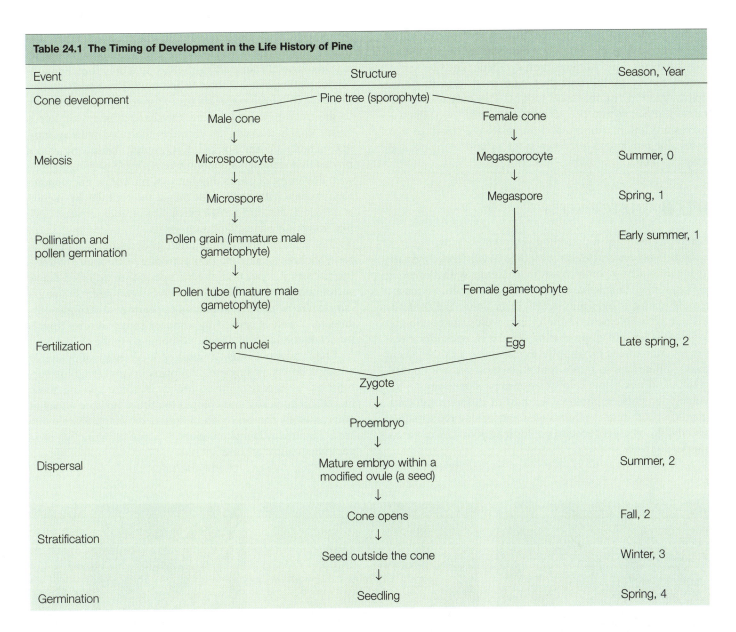

While the embryo develops, the female gametophyte also continues to grow by digesting the remaining nucellus, and its cells become packed with food. The integument hardens and becomes a seed coat. The micropyle has closed. The ovulate cone enlarges still more, becomes woody, and loses its green color.

When the seeds are mature, the cone scales open and the seeds fall out. Pine seeds usually have a wing attached to them, which catches the wind and aids in the dispersal of seeds from the parent tree (Fig. 24.19n). The wing is formed from the cone scale. Seeds are usually shed immediately on ripening, in late summer or early fall. The mature, open cone is two years old when seeds are dispersed (Table 24.1).

Seeds are dispersal packages containing a dormant embryo, stored food, and a protective outer coat. Their richness in carbohydrates has made pine seeds an important food source for human and animal populations. Native Americans ground the seeds into a meal and used the flour, much as other cultures have used grains. Many birds and mammals eat and cache pine seeds.

Seeds that escape predation by herbivores usually lie dormant until the next spring. The dormant embryos have a low metabolic rate, and their water content is low. They are able to tolerate aridity, anaerobic conditions, and temperature extremes while in this state. Their dormancy is broken by exposure to cold, wet winter conditions. Dormancy can be artificially broken by storing seeds in wet cheesecloth at temperatures a few degrees greater than freezing for 4 to 6 weeks, a process called **stratification.** Dormancy prevents pines from germinating on warm fall days, because that would force seedlings to endure inhospitable winter growing conditions.

When the seed does germinate, the radicle emerges first. Then the hypocotyl elongates, taking the cotyledons and epicotyl above the surface while still enclosed in the seed coat. The cotyledons absorb nutrients stored in the female gametophyte and then enlarge, pulling themselves out of the seed coat (Fig. 24.19o). When light strikes the cotyledons, they turn green and become photosynthetic. The young pine grows slowly in the first year, and most growth occurs below ground, in the root system. The cotyledons may remain attached into the second year; until then, the true leaves are few in number and are relatively small. In such cases, the cotyledons continue to grow and elongate until they are shed.

24.8 GNETOPHYTES

Gnetophytes are represented by three living genera: *Ephedra, Gnetum,* and *Welwitschia* **(Fig. 24.23).** This enigmatic group has a number of traits in common with flowering plants, and botanists have speculated that gnetophytes are closely related or even ancestral to them. Some of these traits include vessel elements in the xylem, ovules surrounded by a fleshy layer, pollen-producing structures resembling stamens, and reduced gametophytes. However, a closer look at these similarities suggests that these features are convergences and not evidence of relationships. For example, the vessel elements of gnetophytes have a different developmental origin than the vessel elements of flowering plants, and the fleshy layers around gnetophyte seeds are not at all similar to fruits.

Despite great dissimilarity in their appearances, the three genera of gnetophytes share a series of morphological features strongly supporting the idea that they form a monophyletic group. In addition to the traits mentioned earlier, gnetophytes share opposite (or whorled) leaves, extension of the micropyle into a long tube, and a series of sterile bracts (bracts that are not directly associated with sporangia) making up the strobili. One shared feature of gnetophytes that represents an amazing example of convergence with flowering plants is that both sperm from a microgametophyte fuse with a cell in the megagametophyte. In *Gnetum* and *Welwitschia*, the gametophytes are highly reduced and, like flowering plants, lack archegonia.

The basic units of a gnetophyte pollen strobilus are similar in all three genera and bear a striking resemblance to flowers. Each unit consists of several opposite or whorled bracts surrounding stamen-like structures. Several microsporangia dangle from these structures, appearing amazingly like anthers. Pollen is carried by wind or sometimes by insects (typically moths) from the microsporangia to the ovules.

Ovulate strobili also consist of a series of opposite or whorled bracts. A single ovule forms in each strobilus surrounded by a pair of fused bracts. In some species, these bracts can be fleshy and colored to attract birds, whereas in other species, the bracts form collars or wings around the seed. The integuments are extended into a distinctive long papery, micropylar tube.

Figure 24.23 Two examples of gnetophytes. (**a**) Branches of the shrub *Ephedra viridis* (Mormon tea) about 1 m in height, growing in the White Mountains of California. (**b**) Detail of a stem node and the small, scalelike leaves of *E. viridis*. Most photosynthesis occurs in the green stems. (**c**) Seed nearly surrounded by a red aril. (**d**) *Gnetum leyboldii*, a tropical vine.

gnetum

Ephedra (also called joint fir or Mormon tea) is the only gnetophyte found in North America. Its 40 species are distributed through the warm-temperate Mediterranean rim, India, China, the southwestern deserts of the United States, and mountainous parts of South America. It is a vine or shrub with opposite or whorled leaves and prominent joints (Fig. 24.23a,b). The leaves are little more than scales; therefore, most photosynthesis is conducted by the green stems. Joint fir is the source of the drug ephedrine, an alkaloid that constricts swollen blood vessels and also is a mild stimulant. An overdose can cause death. The Asian species contain significantly more ephedrine and are important in Chinese herbal medicine, where it is called *ma huang*. Some Native American groups used the tea to cure venereal diseases, and Europeans emigrating to the southwest adopted this remedy. Strong *Ephedra* tea was once a common offering at brothels in Nevada and California.

Gnetum is a tropical genus of 30 species, which include lianas (climbing vines), shrubs, or trees. The leaves are nearly indistinguishable from those of a broad-leafed flowering plant (Fig. 24.23c), and unless reproductive structures are present, *Gnetum* is difficult to identify.

Welwitschia is found in the Namib Desert, along the arid southwest coast of Africa. It has two (rarely four) long, leathery, straplike leaves that trail along the soil surface **(Fig. 24.24).** When Austrian naturalist Friedrich Welwitsch first saw this plant on a field trip in 1859 near the Angolan coast, he fell to his knees in disbelief. Specimens taken to Europe caused Darwin to describe it as the platypus of the plant kingdom. Local Africans called it *otjitumbo* ("stump"), but the English taxonomist Joseph Hooker named it *Welwitschia mirabilis*—the genus honors its European discoverer, and the species epithet is the Latin word for "wonderful."

The plants are slow growing, with most photosynthate going into an exceptionally well-developed tap root. The aboveground portion consists of a thick woody crown and two broad leaves. These grow continuously from their base and fray at the ends, eventually becoming tangled into a large mound of what appears like many leaves. The leaf tissue has many sclereids, a considerable concentration of

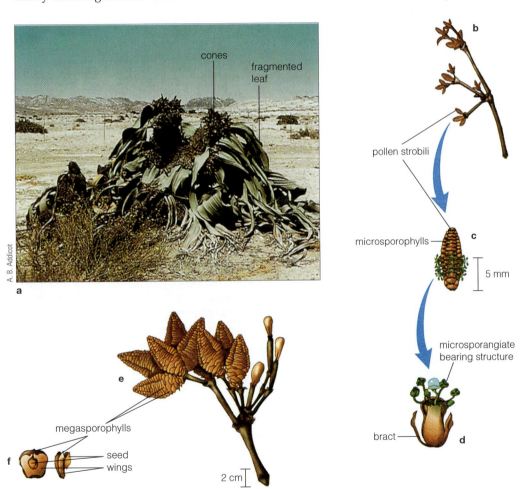

Figure 24.24 *Welwitschia mirabilis*, a plant of the Namib Desert in southwestern Africa. (**a**) The two old foliage leaves of this plant have become fragmented into ribbonlike strips. Strobili are visible at the juncture of the stem crown and leaves. A plant this size, about 1 m in height, is about 1,000 years old. (**b**) A cluster of microsporophylls resembling a conifer pollen strobilus. (**c,d**) The strobilus of microsporophylls at the time of pollen release. Note how flower-like each sporophyll is, with structures bearing microsporangia, which resemble anthers. (**e**) A cluster of megasporangia strobili, which resemble ovulate strobili of conifers. (**f**) Each megasporophyll has a bract that encloses an ovule, and when the ovule has matured, outgrowths of the integument form wings.

lignin, and numerous crystals of calcium oxalate—all combining to make them durable and inedible. Stomata are sunken, probably an adaptation to the arid habitat. (Annual rainfall in its growing region is only 2 cm although condensation from fog can add the equivalent of another 5 cm.) Individual plants can attain ages of 1,000 to 2,000 years.

THE ECOLOGICAL AND ECONOMIC IMPORTANCE OF GYMNOSPERMS

Gymnosperms have great ecological importance. Although their habitats range from tropical forests to deserts, their centers of dominance are the cool-temperate zones of the Northern and Southern Hemispheres. In North America, they are part of low-elevation forests across Alaska, Canada, and New England and down the Pacific coast; in addition, they are part of high-elevation forests in the Appalachian, Rocky, and Cascade-Sierra mountain chains. They also occur in more localized habitats within the southeastern coastal plain. The gymnosperm landscape covers more than one third of the North American landmass. Within this landscape, ecosystems are created largely by conifers. Conifer foliage is rich in organic acid, so its decomposition, in turn, makes the soil acidic and relatively low in nutrients. Only those shrub and herb species that can tolerate such soil conditions, and the low level of light beneath the dense conifer overstory, are able to grow here. The acidity also hinders bacteria but favors fungi, so the decomposer microflora is strongly affected. Conifer foliage and wood are high in secondary compounds that inhibit grazing animals; therefore, mammal and insect diversity is low, as is that of insectivorous birds.

Gymnosperms also have great economic importance. They are a major source of lumber, paper pulp, turpentine, and resins, and they are used as fuel for heat. More than one historian has pointed out that humans have always lived in a Wood Age, even if we talk about the Stone Age, Bronze Age, and Iron Age. The fact is that wood has provided energy for smelting, heating, and cooking, as well as raw building material for habitations and vehicles throughout human history. Minoan, Greek, and Roman civilizations rose and fell depending on their access to forests, often forests made up of gymnosperm trees. In North America, economic growth has depended particularly on lumber of ponderosa pine (*Pinus ponderosa*), eastern white pine (*Pinus strobus*), and Douglas fir (*Pseudotsuga menziesii*). A significant amount of raw or sawn logs is exported to countries that no longer have this natural resource.

The magnitude of forest exploitation on the North American continent rivals modern deforestation in the tropics. Enormous lowland forests of pine from the states bordering the Great Lakes and the Southeast United States, as well as equally extensive upland conifer forests in the West, have been clear-cut in the last 100 years. A cumulative area greater than 40 million hectares (100 million acres) has been clear-cut in this relatively short time. Clear-cuts in some Canadian provinces are so large that they are one of the few human artifacts visible from orbiting spacecraft. The white pine forests of the Great Lakes area have never recovered from harvesting, and the second-growth forests of the West lack the biotic diversity and ecological stability the pristine forests had acquired over centuries of slow development.

Gymnosperms are widely used in landscaping because of their evergreen habit and the diversity of their growth forms. They can be pruned into low ground covers, hedges, and dramatically branched trees. Many have rapid growth rates. Some, such as the Norfolk Island pine (*Araucaria heterophylla*), cycads (*Cycas*), or the Monterey cypress (*Cupressus macrocarpa*), have unique canopy architectures; many people find these trees pleasing, and therefore have brought them to countries far outside their original ranges.

KEY TERMS

closed cone conifer	Pangaea
conifer	pollen, pollen grain
cycad	pollination
fascicle	pollination drop
gnetophyte	proembryo
gymnosperm	progymnosperm
integument	resin
micropylar chamber	resin duct
micropyle	sieve cell
nucellus	strobilus
ovule	tapetum

SUMMARY

1. Gymnosperms are tracheophytes whose life cycles have novel features. The new traits include the ovule, the seed, pollination, and wood, all of which are adaptations for survival on land. *Gymnosperm* means "naked seed."

2. Living gymnosperms number approximately 900 species and belong to four lineages: cycads, ginkgo, conifers, and gnetophytes. They have great economic and ecological importance, mainly in the temperate zones of the world.

3. Pine provides a typical example of a gymnosperm life cycle. Pine is monoecious; an individual pine tree will produce both pollen and ovules. Some other gymnosperms are dioecious.

4. Pollen strobili are small and seasonal. They consist of sporophylls attached to a central axis. Two microsporangia are attached to the underside of each sporophyll. Microspores develop into pollen grains within the microsporangium. The pollen grains (immature male gametophytes) are liberated from the microsporangium and carried passively on the wind to ovules, generally on another tree.

5. Ovulate cones are the cones commonly associated with conifers. When young, a cone is as small as a pollen strobilus, but it ultimately becomes larger. It consists of scales with a bract beneath each scale. Two integumented megasporangia (ovules) are attached to the upper surface of each scale. An ovule consists of an integument with a micropyle and micropylar chamber at one end and nucellus (megasporangium tissue) with a single megasporocyte. This is the stage of development at the time of pollination.

6. The megasporocyte divides by meiosis into four megaspores, but only one develops into a megagametophyte. At maturity, the megagametophyte is a relatively small mass of cells with several archegonia at the micropylar end. In the meantime, the pollen grain has been pulled through the micropyle by a shrinking pollination drop until it lies on the surface of the nucellus. It germinates into a pollen tube that grows slowly through the nucellus toward an archegonium and egg. Cell division in the tube produces two nonmotile sperm nuclei, one of which fertilizes an egg, forming a zygote.

7. The zygote divides to produce a proembryo, some of whose cells divide to produce an embryo with cotyledons and epicotyl, hypocotyl, and radicle regions. Surrounding the embryo is the megagametophyte, itself surrounded by a seed coat (which forms from the integument). Embryos are dormant when seeds are shed. Germination is possible after the seeds are exposed to cold, wet conditions for several weeks—a process called stratification. Thus, a seed is a dormant embryo with stored food and a protective outer coat; its function is to disperse the embryo away from the parent plant over time and space.

8. Cycads comprise 100 species of woody tropical plants that possess some ancestral features such as large, compound leaves and swimming sperm. Plants are dioecious. Strobili are large. Pollination involves beetles as vectors in some species.

9. *Ginkgo biloba* is the sole surviving species of its lineage, and although native to China, it is widely used as a street tree and ornamental. Its fan-shaped leaves are deciduous. Ginkgo is dioecious and has swimming sperm, like the cycads, but has wood similar to the conifers.

10. The conifers are the largest group of gymnosperms and consist of the familiar pines, spruces, firs, junipers, redwoods, yews, cedars, and cypresses. Other members of the group include the monkey puzzle tree, plum yew, and podocarps. All of them are distinguished by the presence of ovulate cones except for the yews which have lost them. Cones are typically woody structures composed of a branch system, but in some cases they are fleshy.

11. Gnetophytes consist of three dissimilar genera. All have a series of features that unite them, including vessels, fleshy coverings on the seeds, a micropylar tube, and fusion by both sperm. Only the genus *Ephedra* is found in North America. *Gnetum* is a tropical genus, and most species are lianas. *Welwitschia*, native to the arid coast of southwest Africa, is a bizarre plant with only two leaves.

Questions

1. How do the novel life cycle features of ovule, pollen, and seed adapt conifers to life on land?

2. In what ways does conifer wood play an important role in technological economies? Is it going too far to say that the human species has always lived in—and still does live in—a Wood Age?

3. What are the key differences between the megasporangium of a seed plant and the megasporangium of *Selaginella*? Compare the products of the megasporangia in both groups.

4. Why is a pollen grain an *immature* male gametophyte? When does it become completely developed?

5. Can you see any trends in gametophyte–sporophyte relationships in the moss-to-fern-to-conifer sequence?

6. In what ways are conifers different from cycads and ginkgo?

 InfoTrac® College Edition

http://infotrac.thomsonlearning.com

Gower, S.T., Richards, J.H. 1990. Larches: deciduous conifers in evergreen world. *BioScience* 40:818. (Keywords: "larches" and "conifers")

Rydin, C., Kallersjo, M., Friist, E.M. 2002. Seed plant relationships and the systematic position of Gnetales based on nuclear and chloroplast DNA: conflicting data, rooting problems, and the monophyly of conifers. *International Journal of Plant Sciences* 163:197. (Keywords: "Gnetales" and "monophyly")

Angiosperms

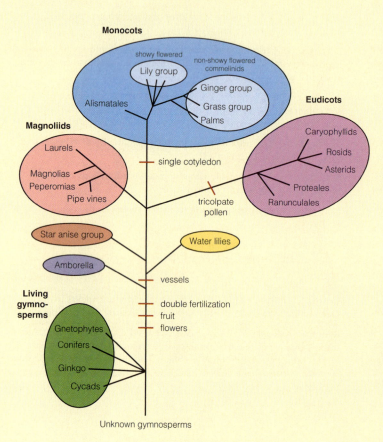

Monocots

showy flowered
Lily group

non-showy flowered
commelinids
Ginger group
Grass group
Palms

Alismatales

Magnoliids

Laurels

Magnolias
Peperomias
Pipe vines

— single cotyledon

Eudicots

Caryophyllids
Rosids
Asterids
Proteales
Ranunculales

tricolpate
pollen

Star anise group

Water lilies

Amborella

**Living
gymno-
sperms**

— vessels

— double fertilization
— fruit
— flowers

Gnetophytes
Conifers
Ginkgo
Cycads

Unknown gymnosperms

Visit us on the web at http://biology.brookscole.com/plantbio2
for additional resources, such as flashcards, tutorial quizzes,
InfoTrac exercises, further readings, and web links.

KEY CONCEPTS

1. Angiosperms, or flowering plants, are unique in having ovules borne inside carpels, rather than on naked scales or leaves, as in gymnosperms. The ovule is fertilized by a pollen grain that is transferred from a stamen to a carpel (by wind, water, or animals), and then germinates into a tube that grows through ovary tissue, ultimately reaching the ovule and its haploid egg cell. The tube bursts, releasing two sperm nuclei. After one nucleus fertilizes the egg, the egg becomes an embryo, the ovule becomes a seed, and the ovary becomes a fruit. The new features—ovary and fruit—increase the protection of the gametophyte generation and of the embryo; they also widen the dispersal of the seed over time and space.

2. Ovaries are part of a new structure called the carpel. Ovule-containing carpel(s) plus pollen-containing stamens and variously colored and shaped petals and sepals collectively comprise a new structure called a flower.

3. Double fertilization is an innovation of the angiosperm life cycle. When a pollen grain becomes a mature microgametophyte, it produces a long pollen tube that contains two sperm nuclei. One nucleus unites with an egg to produce a zygote, whereas the other unites with two other nuclei in the gametophyte (the polar nuclei) to produce an endosperm nucleus. As the embryo develops from the zygote, the endosperm nucleus divides repeatedly and surrounds the embryo with stored food. Double fertilization conserves energy, because food for the embryo does not accumulate until after fertilization.

4. The first angiosperm fossils date back approximately to 135 million years ago. However, the initial split of the lineage leading to angiosperms from other seed plants may have occurred during the time of the early seed plants, more than 250 million years ago. Currently, it is not possible to say which seed plants are the closest relatives, and it may be that no *living* group of seed plants is closely related to the angiosperms. During the Cretaceous period, angiosperms speciated rapidly, and by the early Cenozoic era, they had become the dominant terrestrial plant group.

5. Angiosperms form a monophyletic group. A basal grade of lineages includes *Amborella,* water lilies, and star anise and its relatives. The remaining taxa—that is, the core angiosperms—comprise three major clades: magnoliids, monocots, and eudicots. All together, angiosperms number 257,000 species in 14,000 genera.

6. Magnoliids vary from woody to herbaceous. Magnoliid flowers vary from large and showy, with numerous spirally arranged parts, to small and inconspicuous, with parts in threes. Many important foods, timber, spices, and medicinal and ornamental plants are magnoliids, such as cinnamon, camphor, avocado, bay, sassafras, magnolia, and tulip tree.

7. Monocots typically have parallel-veined leaves, flower parts in threes, embryos with a single cotyledon, sieve-tube members having plastids with protein crystals (lacking starch grains), scattered vascular bundles, and prominent adventitious roots. Monocots consist of such economically and ecologically impor-

tant plants as agaves, bananas, grasses, irises, lilies, onions, orchids, palms, rushes, sedges, yams, and yuccas.

8. Eudicots typically have net-veined leaves, flower parts in fours or fives, embryos with two cotyledons, sieve-tube members having plastids with starch grains, stem vascular bundles in a ring, stamens with slender filaments, and three-apertured (tricolpate) pollen. Basal eudicots include Ranunculales and Proteales, whereas the main group contains three major lineages: caryophyllids, rosids, and asterids. Eudicots include such economically and ecologically important plants as blueberries, buckwheat, cacti, carrots, coffee, grape vines, hemp, legumes, melons, poppies, potatoes, roses, sandalwood, stone fruits, strawberries, sunflowers, tea, teak, tomatoes, and walnuts.

9. Plant geography is the study of plant distribution throughout the world. Some clades are widely distributed, whereas others are narrowly restricted to one part of the world and one type of environment. In the last thousand years, human populations have purposely and accidentally carried plants into regions those plants had not yet reached on their own, creating a more homogeneous global landscape. Angiosperms are of exceptional importance to people for food, fiber, pharmaceuticals, building materials, ornaments, and beverages.

25.1 PLANTS WITH AN ENCLOSED SEED

Flowering plants dominate the earth's vegetation (Fig. 25.1). Hardwood forests and woodlands, shrublands, grasslands, wetlands, deserts, cold high-elevation tundras, and warm low-elevation rain forests have a biomass (weight of organic matter) composed chiefly of flowering plants. Nearly all of the earth's crop plants, orchard trees, garden plants, and ornamentals are flowering plants. Coffee, tea, and cocoa beverages are made from flowering plants. Cotton and linen are fabrics made from flowering plants. Many pharmaceuticals—from aspirin to morphine—come from metabolic chemicals in flowering plants, or are patterned after these chemicals.

Angiosperm is a synonym for flowering plant. It means "seed within a vessel" or "enclosed seed." The defining angiosperm feature is the enclosure of the ovules within surrounding tissue called an **ovary.** The ovary is part of a flower, a structure that occurs only in angiosperms. The ovary (and sometimes associated tissues) eventually forms a fruit, another unique angiosperm structure.

This chapter brings us full circle within this textbook. The initial chapters focused on the development, structure, physiology, and genetics of flowering plants. Then the key features of fungi, protists, algae, bryophytes, seedless vascular plants, and gymnosperms were examined in a sequence that roughly parallels the evolutionary appearance of these groups in the fossil record. Within the green plant lineage, this sequence shows three trends: (1) increasing prominence and complexity of the diploid sporophyte generation; (2) decreasing prominence and complexity of the haploid

Figure 25.1 Vernal pool vegetation in full flower, springtime, Sacramento Valley, CA. (**a**) Aspect, the dominant color coming from gold fields (*Lasthenia* sp.). (**b**) A close-up of *Downingia*, a smaller vernal pool plant.

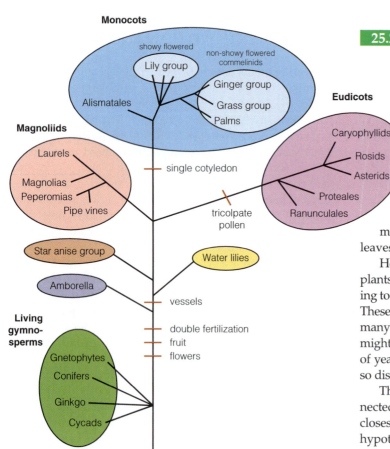

Figure 25.2 Phylogenetic relationships among major angiosperm groups and clades.

gametophyte generation; and (3) ever more elaborate adaptations to life on land. The survey, the sequence, and the trends all peak with the angiosperms, one of the most successful groups of terrestrial organisms. When did this group evolve, and how did it come to dominate the land?

25.2 THE MYSTERIOUS ORIGIN OF THE ANGIOSPERMS

Charles Darwin called the origin of flowering plants "an abominable mystery," because they appear suddenly in the geologic record without a clear fossil history showing a transition from some other plant group (**Fig. 25.2**). The first angiosperm fossils—from the early Cretaceous period, about 135 million years ago—are limited to microfossils such as pollen grains and bits of stem or leaf cuticle. By the mid-Cretaceous period, macrofossils are numerous, diverse, and complex. Fossil leaves, stems, flowers, fruits, and seeds are common.

How can we explain the sudden appearance of flowering plants? Some paleobotanists suggest that the initial line leading to angiosperms diverged from other seed plants very early. These early ancestors of angiosperms would have lacked many of the traits typical of angiosperms today, and they might be difficult to identify in the fossil record. Over millions of years, the full set of characteristics that make angiosperms so distinctive and successful evolved one by one.

The question of when angiosperms originated is connected to the equally difficult problem of identifying their closest relatives. Only a few years ago, many evolutionists hypothesized that the angiosperms were most closely related to gnetophytes, because they share such striking morphological traits as insect pollination, flowerlike structures, vessels, reduced gametophytes, and double fertilization (both sperm fuse with a cell in the megagametophyte). However, when

PLANTS, PEOPLE, AND THE ENVIRONMENT:

Molecular Clocks and the Age of the Angiosperms

Scientists have been developing a method of dating the origin of angiosperms (and other groups) that does not depend on fossils. This technique uses the idea of a **molecular clock**, first proposed by Linus Pauling and Emile Zuckerkandl in the early 1960s and much improved with the advent of recent phylogenetic techniques. Pauling and Zuckerkandl had determined the sequences of amino acids that make up the protein hemoglobin (the molecule that carries oxygen in the blood of many animals) for several species, including human, mouse, horse, bird, frog, and shark species. The number of amino acid differences among the various species appeared to be proportional to how closely related the species were to each other. Human and mouse hemoglobin had 16 amino acid differences, whereas human and horse hemoglobin had about 18. Bird, frog, and shark hemoglobin differed from human by 32, 65, and 79 amino acids, respectively.

Pauling and Zuckerkandl's great insight was to compare these calculated differences in hemoglobin structure with paleontological data concerning the age of divergence of the various lineages to which these species belong. The fossil record provides a rough idea of when each of the groups represented by these species originated (that is, diverged from the vertebrate lineage). When they compared the relative age of the lineages with the differences in their hemoglobin structure, they found a strong correspondence: Hemoglobin seemed to be evolving at a regular, steady rate. This was a surprising result, because few people expected that molecules would change at a steady rate over hundreds of millions of years.

If certain protein or DNA molecules could be shown to evolve in this constant fashion, lineages could be dated by simply assessing the number of differences among them, then figuring out what the rate of change was for that molecule. The time since any two species last shared a common ancestor could, in theory, be determined simply by calculating how much time was necessary for the accumulation of the differences in their amino acid or nucleotide sequences. Unfortunately, it is now known that many molecules do not seem to change in a clocklike fashion, and that different lineages, even sister lineages, may change at different rates. New techniques are being developed that account for rate variation in a molecule's evolution over time.

This technique has been applied to plants. Some molecular clock estimates suggest that land plants originated about 1 billion years ago and that vascular plants separated from bryophytes as early as 700 million years ago. These estimates are controversial because they do not correspond with estimates derived from the fossil record.

Molecular clock estimates of the origin of angiosperms also have been proposed. Research using chloroplast genes has yielded dates ranging from 79 to 220 million years ago. Some of the dates determined in molecular clock studies agree with those determined by dating fossils. New methods of analysis, expanded data sets, and better calibration with known fossils promise to yield molecular clock ages that we can be confident about soon.

 Websites for further study:

Estimate of the Age of the Angiosperm Crown Group:
http://www.2000.botanyconference.org/section13/abstracts/156.shtml

Paleobotany—Angiosperm Evolution:
http://www.monmouth.com/~bcornet/paleobot.htm

botanists began constructing cladograms using molecular data from the relatively unexplored mitochondrion, and then combined these data with sequences from the chloroplast and nucleus, a different picture emerged **(Fig. 25.3).** As these ongoing studies have expanded in scope and scale, they have increasingly supported the idea that angiosperms diverged early in the history of seed plants, not more recently, as most botanists had assumed.

An early divergence, if true, could mean that the closest relatives of angiosperms are now extinct. According to **paleobotanists** Michael Donoghue and James Doyle, it may well be that there are no *living* seed plants closely related to angiosperms. In other words, Darwin's famous statement about the angiosperms' origin being an abominable mystery is still appropriate today.

Well-preserved, early angiosperm macrofossils have been found in China and have been given the name *Archaefructus*. These small, herbaceous plants had no petals or sepals, but they did bear prominent ovaries with enclosed seeds and paired stamens. They flourished in aquatic habitats 125 million years ago to perhaps as early as 140 million years ago, although an origin preceding the Cretaceous period is possible. They do not resemble any known modern group and may actually be submerged aquatic plants that are reduced and highly specialized; they may not preserve many primitive features of the earliest angiosperms.

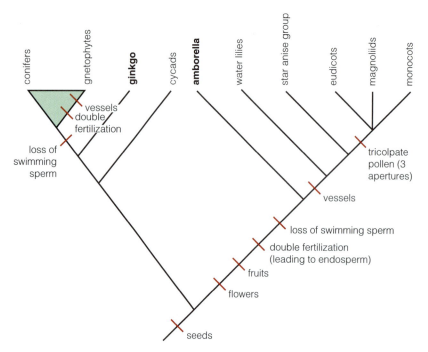

Figure 25.3 A phylogeny for the seed plants, showing the relationships between angiosperms (six groups on the right) and gymnosperms (four groups on the left) and their major lineages.

Key Innovations of Angiosperms Include Both Vegetative and Reproductive Features

Why were flowering plants able to supplant gymnosperms, as well as many fern groups? Probably because they had new vegetative and reproductive features that promoted survival and gave them a competitive advantage. One such feature is an improved vascular system. Angiosperm xylem typically contains large, relatively thin-walled **vessels** in addition to tracheids. The movement of water is much more efficient through these vessels. Angiosperm phloem contains **sieve-tube members** in association with companion cells, in contrast to the sieve cells of gymnosperms. Sieve-tube members have a larger diameter and larger sieve pores, increasing the efficiency of sugar transport.

Another novel feature is the **fruit,** the ripened ovary with enclosed seeds. Fruits aid in the dispersal of seeds by catching on the wind, adding buoyancy in water, or being moved by animals. Some gymnosperms have fleshy arils or scales that accomplish the same end, but not to the same degree of elaboration as in angiosperms.

William Bond (Cape Town University, South Africa) combined these features and others to formulate a **seedling hypothesis** as an explanation of angiosperm dominance. His hypothesis compares gymnosperms to a tortoise and angiosperms to a hare. Gymnosperms are woody and slow-growing, and have lengthy reproductive cycles. The juvenile stage is long. Cotyledons and young leaves are thick and

evergreen, energetically expensive to manufacture, and not changeable in shape. Gymnosperm tracheids and sieve cells are relatively inefficient. All this leads to a slow seedling growth rate.

In contrast, many angiosperms are herbaceous and fast-growing and have short reproductive cycles. The juvenile stage can be short. Cotyledons and young leaves often are thin, deciduous, energetically cheap to make, and variable in shape. Vessels and sieve-tube members are highly efficient pipelines. All this leads to a rapid seedling growth rate. Bond's theory predicts that gymnosperms will be outcompeted everywhere except where angiosperm seedling competition is reduced, as in cold-temperate regions with nutrient-poor soils.

Angiosperms produce a novel structure: the **flower.** The ovary, with its enclosed seeds, is part of this new structure, which serves to aid pollination, protect the developing seeds, and disperse the mature seeds. How could the flower have developed from preexisting organs, such as stems and leaves? The flower is thought to be a modified branch whose leaves have become sepals, petals, stamens, and carpels. Several types of indirect evidence support this hypothesis. The early developmental phases resemble those of leaves, and they are found in the same place that a leaf would be found. The stamens or carpels of some plants, such as *Firmiana simplex,* bear a striking resemblance to leaves. When the ovary of this tree matures, it splits open to show five leaflike carpels that bear seeds along their margins **(Fig. 25.4).**

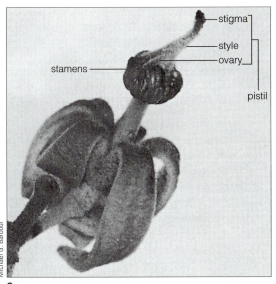

Michael G. Barbour

stigma
style
ovary
pistil
stamens

a

style
expanding
carpels

Michael G. Barbour

b

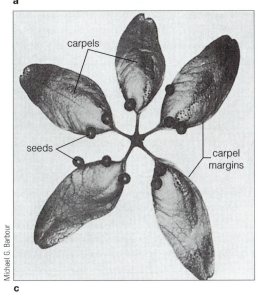

carpels
seeds
carpel
margins

Michael G. Barbour

c

Figure 25.4 *Firmiana plantanifolia.* Note that the pistils (carpels) resemble leaves. (**a**) A single pistil of the flower. (**b**) After pollination, the pistils separate, except along the style and stigma. (**c**) At maturity, each ovary splits open, exposing seeds attached to a leaflike surface.

25.3 **THE RISE OF ANGIOSPERMS TO DOMINANCE**

Angiosperms diversified and became so abundant in the fossil record of the late Cretaceous period that we can conclude they were the dominant plant life on land. As the Mesozoic era ended, so did dominance by gymnosperms and dinosaurs. As far as can be inferred from fossils, approximately 75% of all species on Earth at the time went extinct. There also was a rather sudden cooling of the climate at the boundary of the Mesozoic and Cenozoic eras—so sudden that some paleobotanists turn to theories of catastrophic events to explain the massive environmental change and extinctions that occurred in such a brief time. There is overwhelming evidence that a relatively large meteorite slammed into what has become the Gulf of Mexico, along the Yucatan Peninsula, creating disastrous fires and throwing debris into the atmosphere. During the months or years the debris took to settle back to Earth, it reflected solar radiation back to space; consequently, temperatures on Earth plummeted, extinguishing entire groups of organisms. One assumption of this theory is that such disturbances of extraterrestrial origin have occurred many times—not just at the end of the Mesozoic era—and will occur again in the future.

It is unknown whether such an event alone ushered in the Cenozoic era or whether some other factor contributed, such as the extensive volcanic activity of this period. It is known, however, that the following 65-million-year span of the Cenozoic era has been one of continuous climatic change **(Fig. 25.5),** and that through it all, angiosperms have held dominance.

Figure 25.5 Average temperatures over geologic time for the earth at 40 to 90 degrees North latitude. The time scale is distorted to show more detail for the last 1 million years. The current mean temperature (extreme right) is approximately 0°C. In the Cretaceous period, when angiosperms first appeared in the fossil record, the mean temperature was about 13°C warmer than the current temperature. (Adapted from Dorf, E. 1960. Climatic Changes of the Past and Present. *American Scientist* 48:341, with permission.)

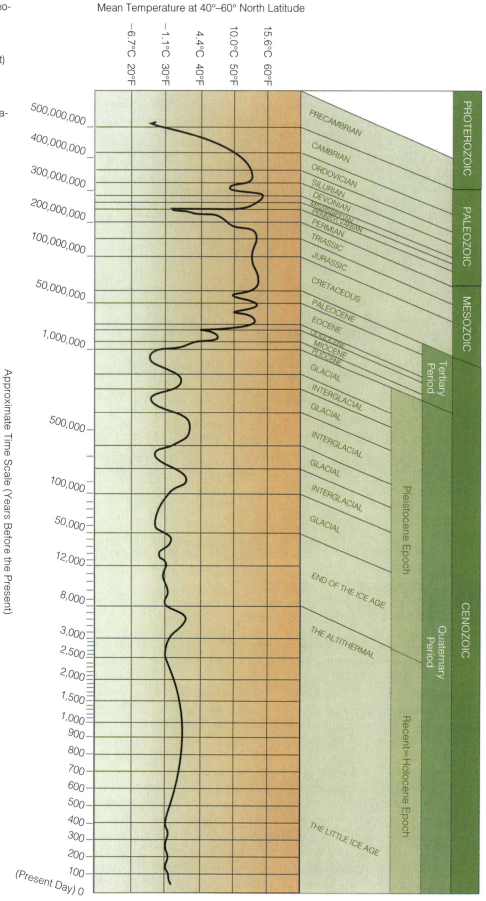

PLANTS, PEOPLE, AND THE ENVIRONMENT:

We Live on a Changing Surface

Currently, the climate of the John Day Basin is semiarid. But in geological terms, this climate has only just arrived, and in earlier times, the climate was wetter and the vegetation more lush.

In the Eocene epoch, which began about 57 million years ago (Fig. 25.5), the dominant vegetation of the John Day Basin consisted of tropical vegetation of the type currently found in the mountain forests of Central America, where annual rainfall is more than 150 cm and there is no winter frost.

Twenty million years later, in the Oligocene epoch, rainfall decreased to perhaps 120 cm per year, summers were wet and mild, and winters were drier and cold (but with little snow). The fossil record shows a change to a rich, mixed conifer–hardwood forest. Leaves had become smaller, with dentate (toothed) or convoluted margins, indicating a drier habitat (because smaller leaves with lobed or uneven margins are able to be cooled directly by wind instead of by

transpiration). Deciduousness implies a colder winter. This type of mixed forest is not found anywhere on Earth today because the component species have been fragmented and separated. The closest approximations are along the cool, wet Oregon–California coast, in the Great Smoky Mountains of Tennessee, and in parts of China and Japan.

Twelve million years later, in the Miocene epoch, the shift to deciduous trees was even more pronounced, and conifers became less abundant. Species richness declined. Paleobotanists imagine that the climate was much like that in modern Ohio: 100 cm of annual precipita-

tion, with half of that falling as snow in winter, when freezing temperatures were common.

By the Pliocene epoch, fir and spruce—conifers that we associate today with cold climates—were common, along with winter-deciduous hardwoods. Some leaves were quite small and hard, indicating adaptation to a cold and arid climate.

Currently, the John Day Basin is dominated by sagebrush (*Artemisia tridentata*). Trees are absent except along watercourses. Annual precipitation is only 25 cm per year, most of that falling as snow in winter. The area is a semiarid cold desert.

Angiosperm Fossils Show Climatic Change during the Tertiary Period

The Cenozoic era is traditionally divided into two periods: the Tertiary and the Quaternary. The Tertiary period extended from approximately 65 million to 2 million years ago. During this time, continents continued to break apart, increasing the diversity of climate and vegetation types (Fig. 25.6). An excellent case study example of climate and vegetation change during that time comes from the fossil record of the John Day Basin in northeastern Oregon (see "PLANTS, PEOPLE, AND THE ENVIRONMENT: We Live on a Changing Surface" sidebar).

The Ice Age Affected the Diversity of Plants in Temperate Zones

The Quaternary period began about 2 million years ago, when the earth's climate cooled by a few degrees—just enough to alter hydrologic balances near the poles and at

high elevations. Summer melting no longer kept up with winter snowfall, resulting in snow accumulation year after year. Its layers compressed into ice, and great ice sheets called *glaciers* coalesced, moved, and came to cover much of the cool-temperate zone. The balance of water shifted from the oceans toward bodies of freshwater and ice; sea level fell. This **Ice Age,** which has occupied most of the Quaternary period, is known as the Pleistocene epoch (Fig. 25.5).

Periodically during the Pleistocene, the climate warmed and glaciers retreated; then the climate cooled and glaciers readvanced. Glacial–interglacial cycles occurred approximately every 41,000 years until about 800,000 years ago. Since then, the cycle has been on the order of 100,000 years. During glacial advance, almost all of Canada, the northern third of the United States, most of Europe, all of Scandinavia, and large parts of Siberia were covered by ice. Mountain chains well into the tropics were all capped by glaciers that extended downslope, filling wide canyons and moun-

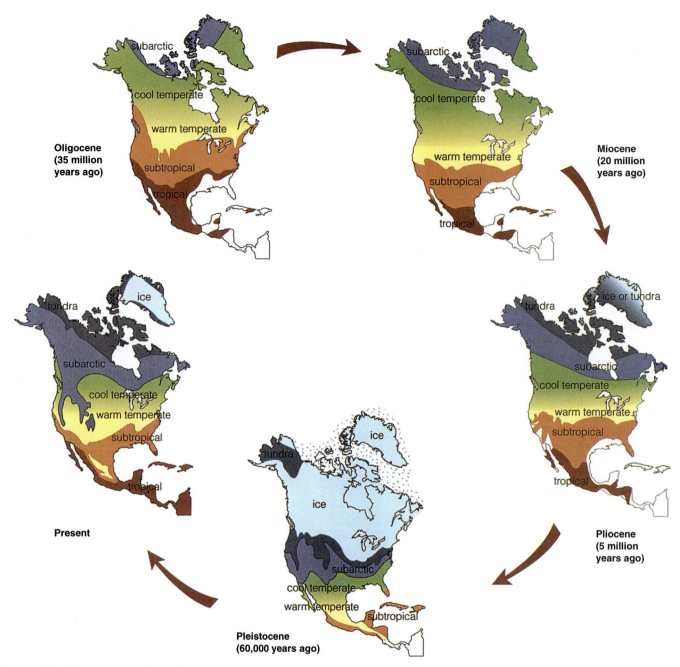

Figure 25.6 Climate zones of North America during the last 35 million years. (Redrawn from Dorf, E. 1960. Climatic Changes of the Past and Present. *American Scientist* 48:341, with permission.)

tain valleys. Ice covered 30% of the earth's land surface. Sea level was 100 m (325 feet) lower than current levels because so much water was locked in glacial ice.

When glaciers advanced, vegetation zones were pushed lower in elevation or lower in latitude. For example, at full glacial advance 18,000 years ago (Fig. 25.7), a spruce–pine boreal forest covered the eastern United States as far south as North Carolina. Treeless tundra covered the Appalachian Mountains. A thin strip of mixed conifer–hardwood forest formed a northern cap to oak–hickory deciduous forest, and

Florida (much wider then) was covered by sand dune scrub. Only 800 km (~495) of north-to-south distance separated the front of an ice sheet from deciduous forest, indicating a sharp gradient in climate.

When glaciers retreated, vegetation zones rebounded, the climatic gradient softened, and landscape diversity increased (Fig. 25.7). Each cycle of advance and retreat caused some extinctions, however, so that the biota of the temperate zone became simpler during the Pleistocene epoch.

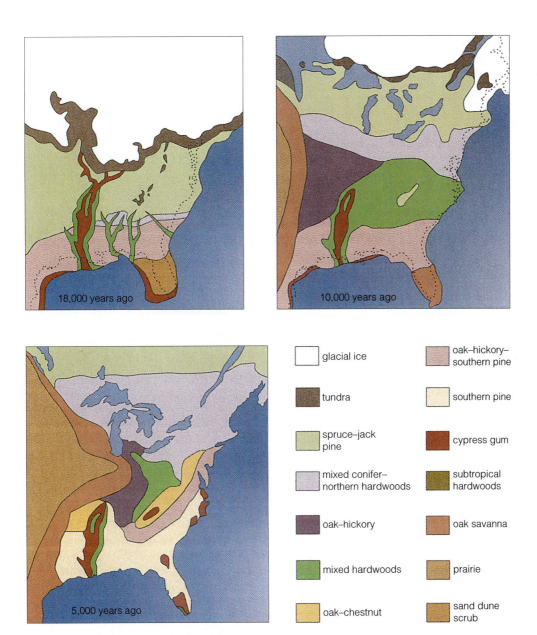

18,000 years ago

10,000 years ago

5,000 years ago

glacial ice

tundra

spruce–jack pine

mixed conifer–northern hardwoods

oak–hickory

mixed hardwoods

oak–chestnut

oak–hickory–southern pine

southern pine

cypress gum

subtropical hardwoods

oak savanna

prairie

sand dune scrub

Figure 25.7 Reconstruction of major vegetation zones in eastern North America since the last glacial maximum. (Redrawn from Delcourt, H.R., et al. 1983. *Quarterly Science Review* 1:153–175, with permission.)

Angiosperm Evolution Was Affected by Humans in the Quaternary Period

Humans also played a role in plant distribution and evolution during the Quaternary period. Initially, humans in hunter–gatherer cultures might merely have harvested wild plants as they found them. However, they gradually began to cultivate and select some of those species for convenience and greater yield. They did not sow seeds in geometric patterns and till the land, but they did use fire, pruning, selective harvesting, and sowing (of rhizomes, bulbs, or seeds) *without* cultivation to favor the abundance of certain food plants. Anthropologists call this stage **protoagriculture.** Protoagriculture might have gone on for thousands of years before agriculture and the domestication of crops were fully established.

The "root" crops cassava (*Manihot esculenta*) and taro (*Colocasia esculenta*) were cultivated in southeastern Asia as early as 15,000 years ago. Human-induced selection in these species has gone on for so long that some varieties have lost, or nearly lost, their capacity for sexual reproduction. It is doubtful that they could survive in nature. The earliest archaeological evidence for seed agriculture (cultivation of annuals such as rice, wheat, beans, or squash) goes back 11,000 years. Wild annuals that demonstrated exceptional productivity were valued and propagated. Over time, such artificial selection pressure on the genetic makeup of plants resulted in so much change from their wild relatives that it is now difficult to determine where, and from what wild stock, they were first domesticated.

Humans also have accidentally domesticated and favored the evolution of weeds. Weeds are plants that grow well in disturbed or trampled soil, in waste areas rich in nitrogen, or interspersed with crop plants. Some weeds have evolved seeds that are the same size as crop seeds; therefore, they are not easily separated during threshing or sieving. Thus, when the next season's crop is sown, the weeds are sown inadvertently right along with it.

As humans explored new lands, they brought along not only their culture but also their domesticated plants and companion weeds. Sometimes ornamental plants were brought into new regions. The weeds and ornamentals often turned out to be aggressive competitors in their new homes, able to displace native species from the landscape. As much as one third of the flora of some parts of the United States is composed of weeds imported from various parts of the world, including Europe, Australia, Asia, and Africa. Introduced plants have become common in forests, grasslands, woodlands, wetlands, coastal strands, and deserts. Only the most stressful habitats, such as alpine tundra, salt marshes, and rocky outcrops, seem free, or almost free, of introduced species. The accelerated pace of land disturbance, single-species cultivation, and travel are making the flora of the world increasingly homogeneous and lower in biological diversity.

25.4 NOVEL FEATURES OF THE ANGIOSPERM LIFE CYCLE

Angiosperms share with gymnosperms the characteristic of producing a seed, but the life cycles of the two groups are different in several important features. In flowering plants, the size and complexity of the gametophyte generation become reduced, the location of the ovule becomes hidden, there are two fertilization events (double fertilization), and the dispersal of the seed is improved by its enclosure within a fruit. These novel life cycle features help flowering plants adapt to life on land and conserve their food reserves.

Our survey of plants has revealed a trend of decreasing gametophyte size and increasing dependence on the sporophyte. Angiosperms carry the reduction further than gymnosperms, for both male and female gametophytes. In angiosperms, the male gametophyte (in the form of a pollen grain) has only two or three nuclei. The female gametophyte has been reduced to seven cells and eight nuclei, and an archegonium is no longer recognizable. This reduction probably leads to more efficient reproduction: When an egg is not fertilized or when a pollen grain does not land in a suitable flower, less energy is wasted.

Additional savings in energy result when animal vectors transfer pollen grains to the stigmas of other flowers (Fig. 25.8). Many of these living pollen carriers move selectively, not randomly, visiting sequential flowers of the same species. Thus, fewer pollen grains are needed to make seeds.

The enclosure of an ovule within an ovary shelters the ovule against drying and attack by herbivores or pathogens. Enclosure also allows selectivity for appropriate pollen because the pollen has to germinate on a receptive stigma

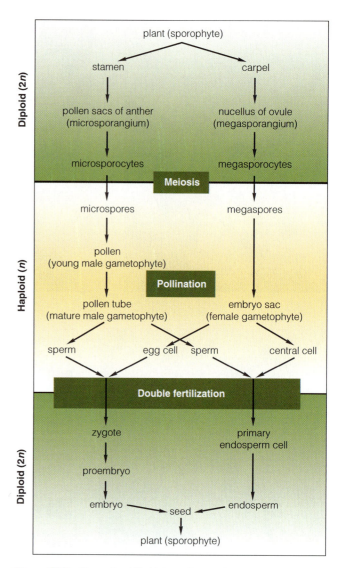

Figure 25.8 Generalized life history of an angiosperm.

and then grow through an accepting style. Finally, the ovary wall later matures into the fruit. Fruits protect the seeds, enhance seed dispersal, and can control seed germination. The consequences are that seedlings will emerge when and where conditions are most favorable.

Double fertilization (Fig. 25.8) also conserves energy because food storage tissue does not accumulate until after fertilization. Energy-rich endosperm tissue is not part of a preexisting female gametophyte; it is a new tissue that is created only by the act of fertilization. If unfertilized ovules contained as much energy-rich tissue around them as fertilized ovules, a significant amount of food reserves would be wasted.

About 257,000 species of flowering plants have been described, and taxonomists estimate that many more thousands of unknown species exist in poorly explored regions. Certainly the number of described species increases each year. When Linnaeus published *Species Plantarum*, his compendium of the plant world in the mid-eighteenth century, he listed less than 8,000 species. The novel life cycle traits described above are possibly a major reason for the diversity, success, and dominance of this group.

25.5 ANGIOSPERM DIVERSITY

Our understanding of angiosperm relationships has undergone a revolution in the last several years. The advent of cladistic techniques (bolstered by the development of powerful computers, software, and a rapidly growing wealth of molecular characters) has been the driving force behind this revolution.

Early taxonomists created arbitrary (nonphylogenetic) classifications that were based on ease of use or similarity of form. In the twentieth and twenty-first centuries, the trend has been to classify plants on the basis of evolutionary relationships—phylogeny. A great variety of phylogenetic classification systems have been proposed. The system of Arthur Cronquist has been particularly influential and remains standard in many places. However, Cronquist's system is based on his great experience and intuition. It has been contradicted by recent cladistic studies. Some researchers have proposed that a completely new classification system, based on naming nodes on a cladogram rather than on assigning each taxon a rank with a standard suffix (for example, family, order, class) in the manner of Linnaeus and Cronquist, is desirable. Many articles are currently being published that name groups on the basis of phylogeny rather than on hierarchical rank.

Currently, we have a clear picture of the basic phylogenetic structure within the angiosperms. A variety of molecular studies have pointed to a series of well-supported clades and subclades. Several distinctive lineages of angiosperms form a basal grade. These include: the shrub *Amborella*; the shrubs, vines, or trees of star anise and its relatives; and the aquatic, herbaceous water lilies. These three lineages comprise the basal-most groups in the angiosperms, and, therefore, are sister taxa to all other flowering plants (see Figs. 25.2 and 25.3).

The remaining angiosperms comprise three large, diverse groups: magnoliids, monocots, and eudicots.

Basal Angiosperm Groups Include the Water Lilies

The currently identified **basal angiosperms** consist of about 170 species of herbs, shrubs, and trees widely distributed throughout tropical and temperate zones. Shared traits include elongate vessels with slanted perforation plates (or else no vessels); radially symmetrical flowers with several to many free carpels and stamens; stamens with broad, short, petal-like, or poorly differentiated filaments; carpels with short or missing styles but with an elongated stigmatic region; pollen with a single aperture; and seeds with small embryos but with a significant amount of endosperm.

The living lineage that may have been the first to diverge is represented today by a single species, a shrub called *Amborella trichopoda*, which is found only on New Caledonia. *Amborella* lacks vessels in its wood, and although the plants are dioecious, the flowers have vestigial structures that suggest they evolved from plants that produced both pollen and ovules in the same flower.

The water lilies are another group of basal angiosperms. They consist of 70 aquatic, rhizomatous wetland herb species. Their leaves and flowers float, because oxygen is carried from the leaves down through the petioles to rhizomes and roots through conspicuous air canals. Flowers are large, with numerous **tepals** (colored flower parts not differentiated into petals and sepals), stamens, and carpels **(Fig. 25.9).** Many wild species occur in North American

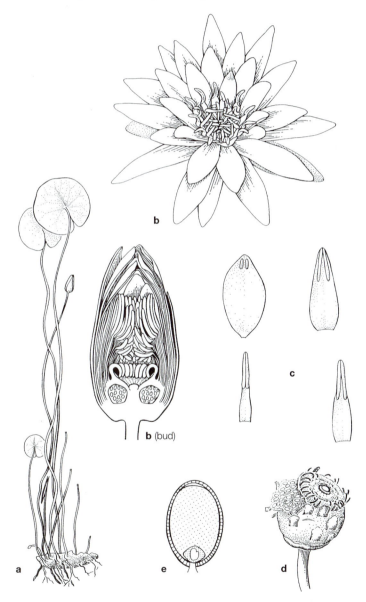

Figure 25.9 The water lily *Nymphaea odorata*. (**a**) Leaves on long petioles, growing upward through water from a rhizome anchored to the pond's bottom sediments. (**b**) A flower with numerous parts. (**c**) Petal-like stamens. (**d**) An opening fruit and many small seeds. (**e**) A longitudinal section of a seed, showing a small embryo and abundant endosperm. (Adapted from Judd, W. S., et al. 2002. *Plant Systematics*, 2nd ed. Sunderland, MA: Sinauer Associates, p. 227, with permission.)

ponds and lakes, but they also are common ornamentals in garden pools.

The star anise group contains 100 species of plants, some with medicinal value. Star anise (*Illicium verum*) is the most economically important, as a source for the spice and anise oil. These plants are vines, shrubs, or trees mostly of warmer climates. Few morphological features identify these plants as a clade, but DNA has strongly supported the group as monophyletic and as a basal angiosperm lineage.

Core Angiosperm Groups Include Most of the Flowering Plants

Most angiosperm species are classified in a clade called **core angiosperms.** Phylogenetic data support dividing the core angiosperms into three subclades: **magnoliids, monocots, and eudicots.** Although each of these contains a great variety of forms and characteristics, two of them can be distinguished by single characters: monocots by a single cotyledon, and eudicots by three-apertured pollen (or evolutionary derivatives of three-apertured pollen). The exact relationships among the core angiosperm groups currently are unknown, but we believe each of the groups is monophyletic.

MAGNOLIIDS Before the advent of molecular cladistic evidence, the magnoliid group often was considered typical of the earliest angiosperms. However, because the group ranges from herbs to trees and has a great variety of morphological, anatomical, biochemical, and cellular variety, reconstructing the features of the earliest angiosperms was a complex task. Now that it is known that the magnoliid group does not represent a basal angiosperm lineage, the characteristics can be assessed more appropriately. For example, some of the angiosperms that lack vessels are nested within the magnoliids, suggesting that vessels have been lost several times.

Magnoliids typically are tropical and warm-temperate, although some occur in temperate zones. Many are woody plants with simple leaves and pinnate venation. Some have large flowers with many spirally arranged tepals, stamens, and carpels, such as *Magnolia* **(Fig. 25.10).** Others have small flowers with parts in threes, such as cinnamon and camphor (*Cinnamomum*). Magnoliids include important spices and fruits, such as nutmeg (*Myristica fragrans*), sassafras (*Sassafras*), avocado (*Persea americana*), bay laurel (*Laurus nobilis*), black pepper (*Piper nigrum*), and pawpaw (*Asimina triloba*). Some magnoliids are medicinal and ornamental as well, such as peperomia (*Peperomia*), betel pepper (*Piper betle*), wild ginger (*Asarum*), and pipe vine (*Aristolochia*).

MONOCOTS In addition to the single cotyledon, monocots typically have parallel-veined leaves, flower parts in threes, sieve-tube members having plastids containing protein crystals, stems with scattered vascular bundles, an absence of secondary growth, and primary roots that abort early and are replaced by an adventitious root system.

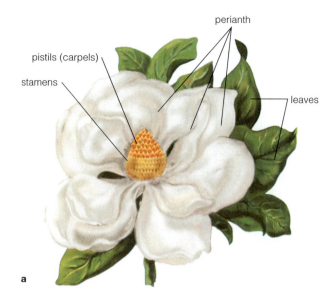

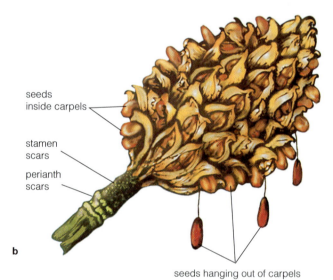

Figure 25.10 The southern magnolia (*Magnolia grandiflora*), a member of the magnoliid group. (**a**) A flower. (**b**) The fruit and scars showing where the stamens and petals had once been attached.

Monocots number about 65,000 species and include such economically and ecologically important plants as the grains, known as the "staff of life" because so many humans depend on them as a food staple—for example, barley, corn, millet, oats, rice, rye, and wheat. Domestic animals also depend on grain and other grasses for forage. Other important monocots are the agaves, bananas, irises, lilies, onions, orchids, palms, rushes, sedges, yams, and yuccas.

Relationships within monocots are poorly known. A variety of analyses suggest that a clade called **Alismatales** forms the basal lineage **(Fig. 25.11).** This basal group of some 3,000 species includes a number of important plants such as the aroids, many of which are well-known plants with distinctive flowers. Members include *Philodendron*, the calla lily (*Zantedeschia*), *Anthurium*, and taro (*Colocasia*). Other Alismatales include a widespread assemblage of aquatic plants,

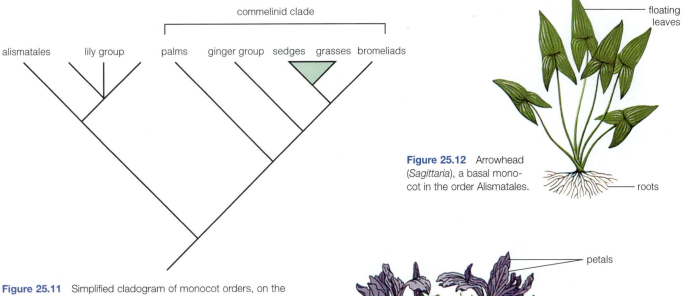

Figure 25.11 Simplified cladogram of monocot orders, on the basis of DNA analysis. The lily group is shown as a polytomy, but it could be a grade of lineages leading to the commelinid clade.

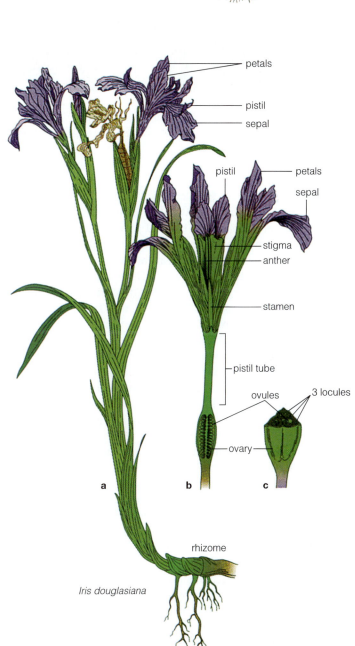

Figure 25.12 Arrowhead (*Sagittaria*), a basal monocot in the order Alismatales.

Figure 25.13 The monocot Douglas iris (*Iris douglasiana*). (**a**) A whole plant. (**b**) One flower with the ovary sectioned lengthwise to show many ovules. (**c**) Cross section of the ovary, showing three locules that represent three fused carpels.

such as the pernicious aquatic weeds *Hydrilla* and *Elodea*, and the ecologically important pondweeds. Arrowhead species, such as *Sagittaria* **(Fig. 25.12),** have been cultivated for their edible tubers in Hawaii, Malaya, Kashmir, and by Indians in North America. Arrowhead is a perennial herb commonly found in freshwater marshes, ponds, swamps, and lakes. It has a rhizomatous stem rooted in bottom sediments and arrow-shaped leaves on long petioles, which may float on the surface or may be submerged. Sepals and petals are distinct and each whorl numbers three; stamens and carpels can be numerous; nectaries are present, and pollinators include bees and flies. Seeds either disperse by floating on water or they are eaten and spread by birds.

The remaining monocots (besides Alismatales) include two large groups: one with typically showy flowers and another primarily composed of plants lacking showy flowers. There is strong evidence that the nonshowy group, the **commelinids,** forms a clade. The showy flowered (or lily) group may or may not be a clade. More research is needed.

Among the showy-flowered group are yams (*Dioscorea*) and a large clade of lilies and their relatives, including irises **(Fig. 25.13),** amaryllis, hyacinth, daffodil, tulip, agave, asparagus, onion, and one of the largest of all plant groups, the orchids, with 17,500 species.

The palms, with 2,700 species distributed throughout the tropics, comprise the basal lineage in the nonshowy-flowered commelinid clade. A few species enter the warm-temperate zone, such as California fan palm and saw palmetto **(Fig. 25.14).** Palms have a distinctive growth form of a typically unbranched trunk, a terminal tuft of compound or dissected leaves, and fruits called drupes, which have a berrylike, fleshy outer layer, but contain one to several hard pits. The coconut is a large drupe. Most palms are sensitive to frost, and if their single apical meristem is killed, the entire plant dies. One advantage of an unbranched mor-

(a–b) Michael G. Barbour

Figure 25.14 Palms that grow in warm-temperate latitudes. (**a**) A California fan palm (*Washintonia filifera*) in a southern California desert oasis. (**b**) A saw palmetto (*Serenoa repens*) in the understory beneath pine trees in north-central Florida.

phology is that it blocks a minimal amount of damaging wind during hurricane or typhoon conditions. Palms are among the most important tropical groups. Coconut palm (*Cocos nucifera*) yields edible endosperm from its huge seeds, oil pressed from dried coconut meal (copra), and cordage (coir) from its stringy outer husk. Dates are produced by *Phoenix dactylifera,* carnauba wax is extracted from *Copernicia,* cooking oil comes from *Elaeis,* betel nuts from *Areca,* and basketry material from *Raffia.* Many species are popular ornamentals in warm climates.

Other commelinids include a subclade of largely wind-pollinated plants such as bamboo and other grasses, cattails, rushes, sedges, and tules. This subclade also includes, as one of its basal lineages, the bromeliads, of which pineapple (*Ananas*) is a member. This group is one of the most species-rich among the monocots, the grasses alone comprising more than 8,000 species and the sedges comprising 3,600 species. Another subclade of commelinids, the ginger group, is not typical because it has showy, insect-pollinated flowers, some of economic importance. Members include ginger (*Zingiber*), cardamom (*Elettaria*), turmeric (*Curcuma*), banana (*Musa*), canna (*Canna*), maranta (*Maranta*), bird-of-paradise (*Strelitzia*), and the large tropical genus *Heliconia.*

EUDICOTS The third major lineage of core angiosperms, eudicots, is defined by a single feature: pollen with three apertures (**Fig. 25.15**). The simplest type of three-apertured pollen is called tricolpate. These pollen grains have three slits (or colpi) that divide the grain into three longitudinal wedges. Some members of this group have modified their tricolpate pollen to other forms. Some, for example, may have pores located in the slits (tricolporate; Fig. 25.15), and others have only pores (triporate). Some types of pollen may

have more than three apertures, but all of these variations are clearly derived from a tricolpate ancestor.

Eudicots typically have net-veined leaves, flower parts in fours or fives, embryos with two cotyledons, sieve-tube members usually having plastids with starch grains, stem vascular bundles arranged in a ring, and stamens with slender filaments. This group includes many economically and ecologically important plants such as blueberries, buckwheat, cacti, carrots, coffee, grapes, hemp, legumes, melons, poppies, potatoes, roses, sandalwood, stone fruits, strawberries, sunflowers, tea, teak, tomatoes, and walnuts.

A cladogram of eudicots based on Rubisco genes and ribosomal genes from the nucleus shows several basal lin-

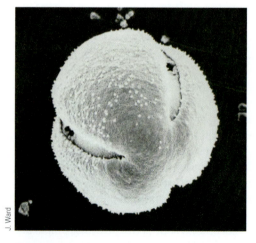

J. Ward

Figure 25.15 Pollen with three apertures, characteristic of eudicots. Note the longitudinal slits, two of the three being visible.

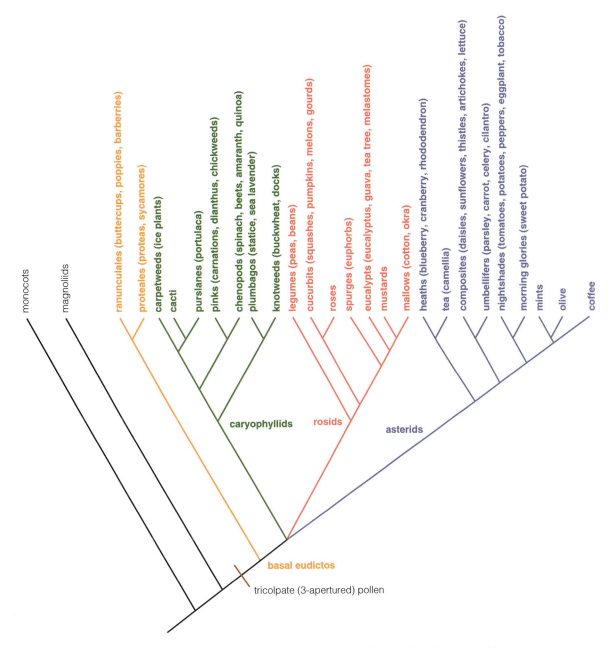

monocots

magnoliids

ranunculales (buttercups, poppies, barberries)

proteales (proteas, sycamores)

carpetweeds (ice plants)

cacti

purslanes (portulaca)

pinks (carnations, dianthus, chickweeds)

chenopods (spinach, beets, amaranth, quinoa)

plumbagos (statice, sea lavender)

knotweeds (buckwheat, docks)

legumes (peas, beans)

cucurbits (squashes, pumpkins, melons, gourds)

roses

spurges (euphorbs)

eucalypts (eucalyptus, guava, tea tree, melastomes)

mustards

mallows (cotton, okra)

heaths (blueberry, cranberry, rhododendron)

tea (camellia)

composites (daisies, sunflowers, thistles, artichokes, lettuce)

umbellifers (parsley, carrot, celery, cilantro)

nightshades (tomatoes, potatoes, peppers, eggplant, tobacco)

morning glories (sweet potato)

mints

olive

coffee

caryophyllids

rosids

asterids

basal eudicots

tricolpate (3-apertured) pollen

Figure 25.16 Simplified cladogram of eudicots, based on several data sets: ribosomal nuclear genes, chloroplast photosynthetic genes, morphology, and others.

eages and three major clades: the rosids, the asterids, and the caryophyllids **(Fig. 25.16).** The basal lineages include Ranunculales and Proteales. Ranunculales consists of about 3,500 species. They are mainly herbs in temperate latitudes with lobed leaves, numerous flower parts, superior ovary position, and seeds with small embryos. Buttercups (*Ranunculus*; **Fig. 25.17**) are a good example, but other genera in this lineage are *Anemone, Aquilegia* (columbine), *Delphinium* (larkspur), *Berberis* (barberry), and *Papaver somniferum* (opium poppy). Edible species are uncommon because most contain poisonous alkaloids.

Proteales are especially abundant in Africa and Australia, but the widespread Northern Hemisphere sycamores (*Platanus*) also are members. Proteales typically are trees or

shrubs with highly reduced, wind-pollinated flowers. They frequently are grown as ornamental shrubs (for example, the genera *Banksia, Grevillea, Hakea*) or street trees (sycamore).

The **caryophyllid** clade contains ice plants, carpetweeds, cacti **(Fig. 25.18),** pinks, and amaranths. Plants important to humans include sugar beet (*Beta*), spinach (*Spinacea*), purslane (*Portulaca*), rhubarb (*Rheum*), buckwheat (*Fagopyrum*), amaranth (*Amaranthus*) that produces a kind of grain, and the South American quinoa plant (*Chenopodium quinoa*). Landscape ornamentals include tropical *Bougainvillea*, temperate carnations (*Dianthus*), sea lavenders (*Limonium*), and many succulents.

The **rosid** clade is a diverse group, the largest and most familiar members of which are the legumes (16,400 species),

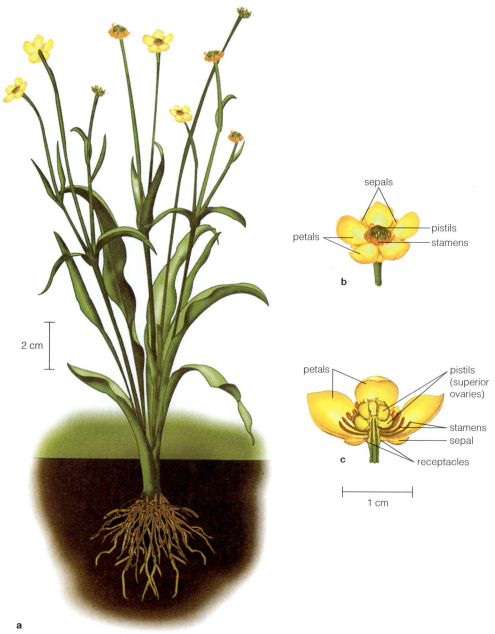

Figure 25.17 The buttercup (*Ranunculus alismaefolius*), a basal eudicot. (**a**) The whole plant. (**b**) One flower, showing regular symmetry. (**c**) A section through one flower, showing multiple, separate parts and superior ovary position. Primitive traits include regular symmetry, many flower parts on an elongated receptacle, a superior ovary position, and simple leaves.

Figure 25.18 A cactus flower with multiple parts.

drugs. Botanical oddities range from the small insectivorous Venus flytrap, to parasitic mistletoes, to giant ground-hugging *Rafflesia* flowers 1 m across.

The flowers of a typical rosid are illustrated in **Figure 25.19.** Sepals and petals often are fused into a ring, which can either be fused to the ovary or free from it. Sepals and petals typically occur in fives, and the two whorls are distinctly different in texture and color. Stamens are numerous, and sometimes so are the carpels. Flowers are moderately large, usually radially symmetrical, and present such rewards as nectar or pollen to bee pollinators. Two other representatives of the rosid group are gourds **(Fig. 25.20)** and cotton plants **(Fig. 25.21).** Gourds usually are monoecious (separate pollen and seed-producing flowers, both of which are found on the same plant); petals are fused (unusual for rosids) into an elongate funnel shape; flower parts are reduced in number; the ovary is inferior; and plants are herbaceous vines. Cotton flowers have separate petals, but the numerous stamens are characteristically fused together in a tube that surrounds the elongated style. When the seeds are mature, the dry, papery capsule splits open, revealing seeds whose seed coats have long white hairs; these are the hairs that are mechanically harvested and separated from the seed to become cotton fabric.

spurges (7,800 species), tropical melastomes (4,800 species), Australasian eucalypti (3,900 species), roses (3,000 species), and mustards (3,000 species). Economically important products from rosids include fruits, nuts, vegetables, ornamentals, timber trees, spices and flavorings, fibers, dyes, and

compound leaf

young fruit

flower

stigma

ovary (future achene)

receptacle

stamen

petal

sepal

a

b

Figure 25.19 The rosid group includes many species in the family Rosaceae with edible fruits, such as boysenberry (*Rubus ursinus* var. *loganbaccus*). (**a**) A portion of a stem, showing compound leaves, a mature flower, and young fruits. (**b**) A section of one flower. Primitive traits include regular symmetry and many flower parts; derived traits include compound, toothed leaves and some degree of fusion of petals and sepals at their bases.

Figure 25.20 The gourd family (Cucurbitaceae), another member of the rosid group. (**a**) A *Cucurbita pepo* plant, showing creeping stem, leaves, dioecious flowers, and a fruit. (**b**) A staminate flower. (**c**) A pistillate flower with the ovary sectioned to show ovules. (**d**) A young fruit in cross section, showing that the pistil is made from several fused carpels. Derived traits include fusion of flower parts, inferior ovary position, reduction in number of flower parts, unisexual flowers, and lobed, toothed leaves.

Figure 25.21 Cotton (*Gossypium hirsutum*), an economically important member of the rosid group. (**a**) Upper portion of one plant, showing leaves, a mature flower, and developing fruit. (**b**) A longitudinal section of one flower. (**c,d**) A mature five-section fruit, within which are many seeds densely covered with epidermal fibers.

Figure 25.22 Nightshade (*Solanum parishii*), of the Solanaceae or tomato family, in the asterid group. (**a**) Upper portion of one plant showing flower buds, mature flowers, and developing fruit. (**b**) One flower with some stamens removed to show the superior ovary position. (**c**) Cross section of a young berry fruit showing two locules and developing seeds. Derived traits include small flower size and fused petals, sepals, and stamens.

Figure 25.23 Sage (*Salvia mellifera*), of the Lamiaceae or mint family, in the asterid group. (**a**) Upper part of one plant showing opposite leaves, square stems, and lateral clusters of irregular flowers. (**b**) One flower; note the fused petals and sepals. (**c**) A longitudinal section of a flower; note that the stamens are fused to the petals. In this species, some stamens do not fully develop and are small pegs attached to the corolla near the throat. Derived traits include irregular symmetry, small flower size, fusion of parts within and between whorls, and a nonterminal inflorescence.

chicory (*Cichorium*); lettuce (*Lactuca*); mint (*Mentha*); olive (*Olea*); oregano (*Oreganum*); peppers, paprika, and chilis (all species of *Capsicum*); potato (*Solanum*); safflower (*Carthamnus*); sage (*Salvia*); sesame (*Sesamum*); sunflower oil (*Helianthus*); sweet potato (*Ipomoea*); thyme (*Thymus*); tomato (*Lycopersicon*); tomatillo (*Physalis*); and tobacco (*Nicotiana*).

25.6 **PLANT GEOGRAPHY**

Plant geography is the branch of plant biology that describes the distribution of plants over the surface of the earth. It also studies possible explanations for how, why, or when these patterns of distribution occurred. Generally, seed plants—and particularly flowering plants—are the focus. Rather than relying on classification theories or technological breakthroughs, studies of plant distribution require only field trips conducted by energetic botanists with keen observational skills.

Plant geography was stimulated by voyages of exploration in the seventeenth, eighteenth, and nineteenth centuries. Plants new to science were brought back to botanical gardens in Europe alive and whole, as seeds, or as pressed

The **asterid** clade (Fig. 25.16) contains some of the most highly specialized core angiosperms. Fused petals predominate. The heaths (2,700 species); tomatoes (**Fig. 25.22**); potatoes; peppers and relatives (2,900 species); mints (7,000 species; **Fig. 25.23**); carrot (**Fig. 25.24**); parsley and relatives (3,100 species); and sunflowers, daisies, and related genera (21,000 species) are included in this group. Flower parts typically are fewer, ovary position often is inferior, and the majority of species have irregular symmetry. Asterids typically are herbaceous.

No major global food plants are found in the asterid clade. However, components and flavors of regional diets are provided by artichoke (*Cynara*); basil (*Ocimum*); coffee (*Coffea*); tea (*Camellia*); elderberry (*Sambucus*); endive and

umbellet

compound umbel

umbellet

b

anthers

stamens

filament

petal

2 styles

pistil
(2 carpels)

ovule

c

bracts
of umbel

1 cm

petiole
base

a

Figure 25.24 The carrot family (Apiaceae), an asterid, represented here by the large herb cow parsnip (*Heracleum lanatum*). (**a**) Upper portion of a large plant, about 1 m in height, showing deeply lobed leaves and several inflorescences called umbels or compound umbels. (**b**) One portion of a compound umbel (an umbellet) made up of two dozen flowers. (**c**) One flower, sectioned to show the inferior ovary position. Advanced traits include reduced flower size, reduced number of flower parts, inferior ovary, fusion of two carpels, and deeply lobed and toothed leaves.

herbarium specimens (see "ECONOMIC BOTANY: Botanical Gardens, Smuggling, and Colonialism" sidebar). Important botanists who tried to make sense of this cascading mountain of information included Carl von Willdenow, Alexander von Humboldt, Johannes Schouw, August Grisebach, Alphonse de Condolle, Oscar Drude, Adolf Engler, George Marsh, Asa Gray, and Charles Darwin.

Humboldt, for example, led an expedition to Central and South America at the start of the nineteenth century. For 5 years he traveled in Cuba, Venezuela, Peru, Mexico, and the Orinoco and Amazon River basins. He walked through steamy lowland rain forest, semiarid thorn scrub, dry deserts, and cold alpine elevations as high as 6,000 m atop Mt. Chimborazo. His expedition took along the best equipment at that time for measuring elevation, location, and weather. More than 60,000 plant specimens were collected. On his return to Europe, Humboldt wrote a monumental 30-volume work summarizing the expedition; the first 14 volumes were devoted to botany. He is generally credited with being the founder of the science of plant geography. In addition, his keen observations, records, and interpretations were important contributions to the new discipline of ecology.

It has become clear that the world's flora is not uniform. Although a dozen or so large families of flowering plants are cosmopolitan (Table 25.1), that is, commonly found on every continent except Antarctica, most families have a regional flavor or habitat bias in their distribution. Entire clades or subclades are restricted to certain continents or regions. As a result, plant geographers have divided the world into floristically homogeneous units. Ronald Good's scheme recognizes more than 30 global units (Fig. 25.25). Each unit is characterized by its own endemic plants in addition to unique mixes of the more cosmopolitan lineages.

Studies by plant geographers also have shown that the vegetation of one particular kind of climate sometimes looks similar wherever that climate recurs around the world— even though plants at each location may belong to taxa that are not closely related. For example, desert vegetation, with cactus-like growth forms, exists in the New and Old Worlds. The New World species, however, are in the cactus family (caryophyllids), whereas the Old World species are in the *Euphorbiaceae*, a family of the rosid lineage. Similar shrubby vegetation with a dense, stiff, small-leaved canopy occurs in five different places with a Mediterranean-type climate, yet the plants are not closely related. Tropical rain forests of Africa, Australasia, and South America share a similar climate, a similar array of plant growth forms, and a similar forest architecture, but in each continent the assemblage of important plants is different.

Table 25.1 The Twenty Largest Families of Flowering Plants

Rank	Family	Common Name	No. Species	Distribution
1	Asteraceae	Sunflower family	21,000	Cosmopolitan
2	Orchidaceae	Orchid	17,500	Cosmopolitan
3	Fabaceae	Bean	16,400	Cosmopolitan
4	Rubiaceae	Madder	10,700	Cosmopolitan
5	Poaceae	Grass	7,950	Cosmopolitan
6	Euphorbiaceae	Spurge	7,750	Cosmopolitan
7	Lamiaceae	Mint	5,600	Cosmopolitan
8	Melastomataceae	Melastoma	4,750	Tropical, especially South America
9	Liliaceae	Lily	4,550	Cosmopolitan
10	Scrophulariaceae	Snapdragon	4,500	Cosmopolitan
11	Acanthaceae	Acanthus	4,350	Tropical
12	Myrtaceae	Eucalyptus	3,850	Tropical, especially Australia
13	Cyperaceae	Sedge	3,600	Cosmopolitan
14	Ericaceae	Heath	3,350	Cosmopolitan
15	Apiaceae	Carrot	3,100	Temperate
16	Rosaceae	Rose	3,050	Temperate, especially Northern hemisphere
17	Brassicaceae	Mustard	3,000	Cosmopolitan
18	Moraceae	Mulberry	2,975	Cosmopolitan
19	Araceae	Arum	2,950	Tropical and subtropical
20	Asclepiadaceae	Milkweed	2,900	Tropical, especially Africa

From Mabberley, D.J. 1987. *The Plant Book.* Cambridge, UK: Cambridge University Press, with permission.

1. Arctic and Sub-arctic
2. Euro–Siberian
   A. Europe
   B. Asia
3. Sino–Japanese
4. W. and C. Asiatic
5. Mediterranean
6. Macaronesian
7. Atlantic North American
   A. Northern
   B. Southern
8. Pacific North American
9. African–Indian Desert
10. Sudanese Park Steppe

11. N.E. African Highland
12. W. African Rain Forest
13. E. African Steppe
14. South African
15. Madagascar
16. Ascension and St. Helena
17. Indian
18. Continental S.E. Asiatic
19. Malaysian
20. Hawaiian
21. New Caledonia
22. Melanesia and Micronesia
23. Polynesia
24. Caribbean

25. Venezuela and Guiana
26. Amazon
27. South Brazilian
28. Andean
29. Pampas
30. Juan Fernandez
31. Cape
32. N. and E. Australian
33. S.W. Australian
34. C. Australian
35. New Zealand
36. Patagonian
37. S. Temp. Oceanic Islands

Figure 25.25 Floristic regions of the world according to Good. (Redrawn from Good, R. 1961. *The Geography of the Flowering Plants.* New York: John Wiley, with permission.)

Figure 25.26 Alpine vegetation in African mountains. (**a**) A giant lobelia (*Lobelia rhynchorpetalum*). (**b**) A giant senecio (*Senecio keniodendron*).

(a–b) G.J. James/Biological Photo Service

Sometimes, however, the environmental conditions, the isolation of a place, and the genetic potential of plants growing there combine to create unique and bizarre vegetation found nowhere else in the world. An excellent example is the flora and vegetation of alpine zones on tropical African mountains such as Mt. Kilimanjaro in Kenya. Tree-sized plants in the normally herbaceous genera *Senecio* and *Lobelia* dominate the landscape **(Fig. 25.26)**. The plants have clustered leaves at the ends of irregularly formed branches, and they may live for a century.

The desert of Baja California is another place where unique taxa and growth forms exist. The boojum tree (*Fouquieria columnaris*), giant cacti, arborescent yuccas, and trees with succulent stems (*Bursera, Pachycormus*) give the horizon an otherworldly aspect **(Fig. 25.27)**.

The process of evolution has not been equal for all taxa of flowering plants. Some groups have many species throughout the world, but nevertheless have a limited range of growth forms. Other groups have fewer species that are less wide-

ECONOMIC BOTANY:

Botanical Gardens, Smuggling, and Colonialism

Botanical gardens are collections of living plants. They began as places where plants that were new to science (collected on voyages of discovery in the sixteenth and seventeenth centuries) could be propagated and studied. The first established botanical garden, in 1545, is the Royal Botanic Garden at Padua, Italy. Other Italian gardens at Pisa, Florence, and Bologna soon followed. In the seventeenth century, gardens were constructed in France, the Netherlands, England, Germany, Sweden, Scotland, and Japan. Early botanical gardens were small and formally laid out; the English were the first to plant larger, more naturally landscaped gardens.

The Royal Botanic Gardens at Kew, England, are the finest botanical gardens in the world. They began on 4 hectares (10 acres) in 1759, but they have expanded to occupy more than 100 hectares (250 acres)

today. More than 28,000 species of plants from around the world are growing in the garden **(Fig. 1)**. Major sections include an arboretum (trees), tropical plants maintained in warm greenhouses, a rock garden, alpine plants in refrigerated greenhouses, and a grass collection. Libraries, research laboratories, collections of botanical art, and a herbarium of 6 million pressed specimens also are on the grounds. Kew Gardens exemplifies several objectives common to most botanical gardens. They conduct basic taxonomic research on living specimens, investigate plants for potential economic value, propagate horticultural plants that have ornamental value, educate the public, provide the public with a place to enjoy nature, and provide a secure location to maintain rare plants.

Sometimes the collection of plants for botanical gardens has involved question-

able activities. For example, by the mid-nineteenth century, rubber had become a valuable, desired plant product. The industrial process of vulcanization had enormously expanded the use of rubber. Rubber comes from tropical plants of the New World, and the best rubber trees (*Hevea brasiliensis*) were native to Brazil. The Brazilian government wished to maintain its monopoly and prohibited the export of rubber tree cuttings or seeds.

The British India Office, however, was determined to start rubber plantations in the new British colony of Ceylon. The India Office believed that Ceylon soils and climate were similar enough to those of Brazil to provide a suitable new home for rubber production. In 1876, the India Office secretly commissioned the English rubber worker H. A. Wickham to smuggle out rubber tree seeds, agreeing to pay him the equivalent of $50 per 1,000 seeds.

Figure 25.27 Desert vegetation in Baja California, showing boojum trees (*Fouquieria*), century plants (Yucca), and cardon cactus (*Pachycereus*). No other desert in the world has a similar collection of growth forms.

spread, yet they exhibit a wide range of growth forms. Still others are narrowly restricted and have few species. Apparently, some groups have the genetic potential to be numerous, aquatic, succulent, woody, frost tolerant, or salt tolerant; to evolve flowers pollinated by bats, or fruits dispersed by birds; or to accumulate herbivore-inhibiting chemicals. Others are not so genetically equipped. No group of plants has the potential to fill every terrestrial niche, but some groups can fill more niches than others. We also see that environmental stresses or problems are solved by different plants in different ways.

Plant **biogeographers,** like systematists, have benefited greatly from phylogenetic analyses. Cladograms can be used to infer centers of origin, to identify long-distance dispersal events, and to investigate the process of domestication in many groups. In addition, the ecologic value of various characteristics can be studied by reference to a cladogram.

Perhaps future plant biologists will be able to explain why evolution has proceeded along the exact route it has taken among the flowering plants. For now, we can only marvel at the diverse results, treasure this biotic heritage of the past, and protect it for the future.

Wickham proved to be efficient; he collected 70,000 seeds, packing them in covered baskets, labeled simply "Botanical Specimens for Her Majesty's Gardens at Kew." He eventually found a willing ship captain with available space who slipped the seeds past the customs office. Once in England, the seeds were turned over to botanists at Kew Gardens, who carefully germinated and nurtured 2,400 seedlings from the 70,000 seeds. Most of these plants were shipped within a year to plantations that had been made ready in Ceylon. By 1912, these plantations were producing a princely income of $45,000 per hectare (2.5 acres) per year. Not surprisingly, Wickham was knighted in 1920. Currently, 95% of the world's rubber comes from Southeast Asia, and plantations also have been established in Africa.

Many other economically valuable plants were moved to new homes during the imperialist and colonialist times of the seventeenth to nineteenth centuries. Trade in plants was one of the driving economic forces during those centuries. For example, profits in brokering spice between Southeast Asia and Europe were every bit as attractive to investors as trade in gold, cotton, or slaves. Cornering the market of a spice or any other important plant species was a reward for aggressive imperialism. Consequently, coffee (*Coffea*) was

Figure 1 Kew Gardens, United Kingdom. This glasshouse features tropical vegetation. In the foreground are large floating leaves of *Victoria*, a water lily.

taken from Africa to Central and South America, to Indonesia, and to Hawaii; tea (*Thea*) was moved from China to India; chili peppers (*Capsicum*) were transplanted from the Caribbean to Europe; pineapples (*Ananas*) were hijacked from South America to Hawaii; rice (*Oryza*) was taken from Asia to Texas and California; and oranges (*Citrus*) were brought from China to Florida. Sugarcane (*Saccharum*) originated on the Pacific islands. In all these cases, modern centers of production are a far distance from the places where the plants were once native and restricted. Truly, humans are homogenizing the earth's edible landscape.

KEY TERMS

Alismatales	flowering plants
angiosperm	fruit
asterid clade	Ice Age
basal angiosperms	magnoliids
biogeography	monocots
caryophyllid clade	ovary
commelinids	paleobotany
core angiosperms	plant geography
dicots	protoagriculture
double fertilization	rosid clade
endosperm	seedling hypothesis
eudicots	tepals
flower	

SUMMARY

1. Angiosperms are seed plants. The ovule, and the seed that develops from the ovule, are enclosed within an ovary, rather than born naked, as in gymnosperms. The ovary is part of a new organ called a carpel, and the carpel is part of a new complex structure called the flower. The flower is thought to represent a modified leafy shoot.

2. The fossil record of angiosperms extends back to the early Cretaceous period, about 140 million years ago. Paleobotanists theorize that the first appearance of angiosperms on Earth may have been as early as 190 million years ago, and that the initial split of the line that became angiosperms diverged from that of vascular plants and gymnosperms more than 300 million years ago during the Carboniferous period. It is not clear which living plants are the closest relatives to angiosperms, and it may be that no *living* group is closely related.

3. Flowering plants have dominated Earth's vegetation throughout the Cenozoic era, despite major geographic, climatic, and vegetational changes. The temperate zone of North America showed a trend toward a progressively cooler and drier climate. The last 2 million years of the Cenozoic era form the Quaternary period, which includes the Ice Age (Pleistocene) and Recent (Holocene) epochs. During the Pleistocene epoch, ice sheets advanced and retreated several times. Temperate-zone vegetation is still recovering from the last glacial retreat, which ended 10,000 years ago. The last 10,000 years—the Holocene epoch—may merely be another interglacial stage.

4. Humans have affected plant evolution during the Pleistocene and Holocene epochs by selecting plants for food, managing the landscape with fire, and introducing exotic species to new locations.

5. Novel aspects of the angiosperm life cycle include further reduction (when compared with most gymnosperms) in the size and complexity of the gametophyte generation; location of the ovule within another structure (the ovary); double fertilization, where the second sperm produces endosperm tissue; and dispersal of the seed within a fruit. Vegetative innovations include an improved vascular system (with vessels, sieve-tube members, and companion cells), and an accelerated seedling phase. In comparison with gymnosperms, these innovations further adapt angiosperms to life on land, and they help conserve plant food reserves. Thus, these adaptations have ecological consequences and may explain why angiosperms supplanted gymnosperms as the dominant terrestrial plant life form.

6. About 257,000 named species of flowering plants exist on Earth today. Phylogenetic classification of these plants has recently become less subjective because of research using DNA and cladistic methods. Angiosperms consist of a basal group with three lineages: magnoliids (a diverse clade of both woody and herbaceous plants), monocots, and eudicots.

7. Basal angiosperms number 170 species among three lineages (*Amborella*, water lilies, and star anise and relatives), with the following shared traits: vessels lacking or, if present, elongated with slanted perforation plates; radially symmetrical flowers with numerous free carpels and stamens; stamens with broad, short, petaloid, or poorly differentiated filaments; carpels with short or missing styles, but with an elongated stigmatic region; pollen with a single aperture; and seeds with small embryos and copious endosperm.

8. Core angiosperms include magnoliids, monocots, and eudicots. The magnoliids were long considered to be basal angiosperms, but more recent research considers them to be derived and part of the core group. They are mostly tropical and warm-temperate trees, shrubs, vines, and herbs with simple leaves, pinnate venation, large flowers with many stamens and pistils, and relatively simple wood anatomy (some lack vessels). Example genera and products are avocado, black pepper, cinnamon, camphor, laurel, nutmeg, *Magnolia,* and pawpaw.

9. Monocots typically have parallel-veined leaves, flower parts in threes, embryos with a single cotyledon, sieve-tube members having plastids containing protein crystals, and stems with scattered vascular bundles, and they are capable of producing adventitious roots. Monocots comprise three groups: a basal Alismatales lineage; the commelinid lineage with typically reduced, wind-pollinated flowers; and a group consisting mostly of monocots with large, showy flowers. Monocots include such economically and ecologically important plants as the cereal grains and other grasses, agaves, banana, irises, lilies, onion, orchids, palms, rushes, sedges, yam, and yuccas.

10. The Alismatales include such well-known plants as *Philodendron,* the calla lily *Anthurium,* taro (*Colocasia*), and the pernicious aquatic weeds *Hydrilla* and *Elodea.* The showy-flowered clade includes several large groups (orchids and lilies and their relatives). The commelinid group includes grasses, sedges, rushes, and similar plants, as well as the palms and ginger.

11. The last core angiosperm clade, eudicots, typically has net-veined leaves, flower parts in fours or fives, embryos with two cotyledons, sieve-tube members having plastids with starch grains, vascular bundles in a circle, stamens with slender filaments, and tricolpate pollen (or pollen derived from a tricolpate ancestor).

12. Eudicots include many economically and ecologically important plants. The Ranunculales and Proteales are the major basal eudicot lineages. The Ranunculales are prominent in North America and include herbs of temperate latitudes with lobed leaves, numerous flower parts, a superior ovary position, and seeds with small embryos. The eudicots consist of three major lineages: the caryophyllids (cacti, chenopods, pinks), the rosids (stone fruits, almonds, roses, legumes, mustards), and the asterids (blueberries, cranberries and relatives, tomato, potato and relatives, the mints, carrot, parsley and relatives, and the sunflowers—the latter being the largest family of angiosperms).

13. Plant geography is the study of distribution patterns of plants over the earth's surface and the processes that caused those patterns. Flowering plants usually are the focus. Some plant groups are cosmopolitan, but others are more narrowly distributed and characterize floristic regions of the earth. Each plant group has its own genetic limitations and cannot evolve to fit into all available niches. The same environmental stresses or problems are solved by different plant groups in different ways.

Questions

1. Diagram the evolutionary steps necessary to create an ovary of a flowering plant. According to the fossil record as currently understood, when did the first flowering plant appear?

2. Compare the size, anatomical complexity, and degree of independence of a fern gametophyte, a pine female gametophyte, and a magnolia female gametophyte. Which one is the most insulated and protected from the environment?

3. The megagametophyte of angiosperms is greatly reduced, but may contain vestiges of the structures present in larger gametophytes of seed plants. Speculate about what the synergids and the antipodals represent in an angiosperm megagametophyte.

4. How can a fossil assemblage of flowering plants in a given location be used to estimate the climate at that location at that time? What assumptions are made to reconstruct the previous climate?

5. In your opinion, which group or family of angiosperms is most important to the most people? Give some examples of which plants within that group you consider to be important. Are they rich in species or widely distributed?

6. How can widely distanced places in the world that share similar environments have similar vegetation when the taxa of flowering plants that make up the vegetation are completely different from place to place?

InfoTrac® College Edition

http://infotrac.thomsonlearning.com

Angiosperm Origin

Ge, S., Dilcher, D.L., Zheng, S., Zhou, Z. 1998. In search of the first flower: a jurassic angiosperm, *Archaefructus*, from Northeast China. *Science* 282:1692. (Keywords: "jurassic" and "angiosperm")

Mathews, S., Donoghue, M.J. 1999. The root of angiosperm phylogeny inferred from duplicate phytochrome genes. *Science* 286:947. (Keywords: "phylogeny" and "phytochrome")

Roush, W. 1996. Probing flowers' genetic past (living weed gene that controls both ovule formation and flower growth may be evidence of the origin of flowering plants). *Science* 273:1339. (Keywords: "ovule" and "origin")

Soltis, P.S., Soltis, D.E., Zanis, M.J., Kim, S. 2000. Basal lineages of angiosperms: relationships and implications for floral evolution. *International Journal of Plant Sciences* 161:S97. (Keywords: "Soltis" and "angiosperms" and "lineages")

Plant Geography

Holmes, B. 1998. The coming plagues? (non-native species on the move due to global warming). *World Press Review* Nov 1, p. 36. (Keywords: "plagues" and "global warming")

Pitelka, L.F. 1997. Plant migration and climate change. *American Scientist* 85:464. (Keywords: "plant" and "migration")

Ecology, Ecosystems, and Plant Populations

Visit us on the web at http://biology.brookscole.com/plantbio2 for additional resources, such as flashcards, tutorial quizzes, InfoTrac exercises, further readings, and web links.

1. Each organism has one of three roles in any ecosystem: producer, consumer, or decomposer. (a) Producers are green plants or protists that manufacture their own carbohydrate food from inorganic water and carbon dioxide. (b) Consumers are animals, pathogens, or parasites that obtain food by ingesting other organisms. (c) Decomposers are nongreen protists or prokaryotes that digest dead organic remains of producers and consumers.

2. Only about 1% of the solar radiation that reaches vegetation is absorbed and converted into metabolic energy, which can be measured as the caloric content of tissue. The transfer of metabolic energy by herbivores from plant tissue consumed to herbivore tissue is incomplete; only about 10% of available plant tissue is browsed and the rest remains uneaten. A similar small fraction of energy is transferred from herbivores to predators. Because only 1% to 10% of available energy is transferred at each level up the food chain, the maximum biomass at each level declines steeply, from producers to herbivores to predators.

3. A population is a local group of organisms belonging to the same species. Each population is genetically distinct. Most species have hundreds to thousands of unique populations, with each population subtly adapted to variations in the local environment. Thus, the population is the basic ecological unit of any species.

4. Every species can be assigned to one of a relatively small number of life history patterns. Each pattern represents a unique budget of time, activities, and resources that allows a population to continue in existence from generation to generation. Some of the activities budgeted are germination, growth, and reproduction.

5. The distribution of a population is affected by abiotic factors such as soil nutrient level, soil moisture availability, intensity of solar radiation, or the incidence of wildfire. Every plant must solve a zero-sum budget for dissipating incoming solar radiation energy into reflection, reradiation, convection, metabolism, storage, and transpiration.

6. Distribution also is determined by biotic interactions with other species, such as competition (in which a substance is removed from the environment), amensalism (in which a substance is added to the environment), herbivory (consumption by an animal), and mutualism (cooperative behavior in which the probability of survival is increased for both interacting populations).

26.1 ECOSYSTEMS AND BIOMASS PYRAMIDS

Plant ecologists seek to find an underlying order to the pattern of plant distribution over the earth's surface.

What threads link plants to each other and to their environment? How flexible are those threads, and how intertwined? How has the process of evolution selected plant behaviors that permit plants to acquire enough energy and nutrients to grow, compete, deter herbivores, reproduce, and disseminate seeds to appropriate habitats? Are there as many ways to budget time and energy during a life cycle as there are species, or are there instead only a small number of successful "strategies" that many species share (Fig. 26.1)? How do plants either resist episodic stresses and disturbances or recover from them? What can certain species tell us, by their presence, vigor, or abundance, about the past and the future of a habitat? Can plants be used as scientific probes to analyze the environment, or as tools to restore degraded landscapes?

All these questions, and more, are being investigated by plant ecologists. The word **ecology** (derived from the Greek roots *oikos* ["home"] and *logos* ["study of"]) was coined more than a century ago by the German zoologist Ernst Haeckel. Ecology is the study of organisms in their home, the environment. More generally, ecology is the study of organisms in relation to their natural environment.

Environment is the sum of all biotic (living) and abiotic (nonliving) elements that surround and influence an organism. Environment is synonymous with the terms *habitat* and *ecosystem*. Examples of biotic elements in an ecosystem are neighboring plants, animals, and soil microbes; examples of abiotic elements in an ecosystem are temperature, moisture, wind, sunlight, soil nutrients, and episodic fires. Depending on the organisms being studied, a habitat or ecosystem can be a few square centimeters of bare rock being colonized by

Figure 26.1 A 1-year-old conifer seedling. During its life span of several hundred years, the plant that grows from this seedling must successfully budget time and resources to satisfy the demands of growth, maintenance, reproduction, competition, and herbivore defense.

mosses, a small lake filling with sediment, or an entire mountain range.

All the hundreds of species within any one ecosystem can be conveniently placed in a few **trophic** (nutritional) categories. Photosynthetic plants, whether seed plants, algal protists, or prokaryotes, are considered **producers** because they generate hydrocarbon food from sunlight and inorganic chemicals alone. Herbivorous animals and plant parasites are **primary consumers.** Carnivorous animals are **secondary consumers,** and carnivores that consume secondary consumers are **tertiary consumers.** Carrion feeders, nonphotosynthetic protists, and microbes are **decomposers.** Consumers and decomposers need organic hydrocarbons to survive, and they are ultimately manufactured by producers.

Each trophic level can be described in terms of the number of species that belong to it, in terms of the weight of all individuals belonging to it (biomass), or in terms of the total caloric energy contained in the individuals. As a general rule, 1 g dry weight of plant or animal tissue contains 5,000 calories.

The biomass or energy of trophic levels decreases sharply from solar energy to producers to herbivores to predators. For example **(Fig. 26.2),** 1 acre of grassland in southern Michigan receives 4,700,000,000 calories (four billion seven hundred million, or 47×10^8) over the course of 1 year. The vegetation, however, is capable of absorbing and transforming no more than 1% of that energy into plant tissue. One acre of vegetation, at the end of a year's growth, has added only 47,000,000 calories (4.7×10^7). Herbivores, in turn, do not consume all that vegetation, but rather only about 10% of it, and most of that energy is burned in respiration. One acre of mice will consume about 5,000,000 calories of plant tissue, but their weight gain during the year will translate into only 50,000 calories (5×10^4). Finally, the primary predator on this particular acre is a weasel. It can consume 10% of the mice, but only a fraction of that energy can be measured as growth by 1,000 calories (1×10^3). If a mountain lion, bobcat, or wolf was a secondary predator in this system, it would have to forage over dozens of acres to find and consume enough weasels to show any gain in weight. Figure 26.2 is an example of a biomass pyramid (energy pyramid). Pyramids for an acre of rain forest, salt marsh, desert scrub, or alpine lake typically show the same steep reduction in biomass or energy through their trophic levels.

26.2 THE POPULATION: THE BASIC ECOLOGICAL UNIT

Organisms usually are not studied as individuals by ecologists; rather, they are grouped together into populations. A **population** is a group of freely interbreeding individuals belonging to the same species and occupying the same habitat. A plant population might consist of thousands of individuals, especially if they are small; or it might include only a few individuals, if they are rare. Not all individuals in a population actually interbreed every year, sharing pollen and genes, but they all grow close enough together so that they could *potentially* interbreed. Enough of them do interbreed to maintain a relatively homogeneous mix of genes throughout the population.

In special circumstances, a population can consist of many genetically identical individuals. Aspen (*Populus tremuloides*) reproduces not only sexually but also by underground roots and sucker shoots. A grove of aspen **(Fig. 26.3)** may have hundreds of trees, all connected underground, representing a single clone.

Although a population consists of only a single species, most species are made up of many populations. Each population occupies the same kind of habitat, scattered over a landscape or region. Some widespread species have thousands of populations, which range over many degrees of latitude or longitude and through hundreds of meters of elevation. In such a case, populations at the extremes of the species' territory are isolated from each other and have little opportunity to share genes. Over time, these outlier populations may evolve different structural or functional traits that allow the species to exploit new habitats.

Red maple (*Acer rubrum*) is a good example of a widespread species: It ranges throughout the entire eastern United States **(Fig. 26.4)** into habitats that vary from wet to dry. An uncommon species, such as California coast redwood (*Sequoia sempervirens*), has narrower habitat requirements than red maple. As a result, it has a smaller range and fewer populations (Fig. 26.4), although the number of individuals in each population is large. The rarest species with

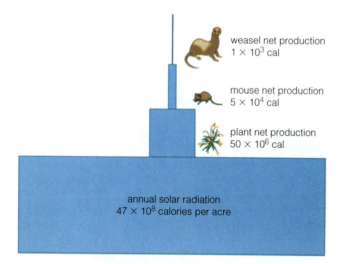

weasel net production
1×10^3 cal

mouse net production
5×10^4 cal

plant net production
50×10^6 cal

annual solar radiation
47×10^8 calories per acre

Figure 26.2 A biomass/energy pyramid for 1 acre of grassland in southern Michigan. Notice how steeply each trophic level's caloric content declines. In this particular ecosystem, solar energy reaching the vegetation over the course of 1 year totals 47×10^8 calories per acre. Green plants (the producer trophic level) transformed only a small fraction of solar energy into growth, net growth (in calories) being about 1% of the solar energy received. Mice (herbivore trophic level) were, in turn, capable of transforming only about 10% of the vegetation into weight gain and reproduction. Weasels (primary predator trophic level) similarly transformed only 10% of mouse tissue into their own weight gain. Net weasel growth thus captured only 0.0002% of the solar radiation that fueled the base of the pyramid.

Figure 26.3 A population of aspen (*Populus tremuloides*). Aspen trees are capable of spreading asexually. This group of individuals could all belong to the same genotype, but most aspen stands consist of a mixture of genotypes.

the narrowest habitat requirements may have only one or a few small populations. For instance, the closed-cone conifer Tecate cypress (*Cupressus forbesii*) occurs in only four little groves, all in southern California (Fig. 26.4).

Ecologists who study single populations and environmental factors that affect the population are engaged in population ecology. In contrast, community ecology or ecosystem ecology is the study of groups of different populations that coexist in the same habitat and of the environmental factors that affect them. This chapter discusses population ecology, and Chapter 27 discusses community and ecosystem ecology.

26.3 LIFE HISTORY PATTERNS

Every plant population has a problem to solve: how to allocate time and energy during the life cycle so as to maximize the probability of successful reproduction. Every plant has limits to its life span, its size, and the resources available to it; therefore, a plant's budget of time and energy is limited and finite.

Ecological research now indicates that there are only a few basic kinds of life cycle budgets and that each kind is shared by many species. We can think of each budget as a life history pattern: a collection of inherited traits and behav-

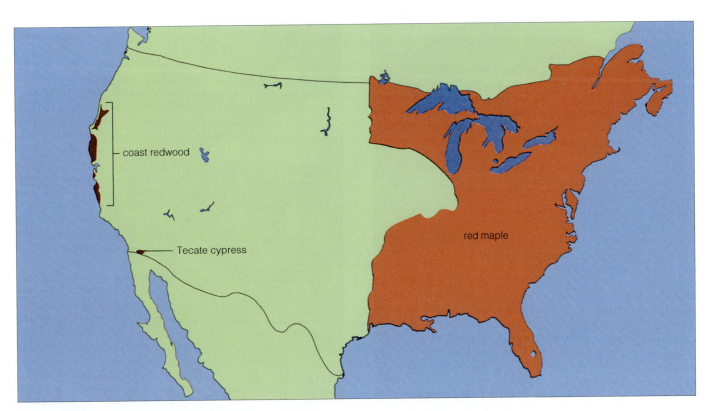

Figure 26.4 Distribution limits of a common species (red maple, *Acer rubrum*), an uncommon species (coast redwood, *Sequoia sempervirens*), and a rare species (Tecate cypress, *Cupressus forbesii*). (Redrawn from Fowells, H.A. 1965. Silvics of forest trees in the United States. *Agriculture handbook no. 271*. Washington, DC: U.S. Department of Agriculture.)

iors that permits a plant population to survive. Life spans, growth forms, timing of reproduction, and sexuality are the major components of life history patterns.

Life Spans Are Annual, Biennial, or Perennial

Annual plants go from seed to seed in less than 1 year. They usually are small and inhabit open areas with short, unpredictable growing seasons. Desert annuals are good examples **(Fig. 26.5)**. *Boerhaavia repens* (desert four o'clock) of the Sahara is a small annual that can go from seed to seed in only 10 to 14 days, but most annuals have a life span of 3 to 8 months. Perhaps the term **ephemeral** is more appropriate than annual, because it does not imply a full year of life.

Winter annuals in the Mojave Desert of California and Nevada usually germinate in November, with the onset of winter rains. They grow slowly during the winter, mainly developing a root system and a whorl of basal leaves. Warming temperatures in March and April bring rapid stem elongation and flower production. Seed maturity occurs by May, and the parent plant dies in June, killed by both the exhaustion of energy reserves that went into reproduction and the stresses of summer heat and drought. The total life span is about 8 months. However, the same winter annuals can complete their life cycle in only 3 to 4 months if winter rains are delayed until March. The plants then skip the winter phase of growth and go on to the reproductive stage; they flower as much smaller individuals but nevertheless complete seed set before dying on schedule in June.

Biennials live for 2 years. In the first year, growth takes place in the root system and a basal rosette of leaves. Cold temperatures during the first winter induce hormonal changes, so that during the second spring the shoot elongates and produces flowers. After setting seed, the plant dies. Carrot (*Daucus*) and lettuce (*Lactuca*) are examples of biennials. Most biennials can be induced experimentally to live longer than 2 years; therefore, we know that they are not genetically programmed to die in that span of time. If a biennial is kept warm over the winter, it remains in the vegetative state longer. Alternatively, biennials can be sprayed with gibberellin and experimentally induced to flower during the first year.

Perennials live for many years and usually flower repeatedly. Their growth form may be herbaceous or woody. **Herbaceous perennials** die back to underground parts each winter. Only their bulbs, corms, roots, or rhizomes are perennial; the aboveground leaves and stems are annual. *Iris*, *Gladiolus*, onion (*Allium*), potato (*Solanum*), the ferns, many wild grasses, and most spring-flowering herbs of the eastern deciduous forest **(Fig. 26.6)** are herbaceous perennials. They often occupy moderately harsh environments such as desert, seasonally dry grassland, shaded forest floors, or alpine regions.

It is difficult to determine the age of a herbaceous perennial because a woody stem with growth rings is absent. The

Michael G. Barbour

Figure 26.5 Desert ephemerals in flower. These annuals have a flexible life span, as short as 3 months or as long as 8 months, depending on temperature and rainfall.

Michael G. Barbour

Figure 26.6 Jack-in-the-pulpit (*Arisaema triphylla*), a dioecious perennial herb. Plant sex depends on plant size, male plants being smaller than females, and juvenile (nonsexual) plants being smallest of all.

perennial organs may retain a scar from each year's leaf system, however, and the scars can be counted to determine plant age. Other perennials spread outward vegetatively from rhizomes or runners, and their age can be estimated by knowing the annual rate of spread. From these types of evidence, we think that many herbaceous perennials live 20 to 30 years.

Woody perennials accumulate aboveground stem tissue year after year. What makes them woody is that the stems have secondary growth. The size and shape of the stems determine whether the plants are subshrubs, shrubs, vines, or trees. **Subshrubs** are multibranched and genetically dwarfed, seldom becoming taller than 30 cm. At the end of each growing season, their stems die back partially, but not all the way to the soil surface. They have short life spans, similar in length to those of herbaceous perennials. They often invade disturbed sites and are later overtopped and outcompeted by taller, longer living shrubs and trees. Rabbitbrush (several species of *Chrysothamnus*) is a common subshrub of the intermountain Great Basin.

Shrubs also are multibranched, but they do not die back annually. Sometimes an entire branch may die, but the tips of other, living branches continue to grow taller. Shrubs have life spans that may exceed a century. Some reproduce asexually with underground stems, and an entire clone can attain a very great age. Creosote bush (*Larrea tridentata*), for example, is a desert shrub capable of spreading vegetatively from the base. Some clones are old enough to form a ring several meters in diameter (Fig. 26.7), estimated to be several thousand years old.

Vines (called **lianas** in the tropics) have weak, single trunks. They require additional support from neighboring shrubs or trees, obtaining that support by twining about the host's trunk or literally sprawling on top of the host's leafy canopy. Common vines in the eastern deciduous forest include Virginia creeper (*Parthenocissus quinquefolia*) and

Dutchman's pipe (*Aristolochia durior*). Vines have secondary wood, but typically that wood is light because it contains many more vessels than tracheids and fibers. Such wood is metabolically inexpensive to build, leaving more energy to be allocated to the plant's growth in height. Consequently, vines have a relatively rapid growth rate. The life span of a vine is similar to that for subshrubs and shrubs.

Trees have strong, single trunks and an elevated branch system that does not die back annually. As with shrubs, individual branches or even a portion of the trunk may die, but the rest of the tree continues to grow in height and girth. Broadleaf trees, such as oak (*Quercus* species), are typically angiosperms; they also are called **hardwoods** because of the density of their secondary wood. Needleleaf trees, such as pine (*Pinus* species), are gymnosperms; they are called **softwoods** because the uniform size and arrangement of tracheids in their secondary zylem make the wood relatively easy to cut through and work.

Hardwoods and softwoods can be either evergreen or deciduous. If they are evergreen, each individual leaf has a life span longer than 1 year; if they are deciduous, all leaves have a life span less than 1 year and fall synchronously (at the same time), leaving the tree bare for part of the year. Leaf life spans mirror the pattern for tree life spans, being longer for softwood trees than for hardwood trees. **Broadleaf evergreens,** such as some southern and western oaks, have leaf life spans of only 2 years; but needle-leaf evergreens, such as some mountain pines, have leaf life spans of 5 to 30 years.

Evergreen leaves, also called sclerophylls (hard leaves), are energetically more expensive to manufacture than deciduous leaves. They are thick, tough in texture, and bounded by a well-developed cuticle; they also have few air spaces in the mesophyll, exhibit a low surface-to-volume ratio, and contain high concentrations of metabolic byproducts. Collectively, these features mean that it takes more carbohydrates to build an evergreen leaf than a deciduous leaf.

Figure 26.7 Creosote bush (*Larrea tridentata*), a common shrub of the warm deserts of North America. Individuals can live for more than 100 years. They are capable of reproducing vegetatively, producing an ever-widening circle of offspring. Some of these clones are estimated to be several thousands of years old.

Evergreen leaves are more drought tolerant than deciduous leaves, but at the same time they have a lower maximum rate of photosynthesis. Ecologists think that the long life of sclerophylls and their efficient use of water are compensations for their high expense and low photosynthetic rates.

Reproduction Can Be Semelparous or Iteroparous

Timing of reproduction is another component of a life history pattern.

Annuals, biennials, and some herbaceous perennials reproduce sexually only once, at the end of their life spans. This is called **semelparous** reproduction (derived from the Latin *semel* ["once"] and *parere* [to "bear"]). An advantage of semelparous reproduction is that a single burst of reproduction, delayed to the time of maximum plant size, yields so many seeds that some are likely to germinate and live long enough to produce the next generation. The California chaparral plant, our Lord's candle (*Yucca whipplei*, **Fig. 26.8**), is a semelparous perennial that flowers at the end of its 20-year life span. Semelparous plants spend 25% of their stored caloric resources on their single massive reproductive event. It is possible that they die by exhausting their food reserves in setting so many flowers and seeds in such a short time.

In contrast, **iteroparous** plants reproduce many times during their lives (*itero-* means "repeat"). Each reproductive event uses 5% to 15% of the plant's food reserves. Iteroparous plants go through a juvenile period before the first reproductive event. The length of the juvenile period depends on the species; in general, it accounts for 10% of the

entire life span. In commercial fruit trees, the juvenile phase lasts 4 to 6 years, but wild trees often have a much longer juvenile period. Some iteroparous species reproduce at a constant rate once the juvenile period is over. Others, such as oak, pine, and fir, reproduce more abundantly some years than others. The years of high reproduction are called **mast** years.

Sexual Identity Is Not Always Fixed

The distribution of sexual identity among individuals is yet another component of life history patterns. Most species of angiosperms are **monoecious,** with male and female flower parts (stamens and pistils) on the same individual. Other species that have only male or only female flower parts on one individual are **dioecious.** We can imagine that the advantage of being dioecious is that cross-pollination is absolutely required; thus, genetic recombination and variation are high. The advantage of being monoecious is that a single isolated individual can reproduce successfully. Most weedy, invasive species are monoecious.

In dioecious species, the sex of the individual plant is usually genetically determined, but not by X and Y chromosomes, as in humans. Sometimes sex is not under any genetic control but is a consequence of plant size or habitat conditions. Jack-in-the-pulpit (see Fig. 26.6) is a herbaceous perennial in which the sex of an individual is environmentally controlled. Very young individuals go through a juvenile phase during which they grow vegetatively and become larger but do not reproduce. Plants exhibiting sexuality are older and larger in size (as measured by total area of leaves produced and volume of perennial underground parts) than the juveniles. Of these adult plants, the largest—but not necessarily the oldest—are female. Sexuality in this case is actually a function of size, not of age. If a female plant is manipulated so that it does not accumulate much stored food one year, it will produce fewer leaves the next year, and the size of its underground system will shrink. If the loss in size is significant, the plant will become male. As it accumulates stored food over the years, it will become female again. If male plants are similarly starved, the following year they may become nonreproductive, like juveniles. As they grow, they will become male again.

Some species segregate sex by habitat. Female plants tend to occupy moister or nutritionally richer soils, whereas male plants tend to occupy drier, sterile sites. Female plants require better sites because so much stored food must go into seeds and fruits, but male plants can occupy marginal sites because less energy is needed to produce pollen. Dioecious species thus subdivide the habitat and take advantage of all sites.

Another advantage of the dioecious state is that seed predation by animals is reduced. Female plants—with their energy-rich seeds and fruits—are mixed with male plants across the landscape; thus, an herbivore that finds a female plant may not encounter another female plant nearby, making it unlikely that the herbivore will devour all the seeds in

Michael G. Barbour

Figure 26.8 Our Lord's candle (*Yucca whipplei*), a semelparous perennial plant. It flowers only once, at the end of its 20-year life span. This species grows in coastal hills of central and southern California.

a locale. Patterns of sexuality in other species are less obvious. Some red maple (*Acer rubrum*) individuals, for example, seem to change sex randomly from year to year, with few remaining constantly male or female. We still have much to learn about the allocation of sexuality in plants of dioecious species.

Life History Patterns Range from r to K

When life span, growth form traits, timing of reproduction, and allocation of sexuality are put together, some concrete life history patterns emerge. It is useful to recognize two extremes because all other types can be arranged between the extremes. Those extremes are called r and K life history patterns **(Table 26.1).**

Typically, **r-selected** species are annuals or biennials that grow in open habitats. They have a rapid growth rate, deciduous leaves, and a short juvenile period. They are semelparous and allocate considerable caloric energy to reproduction. All of the common weeds that inhabit frequently disturbed locations are r-selected species: ragweed, wild radish, vetch, pigweed, dock, wild oats, and bindweed.

K-selected species usually are perennials with a larger body size. These plants occupy more stable habitats that may be buffered (from wind, sun, and frost) by an overstory leaf canopy. Herbs, shrubs, and trees in old-growth forests are good examples, such as hemlock, red oak, sugar maple, viburnum, false Solomon's seal, and Virginia creeper. They may be evergreen or deciduous. Compared with r-selected species, they have a slower growth rate and a longer juvenile stage; they expend less energy on reproductive events and are iteroparous. Basically, r-selected species maximize reproduction in unstable habitats at the expense of long life and large body size, whereas K-selected species maximize

long-term occupation of a site and competitive abilities at the expense of rapid growth and early reproduction. Most species lie somewhere along a continuum between these extremes.

It is important to remember that r and K strategies are relative, not absolute, categories. For example, one might think that all annual herbs are r-selected and that all trees are K-selected. But tree species include some that are r-selected and others that are K-selected. Cottonwood trees (*Populus* species) grow rapidly, produce very light and easily decayed wood, have thin deciduous leaves with low concentrations of herbivore-deterring metabolic byproducts; they generate many seeds every year that are distributed randomly by the wind; and the trees favor disturbed, open habitats. All of these characteristics are r-type traits. Red oak trees (*Quercus rubra*), in contrast, grow slowly, produce dense and decay-resistant wood, have tougher leaves rich in tannins, produce large numbers of acorns only every several years, use animals to disperse the large seeds nonrandomly, and favor undisturbed, shaded habitats. All of these characteristics are K-type traits. In a similar manner, some herbaceous plants are K-selected, whereas others are r-selected.

Another classification of life history patterns recognizes three extremes **(Fig. 26.9)** instead of two. In this system, the **R type** (abbreviation for ruderal, or roadside) is equivalent to the r in the r–K continuum described above. The **C type** (for competitive) is equivalent to the K. The **S type** (stress-tolerant) is a new category, representing species that grow in stressful habitats such as salt marshes, rock outcrops, moving sand, and deserts. Plant density in such habitats is low; thus, a tolerance of biotic stresses (such as competition) is less important than a tolerance of abiotic stresses.

R, C, and S species can sometimes grow in close association. For example, rock outcrops in the Appalachian Moun-

Table 26.1 Some Traits of r- and K-selected Species

Trait	r-selected	K-selected
Habitat	Often disturbed	More constant
Population	Variable size	More constant, often near the carrying capacity
Survivorship	Type I or II	Typically type III
Competition	Variable; not an important factor	Usually keen
Life span and reproductive effort	Usually short; semelparous; high reproductive effort	Long; iteroparous; low reproductive effort (except in mast years)
Growth pattern	Rapid growth and development	Slower growth and development
Body size	Small	Large
Juvenile period	Short; several months for herbs, 5 years for trees	Long (10% or more of entire life span)

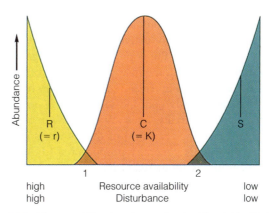

Figure 26.9 Diagram of the abundance of ruderal (roadside, R) plants, stress-tolerant (S) plants, and competitive (C) plants along environmental gradients of resource availability and frequency of disturbance. At *point 1*, disturbance is rare enough to favor C types over R types. At *point 2*, resources become so limiting that S types are favored over C types. (Redrawn from Grime, J.P. 1977. Evidence for the existence of three primary strategies in plants and its relevance to ecological and evolutionary theory. *American Naturalist* 11:1169–1194, with permission.)

Michael G. Barbour

Figure 26.10 C-, S-, and r-selected plants may occur in the same local habitat, separated only by microenvironmental gradients. Rock outcrops in the Appalachian Mountains show a concentric zonation of plants, with S types at the outermost fringe, R types in a middle zone, and C types in the very center.

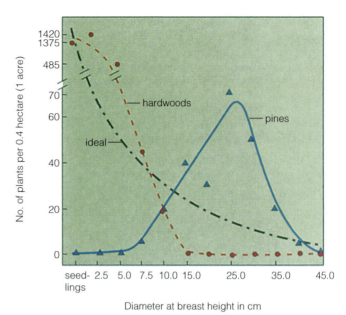

Figure 26.11 Age distribution (on the basis of trunk diameters at breast height) for pines and hardwoods in a forest near Gainesville, FL. The pine population is senescent, and the hardwood population is invading. The *green line (dashes and dots)* shows the expected age distribution of a stable population, in contrast to the pines and hardwoods.

tains often are occupied by islands of vegetation (Fig. 26.10). At the outermost edge, growing on bare rock, are the mosses *Grimmia*, *Polytrichum*, and *Bryum*. These slow-growing plants are S types. Farther into the island of vegetation, growing on top of a thin layer of soil that mosses have accumulated, occur herbaceous R-type plants such as hair grass (*Deschampsia flexuosa*). In the center, with deepest soil, grow C-type shrubs and trees such as *Rhododendron* and chestnut oak (*Quercus prinus*). Most ecologists agree that this tripartite division is a concept that has improved our understanding of life history patterns.

26.4 PLANT DEMOGRAPHY: POPULATION AGE STRUCTURE OVER TIME

A population usually contains individuals of all ages: seedlings, juveniles, reproductive adults, and senescent oldsters. The proportion of the population in each age category tells us something about the history and future of that population. For example, Figure 26.11 is a summary of the ages of all pine and hardwood individuals in a forest near Gainesville, FL. (Actually, ages are not shown; rather, the diameter at breast height [dbh] of the trunks is indicated. However, tree age is generally proportional to dbh. A tree with a 25-cm dbh is about 100 years old in this forest; a tree with a 45-cm dbh is about 200 years old.)

Notice that the pine population in the figure has many mature trees 15 to 30 cm in dbh, but it has no seedlings, saplings, or young trees smaller than 5 cm in dbh. From this, we conclude that the pine population has not been reproducing itself for some decades. The hardwood population, in contrast, has only seedlings and saplings, with no large, mature individuals. This suggests that the hardwood population is a recent invader of the forest.

Putting the two population trends together, we can predict that the pine forest will be replaced by a hardwood forest. The current overstory pines eventually will reach their natural life span limits, about 200 years, and die. As they leave the canopy, they will be replaced by the young hardwoods growing beneath them.

In time, the forest will become a hardwood forest, and pines will be rare or absent. When that happens, we can predict that the distribution of ages is likely to follow the ideal curve shown in Figure 26.11: There will be many young hardwoods and less and less old ones. Such an age structure will be stable over time because as each old tree dies, the probability is great that youngsters of the same population are already growing beneath it and they will grow up to maintain the same overstory composition.

Plant demography is the study of changes in population age structure over time. By examining age structure, a demographer can create a mathematical model for predicting how long an average individual will live, when it will enter and leave reproductive age, how many seeds it will shed in its lifetime, and how many of those seeds will survive to germinate.

Figure 26.12 is a diagram summarizing the demography of one particular species. Let us tease apart that model. On the left (represented by small orange circles) are all the seeds currently in the ground; this is the seed pool. These seeds are 26 in number (N = 26). Environmental and physiological stresses (such as being eaten by animals, parasitized by soil microbes, or kept dormant by some internal mechanism) take their toll, so that only a fraction of those seeds (9 of 26) germinate this year. The nine that germinate together are part of a single **cohort,** a segment of the population that is

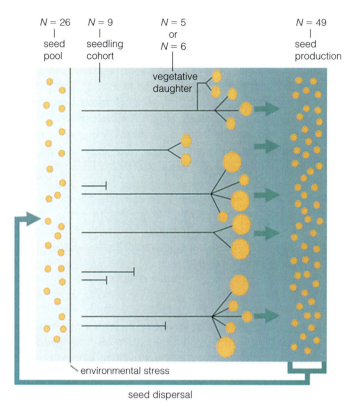

N = 26 | seed pool
N = 9 | seedling cohort
N = 5 or N = 6 | vegetative daughter
N = 49 | seed production

environmental stress

seed dispersal

Figure 26.12 A model of one population's life history. Time increases along the *horizontal line,* from left to right. This species is semelparous and flowers only once at the end of its life span. It is capable of vegetative production during this life span. Note that not all seeds germinate and that not all seedlings reach maturity. *N* is the number of living individuals in the cohort at the times shown. Because seed output is larger than the initial seed pool, this population is growing. (Redrawn from Harper, J.L. 1977. *Population Biology of Plants.* New York: Academic Press, with permission.)

the same age. Of the nine germinated seeds, environmental stresses—such as heat, drought, or disease—kill four before they can reproduce (represented in Fig. 26.12 by short, truncated lines). The remaining five plants of the cohort all reach sexual maturity and produce flowers (indicated as large orange circles in Fig. 26.12). One member of the cohort reproduces vegetatively before this event, producing an additional sixth plant, which also flowers. Notice that not all the plants produce the same number or sizes of flowers. Their reproductive effort probably correlates with the particular patch of ground each plant occupies.

This particular model is for a semelparous species because all six plants die on release of seeds (the horizontal lines stop). Notice that the number of seeds released (49) is much greater than the number at the start (26). Therefore, we might predict that this population will grow in number. Because about 35% of the original seed pool germinated (9/26 × 100 = 35%), we can predict that next year's cohort of seedlings should total about 17 (35% of 49 seeds).

Another way of presenting changes in population age structure over time is in the form of a **survivorship curve (Fig. 26.13).** Plant and animal populations have three basic kinds of survivorship curves. A **type I** curve is characteristic

Michael G. Barbour

John Shaw/NHPA

George Ranalli/Photo Researchers

a

b

c

Type I

Type II

Type III

No. of survivors (log scale)

1000
100
10
1
0.1

Age ⟶

Figure 26.13 Three different survivorship curves. (**a**) A type I curve is characteristic of annual plants. (**b**) A type II curve is characteristic of plants in extreme environments. (**c**) A type III curve is characteristic of large perennial plants. Note that the vertical axis of the graph is logarithmic, meaning that each population unit changes by an order of magnitude. (Redrawn from Deevey, E.S., Jr. 1947. Life tables for natural populations. *Quarterly Review of Biology* 22:283–314, with permission.)

of r-selected (or R-type) populations, such as annual plants. There is high survivorship (low mortality) for most of the life span; then all the individuals die within a relatively short time. A **type III** curve is characteristic of K-selected (or C-type) populations, such as forest trees. There is high mortality early in life, but then very low mortality once the individuals reach maturity. A **type II** curve is characteristic of S-type populations such as alpine herbaceous perennials. There is constant mortality throughout the life span because of the overwhelming importance of abiotic environmental stress at all stages of the life cycle.

26.5 POPULATION INTERACTIONS WITH THE ENVIRONMENT

We have already divided the environment into biotic and abiotic components, but an additional division is useful. The **macroenvironment** is that part of the environment determined by the general climate, elevation, and latitude of the region. Weather bureau data on rainfall, wind speed, and temperature are measurements of the macroenvironment. Such measurements are taken at a standard height above the ground, in cleared areas well away from buildings or trees.

The **microenvironment** is that part of the environment close to the surface of a plant, animal, structure, or the ground. The microenvironment is modified by nearby surfaces. For example, bare soil tends to absorb heat; consequently, on a clear day the temperature just above or below the soil surface is much greater than air temperature. Prostrate plants with basal leaves will experience microenvironmental temperatures much warmer than those recorded by the weather bureau 1.5 m above the ground. Microenvironmental air as far as 10 mm from the surface of a leaf on a still day is less turbulent, greater in humidity, and warmer than free air farther from the leaf.

Light quality and quantity are much different for herbs in the microenvironment beneath a forest canopy than for the leaves of the canopy itself. Herbs on the forest floor sense perhaps only 1% to 5% of the light received at the top of tree canopies. Moreover, the light that passes through the canopy is relatively enriched in green and far-red wavelengths because photosynthetic pigments in the canopy leaves absorb in the blue and red portions of the spectrum.

Plants growing in shallow depressions experience frost more frequently than those growing on higher ground because cold air flows downhill. Soils may also be wetter and more anaerobic in depressions.

Water and Soil Are Important Environmental Factors

Probably no single abiotic factor is as significant in the distribution of plant populations as the supply of water. So important are plant–water relationships that species have been formally classified into **xerophytes** (plants able to grow in very dry places), **mesophytes** (those that grow best in moist soil), **helophytes** (those able to grow in saturated soil

such as marshes and bogs), and **hydrophytes** (those able to grow rooted, submerged, emergent, or floating in standing water).

Plants in each category have unique morphological, anatomical, and physiological traits that adapt them to a particular moisture regimen. Xerophytes, for example, may have small, hard leaves; an epidermis with thick cuticle, light color, and hairlike trichomes; stomata that are sunken in epidermal depressions and that close early during the day; and, sometimes, **succulent** tissue (the cells of which are large and contain water-filled vacuoles). Leaves may be permanently absent or seasonally absent. If they are permanently absent, green stems conduct photosynthesis. If leaves are seasonally absent and they fall off during hot rainless periods (rather than during cold winter periods), the plants are called **drought-deciduous.**

Xerophytes can live in humid environments, provided that their roots are in dry microenvironments. Some xerophytes live in tropical rain forests as **epiphytes**—plants that grow on tree trunks or branches with their roots twining through relatively dry bark; they obtain all their water by trapping rainfall in leaf axils. Many of them have succulent leaves and stems and specialized water-storing tissue on root surfaces. One common group of tropical epiphytes belong to the Cactaceae, or cactus, family **(Fig. 26.14).** Cacti probably evolved in the tropics, later moving into dry deserts as those habitats appeared in the late Cenozoic. Other important epiphytic families are the Bromeliaceae and

Figure 26.14 An epiphytic cactus in a semitropical forest.

the Orchidaceae, to which the pineapple (*Ananas*) and the orchid, respectively, belong.

Other xerophytes live on soil in which the physical or chemical properties limit the amount of free water available to plant roots. Very coarse soils, high in sand or gravel, permit most rainwater to percolate rapidly, leaving an island of aridity within a climate that might seem wet and adequate for luxurious plant growth. Similarly, shallow soils or rock outcrops store little water in the root zone. Soils high in clay can potentially store much water, but the rate at which they can absorb water is slow; if rains come in torrential bursts, much of the precipitation will run off, rather than percolate. Saline soils are said to be physiologically arid because dissolved salts decrease the water potential of soil water, making the water less available. Again, locally dry islands of vegetation are the result.

The seasonal timing of precipitation can be as critical to vegetation as the annual amount. In the tropical latitudes, annual precipitation is generally more than 200 cm. When that precipitation is evenly divided among 12 months, a three-storied, evergreen forest results; but if the same annual precipitation is divided into wet and dry seasons, a less complex forest with many deciduous trees is the result. The form of precipitation also is important. Through the mid-elevations of California mountains, annual precipitation changes little, averaging 150 cm. Below an elevation of 2,000 m, however, most of that 150 cm falls as rain. Above 2,000 m elevation, most of it falls as snow, building a winter snowpack 3 to 4 m deep. As a result, the species of trees, shrubs, and herbs are dramatically different in the two zones, even though total and seasonal precipitation are the same.

Solar Radiation Is another Environmental Factor

Solar radiation strikes the outer limits of our atmosphere with an energy content of about 2 calories per square centimeter per minute, a value called the *solar constant*. Solar wavelengths of the electromagnetic spectrum range from 300 to more than 10,000 nm. Technically, light is the portion of the solar spectrum visible to the human eye: wavelengths from 400 (violet light) to 740 nm (red light). Plants can use the same range of wavelengths in photosynthesis.

Only half of the solar constant reaches the lower atmosphere and vegetation within it because certain wavelengths are absorbed or reflected back to space. In particular, a significant amount of shortwave, ultraviolet radiation is removed by ozone in the stratosphere.

Solar radiation is further depleted as it passes through foliage. Below a forest canopy, sunlight is reduced to 5% of its level just above the foliage. Many shade-tolerant herbs, shrubs, and tree saplings can still grow in this level of sunlight.

Long-living plants must be flexible in their light requirements because, as they grow, they experience different levels of light intensity. As saplings they may grow in dense shade, but as adults they are exposed to full sun. Furthermore, adult trees are so large that not all the leaves experience the same amount of sunlight. Leaves high in the

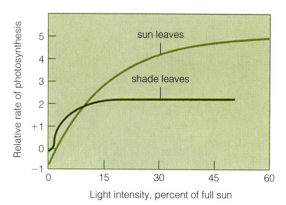

Figure 26.15 The rate of photosynthesis in sun leaves and in shade leaves at different light intensities. Leaves are from beech (*Fagus sylvatica*). (Redrawn from Boysen-Jensen, P., Muller, D. 1929. *Jahrbuch fuer Wissenschaftliche Botanische* 70:493, with permission.)

canopy, **sun leaves,** receive high amounts of solar radiation, whereas those lower in the canopy, **shade leaves,** receive less. Shade leaves exhibit a different morphology, anatomy, and physiology from those that develop in sunlight, even when both are attached to the same plant. Shade leaves are larger, thinner, contain less chlorophyll per gram of tissue, and have less well-defined palisade and spongy mesophyll layers. They reach their maximum rate of photosynthesis at much lower light intensities than sun leaves **(Fig. 26.15).**

Eastern hemlock (*Tsuga canadensis*) is a good example of a long-living, shade-tolerant tree species: It can grow in deep shade as a sapling but continues growth in full sun as an adult. Mature trees live for 1,000 years. Saplings grow slowly while shaded. Saplings only 2 m in height and 2 to 3 cm in dbh may have a ring count indicating an age of 60 years. Juveniles can remain alive in deep shade for as long as 400 years. If the canopy opens up because of the death of a mature tree, the increased light releases a juvenile into a fast growth mode, and it soon fills the canopy gap. Many other trees, especially pines, are not shade tolerant, and their juveniles can grow only in open habitats.

Solar radiation also is depleted when it passes through water. At about 170 m below the surface of clear ocean water, sunlight is reduced to 0.5% of its level at the surface. Light level sets the **compensation depth** for most aquatic plants—the depth at which they can just maintain positive net photosynthesis.

BALANCING INCOMING AND OUTGOING SOLAR RADIATION Currently, the earth is neither cooling down nor heating up. Therefore, we can conclude that what comes in also goes out; that is, the solar energy budget equals zero. This zero-sum budget holds not only for the earth itself, but for every organism and for every part of an organism on Earth. Let us examine the energy budget of an imaginary plant leaf as an example **(Fig. 26.16).**

Some of the solar radiation that reaches the leaf will be reflected. Perfectly white leaves reflect all radiation, but typical green leaves reflect only 30% of the energy. The energy

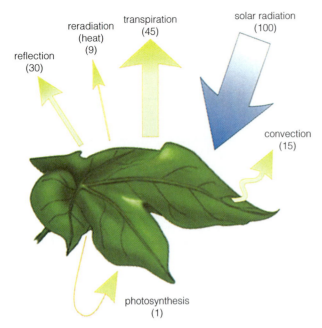

reflection
(30)

reradiation
(heat)
(9)

transpiration
(45)

solar radiation
(100)

convection
(15)

photosynthesis
(1)

Figure 26.16 Energy budget of a leaf. Incoming solar radiation is given a relative value of 100 energy units. The leaf gives back 99 units in the forms of reflected light, reradiation, transpiration, and convection (heating the air). Only 1 unit was assimilated in the process of photosynthesis.

that is absorbed is converted to long-wave energy (wavelengths greater than 740 nm), which is felt as heat—that is, the leaf warms up. This heat is dissipated in four ways: transpiration, convection, reradiation, and metabolism (photosynthesis).

Transpiration is the conversion of water into gas and its loss from the leaf. When water changes state to gas, it uses up a significant amount of energy. For our imaginary leaf, with well-hydrated tissues and open stomata, transpiration accounts for 45% of the energy budget. In a desert, transpiration would account for less of the energy budget. **Convection** is the transfer of heat energy back to air. For our imaginary large, entire leaf, convection accounts for only 15% of the energy budget. In a desert plant, with smaller leaves, convection would account for more of the energy budget. **Reradiation** is the loss of long-wave radiation back to space; its effect is best seen at night, when the temperature of the leaf decreases. Under the cover of a forest canopy, reradiation accounts for only 9% of our imaginary leaf's heat budget. In a desert, under a cloudless sky, reradiation would account for more of the energy budget. The overall remainder of the solar radiation, 1%, is converted into chemical energy by photosynthesis. Photosynthesis does not vary much, regardless of the plant or habitat.

This particular energy budget formula is:

incoming solar radiation
(100 units)

= [reflection + transpiration + convection + reradiation + photosynthesis]
= (30 units) + (45 units) + (15 units) + (9 units) + (1 unit)

The equation is balanced: 100 incoming units = 100 outgoing units, and there is zero gain or loss.

THE EFFECT OF SOLAR RADIATION ON TEMPERATURE   As noted earlier, heat is a consequence of solar radiation. Long-wave radiation is experienced as heat, and it is an important environmental factor. Heat can be measured with thermometers and expressed in many ways, temperature being just one.

Latitude and topography greatly influence solar radiation and heat. As we move north or south from the equator, the angle at which solar radiation strikes the ground becomes less direct and more oblique; this means that radiation travels through more of the earth's atmosphere and that more of it is absorbed or reflected. Light at the poles in summer may last 24 hours a day, but it is a weak light compared with the intensity at the equator. One consequence is that polar summer days are not very warm.

Because temperature tolerance differs among plant species, different species prosper at different latitudes and elevations. For instance, some plants are restricted to certain belts of elevation in mountains, and the elevation of those belts changes with latitude. Plants that grow in mountains respond to the diminishing amount of heat away from the equator by shifting their habitat lower and lower. Mountain hemlock (*Tsuga mertensiana*), for example, grows at high subalpine elevations (2,700 m) in the Sierra Nevada of California, but its habitat drops toward the north until it reaches sea level along the Alaskan coast (**Fig. 26.17**).

The direction a slope face also affects solar radiation and heat. In the northern hemisphere, south-facing slopes receive more sunlight than north-facing slopes; consequently, they have greater temperatures and drier soils. These microenvironmental differences affect plant population distribution and behavior. A canyon gorge running east–west through an Indiana forest illustrates the magnitude of this effect. The gorge is 65 m in width at the top and 45 m in depth; its sides support scattered trees, shrubs, and herbs. Microenvironmental instruments were placed 15 cm above and below the soil surface midway down each side. In spring, the south-facing slope had a larger daily range of soil and air temperature, a greater average air temperature, a greater rate of soil water evaporation, lower soil moisture, and lower relative air humidity than the north-facing slope. Of nine spring-flowering species present on both slopes, the flowering time averaged 6 days earlier on the south-facing side. To gain a similar difference in flowering time on level land running north-to-south, one would have to pick sites 180 km apart.

Some species are sensitive to brief extremes of high or low temperature, and their distributions are determined more by the extremes of temperature than by long-term average temperatures. Most species of palms and cacti, for example, are sensitive to frost and do not occur in regions that experience yearly frost, even if the duration of freezing temperature is short.

Other species are sensitive to variation in temperature over seasons or days, and it is this amplitude of temperature that is more important than the average temperature. The difference between maximum daytime temperature and

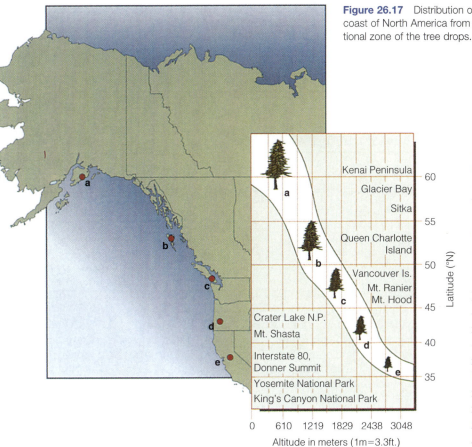

Figure 26.17 Distribution of mountain hemlock (*Tsuga mertensiana*) along the west coast of North America from latitude 62° N to 38° N. As latitude increases, the elevational zone of the tree drops.

Kenai Peninsula
Glacier Bay
Sitka
Queen Charlotte Island
Vancouver Is.
Mt. Ranier
Mt. Hood
Crater Lake N.P.
Mt. Shasta
Interstate 80, Donner Summit
Yosemite National Park
King's Canyon National Park

Latitude (°N)

0   610   1219   1829   2438   3048
Altitude in meters (1m=3.3ft.)

chaparral scrub, the northern boreal forest, southeastern forests (Fig. 26.18), some mountain conifer forests, and even desert palm oases owe their presence in large measure to climates with episodic natural fires.

When humans suppress fires in the mistaken notion that all fires are unnatural and catastrophic, they begin to change the mix of species and the shape of the landscape (see the "PLANTS, PEOPLE, AND THE ENVIRONMENT: The Natural Fire Cycle in the Southeastern Pine Savanna" sidebar). This is because many species not only tolerate fire but depend on it, either to complete their life cycle or to maintain dominance over other species. Despite our current understanding of fire's importance and the creation of programs to use it in national parks and forests, only a small fraction of the land that should burn each year actually does burn. Thus, the accumulated magnitude of environmental change gets larger year by year, and we have not returned North America to the landscapes seen by the first explorers.

Each Population Is Ecologically and Genetically Unique

We now realize that each population interacts with its local environment so that over evolutionary time it becomes genetically better adapted to its microenvironment. Each

minimum nighttime temperature within a 24-hour cycle is called a **thermoperiod.** California coast redwood (*Sequoia sempervirens*) grows best under the almost constant day–night and winter–summer temperatures of its habitat, near the temperature-buffering Pacific Ocean and bathed by summer fogs (Fig. 26.4). Experiments with seedling redwoods in growth chambers have shown that the optimum thermoperiod for coast redwood is close to zero. In contrast, conifers that grow inland in the Sierra Nevada have optimum thermoperiods greater than 13°C.

Fire Can Be a Natural Part of the Environment

In the last several decades, fire has been rediscovered as a natural environmental factor in North America. We say "rediscovered" because Native Americans were well aware of its occurrence and its effects, and they learned how to use it for their own purposes. Early Euro-American explorers and colonists invariably commented on the frequency of fires in forests and grasslands, but the pervasiveness of natural fire was ignored by land use managers until recently.

Most natural fires are started by lightning strikes unaccompanied by rain, which occur after a prolonged spell of dry weather. Climates that have such conditions are said to be fire-type climates, and they are widespread throughout North America. Lightning accounts for 5,000 fires in the United States each year, which is two thirds of all our wildfires. Human-caused fires are in the minority. Grasslands,

Paige Martin/Archbold Biological Station

Figure 26.18 Fire in the southeastern coastal plain.

PLANTS, PEOPLE, AND THE ENVIRONMENT:

The Natural Fire Cycle in the Southeastern Pine Savanna

One of the first vegetation types in the United States shown to be dependent on fire for its maintenance was the pine savanna of the southeastern coastal plain **(Fig. 1).** Tall, scattered loblolly, slash, shortleaf, and longleaf pines dominate the region, and a thick growth of grasses (mainly *Andropogon* and *Aristida*), with some broad-leaved herbs, cover the ground beneath. Euro-American settlers found the pines to be valuable for lumber and turpentine, and the grass to be valuable for grazing livestock. Wildfires were thought to be dangerous; therefore, during the nineteenth century some of this vegetation was protected from fire. Gradually, however, the pines gave way to oaks, and in time a dense oak forest with little grass replaced the pine savanna.

Ecological studies in the 1930s revealed the importance of fire to the pines, especially to longleaf pine (*Pinus palustris*). Longleaf pine is not only tolerant of fire, it is dependent on it. Its seeds germinate in the fall, soon after dropping to the ground from cones. During their first year of growth, the seedlings are very sensitive to even the lightest fire. However, during the next several years, longleaf pine seedlings are in a "grass" stage (Fig. 1). Food reserves are shunted to the root system, the stem remains stunted, and a dense cluster of long needles surrounds the apical meristem, which is located at the ground surface. If a fire sweeps through the area at this time, the apical bud is insulated, and the seedling remains alive. A fire actually improves seedling survival because the high temperatures kill a common fungus that otherwise parasitizes the needles.

After the grass stage, longleaf pine saplings enter a 4- to 5-year period of rapid stem elongation. By the time the sapling is 8 to 9 years old, its canopy is high enough above the ground that it is beyond the reach of surface fire flames. In addition, the bark is thick enough to insulate the cambium from damage.

Pine savanna systems maintain themselves naturally so long as ground fires sweep through every several years. However, if fires are suppressed for as long as 15 years, hardwood seeds carried into the area by animals will germinate and produce young saplings. Hardwood saplings are very sensitive to ground fire; they would not be able to invade an area where burns occurred every few years. In the absence of fire, however, they compete well with grasses and pines, and they are tolerant of shade from the overstory pines. As the oaks grow, they create more shade, which inhibits pine seedling establishment, growth, and survival. When overstory pines die, they are replaced by hardwoods, not by pines. In time, an oak forest exists where a pine savanna used to be.

In addition to pines, many herbaceous plant species decline in abundance because fire no longer keeps down competing grass and smothering litter. Other species disappear because their life cycles somehow require fire: Their seeds can germinate only after the seed coat is scarified by high temperatures, or they flower profusely only after fire has released a pulse of nutrients into the soil. In addition, certain grazing animals, birds, and insects that live best in open savanna will be absent from this oak forest. In other

population within a given species, then, is probably genetically different from all others. The genetic differences can be expressed as plant morphology, timing of life cycle events (such as flowering or leaf drop), or plant metabolism.

Not so long ago, botanists did not understand that species could contain such genetic diversity. In the 1920s, the botanist Gote Turesson collected seeds of species that had a wide range of habitats—from lowland, southern, and central Europe to northeastern Russia in the Ural Mountains—and he germinated them all in a garden in Åkarp, Sweden. He was the first to demonstrate that members of a widespread species were not genetically homogeneous. Despite the uniform environment of the common garden, plants that grew from seeds collected in warm, lowland sites often were taller and flowered later in the year than those that grew from seeds collected in cold, northern sites. They differed in frost tolerance, leaf traits, and onset of dormancy. Yet, these variants all had the same flower traits and all could be cross-pollinated; therefore, they are members of the same species.

Turesson called these variants within species **ecotypes.** Today, after more research, we now understand that every population is its own ecotype.

words, a whole community of organisms will change in the absence of fire.

Standard land management practice currently is to set a prescribed surface fire purposely every 4 years if a natural fire has not visited the area. Under this system, the pine savanna maintains itself indefinitely, just as it would with natural wildfires.

The use of natural or prescribed fire as a management tool is still not widely practiced because it affects air quality, is expensive, and can be conducted only during narrow windows of time when there is a low probability that a fire will escape. The infamous Yellowstone fire in 1988 is a good example of a natural fire that was allowed to burn but that got out of control when the weather changed. It caused a great deal of economic damage to surrounding communities and resulted in a congressional investigation of the whole concept of fire management. The congressional report ratified the concept that natural fire is beneficial, but it recommended additional restrictions as to when natural fires would be allowed to burn.

Michael G. Barbour

a

Figure 1 Longleaf pine (*Pinus palustris*) savanna in North Carolina. (**a**) Aspect of the savanna. Overstory trees are widely spaced and average 50 years of age. Beneath them is a well-developed understory of grasses and broad-leaved herbs. (**b**) Close-up view of a young longleaf pine in the "grass" stage.

Michael G. Barbour

b

🌐 Websites for further study:

Floridata—*Pinus palustris*: http://www.floridata.com/ref/p/pinu_pal.cfm

Western Fire Ecology Center: http://www.fire-ecology.org/

USGS Fire Ecology Research: http://www.werc.usgs.gov/fire/

Yellowstone National Park—Fire Ecology:
http://www.nps.gov/yell/technical/fire/ecology.htm

One example of ecotypes involves alpine sorrel (*Oxyria digyna*), a small perennial herb that grows in rocky places above timberline. Some populations grow at high elevations in mountains; these are alpine populations (Fig. 26.19a). Others grow in the far north at low elevations; these are the arctic populations (Fig. 26.19b). Both populations experience certain environmental stresses in common, but they differ in other stresses. For example, they both are exposed to long periods of freezing winter weather, but during summer days alpine plants receive more solar radiation. Researchers have demonstrated that the alpine and arctic populations are separate ecotypes and that they differ in such physiological ways as the amount of light that is optimum for photosynthesis (Fig. 26.19c). Alpine plants photosynthesize best at 50% full sun, whereas arctic plants do best in lower light at 25% full sun.

The ecotype concept is important in forestry. After logging, sites may be planted with tree seedlings. The seedlings come from seeds that were collected as close (in distance and elevation) to the logged area as possible. In mountainous terrain, seed sources must be within 100 m of elevation of the planting site. The Forest Service recognizes that the

a

b

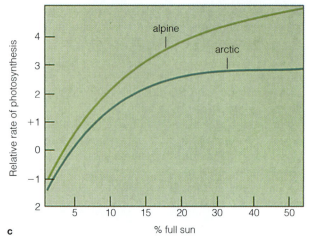

c

Figure 26.19 Ecotypes of alpine sorrel (*Oxyria digyna*). (**a**) Alpine ecotype. (**b**) Arctic ecotype. (**c**) Photosynthesis of each at different light intensities. (a, Reproduced from Mooney, H.A., Billings, W.D. 1961. Comparative physiological ecology of arctic and alpine populations of *Oxyria digyna. Ecological Monographs* 31:1–29, with permission.)

probability of seedling survival will be greatest for a population that is already genetically adapted to the local region.

<table>
<tr><td>26.6</td><td></td></tr>
</table>

26.6 INTERACTIONS AMONG NEIGHBORING POPULATIONS

Rarely does a single population, to the exclusion of all others, occupy a habitat. Usually, several plant, animal, and microbe populations—representing as many different species—coexist. Individuals of the various populations intermingle, growing next to each other, and this proximity allows them to affect each other. Sometimes the results are positive, with the organisms showing enhanced growth or survival; sometimes the results are negative, with the organisms suffering a decline. There are four basic kinds of interactions: competition, amensalism, herbivory, and mutualism.

Competition Creates Stress by Reducing the Amount of a Commonly Required Resource

Competition may be defined as the decreased growth of two interacting populations because of an insufficient supply of a necessary resource such as light, moisture, space, nutrients, or pollinators. Sometimes the limitation lies with a single resource, but most often several resources are lacking. Competition may be equal, allowing the two populations to coexist indefinitely, or it may be unequal, eventually resulting in the displacement of one population and the occupation of its space by the other population.

Competition may be the most important biotic factor affecting plant distribution. Many populations restricted to saline, dry, or nutritionally poor soils, for example, would actually grow better on normal soil if other populations were first removed. These restricted populations are more tolerant of stress but are poor competitors in comparison with the plants that populate normal soil. When planted together, the restricted populations have slower root and shoot growth rates; consequently, they obtain less soil water and sunlight, and fewer of them survive to produce seeds. In time, they are completely eliminated from all but the most stressful sites, where plants from normal soils cannot maintain themselves.

The intensity of competition lessens in a process called divergent evolution. Over time, two populations of different species become more different in such traits as the time of germination or flowering, tolerances for soil aridity or depth, nature of pollination or seed-dispersal vectors, or degree of shade tolerance. The portion of the microenvironment each population uses is called its niche. Over time, each population's niche becomes more distinct and separate from the niches of nearby populations. It is theoretically impossible for two populations to have the same niche because competition would be too intense; hence, the expression "One niche, one population."

Amensalism Creates Stress by Adding Something to the Environment

Amensalism (sometimes the word **allelopathy** is substituted) may be defined as the inhibition of one population by another through the addition of something to the environment. The added material can be a metabolic byproduct exuded from a living root or the decomposition products from dead litter. It can be a solid that accumulates beneath the parent plant, a liquid carried into the soil by percolating rainwater, or a volatile molecule carried off by the wind. A striking pattern of avoidance by two species in nature often is taken as evidence of amensalism (Fig. 26.20).

Amensalism may be common because plants are leaky systems, passively contributing all sorts of chemicals into the environment. One investigator grew seedlings representing 150 different flowering plant species in a nutrient culture. Water around the roots contained several radioactive elements

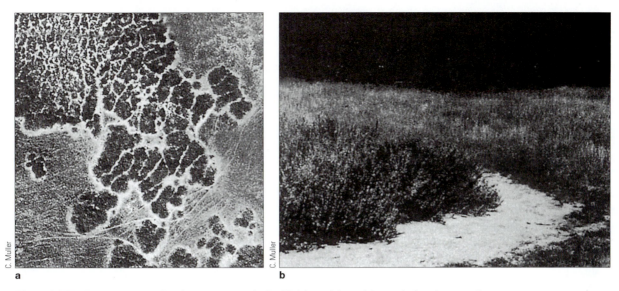

Figure 26.20 Possible amensalism between sage shrubs (*Salvia* and *Artemisia* species) and grasses in coastal areas of southern California. (**a**) Aerial photograph of a patch of sage. Note the bare soil beneath the shrubs and for some distance away from them. (**b**) View along the edge of a patch of shrubs, showing a zone of bare soil and stunted grasses several meters in width. Some researchers speculate that a volatile chemical exuded by shrub leaves prevents the grasses from germinating or growing normally.

as markers. The plants took up the isotopes through their roots and transported them to all organs, including the leaves. The plants then were exposed to a mist, and the water that condensed and ran off the leaves was collected for analysis. The investigator found that 14 elements, 7 sugars, 23 amino acids, and 15 organic acids—all radioactive—had been leached from the plants.

Another study showed that roots are similarly leaky. Root tips were collected from mature trees in a New Hampshire forest and rinsed in distilled water. Analysis of the water revealed sugars, amino acids, organic acids, and various cations and anions. Birch (*Betula alleghaniensis*) exuded five times the amount of substances as maple (*Acer saccharum*). In nature, such substances from leaves, roots, or litter would accumulate in the soil, where they might affect microbial decomposers, soil pH, soil physical structure, and the growth of nearby plants.

Herbivory Is the Consumption of Plant Biomass by Animals

Herbivory is the harvesting by an animal of vegetative or reproductive parts of a plant; it may leave the plant still alive or kill it. Herbivory can play a striking role in plant distribution. A glance along a pasture fence shows that certain palatable species are absent inside the pasture yet common outside it. Grazing animals are responsible for the difference.

Although plants often are thought of as passive organisms, unable to move as easily as animals, they are able to defend themselves against herbivores. The defenses are slowly elaborated over evolutionary time. One defense is the dispersal of population members over space, making it

difficult for herbivores to locate and damage all the individuals. Dispersal can be achieved by vegetative runners or rhizomes, by explosive fruits, or by the use of wind as a vector. Another defense is the dispersal of population members over time. This can be accomplished by seeds with complicated dormancies or by reproduction that is irregular from year to year (as with masting or semelparous flowering). A third defense is the development of physical barriers such as sclerophylls, spines, thick bark, and hard fruits or seed coats.

A fourth defense is amensal: the manufacture and accumulation of metabolic byproducts that are distasteful or that otherwise inhibit herbivores from feeding. There has been much recent research into chemical herbivore defenses. The type of chemical a plant manufactures varies with its life history pattern: r-selected plants tend to manufacture toxins, whereas C-type plants tend to accumulate tannins or terpenes. Toxins repel herbivores by interfering with nerve and muscle activity, hormone function, or liver and kidney metabolism. Toxins are small, relatively energetically inexpensive molecules of less than 500 molecular weight. They are effective in small concentrations, and they account for less than 2% of leaf dry weight. In contrast, tannins and terpenes repel herbivores by their bitter, unpleasant taste. They are large, complex molecules of 500 to 3,000 molecular weight and may account for 6% of leaf dry weight. Animals attempting to feed on plant tissue that is rich in tannins or terpenes move away to other plants.

Some chemical herbivore defenses are induced in response to an attack by an herbivore, parasite, or pathogen. Such chemicals are called **phytoalexins.** The plant's chemical defense is analogous to antibody formation in animals. Humans are herbivores, too, and therefore plant toxins can and do poison humans, as well as grazing animals or insects. Humans, however, have learned to use some poisons as medicines or psychotropic drugs, and we cultivate the plants that manufacture them. The nicotine that accumulates in tobacco (*Nicotiana*) leaves inhibits herbivores, but humans use nicotine as a drug. Cocaine functions as an herbivore deterrent in coca shrubs (*Erythroxylon*), but Andean Indians learned to chew the leaves for stimulation while doing strenuous activity at high elevations.

Mutualism Increases the Success of Both Populations

The previous examples of interactions were all negative, causing depressed growth or reproductive success in one or both interacting populations. **Mutualism** is an interaction that benefits both partners; furthermore, it is essential in the sense that the success of both partners is reduced when one of the partners is absent.

Several examples of mutualism already have been described earlier in this textbook. Animal partners form mutualistic relationships with plants. Bees, moths, and beetles—which cross-pollinate flowers while themselves feeding on nectar, pollen, or flower parts—are familiar examples. Ants, birds, and mammals that consume fruits while dispersing seeds are other examples. Sea anemones, which mutualistically harbor algae in their cells, provide yet another example. Lichens are mutualistic associations of fungi and algae, mycorrhizae are mutualistic associations of fungi and higher plants, and nitrogen-fixing bacteria live symbiotically in the root nodules of legumes.

If the benefits are not shared by both interacting populations (that is, the interaction is not obligate), the relationship is not mutualistic but **commensal.** Epiphytes growing in a tropical tree, for example, form a commensal relationship with the tree. The host tree gains no benefit from the epiphyte. At the same time, the host tree is not usually hindered by the epiphyte, because the epiphyte is not a parasite and it is not so large that it shades the tree or breaks off limbs. The tree neither benefits nor loses by the association, whereas the epiphyte is promoted.

KEY TERMS

amensalism (allelopathy)	drought-deciduous
annual	ecology
biennials	ecotype
broadleaf evergreen	environment
C type	ephemeral
cohort	epiphyte
commensal	hardwood
compensation depth	helophyte
competition	herbaceous perennial
convection	herbivory
decomposer	hydrophyte
dioecious	iteroparous

K-selected	secondary consumer
macroenvironment	semelparous
mast	shade leaf
mesophyte	shrub
microenvironment	softwood
monoecious	subshrub
mutualism	succulent
perennial	sun leaf
phytoalexin	survivorship curve
plant demography	thermoperiod
population	transpiration
primary consumer	tree
producer	trophic
R type	vine (lianas)
reradiation	woody perennial
r-selected	xerophyte
S type	

SUMMARY

1. Ecology is the study of organisms in relation to their natural environment. The environment contains all the biotic and abiotic elements that surround and influence an organism. The macroenvironment reflects the general, regional climate; the microenvironment reflects conditions near the surface of organisms or objects. Ecosystems consist of organisms and their environment. These organisms participate in trophic functions within the ecosystem: They produce food, they consume it, or they decompose it.

2. The basic ecological unit is the population. Every species typically has many populations, each one genetically distinct and adapted to its particular habitat. A population may be equivalent to an ecotype, unique in its morphology, physiology, or behavior. Population ecology is the study of a population, together with the environmental factors affecting that population.

3. A life history pattern is the budget of time and energy that carries a population through all phases of its life span. Plant life spans are annual (ephemeral), biennial, or perennial. Herbaceous perennials die back to below-ground organs each year. They occupy stressful habitats and commonly live for 20 to 30 years. Woody perennials include subshrubs, shrubs, vines, and trees. Trees include broadleaf angiosperm and needle-leaf gymnosperm categories. Broadleaf trees are shorter lived, have shorter-living leaves, and build harder wood than needle-leaf trees.

4. Semelparous populations reproduce only once, at the end of a plant's life span. Iteroparous populations reproduce repeatedly. Each pattern has its own advantages. Sexuality is sometimes determined by environmental factors or plant size; it is not necessarily under genetic control. The dioecious condition promotes cross-pollination, allows a popula-

tion to occupy a wider array of sites, and reduces the loss of seeds to herbivores. The monoecious condition permits weedy, aggressive plants to disperse into new habitats, even when only a single individual has arrived.

5. Species with an r-selected life history pattern typically are small, fast-growing annuals adapted for open habitats; they are semelparous and allocate a great deal of caloric energy to reproduction. K-selected species typically are slow-growing but ultimately large and long-lived forest trees in more stable habitats; they are iteroparous and allocate less energy to reproductive events. Most species lie somewhere along a continuum between these extremes.

6. Plant demography is the study of population age structure over time. It uses mathematical models to summarize the life span of individuals and cohorts in a population, their reproductive potential, their mortality rate over time, and future changes in the size of the population.

7. The intensity and quality of solar radiation change as the radiation passes through the atmosphere, plant leaves, and water. Light (the wavelengths of solar radiation visible to humans) is used by plants in photosynthesis. Large, long-living, K-selected plants successively live through different light environments, and therefore must be flexible in their light requirements. This flexibility is enhanced by an ability to produce sun leaves and shade leaves and to be shade tolerant when young.

8. Every organism and organ has an energy budget with a zero sum. Solar radiation striking a leaf, for example, is first partitioned into reflected and absorbed energy. Absorbed energy is dissipated as transpiration, convection, reradiation, and photosynthesis. Mesophytes expend most solar radiation by transpiration; xerophytes expend solar radiation by convection, reradiation, and reflection.

9. Fire caused by dry lightning strikes is a natural occurrence in many parts of North America. When natural fires are suppressed, the distribution of many plant populations is affected. Open pine savannas of the southeastern coastal plain, for example, change into dense oak forests if fires are not allowed to burn every 3 to 15 years. Currently, our use of prescribed fire is too limited to restore all North American landscapes to conditions before fire suppression.

10. Neighboring populations in the same habitat interact and affect each other through competition, amensalism, herbivory, and mutualism. Competition, amensalism, and herbivory are negative interactions, with one or both partners suffering a decline. Competition is caused by a deficient amount of an essential resource. Amensalism is caused by the addition of a deleterious substance into the microenvironment. Herbivory is the consumption of vegetative or reproductive plant material by an animal. Plants have evolved herbivore defenses in the following forms: (a) dispersal over space and time; (b) repulsive physical structures; and (c) toxic or unpalatable metabolic

byproducts. The latter mechanism is a form of amensalism. Mutualism is a positive interaction, with the partners showing enhanced growth or survival.

Questions

1. A grassy acre in southern Michigan receives 47×10^8 calories of solar radiation over the course of one year. Plants fix, in photosynthesis, about 58×10^6 calories of that radiation. They retain 50×10^6 calories, and of these, mice consume 25×10^4 calories. The mice retain 5×10^4 calories, and of these, weasels consume 5×10^3 calories. Weasel net growth is 1×10^3 calories. If a young weasel must gain 2,000 g of weight in that year, and each gram is equivalent to 5×10^3 calories, how many acres of grassland must be its range?

2. Can a species be rare and in danger of extinction if its only population has thousands of individuals? Can a species be rare and in danger of extinction if it has thousands of populations, each population of which has only a few individuals?

3. Why do semelparous plants die after their single reproductive event? What are possible advantages to reproduction in this way? Shrubs and trees are rarely semelparous. Do you know of any in your region?

4. Is a single habitat inhabited by only r-selected or only K-selected species, or can they coexist in the same habitat?

5. What is the definition of plant demography? Try to diagram the demography of a human population using the same model shown for a plant population in Figure 26.12.

6. In an irrigated, dark green pasture in Texas, what fraction of annual solar radiation do you guess would be dissipated by transpiration, by reflection, by convection, and by photosynthesis? If that same pasture were allowed to revert to semiarid, nonirrigated natural grassland, what might those figures be?

7. How did Turesson show that species were not ecologically and genetically homogeneous? Is ecotype a synonym for population?

8. Describe the difference between competition and amensalism. Why are both of these interactions called negative, whereas mutualism is said to be positive?

InfoTrac® College Edition

http://infotrac.thomsonlearning.com

Ecosystem Functions

Bergelson, J. 1996. Competition between two weeds (common groundsel and annual bluegrass). *American Scientist* 84:579. (Key words: "groundsel" and "bluegrass")

Gomez, J.M. 2003. Herbivory reduces the strength of pollinator-mediated selection in the Mediterranean herb *Erysimum mediohispanicum*: Consequences for plant specialization. *The American Naturalist* 162:242. (Keywords: "herbivory" and "Mediterranean")

Uttley, A. 2002. Energy transfer: You may have wondered why the real-life food chains you have studied aren't very long. In this article we show how the loss of energy from food chains limits the number of animals that can survive on the energy fixed by a patch of vegetation. *Catalyst* 13:9. (Keywords: "energy" and "food chains")

Ecology and Plant Communities

Visit us on the web at http://biology.brookscole.com/plantbio2 for additional resources, such as flashcards, tutorial quizzes, InfoTrac exercises, further readings, and web links.

1. The plant community is a group of recurring species that: share a characteristic habitat; collectively create a unique physiognomy; attain a typical range of species richness, annual productivity, and standing biomass; and through which nutrients and energy pass at predictable rates and with predictable efficiency. North America contains thousands of plant communities.

2. A vegetation type is composed of many communities that differ only in the identity of dominant or associated species, or both, but that otherwise share a similar physiognomy and environment. Two-thirds of North America is covered by only three major vegetation types: boreal forest, grassland, and tundra.

3. Successional plant communities change over time, whereas climax plant communities do not show any directional change, although they may fluctuate from year to year.

4. Conservation biology is a relatively new science that investigates ways to preserve, restore, and maintain biotic diversity in the face of human exploitation of natural ecosystems.

Figure 27.1 A red spruce (*Picea rubens*)–Fraser fir (*Abies fraseri*) forest community in the Appalachian Mountains of North Carolina.

27.1 THE NATURE OF PLANT COMMUNITIES

Most plant populations do not grow in isolation. Single populations do not usually monopolize a habitat to the exclusion of all other organisms. Normally, there is a mixture of coexisting plant and animal populations, as well as of many less visible fungal, algal, and bacterial populations.

Wherever a particular habitat repeats itself within a region, many of the same species recur. The species composition does not replicate itself completely, but there is a nucleus of species that do repeat. These clustered species are said to be associated with each other and to be members of a biotic **community.** For simplicity, we can break down any community into its complement of animal, plant, or microbial species. Because this is a plant biology textbook, this chapter discusses communities from the standpoint of the plants they contain—that is, the plant community is discussed.

Every plant community is named after its dominant species and is characterized by its own roster of associated species and their combined architecture. For example, high peaks in the Great Smoky Mountains of North Carolina all have red spruce (*Picea rubens*) and Fraser fir (*Abies fraseri*) trees as the tallest layer of vegetation. The conifers are rather dense, and their canopies usually touch, creating a deep shade **(Fig. 27.1).** Beneath this needle-leaf canopy is a second, much more open layer of vegetation of scattered mountain ash trees (*Sorbus americana*). Mountain ash is a deciduous broadleaf species that sheds its leaves in the fall. A final layer of vegetation carpets the ground and consists mainly of ferns, mosses, and broad-leaved herbaceous perennials.

The spruce and fir populations are said to **dominate** the community because they contain the largest individuals and contribute the most biomass. By being the largest, they create the microenvironment within which smaller associated species live.

The architecture of this red spruce–Fraser fir community can be summarized as having three canopy layers: The uppermost is almost continuous and is made up of needle-leaf evergreens; the second is open and is made up of shorter, broadleaf trees; and the third is a continuous ground layer of herbs. The habitat is high-elevation, north-temperate, winter-cold, with well-drained soils. Wherever this habitat appears in the Appalachian Mountains, the spruce–fir community recurs.

There are many other plant communities in the Appalachian Mountains. They are dominated by different species with different growth forms—such as shrubs, grasses, broad-leaved evergreen trees, or broad-leaved herbs. If we expand our vision to the entire North American continent, we could identify thousands of different plant communities. In addition, we could discern blendings or mixtures in **ecotones,** where two or more communities and environments grade into each other.

Each Plant Community Has Unique Attributes

Each plant community has features that transcend a mere list of member species. These features, or **attributes,** have to do with community architecture, species richness, the spatial patterns in which individuals are arranged, the efficiency with which they trap sunlight and cycle energy or nutrients through the community, and the stability of the associated species in the face of environmental stress or change. These attributes are summarized in **Table 27.1;** a few of the attributes also are discussed in the following paragraphs.

PHYSIOGNOMY The architecture of an Appalachian spruce–fir community was described earlier in this chapter. A technical synonym for community architecture is physiognomy. **Physiognomy** is the external appearance of the community, its vertical structure, and the growth forms that dominate each canopy layer. A desert community, consisting of only a single canopy layer of widely scattered shrubs **(Fig. 27.2a),** might have 10% canopy cover or less—that is, 10% of the ground is directly beneath the foliage of the shrubs; 90% is open and unshaded. A tropical rain forest, with several overlapping tree layers **(Fig. 27.2b),** has 100% canopy cover.

Another attribute of the canopy has to do with its thickness—that is, the number of layers of overlapping leaves through which light must pass on its way to the ground. Canopy cover does not measure this attribute, but *leaf area index* (LAI) does. LAI is the total area of leaf surface (one side only) for all leaves that project over a given area of ground. If the leaf and ground areas are measured in the same units (so many square meters of leaf surface per square meter of ground), then LAI becomes a dimensionless number. The desert community mentioned earlier in this chapter has an LAI of 1, and the tropical rain forest has an LAI of 10. The optimal LAI for crops such as corn is 4. Planting corn so densely that the LAI is greater creates too much shading for the lowest leaves.

SPECIES RICHNESS Biotic diversity—a term often used by the news media—simply means the total number of species that occur in a given community or region. In this sense, **diversity** is synonymous with species **richness.** Communities differ in the number of associated species they contain. Tropical rain forests appear to have the greatest diversity of plant species, up to 365 per 10,000 m². Temperate forest and woodland communities have more moderate diversities of 50 to 100 species in a similar area, and desert communities have less than 50 species in the same area.

Not every species in a community is equally important. A few species are represented by many individuals (such species are common), a few other species are rare, and most are intermediate in abundance. It is the relative scale of importance of each species that collectively gives the community its unique physiognomy. The importance of each species can be quantified by counting the **density** of individuals (the number per unit area), its canopy cover, its biomass, or its **frequency** of occurrence in the community. A widely dispersed population has individuals that would be frequently encountered, whereas individuals that are clumped together would be encountered less often **(Fig. 27.3).**

BIOMASS AND PRODUCTIVITY Communities differ in the amount of biomass they have above- and belowground. Tropical rain forests, for example, may have up to 500,000 kg of aboveground biomass per hectare, whereas desert communities have only 1% of that amount. Allocation of bio-

Figure 27.2 Communities with different physiognomies. (**a**) Desert scrub in southern Nevada, dominated by creosote bush (*Larrea tridentata*), has a single open canopy layer. (**b**) A tropical rain forest in northern Argentina has several tree canopy layers.

Michael G. Barbour

a

Michael G. Barbour

b

mass above- and belowground also varies with the community. Forest communities typically have five times as much biomass above as belowground. Another way of saying this is that the **root-to-shoot biomass ratio** of a forest is 0.2. In contrast, scrub communities have equal amounts of biomass above- and belowground (the ratio is 1.0), and grassland communities have more biomass belowground than aboveground (the ratio is 3.0).

Communities also differ in the amount of biomass they produce each year, their **productivity.** Those communities with the greatest biomass usually are the most productive because they have a greater LAI with which to trap solar radiation. High-biomass communities also tend to live where the growing season is longest, another factor contributing to productivity. Thus, tropical rain forest communities have an annual productivity of 20,000 kg per hectare, compared with only 2,000 kg per hectare for deserts.

NUTRIENT CYCLING Communities differ in their demand for mineral nutrients from the soil, in the locations where such nutrients travel and accumulate, and in the rate at which they are returned to the soil or are lost from the ecosystem by erosion. This process of nutrient uptake, use, and return is called a **nutrient cycle.**

The nitrogen cycle **(Fig. 27.4),** for example, moves nitrogen from inorganic to organic forms and back again. Nitro-

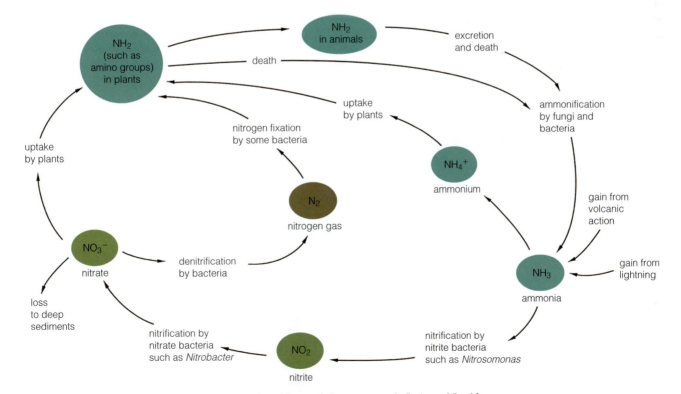

Figure 27.4 The nitrogen cycle. Blue color indicates reduced forms of nitrogen; green indicates oxidized forms.

gen gas (N_2) in the air is ordinarily not available to plants, which can absorb nitrogen only as ammonium (NH^{4+}) or nitrate (NO_3^-). The biological process of nitrogen fixation accomplished by certain bacteria converts nitrogen gas into ammonium. Volcanic action also creates ammonium, as does lightning; this ammonium is then brought to the ground (and to plant roots) as ions dissolved in rainwater. Certain bacteria are capable of converting ammonium into nitrate.

Once in plants, the ammonium or nitrate is metabolically altered, moved onto amino acids, and then incorporated into proteins, nucleic acids, alkaloids, and many other important molecules. Some of the nitrogen returns to the soil when leaves are shed or when the plant dies. Certain soil microbes convert the organic forms of nitrogen in litter back into ammonium, and others change the ammonium into nitrite (NO_2) and then into nitrate. About 10% of the nitrogen in a plant is passed on to the next generation in the form of stored food in fruits and seeds; another 10% is ingested by grazing animals and becomes part of their metabolism. When they excrete waste or die, their nitrogen is converted into ammonium, nitrite, and nitrate by the same microbes that decompose plant remains.

Similar nutrient cycle diagrams could be prepared for sulfur, carbon (Fig. 27.5), water, and phosphorus. The details

of each cycle, the rates of nutrient movement within it, and its overall efficiency are different from community to community.

Tropical forest communities, for example, contain broadleaf trees with a high demand for nitrogen and other nutrients. The forest trees store nutrients in trunks for long periods, and they return only a small fraction of the nutrients to the soil each year through leaf shedding and decomposition. The largest pool of nutrients in this particular ecosystem (80%) is in the wood. Decomposition, from litter to humus to soil, is rapid, the half-life of litter decomposition being less than 1 year because of continuously warm temperatures and adequate moisture. Most nutrients released to the soil are rapidly taken back up by a dense network of fine roots in the topsoil. Consequently, tropical soils tend to be low in residual nutrients. There is little loss of nutrients from erosion because the soil is covered with vegetation and anchored in place by dense roots. Most minerals are said to have a long *residence time* in the ecosystem, and the cycling of nutrients is "tight."

Northern forest communities, in contrast, contain conifer trees with a low demand for nutrients. These trees have a smaller pool of nutrients in their wood (33%), and they return more nutrients to the soil in an annual rain of litter.

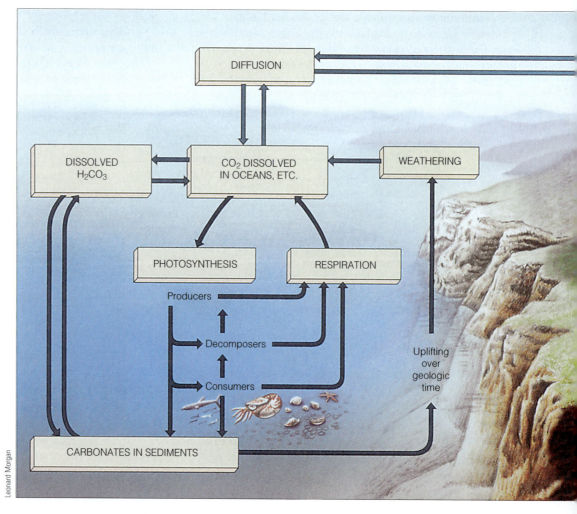

Figure 27.5 Diagram of the global carbon cycle. The portion on the left shows movements of carbon through the marine ecosystem; on the right is shown movements of carbon through land ecosystems. Carbon is shown being released from respiration, by decay, from sediments, and from volcanic eruptions. Methane (CH_4) is also released by decay and it is moved through the atmosphere. Although not as abundant as CO_2, it has a longer residence time and retains heat more effectively than CO_2.

The half-life of litter decomposition is slow (10 years) because of cold temperatures; therefore, litter accumulates in a thick layer. The largest pool of nutrients in this ecosystem (60%) is in semi-decomposed litter called humus. Soils tend to be low in nutrients because they are acidic. Conifer foliage contains high amounts of organic acids, which dissolve cations in the soil, allowing them to be leached from the root zone. The cycling of nutrients in such a community is "loose."

Plant Communities Change over Time

The microenvironment within a plant community is very different from that outside. Temperature, humidity, soil moisture, and light are all affected and modified. A stable community consists of K-selected species whose seedlings can survive to maturity in this unique microenvironment (see Chapter 26). Seedlings of species from other communities are at a disadvantage and cannot normally invade.

CHARACTERISTICS OF SUCCESSION If the stable community is removed by some disturbance—such as landslide, storm, canopy fire, or logging—then the soil surface is exposed, and the microenvironment has changed. The first species to colonize the site after the disturbance usually are not those of the old community. They are seedlings of r-selected species adapted to open sites. Only after a passage of time, when the biomass of invasive species has altered the microenvironment, will species of the old, stable community gradually return. The process of community change at one place over time is called **plant succession.**

Succession has limiting characteristics at population and community scales. For example, the death of one individual and its replacement by another is not succession. We must instead look to change over a larger area than just the space taken up by one individual for evidence of succession. Succession also has a *maximum* area limit because it must occur within a uniform macroenvironment. When we begin to cross soil boundaries or zones of elevation or latitude, the area is no longer homogeneous, so the changes that we see from place to place are not the result of succession. Succession generally occurs over an area from 1 hectare to several square kilometers.

Succession usually is measured over the course of several years to several hundred years. Thus, plant changes from season to season are not succession because they merely reflect life cycle phases. Succession does not occur over time periods shorter than a year. There also is a maxi-

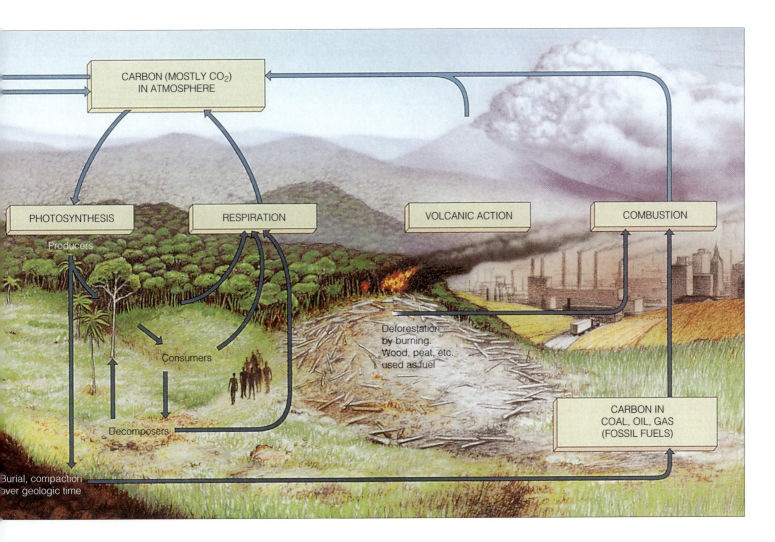

mum time—certainly shorter than 1,000 years—because succession must occur within a constant macroenvironment. Once we reach scales of time that incorporate global climate change, the community changes we see are not part of succession. Furthermore, we all know that the environment is not constant from year to year. In good years, the biomass, productivity, and species richness of a community will be different from those in poor years. These random fluctuations in a community over relatively short periods are not part of succession.

STAGES OF SUCCESSION Succession proceeds from pioneer to climax phases. For example, in parts of the southeastern United States, oak–hickory forests seem to be stable. If such a forest is cleared and then abandoned, the path of succession is as follows: The first plants to invade are annual and perennial herbs that grow well in high light intensity, such as horseweed (*Conyza canadensis*), *Aster pilosus*, and broomsedge (*Andropogon virginicus*). These make up a **pioneer stage (Fig. 27.6a).**

Pine seeds blow in from large, old pines that grow as scattered individuals in the surrounding oak–hickory forest. They are more readily dispersed than the heavier seeds of oak (*Quercus*) and hickory (*Carya*). They also germinate and grow well in semi-open habitats. Within 10 years of the disturbance, many short pine saplings are visible **(Fig. 27.6b).** As the pines grow, their emerging canopies begin to alter the microenvironment. Some of the pioneer herbs do not maintain their populations and disappear from the community or become rare. Within 30 years of the disturbance, a pine savanna has grown up **(Fig. 27.6c).**

Examination of the herb layer beneath the pines shows many oak and hickory seedlings and few pine seedlings. Pine seedlings grow poorly in shade and under conditions of root competition for moisture from other trees, but those of oak and hickory are more tolerant of shade and competition.

As seen in Chapter 26, if fire sweeps through the pine savanna community frequently enough, the oaks and hickories are selectively eliminated. But if fire does not visit this community, succession continues. In this case, within 50 years of disturbance, a well-defined understory of hardwood saplings and young trees exists beneath the pines (Fig. 27.6b). Whenever a pine dies, it is replaced in the upper canopy by a maturing oak or hickory. Within 200 years of the disturbance, the forest again consists of mature oak and hickory trees with an occasional old pine. The shrubs and herbs associated with oak and hickory also are present, and the dense grass understory of the pine savanna has disappeared **(Fig. 27.6d).** The stable oak–hickory forest community is the end point of succession and is called the **climax stage.**

The succession just described is an example of **secondary succession,** which takes place on vegetated land. **Primary succession** occurs on newly exposed ground not previously occupied by plants. Examples of new land include mobile sand dunes, volcanic lava flows, mud exposed by a drop in lake water level, bare rock scraped clean by a retreating glacier, or the infilling of a small lake. The rate of succession is slower in primary succession for several reasons. First, the parent materials may be bedrock, coarse sand, or windblown silt—all of which lack clay particles, and thus essential nutrients. Second, there is no bank of plant seeds, bulbs, or rhizomes already in the soil, left over from a predistur-

a

b

d

Figure 27.6 Secondary succession in abandoned fields in the Piedmont region of North Carolina. (**a**) Field abandoned for 1 year. Dominant plants include short-lived herbs and grasses. (**b**) A 10-year-old field, showing perennial herbs and grasses and 7- to 8-year-old pine saplings are in the background. (**c**) Pine stand on field abandoned for 50 years. Old furrows are still obvious in the picture. Understory is dominated by broad-leaved tree seedlings and samplings. (**d**) Mature deciduous forest that has developed on a field abandoned for 150 years.

Norm Christensen, Duke University (a–d)

c

Table 27.2 Comparison of Some Community Traits during Early and Late Stages of Progressive Succession

Trait	Early Stages	Later Stages
Biomass	Small	Large
Architecture	Simple	Complex
Nutrient pool	Soil	Vegetation
Mineral cycling	Loose	Tight
Productivity	High	Low
Stability	Low	High
Species diversity	Low	High
Life history	r-selected	K-selected
Site quality (microenvironment)	Extreme and not well-developed	Moderate and well-developed

bance community. Primary successions often take several hundred years, whereas secondary successions may take only decades.

Many community traits change during succession, and these are summarized in **Table 27.2.** Basically, these changes make the community more complex and massive, the cycles of energy and nutrients more efficient, and the microenvironment less stressful. These changes are the result of **progressive succession.**

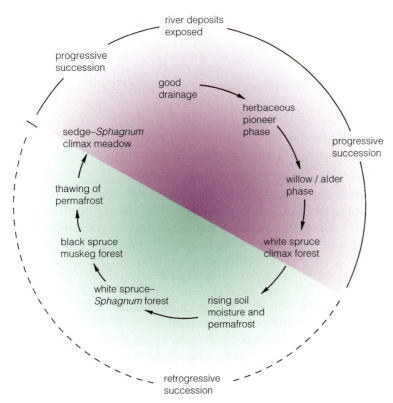

Figure 27.7 Progressive (*purple*) and retrogressive (*blue*) succession on an Alaskan floodplain. (Redrawn from Drury, W.H. Jr. 1956. Bog flats and physiograph processes in the upper Kuskokwin River region, Alaska. *Contributions from the Gray Herbarium* 178:1–30, with permission.)

But not all succession is progressive. For example, **Figure 27.7** is a summary of primary succession on Alaskan floodplains left dry when a river meanders and changes form. At first, progressive succession (clockwise in Fig. 27.7) leads to increasing herb cover and then to invasion by willow (*Salix*) and alder (*Alnus*) shrubs. Alders can fix nitrogen in their roots, so soils become richer. Biomass increases, shading the site. K-selected white spruce (*Picea glauca*) then invades the shade and slowly grows through the shrub canopy. As the spruce creates an overarching tree canopy, the r-selected alders decline in the deepening shade.

Now **retrogressive succession** begins, reversing the process, so that the community becomes simpler and less massive, cycles of energy and nutrients less efficient, and the microenvironment more severe. The mature spruce canopy creates a cold, dense shade. One consequence is that *Sphagnum* moss invades the ground surface; this moss has the capacity to retain large amounts of water, and it literally raises the water table. The shade and thick layer of *Sphagnum* cools the soil, allowing a permanently frozen layer (**permafrost**) to rise toward the surface. White spruce requires well-drained soil and a deep water table, so it begins to die. Scattered, stunted black spruce (*Picea mariana*) trees replace the declining white spruce. Black spruce woodland permits more light to reach the ground, heating the soil and melting back the permafrost, creating an even wetter habitat. Consequently, the soil becomes shallow and soggy, and an open bog or muskeg replaces the spruce forest. Retrogressive succession ends in a meadow with dwarf shrubs, grasses, and sedges as the dominant growth forms.

27.2 VEGETATION TYPES

Plant communities blend into each other. Every species has its range limits, and at those limits each can be replaced by some other species that may have a similar growth form and niche. Therefore, the name of the plant community will change as the species range limits are crossed, but the architecture of the vegetation and the general habitat might stay the same. So long as the architecture and environment remain constant, several sequential communities all belong to a single vegetation type.

The term **vegetation** refers to the dominant growth form, not to the dominant species. Each vegetation type has a two-part name that describes the dominant growth form and the habitat; the name does not include any information about species. Thus, we have upland conifer forest, tropical rain forest, alpine tundra, eastern deciduous forest, tidal marsh, and desert scrub (scrub = shrub) vegetation types, among many others. The Appalachian red spruce–Fraser fir community fits within a more widespread vegetation type called subalpine conifer forest, which includes spruce–fir communities from the Rocky Mountains and the Cascade Range, as well as pine–hemlock communities from the Sierra Nevada.

There are, then, many fewer vegetation types than communities. North America has thousands of communities but

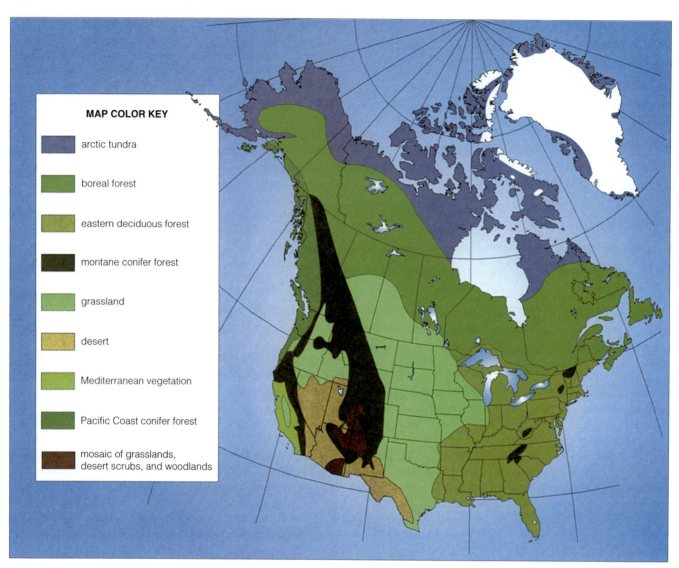

Figure 27.8 Major vegetation types of North America. (Redrawn from Barbour, M.G., Christensen, N. Vegetation of North America. In: *Flora of North America*, *Vol. 1*. New York: Oxford University Press, with permission.)

MAP COLOR KEY

- arctic tundra
- boreal forest
- eastern deciduous forest
- montane conifer forest
- grassland
- desert
- Mediterranean vegetation
- Pacific Coast conifer forest
- mosaic of grasslands, desert scrubs, and woodlands

fewer than a dozen major vegetation types **(Fig. 27.8).** Most of the world's major vegetation types are represented in North America, except for extremely arid deserts and several tropical grassland, savanna, and forest types. This chapter describes nine of North America's major vegetation types, beginning in the north and moving clockwise through the continent. The vegetation, of course, is simply one component of the regional ecosystem, but because of limited space, our descriptions will not dwell on such other ecosystem components as animals, microbes, and certain environmental factors. These large, regional, climatically controlled ecosystems often are called **biomes.**

Tundra Vegetation Occurs Beyond Timberline

As one travels north toward the pole, trees gradually become scattered, stunted, and less common. They finally disappear completely in a zone called timberline. At the same time, low

shrubs, perennial herbs, grasses, and sedges become dominant. Most of the biomass is below the soil surface and the root-to-shoot ratio is high (about 4). Shrubs are dwarfed, gaining normal height only in the protection of boulders or small hills. The winter wind, carrying drying air and ice, acts like a sandblast, pruning back stems wherever they project above the top of snow or beyond a protecting object. Perennials produce many large flowers, but the most successful reproduction is by rhizomes. Annuals are rare. This is **tundra,** a word derived from Finnish or Lapp, meaning a "marshy hill." Vegetation ecologists, however, use the word to describe vegetation **(Fig. 27.9a),** not topography. This vegetation type covers 19% of North America's land area.

Tundra is one-layered vegetation with nearly 100% cover at its southern extent, but decreasing to 5% to 50% cover at its northern limits. Similar vegetation occurs at lower latitudes in mountains, at elevations above timberline **(Fig. 27.9b).** The climates of the two tundra regions—arctic and alpine—have

James McGraw

Michael G. Barbour

a

b

Figure 27.9 Tundra vegetation. (**a**) Arctic tundra at high latitudes in Alaska, north of timberline. The antlers are from a caribou. (**b**) Alpine tundra at high elevation above timberline at 3,500 m in southern Colorado.

both similarities and differences. In both regions the warmest month has an average daily mean temperature of 10°C or less. The growing season is short, only 2 to 3 months of the year having average daily temperatures above freezing. During the growing season, the top 30 cm of soil thaws, and roots may freely penetrate it; below this depth, soil and water are frozen year-round in an impermeable, meters-thick permafrost layer. Annual precipitation is less than 25 cm; thus, winter snowpacks are shallow and only deep enough to cover and insulate the low-growing vegetation. Few plants have evolved to tolerate this environment. Only 700 species in North America have ranges that extend into the arctic tundra.

The **alpine tundra** differs from the **arctic tundra** in that summertime solar radiation and temperatures near the soil surface are greater; also, **thermoperiods** have a broader range than in arctic tundra. An alpine plant could experience leaf temperatures of 30°C at midday and −5°C at night, for a 35°C thermoperiod. During the same 24-hour period, an arctic plant would experience a weak sun that never sets and temperature extremes of only 15°C maximum and 5°C minimum, for only a 10°C thermoperiod. Alpine tundra plants share many of the same environmental stresses felt by desert shrubs thousands of meters lower in elevation: strong winds, high solar radiation, short growing season, low soil moisture, and widely fluctuating temperatures.

Boreal Forest Is the Taiga of North America

South of the arctic tundra lies a broad belt of low-elevation conifer forest, the **boreal forest.** It covers 28% of North America. At the forest's northern limit, trees meet and mingle with the tundra, and they can be shrublike and slow growing. The critical environmental factor determining the location of timberline is the amount of heat received during the growing season. The growing season in the boreal forest is 3 to 4 months in duration, and temperatures can be much warmer than in the tundra. Annual precipitation is 30 to 90 cm, most of it falling in summer, so water is not limiting. Winter temperatures, however, can be even lower than they are in the tundra because oceans—with their moderating effect on temperature—are more distant. Soils are relatively young, acidic, and leached of nutrients. Some soils beneath the northern portion of the boreal forest have a permafrost layer.

This vegetation has a two-layered architecture. Trees are slender, short (15–20 m in height), and relatively short-lived (less than 300 years), but they are densely packed (Fig. 27.10). The Eurasian term for this vegetation, **taiga,** means "dense forest." Carpeting the ground beneath the overstory canopy is a continuous layer of bryophytes, seedless vascular plants, and herbaceous perennial angiosperms.

Michael G. Barbour

Figure 27.10 Boreal forest in northern Canada, adjacent to a clear-cut.

Disturbance by storms and wildfires recurs every 200 or more years, affecting large areas and setting secondary succession in motion. Black spruce (*Picea mariana*), one of the most widespread trees in the boreal forest, is a **closed-cone conifer** well-adapted to fire. As an intense fire sweeps through a black spruce forest, killing all aboveground vegetation, the millions of sealed cones are melted open, and for days after the burn a quiet rain of seeds falls on an ash-rich seed bed. A new black spruce forest will replace the old one.

Eastern Deciduous Forest Has a Complex Physiognomy

Boreal softwoods give way, at about 45 degrees N latitude, to deciduous hardwoods as the growing season increases to 6 months and annual precipitation climbs above 100 cm. Winters are cold, but they are neither as intensely cold nor as long as in the boreal forest. Soils are richer in nutrients and less acidic. This is the **eastern deciduous forest,** a vegetation type dominated by a diversity of broadleaf, winter-deciduous tree species **(Fig. 27.11a).** Along the southern limits of the forest, at about 30 degrees N latitude, winters are mild enough for evergreen broadleaf trees to mix, as a minor element, with deciduous trees. Precipitation declines to the west, where arid grassland exists as a boundary to the forest. Sometimes the transition from forest to grassland is abrupt, and sometimes it is many kilometers wide. This important ecotone between woody and herbaceous vegetation meanders north to south along the line of 95 degrees W longitude.

The eastern deciduous forest, which covers 11% of North America, is a complex mix of communities. Little of this forest has escaped human impact. Native Americans influenced its structure for thousands of years by burning the understory. They maintained an open forest to encourage populations of birds and grazing animals used for food. Before the forest changes brought about by European colonists, forest buffalo and grouse grazed as far east as New York State. In 200 to 300 years, Euro-Americans modified the forest further by clearing, selectively cutting, grazing, or accidentally introducing foreign pathogens and weeds. The modern forest no longer resembles the descriptions by early explorers. The forest overstory of Ohio and Pennsylvania, for example, was once dominated by 400-year-old oaks (*Quercus*), sugar maples (*Acer saccharum*), and chestnuts (*Castanea*) 1 to 2 m in diameter at breast height, with straight trunks rising 25 m before the first side branch. Black walnuts (*Juglans*), shagbark hickories (*Carya*), and cottonwoods (*Populus*) grew in flood plains near rivers and were big enough to be made into dugout canoes 20 m long and more than 1 m across. Currently, only a few forest reserves contain remnants of this ancient forest.

Seasonality is a striking feature of eastern deciduous forests. Overstory leaves turn brilliant yellow, red, and orange in autumn. After winter dormancy, the first green of spring comes from a diversity of herbaceous perennials on the forest floor **(Fig. 27.11b).** These species take advantage of a narrow time window when light and temperature permit rapid growth. As the overstory canopy completes its expansion of new leaves, these spring-flowering herbs shed seeds

a

b

Figure 27.11 The eastern deciduous forest. (**a**) The leafy overstory canopy changes color dramatically in autumn, just before leaves drop. (**b**) Active spring growth of perennial herbs occurs before the new overstory completely leafs out.

and enter dormancy. Two intermediate canopy layers—between the ground herbs and the overstory trees—consist of scattered shrubs (many in the heath family) and small trees such as dogwood (*Cornus*). In addition, many vines (such as Virginia creeper, *Parthenocissus quinquefolia*), grow up through all of the tree and shrub canopies.

Grasslands Cover One-Fifth of North America

Across an east–west expanse of perhaps 100 km, we pass along a gradient of increasing aridity and a series of vegetation types. At the western edge of the deciduous forest, the overstory tree canopy becomes discontinuous and herbaceous plants begin to form a continuous ground cover. Forest becomes woodland and then savanna, steppe, and finally grassland. **Woodland** is grassland with overtopping trees whose canopies cover 30% to 60% of the ground; **savanna** is grassland with overtopping trees that are regularly present but whose canopies cover less than 30% of the ground; **steppe** is grassland interspersed with shrubs. **Grassland** (synonym, **prairie**), finally, is vegetation dominated by herbaceous plants growing in a climate too dry for trees **(Fig. 27.12).** Trees may be present, but they are restricted to special, localized topography, such as along waterways or on rocky ridgelines with thin soil. Perennial and annual grasses dominate the biomass, but broad-leaved dicot herbs (forbs) dominate in terms of numbers of species.

Some plant geographers have concluded that there is no such thing as a grassland climate. Their argument is that grasslands replace woodlands for reasons other than gradients in climate. A certain annual precipitation—for example, 50 cm—could support grassland, savanna, woodland, or forest. Grassland is favored in those places where variation in rainfall from year to year is high, creating years of drought that alternate with years of above-average precipitation. Grassland also is favored in fire-type climates, where the probability of wildfire revisiting the same hectare of land every 1 to 3 years is high. In addition, grassland is favored where growing temperatures are high enough and air humidity low enough to promote transpiration. In forests, the ratio of incoming precipitation to outgoing transpiration is greater than 1; but in grasslands, the ratio decreases to 1 or just below. Finally, grassland is favored where soils are deep and texture is loamy to clayey.

Grasslands once covered 21% of North America. The largest area of grassland occupies the center of North America, from Manitoba to Texas on the east and from Iowa to the Rocky Mountains on the west. It consists of tall-grass, mixed-grass, and short-grass prairies. Other major grasslands include: (1) along the edge of warm deserts in Texas, New Mexico, and Arizona; (2) scattered through the intermountain Great Basin (with a finger extending into the Palouse area of southeastern Washington and outliers in the Willamette Valley of Oregon and north coastal California); and (3) within the Central Valley of California.

All of our grasslands have been significantly modified by Euro-Americans in the last 200 years. Most of the central grasslands have been cleared and plowed, converted into farmland. The desert grasslands have been overgrazed and fire has been suppressed; as a result, desert shrubs have invaded. The intermountain, Palouse, and California grasslands have been overgrazed and invaded by aggressive annuals from Eurasia (for example, cheatgrass, *Bromus tectorum*), as well as having had many hectares converted to farmland, pasture, and urban sprawl.

Desert Scrub Is Dominated by Shrubs

Scrub is any vegetation dominated by shrubs. It occurs where either precipitation or the water storage capacity of the soil is too low to support grassland. Thorn scrub, chaparral, and desert scrub are examples. **Desert scrub** occurs where annual rainfall is less than 25 cm and a pronounced dry season exists every year. The annual precipitation-to-transpiration ratio decreases to 0.1. Another feature of desert climate is the high variation in rainfall from year to year, so that the concept of an average rainfall is meaningless. Desert vegetation covers only 5% of North America.

Deserts generally are warm to hot during the summer, but they may be quite cold in winter. The high-elevation intermountain desert, for example, regularly receives snow and hard frost every winter; this stress reduces species and growth form diversity **(Fig. 27.13).** Low-elevation and more southerly Sonoran desert scrub has almost no frost; it has a greater diversity of species and growth forms. The two other North American deserts—the Mojave of southern California and Nevada, and the Chihuahua of Texas, New Mexico, and Mexico—are intermediate in elevation and temperature regimens.

Associated with shrubs in the warm deserts may be succulent cacti, green-stemmed trees, subshrubs, herbaceous perennials, and ephemerals. Total ground cover may be 50% at a maximum and 5% at a minimum. Other parts of the world have more extremely arid climates than North America, and in those places plant cover may be less than 1%.

Michael G. Barbour

Figure 27.12 Grassland on rolling hills in North Dakota.

Figure 27.13 The Great Basin desert vegetation has a uniform one-layered scrub architecture dominated by sagebrush shrubs (*Artemisia tridentata*). Warm desert vegetation is more diverse and includes succulents, trees, and drought-deciduous shrubs.

All plants in desert scrub must be adapted to survive extended droughts. There are five basic techniques for drought tolerance or avoidance. One is the **phreatophyte** syndrome, a suite of traits shared by woody plants that have deep roots in permanent contact with groundwater. These plants have green stems and leaves that are winter deciduous. Their leaves are well supplied with water during the hot summer, and therefore they are under stress (of cold temperature) only in wintertime, which is when the leaves are shed.

Other shrubs with shallower roots retain their leaves only during the wet season and drop them during dry seasons; this is the **drought-deciduous** syndrome. Drought-deciduous leaves are thin and energetically inexpensive; they can be cast off and remade several times a year (Fig. 27.14).

A few shrubs, the true xerophytes, have evergreen leaves. They are able to tolerate the desert dry season because their metabolism proceeds at a slow rate the entire year. Under prolonged drought, some of the leaves may be shed; but if all fall, the plant dies. Their leaves are typically small and have many anatomical features that retard transpiration.

Succulents, such as cacti, store water in the vacuoles of large cells. They typically exhibit **crassulacean acid metabolism,** which features stomata that open by night and close by day, thereby lowering transpiration. Their leaves and bodies also minimize the amount of surface for a given volume or mass, another feature that reduces transpiration. Cacti have shallow root systems, capable of absorbing moisture from even light rains. Succulent plants appear to avoid drought, rather than tolerate it.

Ephemerals exhibit a fifth syndrome: They live for 6 weeks to 6 months and complete their life cycle during the wettest, least stressful part of the year. They avoid the drought season by passing through it in a state of dormancy, as seeds. Although woody plants dominate the desert biomass, ephemerals contribute the most species.

Chaparral and Woodland Are Mediterranean Vegetation Types

Mediterranean climates and vegetation are found in five locations throughout the world: (1) the Mediterranean rim of southern Europe, the Middle East, and northern Africa; (2) the Cape region of South Africa; (3) southern and southwestern Australia; (4) central Chile; and (5) California. All these regions lie between 40 and 32 degrees N or S latitude. They occupy western or southwestern edges of continents, receive 27 to 90 cm of annual precipitation (mainly winter rainfall), have minimal frost, and experience episodic wildfire. Mediterranean climates are fire-type climates, basically with hot, dry summers and cool, wet winters.

Vegetation is remarkably similar in these five areas, even though different families of plants predominate in each. Mediterranean vegetation ranges from forest (in the wettest locations) to woodland to scrub (in the driest locations). All these vegetation types are dominated by broad-leaved, evergreen, woody flowering plants, with a relatively high percentage of canopy cover.

Figure 27.14 Coachwhip (*Fouquieria splendens*), a drought-deciduous plant that is able to shed and remake several crops of leaves each year. (**a**) Plants with leaves, soon after heavy rains. (**b**) Same plants, leafless during a later dry period.

a

b

Mediterranean scrub is called **chaparral,** derived from the Spanish as a term that generally means "low-growing" woody vegetation (in contrast to *forestal,* tall vegetation dominated by trees). Chaparral is dense, one-layered, and about 1 to 3 m in height; it is composed of rigidly branched shrubs with small, hard leaves and an extensive root system (Fig. 27.15a). In contrast to open desert scrub, chaparral has 100% cover and an LAI twice that of desert scrub. Herbs are uncommon.

Wildfire recurs every 20 to 50 years, consuming all the aboveground vegetation and producing very high temperatures. Chaparral shrubs respond to fire in several ways. Some are capable of sprouting from the root crown, which is buried beneath the soil surface, and thus insulated from high temperatures during the fire. Others have hard-coated seeds that lie dormant until cracked by moderately high temperatures. Some have seeds that are stimulated to germinate by some active ingredient in smoke. By both resprouts and seedlings, the chaparral community recovers its preburn cover and species composition within half a dozen years.

Figure 27.15 Mediterranean vegetation types. (**a**) Chaparral. (**b**) Foothill woodland.

Chaparral grows on steep slopes with coarse, shallow soils, at elevations below 1,000 m. Nearby, within the same macroenvironment, are foothill woodlands, dominated by oak (*Quercus*) trees and underlain by grasses and forbs (Fig. 27.15b). Tree canopy cover is 30% to 60%. Woodland and chaparral vegetation types are spatially separated according to microenvironmental factors, not macroenvironmental ones. The woodland occupies gentler slopes (with deeper, less coarse soil), as well as moist north-facing or east-facing slopes (which burn less frequently than the drier chaparral sites).

Before the arrival of Euro-Americans, the woodland understory consisted mainly of perennial bunchgrasses. Because of the introduction of aggressive, weedy annuals in the last 150 years, the understory today consists mainly of annual grasses. These woodlands once supported the greatest population densities of Native Americans in all of North America. Oak acorns were the major resource that supported them. Acorns were easily collected every fall, leached of tannin, ground into flour, stored, and used the entire year, much as grains are used in other cultures.

Mediterranean vegetation covers only 1% of North America, but many people are familiar with it because California landscapes often are in the news—witness the ever-recurring catastrophes of wildfires, floods, landslides, earthquakes, and droughts—and because they serve as backdrops to many popular films and television programs.

Pacific Coast Conifer Forests Are the Most Massive in the World

The conifer forests of the Pacific coast are the most luxuriant, most productive, most massive vegetation type in the world. They are dominated by a rich diversity of big, long-living tree species underlain by equally rich canopies of shrubs, herbs, bryophytes, and epiphytes (Fig. 27.16). The Pacific Coast conifer forests are situated in a low elevation strip of coastline that extends from Cook Inlet, Alaska, south to Monterey, California. The climate is mild, buffered by the nearby ocean and summer fog banks; therefore, thermoperiods are only 6 to 10°C. Hard frosts are uncommon. Annual precipitation is very high, 80 to 300 cm, most of which falls in wintertime. Because of the climate, this type of forest often is called a **temperate rain forest.**

Dominant tree species include coast redwood (*Sequoia sempervirens*), Douglas fir (*Pseudotsuga menziesii*), lowland white fir (*Abies grandis*), Sitka spruce (*Picea sitchensis*), western hemlock (*Tsuga heterophylla*), and western red cedar (*Thuja plicata*). These trees all commonly live for 400 to 1,200 years and attain heights of 100 m. Annual productivity is as high as 25 metric tons per hectare, and standing biomass is typically 85 metric tons per hectare; in one particular location, it is as high as 230 metric tons per hectare. Tropical rain forest vegetation has a similar productivity, but its biomass is significantly smaller.

Although this vegetation type covers only 3% of North America, its timber volume and value have been highly sig-

Michael G. Barbour

a

Michael G. Barbour

b

Figure 27.16 Pacific Coast conifer forest (also called a temperate rain forest). (**a**) Cathedral Grove Provincial Park, Vancouver Island, BC, Canada. The dominant trees are 2 m in diameter breast height and more than 65 m in height. Note person in orange jacket near the bottom of the photograph. (**b**) Because of the high rainfall and humidity, tree trunks and branches in this forest support a rich assortment of epiphytic mosses and lichens.

nificant to both local human communities and distant corporations during the past century. Logging has had a heavy impact on the region. Conservationists, using the endangered northern spotted owl as a symbol of logging threats to biotic diversity, have managed to decrease the intensity of logging in the United States in recent years. Canada has not yet reduced its harvest of coastal forests in British Columbia, however, and clear-cuts in British Columbia are said to be among the few human artifacts visible from orbiting spacecraft.

Upland Conifer Forests Have a Wide Distribution

Coniferous forests clothe the slopes of the higher Appalachian peaks, the Rocky Mountains, the Cascade-Sierra Nevada axis, the Coast Ranges of Washington and Oregon, and the Transverse and Peninsular Ranges of southern California. These upland forests are called **montane conifer forests.** They range from 65 to 19 degrees N latitude (well south, into the Sierra Madre of Mexico), and they cover 7% of

Greg Vaughn/Tom Stack & Associates

Figure 27.17 Mid-montane conifer forest, Yosemite National Park, CA, about 2,000 m elevation.

North America. Annual precipitation increases from 60 cm at lowest elevations to more than 200 cm at highest elevations. Precipitation in winter is in the form of snow, and deep snowpacks can accumulate. Summers are warm and relatively dry. Montane conifer forests of the West are in a fire-type climate.

The structure, diversity, and productivity of montane conifer forests are intermediate between the two other conifer-dominated vegetation types described earlier, the boreal forest and the Pacific Coast forest. Many of our most popular national parks, the jewels of the park system, are located in this vegetation type: Glacier, Great Basin, Great Smoky Mountains (uppermost elevations), North Cascades, Rocky Mountains, Yellowstone, and Yosemite National Parks (Fig. 27.17).

Zonation of forest communities along elevation gradients is a common phenomenon. Lower montane (low-elevation) forests tend to be rather open savannas or woodlands, intermingled with species from adjacent grasslands, Mediterranean woodlands and chaparral, or deserts. Frequent wildfires seem to be essential to the maintenance of some of these communities, and they have been significantly degraded by overgrazing and changes in fire frequency over the last 150 years. Pinyon pines (*Pinus monophylla* and others), ponderosa pine (*Pinus ponderosa*), and junipers (*Juniperus scopulorum* and others) are common woodland trees.

Mid-montane (intermediate-elevation) forests are typically rich in overstory species, such as Douglas fir (*Pseudotsuga menziesii*), white fir (*Abies concolor*), ponderosa pine (*P. ponderosa*), and many regionally limited tree species. A variety of shrubs in the heath and rose families also are common, as are seasonally present herbaceous perennials.

These are complex, four-layered forests that often require wildfire to maintain their structure. They need surface fires that burn relatively cool, consuming only litter, shrubs, and young trees. When fire is kept out by fire suppression management, these forests change their physiog-

Figure 27.18 Forest fires can be surface fires (*left*) or crown fires (*right*). Recurrent natural wildfires keep many montane forests open, so fires remain on the ground and consume only litter, shrubs, and young trees. Fire temperatures are moderate, and mature trees usually survive. Fire suppression, however, allows more trees of all sizes to coexist; thus, a surface fire can ladder up to the crowns of the tallest trees, becoming hotter, and killing all above-ground vegetation. (Drawing by M. Yuval from Barbour, M., et al. 1993. *California's Changing Landscapes*. Sacramento: California Native Plant Society, with permission.)

ground fire   crown fire

nomy, becoming denser. Then, when a surface fire does start (and the presence of a fire-type climate assures us that it will start), the surface fire will "ladder" up the younger trees and become a raging, destructive crown fire (Fig. 27.18).

The mid-montane elevation belt contains many other vegetation types besides forest. Perhaps half the area is interrupted by meadow or scrub vegetation. **Meadow** is grassland that occurs within a climate capable of supporting forest vegetation. Forbs often dominate meadows, grasses being reduced to associate status. Forest is absent here because of local topography or soil that creates seasonally wet conditions. **Montane scrub** occurs on rocky ridges, on south-facing slopes, or as a temporary successional stage after wildfire. In the latter case, forests will slowly reclaim the site by growing through the scrub and shading it out.

Upper montane and **subalpine** (highest-elevation) forests are densest and simplest and experience the deepest snow packs. Fir (*Abies*), hemlock (*Tsuga*), pine (*Pinus*), and spruce (*Picea*) are the dominant genera. This is the elevation zone where the oldest individual plants in the world exist, the bristlecone pines (*Pinus longaeva*; Fig. 27.19).

Above the subalpine forest, the amount of heat during the growing season is too little to support the growth of trees, and we enter an alpine, high-elevation type of tundra.

Michael G. Barbour

Figure 27.19 Bristlecone pine (*Pinus longaeva*) in the subalpine zone of the White Mountains of California. This species occurs on high peaks throughout the Great Basin, and individuals can reach ages of 5,000 years.

Wetlands and Aquatic Ecosystems Are Productive

SALINE WETLANDS Wetlands are terrestrial sites where the upper soil is saturated by saline or freshwater for at least a few weeks of the year. **Tidal wetlands,** or **salt marshes,** are coastal meadows subject to periodic flooding by the sea (Fig. 27.20). They occupy nearly level shores that receive only low-energy waves, such as are found along estuaries or tidal flats behind barrier islands or sandy peninsulas. Environmental stresses include flooding—bringing mechanical disturbance and anaerobic conditions—and salinity. As the ground slopes up toward land, however, these two stresses decline in frequency and intensity.

The vegetation is usually a single, low-growing, nearly closed layer of perennial herbs. The soil beneath is crowded

Michael G. Barbour

Figure 27.20 Tidal wetland (salt marsh) vegetation.

with rhizomes and roots. The flora is rather simple, with only a handful of species coexisting in a given local wetland. Tidal wetlands along the entire Atlantic and Gulf coasts of the United States support less than 350 vascular plant species, and those along the entire Pacific coast, from Point Barrow to the tip of the Baja California peninsula, contain less than 80 species. Apparently, not many species have the genetic capacity to tolerate tidal marsh stresses. A few have enormous ranges that extend through nearly all coastlines of North America: arrow-grass (*Triglochin maritimum*), cord-grass (*Spartina alterniflora*), pickleweed (*Salicornia virginica*), saltbush (*Atriplex patula*), and salt grass (*Distichlis spicata*).

Some of the traits these successful species exhibit are succulence, asexual reproduction by rhizomes, and **aerenchymna** tissue in stems and roots (which channels oxygen down into the anaerobic soil).

Enormous losses of acreage in tidal wetlands have occurred during the twentieth century. Tidal wetlands have been diked, drained, filled, and converted to farmland, ports, and cities. We have discovered their ecological value rather late. One ecological function these wetlands serve is as a biological filter for runoffs from the land, which contain pollutants and excessive nutrients. The marsh and its soil act as a sieve, cleaning the water before passing it on to the ocean. Another ecological function is service as a nursery for the young of many aquatic animals, including commercially

valuable fish. Although salt marsh vegetation has a low profile, its annual productivity is as great as that of a tropical rain forest. The plants are herbaceous, however; therefore, this tremendous amount of new plant biomass does not accumulate as woody tissue. Instead, it is shed each year into the water, where it fuels an extensive food chain through microbes, algae, plankton, invertebrates, fish, and humans.

Along exposed coasts that receive the full brunt of wave action, a **rocky intertidal** ecosystem replaces the salt marsh. Flowering plants are few in the rocky intertidal zone, but they include surf grass (*Phyllospadix*), a relative of other sea grasses more commonly found in quieter harbor waters. Many seaweeds and a few kelps also are attached to rocks in this habitat, each at its own particular depth within the intertidal zone. The different depths at which they are distributed probably reflect different degrees of adaptation to the stresses of exposure.

Below the intertidal zone is the **neritic zone.** This aquatic ecosystem is a rocky shelf always covered by water but shallow enough to admit adequate sunlight for attached algae to grow along the bottom. Many kelps grow in the neritic zone, especially in cool-temperate oceans. As in the intertidal zone, different species of algae dominate at different depths **(Fig. 27.21).** The pattern of distribution probably reflects differing tolerances among algal species to low light. Both the

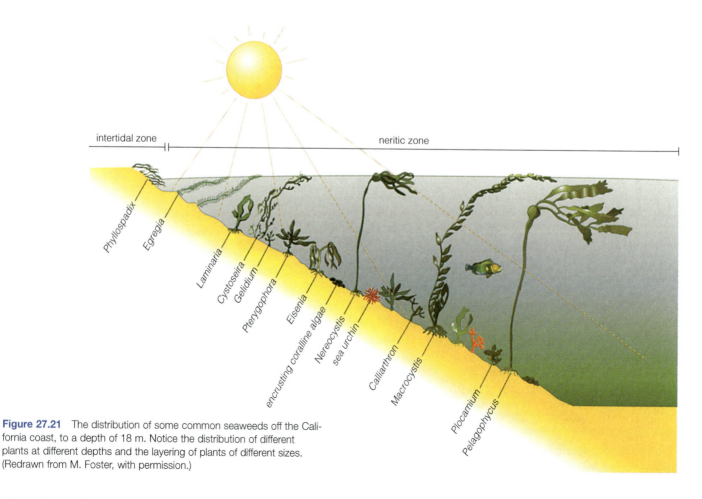

Figure 27.21 The distribution of some common seaweeds off the California coast, to a depth of 18 m. Notice the distribution of different plants at different depths and the layering of plants of different sizes. (Redrawn from M. Foster, with permission.)

quality and quantity of light change with passage through water. The compensation depth, where positive growth is no longer possible, is usually shallower than 170 m.

In deep oceanic bodies of water, the habitat between the surface and the compensation depth is dominated by phytoplankton. One exception is the Sargasso Sea off Bermuda, where large, floating masses of *Sargassum* seaweed are present. This seaweed may play an important role in the survival of young sea turtles, giving them a place to feed and hide from predators until they become large adults.

FRESHWATER WETLANDS Freshwater wetlands are found along the shorelines of lakes, rivers, sinks, seeps, and springs. Trees, shrubs, and herbs that occupy these habitats must be tolerant of occasional flooding, and they must be resilient to the physical disturbance of floodwaters. They often have fast growth rates, produce abundant wind-distributed seed, and are capable of vegetative reproduction. This fringe of vegetation is ecologically important as a filter of eroded soil and nutrients that otherwise would enter adjacent aquatic ecosystems and degrade them. The filtering function is exceptionally important in the modern landscape because intensive agricultural practices add fertilizers and pesticides to the land.

At shallow margins of freshwater aquatic ecosystems (Fig. 27.22) are flowering plants such as sedge (*Scirpus, Carex, Cyperus*), cattail (*Typha*), pondweed (*Potamogeton*), and water lily (*Nymphaea*). These are called emergent aquatic plants because, although rooted under water, some part of the plant body extends above the water surface. Associated with them are algae (diatoms, green algae, golden algae) that grow attached to the muddy surface or epiphytically to other plants.

In somewhat deeper water, submerged or floating flowering plants are common, associated with larger algae such as stonewort (*Chara*). If lakes are deep enough, their water may stagnate into zones. An upper zone, the **epilimnion,** is relatively warm, and sunlight is intense enough to support large phytoplankton populations. Oxygen levels in the epilimnion are high. Below the epilimnion is a narrow transition zone where temperature declines rapidly with depth (on the order of 4°C per meter). This is the **thermocline,** and it serves as a barrier to any mixing between the upper epilimnion and the lower **hypolimnion,** which extends to the bottom of the lake. Phytoplankton may occur in the hypolimnion, but only at low densities.

This chapter concludes with some thoughts about how this diversity of vegetation may be maintained into the future.

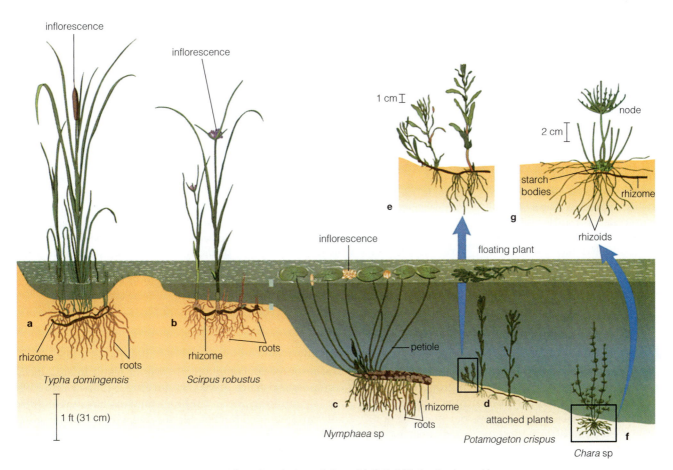

Figure 27.22 Freshwater ecosystems support flowering plants and algae. (**a**) Cattail (*Typha domingensis*). (**b**) Sedge (*Scirpus robustus*), more tolerant of flooding than cattail. (**c**) Water lily (*Nymphaea*), with floating leaves and flowers. (**d,e**) Submerged and floating segments of pondweed (*Potamogeton crispus*). (**f,g**) The stonewort *Chara*, a complex alga differentiated into rootlike rhizoids and rhizome-like and stemlike regions.

PLANTS, PEOPLE, AND THE ENVIRONMENT:

Nature in Flux or Nature in Balance?

A **paradigm** is any widely accepted viewpoint. In science, a paradigm is made up of a family of related theories that seem to explain some behavior in nature. Until recently, one classical paradigm in ecology was the **equilibrium paradigm,** which defined nature as dominated by stable, long-undisturbed climax ecosystems. An old-growth montane forest, 1 acre of central prairie, or a strip of undisturbed salt marsh was each seen to be in equilibrium, or in balance, with its environment. This paradigm and its metaphor, "the balance of nature," suggest that any climax unit in a landscape will maintain itself, especially if isolated from direct human interference.

An example of the application of this paradigm is Mettler's Woods, the last remaining uncut upland forest in central New Jersey. The forest, which had been protected since colonial days and eventually became a part of Rutgers University, is now known as the Hutcheson Memorial Forest Center (HMFC). The forest was considered to be climax, and thus self-perpetuating. No human disturbance was to be permitted. Visitors were only allowed entrance when accompanied by a qualified guide and restricted to a single trail.

The last 30 years of research have tended to show that this paradigm is incorrect. In the case of Mettler's Woods, merely protecting the forest from humans has not been sufficient to maintain it. The

old oaks (*Quercus* sp.), the average age of which is 250 years, are senescing, falling, and leaving gaps in the canopy. The overstory is quite open, the mid-layer dogwoods (*Cornus florida*) have succumbed to a regional epidemic of the fungal disease anthracnose, and many species of shade-tolerant wildflowers have become rare. After severe defoliation of the oaks caused by gypsy moth caterpillars in the early 1980s, several sun-loving shrubs and herbs invaded the forest understory and today they continue to maintain large populations. Using insecticides to manage the gypsy moth population was considered to be inappropriate in a nature reserve, and so was not done. Although the oaks do produce large seed crops, there is an almost complete failure of seedling oaks to survive into the sapling category.

It is now understood that surface fires had been a regular part of the history of the forest before fire protection policies were adopted in 1711. Fire scars preserved in the trunks of the oldest oaks document that light surface fires used to occur every 10 years. Perhaps these fires cleared away competing herbs and shrubs, thereby permitting seedlings to survive. Thus, the paradigm—and the conservation strategy derived from it—have failed to preserve the forest.

Failures of the equilibrium paradigm became so numerous and well-known by 1980 that ecologists developed a new par-

adigm, the **nonequilibrium paradigm.** First, it accepts natural systems as being open, meaning that they must be put into the context of their surroundings. Second, it recognizes that natural systems are subject to physical disruption from a wide range of natural forces and events such as fire, drought, windstorm, earth movement, volcanism, herbivore outbreak, disease epidemics, and so on. Third, the new paradigm permits the inclusion of humans in the scope of conservation. Once the importance of natural disturbances is recognized, it is a short logical step to include humans as just another agent of change. A metaphor for this new viewpoint is "nature in flux."

The simplest translation of the contemporary paradigm into conservation biology would state that conservation cannot always be passive, but must often involve active management. If natural disturbance had been a characteristic that preserved a particular ecosystem, then managed disturbances that mimic natural ones must be created. Managed disturbances might include setting a prescribed fire, thinning and mulching young trees, biologically controlling certain insects or pathogens, weeding out invasive plants, or permitting seasonal flooding. The size of the reserve must be large enough to permit such episodic disturbances to run their course within at least a portion of the reserve.

27.3 CONSERVATION BIOLOGY

Each year, the extent of natural vegetation grows smaller. Vast areas of some vegetation types, such as tundra and boreal forest, still remain as wilderness, but many of our deciduous forests, grasslands, deserts, woodlands, montane conifer forests, and tidal wetlands are endangered. They are threatened by timber harvest, land conversion, water diversion, the spread of weeds and pathogens, and even by the recreational impact of too many people in too small a place.

Only a generation ago, conservation meant a rate of natural resource consumption that would result in sustained, continued existence of that resource far into the future. We are beginning to understand that humans have great difficulty in staying within boundaries that would sustain a particular resource. Our species has had a sad, consistent history of overshooting these balance points. Our farming, fishing, and forestry activities have consistently led to overexploitation and decline of natural resources. Consequently, **conservation** now usually means the restricted use, nonuse, or preservation of some natural resources.

The location of a reserve must be chosen in appropriate relation to the land use practices, politics, and social attitudes within adjoining parcels. All pieces of an ecosystem are not equally capable of preservation, and active management cannot be practiced safely everywhere. Preserves must be large enough to permit episodic wildfires to burn some patches and to permit the movement of displaced animals to adjacent, unburned areas. In addition, adjacent parcels must be compatible with a "let-burn" policy for wildfires. We need to demarcate entire landscapes as parks, preferably with surrounding buffers and with corridors that connect parks so that biota can move more widely. The complexities of compromise and management required for preserve management today are very challenging.

A modern approach to regional preserve planning is being attempted in southern California, where decades of residential development have displaced and fragmented a major regional vegetation type called southern coastal scrub (**Fig. 1**). A number of rare plant and animal species are at risk for extinction if development continues on the scale of the past several decades. As an alternative to continued future habitat and biotic diversity loss, a consortium of federal, state, and local government agency planners, developers, environmentalists, and citizens of several counties joined together. Their objective is to locate the best and largest remnants of the scrub, then to suggest optimal boundary shapes, to design corridors between preserves, to designate appropriate human activities within buffer lands next to the corridors and preserves, and to identify developable areas and transportation routes within the context of this mosaic of regional open space. Will this approach be successful? At the present time, we still do not know. But

Michael G. Barbour

Figure 1

conservationists are hopeful that it will work because the old method of conservation has not been successful, when the ecosystems of focus have been close to urban centers.

Portions of this essay were condensed from Pickett, S.T.A., Parker, V.T., Fiedler, P.L. 1992. The new paradigm in ecology: Implications for conservation biology above the species level. In: Fiedler, P.L., Jain, S.K., eds. *Conservation biology.* New York: Chapman and Hall, pp. 65–88.

Conservation biology is a relatively new science that studies the impact of human societies on the nonhuman landscape. Conservation biologists ask the question: Can a growth-oriented, technological culture coexist with its surrounding natural systems? We know that nontechnological cultures, such as those of Native Americans, did coexist with their surrounding natural systems for thousands of years, but we recognize that their demands on nature were much different from ours.

Conservation biologists are investigating ways to measure sustainability: How do we know when a plant or animal population is maintaining itself? How do we measure biotic diversity? How do we design parks so that the probability of extinction for any rare population is as low as possible (see "PLANTS, PEOPLE, AND THE ENVIRONMENT: Nature in Flux or Nature in Balance?" sidebar)? How do we restore degraded habitats and their plant and animal communities? How do our technological activities interweave with the biosphere in unexpected ways to magnify into global stresses (acid rain, ozone depletion, climate change, pollutants carried through food chains), and how might we best modify these technological activities to reduce the stress?

Conservation biology is an exciting field, but its contributions will be limited if our human population continues to increase. In the middle of the twentieth century we numbered 3 billion, and we reached 6 billion by the start of the twenty-first century. Sustainability is an unattainable myth in the face of such climbing population pressure on the earth. Without a doubt, the control of human population growth has been and remains the greatest challenge our species must solve.

Ecosystem Restoration

Some vegetation types and entire ecosystems cannot be conserved because they have been largely changed by recent human activities. Exotic weedy plants and animals have invaded and become widely established, reducing the abundance or even eliminating some native species. Domesticated livestock have caused severe surface erosion. Dams no longer permit the seasonal fluctuation in the volume of water that previously characterized a river. The program of fire suppression management has led to forest thickening, loss of species richness, and epidemic tree mortality during period droughts. Perhaps relictual, unchanged examples of a particular ecosystem exist in small, scattered, remote locations, but for all practical purposes, these ecosystems have been completely degraded.

In such cases, active management techniques have to be used to reverse the human-caused changes and to bring back the previous ecosystem. Such a reversal is called **ecosystem restoration,** and it is a rapidly developing area of research and practice. One advantage of restoring a habitat, vegetation type, or ecosystem is that the restored system tends to be self-maintaining, requiring less investment in terms of human attention. Another advantage of restoration is that **ecosystem services** increase; such valuable attributes of an ecosystem include soil stabilization, protection against catastrophic fire, the filtering of contaminants from agricultural runoff, the provision of maximum biodiversity, and the moderation of temperature.

Restoration techniques imitate natural successional or disturbance processes. For example, if a forest type has become too dense because of decades of fire suppression management, then a fraction of the trees are cut and removed (the stand is thinned), followed by the setting of a prescribed fire that imitates a natural (lightning-started) fire. The prescribed fire is started during weather conditions that keep the flame front short and moving slowly. Thereafter, at set intervals of years that imitate the natural fire return interval, prescribed fires continue to be set **(Fig. 27.23).** Grasslands can be similarly restored to their previous species richness and composition by purposely setting fires every several years, or instead by instituting livestock grazing practices or mowing that imitates the seasonality and intensity of native animal herds—herds now depleted or even extinct.

Restoration has been somewhat successful in wetlands, grasslands, and certain forest types. It has been much less successful in such stressful settings as deserts, subalpine meadows, on mine spoil, and in areas damaged by smog or acid deposition. Restoration is a young science, and it will require decades more experimentation and accumulated wisdom before we can confidently presume that a given degraded ecosystem can be restored. Until that time, the wiser course of action is to conserve what still remains.

a   b   c

Figure 27.23 Paired photographs taken before and after 64 years of fire-suppression management near Ebbetts Pass in the Sierra Nevada of California (**a**, in 1929; **b**, in 1993). Thinning, followed by prescribed fire, has restored a similar forest (**c**) back to a condition that resembles the natural vegetation.

KEY TERMS

aerenchymna

alpine tundra

arctic tundra

attributes

biome

boreal forest

chaparral

climax stage

closed-cone conifer

community

conservation

density

desert scrub

diversity

dominate

drought-deciduous

eastern deciduous forest

ecosystem restoration

ecosystem services

ecotone

ephemeral

epilimnion

equilibrium paradigm

frequency

freshwater wetland

grassland (prairie)

hypolimnion

leaf area index (LAI)

meadow

Mediterranean climate

montane conifer forest

montane scrub

neritic zone

nonequilibrium paradigm

nutrient cycle

permafrost

phreatophyte

physiognomy

pioneer stage

primary succession

productivity

progressive succession

retrogressive succession

richness

rocky intertidal

root-to-shoot biomass ratio

saline wetland

savanna

scrub

secondary succession

steppe

subalpine

succulents

taiga

temperate rain forest

thermocline

thermoperiod

tidal wetlands (salt marshes)

tundra

upper montane

vegetation

woodland

SUMMARY

1. A plant community consists of a cluster of associated species that repeats wherever a particular habitat repeats itself. Plant communities are named after their dominant or characteristic species.

2. Plant communities have a characteristic architecture, or physiognomy, which is a combination of the external appearance of the community, its vertical structure, and the growth forms of each canopy layer.

3. Other community attributes include percent cover, LAI, species richness, productivity, biomass, allocation of biomass (to roots, woody tissue, leaves, reproductive organs), rate of nutrient cycling, and relative stability.

4. Plant communities change over time in a process called succession. Technically, succession is cumulative, directional change in a homogeneous area over several years to several hundred years. Secondary succession takes place on already vegetated land that is disturbed; primary succession takes place on new land not previously occupied by plants. Primary succession is much slower than secondary succession.

5. Progressive succession leads to increasingly complex and massive communities, in which the cycles of energy and nutrient flow become tighter and more efficient and the microenvironment becomes less stressful and more buffered. Retrogressive succession exhibits the reverse trends.

6. Plant communities blend gradually into each other over space because every species has its own range limits. However, the architecture and habitat may remain constant. All plant communities that share the same architecture and habitat belong to the same vegetation type. Vegetation types are named after location and dominant growth form, not after the dominant species. North America has thousands of plant communities, but many less vegetation types.

7. The major vegetation types of North America differ profoundly in their productivity, biomass, physiognomy, habitat traits, and in the portion of North America that they occupy.

8. Two-thirds of North America is covered by boreal forest, grassland, and tundra. Tundra is herbaceous vegetation that occurs where the growing season is too cool and short for trees. Arctic tundra is at low elevations in far northern latitudes, whereas alpine tundra is at high elevations in mountain chains at more southern latitudes. Grassland vegetation lacks trees because of periodic droughts, a low precipitation-to-evaporation ratio, recurring wildfire, and a fine soil texture. Boreal forest (taiga) is dense, short, and relatively low in biotic diversity. It occurs in a wide band just south of arctic tundra, where the growing season is 3 to 4 months long and the difference between winter and summer temperatures is relatively large.

9. Other North American forest types include the eastern deciduous forest, dominated by winter-deciduous broadleaf trees; the Pacific Coast temperate rain forest, the most massive vegetation type in the world; and montane conifer forests, home to famous national parks in the Rocky Mountains, the Cascade Range, and the Sierra Nevada.

10. Nonforest vegetation types include Mediterranean scrub and woodlands, desert scrub, and wetlands. Mediterranean climate occurs in only five small parts of the world; in North America, it dominates California. Characteristic vegetation includes a dense scrub called chaparral and an open woodland of evergreen and drought-deciduous small trees.

11. Desert scrub plants are more dispersed than chaparral shrubs, and they grow in a more arid environment. All desert plants either tolerate or avoid drought, and they do so by being phreatophytes, drought deciduous, true xerophytes, succulents, or ephemerals.

12. Wetlands are very productive vegetation types, and they play important filtering roles as a buffer between terrestrial and aquatic ecosystems. Wetlands may be saline or freshwater. Saline wetlands include salt marshes and rocky intertidal habitats; freshwater wetlands include riparian vegetation along the banks of rivers, lakes, springs, and sinks.

13. Conservation biology is a relatively new science that investigates ways to preserve, restore, and maintain biotic diversity in the face of human exploitation of natural ecosystems. Given the current global population (projected by the U.S. Census Bureau to hit 7 billion by 2013), it will be difficult to develop a resource management plan capable of sustaining our natural ecosystems.

14. Restoration ecology seems to offer great promise in our ability to restore degraded ecosystems to health, but, in fact, most successful restorations have been limited to wetlands or grasslands, both dominated by herbs. The restoration of scrub and forest vegetation types on stressful substrates has been far less successful. It will be several decades before the science and technology of restoration will be more useful.

Questions

1. Some ecologists believe that plant communities are very tightly organized, with the associated species being somehow interdependent on one another. How could you test this idea, either by observation or experimentation?

2. What traits do communities exhibit that individual plants or plant populations do not?

3. When Mount St. Helens erupted in the 1980s, the explosion blew down forest trees and covered the ground with a deep deposit of volcanic ash. Succession has been studied since the eruption, and it is proceeding slowly. Is this primary or secondary succession, or both?

4. In the boreal forest, storms often uproot and blow down every tree in a stand. Over time, herbs and shrubs invade and cover the site; then aspen may dominate, and conifers reinvade the site under the shade of the aspen. Later, the conifers overtop the aspen, and a boreal forest is re-established. Explain why this succession is progressive, rather than retrogressive.

5. Why is it critically important for tundra plants to be low-growing perennial herbs?

6. Describe the physiognomy (architecture and dominant growth forms) of the eastern deciduous forest.

7. Consider a desert phreatophyte, a cactus, and an evergreen shrub all growing near each other in a hectare of desert. Which one do you think would grow the least in 1 year, and why?

8. Why is wetland vegetation important to an adjacent aquatic ecosystem?

9. Do you think it is possible for humans to do all four of the following simultaneously: continue economic growth, continue population growth, maintain biological diversity, and attain sustainability in natural resource use? If not, how many of these four could be simultaneously achieved?

InfoTrac® College Edition

http://infotrac.thomsonlearning.com

Community Succession

Bardgett, R. 2001. Plant succession: Succession is the term used to describe the changes in species of organisms that occupy an area over time. This article focuses on plants, to demonstrate how succession generates biologically diverse ecosystems from simple and often harsh beginnings. *Biological Sciences Review* 14:2. (Keywords: "Bardgett" and "succession")

Culotta, E. 1996. Exploring biodiversity's benefits (effect of biodiversity on ecosystem productivity and stability). *Science* 273:1045. (Keyword: "biodiversity's benefits")

Rees, M., Condit, R., Crawley, M., Pacala, S., Tilman, D. 2001. Long-term studies of vegetation dynamics. *Science* 293:650. (Keywords: "long-term" and "vegetation dynamics")

Zimmer, C. 1994. More productive, less diverse (too much biological productivity reduces biodiversity). *Discover* 15:24. (Keywords: "biodiversity" and "productivity")

Forest Ecology

Hansen, A.J., Neilson, R.R., Dale, V.H., Flather, C.H., Iverson, L.R., Currie, D.J., Shafer, S., Cook, R., Bartlein, P.J. 2001. Global change in forests: Responses of species, communities, and biomes. *BioScience* 51:765. (Keywords: "global change" and "forests")

Mcnulty, S.G., Aber, J.D. 2001. US national climate change assessment on forest ecosystems: An introduction. *BioScience* 51:720. (Keywords: "assessment" and "forest ecosystems")

Pike, J. 2002. Good news is that bad news is wrong: Despite the almost-constant warnings of an impending ecological doom, forests in the eastern United States are flourishing and once-endangered animals are abundant. *Insight on the News* 18:27. (Keywords: "forests" and "flourishing")

Appendix 1: Units of Measure

Metric-English Conversions

Length

1 inch	=	2.54 centimeters
1 foot	=	0.30 meters
1 yard	=	0.91 meters
1 mile	=	1.61 kilometers
1 centimeter	=	0.39 inches
1 meter	=	1.09 yards
1 kilometer	=	0.62 miles

Area

1 acre	=	0.40 hectares
1 square mile	=	2.61 square kilometers
1 hectare	=	2.47 acres
1 square kilometer	=	0.38 square miles

Volume

1 cubic inch	=	16.39 milliliters
1 quart (U.S., liq.)	=	0.95 liters
1 gallon (U.S., liq.)	=	3.79 liters
1 milliliter	=	0.034 fluid ounces (U.S.)
1 liter	=	1.06 quarts (U.S., liq.)

Weight

1 ounce	=	28.6 grams
1 pound	=	0.45 kilograms
1 ton (short: 2000 lbs)	=	907 kilograms
1 gram	=	0.035 ounces
1 kilogram	=	2.20 pounds
1 metric ton	=	2204 pounds

Metric Prefixes

giga	G	10^9
mega	M	10^6
kilo	k	10^3
hecto	h	10^2
deka	da	10^1
deci	d	10^{-1}
centi	c	10^{-2}
milli	m	10^{-3}
micro	μ	10^{-6}
nano	n	10^{-9}
pico	p	10^{-12}
femto	f	10^{-15}

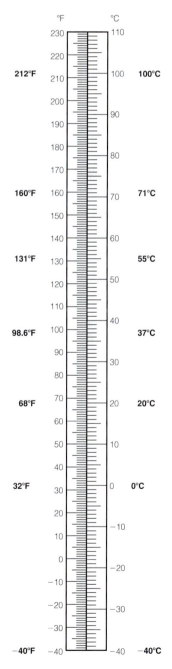

Appendix 2: Periodic Table of the Elements

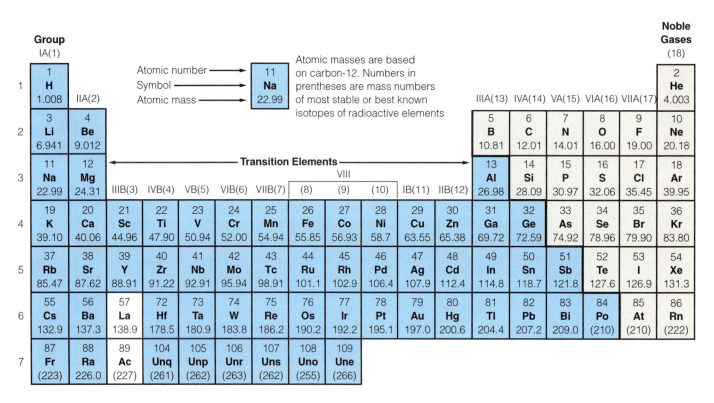

Appendix 3: Geologic Time Scale

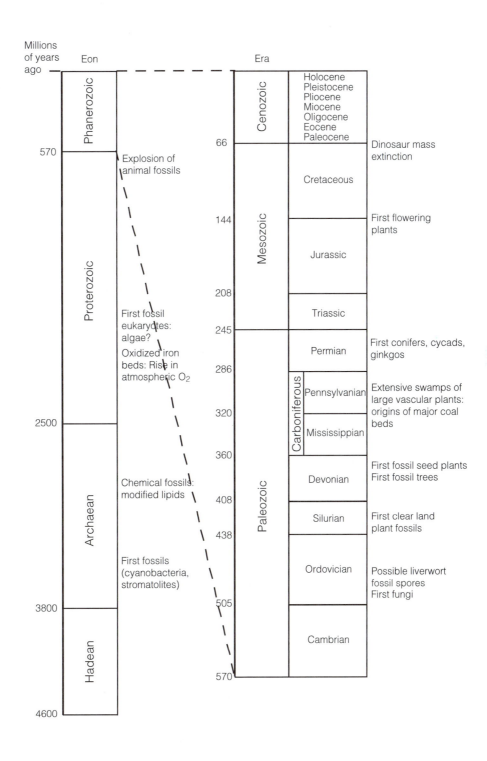

Appendix 4: Answers to Genetics Problems

1. Two plants with red flowers may produce 25% white-flowered progeny if both the parents are heterozygous. Under the assumptions of the problem, white-flowered plants are homozygous and cannot produce red-flowered progeny.

2. If *ein* is recessive and the tall plant is mutant in *ein,* then the tall plant must be homozygous.

3. 50% of the seeds will be yellow; 50% will be green.

4. She expected 25 of the 100 seeds to be green. She would not necessarily expect a second plant to have green seeds because it might be homozygous for yellow seeds.

5. The genotypes of the parents were *ccPP* and *CCpp.*

6. Red, normal *(RrShsh);* yellow, normal *(rrShsh);* red, shrunken *(Rrshsh);* and yellow, shrunken *(rrshsh)*—all in equal proportions (25%).

7. Alleles B and C (and *b* and *c*) are linked. Gene *A(a)* is not linked to either gene *B(b)* or *C(c).*

8. The homologous chromosomes containing the *R(r)* and *W2(w2)* genes looked like this: *R____w2* and *r____W2*. The proportion of recombinant pollen was $(20 + 28)/(20 + 172 + 180 + 28) = 0.12$.

9. Assuming that the starting strains were homozygous, 100% of the progeny of cross 1 would have the *Rv* allele and so be resistant. All the progeny from this cross (and further crosses) that are resistant would be heterozygous; so 50% of the progeny of cross 2 would be resistant, as would 50% of progeny from all further crosses (including cross 10). 100% of the progeny of cross 1 would have unlinked wild tomato alleles. For cross 2, 50% would have wild tomato alleles for any gene unlinked to *Rv* (the other 50% would have domestic alleles). For cross 3, because 50% of the parents (progeny from cross 2) have wild tomato alleles, $50\% \times 50\% = (0.5)^2 \times 100 = 25\%$ of the progeny would have unlinked alleles; by extension, for cross 10, the number would be $(0.5)^9 \times 100\%$.

10. If the trait was controlled by a nuclear gene, it would be inherited according to Mendelian laws. Taking this as a hypothesis, the fact that all progeny from the first cross were sterile implies that the male-sterile allele must be dominant and that the progeny from the first cross must be heterozygous. But if that was true, some of the progeny from the second cross should be male-fertile. Because they were not, the trait could not be controlled by a single Mendelian gene. The results are consistent with a maternally inherited gene. (There are other possibilities: What if the male-sterile mutant had dominant alleles in many genes, any one of which could cause male sterility? The probability of finding a male-fertile progeny in the second cross could become very small.)

Appendix 5: House Plants

Plants can make an otherwise sterile living space look friendly and comfortable. While anyone can buy plants at a local garden store (or, often, supermarket), choosing and caring for plants so that they thrive indoors takes a bit more knowledge and planning. Plants differ in their favored environments—the amounts of light, water, and minerals they need. The table below gives a selection of common house plants, together with some suggestions for their use and care. Although in general we have given only one entry for a species (and sometimes for a genus), you should recognize that varieties may differ in their requirements. Photographs of some of these plants and their flowers can be found at the "Photogallery" of http://ohric.ucdavis.edu/.

Although different species have very different preferences, there are some rules of thumb that can help one maintain many house plants successfully.

Indoor light is generally low in intensity—lower than is apparent, because human eyes adjust to intensities that differ by several orders of magnitude. Plants that can live with low or medium light may find artificial (especially fluorescent) light sufficient; plants that require high or very high light should be placed near windows. However, windows that receive indirect skylight are best, because direct sunlight through a window can often scorch leaves that would survive outside.

The relative humidity in a house or apartment is often very low, especially in the winter when the living space is heated. (Relative humidity drops as air is warmed, unless more water vapor is provided.) A terrarium provides a humid atmosphere. Some plants appreciate a daily misting of their leaves, but misting makes some plants susceptible to fungal or bacterial infections. Overwatering does *not* help.

Indoor plants are generally restricted in the growth of their roots. A porous potting mix with a considerable amount of organic matter allows airflow for good respiration and maintains inorganic nutrients in the soil solution at an appropriate level. It is good to water a pot periodically to saturation and allow any excess water to drain away. This is especially important in areas with hard water or water with problem substances such as borate. Elevating a pot in a tray with gravel or some other spacer to prevent excess water from being drawn back into the pot will prevent salt buildup. The frequency of watering depends on the soil, the plant, the amount of light, and the humidity, and the optimum schedule must be determined empirically for each situation.

Keys—Light: VH, very high; H, high; M, medium; L, low. Water: W, wet; M, moist; D, dry. Container suggestions: HB, hanging basket; T, terrarium; DD, desert dish; LT, large tub; GC, creeping ground cover for planting box; CP, climbing pole.

Scientific Name	Common Name	Light	Water	Container	Notes
Abutilon spp.	flowering maple	H	M		
Acalypha hispida (*A. wilkesiana*)	chenille plant	H	M		
Achimenes spp.	magic glower	H	M	HB	
Adiantum cuneatum	maidenhair fern	M	W		
Aechmea fasciata	bromeliad	M	M	T	Prefers moderate to high humity, but not overwatering
Aeschynanthus pulcher	lipstick plant	M	D	HB	
Agave americana	century plant	M	M		
Aglaonema modestum	Chinese evergreen	L	M	T	Durable house plant that tolerates poor light, dry air, and air-conditioning
Aglaonema x pseudo-bracteatum	golden aglaonema	L	M		Also easy to grow

Scientific Name	Common Name	Light	Water	Container	Notes
Aglaonema roebelenii	pewter plant	L	M		Also easy to grow
Aloe variegata	aloe	VH	D	DD	
Alternanthera bettzickiana		M to H	M		
Ananas comosus	pineapple	H	M		
Anthurium andreanum	anthurium	M	M		Sensitive to cold, fluctuating temperature, hard water, and dry air
Aphelandra squarrosa	zebra plant	M	M		Sensitive to cold, dry air; give even moisture and medium light
Araucaria heterophylla	Norfolk island pine	H	M		Hardy: tolerates dim and bright light and some drought
Ardisia crispa	coral ardisia	M	M		
Asparagus plumosus (*A. setaceus*)	bride's bouquet fern	H	M	HB	
Asparagus sprengeri (*A. densiflora*)	asparagus fern	M	M	HB	
Aspidistra elatior	cast-iron plant	L	M		Easy to grow: tolerates heat, dust, poor light, and dryness
Asplenium nidus	bird's nest fern	M	W		
Aucuba japonica	gold-dust plant	M	D		
Beaucarnea recurvata	ponytail palm	M	D		
Begonia spp.	begonia	H	M		Grow in filtered sunlight; avoid standing water (susceptible to rot)
Beloperone guttata	shrimp plant	H	D		
Billbergia zebrina	billberia	M	D to M		
Bougainvillea glabra	bougainvillea	VH	D		Thorns require caution
Browallia speciosa	bush violet	M to H	M		
Caladium spp.	caladium	H	M		
Calathea spp.	zebra plant, peacock plant	M	M		Needs high humidity: good for terrariums; metallic-hued leaves
Calceolaria herbeahybrida	pocketbook plant	H	M		
Campanula isophylla	star-of-Bethlehem	H	M		Flowers indoors for short period, but not sustainable
Capsicum annuum	Christmas pepper	H	M		
Carissa grandiflora	Natal plum	H	D		
Cattleya hybrids	cattleya orchid	M	M		
Chamaedorea elegans	parlor palm	L	M	T	
Chamaedorea erumpens	bamboo palm	L	M		
Chamaerops humilis	European fan palm	H	M		
Chlorophytum comosum	spider plant	M to H	D to M	HB	Variegated varieties need more light than others
Chrysalidocarpus lutescens	areca palm	M	M		
Chrysanthemum morifolium	chrysanthemum	VH	M	HB	Flowers indoors for short period, but not sustainable
Cissus antarctica	kangaroo vine	H	M	HB, CP, T	Tolerates sun, shade, and dry air
Cissus discolor	rex begonia vine	M	D to M	HB, CP	
Cissus rhombifolia	grape ivy	M	D to M	HB, CP	Very easy to grow
Cissus rotundifolia	Arabian wax cissus	M	D to M	HB	

Scientific Name	Common Name	Light	Water	Container	Notes
Citrus spp.	calemondin orange, Meyer lemon, kumquat, et al.	H	D		Needs high light, constant temperature, acid soil, and iron fertilizers
Clerodendrum thomsoniae	bleeding heart vine	M to H	M	CP	
Clivia miniata	Kaffir lily	M	D		
Clusia rosea	autograph tree	M	D to M		
Codiaeum variegatum	croton	VH	D	LT	Needs bright sun, high humidity; avoid drafts
Coffea arabica	coffee tree	M	M	T	
Coleus blumei	coleus	H to VH	M		Bright light helps develop color; short-lived indoors
Columnea spp.	columnea	M	M		*C. microphylla* good for hanging baskets
Convallaria majalis	lily-of-the-valley	H	M		
Cordyline terminalis	Hawaiian ti plant	M	M		*C.t. minima* "baby ti" good for terraria
Crassula spp.	succulents	VH	D		*C. lycopodioides, C. rupestris* good for terraria
Crassula argentea	jade plant	H	D	DD	
Crocus spp.	crocus	H	M		Bulbs can be forced to bloom indoors
Crossandra infundibuliformis	crossandra	M	M		
Cryptanthus spp.	dwarf bromeliad	M	D to M	T	Prefers moderate to high humity, but not overwatering
Cyanotis kewensis	teddy bear vine	M	M	HB	
Cycas revoluta	sago palm	M	D		
Cyclamen spp.	cyclamen	H	M		Prefers cool conditions, less than 70°F
Cymbidium hybrids	cymbidium orchid	M	M		
Cyperus alternifolius	umbrella plant	M	W		
Cyrtomium falcatum	Japanese holly fern	L to M	M		
Davallia fejeensis	rabbit's foot fern	L	M	HB	
Dichorisandra reginae	queen's spiderwort	M	M		
Diffenbachia amoena	dumbcane	M	D	LT	Especially easy: grow in indirect sunlight; avoid overwatering
Dionaea muscipula	venus fly trap	M	M to W		Special requirements: boggy, acid, warm, high humidity
Dizygotheca elegantissima	false aralia	H	M	T	
Dracaena deremensis	green dracaena	M	M to W		
Dracaena deremensis "Warneckii"	white-striped dracaena	M	D to M		
Dracaena fragrans massangeana	corn plant	L	M to W		
Dracaena marginata	dragon tree	M	M		Dracaenas tolerate low light, low humidity
Dracaena sanderiana	ribbon plant	M	M	T	
Dracaena surculosa (D. godseffiana)	gold-dust dracaena	M	M to W	T	

Scientific Name	Common Name	Light	Water	Container	Notes
Echeveria spp.	hen-and-chickens	M	D	DD	
Epiphyllum hybrids	orchid cactus	H	M		Tropical spineless cactus
Epipremnum aureum	pothos	L	M	HB	
Episcia spp.	flame violet	M	M	GC	*E. cupreata* good for hanging baskets
Eranthemum nervosum	blue sage	M	M		
Erica spp.	heather	H	M		Flowers indoors for short period, but not sustainable
Euphorbia lactea	candelabra cactus	M	D	DD	
Euphorbia milii	crown-of-thorns	H	D		
Euphorbia pulcherrima	poinsettia	VH	D to M		
Exacum affine	Persian violet, exacum	M	M		
Fatshedera lizei	botanical wonder	M	M	LT	
Fatsia japonica (Aralia japonica)	Japanese aralia	M	M	LT	
Ficus benjamina "Exotica"	weeping java fig	H	M		Give bright light; avoid overwatering and over-handling
Ficus elastica "Decora"	rubber plant	M	M	LT	Easy to grow!
Ficus lyrata	fiddleleaf fig	M	M	LT	Easy to grow!
Ficus repens var. *pumila*	creeping fig	L to M	M	GC, CP	
Ficus retusa nitida	India laurel	L	D		Easy to grow!
Ficus triangularis	triangleleaf fig	M	M		
Fittonia verschaffeltii	silver nerve plant	M	M	GC, T	Needs high humidity: good for terrariums
Fuchsia spp.	fuchsia	VH	M		Give diffuse sunlight, cool temperatures, and high humidity
Gardenia jasminoides	gardenia	VH	M		Needs bright, warm conditions, acid soil
Guzmania lingulata	scarlet star	M	D to M		
Gynura aurantiaca	velvet plant	M	M		
Haemanthus coccineus	blood lily	VH	M		
Haworthia spp.	zebra haworthia, wart plant	M	D	DD	
Hedera helix	English ivy	L to M	M	GC	Easy to grow; some varieties good for hanging baskets
Helxine soleirolii	baby's tears	M	M		Needs high humidity: good for terrariums
Hemigraphis exotica	waffle plant	M	M	GC, HB	
Hibiscus rosa-sinensis	Chinese hibiscus	VH	M		Needs bright, warm conditions
Hippeastrum vittatum	amaryllis	H	M		Bulbs can be forced to bloom indoors
Howea forsterana	kentia palm	L to M	M		
Hoya carnosa	wax plant, Hindu rope plant	M	D to M	HB	Avoid overwatering; allow winter resting in cool, dry conditions
Hyacinthus orientalis	hyacinth	H	M		Bulbs can be forced to bloom indoors
Hydrangea macrophylla	hydrangea	H	M		Flowers indoors for short period, but not sustainable
Hypocyrta nummularia	goldfish plant	M	M	HB	

Scientific Name	Common Name	Light	Water	Container	Notes
Hypoestes sanguinolenta	polka-dot plant	M	M		
Impatiens spp.	impatiens	H to VH	M		Flowers indoors for short period, but not sustainable
Ipomoea batatas	sweet potato	M to H	M to D	HB	
Kalanchoe spp.	kalanchoe	H	D to M		
Kalanchoe blossfeldiana	kalanchoe, panda plant	M to H	D to M		
Lantana camara	lantana, yellow sage	H	D to M		
Lantana montevidensis	trailing lantana	H	D to M		
Leea coccinea	leea	M	M		
Lilium longiflorum	Easter lily	H	M		Bulbs can be forced to bloom indoors
Maranta leuconeura	prayer plant	M	M	T	Needs high humidity: good for terrariums
Mikania ternata	plush vine	M	M		
Mimosa pudica	sensitive plant	H	M		Novelty: fun to grow, but short-lived
Monstera spp.	swiss cheese plant, split-leaf philodendron	L to M	M	LT	Easy to grow: tolerates low light, variable temperatures, low humidity
Muscari spp.	grape hyacinth	H	M		Bulbs can be forced to bloom indoors
Narcissus pseudonarcissus	daffodil	L	D		Bulbs can be forced to bloom indoors
Nautilocalyx lunchii	coral plant	M	M		
Neoregelia carolinae tricolor	tricolor bromeliad	M	D to M		Prefers moderate to high humity, but not overwatering
N. spectabilis	fingernail plant	M	M to W		
Nephrolepis exaltata bostoniensis	Boston fern	M	M	HB, T	
Nerium oleander	oleander	H	M		
Oxalis spp.	oxalis	H	M		
Pachystachys coccinea	lollipop pant	M	M		
Pandanus veitchii	screw pine	M	M	LT	
Paphiopedilum hybrids	lady-slipper orchid	M	M		Fussy: has special requirements
Passiflora spp.	passion flower	VH	M		
Pelargonium spp.	geranium	VH	D to M		Prefers at least 4 hr of direct sunlight, but cool temperatures
Pellionia caperata	emerald ripple	L	D		
Pellionia daveauana	trailing watermelon vine	M	M	HB	
Pellionia glabella "Variegata"	variegated wax privet peperomia	H	W	HB	
Pellionia metallica	metallic peperomia	M	M		Good terrarium plant: likes high humidity (but stems rot in overwet soil)
Pellionia obtusifolia	oval leaf peperomia	M	M		
Pellionia pulchra	satin pellionia	M	M	HB	
Pellionia scandens	philodendron peperomia	M	M		
Phalaenopsis hybrids	phalaenopsis orchid	M	M		
Philodendron domesticum "Hastatum"	elephant ear philodendron	M	M	LT	Easy to grow: most philodendrons tolerate dim light, drought, dust, dry air, neglect

Scientific Name	Common Name	Light	Water	Container	Notes
Philodendron micans	velvetleaf philodendron	M	M	HB	Easy to grow!
Philodendron oxycardium	common philodendron	L	D	HB	Easy to grow!
Philodendron panduriforme	fiddleleaf philodendron	M	M		Easy to grow!
Philodendron selloum	selloum philodendron	M	M	LT	Easy to grow!
Phoenix roebelenii	dwarf date palm	M	W		
Pilea cadierei	aluminum plant	M	M	T	Needs high humidity
Pilea microphylla	artillery plant	M	M	T	
Pilea nummariifolia	creeping charley	M	M	GC, HB, T	
Pisonia grandis "Tricolor"	bird catcher tree	H	M		
Pittosporum tobira	mock orange	H	D		
Platycerium spp.	staghorn fern	M	M to W	HB	
Plectranthus australis	Swedish ivy	M	M	HB	
Pleomele reflexa	green pleomele	M	M to W		
Pleomele thallioides	lance dracaena	M	M		
Plumeria rubra	frangipani	M to D	H		
Podocarpus macrophyllus	podocarpus	H	M		
Polyscias guilfoylei	parsley aralia	M	M		
Primula malacoides	fairy primrose	H	M		Flowers indoors for short period, but not sustainable
Primula obconica	German primrose	H	M		Flowers indoors for short period, but not sustainable
Pteris ensiformis	silver table fern	M	M		
Rhapis excelsa	lady palm	M	M to W		
Rhipsalis spp.	mistletoe cactus	H	D to M	HB	
Rhododendron spp.	azalea	H	M		Flowers indoors for short period, but not sustainable
Rhoeo discolor	Moses-in-the-cradle	M	M		
Rosa chinensis v. *minima*	miniature rose	VH	M		Flowers indoors for short period, but not sustainable
Ruellia makoyana	ruellia	M	M	HB	
Saintpaulia spp.	African violet	H to VH	M	T	Sensitive to hot sun, temperature fluctuations, dry air, cold water, and rotting in overwet soil
Sansevieria trifasciata, S. hahuii	snake plant	L	D		Easy to grow: tolerates poor light, drought, drafts, and low humidity; avoid overwatering
Saxifraga sarmentosa	strawberry begonia, strawberry geranium	VH	D to M	GC	Good hanging basket plant
Schefflera actinophyla	schefflera, umbrella tree	M	M	LT	Aggressive nuisance outdoors
Schefflera arborcola		M	M		Easy to grow shrub
Schlumbergera bridgesii	Christmas cactus	H	M	HB	Tropical cactus: for Christmas flowers, give >12 hr per day of complete darkness starting in September
Sedum morganianum	burro's tail	M to VH		DD	
Selaginella lepidophylla	resurrection plant	L to M	D		Needs high humidity
Senecio cruentus	cineraria	H	M		Flowers indoors for short period, but not sustainable

Scientific Name	Common Name	Light	Water	Container	Notes
Senecio macroglossus "Variegatum"	variegated wax ivy	M	M		
Senecio mikanioides	German ivy	M	M		
Senecio rowleyanus	string of pearls	M	D	HB	Great conversation piece
Setcreasea purpurea	purple heart	M	D to M	HB	
Sinningia spp.	gloxinia	VH	M to W		Miniature varieties good for terraria
Solanum pseudocapsicum	Jerusalem cherry	H	D to M		Fruit is toxic
Spathiphyllum "Mauna Loa"	white flag	M	D to M		Easy to grow!
Strelitzia reginae	bird of paradise	M to H	M	LT	
Streptocarpus spp.	Cape primrose	H	M		*S. saxorum* good for hanging baskets
Strobilanthes dyerianus	Persian shield	M	M		
Syngonium podophyllum	arrowhead vine, nephthytis	L	M	T	Easy to grow: tolerates neglect, low light
Thunbergia alata	black-eyed susan vine	H	M		
Tolmiea menziesii	piggyback plant	H	M		Sensitive to salt in water
Tradescantia spp.	wandering Jew	M	D to M	GC	*T. albiflora* and *T. sillamontana* good hanging basket plants
Tulipa spp.	tulip	H	M		Bulbs can be forced to bloom indoors
Vinca major "Variegata"	periwinkle		D	GC	
Vriesea splendens	flaming sword	M	M		
Yucca spp.	yucca	L	D		
Zantedeschia spp.	calla lily	H	M to W		Flowers indoors for short period, but not sustainable
Zebrina spp.	wandering Jew	M	M	HB	Easy to grow!

References

1. Ralph C. Gay and Dennis R. Pittenger, "House Plants", in Pittenger, D.R., ed., *California Master Gardener Handbook*, University of California Division of Agriculture and Natural Resources, 2002, Chapter 11.

2. Alberta Government, "House Plants: List of Species," http://www1.agric.gov.ab.ca/$department/deptdocs.nsf/all/web-doc1374?opendocument, March 17, 2003.

3. Tim Metcalf and Ernesto Sandoval, University of California, Davis Botanical Conservatory; http://greenhouse.ucdavis.edu/conservatory.htm

4. Don Shor, Owner, Redwood Barn Nursery, Inc., Davis, CA; www.redwoodbarn.com

Appendix 6: Toxic Plants

Many plants protect themselves from herbivory and parasitism by producing disagreeable or poisonous compounds. In fact, it is astonishing how many familiar plants can be extremely dangerous. A knowledge of these plants can be important in choosing plants for home or garden. On the one hand, one should avoid certain plants in the home if there are curious and adventuresome pets or young children present; on the other hand, it is possible to choose plants that will ward off or discourage garden pests. Information about toxic plants can also be a valuable clue to possible new medicinal products. The table below presents a selection of toxic plants and their dangers. In case of ingestion or contact with these plants, or in case symptoms of poisoning or irritation occur, call the local Poison Control Center or a doctor.

KEY TO "TOXICITY CLASS"

1. **Major toxicity**. These plants can cause serious illness or death if the poisonous parts are ingested. Note that some plants that produce common, nutritious foods have very poisonous components (e.g., apple seeds, tomato leaves).

2. **Minor toxicity.** Ingestion of these plants may cause minor illness such as vomiting or diarrhea.

3. **Oxalates**. The juice or sap of these plants contains crystals of oxalic acid, which bind calcium and irritate the skin, mouth, tongue, and throat. This can result in throat swelling and serious breathing difficulties, as well as pain and nausea.

4. **Dermatitis**. The juice, sap, or thorns of these plants can cause a skin rash or irritation. Wash the affected area with soap and water as soon as possible following contact with these plants.

KEY TO "REFERENCES"

1. This table is based on a website authored by Dr. Ann King Filmer, University of California, Davis, http://envhort.ucdavis.edu/ce/king/PoisPlant/Tox-SCI.htm, and a publication "Know Your Plants . . . Safe or Poisonous?" by E. Stone and A. King, University of California Cooperative Extension, 1997. The website has lists of toxic plants by both scientific or common name. It also has lists of plants generally recognized as safe. If you have difficulty locating it, try http://ohric.ucdavis.edu.

2. Further information, including photographs, is available on a website hosted by Cornell University: http://www.ansci.cornell.edu/plants/comlist.html
   This site has lists of toxic plants by both scientific or common name.

3. Further information, including photographs, is available on a website hosted by North Carolina State University: http://www.ces.ncsu.edu/depts/hort/consumer/poison/indcoa_e.htm
   This site has lists of toxic plants by both scientific or common name.

4. Further information, including a description of poisoning symptoms, is available on a website hosted by Texas A&M University: http://aggie-horticulture.tamu.edu/plantanswers/publications/poison/poison.html

Scientific Name	Common Name	Toxicity Class	References (1)
Abies balsamea	Balsam fir	4	
Abrus precatorius	Jequirity bean, Rosary bean, Rosary pea	1	2,3
Acalypha spp.	Chenille plant, Copperleaf, Firetail	2,4	3
Acer spp.	Maple	4	
Achillea millefolium	Achillea, Yarrow	2,4	3
Acokanthera spp.	Bushman's poison, Wintersweet	1	
Aconitum spp.	Aconite, Monkshood	1	2,3,4
Aesculus spp.	California buckeye, Horsechestnut	2	2,3
Agapanthus spp.	African lily, Agapanthus, Lily-of-the-Nile	2,4	3
Agave spp.	Agave, Century plant	2,3,4	3
Aglaonema spp.	Aglaonema, Chinese evergreen	3,4	3
Ailanthus altissima	Ailanthus, Tree-of-heaven	2,4	
Alcea rosea	Hollyhock	4	
Allium spp.	Allium, Wild onion	2	2,3
Alnus spp.	Alder	4	
Alocasia spp.	Alocasia, Elephant's ear	3,4	3
Alstroemeria spp.	Alstroemeria, Peruvian lily	2,4	3
Amaranthus caudatus	Love-lies-bleeding, Tassel flower	1	
Amaryllis belladonna	Amaryllis, Belladonna lily, Naked lady	2,4	
Ammi majus	Bishop's weed, False Queen Anne's lace	4	
Anagallis arvensis	Scarlet pimpernel	2,4	
Ananas comosus	Pineapple	4	3
Anemone spp.	Anemone, Pasque flower, Windflower	2,4	3
Anthurium spp.	Anthurium	3,4	3
Aquilegia spp.	Columbine	2	
Arisaema triphyllum	Indian turnip, Jack-in-the-pulpit	3,4	2,3,4
Artemisia spp.	Sagebrush, Wormwood	4	3
Arum spp.	Arum, Black calla, Italian arum	3,4	3
Asclepias spp.	Butterfly weed, Milkweed	2,4	2,3
Asparagus officinalis	Garden asparagus	4	3
Asparagus densiflorus	Sprenger asparagus	4	3
Aster spp.	Aster	4	
Atropa belladonna	Belladonna, Deadly nightshade	1	2,3
Aucuba japonica	Japanese aucuba	2	3
Begonia spp. (some spp.)	Begonia	2,3	3
Bellis perennis	English daisy	4	
Berberis spp.	Barberry	2,4	
Betula spp.	Birch tree	2,4	
Bougainvillea spp.	Bougainvillea (thorns)	4	
Brachychiton populneus	Bottle tree	4	
Brugmansia spp.	Angel's trumpet, Jimson weed	1	3,4
Buxus sempervirens	Boxwood	2,4	3
Cactus spp.	Cactus (thorns and sap)	4	
Caesalpina gilliesii	Bird-of-paradise shrub, Poinciana	2	3

Scientific Name	Common Name	Toxicity Class	References (1)
Caladium bicolor	Caladium	3,4	3
Calluna vulgaris	Heather	1	
Caltha palustris	Marsh marigold	2	3
Campsis radicans	Trumpet creeper, Trumpet vine	4	3
Carissa macrocarpa	Natal plum	2	
Caryota spp.	Fishtail palm	3,4	3
Celastrus scandens	Bittersweet	2	3
Cestrum spp.	Cestrum, Jessamine, Night-blooming jessamine	1	
Chamaemelum nobile	Chamomile	4	
Chrysanthemum spp.	Chrysanthemum, Daisy, Marguerite daisy	2,4	3
Chrysanthemum maximum	Shasta daisy	2,4	
Cicuta spp.	Water hemlock	1	2,3,4
Cinnamomum camphora	Camphor tree	2	
Cissus rhombifolia	Grape ivy	4	
Clematis spp.	Clematis	2,4	3
Clivia spp.	Clivia, Kaffir lily	2,4	3
Codiaeum variegatum	Croton	2,4	3
Coffea arabica	Coffee plant	2	
Colchicum autumnale	Autumn crocus, Meadow saffron	1,4	3,4
Colocasia spp.	Elephant's ear, Taro	3,4	3
Conium maculatum	Poison hemlock	1	2,3,4
Convallaria majalis	Lily-of-the-valley	1,4	2,3,4
Cornus spp.	Dogwood	4	
Cortaderia selloana	Pampas grass	1	
Corynocarpus laevigata	New Zealand laurel	2	
Cotinus coggygria	Smoke bush	4	
Cotoneaster spp.	Cotoneaster	2	
Crassula argentea	Jade plant	2,4	
Crinum spp.	Crinum lily	2,4	3
Cuphea hyssopifolia	False heather	4	
Cycas revoluta	Sago palm	2	3
Cyclamen spp.	Cyclamen	2,4	3
Cynodon dactylon	Bermudagrass	4	
Cyperus alternifolius	Umbrella plant	2	
Cypripedium spp.	Lady slipper orchid	4	
Cytisus spp.	Broom, Scotch broom	2	3
Daphne spp.	Daphne	1	2,3,4
Daucus carota	Queen Anne's lace	4	3
Delphinium spp.	Delphinium, Larkspur	1	2,3,4
Dendranthema spp.	Chrysanthemum, Daisy, Marguerite daisy	2,4	
Dianthus spp.	Carnation, Pink, Sweet William	2,4	3
Dicentra spp.	Bleeding heart	4	2,3,4
Dichondra micrantha	Dichondra	4	
Dieffenbachia spp.	Dieffenbachia, Dumb cane	3	3,4

Scientific Name	Common Name	Toxicity Class	References (1)
Digitalis purpurea	Foxglove	1	2,3,4
Echium vulgare	Echium Blue weed, Blue devil	1,4	3
Erigeron spp.	Fleabane	4	
Epipremnum aureum	Pothos	3,4	3
Eriobotrya japonica	Loquat (seeds)	1	3
Erythrina spp.	Coral tree	1	3
Eucalyptus spp.	Blue gum, Eucalyptus	2,4	3
Euonymus spp.	Burning bush, Euonymus	2	3
Euphorbia spp.	Crown of thorns, Euphorbia, Gopher plant, Pencil tree, Poinsettia, Snow-on-the-mountain	2,4	2,3
Festuca spp.	Fescue (grass)	4	2
Ficus spp.	Fig, Weeping fig, Fiddleleaf fig	4	3
Ficus elastica	Rubber plant	4	
Fraxinus spp.	Ash	4	
Fritillaria meleagris	Checkered lily, Snakeshead	1	
Gaillardia spp.	Blanket flower	4	
Galanthus spp.	Snowdrop	2,4	3
Gelsemium sempervirens	Carolina jessamine	1,4	2,3
Ginkgo biloba	Ginkgo, Maidenhair tree	4	3
Gladiolus spp.	Gladiolus	2,4	
Glechoma hederacea	Creeping Charlie, Ground ivy	2	2
Gloriosa spp.	Climbing lily, Glory lily	1	3
Grevillea spp.	Grevillea, Silk oak	4	3
Gypsophila paniculata	Baby's breath	4	3
Haemanthus spp.	Blood lily	2,4	
Hedera spp.	English ivy; Ivy	2,4	3
Heliotropum arborescens	Heliotrope	1	3
Helleborus spp.	Christmas rose, Hellebore, Lenten rose	1,4	2,3
Heteromeles arbutifolia	Toyon	1	
Hippeastrum spp.	Amaryllis	2	3
Hyacinthus orientalis	Hyacinth	2,4	3,4
Hydrangea spp.	Hydrangea	1,4	3
Hymenocallis spp.	Spider lily, Sea daffodil	2,4	3
Hyoscyamus niger	Black henbane, Deadly nightshade	1	2,3
Hypericum spp.	St. Johnswort, Klamath weed	1,4	2
Iberis sempervirens	Evergreen candytuft	4	
Ilex spp.	Holly (berries)	2	3
Ipomoea spp.	Morning glory (seeds)	1	3
Iris spp.	Iris	2,4	2,3,4
Jatropha spp.	Coral plant, Jatropha	2,4	
Juglans spp.	Walnut	4	
Juniperus spp.	Juniper	2	3
Kalmia latifolia	Mountain laurel	2	3
Laburnum anagyroides	Goldenchain tree, Laburnum	2	2,3,4
Lantana camara	Lantana, Red sage, Shrub verbena	1	2,3,4

Scientific Name	Common Name	Toxicity Class	References (1)
Lathyrus odoratus	Sweet pea (seeds)	2	2,3
Leucojum spp.	Snowflake	2,4	
Ligustrum spp.	Ligustrum, Privet	2,4	3
Lilium spp. (some spp.)	Lily	2,4	
Linum usitatissimum	Flax	4	2,3
Lobelia spp.	Cardinal flower, Lobelia	1,4	2,3
Lupinus spp.	Lupine	1	2,3
Lycopersicon esculentum	Tomato (non-fruit parts)	1,4	3
Lycoris spp.	Spider lily	2	3
Malus spp.	Apple, Crabapple (seeds)	1	3
Melaleuca quinquenervia	Cajeput tree	4	
Melia azedarach	Chinaberry	1	3
Melianthus spp.	Honey bush	1	
Mirabilis jalapa	Four-o'clock	2,4	3
Monstera deliciosa	Split-leaf philodendron	3,4	3
Myoporum laetum	Myoporum	1	
Myrsine africana	African boxwood	2	
Myrtus communis	Myrtle	2	
Narcissus spp.	Daffodil, Jonquil, Narcissus (bulb)	2,4	3,4
Nerine spp.	Guernsey lily, Nerine	2,4	
Nerium oleander	Oleander	1,4	2,3,4
Nicotiana glauca	Flowering tobacco, Tree tobacco	1	2,3
Nigella damascena	Love-in-a-mist	2	
Ornithogalum spp.	Ornithogalum, Pregnant onion, Star-of-Bethelehem	1	3
Papaver spp.	Poppy	2,4	2,3
Papaver nudicaule	Iceland poppy	3,4	
Parthenocissus spp.	Boston ivy, Virginia creeper	3,4	3
Phacelia spp.	Desert bluebells	4	
Philodendron spp.	Heart leaf, Philodendron	3,4	3
Phoradendron spp.	Mistletoe	2,4	3,4
Physalis spp.	Chinese lantern, Groundcherry	1	3
Phytolacca americana	Pokeweed	2	3
Pieris japonica	Japanese pieris	1	2,3
Pittosporum spp.	Pittosporum	1	
Plumbago auriculata	Cape plumbago	4	3
Plumeria rubra	Frangipani, Plumeria	4	
Podocarpus macrophyllus	Yew pine	2	3
Podophyllum peltatum	May apple, Mandrake	1	2,3
Polianthes tuberosa	Tuberose	2	
Polyscias spp.	Ming aralia	2,4	3
Primula spp.	Primrose, Primula	4	3
Prunus spp.	Almond, Apricot, Black cherry, Cherry, Nectarine, Peach, Plum (seeds)	1	2,3,4
Prunus caroliniana	Carolina laurel cherry	1	3
Prunus laurocerasus	English laurel	1	3
Prunus virginiana	Chokecherry	1	

Scientific Name	Common Name	Toxicity Class	References (1)
Pyracantha spp.	Firethorn, Pyracantha	2,4	
Pyrus spp.	Pear (seeds)	1	
Quercus spp.	Oak tree (acorns)	2,4	2,3,4
Ranunculus spp.	Buttercup, Ranunculus	2,4	2,3,4
Rhamnus spp.	Buckthorn, Coffeeberry	2,4	3
Rheum rhabarbarum	Rhubarb (leaves)	3	3,4
Rhododendron spp.	Azalea, Rhododendron	1	3
Rhoeo spathacea	Moses-in-the-cradle, Oyster plant	4	3
Ricinus communis	Castor bean	1	2,3,4
Robinia pseudoacacia	Black locust (seeds)	1	2,3,4
Ruta graveolens	Rue	4	3
Sambucus spp.	Elderberry (ripe fruit is nontoxic)	1	2,3,4
Sansevieria spp.	Snake plant	2,4	3
Sapium sebiferum	Chinese tallow tree, Popcorn tree	4	3
Schefflera actinophylla	Schefflera, Umbrella tree	2,4	3
Schinus molle	California pepper tree	4	
Schinus terebinthifolius	Brazilian pepper tree	2,4	
Scilla spp.	Peruvian scilla, Squill	1	3
Senecio spp. (some spp.)	Dusty miller, String of beads, Cineraria	2,4	2,3
Sequoia sempervirens	Coast redwood	2,4	
Solanum spp.	Nightshade, Black or Deadly, Horse nettle, Buffalo bur, Potato, Jerusalem cherry	1	2,3,4
Solanum tuberosum	Potato plant (green parts)	1	
Spathiphyllum spp.	Spathiphyllum	3,4	3
Symphoricarpos spp.	Coralberry, Indian currant, Snowberry	2	2,3
Syngonium podophyllum	Arrowhead plant, Nephthytis	3	3
Tagetes spp.	Marigold	4	3
Tanacetum spp.	Tansy	4	3
Taxus spp.	Yew, English and Japanese (seeds)	1	2,3,4
Thevetia peruviana	Yellow oleander	1,4	
Thuja spp.	Arborvitae	2,4	
Toxicodendron spp.	Poison oak, Poison ivy, Poison sumac	4	2,3
Tradescantia spp.	Wandering jew	4	3
Tulipa spp.	Tulip (bulb)	2,4	3
Ulmus spp.	Elm tree	4	
Umbellularia californica	California bay, California laurel	4	
Urtica spp.	Stinging nettles	4	2,3
Veratrum spp.	Skunk cabbage, Corn lily, False hellbore	2	2,3
Vinca spp.	Myrtle, Periwinkle, Vinca	1	
Viola spp.	Pansy, Violet (seeds)	2	
Viscum album	European mistletoe	1	
Wisteria spp.	Wisteria	2	2,3,4
Xanthosoma spp.	Elephant's ear	3,4	
Zantedeschia aethiopica	Calla lily	3,4	3
Zigadenus spp.	Death camas	1	2,3

Glossary

Technical Terms Including Their Origin

Abbreviation	Meaning
A.S.	Anglo-Saxon
D.	Dutch
dim.	diminutive
F.	French
Gr.	Greek
It.	Italian
L.	Latin
Lapp.	Lapland
M.E.	Medieval English
M.L.	Medieval Latin
N.L.	New Latin
O.F.	Old French
O.E.	Old English
R.	Russian
Sp.	Spanish

The glossary serves two purposes: (a) to define terms used in the text, and (b) to give their derivation. Note that in carrying out the first purpose, with few exceptions, only the meanings actually used in the text are given. Reference to dictionaries will give additional meanings, and other textbooks may define the same words in slightly different ways.

Scientific language and slang have much in common. Both are vital, growing, and changing phases of modern English. The same words can have different meanings in different parts of the country.

Scientific language grows in a number of ways. Some botanical terms, such as "seed," are used by scientist and layman alike and can be traced back to Anglo-Saxon days, when the word *sed* was used to designate anything sown in the ground, or, in our terms, a seed. In other instances, words from Greek or Latin have been taken into the scientific language with their identical original meaning; for instance, the Greeks called wood xylon, and we use this term, changed to xylem, to designate the woody conducting tissue of plants. Other words have a more complex history. *Metabolos* is a Greek word meaning "to change." When the scientists of Europe wrote in Latin they took this Greek word, made a Medieval Latin word from it, and used it to describe the changes undergone by some insects. It was used in 1639 to mean changes in health, and in 1845, the German scientist who elaborated the cell doctrine made a German word of it and first used it as this textbook uses it, that is, designating the changes taking place within a cell.

In using the glossary, pay as much attention to derivation as you do to the definition. To memorize the definition alone is to learn, parrot-fashion, only one word. To learn the derivation is to understand the word and possibly to introduce yourself to a whole family of new words. Pay particular attention to such combining forms as *hetero-, auto-, micro-, -phyll, angio-, -plast* or *-plasm, -spore,* and the like.

ABA abscisic acid

Abiotic nonliving components of the environment that influence plant life, such as wind pollination

Abscisic acid a plant hormone variously inducing abscission, dormancy, stomatal closure, growth inhibition, and other responses in plants

Abscission the process of leaves separating from a stem

Abscission zone (L. *abscissus,* cut off) zone of delicate, thin-walled cells extending across the base of a petiole, the breakdown of which disjoins the leaf or fruit from the stem

Absorb (L. *ab,* away + *sorbere,* to suck in) to suck up, to drink up, or to take in; in plant cells, materials are taken (absorbed) in solution

Absorption spectrum a graph relating the absorption of a beam of light by a substance to the wavelength of the light

Absorptive nutrition a feeding system characteristic of fungi, in which large molecules are broken into small molecules outside the feeder's body, and the small molecules are taken into the body one by one; contrast with ingestion of foods by animals

Accessory bud a bud located above or on either side of the main axillary bud

Accessory pigment a pigment that absorbs light energy and transfers energy to chlorophyll *a*

Achene simple, dry, one-seeded indehiscent fruit, with seed attached to ovary wall at one point only

Acid (F. *acide*, from L. *acidus*, sharp) a substance that can donate a hydrogen ion; most typical acids are sour and are compounds of hydrogen with another element or elements

Acid-growth hypothesis a hypothesis to explain the stimulation of growth of plant cells by auxin; states that the main effect of auxin is to cause cells to secrete acid (H^+ ions) and that the acid stimulates the changes in plasticity

Actinomorphic (Gr. *aktis*, ray + *morphe*, form) said of flowers of a regular or star pattern, capable of bisection in two or more planes into similar halves

Action spectrum (F. *acte*, a thing done) a graph relating the degree of physiologic response (for example, phototropism, photosynthesis) caused by different wavelengths of light

Activation energy the increase in potential energy of a complex of two or more molecules required for a chemical reaction

Active site a region of an enzyme whose shape permits the binding of substrates and where catalytic activity takes place

Active solute uptake accumulation of solutes in a cell by means of active transport

Active transport the transport of a substance across a cellular membrane by a mechanism that expends cellular energy to control the net direction of transport.

Adaptation (L. *ad*, to + *aptare*, to fit) in evolution, a trait that evolved in response to directional selection and that enhances the success of the species in the selecting environment; also, the process of generating an adaptation; in physiology, adjustment of an organism to its environment

Adenine a purine base present in nucleic acids and nucleotides

Adenosine triphosphate (ATP) a substance formed in metabolism from ADP and inorganic phosphate; the most prominent and universal molecule that acts as a carrier of energy in metabolism

Adhesion (L. *adhaerere*, to stick to) a sticking together of unlike things or materials

Adnation (L. *adnasci*, to grow to) in flowers, the growing together of two or more whorls to a greater or lesser extent; compare Adhesion

ADP adenosine diphosphate

Adsorption (L. *ad*, to + *sorbere*, to suck in) the concentration of molecules or ions of a substance at a surface or an interface (boundary) between two substances

Advanced (M.E. *advaunce*, to forward) said of a taxonomic trait thought to have evolved late in time from some more primitive trait

Adventitious (L. *adventicius*, not properly belonging to) referring to a structure arising from an unusual place: buds at other places than leaf axils, roots growing from stems or leaves

Aerate to supply or impregnate with common air, such as by bubbling air through a culture solution

Aerenchyma (Gr. *aer*, air + *en*, in + *khein*, to pour) stem or root tissue that contains irregular, large pores capable of diffusing air from aboveground organs to belowground organs

Aerobe (Gr. *aer*, air + *bios*, life) an organism living in the presence of molecular oxygen and using it in its respiratory process

Aerobic respiration the biochemical pathways through which organic molecules (for example, carbohydrate) are converted into simpler components and oxidized using molecular oxygen as a terminal electron acceptor

Aflatoxin a class of cancer-causing compounds produced by the fungus *Aspergillus flavus* and deposited in infected nuts and grains

Agar (Malay *agaragar*) a gelatinous substance obtained mainly from certain species of red algae

Aggregate fruit (L. *ad*, to + *gregare*, to collect; to bring together) a fruit developing from the several separate carpels of a single flower; for example, a strawberry

Akinete (Gr. *a*, not + *kinein*, to move) enlarged, thick-walled, nonmotile reproductive cell produced by some cyanobacteria

Albuminous cell parenchyma-like cell associated with sieve cells in the phloem of gymnosperms and some ferns

Alcohol (M.L. from Arabic *al-kuhl*, a powder for painting eyelids; later applied, in Europe, to distilled spirits that were unknown in Arabia) a product of the distillation of wine or malt; any one of a class of compounds analogous to common alcohol; the ending designates a member of this class of compounds

Alcoholic fermentation the biochemical process by which organic molecules are used as substrates to produce alcohol (generally, ethanol)

Aleurone layer (Gr. *aleurone*, flour) the outermost cell layer of the endosperm of wheat and other grains

Alfisol modified podzol soil, typical of the northern part of the deciduous forest

Alga (plural, algae) (L. *alga*, seaweed) informal name for a large group of species of simple, photosynthetic, mainly aquatic plants that lack vascular tissue (and therefore lack true stems, leaves, and roots); in this textbook, some algal groups (green, red) are classified in the plant kingdom, whereas most (brown, golden, diatoms) are classified as protists

Algin a long-chain polymer of mannuronic acid found in the cell walls of the brown algae

Alismatales an order of plants that is a basal clade or group to the rest of the monocots; includes a variety of aroids and aquatic plants

Alkali (Arabic *alqili,* the ashes of the plant saltwort) a substance with marked basic properties

Allele (Gr. *allelon,* of one another, mutually, each other) variant form of a gene

Allelopathy (Gr. *allelon,* mutually + *pathos,* suffering) synonym for amensalism; also, the special case of amensalism among plants

Alpine (L. *Alpes,* the Alps Mountains) meadowlike vegetation at high elevation, above tree line

Alternate referring to bud or leaf arrangement in which there is one bud or one leaf at a node

Alternation of generations the alternation of haploid (gametophytic) and diploid (sporophytic) phases in the life cycle of many organisms; the phases (generations) may be morphologically quite similar or very distinct, depending on the organism

Alveolates (L. *alveus,* small cavity) a group of protists that have membrane-bounded sacs lying beneath the plasma membrane; includes dinoflagellates, foraminifera, ciliates, and the apicomplexa

Amensalism (L. *a,* not + *mensa,* table) a form of biological interaction in which one organism is inhibited by another, but the other is neither inhibited nor stimulated

Amino acid (Gr. *Ammon,* from the Egyptian sun god, in N.L. used in connection with ammonium salts) an acid containing the group NH_2; one of the building blocks of a protein

Ammonification (*Ammon,* Egyptian sun god, near whose temple ammonium salts were first prepared from camel dung + L. *facere,* to make) decomposition of amino acids, resulting in the production of ammonia

Amoeboid (Gr. *amoibe,* change) eating or moving by means of temporary cytoplasmic extensions from the cell body

Amyloplast (L. *amylum,* starch + *plastos,* formed) cytoplasmic organelle specialized to store starch; abundant in roots and in storage organs such as tubers

Anabolism (Gr. *ana,* up + metabolism) the constructive phase of metabolism, in which more complex molecules are built from simpler substances

Anaerobe (Gr. *a,* without + *aer,* air + *bios,* life) an organism able to respire in the absence of free oxygen, or in greatly reduced concentrations of free oxygen

Anaerobic respiration the biochemical pathways through which organic molecules (for example, carbohydrate) are converted into simpler components in the absence of molecular oxygen

Analogous traits traits that have a similar form, function, or both, but were not evolved from the same ancestral trait; contrast with homologous traits

Anaphase (Gr. *ana,* up + *phais,* appearance) that stage in mitosis in which half chromosomes or sister chromatids move to opposite poles of the cell

Anatomy (Gr. *anatome,* dissection), the study of the internal structure of organisms

Ancestral character state a character state that a clade inherited from its immediate ancestor; contrast with derived character state

Androecium (Gr. *andros,* man + *oikos,* house) the aggregate of stamens in the flower of a seed plant

Aneuploid (Gr. *aneu,* without + *ploid*) the condition in which the number of chromosomes differs from the normal by less than a full set (*n*), for example, $3n + 1$, $2n - 1$, $2n + 4$; compare Polyploid

Angiosperm (Gr. *angion,* a vessel + *sperma* from *speirein,* to sow, hence a seed or germ) literally a seed borne in a vessel, thus a group of plants whose seeds are borne within a matured ovary

Angstrom (Å) (after A. J. Ångstrom, a Swiss physicist, 1814–1874) a unit of length equal to 0.0001 of a micron (0.1 nanometer), 10,000 angstroms = 1 micron = 0.001 millimeter

Animalia (L. *animalis,* living) the animal kingdom

Anions (Gr.) a negatively charged ion, as hydroxide, chloride, or a protein with a net excess of negatively charged carboxylate functional groups

Anisogamy (Gr. *an,* prefix meaning not + *isos,* equal + *gamete,* spouse) the condition in which the gametes, although similar in appearance, are not identical

Annual (L. *annualis,* within a year) a plant that completes its life cycle in one year and then dies

Annual ring in wood, a layer of growth formed during one year and consisting of springwood and summerwood

Annular vessels (L. *annularis,* a ring) vessels with lignified rings of secondary wall material

Annulus (L. *anulus* or *annulus,* a ring) in ferns, a row of specialized cells in a sporangium, of importance in opening of the sporangium; in mosses, thick-walled cells along the rim of the sporangium to which the peristome is attached

Anther (M.L. *anthera-* from the Gr. *anthros,* meaning flower) pollen-bearing portion of stamen

Antheridiophore a stalked structure on which male gametangia are produced, as with the liverwort *Marchantia*

Antheridium (plural, antheridia) (anther + Gr. *idion,* dim. ending, thus a little anther) in plants, the male gametangium or sperm-bearing organ of plants other than seed plants; in ascomycete fungi, a specialized hyphal tip that fuses with an ascogonium to deliver gamete nuclei

Anthocyanin (Gr. *anthros,* a flower + *kyanos,* dark blue) a blue, purple, or red vacuolar pigment

Antibiotic (Gr. *anti,* against or opposite + *biotikos,* pertaining to life) a natural organic substance that retards or prevents the growth of organisms; generally used to designate substances formed by microorganisms that prevent growth of other microorganisms

Antibody (Gr. *anti,* against + body) a protein produced in an organism, in response by the organism to a contact with a foreign substance, and having the ability of specifically reacting with the foreign substance

Anticlinal cell division (Gr. *anti,* against + *klinein,* incline) cell division where the newly formed cell wall is perpendicular to the axis of the organ surface

Antipodal (Gr. *anti,* opposite + *pous,* foot) cells or nuclei at the end of the embryo sac opposite that of the egg apparatus

Apex (L. *apex,* a tip, point, or extremity) the tip, point, or angular summit of anything: the tip of a leaf, that portion of a root or shoot containing apical and primary meristems

Apical dominance the inhibition of lateral buds or meristems by the apical meristem

Apical meristem a mass of dividing cells at the very tip of a shoot or root

Apicomplexa one group of pathogenic protists in the alveolate clade; contains some organisms previously classified as protozoa, for example, *Plasmodium,* the cause of malaria

Apogamy (Gr. *apo,* away from + gamete) the process in which haploid gametophytes produce haploid sporophytes without the fusion of gametes

Apomixis (Gr. *apo,* away from + *mixis,* a mingling) the production of offspring in the usual sexual structures without the mingling and segregation of chromosomes

Apoplast (Gr. *apo,* away from + *plastides,* formed) the region of a plant tissue between the cells, outside of the plasma membrane of the cells that form it; contrast to Symplast

Apospory (Gr. *apo,* away from + spore) the process in which diploid sporophytes produce diploid gametophytes without the occurrence of meiosis

Apothecium (Gr. *apotheke,* a storehouse) a cup- or saucer-shaped open ascoma

Arbuscule a nutrient-transferring mycelial structure made by a glomeromycete fungus in endomycorrhizal association with a host plant; consisting of highly branched hyphae, it forms in the space enclosed by the walls of a single host plant cell

Archaea (Gr. *arckhaios,* ancient) a domain of prokaryotic organisms, including methane-producing, halophilic, and hot acid–dwelling forms

Archaebacteria see Archaea

Archegoniophore a stalked structure on which female gametangia are produced, as with the liverwort *Marchantia*

Archegonium (L. dim. of Gr. *archegonos,* literally a little founder of a race) female gametangium or egg-bearing organ, in which the egg is protected by a jacket of sterile cells

Arctic zone a vegetational or climatic zone at high latitudes, where conditions during the growing season are too cool for trees to exist

Aril (M.L. *arillus,* a wrapper for a seed) an accessory seed covering formed by an outgrowth at the base of the ovule in *Taxus*

Ascocarp (Gr. *askos,* a bag + *karpos,* fruit) a fruiting body of the Ascomycetes, generally either an open cup, a vessel, or closed sphere lined with special cells called asci; see Ascus

Ascogenous hyphae hyphae arising from the ascogonium, after the formation of *n* + *n* paired nuclei; the hymenial layer of the ascocarp develops from the ascogenoushyphae

Ascogonium (plural, ascogonia) in ascomycete fungi, an enlarged hyphal tip that accepts gamete nuclei from an antheridium and initiates the dikaryotic stage of the sexual life cycle

Ascoma (plural, ascomata) a sexual fruiting body made by an ascomycete fungus

Ascomycota (Gr. *askos,* a bag + *mykes,* fungus) a monophyletic phylum in kingdom Fungi, consisting of fungi that produce sexual spores in asci; mycelial members have septate hyphae, a dikaryotic stage in the sexual life cycle, and spread primarily by releasing conidia (mitospores)

Ascospore (Gr. *askos,* a bag + spore) meiospore produced within an ascus by ascomycete fungi

Ascus (plural, asci) (Gr. *askos,* a bag) a saclike cell in which ascomycete fungi produce meiospores

Asepsis (Gr. *a,* not + *septos,* putrid) the condition of being germ free

Aseptate characteristic of a hypha that is not divided into cell-like compartments by regularly spaced cross walls

Asexual (Gr. *a,* without + L. *sexualis,* sexual) any type of reproduction not involving the union of gametes or meiosis

Aspect (L. *aspectus,* appearance) the direction of slope of a surface, as a hillside with a south-facing aspect

Assimilation (L. *assimilare,* to make like) the transformation of food into protoplasm

Asterid clade a large number of herbaceous species (in many flowering plant families) that are the most highly specialized eudicots, including heaths, tomatoes and potatoes, mints, umbellifers, and sunflowers

Atoms (F. *atome,* from the Gr. *atomos,* indivisible) the smallest particles in which the elements combine either with themselves or with other elements, and thus the smallest quantity of matter known to possess the properties of a particular element; a unit of matter consisting of a dense, central nucleus, consisting of several positively charged protons and uncharged neutrons, surrounded by a number of negatively charged electrons

ATP see Adenosine triphosphate

ATP synthetase an enzyme that produces ATP from ADP and phosphate, using as an energy source the flow of protons across a membrane in response to a concentration gradient or electrical potential difference

Auricles (L. *auricula,* dim. of *auris,* ear) earlike structures; in grasses, small projections that grow out from the opposite side of the sheath at its upper end where it joins the blade

Autoecious (Gr. *auto,* self + *oikia,* dwelling) having a complete life cycle on the same host

Autoradiograph (Gr. *auto,* self + L. *radio*Ins, a ray + Gr. *graphe,* a painting) a photographic print made by a radioactive substance acting on a sensitive photographic film

Autotetraploidy the condition in which the doubling of the chromosome number occurs in one cell or between cells on the same plant

Autotrophic (Gr. *auto,* self + *trophein,* to nourish with food) pertaining to an organism that is able to manufacture its own food

Auxin (Gr. *auxein,* to increase) a plant hormone regulating cell elongation and various other aspects of development

Axial system the secondary vascular cells oriented parallel to the long axis of a stem or root

Axil (Gr. *axilla,* armpit) the upper angle between a petiole of a leaf and the stem from which it grows

Axile form of placentation where the ovules are attached along the central axis of the ovary

Axillary bud a bud formed in the axil of a leaf

Bacillus (L. *baculum,* a stick) a rod-shaped bacterium

Backcross a mating in which a hybrid organism mates with a member of one of the hybrid's parental types

Bacteria (Gr. *bakterion,* a stick) a domain of prokaryotic organisms; also, common name for prokaryotes

Bacteriochlorophyll a light-harvesting pigment found in certain bacteria; the molecular structure is similar to that of chlorophyll *a,* but certain side groups and the absorption spectrum are unique

Bacteriology (bacteria + Gr. *logos,* discourse) the science of bacteria

Bacteriophage (bacteria + Gr. *phagein,* to eat) literally, an eater of bacteria; a virus that infects specific bacteria, multiples therein, and usually destroys the bacterial cells

Ballistospore release a mechanism of active meiospore release found only in basidiomycete fungi, based on physical forces that arise from merging a water droplet with a water film

Bark (Swedish *bark,* rind) the external group of tissues, from the cambium outward, of a woody stem or root

Basal angiosperms a small group of species (<200) thought to be ancestral to all other monocots and dicots; includes water lilies, star anise, and *Amborella*

Base a substance that can accept a proton (H$^+$); also, the purine and pyrimidine groups in nucleic acids and nucleotides are collectively called bases

Base pair the nitrogen bases that pair in the DNA molecule, adenine with thymine and guanine with cytosine

Base sequence a sequence of nucleotide bases in DNA or RNA potentially containing information used to direct the synthesis of a protein

Basidioma (plural, basidiomata) a sexual fruiting body of a basidiomycete fungus, composed of branching dikaryotic hyphae, basidia, and basidiospores

Basidiomycota (M.L. *basidium,* a little pedestal + Gr. *mykes,* fungus) a monophyletic phylum in kingdom Fungi, in which all members extrude meiospores from a basidium; mycelial members have septate hyphae, and a dikaryotic stage in the sexual life cycle

Basidiospore (M.L. *basidium,* a little pedestal + spore) meiospore that is extruded from a basidium in basidiomycete fungi

Basidium (plural, basidia) (M.L. *basidium,* a little pedestal) a thickened, club-shaped or elongate cell that extrudes meiospores in basidiomycete fungi

Berry a simple fleshy fruit, the ovary wall fleshy and including one or more carpels and seeds

Biennial (L. *biennium,* a period of two years) a plant that requires two years to complete its life cycle; flowering is normally delayed until the second year

Bifacial leaf (L. *bis,* twice + *facies,* face) a leaf having distinctly different upper and lower surfaces

Binary fission the process of diving in two; the reproductive process of single-celled prokaryotes

Binomial (L. *binominis,* two names) the formal name given to a species, consisting of the genus name followed by a species epithet; also called the species name

Bioassay (Gr. *bios,* life + L. *exagere,* to weigh or test) to test for the presence or quantity of a substance by using an organism's response as an indicator

Biodiversity the diversity of life forms found in a region; measures include the number of species and their relative abundance

Biogeography (Gr. *bios,* life + *ge,* Earth + *graphein,* to write) the study of the geographic distribution of organisms and the reasons for the limits to their distribution, often including distributions over geologic time and with a focus on taxonomically related organisms (families, genera); see Plant geography

Biological barrier a barrier to crossing (hybridization) of plants caused by differences in pollination vector or timing in flower opening, in contrast to physiologic barriers (incompatibility of pollen with stigma or style) or ecologic barriers (habitats too far apart)

Biological clock an endogenous timing mechanism inferred to exist in cells to explain various cyclical behaviors

Biological species a group of organisms, the members of which reproduce successfully only with other members of the same group and not with members of other groups

Biology (Gr. *bios,* life + *logos,* word, speech, discourse) the science that deals with living things

Biomass pyramid the diagrammatic summary of biomass of each trophic level, starting with producer biomass on the bottom layer of the diagram and ending with tertiary consumers on the top layer; each higher layer is smaller than the one below, so that the combination of layers look like a stepped pyramid

Biosystematics (Gr. *bios,* life + *synistanai,* to place together) a field of taxonomy that emphasizes breeding behavior and chromosome characteristics

Biotic (Gr. *biokitos,* relating to life) referring to living components of the environment that affect plant life, for example, insect pollination

Biotin a vitamin of the B complex

Bird's nest fungus a basidiomycete that makes a basidioma resembling a miniature nest with eggs

Bladder (O.E. *bladre,* a blister) a gas-filled sac with buoyancy that keeps some aquatic plants upright

Blade typically the thin, expanded portion of a leaf; in some algae, the leaflike frond

Bloom (Gr. *blume,* flower; also IndoEuropean *bhlo,* to spring up) an increase in phytoplankton density sufficient to color bodies of freshwater

Blue–green algae see Cyanobacteria

Bolting the rapid formation of a flowering shoot from the apex of a plant with a rosette habit

Bordered pit a pit in a tracheid, vessel member, or sometimes a fiber having a distinct rim of the cell wall overarching the pit membrane

Boreal forest any forest at low elevations in the boreal zone, a north-temperate belt between the arctic and temperate zones; these forests are usually dominated by conifers

Botany (Gr. *botane,* plant, herb) the study of plant and plantlike organisms; synonym, plant biology

Bracket fungus see Shelf fungus

Bract (L. *bractea,* a thin plate of precious metal) a modified leaf, from the axil of which arises a flower or an inflorescence

Brown algae multicellular, cool-water, marine algae; includes the large kelps; placed among other heterokont protists in this text; also classified in its own division, the Phaeophyta

Bryophytes (Gr. *bruon,* moss) a group of small, mainly terrestrial plants that have an embryo life cycle phase but lack vascular tissue; contains mosses, liverworts, and hornworts

Bud (M.E. *budde,* bud) an undeveloped shoot, largely meristematic tissue, generally protected by modified scale-leaves; also a swelling on a yeast cell that will become a new yeast cell when released

Bud scale a modified protective leaf of a bud

Bud scar a scar left on a twig when the bud or bud scales fall away

Bulb (L. *bulbus,* a modified bud, usually underground) a short, flattened, or disk-shaped underground stem, with many fleshy scale-leaves filled with stored food

Bundle scar scar left where conducting strands passing out of the stem into the leaf stalk were broken off when the leaf fell

Bundle sheath sheath of parenchyma or sclerenchyma cells that surround the vascular bundles of leaves, sometimes called border parenchyma

Bundle sheath sheath of parenchyma cells that surround the vascular bundles of leaves, sometimes called border parenchyma

^{13}**C,**^{14}**C** isotopes of carbon that have additional neutrons; only ^{14}C is radioactive

C_3 **cycle** the Calvin Benson cycle of photosynthesis, in which the first products after CO_2 fixation are three-carbon molecules

C_4 **cycle** the Hatch–Slack cycle of photosynthesis, in which the first products after CO_2 fixation are four-carbon molecules

C type life history a life history pattern that describes long-lived, large, competitive plants; similar to k-selected plants

Callose (L. *callum,* thick skin + *ose,* a suffix indicating a carbohydrate) an amorphous carbohydrate deposited around pores in sieve-tube members and in other areas in cell walls

Callus (L. *callum,* thick skin) mass of large, thin-walled cells, usually developed as the result of wounding

Calorie (L. *calor,* heat) the amount of heat needed to increase the temperature of 1 g water 1°C (usually from 14.5 to 15.5°C), also called gram-calorie; 1,000 calories = 1 kilocalorie

Calyptra (Gr. *kalyptra,* a veil, covering) in bryophytes, an envelope covering the developing sporophyte, formed by growth of the venter of the archegonium

Calyx (Gr. *kalyx,* a husk, cup) sepals collectively; outermost flower whorl

Cambium (L. *cambium,* one of the alimentary body fluids supposed to nourish the body organs) a layer, usually regarded as one or two cells in thickness, of persistently meristematic tissues, giving rise to secondary tissues, resulting in growth in diameter

Canopy (Gr. *kanopeion,* a cover over a bed to keep off gnats) the leafy portion of a tree or shrub

Cap in mushrooms, the umbrella-like portion in which meiospores are made

Capillaries (L. *capillus,* hair) very small spaces, or very fine bores in a tube

Capillary forces forces that pull water into the narrow space between hydrophilic surfaces; includes adhesive force beween water and the surfaces and cohesive force between water molecules

Capsule (L. *capsula,* dim. of *capsa,* a case) in angiosperms, a simple, dry, dehiscent fruit, with two or more carpels; in mosses, a spore-producing structure with diploid photosynthetic tissue, capable of splitting open along slits or at an apical pore

Carbohydrate (chemical combining forms, *carbo,* carbon + *hydrate,* containing water) a food composed of carbon, hydrogen, and oxygen, with the general formula $C_nH_{2n}O_n$

Carbon fixation the enzymatic reaction in which CO_2 is attached to a receiver compound such as ribulose bisphosphate, thereby adding to the supply of organic carbon; occurs chiefly in photosynthesis

Carboxylase/oxygenase (rubisco) see Ribulose bisphosphate carboxylase

Carboxysome (F. *carbboe,* carbon + Gr. *oxys,* acidic + Gr. *soma,* body) polyhedral bodies rich in ribulose bisphosphate carboxylase, found in some photosynthetic bacteria and cyanobacteria

Carcinogenic (Gr. *karkinoma,* cancer + L. *genitalis,* to beget) that which induces cancer

Carinal canal an open canal, devoid of cells, which may function to transport air down through the cortex to underground rhizomes and roots; characteristic of horsetail stems

Carnivore (L. *caro,* flesh + *vorare,* to swallow up) an organism that obtains food by killing and consuming other organisms; normally refers to an animal that eats other animals, but the term also can be used for microbial protozoans or for an herbivorous animal that consumes entire plants or their seeds

Carotenoid (L. *carota,* carrot) a reddish orange plastid pigment

Carpel (Gr. *karpos*, fruit) a floral leaf bearing ovules along the margins

Carrageenan a polysaccharide found in the walls of red algae that reacts with milk proteins to make a stable, creamy, thick solution or gel

Caruncle (L. *caruncula*, dim. of *caro*, flesh, wart) a spongy outgrowth of the seed coat, especially prominent in the castor bean seed

Caryophyllid clade a group of eudicots with small flowers that have superior ovaries; includes many succulents (ice plants, cacti) and also carnations, amaranths, buckwheat, spinach, and sugar beet

Caryopsis (Gr. *karyon*, a nut + *opsis*, appearance) a simple, dry, one-seeded, indehiscent fruit, with pericarp firmly united all around to the seed coat; also called a grain

Casparian strip suberized strip that impregnates the radial and transverse wall of endodermal cells

Catabolism (Gr. *kata-*, down + *bole*, a throw) the destructive phase of metabolism, in which complex molecules are broken down to simpler products

Catalyst (Gr. *katelyein*, to dissolve) a substance that accelerates a chemical reaction but that is not used up in the reaction

Cation exchange (Gr. *kata*, downward) the replacement of one positive ion (cation) by another, as on a negatively charged clay particle

Cations (Gr.) positively charged ions, such as sodium, potassium, and hydronium ions, and proteins with a net positive charge

Catkin (literally a kitten, apparently first used in 1578 to describe the inflorescence of the pussy willow) a type of inflorescence, really a spike, generally bearing only pistillate flowers or only staminate flowers, which eventually fall from the plant entirely

Caulescent (Gr. *kaulos*, a plant stem) a plant with a stem that bears leaves separated by visibly elongated internodes, as opposed to a rosette plant

Cell (L. *cella*, small room) a structural and physiologic unit composed of living organisms, in which take place the majority of complicated reactions characteristic of life; it is surrounded by a plasma membrane, contains a metabolic system, and has a store of DNA

Cell cycle the repeating sequence of events involved in the reproduction of a eukaryotic cell, including G1, S (DNA synthesis), G2, and M (mitosis and cytokinesis) phases

Cell differentiation processes of cell growth and other changes in a cell that lead to it becoming specialized

Cell plate a structure that forms at the equatorial plane of the cell at right angles to the spindle fibers during cytokinesis; the precursor of the middle lamella

Cell theory the theory that states that the cell is the fundamental unit of living matter and that organisms are formed from one or more cells

Cellulose (cell + *ose*, a suffix indicating a carbohydrate) a complex carbohydrate occurring in the cell walls of the majority of plants; it is composed of hundreds of simple sugar molecules, glucose, linked together in a characteristic manner; cotton fibers are largely cellulose

Cell wall a protective meshwork of molecules enclosing an Archaeal, Bacterial, fungal, protist, or plant cell; usually secreted by the cell

Cenozoic (Gr. *kainos*, recent + *zoe*, life) the geologic era extending from 65 million years ago to the present

Central cell large cell making up the central portion of the embryo sac

Centromere the region of a chromosome that binds to the spindle during mitosis

Chalaza (Gr. *chalaza*, small tubercle) the region on a seed at the upper end of the raphe where the funiculus spreads out and unites with the base of the ovule

Channels proteins in a biological membrane that allow specific molecules to cross the membrane

Chaparral (Sp. *chaparro*, low, woody vegetation) a vegetation type characterized by small-leaved, evergreen shrubs growing together into a nearly impenetrable scrub

Character in genetics and cladistics, a distinctive trait or aspect of the phenotype, controlled by one or more genes; it may be a detail of body form, either physical or chemical, or the organism's function or behavior

Character matrix a table that compares states of shared characters in selected taxa

Character states alternative forms that a given character may take

Charophytes a group of algae closely related to the chlorophytes, classified in this text as plants

Chemiosmotic theory the theory that ATP is synthesized in photosynthesis and mitochondrial respiration using as a direct energy source the flow of protons across a membrane in response to a concentration gradient or electrical potential difference, the gradient and/or potential difference being formed during electron transport reactions within the membrane

Chemoautotroph an organism that oxidizes reduced inorganic compounds such as H_2S to obtain energy and that uses CO_2 as a carbon source

Chemoheterotroph an organism that obtains both energy and carbon from organic sources

Chemotroph (Gr. *chymeia*, to pour, later alchemy, chemistry + *trophein*, to feed) bacteria that oxidize reduced inorganic compounds such as H_2S to obtain energy

Chemotropism (Gr. *chymeia*, to pour + Gr. *tropos*, a turning) an orientation of growth in response to a chemical agent; the organ usually grows toward or away from the source of the chemical

Chiasma (Gr. *chiasma*, two lines placed crosswise) the microscopically visible cross formed between two chromatids of homologous chromosomes during prophase 1 of meiosis; associated with the breaking and rejoining of the broken ends of different chromatids

Chitin (Gr. *chiton*, a coat of mail) a polymer in which the monomer unit is the modified sugar *N*-acetyl glucosamine; the principal stiffening material in cell walls of true fungi and the exoskeleton of insects and crustaceans

Chlamydospore (Gr. *chlamys*, a horseman's or young man's coat + *spore*) a heavy-walled resting asexual spore

Chlorenchyma (Gr. *chloros*, green + *enchyma*, a suffix meaning tissue) parenchyma tissue possessing chloroplasts

Chlorophyll (Gr. *chloros*, green + *phyllon*, leaf) the green pigment found in the chloroplast, important in the absorption of light energy in photosynthesis

Chlorophytes (Gr. *chloros*, green + *phyton*, a plant) a group of mainly freshwater, photosynthetic protists commonly called green algae; they share several complex chemicals with land plants, and in this text they are classified as plants

Chloroplast (Gr. *chloros*, green + *plastos*, formed) specialized cytoplasmic body, containing chlorophyll, in which occur the reactions of photosynthesis, including sugar and starch synthesis

Chlorosis (Gr. *chloros*, green + *osis*, diseased state) failure of chlorophyll development, because of a nutritional disturbance or because of an infection of virus, bacteria, or fungus

Chromatid (chromosome + L. *id*, suffix meaning daughters of) the half chromosome during prophase and metaphase of mitosis, and between prophase I and anaphase II of meiosis

Chromatin (Gr. *chroma*, color) substance in the nucleus that readily takes artificial staining; a complex of DNA and proteins

Chromatin bodies bodies in bacteria that give some of the histochemical reactions that are associated with the chromosomes of higher organisms

Chromatophores (Gr. *chromo*, color + *phorus*, a bearer) in algae, bodies bearing chlorophyll; in bacteria, small bodies, about 100 nm in diameter, containing chlorophyll, protein, and a carbohydrate

Chromoplast (Gr. *chroma*, color + *plastos*, formed) specialized plastid containing yellow or orange pigments

Chromosome (Gr. *chroma*, color + *soma*, body) a nuclear body containing genes in a linear order and undergoing characteristic division stages; one of the units of condensed chromatin visible during cell division

Chromosome set the collection of different chromosomes within a cell; a diploid cell has two chromosome sets; a haploid cell has only one chromosome set

Chrysophyta (Gr. *khrusos*, gold) a small group of freshwater and (mainly) unicellular algae that contain chlorophylls *a* and *c*, have cellulose walls (sometimes with silica), and store carbohydrate as oil droplets; classified in this text with protists in the heterokont group

Chytridiomycota an artificial (nonmonophyletic) phylum in kingdom Fungi, containing all true fungi that have swimming reproductive cells

Cilia (singular, cilium) (F. *cil*, an eyelash) protoplasmic hairs that, by a whiplike motion, propel certain types of unicellular organisms, gametes, and zoospores through water

Ciliates (F. *cil*, an eyelash) one group of motile protists in the alveolate clade; contains some organisms previously classified as protozoa, for example, *Paramecium*, capable of swimming by the synchronized action of many cilia

Cisterna (plural, cisternae) (L. *cistern*, a reservoir) generally referring to sections of the endoplasmic reticulum that appear in electron micrographs as parallel membranes, each about 5 nm in thickness bounding a space about 40 nm in width

Citric acid cycle see Tricarboxylic acid cycle

Clade (Gr. *klados*, twig) a branch in the tree of life, consisting of an originating taxon and all of its descendant taxa

Cladistics (Gr. *klados*, twig) a set of quantitative methods and concepts for generating hypotheses about the evolutionary relationships among taxa; a form of systematics that shows evolutionary relationships among organisms by arranging them in a branching diagram (a cladogram), each branch representing organisms that share certain derived traits

Cladode (Gr. *kladodes*, having many shoots) a cladophyll

Cladogram (Gr. *klados*, twig + *gramma*, letter) a diagram presenting a hypothesis on the sequence of branching events that occurred to generate a given set of present-day taxa from their common ancestor

Cladophyll (Gr. *klados*, a shoot + *phyllon*, leaf) stem or branch resembling a leaf

Clamp connection a hyphal relic of events that maintain the dikaryotic condition in a basidiomycete mycelium, named for the relic's resemblance to a carpenter's clamp

Class (L. *classis*, one of the six divisions of Roman people) in the taxonomic hierarchy, a group that contains one or more orders

Clay soil particles less than 2 microns in diameter, composed mainly of aluminum (Al), oxygen (O), and silicon (S)

Cleistothecium (plural, cleistothecia) (Gr. *kleistos*, closed + *thekion*, a small receptacle) a closed spherical ascoma

Climax community the last stage of a natural succession; a community capable of maintaining itself as long as the climate does not change

Clone (Gr. *klon*, a twig or slip) a group of genetically identical individual organisms produced asexually

Closed bundle a vascular bundle lacking cambium

Closed-cone conifer a conifer (usually in the genera *Pinus* or *Cupressus*) that produces cones that neither open at maturity nor fall from the parent tree; the cone scales open only when exposed to an increased temperature, the exact temperature depending on the species

Coal Age the Carboniferous period, beginning 345 million years ago and ending 280 million years ago

Coalescence (L. *coalescere*, to grow together) a condition in which there is union of separate parts of any one whorl of flower parts; synonyms are Connation and Cohesion

Coccus (plural, cocci) (Gr. *kokkos*, a berry) a spherical bacterium

Codominance concerning alleles of a gene, sharing in influence; the trait produced by two codominant alleles is intermediate between (or different from) that produced by either alone; also incomplete dominance

Codon a sequence of three bases (nucleotides) along the RNA molecule that code for a single amino acid

Coenobium (Gr. *koinois*, common + Gr. *bios*, life) a colony of unicellular organisms surrounded by a common membrane

Coenocytic (Gr. *koinos*, shared in common + *kytos*, a vessel) a condition in which an organ or organism consists of a single cell with many nuclei, rather than being divided into many cells

Coenogamete (Gr. *koinos*, shared in common + gamete) a multinucleate gamete, lacking cross walls

Coenomycetes informal term for all true fungi that lack a dikaryotic stage and have aseptate hyphae

Coenzyme a substance, usually nonprotein and of low molecular weight, necessary for the action of some enzymes

Coevolution coordinated evolution of two or more species, driven by interactions that cause changes in each species to favor changes in the other species

Cohesion (L. *cohaerere*, to stick together) union or holding together of parts of the same materials; the union of floral parts of the same whorl, as petals to petals

Coleoptile (Gr. *koleos*, sheath + *ptilon*, down, feather) the first leaf in germination of grasses that sheaths the succeeding leaves

Coleorhiza (Gr. *koleos*, sheath + *rhiza*, root) sheath that surrounds the radicle of the grass embryo and through which the young root bursts

Collenchyma (Gr. *kolla*, glue + *enchyma*, a suffix, derived from parenchyma and denoting a type of cell tissue) a tissue composed of living cells with thickened cell walls capable of stretching that fit rather closely together; found in young stems and petioles

Colloid (Gr. *kolla*, glue + *eidos*, form) referring to matter composed of particles, ranging in size from 0.0001 to 0.000001 mm, dispersed in some medium; milk and mayonnaise are examples

Colony (L. *colonia*, a settlement) a growth form characterized by a group of closely associated, but poorly differentiated, cells; sometimes filaments can be associated together in a colony (as in *Nostoc*), but more typically unicells are associated in a colony

Commelinids a clade of monocots typically having nonshowy flowers; palms, grasses, sedges, bromeliads, ginger, banana, and the bird-of-paradise are examples

Community (L. *communitas*, a fellowship) all the populations within a given habitat; usually the populations are thought of as being somewhat interdependent

Companion cell cell associated with sieve-tube members

Compensation depth (L. *compensare*, to counterbalance) that depth, in a body of water, at which light intensity is so low that photosynthesis of floating or submerged plants just equals respiration

Compensation point (L. *compensare*, to counterbalance) the light intensity (light compensation point) or the carbon dioxide concentration (CO_2 compensation point) at which photosynthesis just equals respiration

Competition (L. *competere*, to strive together) a form of biological interaction in which both organisms (at least initially) decline in growth or success because of the insufficient supply of some necessary factor(s)

Complete flower a flower having four whorls of floral leaves: sepals, petals, stamens, and carpels

Compound (M.F. *compondre*, put together) a chemically distinct substance produced by the union of two or more elements in definite proportion by weight; formed from specific molecules

Compound leaf a leaf whose blade is divided into several distinct leaflets

Compound pistil a group of fused pistils

Conceptacle (L. *conceptaculum*, a receptacle) a cavity or chamber of a frond (of *Fucus*, for example) in which gametangia are borne

Condensation reaction a chemical or physical reaction from which one product is water

Conduction (L. *conducere*, to bring together) act of moving or conveying a substance through the plant; generally the movement of water through the xylem or food through the phloem

Cone (Gr. *konos*, a pine cone) a fruiting structure composed of modified leaves or branches, which bear sporangia (microsporangia, megasporangia, pollen sacs, or ovules), and frequently arranged in a spiral or four-ranked order; for example, a pine cone

Cone scale the flat, woody parts of pine cones that spiral out from the central axis and bear the ovules (and later seeds) on their upper surfaces; each is subtended by a sterile bract

Confocal microscopy a technique in which the illumination and objective lens of a microscope are focused on only one point of a sample on a microscope stage at a time, eliminating light scattering and increasing resolution and contrst

Conidium (plural, conidia) (Gr. *konis*, dust) mitospores of fungi that form individually (not in a sporangium) by releasing parts of a hypha; conidia may arise by fragmentation of a hypha or by releasing cells from the tip of a hypha

Conidiophore (conidia + Gr. *phoros*, bearing) a hypha that makes and releases conidia

Conidiosporangium (Gr. *konis*, dust + sporangium) sporangium formed by being cut off from the end of a terminal or lateral hypha

Conidiospore (conidia + spore) spore formed as described for conidia

Conifer (cone + L. *ferre*, to carry) a cone-bearing tree

Conjugation (L. *conjugatus*, united) in algae, sexual reproduction accomplished by fusing isogametes; in bacteria, transfer of DNA from one bacterium to another bacterium of the same species

Conk the fruiting body of a bracket fungus

Connation (L. *connatus*, to be born together) condition in a flower where there is a union of similar parts of any one whorl of appendages; synonym of Coalescence

Consensus tree a phylogenetic tree that has all the features shared by the equally parsimonious cladograms in a study, while leaving conflicts unresolved; also called a consensus cladogram

Conservation (L. *conservare*, to keep) in biology, the systematic protection of natural resources and species; also the study of techniques for protecting resources and species

Conservative said of a taxonomic trait whose expression is not modified to any great extent by the external environment; a trait that is constant unless its genetic base is changed

Conserved sequence in cladistics, a segment of DNA that has the same base sequence in many taxa

Convergent evolution the independent evolution of similar traits by two taxa that were originally quite different, resulting from exposure to similar selection pressures

Coral fungus a basidiomycete fungus in which the basidioma resembles a marine coral

Core angiosperms a clade containing the most flowering plants; can be subdivided into the subclades magnoliids, monocots, and eudicots

Cork (L. *quercus*, oak) an external, secondary tissue impermeable to water and gases

Cork cambium a meristem that forms the periderm; synonym for phellogen

Corm (Gr. *kormos*, a trunk) a short, solid, vertical, enlarged underground stem in which food is stored

Corolla (L. *corolla*, dim. of *corona*, a wreath, crown) petals, collectively; usually the conspicuous colored flower whorl

Cortex (L. *cortex*, bark) region of primary tissue in a stem or root bounded externally by the epidermis and internally in the stem by the phloem and in the root by the pericycle; develops from the primary meristem, the ground meristem

Cotyledon (Gr. *kotyledon*, a cup-shaped hollow) seed leaf; there are two in the embryo/seedling of dicotyledonous plants; they generally store food and can expand and become photosynthetic; in monocotyledonous plants, only one is present, generally a digestive organ

Covalent bond a chemical bond between two atoms formed by shared electrons, that is, electrons in bonding orbitals shared by the atoms

Crassulacean acid metabolism (CAM) the biochemical pathways by which the succulent genus *Crassula* and other plants fix carbon at night and release it for photosynthesis during the day

Cristae (L. *crista*, a crest) crests or ridges, used here to designate the infoldings of the inner mitochondrial membrane

Crossing-over the exchange of corresponding segments between chromatids of homologous chromosomes

Cross-pollination the transfer of pollen from a stamen to the stigma of a flower on another plant, except in clones

Crozier (Germanic *crosse*, Bishop's staff) in ascomycete fungi, a hook-shaped hypha that produces an ascus and a new crozier

Crustose said of a lichen that grows firmly attached to the substrate along its entire lower surface

Cuticle (L. *cuticula*, dim. of *cutis*, the skin) waxy layer on outer wall of epidermal cells

Cutin (L. *cutis*, the skin) waxy substance that is but slightly permeable to water, water vapor, and gases

Cutinization impregnation of cell wall with cutin

Cyanobacteria (Gr. *kyanos*, blue) a group of unicellular or filamentous photosynthetic prokaryotes also capable of fixing nitrogen; also called blue–green algae

Cycad (Gr. *kukas* or *koix*, a kind of palm tree) a group of gymnosperms that are palmlike in appearance because of their large, evergreen, compound leaves and an absence of trunk branching; female cones are large and occur only at the tip of the trunk

Cyclic electron transport movement of electrons in thylakoid membranes through a closed pathway that does not result in net oxidation or reduction of any intermediate; the movement is stimulated by light and generates a chemiosmotic gradient that provides energy for the synthesis of ATP

Cyclic photophosphorylation synthesis of ATP through the cyclic electron transport

Cyclosis (Gr. *kyklosis*, circulation) movement of cytoplasm around a cell; synonym is cytoplasmic streaming

Cyme (Gr. *kyma*, a wave, a swelling) a type of inflorescence in which the apex of the main stalk or the axis of the inflorescence ceases to grow quite early, relative to the laterals

Cystocarp (Gr. *kystos*, bladder + *karpos*, fruit) a peculiar diploid spore-bearing structure formed after fertilization in certain red algae

Cytochrome (Gr. *kytos*, a receptacle or cell + *chroma*, color) a class of several electron-transport proteins serving as carriers in mitochondrial oxidations and in photosynthetic electron transport

Cytokinesis (Gr. *kytos*, a hollow vessel + *kinesis*, motion) division of cytoplasmic constituents at cell division

Cytokinin (Gr. *kytos*, a receptacle or cell + *kinetos*, to move) a class of growth hormones important in regulation of cell division, delaying senescence, and organ initiation

Cytology (Gr. *kytos*, a hollow vessel + *logos*, word, speech, discourse) the science dealing with the cell

Cytoplasm (Gr. *kytos*, a hollow vessel + *plasma*, form) all the protoplasm of a protoplast outside the nucleus

Cytoplasmic streaming a controlled flow of cell contents along cytoskeletal elements from one part of a cell to another

Cytosine a pyrimidine base found in DNA and RNA

Cytoskeleton the framework of protein filaments, including microfilaments, microtubules, and intermediate filaments, in the cytoplasm

Day-neutral plants varieties of plants whose flowering does not depend on day length; contrast Long-day plants and Short-day plants

Deciduous (L. *deciduus*, falling) referring to trees and shrubs that lose their leaves in the fall

Decomposer (L. *de,* from + *componere,* to put together) an organism that obtains food by breaking down dead organic matter into simpler molecules

Decomposition (L. *de,* to denote an act undone + *componere,* to put together) a separation or dissolving into simpler compounds; rotting or decaying

Dehiscent (L. *dehiscere,* to split open) opening spontaneously when ripe, splitting into definite parts

Deletion (L. *deletus,* to destroy, to wipe out) used here to designate an area, or region, lacking from a chromosome

Denitrification (L. *de,* to denote an act undone + *nitrum,* nitro, a combining form indicating the presence of nitrogen + *facere,* to make) conversion of nitrates into nitrites, or into gaseous oxides of nitrogen, or even into free nitrogen

Density the number of individuals per unit area; for example, 200 red oak trees per hectare; sometimes a term used to describe the degree of canopy closure, as "canopy density of 35%" would mean that the plant canopy covers 35% of the ground area

Deoxyribonucleic acid (DNA) hereditary material; long, double-stranded polymer of nucleotides A, T, G, and C

Derived character state a character state that evolved after the founding of a clade and was not present in the clade's immediate ancestor

Dermal tissue system consists of the outer tissues of the plant body, epidermis (a primary tissue) and the periderm (a secondary tissue); functions to protect the plant body from drying, pathogens, and insects

Desert shrub (M.E. *schrubbe,* shrub) a vegetation type characterized by evergreen or drought-deciduous shrubs growing together rather openly, generally in an area with annual precipitation less than 25 cm

Determinate (L. *determinare,* to limit) generally, having defined limits; in plant development, a morphogenetic process that ends with a cessation of cell division and growth

Determined in plant development, a tissue having a limited number of developmental possibilities

Detritus (L. *detritus,* worn away) particulate organic matter released in the processes of decomposition of dead organisms or parts of organisms (such as plant litter)

Deuterium or heavy hydrogen a hydrogen atom, the nucleus of which contains one proton and one neutron; it is written as ^2H; the common nucleus of hydrogen consists only of one proton

Deuteromycete see Mitosporic fungus

Development (F. *developper,* to unfold) developmental changes of a cell, tissue, or organ leading to the presence of features that equip that cell, tissue, or organ for performing specialized functions

Diastase (Gr. *diastasis,* a separation) a complex of enzymes that brings about the hydrolysis of starch with the formation of sugar

Diatom (Gr. *diatomos,* cut in two) member of a group of golden brown algae with silicious cell walls fitting together much as do the halves of a pill box

Diatomite (Gr. *diatomos,* cut in two) fossil deposits of diatom cell walls; currently mined for such commercial purposes as filters, extenders, and stabilizers

Dichotomy (Gr. *dicha,* in two) the forking of an axis into two branches

Dicots (Gr. *dis,* twice + *kotyledon,* a cup-shaped hollow) formally, dicotyledonous plants: plants with embryos having two cotyledons; current systematics now breaks up this group into eudicots, magnoliids, and a few basal angiosperms

Dictyosome (Gr. *diktyon,* a net + *some,* body) one of the component parts of the Golgi apparatus; in plant cells, a complex of flattened double lamellae

Differentially permeable referring to a membrane through which different substances diffuse at different rates; some substances may be unable to diffuse through such a membrane

Differentiation (L. *differre,* to carry different ways) development from one cell to many cells, accompanied by a modification of the new cells for the performance of particular functions

Diffuse porous wood with an equal and random distribution of large xylem vessel members throughout the growth season

Diffusion (L. *diffusus,* spread out) the movement of molecules, and thus a substance, from a region of greater concentration of those molecules to a region of smaller concentration

Digestion (L. *digestio,* dividing, or tearing into pieces, an orderly distribution) the processes of rendering food available for metabolism by breaking it down into simpler compounds, chiefly through actions of enzymes

Dihybrid cross (Gr. *dis,* twice + *hybrida,* the offspring of a tame sow and a wild boar, a mongrel) a cross between organisms differing in two characters

Dikaryomycetes informal term for true fungi that have a dikaryotic stage in the sexual life cycle; hyphae of these fungi are septate

Dikaryon see Dikaryotic mycelium

Dikaryotic mycelium (Gr. *di,* two + *karyon,* nut) a mycelium that has paired haploid nuclei of different genotypes in each septate compartment; also called a dikaryon

Dikaryotic stage a stage in the sexual life cycle of a dikaryomycete fungus, consisting of a dikaryotic mycelium

Dimorphism the occurrence of two growth forms at different stages in the life of an organism; an example is the occurrence of yeast and mycelial forms in some fungi

Dinoflagellates (Gr. *dinein,* to whorl + L. *flagellum,* a whip) unicellular algae with two unequal lateral flagella; prominent in marine phytoplankton and capable of causing red tides; classified in the alveolate clade in this text

Dioecious (Gr. *dis*, twice + *oikos*, house) unisexual; having the male and female elements in different individuals

Diploid (Gr. *diploos*, double + *oides*, like) having a double set of chromosomes, or referring to an individual containing a double set of chromosomes per cell; usually a sporophyte generation

Directional selection natural selection that favors phenotypes at one end of a population's range of variation, leading to new adaptations

Disease (L. *dis*, a prefix signifying the opposite + M.E. *aise*, comfort, literally the opposite of ease) any alteration from a state of metabolism necessary for the normal development and functioning of an organism

Divergent evolution the evolution of increasingly greater differences between two taxa, usually resulting from exposure to different selection pressures

Diversifying selection natural selection that increases genetic variation in a population, caused by environmental factors that favor two or more distinct types in a population

Division see Phylum

DNA see Deoxyribonucleic acid

DNA replication the formation of a DNA molecule with the same sequence of nucleotides as that of a pre-existing (template) DNA molecule

Dolipore septum a septum in which the wall around the septal pore is thickened (the dolipore) and parenthesomes occur on both sides of the septum; found only in dikaryotic mycelia of hymenomycete fungi

Domain in the taxonomic hierarchy, a group that contains one or more kingdoms

Dominant (L. *dominari*, to rule) in ecology, referring to species of a community that receive the full force of the macroenvironment; usually the most abundant of such species, and not all of them; in heredity, referring to that allele that, when present in a hybrid with a contrasting allele, completely controls the development of the character; in peas, tall is dominant over dwarf

Donor one who gives

Dormant (L. *dormire*, to sleep) being in a state of reduced physiologic activity such as occurs in seeds, buds, and so on

Dorsiventral (L. *dorsum*, the back + *venter*, the belly) having upper and lower surfaces distinctly different, as a leaf does

Double bond a covalent bond that involves four electrons

Double fertilization in the embryo sac, the fusion of the egg and sperm and the simultaneous fusion of the second male gamete with polar nuclei

Double helix the structure of a DNA molecule, consisting of two complementary polynucleotides chains wound around each other in a right-handed corkscrew shape

Drought-deciduous a plant that loses its leaves during dry periods (as opposed to winter-deciduous)

Drupe (L. *drupa*, an overripe olive) a simple, fleshy fruit, derived from a single carpel, usually one-seeded, in which the exocarp is thin, the mesocarp fleshy, and the endocarp stony

Earlywood wood that forms from vascular cambium early in the growing season; synonym for springwood

Eastern deciduous forest a vegetation type that includes a complex sequence of North American forest communities, all dominated by winter-deciduous trees, and occurring east of longitude 95°W and south of latitude 50°N

Ecologic(al) services products of ecosystems that sustain aquatic and terrestrial life, including the production of oxygen (through photosynthesis), the filtering of pollutants, and the stabilization of soil

Ecology (Gr. *oikos*, home + *logos*, discourse) the study of plant life in relation to the environment

Ecosystem (Gr. *oikos*, house + *synistanai*, to place together) an inclusive term for a living community and all the factors of its nonliving environment

Ecosystem services the byproducts or consequences of biota (usually vegetation) to humans; for example, the removal of pollutants from air and water

Ecotype (Gr. *oikos*, house + *typos*, the mark of a blow) genetic variant within a species that is adapted to a particular environment, yet remains interfertile with all other members of the species

Ectomycorrhiza (Gr. *ektos*, outside + *mykos*, fungus + *riza*, root) a mycorrhiza in which fungal hyphae coat the outside of root tips and grow between root cells but do not penetrate through root cell walls

Ectotrophic mycorrhizae symbiotic relationship between roots and fungi

Edaphic (Gr. *edaphos*, soil) pertaining to soil conditions that influence plant growth

Egg (A.S. *aeg*, egg) a female gamete; in plants, one of eight cells in an embryo sac

Egg apparatus an egg cell and two synergid cells in the embryo sac

Elaiosomes food bodies on certain plants; eaten by insects, especially ants

Elater (G. *elater*, driver) an elongated, sometimes coiled cell capable of dispersing spores from a sporangium; commonly found in liverworts

Electromagnetic energy spectrum a graph showing the intensity of light (in energy units, for example, ergs or Joules) as a function of wavelength

Electron (Gr. *elektron*, gleaming in the sun, by way of L. *electrum*, a bright alloy of gold and silver, and finally amber, from which the first electricity was produced by friction) an elementary particle of matter bearing a unit of negative electrical charge; low in mass, electrons surround the atom's positively charged nucleus; their arrangement defines the size and chemical properties of the atom or molecule

Electronegativity (L. *electrum*, amber + negativity) the power of an atom to attract electrons

Electron microscope a microscope that uses a beam of electrons rather than light to produce a magnified image

Electron transport chain a membrane-bound series of electron carriers that controls the flow of electrons from reduced to oxidized compounds, so that some of the energy carried by the electrons is used to form ATP; the chain consists of several compounds (carriers) that alternately accept and donate electrons; found in mitochondria and chloroplasts

Electrophoresis (Gr. *elektron*, amber + *phora*, motion + *esis*, drive) the process of causing charged molecules (for example, proteins) to move between positively and negatively charged poles

Element (L. *elementa*, the first principles; according to one system of medieval chemistry as recent as 1700, there were four elements composing all material bodies: earth, water, air, and fire) in modern chemistry, a substance that cannot be divided by any known chemical means to a simpler substance; a substance formed entirely from atoms of a distinctive atomic number (number of protons)

Embryo (Gr. *en*, in + *bryein*, to swell) a young sporophytic plant, while still retained in the gametophyte or in the seed

Embryophytes (G. *en*, in + *bryein*, to swell + *phyton*, a plant) all members of the plant kingdom that have an embryo phase in their life cycle

Embryo sac the female gametophyte of the angiosperms; generally a seven-celled structure; the seven cells are two synergids, one egg cell, three antipodal cells (each with a single haploid nucleus), and one endosperm mother cell with two haploid nuclei

Emulsion (L. *emulgere*, to milk out) a suspension of fine particles of a liquid in a liquid

Enation (L. *enasci*, to issue forth) a leaf that originates as an epidermal outgrowth, and thus is not associated with a vascular trace and vascular gap

Endocarp (Gr. *endon*, within + *karpos*, fruit) inner layer of fruit wall (pericarp)

Endodermis (Gr. *endon*, within + *derma*, skin) the layer of living cells, with thickened walls (Casparian strips) and no intercellular spaces, which surrounds the vascular tissue of certain plants and occurs in roots; the innermost layer of the cortex in roots

Endogenous (Gr. *endon*, within + *genos*, race, kind) developed or added from outside the cell

Endomembrane system the organelles in a cell that exchange patches of membranes, including the nuclear envelope, endoplasmic reticulum, and Golgi apparatus

Endomycorrhiza (Gr. *endon*, within + *mykos*, fungus + *riza*, root) a mycorrhiza in which fungal hyphae grow between root cells and penetrate through some root cell walls to share the enclosed space with root cells

Endoplasmic reticulum (Gr. *endon*, within + *plasma*, anything formed or molded; L. *reticulum*, a small net) originally, a cytoplasmic network adjacent to the nucleus; now, a system of closed tubules and flattened sacs in the cytoplasm of a eukaryotic cell; frequently abbreviated to ER

Endosperm (Gr. *endon*, within + *sperma*, seed) the nutritive tissue formed within the embryo sac of seed plants; it often is consumed as the seed matures, but remains in the seeds of corn and other cereals

Endosperm mother cell one of the seven cells of the mature embryo sac, containing the two polar nuclei and, after reception of a sperm cell, giving rise to the primary endosperm cell from which the endosperm develops

Endospore (Gr. *endon*, within + *spora*, seed) a thick-walled resting spore, formed within the cells of certain bacteria after a complex process of nuclear division and cytoplasmic reorganization

Endosymbiosis (Gr. *endon*, within + symbiosis) a symbiotic relationship in which cells of one species live within cells of another species

Energy pyramid a diagram of total caloric energy in all the organisms within each trophic level, starting with producers on the bottom layer of the diagram and ending with tertiary consumers; each higher layer is smaller than the one below, so that the combination of layers look like a stepped pyramid

Environment (L. *in* + *viron*, circle) the totality of biotic and abiotic factors that affect an organism

Enzyme (Gr. *ett*, in + *zyme*, yeast) a protein that possesses a characteristic catalytic activity

Ephemeral (Gr. *ephemeros*, daily) an herbaceous plant whose life span is short, in general living for 6 or fewer months; often synonymous with "annual"

Epicotyl (Gr. *epi*, upon + *kotyledon*, a cup-shaped hollow) the upper portion of the axis of embryo or seedling, above the cotyledons

Epidermal cells the most abundant cell type in the epidermis; typically lack chloroplasts

Epidermis (Gr. *epi*, upon + *derma*, skin) a superficial layer of cells occurring on all parts of the primary plant body: stems, leaves, roots, flowers, fruits, and seeds; it is absent from the root cap and is not differentiated on the apical meristems

Epigeal (Gr. *epi*, upon + *ge*, the earth) type of germination where cotyledons rise above the ground

Epigyny (Gr. *epi*, upon + *gyne*, woman) the arrangement of floral parts in which the ovary is embedded in the receptacle so that the other parts appear to arise from the top of the ovary

Epilimnion (Gr. *epi*, upon + *limne*, marsh) an upper, aerated, warm zone of water that lies above a lower, less aerated, cold zone (the hypolimnion); common in large bodies of freshwater such as deep lakes

Epiphyte (Gr. *epi*, upon + *phyton*, a plant) a plant that grows on another plant, yet is not parasitic

Episome a piece of DNA that can replicate either autonomously or as part of a main chromosome; in bacteria, includes plasmids and certain viral genomes

Equatorial rain forest vegetation with several tree strata; characteristic of warm, wet regions; synonymous with tropical rain forest

Equilibrium paradigm once a widely accepted view of natural vegetation, which held that natural disturbances were rare and that the landscape was largely composed of stands of stable, species-rich climax vegetation

ER see Endoplasmic reticulum

Ergot (F. *argot*, a spur) a fungus disease of cereals and wild grasses in which the grain is replaced by dense masses of purplish hyphae

Erosion (L. *e*, put + *rodere*, to gnaw) the wearing away of land, generally by the action of water

Ethylene C_2H_4, a plant hormone regulating fruit ripening, various aspects of vegetative growth, and the abscission process

Etiolation (F. *etioler*, to blanch) a condition involving increased stem elongation, poor leaf development, and lack of chlorophyll, found in plants growing in the absence, or in a greatly reduced amount, of light

Euglenoids typically freshwater, single-celled, protist organisms that lack a cell wall and move with the help of two flagella; about one third of all species (including those in the genus *Euglena*) are photosynthetic, but the majority ingest food

Eukarya (L. *eu*, true + *karyon*, a nut, referring in modern biology to the nucleus) a taxonomic domain of organisms characterized by having cellular organelles, specifically including the nucleus, bounded by membranes

Eukaryote a member of the Eukarya

Eusporangium (Gr. *eu*, good or true + *spora*, seed + *angeion*, a vessel) the sporangium characteristic of all seedless vascular plants except for the true (leptosporangiate) ferns

Eutrophication (Gr. *eu*, good, well + *trephein*, to nourish) pollution of bodies of water resulting from slow, natural, geologic, or biological processes such as siltation or encroachment of vegetation or accumulation of detritus; also called natural eutrophication

Evapotranspiration (L. *evaporare*, *e*, out of + *vapor*, vapor + F. *transpirer*, to perspire) the process of water loss in vapor form from a unit surface of land both directly and from leaf surfaces

Evolution (L. *evolutio*, an unrolling) in biology, any change in hereditary characteristics of a population or species, or the formation of new species

Exine (L. *exterus*, outside) outer layer of pollen grain wall

Extreme halophiles certain Archaea capable of living in hypersaline habitats

Exocarp (Gr. *exo*, without, outside + *karpos*, fruit) outermost layer of fruit wall (pericarp)

Exodermis the outer layer of the cortex in certain roots; sometimes has Casparian strips

Exogenous (Gr. *exe*, out, beyond + *benos*, race, kind) produced outside of, originating from, or caused by external causes

F1 first filial generation in a cross between any two parents

F2 second filial generation, obtained by crossing two members of the F1, or by self-pollinating the F1

Facultative (L. *facultas*, capability) referring to an organism having the power to live under a number of certain specific conditions; for example, a facultative parasite may be either parasitic or saprophytic

Fairy ring a group of mushrooms growing at the perimeter of a mycelium, arranged in a ring

Family (L. *familia*, family) in the taxonomic hierarchy, a group of genera

Fascicle (L. *fasciculus*, a small bundle) a bundle of pine or other needle-leaves of gymnosperms

Fascicular cambium vascular cambium within vascular bundles

Fat (A.S. *faett*, fatted) one of the three major types of foods (the other two are carbohydrates and proteins); nonpolar, containing carbon, hydrogen, and small amounts of oxygen; rich in energy; used synonymously with lipids

Feedback inhibition the inhibition of an enzyme by its product or the product of the metabolic pathway of which it is a part

Female gametophyte a haploid plant that forms eggs; in the flowering plant life cycle, the embryo sac

Fermentation (L. *fermentum*, a drink made from fermented barley beer) a catabolic process in which molecular oxygen is not involved, such as the production of alcohol from sugar by yeasts

Ferredoxin an electron-transferring protein containing iron, involved in photosynthesis and in the biological production and consumption of hydrogen gas

Fertilization (L. *fertilis*, capable of producing fruit) that state of a sexual life cycle involving the union of egg and sperm and, hence, the doubling of chromosome numbers

Fiber (L. *fibea*, a fiber or filament) an elongated, tapering, thick-walled strengthening cell occurring in various parts of plant bodies; one of the cell types in sclerenchyma tissue

Fiber-tracheid xylem elements found in pine that are structurally intermediate between tracheids and fibers

Fibril (dim. of L. *fibea*, fiber) submicroscopic threadlike units of cellulose found in cell walls

Fibruous root system branched root system lacking a clear, single primary root; common in grasses

Fiddlehead the shape of young fern leaves (fronds) as they push up through the soil: tightly coiled like the head of a violin

Field capacity the amount of water retained in a soil (generally expressed as percentage by weight) after large capillary spaces have been drained by gravity

Filament (L. *fi'lum*, a thread) stalk of stamen bearing the anther at its tip; also, a slender row of cells (certain algae)

Fission (L. *fissilis*, easily split) asexual reproduction involving the division of a single-celled individual into two new single-celled individuals of equal size

Flagellum (plural, flagella) (L. *flagellum*, a whip) a long, slender whip of protoplasm (in eukaryotes) or proteins (in bacteria)

Flora (L. *floris*, a flower) an enumeration of all the species that grow in a region; also, the collective term for all the species that grow in a region

Floret (F. *fleurette*, a dim. of *fleur*, flower) one of the small flowers that make up the composite flower or the spike of the grasses

Flower (F. *fleur*, L. *flos*, a flower) floral leaves grouped together on a stem and adapted for sexual reproduction in the angiosperms

Flowering plants a synonym for angiosperms; plants that produce seeds inside an enlarged ovary, which at maturity is called a fruit

Foliose (L. *folium*, a leaf) said of a lichen that is leaflike, attached to the substrate only along part of its surface

Follicle (L. *folliculus*, dim. of *follis*, bag) a simple, dry, dehiscent fruit, with one carpel, splitting along one suture

Food (A.S. *foda*) any organic substance that directly furnishes energy and building materials for vital processes

Food chain the path along which caloric energy is transferred within a community (from producers to consumers to decomposers)

Foot (O.E. *fot*, foot) that portion of the sporophyte of bryophytes and lower vascular plants that is sunk in gametophyte tissue and absorbs food parasitically from the gametophyte

Foraminifera (L. *foramen*, an opening) nonphotosynthetic, aquatic, unicellular protists that have a calcareous shell with perforations through which filaments of protoplasm (pseudopods) project, capable of trapping or engulfing food; classified within the alveolates

Forb (Gr. *phorbe*, fodder or *pherbein*, to graze) an herbaceous plant that is neither a grass nor a grasslike relative

Forest (L. *foris*, outside) vegetation dominated by trees, in which the canopies of adjacent trees usually touch, providing more than 60% cover

Form in the taxonomic hierarchy, a subgroup within a species, equivalent to race or variety

Form genus a scientific name given to an organism from the fossil record, when only a portion of the entire plant has been recovered and is known

Fossil (L. *fossio*, a digging) any impression, natural or impregnated remains, or other trace of an animal or plant of past geologic ages that has been preserved in the earth's crust

Founder effect a difference in allele ratio among populations that occurs when an offshoot population is founded by so few individuals that chance strongly affects which alleles of the parental population are present in the founders

Free energy the internal thermodynamic potential of a portion of matter; change in free energy during a chemical reaction determines the direction of the reaction

Frequency (L. *frequens*, crowded) the percentage of vegetation samples containing a given species; if 7 of 10 quadrats (samples) contained post oaks, post oak has a 70% frequency

Fret (O.F. *frette*, latticework) flattened membrane sacs that connect grana in chloroplasts

Frond (L. *frons*, branch, leaf) a synonym for a large divided leaf, especially a fern leaf; also the leaflike blades of some algae

Fruit (L. *fructus*, that which is enjoyed, hence product of the soil, trees, cattle, and so on) a matured ovary; in some, seed plants and other parts of the flower may be included; also applied, as fruiting body, to reproductive structures of other groups of plants

Frustule (L. *frustulum*, little piece) a diatom cell, composed of two overlapping halves (valves)

Fruticose (L. *frutex*, a shrub) said of a lichen that is highly branched and erect or pendant

Fucoxanthin (Gr. *phykos*, seaweed + *xanthos*, yellowish brown) a brown pigment found in brown algae

Functional group a portion of a biochemical molecule that participates in a chemical reaction; sulfhydryl (-SH) and amino ($-NH_2$) groups in proteins are functional groups

Fungi the monophyletic kingdom that contains all true fungi

Fungus (plural, fungi) (L. *fungus*, a mushroom) any chemoheterotrophic eukaryote that reproduces with spores and has cell walls at some stage of life

Funiculus (L. *funiculus*, dim. of *funis*, rope or small cord) a stalk of the ovule, containing vascular tissue

Fusiform initials (L. *fusus*, spindle + form) meristematic cells in the vascular cambium that develop into secondary xylem and phloem cells composing the axial system of a stem or root

G1 period of cell cycle preceding DNA synthesis

G2 period of cell cycle preceding mitosis

Gametangium (Gr. *gametes*, a husband, *gamete*, a wife + *angeion*, a vessel) a cell or organ that produces gametes

Gamete (Gr. *gametes*, a husband, *gamete*, a wife) a protoplast that fuses with another protoplast to form the zygote in the process of sexual reproduction

Gametic life cycle a life cycle in which only the gametes are haploid, all other phases being diploid

Gametophyte (gamete + Gr. *phyton*, a plant) the gamete-producing plant

Gel (L. *gelare*, to freeze) jelly-like, colloidal mass

Gemma (plural, gemmae) (L. *gemma*, a bud) a small mass of vegetative tissue; an outgrowth of the thallus

Gene (Gr. *genos*, race, offspring) a group of base pairs in the DNA molecule in the chromosome that determines or conditions one or more hereditary characters

Generation (L. *genus*, birth, race, kind) any phase of a life cycle characterized by a particular chromosome number, as the gametophyte generation and the sporophyte generation

Generative cell one of the two cells making a pollen grain; will divide to form two sperm cells

Genetic code the relationship between codons (sequences of three bases) in DNA or messenger RNA and the amino acids they specify in a protein

Genetic drift random changes in the ratio of alleles that occur in small populations as a result of chance events

Genetics (Gr. *genesis*, origin) the science of heredity

Genome (gene + chromosome) the collection of all genes in an organism

Genotype (gene + type) the assemblage of genes in an organism

Genus (plural, genera) (Gr. *genos*, race, stock) in the taxonomic hierarchy, a group of species

Geographic isolation spatial separation that prevents individuals of two populations from meeting to exchange genes

Germination (L. *germinare,* to sprout) the beginning or resumption of growth by a seed, spore, bud, or other resting structure

Germ sporangium in zygomycete fungi, the sporangium produced a hypha that grows directly from the germinating zygospore

Gibberellins a group of growth hormones, the most characteristic effect of which is to increase the elongation of stems in a number of kinds of higher plants

Gill (M.E. *gile,* a lip, probably because of its resemblance in shape and arrangement to gills of fishes) in certain basidiomata, thin, spore-bearing plates on the underside of the cap

Girdle (O.E. *gyrdel,* enclosure, girdle) that region of a stem from which a ring of bark extending to the cambium has been removed

Glomeromycota a monophyletic phylum in kingdom Fungi, noted for members engaging in arbuscular endomycorrhizae

Glucose (Gr. *glykys,* sweet + *ose,* a suffix indicating a carbohydrate) a simple sugar, grape sugar, $C_6H_{12}O_6$

Glume (L. *gluma,* husk) an outer and lowermost bract of a grass spikelet

Glycogen (Gr. *glykys,* sweet + *gen,* of a kind) a carbohydrate related to starch but found generally in the liver of animals

Glycolipid a molecule consisting of a combination of one or more sugars with a hydrocarbon

Glycolysis (Gr. *glykys,* sweet + *lysis,* a loosening) the biochemical pathway affecting decomposition and oxidation of sugar compounds without involving free oxygen; early steps of respiration

Glyoxisome (Gr. *glykys,* sweet + *soma,* body) a type of microbody that contains enzymes involved in the conversion of fats to carbohydrates during germination of fat-storing seeds

Gnetophytes a small group of seed plants with life cycles having some traits of gymnosperms and some of angiosperms; once thought to be ancestral to flowering plants, they are now considered to be a sister group to angiosperms and gymnosperms

Golgi apparatus (Italian cytologist Camillo Golgi [1844–1926], who first described the organelle) in animal cells, a complex perinuclear region; in plant cells, a series of flattened bladders associated with packaging, secretion, or both

Grain see Caryopsis

Gram's stain (after Hans C. M. Gram) a method for the differential staining of bacteria; distinguishes species on the basis of cell wall composition, which determines whether stain is retained

Grana (singular, granum) (L. *granurn,* a seed) dense collections of thylakoid membranes within chloroplasts, seen as green granules with the light microscope and as a series of apposed lamellae with the electron microscope

Grassland a vegetation type dominated by herbaceous species, including forbs and grasses, and occupying an area too dry to support trees

Gravitropism (L. *gravis,* heavy + Gr. *tropos,* turning) a growth curvature induced by gravity

Green algae a group of (mainly freshwater) aquatic, simple plants lacking embryos and vascular tissue, but similar to more complex land plants by possessing starch, cellulose, and the same photosynthetic pigments; includes the charophytes, from which land plants are thought to have been derived

Ground meristem (Gr. *meristos,* divisible) a primary meristem that gives rise to cortex, mesophyll, and pith

Ground tissue system tissues of the plant body responsible for housekeeping functions, for example, storage, support, photosynthesis, and others; collenchyma, sclerenchyma, and parenchyma

Growth (A.S. *growan,* probably from Old Teutonic *gro,* from which grass also is derived) increasing the size and number of cells in an organ; has two components: cell division and cell enlargement

Growth retardant a chemical (such as cycocel, CCC) that selectively interferes with normal hormonal promotion of growth, but without appreciable toxic effects

Guanine a purine base found in DNA and RNA

Guard cells specialized epidermal cells found on young stems and leaves; between each pair of guard cells is a small pore through which gases enter or leave; a pair of guard cells in addition to the pore constitute a stoma

Guttation (L. *gutta,* drop, exudation of drops) exudation of water from plants, in liquid form

Gymnosperm (Gr. *gummos,* naked + *sperma,* seed) seed plants that produce seeds that are exposed to the environment, rather than being enclosed in a fruit; the seeds may be located on the scales of cones or may be borne singly, covered with colorful aril tissue

Gynoecium (Gr. *gyne,* woman + *oikos,* house) the aggregate of carpels in the flower of a seed plant

Habitat (L. *habitare,* inhabit, dwell) the place or natural environment where an organism naturally grows

Hallucinogenic (L. *hallucinari,* to mentally wander + *genitalis,* to beget) that which induces hallucinations

Halophile an organism adapted to living in a high-salt environment; a group of Archaea are halophiles

Haploid (Gr. *haploos,* single + *oides,* like) having a single complete set of chromosomes, or referring to an individual or generation containing such a single set of chromosomes per cell; usually a gametophyte generation

Haptera (singular, hapteron) finger-like, tubular projections that make up the holdfast of certain kelps

Hardwood an angiosperm tree; having wood that is dense and more difficult to work than the wood of a conifer

Hardy–Weinberg equilibrium within a population, constancy of the ratio of alternative alleles for a particular gene from one generation to the next

Haustorial root modified root of certain parasitic plants, such as dodder, that penetrates into the host plant

Haustorium (plural, haustoria) (M.L. *haustrum,* a pump) a projection of fungal hyphae that acts as a penetrating and absorbing organ

Head an inflorescence; typical of the composite family, in which flowers are grouped closely on a receptacle

Heartwood wood in the center of old secondary stems that is plugged with resins and tyloses and is not active

Helix (Gr. *helix,* anything twisted) anything having a spiral form; here, quite generally refers to the double spiral of the DNA molecule

Hemicellulose (Gr. *hemi,* half + cellulose) a class of polysaccharides of the cell wall, built of several different kinds of simple sugars linked in various combinations

Herb (L. *herba,* grass, green blades) a seed plant that does not develop woody tissues

Herbaceous (L. *herbaceus,* grassy) referring to plants having the characteristics of herbs

Herbaceous perennial a perennial plant without any wooden parts; an herbaceous plant capable of dying aboveground every year, but remaining alive belowground, and thus being able to vegetatively recover every year; lifespans are one to several decades

Herbal (L. *herba,* grass) a book that contains the names and descriptions of plants, especially those that are thought to have medicinal uses

Herbarium (L. *herba,* grass) a collection of dried and pressed plant specimens

Herbicide (L. *herba,* grass or herb + *cidere,* to kill) a chemical used to kill plants, frequently chemically related to a hormone (as the herbicide 2,4-D is related to the hormone IAA); an herbicide may have narrow or wide selectivity (range of target organisms)

Herbivory the ingestion of plants by animals

Heredity (L. *hereditas,* being an heir) the transmission of morphologic and physiologic characters of parents to their offspring

Hermaphrodite flower (Gr. *herinaphroditos,* a person having the attributes of both sexes, represented by Hermes and Aphrodite) a flower having both stamens and pistils

Hesperidium type of berry, such as lemons and oranges, with a thick rind

Heterocyst (Gr. *heteros,* different + *cystis,* a bag) an enlarged colorless cell that may occur in the filaments of certain blue–green algae

Heteroecious (Gr. *heteros,* different + *oikos,* house) referring to fungi that cannot carry through their complete life cycle unless two different host species are present

Heterogametes (Gr. *heteros,* different + gamete) gametes dissimilar from each other in size and behavior, such as egg and sperm

Heterogamy (Gr. *heteros,* different + *gamos,* union or reproduction) reproduction involving two types of gametes

Heterokaryotic having nuclei of more than one genotype in uncontrolled proportions within a single living body

Heterokonts a clade of protists that includes several algal groups with chlorophylls *a* and *c* (golden algae, brown algae, diatoms) and the oomycota (water molds, downy mildews)

Heteromorphic (Gr. *hetero,* other + *morphe,* form) referring to organisms with sporic life cycles in which the sporophyte and gametophyte generations look different

Heterophylly condition where a plant has leaves of more than one shape

Heterophyte (Gr. *heteros,* different + *phyton,* a plant) a plant that must secure its food ready-made

Heterosis (Gr. *heteros,* different + *osis,* suffix indicating a state of) the state of a genotype having a large degree of heterozygosity

Heterospory (Gr. *heteros,* different + spore) the condition of producing microspores and megaspores

Heterothallic (Gr. *heteros,* different + thallus) referring to species in which male gametangia and female gametangia are produced in different filaments or by different individual plant bodies

Heterotrichy (Gr. *heteros,* different + *trichos,* a hair) in the algae, the occurrence of two types of filaments: erect and prostrate

Heterotrophic (Gr. *heteros,* different + *trophein,* to nourish with food) referring to a prokaryote that obtains carbon from the breakdown of organic molecules, and more generally, to an organism that requires organic compounds it cannot make for itself and, hence, must get from the environment

Heterozygous (Gr. *heteros,* different + *zygon,* yoke) having different allels of a Mendelian gene pair present in the same cell or organism; for instance, a tall pea plant with alleles for tallness, *T,* and dwarfness, *t*

Hexose (Gr. *hexa,* six + *ose,* suffix indicating, in this usage, carbohydrate) a carbohydrate with six carbon atoms

Hierarchy a set of categories that has nested levels, such as the Linnaean system of taxonomy

Hilum (L. *hilum,* a trifle) scar on a seed, which marks the place where the seed broke from the stalk

Histology (Gr. *histos,* cloth, tissue + *logos,* discourse) science that deals with the microscopic structure of animal and vegetable tissues

Holdfast an anchoring organ in certain seaweeds; not a true root because it lacks vascular tissue; furthermore, most absorption occurs elsewhere on the thallus

Homologous chromosomes (Gr. *homologos,* the same) members of a chromosome pair; they may be heterozygous or homozygous

Homologous traits traits that evolved from the same ancestral trait; in cladistic terms, homologous traits are alternative states of the same character

Homospory (Gr. *homos,* one and the same + spore) the condition of producing one sort of spore only

Homothallic (Gr. *homos,* one and the same + thallus) referring to species in which male gametangia and female gametangia are produced in the same filament or by the same individual plant body

Homozygous (Gr. *homos,* one and the same + *zygon,* yoke) having identical alleles of a Mendelian gene pair present in the same cell or organism; for instance, a tall pea plant with alleles for tallness *(TT)* only

Hormone (Gr. *hormaein,* to excite) a specific organic product, produced in one part of a plant or animal body and transported to another part where, effective in small amounts, it controls or stimulates another and different process

Hornworts (Gr. *horn,* horn + *wort,* plant) a group of bryophytes with long, pointed sporophytes (resembling horns) that rise from the thalloid gametophytes

Horsetails a small group of 25 species within a single genus, *Equisetum,* which have stems that are hardened by ridges and an epidermis containing silica; branching occurs in whorls, and leaves are reduced; classified in this text among the monilophyte lineage with whisk ferns, ophioglossalean ferns, marattialean ferns, and true ferns

Host (L. *hospes,* host, guest) an organism on or in which another organism lives

Humidity, relative (L. *humidus,* moist) the ratio of the weight of water vapor in a given quantity of air to the total weight of water vapor that quantity of air is capable of holding at the temperature in question, expressed as a percentage

Humus (L. *humus,* the ground) decomposing organic matter in the soil

Hybrid (L. *hybrida,* offspring of a tame sow and a wild boar, a mongrel) in taxonomy, an organism that results from a mating between organisms of different species; in genetics, the offspring of any mating between organisms that carry different alleles for a gene of interest

Hybridization in evolution, the formation of progeny by a mating between organisms of different species

Hybrid vigor (heterosis) (Gr. *heterosis,* alteration fr. *heteros,* other) a greater capacity for growth frequently observed in crossbred animals or plants as compared with those resulting from inbreeding

Hydathode (Gr. *hydro,* water + O.E. *thoden,* stem or *thyddan,* to thrust) a structure, usually on leaves, that releases liquid water during guttation

Hydrocarbon compound a molecule formed from hydrogen and carbon

Hydrogen acceptor a substance capable of accepting hydrogen atoms or electrons in the oxidation–reduction reactions of metabolism

Hydrogen bond a weak bond in which a hydrogen atom interposes between and holds together two strongly electronegative atoms (for example, oxygen, nitrogen)

Hydroid (Gr. *hydro,* water + *eidos,* a shape) a water-conducting cell found in some mosses

Hydrolysis (Gr. *hydro,* water + *lysis,* loosening) the breaking of a covalent bond within a compound through reaction with water; in metabolism, generally controlled by enzymes

Hydrophilic (Gr. *hydro,* water + *philos,* loving) soluble in water, capable of forming weak associations with water molecules

Hydrophobic (Gr. *hydro,* water + *phobos,* fearing) insoluble in water

Hydrophobic bond the tendency of hydrophobic molecules in an aqueous environment to stick together, occurring because of the high free energy required to force them into solution

Hydrophobin a water-repellant protein secreted by certain lichen fungi

Hydrophyte (Gr. *hydro,* water + *phyton,* a plant) a plant that grows wholly or partly submerged in water

Hymenium (Gr. *hymen,* a membrane) spore-bearing tissue in various fungi

Hymenomycetes a monophyletic class of basidiomycete fungi, characterized by dolipore septa and complex basidiomata

Hypantheum (L. *hypo,* under + Gr. *anthos,* flower) fusion of calyx and corolla partway up their length to form a cup, as in many members of the rose family

Hypersensitivity response a reaction of plant cells to the presence of pathogenic agents, involving a loss of ions, the synthesis of reactive oxygen species (superoxide, hydrogen peroxide), and eventual death

Hypertonic (Gr. *hyper,* above, over + *tonos,* to stretch) a solution having a concentration high enough so that water will move into it across a membrane from another solution

Hypertrophy (Gr. *hyper,* over + *trophein,* to nourish with food) a condition of overgrowth or excessive development of an organ or part

Hypha (plural, hyphae) (Gr. *hypee,* a web) a tubular, thread-like eukaryotic cell (or portion of a cell) that bears one or more nuclei and grows at the tip, characteristic of fungi

Hypocotyl (Gr. *hypo,* under + *kotyledon,* a cup-shaped hollow) that portion of an embryo or seedling between the cotyledons and the radicle or young root

Hypogeal (Gr. *hypo,* under + *ge,* the earth) type of germination where cotyledons remain belowground

Hypogyny (Gr. *hypo,* under + *gyne,* female) a condition in which the receptacle is convex or conical, and the flower parts are situated one above another in the following order, beginning with the lowest: sepals, petals, stamens, carpels

Hypolimnion (Gr. *hypo,* under + *limne,* marsh) a lower, cold, relatively nonaerated zone of water that lies below a warmer zone (the epilimnion); common in large bodies of freshwater, such as deep lakes

Hypothesis (Gr. *hypothesis,* foundation) a tentative theory or supposition provisionally adopted to explain certain facts and to guide in the investigation of other facts

Hypotrophy (Gr. *hypo,* under + *trophein,* to nourish with food) an underdevelopment of an organ or part

IAA see Indoleacetic acid

Ice Age a synonym for the Pleistocene epoch, generally defined as the period between 2 million and 10,000 years ago; during the Pleistocene epoch, there were at least five glacial advances, separated by interglacial warm periods

Imbibition (L. *imbibere*, to drink) the absorption of water by a seed, an adsorption phenomenon

Imperfect flower a flower lacking either stamens or pistils

Imperfect fungi fungi reproducing only by asexual means

Inbreeding the production of offspring that originated when the sperm and egg came from genetically similar or genetically identical plants

Inclusion body a body found in the cells of organisms with a virus infection

Incomplete dominance see Codominance

Incomplete flower a flower lacking one or more of the four kinds of flower parts

Indehiscent (L. *in*, not + *dehiscere*, to divide) not opening by valves or along regular lines

Indeterminate generally, having no defined limits; in plant development, a morphogenetic process that produces new organs or sections of an organ while preserving a meristem in which new cells are formed to continue the process indefinitely

Indicator species a species that has a narrow range of tolerance for one or more environmental factors so that, from its occurrence at a site, one can predict these factors at that site (for example, nutrient availability or summer temperatures)

Indoleacetic acid a naturally occurring growth regulator, an auxin

Indusium (plural, indusia) (L. *indusium*, a woman's undergarment) membranous growth of the epidermis of a fern leaf that covers a sorus

Infect (L. *infectus*, to put into, to taint with morbid matter) specifically, to produce disease by such agents as bacteria or viruses

Inferior ovary an ovary partially or completely united with the calyx

Inflorescence (L. *inflorescere*, to begin to bloom) a flower cluster

Ingroup in cladistics, the set of taxa that is the target of study

Inheritance (O.F. *enheritance*, inheritance) the reception or acquisition of characters or qualities by transmission from parent to offspring

Inorganic referring in chemistry to compounds that do not contain carbon

Integument (L. *integumentum*, covering) external layer of ovule that later develops into the seed coat

Inter a prefix, from the Latin preposition *inter*, meaning between, in between, in the midst of

Intercalary (L. *intercalare*, to insert) descriptive of meristematic tissue or growth not restricted to the apex of an organ, that is, growth at nodes

Intercellular (L. *inter*, between + cells) lying between cells

Interfascicular cambium (L. *inter*, between + *fasciculus*, small bundle) cambium that develops between vascular bundles

Intermediary metabolism the collection of all the metabolic pathways in a cell

Intermediate (L. *inter-*, between + *medius*, middle) a compound in a metabolic pathway that occurs before the final product

Internode (L. *inter*, between + *nodus*, a knot) the region of a stem between two successive nodes

Interphase (L. *inter*, between + Gr. *phasis*, appearance) the period of preparation for cell division; state between two mitotic or meiotic cycles

Intertidal zone the strip of coastal land that is alternately inundated and exposed as tides rise and fall

Intine (L. *intus*, within) the innermost coat of a pollen grain

Intra a prefix from the Latin preposition *intra*, meaning on the inside, within

Intracellular (L. *intra*, within + cell) lying within cells

Introgression (L. *intro*, to the inside + *gress*, walk) transfer of genes between two species by means of hybridization followed by back-crossing

Involucre (L. *involucrum*, a wrapper) a whorl or rosette of bracts surrounding an inflorescence

Ion (Gr. *ienai*, to go) a charged particle formed by the breakdown of substances able to conduct an electric current

Ionic bond a chemical bond formed between ions of opposite charge

Ion pump a protein in a cellular membrane that catalyzes the transport of an ion from one side of the membrane to the other; pump implies that the transport is active, requiring the expenditure of metabolic energy and potentially occurring from a smaller concentration of the ion to a greater concentration

Irregular flower flower parts arranged so that only one line can divide the flower into two equal halves or mirror images; zygomorphic flower

Isidium (plural, isidia) a small outgrowth of a lichen that accomplishes asexual reproduction by pinching off to form a new lichen

Isobilateral leaf (Gr. *isos*, equal + L. *bis*, twice, two-fold + *lateralis*, pertaining to the side) a leaf having essentially similar upper and lower surfaces

Isodiametric (Gr. *isos*, equal + diameter) having diameters equal in all directions, as a ball

Isogametes (Gr. *isos*, equal + gametes) gametes similar in size and behavior

Isogamy (Gr. *isos*, equal + *gamete*, spouse) the condition in which the gametes are identical

Isomers (Gr. *isos*, equal + *meros*, part) two or more compounds having the same molecular formula; for example, glucose and fructose are both $C_6H_{12}O_6$

Isomorphic (Gr. *iso*, equal + *morphe*, form) referring to organisms with sporic life cycles in which the sporophyte and gametophyte generations look the same

Isotonic (Gr. *isos*, equal + *tonos*, to stretch) having equal osmotic concentration

Isotope (Gr. *isos*, equal + *topos*, place) any of two or more forms of an element having the same or closely related chemical properties

Iteroparous (L. *iterum*, again + *paritas*, equal) a perennial plant capable of flowering repeatedly throughout its life span

Karyogamy (Gr. *karyon*, nut + *gamos*, marriage) the fusion of two nuclei

Kelp (M.E. *culp*, seaweed) a collective name for any of the large brown algae

Kinetic energy energy associated with moving objects; the energy of a body associated with its motion

Kinetin (Gr. *kinetikos*, causing motion) a purine that acts as a cytokinin in plants but probably does not occur in nature

Kinetochore (Gr. *kinein*, to move + *chorein*, to move apart) specialized component of a chromosome, connecting the centromere to spindle tubules during mitosis or meiosis

Kingdom (O.E. *cyningdom*, territory ruled by a king or queen) in the taxonomic hierarchy, a group of phyla or divisions

Krebs cycle see Citric acid cycle

K selection natural selection that favors long-lived, late-maturing individuals that devote a small fraction of their resources into reproduction; many tree species are K strategists

LAI leaf area index; a ratio that summarizes the surface area of all leaves that shade a given area of ground, a dimensionless number usually between 1 and 11

Lamella (plural, lamellae) (Gr. *lamin*, a thin blade) cellular membranes, frequently those seen in chloroplasts

Lamina (L. *lamina*, a thin plate) blade or expanded part of a leaf

Lateral bud a bud that grows out of the side of a stem

Lateral root branch roots initiated in the pericycle

Lateral root primordium early stage of lateral root

Laterite (L. *later*, a brick) a soil characteristic of rain forest vegetation; color is red from oxidized iron in the A horizon; synonymous with oxisol, a soil order

Latewood secondary xylem that forms in wood late in the growing season; summerwood

Latex (L. *latex*, juice) a milky secretion

Leaf (O.E. *leaf*) lateral outgrowth of stem axis, which is the usual primary photosynthetic organ, and in the axil of which may be a bud

Leaf axil angle formed by the leaf stalk and the stem

Leaf gap the region composed of parenchyma that is located in the primary vascular cylinder above the point of departure of the leaf vascular tissue

Leaflet separate part of the blade of a compound leaf

Leaf primordium (L. *primordium*, a beginning) a lateral outgrowth from the apical meristem, which will become a leaf

Leaf scar characteristic scar on a stem axis made after leaf abscission

Leaf trace the vascular bundle extending from the stem to the base of a leaf

Legume (L. *legumen*, any leguminous plant, particularly bean) a simple, dry dehiscent fruit with one carpel, splitting along two sutures

Lemma (Gr. *lemma*, a husk) lower bract that subtends a grass flower

Lenticel (M.L. *lenticella*, a small lens) a structure of the bark that permits the passage of gas inward and outward

Lepidodendrids (Gr. *lepis*, scale + *dendron*, tree) now-extinct trees with trunks characteristically covered by diamond-shaped pads (scales) where leaves had been attached; early vascular plants related to living lycophytes such as the club moss *Lycopodium*

Leptoid (Gr. *leptos*, small, thin + *eidos*, a shape) a sugar-conducting cell found in some mosses

Leptosporangium (Gr. *leptos*, fine, thin + *spora*, seed + *angeion*, a vessel) the sporangium characteristic of true ferns, originating from a single leaf cell and having a long, thin stalk, a one-cell-layer-thick wall, and an annulus that aids in spore dispersal

Leucoplast (Gr. *leuk-*, white + *plastid*) a colorless plastid

Liana (F. *liane* from *lier*, to bind) a plant that climbs on other plants, depending on them for mechanical support; a plant with climbing shoots

Lichen (Gr. *leichen*, thallus plants growing on rocks and trees) a composite organism consisting of a fungus living symbiotically with an alga

Life cycle the pattern of transfer of genetic information in a species from parent to offspring, or from generation to generation

Life history pattern a collection of traits that describe how a plant allocates time and energy throughout its life span so that it can occupy a habitat and reproduce successfully

Light harvesting complexes groups of chlorophyll-containing proteins in thylakoid membranes that absorb light and transfer the energy to photoreaction centers P_{680} or P_{700}

Lignification (L. *lignum*, wood + *facere*, to make) impregnation of a cell wall with lignin

Lignin (L. *lignum*, wood) an irregular polymer of phenolic molecules impregnating the cellulose framework of certain plant cell walls; adds strength and reduces extensibility of the cell walls

Ligule (L. *ligula*, dim. of *lingua*, tongue) in grass leaves, an outgrowth from the upper and inner side of the leaf blade where it joins the sheath

Line transect a method of sampling vegetation by stretching a tape along a straight line and measuring the canopy cover of plants beneath that line or that cut through a vertical plane described by that line

Linkage the grouping of genes on the same chromosome

Linked characters characters of a plant or animal controlled by genes grouped together on the same chromosome

Lipid (Gr. *lipos*, fat + L. *ides*, suffix meaning son of; now used in sense of having the quality of) any of a group of fats or fatlike compounds insoluble in water and soluble in fat solvents

Lipid body lipid storage organelle found in seeds

Lipopolysaccharide a substance containing lipid and a carbohydrate polymer; —layer, the outer membrane of a gram-negative bacterium

Lithotroph an alternate name for chemoautotroph that derives energy from the oxidation of inorganic compounds

Liverwort (liver + M.E. *zuort*, a plant; literally, a liver plant, so named in medieval times because of its fancied resemblance to the lobes of the liver) common name for the class Hepaticae of the Bryophyta

Loam (O.E. *lam* or Old Teutonic *lai*, to be sticky, clayey) a particular soil texture class, referring to a soil having 30% to 50% sand, 30% to 40% silt, and 10% to 25% clay

Lobed leaf (Gr. *lobos*, lower part of the ear) a leaf divided by clefts or sinuses

Locule (L. *loculus*, dim. of locus, a place) a cavity of the ovary in which ovules occur

Locus in genetics, a site on a chromosome identified by recombinational analysis; the position of a gene

Long-day plants plants that are induced to flower only when the length of the day exceeds a certain value

Longevity (L. *longaevus*, long-lived) length of life

Lumen (L. *lumen*, light, an opening for light) the cavity of the cell within the cell walls

Lycophytes a lineage of seedless vascular plants separate from the monilophytes; once highly diverse and dominant in the Carboniferous period (Coal Age), and now represented by herbs such as *Lycopodium, Selaginella,* and *Isoetes*

Lysis (Gr. *lysis*, a loosening) a process of disintegration and cell destruction

Lysosomes a membrane-bound body within a cell, containing enzymes that when released destroy the cytoplasmic components of the cell

Macroenvironment (Gr. *makros*, large + O.F. *environ*, about) the environment caused by the general, regional climate; traditionally measured some 4 feet above the ground and away from large obstructions

Macroevolution evolution that consists of changes large enough to represent the emergence of a new life form, such as evolution of flowering plants from mosslike ancestors

Macronutrient (Gr. *makros*, large + L. *nutrire*, to nourish) an essential element required by plants in relatively large quantities

Magnoliids a core angiosperm group of (mainly) woody dicots, including magnolia, cinnamon, camphor, nutmeg, sassafras, avocado, betel nut, and ginger

Male gametophyte a haploid plant that forms sperm; in the flowering plant life cycle, the two-celled pollen grain

Map distance on a chromosome, the distance in crossover units between designated genes

Marattialean ferns eusporangiate ferns with upright stems and large leaves (fronds); not classified as true (leptosporangiate) ferns

Mating test a test to determine whether organisms of two populations will spontaneously interbreed and produce fertile offspring when brought together under natural conditions; the basis for defining biological species

Matrix the solution of enzymes in the central chamber of a mitochondrion

Meadow (M.E. *medoue*) a vegetation type dominated by herbaceous plants, including forbs and grasses, but occupying a habitat and climate normally capable of supporting forest vegetation

Mediterranean climate a temperate-zone climate type featuring dry, hot summers and cool, wet winters; frosts are uncommon and annual precipitation is 25 to 75 cm; found in five parts of the world: Australia, South Africa, Chile, California, and around the Mediterranean Sea

Medulla (L. *medulla*, marrow) the filamentous center of certain lichens and kelp blades and stipes

Megafossil (Gr. *megas*, great + L. *fossilis*, dug up) a fossil large enough to be seen without magnification; leaves, flowers, stems, and seeds are examples

Megagametophyte a gametophyte that produces eggs; contrast to microgametophyte, which produces sperm cells

Megapascal (MPa) a measure of pressure and water potential

Megaphyll (Gr. *megas*, great + *phyllon*, leaf) a leaf whose trace is marked with a gap in the stem's vascular system; megaphylls are thought to represent modified branch systems

Megasporangium (Gr. *megas*, large + sporangium) sporangium that bears megaspores

Megaspore (Gr. *megas*, large + spore) the meiospore of vascular plants, which gives rise to a female gametophyte

Megasporocyte (Gr. *megas*, large + *spora*, seed + L. *cyta*, vessel) a diploid cell in which meiosis will occur, resulting in four megaspores; synonymous with megaspore mother cell

Megasporophyll (Gr. *megas*, large + spore + Gr. *phyllon*, leaf) a leaf bearing one or more megasporangium

Meiocyte (meiosis + Gr. *kytos*, currently meaning a cell) any cell in which meiosis occurs

Meiosis (Gr. *meioun*, to make smaller) a type of cell division occurring in sexual reproduction, in which two rounds of division convert one diploid cell to four haploid cells, effecting a segregation of homologous chromosomes

Meiosporangium a structure within which one or more diploid cells undergo meiosis, producing meiospores, each of which can germinate and produce a plant that represents the male gametophyte generation

Meiospores (meiosis + spores) haploid spores made to disperse the recombinant nuclei that result from meiosis

Membrane (L. *membrana*, skin, parchment) generally, a thin, soft, pliable sheet; specifically, a limiting surface, within or surrounding a cell, formed from phospholipids, glycolipids, or other hydrophobic compounds

Meristem (Gr. *meristos*, divisible) sites in the plant body where cells divide and where differentiation into specialized cells and tissues is initiated

Meristoderm (meristem + epidermis) the outer meristematic cell layer (epidermis) of some Phaeophyta

Mesocarp (Gr. *mesos*, middle + *karpos*, fruit) middle layer of fruit wall (pericarp)

Mesophyll (Gr. *mesos*, middle + *phyllon*, leaf) parenchyma tissue of leaf between epidermal layers, where most photosynthesis takes place

Mesophyte (Gr. *mesos*, middle + *phyton*, a plant) a plant that grows best in conditions of moderate moisture and temperature

Mesosome (Gr. *mesos*, middle + *soma*, body) one of a series of paired membranes occurring in many bacteria

Mesozoic era (Gr. *mesos*, middle + *zoe*, life) a geologic era beginning 225 million years ago and ending 65 million years ago

Metabolic pathway a set of compounds related by enzymatically catalyzed chemical reactions, such that the first compound is transformed into the second, the second into the third, and so on

Metabolism (M.L. from the Gr. *metabolos*, to change) the process, in an organism or a single cell, by which nutritive material is built up into living matter, or aids in building living matter, or by which protoplasm is broken down into simple substances to perform special functions

Metabolite (Gr. *metabolos*, changeable + *ites*, one of a group) a chemical that is a normal cell constituent capable of entering into the biochemical transformations within living cells

Metamorphic rock (Gr. *meta*, change + *morphe*, shape or form) one of three major categories of rock; rocks whose original structure or mineral composition has been changed by pressures or temperatures in the earth's crust

Metaphase (Gr. *meta*, after + *phasis*, appearance) stage of mitosis during which the chromosomes, or at least the kinetochores, lie in the central plane of the spindle

Metaxylem (Gr. *meta*, after + *xylon*, wood) last formed primary xylem

Methanogen an Archaeal organism that derives energy by catalyzing the reduction of carbon and oxidation of H_2, forming methane and water

Microcapillary space exceedingly small spaces, such as those found between microfibrils of cellulose

Microenvironment (Gr. *mikros*, small + O.F. *environ*, about) the environment close enough to the surface of a living or nonliving object to be influenced by it

Microevolution evolution that consists of changes too small to alter the fundamental nature of a species, such as alterations in flower color

Microfibrils (Gr. *mikros*, small + fibrils, dim. of fiber; literally, small little fibers) the translation of the name expresses the concept very well; microfibrils are exceedingly small fibers visible only with the high magnifications of the electron microscope

Microfilament an intracellular organelle; threadlike, formed of actin subunits, and often participating in the movement of other organelles around the cell

Micrometer (Gr. *mikros*, small + *metron*, measure) one millionth (10^{-6}) of a meter, or 0.001 millimeter; also called a micron, and abbreviated μm

Micronutrient (Gr. *mikros*, small + L. *nutrire*, to nourish) an essential element required by plants in relatively small quantities

Microphyll (Gr. *mikros*, small + *phyllon*, leaf) a leaf whose trace is not marked with a gap in the stem's vascular system; microphylls are thought to represent epidermal outgrowths

Micropylar chamber a space interior to the micropyle of an ovule, across which pollen grains can be carried to the female gametophyte by a drying pollination drop

Micropyle (Gr. *mikros*, small + *pyle*, orifice, gate) a pore leading from the outer surface of the ovule between the edges of the two integuments down to the surface of the nucellus

Microsporangium (plural, microsporangia) (Gr. *mikros*, little + sporangium) a sporangium that bears microspores

Microspore (Gr. *mikros*, small + spore) a spore that, in vascular plants, gives rise to a male gametophyte

Microsporidia microscopic unicellular eukaryotic parasites of humans and other animals, characterized by lack of mitochondria and by entering host cells through an everted membranous tube

Microsporocyte (Gr. *mikros*, small + *spora*, seed + L. *cyta*, vessel) a diploid cell in which meiosis will occur, resulting in four microspores; synonymous with microspore mother cell

Microsporophyll (Gr. *mikros*, little + spore + Gr. *phyllon*, leaf) a leaf bearing microsporangia

Microsporophyte a sporophyte that produces haploid spores, which, on germinating and cell division, become microgametophytes (gametophytes that produce sperm cells)

Microsurgical dissections (Gr. *mikros*, small + surgical) surgical experiments done on individual cells or groups of cells, such as a shoot apex

Microtubule (Gr. *mikros*, small + *tubule*, dim. of tube) a tubule 25 nm in diameter and of indefinite length, occurring in the cytoplasm of many types of cells; major components of the spindle during cell division

Middle lamella (L. *lamella*, a thin plate or scale) original thin membrane separating two adjacent protoplasts and remaining as a distinct cementing layer between adjacent cell walls

Millimeter the 0.001 part of a meter, equal to 0.0394 inch

Mitochondrion (plural, mitochondria) (Gr. *mitos*, thread + *chondrion*, a grain) a small cytoplasmic particle associated with intracellular respiration

Mitosis (plural, mitoses) (Gr. *mitos*, a thread) nuclear division, involving appearance of chromosomes, their longitudinal duplication, and equal distribution of newly formed parts to daughter nuclei

Mitosporangium an enclosure in which mitospores are made

Mitospore (mitosis + spore) a spore that bears the same genotype as the parental organism because its nucleus was made by mitotic division of a parental nucleus; often functions in asexual reproduction

Mitosporic fungus a fungus that has not been observed to reproduce sexually, but multiplies asexually by producing mitospores; previously called a deuteromycete or an imperfect fungus

Mixed bud a bud containing both rudimentary leaves and flowers

Molecular biology a field of biology that emphasizes the interaction of biochemistry and genetics in the life of an organism

Molecular character a hereditary chemical trait of an organism, such as the detailed structure of a gene, or the ability to make a particular kind of molecule

Molecule (F. *mole*, mass + *cule*, dim.; literally, a little mass) a unit of matter, the smallest portion of an element or a compound that retains chemical identity with the substance in mass; the molecule usually consists of a union of two or more atoms; some organic molecules contain a large number of atoms

Mollisol (L. *mollis*, soft + *solum*, soil, solid) one of the 11 world soil orders, characterized by containing more than 1% organic matter in the top 17.5 cm and associated with grassland vegetation; synonymous with chernozem

Monilophytes a lineage of seedless vascular plants separate from the lycophytes; once more diverse and abundant in the Carboniferous period (Coal Age), and now represented by herbs such as horsetails, whisk ferns, ophioglossalean ferns, marattialean ferns, and true ferns

Monocotyledon (Gr. *monos*, solitary + *kotyledon*, a cup-shaped hollow) a plant whose embryo has one cotyledon

Monoecious (Gr. *monos*, solitary + *oikos*, house) referring to a plant that produces male and female gametes on the same individual; in contrast to dioecious, each individual being one sex only

Monohybrid (Gr. *monos*, solitary + L. *hybrida*, a mongrel) a cross involving one pair of contrasting characters

Monomer (Gr. *monos*, single + *meros*, part) a single subunit that can be used to form a larger complex; for example, glucose, a simple sugar, is a monomer of a cellulose chain

Monophyletic group (Gr. *mono*, single + *phyle*, tribe) a group of organisms or taxa that includes an ancestor, all of its descendants, and nothing else

Monostromatic (Gr. *monos*, single, solitary + *stroma*, a bed, currently meaning a supporting framework) referring to a thallus, one cell in thickness

Montane an adjective, meaning in or of the mountains; montane vegetation grows at significant elevations within mountain chains

Morphogenesis (Gr. *morphe*, form + L. *genitus*, to produce) the structural and physiologic events involved in the development of an entire organism or part of an organism

Morphologic character a hereditary trait that is related to body form, such as flower color or shape

Morphology (Gr. *morphe*, form + *logos*, discourse) the study of form and its development

Moss (L. *muscus*, moss; M.E., O.E. *mos*, moss) one lineage of bryophytes, having gametophytes with leaflike, stemlike, and rootlike organs, and producing sporophytes that grow on and up above the gametophyte, with a stalk and a capsule containing spores

Multiciliate (L. *multus*, many + F. *cil*, an eyelash) having many cilia present on a sperm or spore or other type of ciliated cell

Multiple fruit a cluster of matured ovaries produced by separate flowers; for example, a pineapple

Mutagen an agent that causes changes in DNA, such as ultraviolet radiation or nitrous acid

Mutant (L. *mutare*, to change) an individual containing a gene that differs from those of its parents

Mutation (L. *mutare*, to change) a heritable trait appearing in an individual as the result of a change in DNA sequence or chromosome structure

Mutualism (L. *mutuus*, reciprocal) a symbiotic relationship between two organisms that benefits both organisms

Mycelium (Gr. *mykes*, mushroom) the branching network of hyphae forming the body of most fungi

Mycologist (Gr. *mykes*, mushroom + *logos*, discourse) a scientist who specializes in the study of fungi

Mycorrhiza (plural, mycorrhizae) (Gr. *mykes*, fungus + *riza*, root) a symbiotic association between a fungus and a root of a plant

Myxomycophyta (Gr. *myxa*, mucus + *mykes*, mushroom + *phyton*, plant) a division comprising the "slime fungi"

NAD see Nicotinamide adenine dinucleotide

NADH reduced NAD

NADP see Nicotinamide adenine dinucleotide phosphate

NADPH reduced NADP

Naked bud a bud not protected by bud scales

Nanometer (Gr. *nanos*, small) one billionth (10^{-9}) of a meter, abbreviated nm

Natural selection differential reproductive success of varied genetically determined phenotypes in a population, resulting from interactions with the environment; see also directional, diversifying, and stabilizing selection

Nectar (Gr. *nektar*, drink of the gods) a fluid rich in sugars secreted by nectaries, which often are located near or in flowers

Nectar guide a mark of contrasting color or texture that may serve to guide pollinators to nectaries within the flower

Nectary (Gr. *nektar*, the drink of the gods) a nectar-secreting gland, found in flowers

Neritic zone a subtidal but relatively shallow offshore zone, often dominated by large kelps

Net productivity the arithmetic difference between calories produced in photosynthesis and calories lost in respiration

Net radiation the arithmetic difference between incoming solar radiation and outgoing terrestrial radiation

Netted venation veins of leaf blade visible to the unaided eye, branching frequently and joining again, forming a network

Neutron (L. *neuter,* neither) an uncharged particle found in the atomic nucleus of all elements except hydrogen; the helium nucleus has two protons and two neutrons; mass of a neutron is equal to 1.67×10^{-24} g

Niche (It. *nicchia,* a recess in a wall) the functional position of an organism in its ecosystem

Nicotinamide adenine dinucleotide a coenzyme capable of being reduced (accepting electrons)

Nicotinamide adenine dinucleotide phosphate a coenzyme capable of being reduced

Nitrate reduction the enzymatic reaction that converts nitrate ion, NO_3^-, to nitrite ion, NO_2^-

Nitrification (L. *nitrum,* nitro, a combining form indicating the presence of nitrogen + *facere,* to make) change of ammonium salts into nitrates through the activities of certain bacteria

Nitrogen cycle a continuous series of natural processes by which nitrogen passes through successive stations in air, soil, and organisms, involving principally ammonification, nitrogen fixation, nitrification, nitrate reduction, and denitrification

Nitrogen fixation the process of reducing N_2 gas into ammonia and incorporating it into the protoplast; accomplished only by certain prokaryotes

Node (L. *nodus,* a knot) in plant anatomy, a point on a stem where a leaf and axillary bud occur, and where branches originate; in cladistics, a branch point in a cladogram where one species divides into two new species (speciation)

Nodule (L. *nodulus* dim. of *nodus,* a knot) knot or swelling on a root, especially one containing nitrogen-fixing bacteria

Noncyclic electron transport movement of electrons in thylakoid membranes through a pathway that oxidizes a substrate (for example, H_2O, releasing O_2) and reduces an intermediate (for example, NADP, producing NADPH); the movement is stimulated by light and also generates a chemiosmotic gradient that provides energy for the synthesis of ATP

Noncyclic photophosphorylation synthesis of ATP through noncyclic electron transport

Nonequilibrium paradigm currently, a widely accepted view of natural vegetation, which holds that natural disturbances are so common that the landscape is largely composed of stands of successional vegetation, and one consequence is that the landscape is rich in species

Nonpolar having a uniform distribution of electric charge

Nucellus (L. *nucella,* a small nut) tissue composing the chief part of the young ovule, in which the embryo sac develops; megasporangium

Nuclear envelope the double membrane surrounding the nucleus of a eukaryotic cell

Nucleic acid an acid found in all nuclei, first isolated as part of a protein complex in 1871 and separated from the protein moiety in 1889; all known nucleic acids fall into two classes: DNA and RNA; they differ from each other in the sugar, in one of the nitrogen bases, in many physical properties, and in function

Nucleoid the region in a bacterial cell containing the principal chromosome; differs from a eukaryotic nucleus in not being surrounded by a nuclear envelope

Nucleolus (plural, nucleoli) (L. *nucleolus,* a small nucleus) dense body in the nucleus; site of ribosome synthesis

Nucleosides components of nucleic acids consisting of a nitrogen base and a sugar; in DNA, the sugar is deoxyribose, and in RNA, ribose; adenine, guanine, and cytosine occur in both DNA and RNA, thymine occurs in DNA, and uracil occurs in RNA

Nucleotides components of nucleic acid: nucleoside (nitrogen base + sugar) + phosphoric acid

Nucleus (plural, nuclei) (L. *nucleus,* kernel of a nut) in biology, an organelle characteristic of eukaryotes, containing DNA within an evelope made of two concentric membranes; in chemistry, the positively charged central part of an atom carrying most of the atom's mass

Nut (L. *nux,* nut) a dry, indehiscent, hard, one-seeded fruit, generally produced from a compound ovary

Nutrient cycling the phenomenon of mineral and organic nutrients being taken up by plants, passed on to herbivores and carnivores, and released once again into the soil or the air by decomposers; also the phenomenon of nutrient addition to biota by the weathering of rocks or the eruptions of volcanoes, and the removal of nutrients from biota through erosion and their ultimate deposition into deep-ocean sediments

Nyctinastic movement (Gr. *nyct,* night + *nastos,* close-pressed, firm) a movement of plant parts (for example, leaf petioles) associated with diurnal changes in temperature or light intensity

Obligate anaerobe an organism obliged to live in the absence of oxygen

Obligate parasite an organism obliged to live strictly as a parasite

Obligate saprophyte an organism obliged to live strictly as a saprophyte

Oligosaccharide (Gr. *oligos,* small, few + *sakcharon,* sugar) a molecule consisting of a chain of 2 to 10 simple sugars (for example, glucose)

Ontogeny (Gr. *on,* being + *genesis,* origin) the development of an individual organism or part

Oögamy (Gr. *oion,* egg + *gamete,* spouse) the condition in which the gametes are different in form and activity, that is, sperm and eggs

Oögonium (L. dim. of Gr. *oogonos,* literally, a little egg layer) female gametangium of egg-bearing organ not protected by a jacket of sterile cells, characteristic of the thallophytes

Oömycota (Gr. *oion,* egg + *mukes,* fungus) a group of organisms typically classified as fungi but containing cellulose cell walls and other cellular details shared with several algal groups (golden algae, brown algae, diatoms); classified in this text with the Heterokonts; examples include the causes of downy mildew of grape, potato blight, and sudden oak death

Oöspore (Gr. *oion*, egg + spore) a resistant spore developing from a zygote resulting from the fusion of heterogametes

Open dichotomous venation veins in leaves that repeatedly branch into unequal portions

Operculum (L. *operculum*, a lid) in mosses, cap of sporangium

Ophioglossalean ferns eusporangiate ferns with upright stems and leaves divided into a narrow, fertile (spore-bearing) part and a green, expanded part; not classified as true (leptosporangiate) ferns

Opportunist an organism that is adapted to exploit brief ecologic opportunities, such as a fallen fruit or a patch of ground cleared of competitors by a fire

Opposite referring to bud or leaf arrangement in which there are two buds or two leaves at a node

Orbital a solution of the Schrodinger wave equation describing a possible mode of motion of a single electron in an atom or molecule

Order (L. *ordo*, a row of threads in a loom) in the taxonomic hierarchy, a group of families

Organ (L. *organum*, an instrument or engine of any kind, musical, military, and so on) a part or member of an animal or plant body or cell adapted by its structure for a particular function

Organelle a specialized region within a cell, such as the mitochondrion or dictyosome

Organic referring in chemistry to the carbon compounds, many of which have been in some manner associated with living organisms

Osmometer (Gr. *osmos*, pushing + *meter*, measure) a devise for measuring the magnitude of osmotic force

Osmosis (Gr. *osmos*, a pushing) movement of a solvent through a differentially permeable membrane in response to a difference in solute concentration

Osmotic potential the maximum theoretical suction that can be developed in a solution as a result of osmosis when the solution is placed in an osmometer surrounded by pure water; it is a measure of the concentration of the solute(s)

Osmotic pump a mechanism of forcing solution through a pipe (for example, sieve tube) using osmosis to generate pressure at one end

Outcrossing the production of offspring that originates when the sperm and egg come from genetically different plants

Outgroup in a cladistic study, taxa that are not part of the ingroup but are included in the study to locate the root of the ingroup

Ovary (L. *ovum*, an egg) enlarged basal portion of the pistil, which becomes the fruit

Ovulate referring to a cone, scale, or other structure bearing ovules

Ovule (F. *ovule*, from L. *ovulum*, dim. of *ovum*, egg) a rudimentary seed containing, before fertilization, the female gametophyte, with egg cell, all being surrounded by the nucellus and one or two integuments

Ovuliferous (ovule + L. *ferre*, to bear) referring to a scale or sporophyll bearing ovules

Oxidation (F. *oxide*, oxygen + *tions*, suffix denoting action) to increase the positive valence or decrease the negative valence of an element or ion; loss of an electron by an atom

Oxidation–reduction (redox) any chemical reaction in which electrons are transferred from one molecule to another

Oxidative phosphorylation the synthesis of ATP from ADP plus phosphate as a result of aerobic respiration

P_{680} and P_{700} photoreaction centers of photosystem I (700) and photosystem II (680) in thylakoid membranes of chloroplasts

P_{fr} and P_r abbreviations for the far-red (FR) or red (R) absorbing form of phytochrome (P)

Pi an abbreviation for inorganic orthophosphate (a mixture of $H_2PO_4^-$ and HPO_4^{-2}, depending on the pH)

Palea (L. *palea*, chaff) upper bract that subtends a grass flower

Paleobotany (Gr. *palaios*, ancient, prehistoric + *botane*, a plant) the study of fossil plants

Paleoecology (Gr. *palaios*, ancient) a field of ecology that reconstructs past vegetation and climate from fossil evidence

Paleozoic (Gr. *palaios*, ancient + *zoe*, life) a geologic era beginning 570 million years ago and ending 225 million years ago

Palisade parenchyma elongated photosynthetic cells, containing many chloroplasts, found just beneath the upper epidermis of leaves; also called palisade mesophyll

Palmately veined (L. *paama*, palm of the hand) descriptive of a leaf blade with several principal veins spreading out from the upper end of the petiole

Pangaea (Gr. *pan*, whole + *geo*, Earth) a continuous landmass existing 250 million years ago, which contained all the continents that today are separated; the southern portion of this supercontinent is Gondwanaland and the northern portion is Laurasia

Panicle (L. *panicula*, a tuft) an inflorescence, the main axis of which is branched, and whose branches bear loose racemose flower clusters

Pappus (L. *pappus*, woolly, hairy seed or fruit of certain plants) scales or bristles representing a reduced calyx in composite flowers

Paradermal section (Gr. *para*, beside + *derma*, skin) a section cut parallel to a flat surface, such as a leaf section cut parallel to the surface of the blade

Paradigm (Gr. *paradeigma*, model) a widely accepted opinion or broad viewpoint about some phenomenon

Parallel venation type of venation in leaves in which the vascular bundles are parallel to each other

Paraphysis (plural, paraphyses) (Gr. *para*, beside + *physis*, growth) a slender, multicellular hair (*Fucus*, and others); one of the sterile branches or hyphae growing beside fertile cells in the fruiting body of certain fungi

Parasexual cycle (Gr. *para*, beside) a process by which mitosporic fungi sometimes achieve recombination without meiosis, involving formation of a heterokaryon followed by nuclear fusion and gradual return to the haploid state by loss of chromosomes

Parasite (Gr. *parasitos*, one who eats at the table of another) an organism deriving its food from the living body of another organism

Parasitism a symbiotic relationship in which one organism (the parasite) benefits, whereas the other (the host) is harmed

Parenchyma (Gr. *parenchein*, an ancient Greek medical term meaning to pour beside and expressing the ancient concept that the liver and other internal organs were formed by blood diffusing through the blood vessels and coagulating, thus designating ground tissue) a tissue composed of cells that perform housekeeping functions such as photosynthesis and storage of water and other substances; cells usually have thin walls of cellulose, and often fit rather loosely together, leaving intercellular spaces

Parenthesome (named for resemblance in cross-section view to a parenthesis) in a dolipore septum, a structure made of fused ER cisternae that lies beside the septum and permits passage of molecules but not nuclei through the septal pore

Parent material the original rock or depositional matter from which the soil of a region has been formed

Parietal (F. *parietal*, attached to the wall, from L. *paries*, wall) belonging to, connected with, or attached to the wall of a hollow organ or structure, especially of the ovary or cell

Parietal placentation a type of placentation in which placentae are on the ovary wall

Parsimony (Principle of) in cladistics, the postulate that the cladogram requiring the fewest evolutionary steps is most likely to be correct; that cladogram is said to be the most parsimonious

Parthenocarpy (Gr. *parthenos*, virgin + *karpos*, fruit) the development of fruit without fertilization

Parthenogenesis (Gr. *parthenos*, virgin + *genesis*, origin) the development of a gamete into a new individual without fertilization

Pathogen (Gr. *pathos*, suffering + *genesis*, beginning) an organism that causes a disease

Pathology (Gr. *pathos*, suffering + *logos*, account) the study of diseases, their effects on plants or animals, and their treatment

Peat (M.E. *Pete*, of Celtic origin, a piece of turf used as fuel) any mass of semicarbonized vegetable tissue formed by partial decomposition in water

Pectin (Gr. *pektos*, congealed) a water-soluble polysaccharide, a component of cell walls and the middle lamella between cells; the basis of fruit jellies

Pedicel (L. *pediculus*, a little foot) stalk or stem of the individual flowers of an inflorescence

Peduncle (L. *pedunculus*, a late form of *pediculus*, a little foot) stalk or stem of a flower that is borne singly; or the main stem of an inflorescence

Penicillin an antibiotic derived from the mold *Penicillium*

Pentose (Gr. *pente*, five + *ose*, a suffix indicating a carbohydrate) a five-carbon atom sugar

Pepo type of berry; fruits of plants in the cucumber family

Peptide (Gr. *pepton*, cooked or digested, a substance remaining after the digestion of proteins) two or more amino acids linked end to end

Peptide bond the bond between carbon and nitrogen that unites two amino acid residues in a polypeptide chain or protein

Peptidoglycan a particular macromolecule that makes up the walls of Bacteria, including cyanobacteria

Perennial (L. *perennis*, lasting the whole year through) a plant that lives more than 2 years

Perfect flower a flower having both stamens and pistils; hermaphroditic flower

Perianth (Gr. *peri*, around + *anthos*, flower) the petals and sepals taken together

Pericarp (Gr. *peri*, around + *karpos*, fruit) fruit wall, developed from the ovary wall

Periclinal cell division cell division where the newly formed cell wall is parallel to the axis of the organ

Pericycle (Gr. *peri*, around + *kyklos*, circle) root tissue bound externally by the endodermis and internally by the xylem and phloem; site of initiation of lateral roots, cork cambium, and portions of the vascular cambium in some roots

Periderm (Gr. *peri*, around + *derma*, skin) protective tissue that replaces the epidermis after secondary growth is initiated; consists of cork, cork cambium, and phelloderm

Perigyny (Gr. *peri*, about + *gyne*, a female) a condition in which the receptacle is more or less concave, at the margin of which the sepals, petals, and stamens have their origin, so that these parts seem to be attached around the ovary; also called half-inferior

Periplasmic space the region in gram-negative bacteria between the plasma membrane and the lipopolysaccharide layer

Perisperm nutritive tissue in a seed derived from the nucellus

Peristome (Gr. *peri*, about + *stoma*, a mouth) in mosses, a fringe of teeth about the opening of the sporangium

Perithecium (Gr. *peri*, around + *theke*, a box) a flask-shaped ascoma that opens at the narrow end to release ascospores

Permafrost (L. *permanere*, to remain + A.S. *freosan*, to freeze) soil that is permanently frozen; usually found some distance below a surface layer that thaws during warm weather

Permanent wilting the condition of wilting of a plant when it can no longer obtain moisture from the soil

Permanent wilting percentage the maximum amount of water (expressed as percentage of the dry weight of the soil) that a soil can hold that is unavailable to a plant

Permeable (L. *permeabilis*, that which can be penetrated) said of a membrane, cell, or cell system through which substances may diffuse

Peroxysome an organelle of the microbody class that contains enzymes capable of making and destroying hydrogen peroxide, including glycolic oxidase and catalase

Petal (Gr. *petalon*, a flower leaf) one of the flower parts, usually conspicuously colored

Petiole (L. *petiolus*, a little foot or leg) stalk of leaf

PGA see Phosphoglyceric acid

pH (Fr. *p(ouvoir)*, to be able + *h(ydrogene)*, hydrogen) a symbol for the degree of acidity (values from 0 to 7) or alkalinity (values from 7 to 14); values representing the relative concentration of the hydrogen ion in solution

Phage (Gr. *phago* to eat) a virus-infecting bacteria; originally bacteriophage

Phellem (Gr. *phellos*, cork) outer layers of the periderm consisting of dead cells formed by the phellogen; sometimes called cork

Phelloderm (Gr. *phellos*, cork + *derma*, skin) a layer of cells formed in the stems and roots of some plants from the inner cells of the cork cambium

Phellogen (Gr. *phellos*, cork + *genesis*, birth) cork cambium, a cambium giving rise externally to cork and in some plants internally to phelloderm

Phenetic species a group of organisms that is distinguished from other species on the basis of morphologic or molecular characters, or both, rather than mating tests; contrast with biological species

Phenotype (Gr. *phaneros*, showing + type) the external visible appearance of an organism

Pheromone a small molecule that functions as a chemical signal between organisms, being released by one organism and received by others, where it causes changes in behavior, physiology, or development

Phloem (Gr. *phooos*, bark) food-conducting tissue, consisting of sieve-tube members with companion cells or sieve cells, phloem parenchyma, and fibers

Phosphoenolpyruvate carboxylase the enzyme responsible for the fixation of inorganic CO_2 into oxaloacetic acid in a dark reaction of the C_4 photosynthesis cycle

Phosphoglyceric acid (PGA) a three-carbon compound formed by the interaction of carbon dioxide (CO_2) and a five-carbon compound, ribulose bisphosphate; the reaction yields two molecules of PGA for each molecule of the ribulose bisphosphate; the first step in the C_3 carbon cycle of photosynthesis

Phospholipid a complex lipid compound, composed of fatty acids, glycerol, phosphate, and one additional hydrophilic residue (for example, choline, serine, inositol); a bilayer of phospholipids is a major constituent of most biological membranes

Phosphorylation a reaction in which phosphate is added to a compound, for example, the formation of ATP from ADP and inorganic phosphate; many enzymes are stimulated (or inhibited) by the addition of phosphate

Photoautotroph (Gr. *photos*, light + *autos*, self + *trophein*, to feed) an organism that uses light as an energy source and CO_2 as a carbon source

Photoheterotroph (Gr. *photos*, light + *heteros*, other + *trophein*, to feed) an organism that uses light as an energy source and various organic compounds as a carbon source

Photon a quantum of light; the energy of a photon is proportional to its frequency; $E = hv$, where E is energy; h, Planck's constant, 6.62×10^{-27} erg-second; and v is the frequency

Photoperiod (Gr. *photos*, light + period) the optimum length of day or period of daily illumination required for the normal growth and maturity of a plant

Photophosphorylation a reaction in which light energy is converted into chemical energy in the form of ATP produced from ADP and inorganic phosphate

Photoreceptor (Gr. *photos*, light + L. *receptor*, a receiver) a light-absorbing molecule involved in converting light into some metabolic (chemical energy) form, for example, chlorophyll and phytochrome

Photosynthesis (Gr. *photos*, light + *syn*, together + *tithenai*, to place) a process in which carbon dioxide and water are brought together chemically to form a carbohydrate, the energy for the process being radiant energy

Photosystems I and II components of the noncyclic electron transport system of the light reactions of photosynthesis

Phototroph (Gr. *photos*, light + *trophein*, to feed) an organism that derives energy from light

Phototropism (Gr. *photos*, light + *tropos*, turning) a growth curvature in which light is the stimulus

Phragmoplast a body, consisting of microtubules and vesicles, involved in the synthesis of the cell plate during cytokinesis of plant cells

Phreatophyte (Gr. *phrear*, a well + *phuton*, plant) a tree or shrub that grows with its roots in continuous contact with ground water

Phycobiliproteins pigments found in the red algae and cyanobacteria, similar to bile pigments and always associated with proteins

Phycobilisomes (Gr. *phykos*, seaweed + L. *bilis*, relating to greenish bile + Gr. *soma*, body) rods or discs of accessory pigments that are attached to photosynthetic membranes in cyanobacteria; they absorb light in the green to orange part of the spectrum

Phycobiont (Gr. *phykos*, seaweed + *bios*, life) the algal partner in a lichen

Phycocyanin (Gr. *phykos*, seaweed + *kyanos*, blue) a blue phycobilin pigment occurring in cyanobacteria

Phycoerythrin (Gr. *phykos*, seaweed + *erythros*, red) a red phycobilin pigment occurring in red algae

Phycology (Gr. *phykos*, seaweed + *logos*, word, thought) the science of the study of algae

Phylogenetic systematics the branch of systematics that strives for a taxonomic system based on phylogenetic relationships

Phylogenetic tree a diagrammatic representation of evolutionary relationships for a group of organisms; a cladogram is a type of phylogenetic tree

Phylogeny (Gr. *phylon*, race or tribe + *genesis*, beginning) the series of changes by which a given taxon evolved from its ancestors

Phylum (plural, phyla) (Gr. *phylon*, race or tribe) in the taxonomic hierarchy, a group of classes; also called a division

Physiognomy (Gr. *phusis*, nature + *gnomon*, judge) originally, the art of judging human character from facial features; in plant ecology it is the structure, or architecture, of a vegetation type or a plant community

Physiology (Gr. *physis*, nature + *logos*, discourse) the science of the functions and activities of living organisms

Phytoalexin a flavonoid molecule that accumulates rapidly in a plant undergoing microbial attack; phytoalexins vary among plant families, but each is toxic to a broad spectrum of fungal and bacterial pathogens

Phytobenthon (Gr. *phylon*, a plant + *benthos*, depths of the sea) attached aquatic plants, collectively

Phytochrome a reversible pigment system of protein naturally found in the cytoplasm of green plants; it is associated with the absorption of light that affects growth, development, and differentiation of a plant, independent of photosynthesis, for example, in the photoperiodic response

Phytoplankter (Gr. *phuto*, plant + *plankter*, wanderer) a photosynthetic organism (generally single-celled) that is part of the plankton, a body of floating cells near the surface of fresh or saline bodies of water

Pigment (L. *pingere*, to paint) a substance that absorbs visible light; hence, appears colored

Pilus (L. *pilus*, hair) a structure on the surface of a bacterial cell resembling a hair

Pinna (plural, pinnae) (L. *pinna*, a feather) leaflet or division of a compound leaf (frond)

Pinnately veined (L. *pinna*, a feather + *vena*, a vein) descriptive of a leaf blade with single midrib from which smaller veins branch off, somewhat like the divisions of a feather

Pinnule (L. diminutive of *pinna*, little feather) a leaflet of a leaflet; the ultimate "leaf" on fern fronds that are twice-pinnate, the leaf first being divided into pinnae, then the pinnae into pinnules

Pinocytosis (Gr. *pinein*, to drink + *kytos*, hollow vessel) the process by which a cell may take in food or other material by forming an invagination of the plasma membrane that is pinched off into the cytoplasm

Pioneer (O.F. *pionier*, a foot soldier sent out to clear the way) in succession, an organism capable of occupying newly exposed land or land that has been cleared of previous vegetation by some catastrophic disturbance; generally an r-selected species

Pioneer community the first stage of a succession

Pistil (L. *pistillum*, a pestle) central organ of the flower, typically consisting of ovary, style, and stigma

Pistillate flower a flower having pistils but no stamens

Pit a minute, thin area of a secondary cell wall

Pith the parenchymatous tissue occupying the central portion of a stem

Placenta (plural, placentae) (L. *placenta*, a cake) the tissue within the ovary to which the ovules are attached

Placentation (L. *placenta*, a cake + *tion*, state of) manner in which the placentae are distributed in the ovary

Plankton (Gr. *planktos*, wandering) free-floating aquatic plants and animals, collectively

Plantae the monophyletic kingdom of life that includes all green plants and some algae

Plant biology the study of plant and plantlike organisms; synonymous with botany

Plant demography (Gr. *demos*, people + *graphein*, to write) demography was originally the study of births, deaths, density, and distribution of human populations; plant demography is this study applied to plant populations

Plant geography (L. *planta*, shoot + Gr. *ge*, Earth + *graphein*, to write) a synonym for biogeography; the study of the geographic distribution of plants, often including distributions in geologic time, as well as in the present, and with a focus on taxonomically related plant groups (families, genera)

Plasma membrane (Gr. *plasma*, anything formed + L. *membrana*, parchment) a delicate cytoplasmic membrane found on the outside of the protoplast adjacent to the cell wall

Plasmid (Gr. *plasma*, to form) a piece of extrachromosomal DNA, found in some bacteria

Plasmodesma (plural, plasmodesmata) (Gr. *plasma*, something formed + *desmos*, a bond, a band) fine protoplasmic thread passing through the wall that separates two protoplasts

Plasmodium (Gr. *plasma*, something formed + mod. L. *odium*, something of the nature of) in Myxomycetes, a slimy mass of protoplasm, with no surrounding wall and with numerous free nuclei distributed throughout

Plasmogamy (Gr. *plasma*, anything molded or formed + *gamos*, marriage) the fusion of protoplasts, not accompanied by nuclear fusion

Plasmolysis (Gr. *plasma*, something formed + *lysis*, a loosening) the separation of the cytoplasm from the cell wall caused by removal of water from the protoplast

Plastid (Gr. *plastis*, a builder) a cellular organelle in which carbohydrate metabolism is located

Plastoquinone a quinone, one of a group of compounds involved in the transport of electrons during photosynthesis in chloroplasts

Plumule (L. *plumula*, a small feather) the first bud of an embryo or that portion of the young shoot above the cotyledons

Pneumatophores (Gr. *pneuma*, breath + *phore*, F. *pherein*, to carry) extensions of the root systems of some plants growing in swampy areas; in contrast to most roots, they are negatively geotropic and grow out of the water, thus assuring adequate aeration

Pod see Legume

Polar a term applied to molecules having a nonuniform distribution of electrons and thus negatively and positively charged regions

Polarity (Gr. *pol*, an axis) the observed differentiation of an organism, tissue, or cell into parts having opposed or contrasted properties or form

Polar nuclei the two nuclei of the central cell in the embryo sac of flowering plants

Polar transport the directed movement within plants of compounds (usually hormones) predominantly in one direction; polar transport overcomes the tendency for diffusion in all directions

Pollen, pollen grain (L. *pollen*, fine flour) the immature male gametophytes of seed plants

Pollen mother cell see Microsporocyte

Pollen profile a diagrammatic summary of the sequence and abundance of pollen types that have been chronologically trapped in sediments

Pollen sac a cavity in the anther of a flowering plant, or a globose swelling attached to a cone scale in a conifer that initially contains diploid cells, each of which divides by meiosis and ultimately develops into a pollen grain

Pollen tube a tube of cytoplasm that grows from the pollen grain through the pistil to an egg; it is the mature male gametophyte of flowering plants and contains only two sperm nuclei

Pollination the transfer of pollen from a stamen or staminate cone to a stigma or ovulate cone

Pollination drop a viscous, sticky fluid that exudes out through the micropyle of a gymnosperm ovule, to which pollen grains may adhere and be drawn back through the micropyle as the pollination drop dries and shrinks, or is reabsorbed by tissue within the ovule

Pollination syndrome unique flower and pollen traits that adapt a flower for pollination by a particular vector

Pollinium (L. *pollentis*, powerful or *pollinis*, fine flour + *ium*, group) a mass of pollen that sticks together and is transported by pollinators as a mass; present in orchids and milkweeds

Polygenes many genes influencing the development of a single trait; results in continuous variability; compare Allele

Polymer (Gr. *polys*, many + *meros*, part) a compound formed by repeating structural units; for example, cellulose is a polymer of glucose, a simple sugar

Polymerization the chemical union of monomers, such as glucose, or nucleotides to form starch or nucleic acid

Polynomial (Gr. *polys*, many + L. *nomen*, name) scientific name for an organism, composed of more than two words; see Binomial

Polynucleotides (Gr. *polys*, much, many) long-chain molecules composed of units (monomers) called nucleotides; nucleic acid is a polynucleotide

Polypeptide chain a linear polymer formed from amino acids held together by peptide bonds

Polyploid (Gr. *polys*, many + *ploos*, fold) referring to a plant, tissue, or cell with more than two complete sets of chromosomes, for example, 4*n*, 6*n*

Polyribosome (Gr. *polys*, many + ribosomes) an aggregation of ribosomes, frequently simply called polysome

Polysaccharides (Gr. *polys*, much, many + *sakcharon*, sugar) long-chain molecules composed of units (monomers) of a sugar; starch and cellulose are polysaccharides

Pome (L. *pomum*, apple) a simple, fleshy fruit, the outer portion of which is formed by the floral parts that surround the ovary

Population (L. *populus*, people) a group of closely related, interbreeding organisms

Population genetics a mathematically based discipline that extends the study of genetics to include evolution

Potential energy energy associated with position; concerning a chemical reaction, includes the energy of a set of reactants associated with the position of their constituent atoms and the shape of their bonding orbitals

Powdery mildew a plant disease caused by certain parasitic ascomycete fungi, characterized by a dusty white layer of conidia on infected leaves

p protein a mass of protein material formerly called slime found in sieve-tube members

Prairie (L. *pratum*, meadow) grassland vegetation, with trees essentially absent; often considered to have more rainfall than does the steppe

Predation (L. *predatio*, plundering) a form of biological interaction in which one organism is destroyed (by ingestion); parasitism, carnivory, and seed herbivory are forms of predation

Preprophase band an aggregation of microtubules, occurring in the G2 phase of a plant cell cycle, which marks the plane of cell division after the next mitosis

Primary (L. *primus*, first) first in order of time or development

Primary consumer an herbivorous organism, gaining nutrition from plants (producers)

Primary endosperm nucleus the 3*n* product of the fertilization of one sperm nucleus with the two polar nuclei; the endosperm develops from this

Primary meristems meristems of the shoot or root tip giving rise to the primary tissues—protoderm, ground meristem, and procambium

Primary phloem the part of the phloem differentiated from the procambium during organ elongation

Primary pit field thin areas of primary cell walls

Primary structure the peptide bonds holding the amino acids of a protein together; the sequence of amino acids in a protein

Primary succession a successional path that begins on recently exposed land that has never before supported vegetation; hence, it has no soil profile or buried plant propagules

Primary thickening meristem found in the shoot tip of monocots; generates growth in length and width

Primary tissues those tissues—epidermis, xylem, phloem, and ground tissues—that form from primary meristems

Primary wall the first-formed cell wall layer

Primary xylem the part of the xylem differentiated from the procambium during organ elongation

Primordium (L. *primus*, first + *ordiri*, to begin to weave; literally beginning to weave, or to put things in order) the beginning or origin of any part of an organ

Procambium (L. *pro*, before + *cambium*) a primary meristem that gives rise to primary vascular tissues and, in most woody plants, to the vascular cambium

Producer (L. *producere*, to draw forward) an organism that produces organic matter for itself and other organisms (consumers and decomposers) by photosynthesis

Productivity net biomass produced by green plants through the process of photosynthesis; usually expressed as a rate (for example, 300 kg/hectare per year)

Proembryo (L. *pro*, before + *embryon*, embryo) a group of cells arising from the division of the fertilized egg cell before those cells that are to become the embryo are recognizable

Progressive succession a pathway of succession that results in higher species richness, greater biomass, higher physiognomic complexity, more continuous plant cover, and a microenvironment that is more buffered from the macroenvironment

Prokaryote (Gr. *pro*, before + Gr. *karyon*, a nut, referring in modern biology to the nucleus) member of the domain Archaea or Bacteria; an organism that does not have its DNA in a nucleus, separated from the cytoplasm by an envelope

Prophase (Gr. *pro*, before + *phasis*, appearance) an early stage in nuclear division, characterized by the shortening and thickening of the chromosomes and their movement to the metaphase plate

Proplastid (Gr. *pro*, before + plastid) a type of plastid, occurring generally in meristematic cells, that will develop into a chloroplast

Prop roots adventitious roots from lower nodes of stems; act to support the shoot axis

Protease (protein + ase, a suffix indicating an enzyme) an enzyme breaking down a protein

Protein body protein storage organelle in seeds, sometimes called aleurone grain

Proteins (Gr. *proteios*, holding first place) naturally occurring complex organic substances, composed of one or more polypeptide chains, which through interactions among amino acids in different parts of the chain(s) acquire specific three-dimensional shapes required for their biological functions

Proterozoic (Gr. *protera*, before in time + *zoe*, life) the earliest geologic era, beginning about 4.5 to 5 billion years ago and ending 570 million years ago; also called Precambrian era

Prothallium (Gr. *pro*, before + *thallos*, a sprout) in ferns, the haploid gametophyte generation

Protista (Gr. *protistos*, the very first) an artificial (nonmonophyletic) grouping of eukaryotic organisms that are not included in kingdoms Plantae, Fungi, or Animalia; organisms within it are called protists

Protoagriculture (Gr. *protos*, first + L. *ager*, land + *cultura*, cultivation) a form of plant cultivation thought to precede agriculture, in which desireable plants are tended and selected for in their natural habitats rather than in prepared fields sown to monocultures

Protochlorophyll (Gr. *protos*, first + *chloros*, green + *phyllos*, leaf) one of the precursors of chlorophyll; it accumulates in dark-grown and potentially green tissue

Protochlorophyllide holochrome a light-sensitive compound or complex composed of photochlorophyll and a protein; absorption of light converts the protochlorophyll part to chlorophyll

Protoderm (Gr. *protos*, first + *derma*, skin) a primary meristem that gives rise to epidermis

Proton (Gr. *protos*, first) a positively charged particle found in the atomic nucleus of all elements; the number of protons in a nucleus determines the identity of the element; the mass of a proton is 1.67×10^{-24} g

Proton pump a transmembrane enzyme that catalyzes the movement of protons from one side of the membrane to the other coupled to the expenditure of free energy (for example, hydrolysis of ATP); because energy is expended, protons can be transported against their concentration gradient

Protonema (plural, protonemata) (Gr. *protos*, first + *nema*, a thread) an algal-like filamentous growth; an early stage in development of the gametophyte of mosses

Protoplasm (Gr. *protos*, first + *plasma*, something formed) living substance

Protoplast (Gr. *protoplastos*, formed first) the organized living unit of a single cell; a cell from which the cell wall has been removed

Protoxylem (Gr. *protos*, first + *xylon*, wood) first formed primary xylem

Pseudopodium (Gr. *psuedes*, false + *podion*, a foot) in Myxomycetes, an armlike projection from the body by which the plant creeps over the surface

Psilophytes (Gr. *psilos*, bare) seedless vascular plants mainly consisting of bare, dichotomously branched green stems; the whisk ferns, *Psilotum* and *Tmistris*, which traditionally have been classified in their own division, the Psilophyta, but in this book are placed with horsetails, ophioglossalean ferns, marattialean ferns, and true ferns in the monilophyte lineage

Puffball a type of basidioma consisting of a rounded mass of spore-producing cells surrounded (at least initially) by a leathery covering

Purine a group of organic bases having a double-ring structure, one five-carbon and the other six-carbon

Pyrenoid (Gr. *pyren*, the stone of a fruit + L. *oides*, like) a denser body occurring within the chloroplasts of certain algae and liverworts and apparently associated with starch deposition

Pyrimidine an organic base having a single-ring structural formula

Quadrat (L. *quadrus,* a square) a frame of any shape that, when placed over vegetation, defines a unit sample area within which the plants may be counted or measured

Quantum (L. *quantum,* how much) an elemental unit of energy; its energy value is hv, where h, Planck's constant, is 6.62×10^{-27} erg-second, and v is the frequency of the waves with which the energy is associated

Quaternary structure the combination and arrangement of separate polypeptide chains in a protein

Quiescent center (L. *quiescere,* to rest) disk-shaped region of root apex containing slowly dividing cells

Raceme (L. *racemus,* a bunch of grapes) an inflorescence in which the main axis is elongated but the flowers are born on pedicels that are about equal in length

Rachilla (Gr. *rhachis,* a backbone + L. dim. ending *illa*) shortened axis of spikelet

Rachis (Gr. *rhachis,* a backbone) main axis of spike; axis of fern leaf (frond) from which pinnae arise; in compound leaves, the extension of the petiole corresponding to the midrib of an entire leaf

Radicle (L. *radix,* root) portion of the plant embryo that develops into the primary root

Radioactive decay the disintegration of the nucleus of an atom through the emission of a high-energy α or β particle

Random plant distribution a distribution pattern of a plant population within an area such that the probability of finding an individual at one point is the same for all points

Raphe (Gr. *rhaphe,* seam) ridge on seeds, formed by the stalk of the ovule, in those seeds in which the funiculus is sharply bent at the base of the ovule

Raphides (Gr. *rhachis,* a needle) fine, sharp, needlelike crystals

Ray initials (L. *radius,* a beam or ray) meristematic cells in the vascular cambium that develop into xylem and phloem cells composing the ray system

Ray system system of cells in secondary tissues that are oriented perpendicular to the long axis of the stem, formed from ray initials of the vascular cambium

rDNA DNA that specifies the structure of the RNA molecules found in ribosomes

Receptacle (L. *receptaculum,* a reservoir) enlarged end of the pedicel or peduncle to which other flower parts are attached

Recessive in heredity, referring to an allele that, when present in a hybrid with a contrasting allele, is not expressed in the phenotype

Recessive character that member of a pair of Mendelian characters that, when both members of the pair are present, is subordinated or suppressed by the other, dominant character

Recombinant DNA a molecule of DNA formed by the joining of segments from different sources

Recombination (L. *re,* repeatedly + *combinatus,* joined) a rearrangement of the collection of alleles of different genes through independent segregation or crossing-over, so that the genotype of a meiotic product (that is, gamete, meiospore) of an individual differs from those of its parents

Red algae multicellular, (mostly) marine, warm-water algae, some of which are large and differentiated into seaweeds with holdfast, stipe, and blade regions; closely related to the green algae and included with them in the plant kingdom

Redox reaction (*reduction* + oxidation) a chemical reaction in which one or more reactants is reduced (gains electrons) and one or more reactants is oxidized (loses electrons)

Red tide a coloring of offshore marine water caused by dense phytoplankton populations; often accompanied by toxic byproducts

Reduction (F. *reduction,* L. *reductio,* a bringing back) originally "bringing back" a metal from its oxide, that is, iron from iron rust or ore; any chemical reaction involving the removal of oxygen from or the addition of hydrogen or an electron to a substance; energy is required and may be stored in the process as in photosynthesis

Region of elongation part of the root tip where cells are elongating

Region of maturation part of the root tip basal to the elongation zone where root hairs form and cells reach maturity

Regular flower a flower in which the corolla is made up of similarly shaped petals equally spaced and radiating from the center of the flower; star-shaped flower; actinomorphic flower

Replication the production of a facsimile or a close copy; here used to indicate the production of a second molecule of DNA exactly like the first molecule

Reproduction (L. *re,* repeatedly + *producere,* to give birth to) the process by which plants and animals give rise to offspring

Reproductive isolation a condition in which two populations cannot exchange genes

Residual meristem meristematic region near the shoot tip that remains after differentiation of the pith and cortex

Resin (L. *resina,* resin) a viscose, sticky fluid of plant origin, used by woody plants to repel wood-burrowing insect pests; humans collect resin as copal, rosin, or amber, and they use resin in the manufacture of various lacquers, varnishes, inks, adhesives, and plastics

Resin duct (L. *resina,* resin + *ductus,* led) resin canal; in conifers, continuous tubes lined with secretory cells that run through the sap-wood; they function as repositories for metabolic byproducts

Respiration (L. *re,* repeatedly + *spirare,* to breathe) a chemical oxidation controlled and catalyzed by enzymes that break down carbohydrate and fats, thus releasing energy to be used by the organism in doing work

Restoration (L. *restaure,* brought back) in ecology, the process of modifying the environment and the vegetation of an ecosystem that has been degraded by human-caused disturbances and stresses to return that ecosystem to its predisturbed state

Restriction enzyme an enzyme that hydrolyzes DNA at a particular base sequence (specific to the type of enzyme); also called restriction endonuclease

Reticulum (L. *reticulum,* a small net) a small net

Retrogressive succession a pathway of succession that results in lower species richness, less biomass, lower physiognomic complexity, less plant cover, and a microenvironment less buffered from the macroenvironment

Rhizoid (Gr. *rhiza,* root + L. *oides,* like) in nonvascular plants, a filament of cells that lacks the vascular system of a root but performs rootlike functions (anchorage and absorption); in fungi, a hypha too narrow to house nuclei, functioning in anchorage, absorption, and sometimes mating

Rhizome (Gr. *rhiza,* root) an elongated, underground, horizontal stem

Rhizophores (Gr. *rhiza,* root + *phoros,* bearing) leafless branches that grow downward from the leafy stems of certain Lycophyta and give rise to roots when they come into contact with the soil

Rhizosphere soil zone immediately outside a root containing microorganisms

Ribonucleic acid (RNA) a nucleic acid containing the sugar ribose, phosphorus, and the bases adenine, guanine, cytosine, and uracil; present in all cells and concerned with protein synthesis in the cell

Ribose a five-carbon sugar, a component of RNA

Ribosomes (ribo, from RNA + Gr. *somatos,* body) small particles 10 to 20 nm in diameter, containing RNA and proteins and involved in the synthesis of proteins

Ribulose bisphosphate see Ribulose bisphosphate carboxylase

Ribulose bisphosphate carboxylase (F. *carbboe,* carbon + Gr. *oxys,* acidic) the enzyme of photosynthesis that catalyzes the initial fixation of carbon, that is, the covalent attachment of carbon in CO_2 to an organic molecule, ribulosebisphosphate

Ring porous wood wood with large xylem vessel members mostly in early wood; compare with Diffuse porous wood

Ripening (A.S. *rifi,* perhaps related to reap) changes in a fruit that follow seed maturation and that prepare the fruit for its function of seed dispersal

RNA see Ribonucleic acid

Root (A.S. *rot*) in plant anatomy, a usually underground organ adapted for growth through soil and absorption of minerals and water, containing a central vascular cylinder; in cladistics, the oldest node in a clade or cladogram

Root apical meristem (RAM) region of cells located at the root tip; derives the cells and tissues of the root

Root cap a thimblelike mass of cells covering and protecting the apical meristems of a root; also site of gravity perception

Rooted cladogram a cladogram in which the earliest or root node of the clade of interest is identified, thereby establishing the sequence of evolutionary events throughout the cladogram

Root hairs epidermal projections of root cells in region of maturation that increase the absorptive surface of the root

Root nodule see Nodule

Root pressure pressure developed in the root as the result of osmosis and inducing guttation

Rootstock an elongated, underground, horizontal stem

Root system all of the roots and branches initially derived from the radicle of the embryo

Rosette (dim. of L. *rose,* rose) a shoot with a very short stem, composed of several unelongated internodes but with fully expanded leaves

Rosid clade a diverse group of eudicots, including legumes, eucalypts, roses, mustards, and tropical melastomes

r selection natural selection that favors short-lived, early-maturing individuals that devote a large fraction of their resources to reproduction; annual herbs are r strategists

R type life hisory a life history pattern that describes short-lived, small plants that often live in frequently disturbed habitats; similar to r-selected plants

Runner a stem that grows horizontally along the ground surface

Rust disease a plant disease caused by urediniomycete fungi (phylum Basidiomycota), named for the rust-colored mitospores that often form on infected leaves and stems

Salt a compound formed from a defined mixture of cations and anions

Salt marsh a wetland close enough to a maritime coastline to be influenced by tidal fluctuations, so that at some daily or seasonal times the wetland is submerged and at other times is exposed; dominated by perennial, succulent forbs and salt-tolerant grasses and sedges

Samara (L. *samara,* the fruit of the elm) simple, dry, one- or two-seeded indehiscent fruit with pericarp bearing a winglike outgrowth

Sand soil particles between 50 and 2,000 microns in diameter

Saprobe (Gr. *sapros,* rotten) an organism that derives its food from organic products released by another organism

Sapwood functional xylem at the periphery of secondary stems; contrast to Heartwood

Savannah (Sp. *sabana,* a large plain) vegetation of scattered trees in a grassland matrix; originally limited to tropical regions

Scalariform vessel (L. *scala,* ladder + form) a vessel with secondary thickening resembling a ladder

Scanning electron microscopy a technique by which the surface of a subject is visualized at high resolution by bombarding it with a tighly focused beam of electrons, measuring the rate at which electrons are scattered into a collector as a function of the position of the beam, and displaying the result on a television screen by coordinating the scanning of the electron beam hitting the sample with the electron beam in the television tube

Schizocarp (Gr. *schizein*, to split + *karpos*, fruit) dry fruit with two or more united carpels that split apart at maturity

Sclereids (Gr. *skleros*, hard) cells of different shapes having heavily lignified cell walls

Scientific method an iterative process of stating, testing, and refining hypotheses that best explain natural phenomena

Sclerenchyma (Gr. *skleros*, hard + *enchyma*, a suffix denoting tissue) a strengthening tissue composed of cells with heavily lignified cell walls

Sclerieds cells typically with thick secondary cell walls and irregular shapes, one of two categories of cell types making up sclerenchyma tissue

Sclerophyll (Gr. *scleros*, hard + *phullon*, leaf) evergreen leaves having a hard texture, thick cuticle and blade, closely packed cells in the mesophyll region, low rate of photosynthesis, and a low rate of transpiration

Scrub (A.S. *scrob*, a shrub) vegetation dominated by shrubs; described as thorn forest in areas with moderate rainfall, or as chaparral or desert in areas with low rainfall

Scutellum (L. *scutella*, a dim. of *scutum*, shield) single cotyledon of grass embryo

Seaweed a large alga that is differentiated into holdfast, stipe, and blade regions (except for some floating seaweeds that lack a holdfast); kelps are large seaweeds in the brown algae group

Secondary consumer a carnivorous animal that feeds on herbivorous animals (primary consumers)

Secondary meristem dividing cells responsible for lateral growth of stems and roots of vascular plants; there are two types: vascular cambium and cork cambium

Secondary phloem tissue formed by division of the vascular cambium, usually toward the outside of the organ

Secondary structure the structure of a protein produced by the bending of its backbone into specific shapes (for example, alpha helix)

Secondary succession a successional pathway that begins with bare land that was once-vegetated and ends with vegetation that had occupied the land before some catastrophic disturbance, such as flood, overgrazing, clear-cutting, fire, or conversion to agriculture

Secondary tissue those tissues, xylem, phloem, and periderm, that form from secondary meristems

Secondary wall wall material deposited on the primary wall in some cells after elongation has ceased

Secondary xylem tissue formed by division of the vascular cambium, usually toward the inside of the organ

Sedimentary rock (L. *sedere*, to sit) rock formed from material deposited as sediment, then physically or chemically changed by compaction and hardening while buried in the earth's crust

Seed (A.S. *sed*, anything that may be sown) popularly, as originally used, anything that may be sown; that is, "seed" potatoes, "seeds" of corn, sunflower, and so on; botanically, a seed is the matured ovule without accessory parts

Seed coat outer layers of a seed, derived from the integuments

Seedless vascular plant an organism with an embryo life cycle phase and with xylem and phloem, but which reproduces with spores rather than seeds; includes the lycophytes and the monilophytes

Seedling hypothesis an explanation, first offered by William Bond, for the dominance of angiosperms over gymnosperms; angiosperm seedlings typically are fast growing, contain more efficient tissues for conducting water and nutrients, and build relatively flimsy tissue that requires a low caloric investment

Selection pressure the overall effect of environmental factors in determining the direction of natural selection

Selective agent an environmental factor that causes natural selection

Self-pollination transfer of pollen from the stamens to the stigma of either the same flower or flowers on the same plant

Semelparous (L. *Semele*, a mythical woman who bore a child and then died + *paritas*, equal) a perennial plant that flowers only once, at the end of its life span

Seminal root adventitious roots forming at the base of the seedling stem

Senescence (L. *senescere*, to grow old) the phase of plant growth that extends from full maturity to death and is characterized by a breakdown of functional cellular components, accumulation of metabolic products, and (often) an increase in respiratory rate and synthesis of ethylene

Sepals (M.L. *sepsium*, a covering) outermost flower structures that usually enclose the other flower parts in the bud

Septal pore a hole through a septum in a hypha, permitting material to pass through; also called a perforation

Septate (L. *septum*, fence) divided by cross walls into cells or compartments

Septate hypha a fungal hypha that is divided into cell-like compartments by regularly spaced crosswalls (septa)

Septicidal dehiscence (L. *septum*, fence + *caedere*, to cut; *dehiscere*, to split open) the splitting open of a capsule along the line of union of carpels

Septum (L. *septum*, fence) any dividing wall or partition; frequently, a cross wall in a fungal or algal filament

Serpentine (L. *serpens*, a serpent) referring to soil derived from metamorphic parent material characterized, among other things, by low calcium (Ca), high magnesium (Mg), and a greenish-gray color

Sessile (L. *sessilis*, low, dwarf, from *sedere*, to sit) sitting, referring to a leaf lacking a petiole or a flower or fruit lacking a pedicel

Seta (plural, setae) (L. *seta*, a bristle) in bryophytes, a short stalk of the sporophyte, which connects the foot and the capsule

Sexual reproduction reproduction that requires meiosis and fertilization for a complete life cycle

Sheath part of a leaf that wraps around the stem, as in grasses

Shelf fungus a fungus that forms a shelflike basidioma on wood

Shoot (derivation uncertain, but early referring to new plant growth) a young branch that shoots out from the main stock of a tree, or the young main portion of a plant growing aboveground

Shoot apical meristem (SAM) dividing cells responsible for differentiation of all primary meristems and primary tissues of the shoot

Shoot system all stems, leaves, buds, flowers, and fruits developed from the shoot apical meristem

Shoot tip portion of the shoot containing apical and primary meristems and early stages of differentiation

Short-day plants plants that are induced to flower only when the length of the day decreases to less than a certain value

Shrub (M.E., *schrubbe*) a woody plant arising from many stems, as opposed to a tree, which has a single stem (trunk)

Sibling species species morphologically nearly identical but incapable of producing fertile hybrids

Side chain the portion of an amino acid attached to its amine-(alpha)carbon-carboxylic acid backbone; each type of amino acid has a different side chain

Sieve area specialized portion of sieve cell and sieve-tube member cell walls containing a cluster of small pores where each pore is surrounded by a carbohydrate called callose

Sieve cells a long and slender sieve element found in the phloem of gymnosperms and some ferns, often associated with a specialized parenchyma-like albuminous cell, with relatively unspecialized sieve areas and with tapering end walls that lack sieve plates

Sieve plate wall area in a sieve-tube member containing a region of pores through which pass strands connecting sieve-tube protoplasts

Sieve tube a series of sieve-tube members forming a long, cellular tube specialized for the conduction of food materials; found in flowering plants

Sieve-tube members portion of a sieve tube composed of a single protoplast and separated from other sieve-tube members by sieve plates

Silique (L. *siliqua*, pod) the fruit characteristic of Brassicaceae (mustards); two-celled, the valves splitting from the bottom and leaving the placentae with the false partition stretched between

Silt soil particles between 2 and 50 microns in diameter

Simple leaf single leaf blade; may be lobed or have seriated edges, but is not divided into leaflets

Simple pistil pistil that consists of only one carpel

Simple pit pit in a secondary cell wall not surrounded by an overarching border; contrast to bordered pit

Smog (*sm* from smoke + *og* from fog) polluted air containing volatile and particulate hydrocarbons, oxides of sulfur and/or nitrogen, and ozone; brown-tinted and less transparent than clean air

Smut disease a plant disease caused by ustilagomycete fungi (phylum Basidiomycota), named for greasy black masses of fungal mitospores that form on the host plant

Softwood a gymnosperm tree; having wood that is light and easier to work than the wood of an angiosperm

Soil (L. *solum*, soil, solid) the uppermost stratum of the earth's crust, which has been modified by weathering and organic activity into (typically) three horizons: an upper A horizon that is leached, a middle B horizon in which the leached material accumulates, and a lower C horizon, which is unweathered parent material

Soil texture refers to the amounts of sand, silt, and clay in a soil, as sandy loam, loam, or clay texture

Solute (L. *solutus*, from solvere, to loosen) a dissolved substance

Solution (M.E. *solucion*, from O.T. *solucion*, to loosen) a homogeneous mixture, the molecules of the dissolved substance (for example, sugar), the solute, being dispersed between the molecules of the solvent (for example, water)

Solvent (L. *solvere*, to loosen) a substance, usually a liquid, having the properties of dissolving other substances

Soredium (plural, soredia) (Gr. *soros*, a heap) an asexual reproductive unit released by lichens, consisting of a few algal cells surrounded by fungal hyphae

Sorus (plural, sori) (Gr. *soros*, a heap) a grouping of sporangia; especially characteristic of ferns

Speciation the splitting of one species into two species

Species (L. *species*, appearance, form, kind) a group of organisms that are more closely related to one another than to organisms of any other kind; members of a given species may also look more like one another and interbreed more freely with one another than with organisms outside the group

Species diversity the sum of every species present in an area, each species being multiplied by its relative abundance; compare with species richness

Species epithet the second word in a binomial species name, which distinguishes the named species from other species in the same genus

Species name see Binomial

Species richness the number of species present in an area, rare species being counted as equal to common species

Sperm (Gr. *sperma*, the generative substance or seed of a male animal) a male gamete

Spermatophyte (Gr. *sperma*, seed + *phyton*, plant) a seed plant

Sphenophytes (Gr. *sphen*, wedge) synonymous with horsetails; until recently, horsetails were classified in their own division, the Sphenophyta, but in this text they are classified as monilophytes, together with whisk ferns, ophioglossalean ferns, marattialean ferns, and true ferns

Spike (L. *spica*, an ear of grain) an inflorescence in which the main axis is elongated and the flowers are sessile

Spikelet (L. *spica,* an ear of grain + *let,* dim. ending), the unit of inflorescence in grasses; a small group of grass flowers

Spindle (A.S. *spinel,* an instrument used in spinning thread by hand) referring in mitosis and meiosis to the spindle-shaped intracellular structure in which the chromosomes move

Spine sharp pointed structure, a modified leaf

Spirillum (L. *spira,* a coil) a bacterial cell that has a spiral shape

Spodosol (Gr. *spodos,* wood ashes; R. *pod,* under + *zola,* ashes) one of the 10 world soil orders, characterized by an ashy, sandy, bleached acidic A$_2$ horizon and associated mainly with coniferous forest vegetation; synonymous with podzol

Spongy mesophyll loosely organized photosynthetic cells of a leaf

Spongy parenchyma cells making up the spongy mesophyll

Sporangiophore (sporangium + Gr. *phorein,* to bear) a hypha that bears one or more sporangia

Sporangium (plural, sporangia) (spore + Gr. *angeion,* a vessel) a single cell or multicellular enclosure in which spores are made

Spore (Gr. *spora,* seed) a one-celled or few-celled reproductive unit that develops into a multicellular organism without first fusing with another reproductive unit; contrast with seed and gamete

Sporic life cycle a life cycle in which both gametophyte and sporophyte generations are multicellular organisms

Sporocyte a cell that will divide to generate spores

Sporophyll (spore + Gr. *phyllon,* leaf) a spore-bearing leaf

Sporophyte (spore + Gr. *phyton,* a plant) in alternation of generations, the plant in which meiosis occurs and which thus produces meiospores

Springwood secondary xylem (wood) that forms early in the growing season; earlywood

Stabilizing selection natural selection that favors the current range of phenotypic variation in a population, preventing or limiting evolutionary changes

Stamen (L. *stamen,* the standing-up things or a tuft of thready things) flower structure made up of an anther (pollen-bearing portion) and a stalk or filament

Staminate flower a flower having stamens but no pistils

Starch (M.E. *sterchen,* to stiffen) a complex insoluble carbohydrate, the chief food storage substance of plants, that is composed of several hundred hexose sugar units and that easily breaks down on hydrolysis into these separate units

Statolith (Gr. *statos,* standing + *lithos,* stone) a starch grain that moves to its position in a cell as a result of gravity, thus providing an initial sensing of, or orientation to, gravity by a cell

Stele (Gr. *stele,* a post) the central vascular cylinder, inside the cortex, of roots and stems of vascular plants

Stem (O.E. *stemn*) the main body of the portion aboveground of tree, shrub, herb, or other plant; the ascending axis, whether aboveground or belowground, of a plant, in contradistinction to the descending axis or root

Steppe (R. *step,* a lowland) a vegetation type in which shrubs are intermixed with perennial grasses in a semi-arid region; sometimes defined as a vegetation type dominated by relatively short, perennial grasses without any woody vegetation

Stereid (Gr. *stereos,* solid) thick-walled cells at or near the epidermis of certain mosses

Sterigma (plural, sterigmata) (Gr. *sterigma,* a prop) a projection from a basidium that extrudes a basidiospore at its tip and assists in ballistospore release

Sterilization (L. *sterilis,* barren) the process of making something free of living organisms

Sterols a class of lipids, including (in animals) cholesterol and (in plants) ergosterol, often contributing to the structure of biological membranes

Stigma (L. *stigma,* a prick, a spot, a mark) receptive portion of the style to which pollen adheres

Stipe (L. *stipes,* post, tree trunk) the stalk portion of a kelp or a mushroom

Stipule (L. *stipula,* dim. of stipes, a stock or trunk) a leaflike structure from either side of the leaf base

Stolon (L. *stolo,* a shoot) a stem that grows horizontally along the ground surface

Stoma (plural, stomata) (Gr. *stoma,* mouth) a minute opening plus two guard cells in the epidermis of leaves and stems, through which gases pass

Stomatal apparatus a stoma plus the surrounding subsidiary cells

Stomatal crypt an indentation on the surface of a leaf into which stomata open; regulates transpiration by providing a relatively stable boundary layer of air outside the stomata

Strobilus (Gr. *strobilos,* a cone) a number of modified leaves (sporophylls) or ovule-bearing scales grouped together on an axis

Stroma (Gr. *stroma,* a bed or matress) the proteinaceous solution in the central region of chloroplasts, the location of the reactions of the carbon cycle of photosynthesis

Stroma lamellae the thylakoid membranes in a chloroplast that are relatively dispersed; contrast to Grana

Style (Gr. *stylos,* a column) slender column of tissue that arises from the top of the ovary and through which the pollen tube grows

S type life history a life history pattern that describes long-lived, slow-growing plants of varying adult size, which live in habitats too stressful for R- or C-type plants; plants that have unusual metabolic or morphologic traits that mitigate environmental stresses

Subalpine zone a montane zone in which the tree growth form becomes diminished and scattered; a transition between montane forest and alpine tundra

Suberin (L. *suber,* the cork oak) a waxy material found in the cell walls of cork tissue

Subshrub a shrub that is shorter than 30 cm in height, and which is woody only at the base of its stems; the life span of subshrubs is shorter than that of shrubs, only one to several decades

Subsidiary cell type of epidermal cell in contact with guard cells on leaves

Substomatal chamber air space immediately below the stomatal pore; provides evaporative surfaces for transpiration

Substrates the reactants of an enzymatically catalyzed chemical reaction

Succession (L. *successio*, a coming into the place of another) a sequence of changes in time of the species that inhabit an area, from an initial pioneer community to a final climax community

Succulent (L. *sucus*, juice) a plant having juicy or watery tissues

Sucrose (F. *sucre*, sugar + *ose*, ending designating a sugar), cane sugar ($C_{12}H_{22}O_{11}$)

Summerwood secondary xylem (wood) that forms late in the growing season; latewood

Superior ovary an ovary completely separate and free from the calyx

Survivorship curve a graph relating age of organisms within a cohort (along the horizontal axis) to the proportion of organisms in that cohort still alive (along the vertical axis)

Suspensor (L. *suspendere*, to hang) a cell or chain of cells developed from a zygote, the function of which is to place the embryo cells in an advantageous position to receive food

Suture (L. *sutura*, a sewing together; originally the sewing together of flesh or bone wounds) the junction, or line of junction, of contiguous parts

Symbiotic relationship (Gr. *syn*, with + *bios*, life) a prolonged close association of two or more kinds of organisms, also called symbiosis; see also mutualism and parasitism

Sympetaly (Gr. *syn*, with + *petalon*, leaf) a condition in which petals are united

Symplast (Gr. *syn*, together + *plastides*, formed) the region of a plant tissue inside of the plasma membrane of the cells that form it, especially when the cell interiors are connected by plasmodesmata; contrast to Apoplast

Synandry (Gr. *syn*, with + *andros*, a man) a condition in which stamens are united

Synapsis the stage of meiosis when homologous chromosomes pair up on the spindle apparatus

Syncarpy (Gr. *syn*, with + *karpos*, fruit) a condition in which carpels are united

Synergids (Gr. *synergos*, toiling together) the two cells at the micropylar end of the embryo sac, which, with the third (the egg), constitute the egg apparatus

Synsepaly (Gr. *syn*, with + *sepals*) a condition in which sepals are united

Systematics the discovery and scientific study of biological diversity with the aim of organizing that diversity into a system of named groups that is easy to use, has predictive value, and reflects our understanding of the relationships between the organisms; systematics and taxonomy are closely related, and many scientists use the terms more or less interchangeably

Taiga (Teleut, *taiga*, rocky mountainous terrain) a broad northern belt of vegetation dominated by conifers; also, a similar belt in the mountains just below alpine vegetation

Tannin a substance that has an astringent, bitter taste

Tapetum (Gr. *tapes*, a carpet) nutritive tissue in the sporangium, particularly an anther

Tap root system a single primary root with lateral branches

Taxon (plural, taxa) (Gr. *taxis*, order) a general term for any taxonomic rank; for example, species, orders, and kingdoms are taxa

Taxonomy (Gr. *Taxis*, arrangement + *nomos*, law) the science of naming, describing, and classifying forms of life

Technology (Gr. *techne*, skill + *logos*, reason) a body of knowledge applied to industrial or commercial objectives

Teliospore (Gr. *telos*, completion + spore) a dikaryotic basidiomycete spore that is adapted to survive through winter and to produce basidiospores thereafter

Telophase (Gr. *telos*, completion + phase) the last stage of mitosis, in which daughter nuclei are reorganized

Temperate rain forest in North America, a conifer forest that dominates a coastal strip of land from northern California to the Gulf of Alaska; also called the Pacific Northwest conifer forest

Template a pattern or guide used in manufacturing; a sequence of DNA bases used to specify the sequence of synthesis of a complementary strand of DNA or RNA

Tendril (L. *tendere*, to stretch out, to extend) a slender coiling organ that aids in the support of stems

Tepals sepals and petals that are indistinguishable

Terminal bud a bud at the end of a stem

Terminal electron transport chain the cytochromes in mitochondria that transfer electrons from NADH or $FADH_2$ to molecular oxygen

Tertiary consumer a carnivore that eats other carnivores

Tertiary structure the three-dimensional structure of a protein molecule, stabilized by interactions among the side chains of the polypeptide chain

Testa (L. *testa*, brick, shell) the outer coat of the seed

Test cross the mating between a recessive homozygote and the corresponding dominant to determine whether the latter is homozygous or heterozygous

Tetrad (Gr. *tetradeion*, a set of four) a group of four, usually referring to the meiospores immediately after meiosis

Tetraploid (Gr. *tetra*, four + *ploos*, fold) having four sets of chromosomes per nucleus

Tetraspores (Gr. *tetra*, four + spores) four spores formed by division of the spore mother cell, used particularly for meiospores in certain red algae

Tetrasporine line (tetraspore + L. *ine*, suffix meaning like) a line of evolutionary development in the algae in which mitosis is directly followed by cytokinesis, resulting in a filament, thallus, or complex plant body of varied form

Thallophytes (Gr. *thallos*, a sprout + *phyton*, plant) all plants whose body is a thallus, that is, lacking roots, stems, and leaves: algae, fungi, liverworts

Thallus (plural, thalli) (Gr. *thallos*, a sprout) a plant, fungus, or protist body that is simpler than a vascular plant, lacking lignified vascular tissue, and therefore lacking roots, leaves, and stems

Thermoacidophiles a group of Archaea that is adapted to live in hot, acid environments

Thermocline (Gr. *therme*, heat + *klino*, slope) a zone of rapidly changing temperature that separates the epilimnion from the hypolimnion in large bodies of freshwater

Thermoperiod (Gr. *therme*, heat + *periodos*, circuit) the difference, in degrees, between the greatest and lowest temperatures experienced at one location within a 24-hour period

Thorn sharply pointed woody structure; a modified branch

Thylakoid (Gr. *thylakos*, sack + N.L. *oid*, a thing that is like) the inner membranes of the chloroplasts

Thymidine a nucleoside incorporated in DNA, but not in RNA

Thymine a pyrimidine occurring in DNA, but not in RNA; ^{3}H-thymidine is radioactive thymidine used to indicate and locate DNA synthesis

Tidal wetland a wetland close enough to a maritime coastline to be influenced by tidal fluctuations, so that at some daily or seasonal times the wetland is submerged and at other times is exposed; also called a salt marsh, and dominated by perennial, succulent forbs and salt-tolerant grasses and sedges

Tiller (O.E. *telga*, a branch) a grass stem arising from a lateral bud at a basal node; tillering is the process of tiller formation

Tissue a group of cells of similar structure that performs a special function

Tonoplast (Gr. *tonos*, stretching tension + *plastos*, molded, formed) the cytoplasmic membrane bordering the vacuole; so-called by de Vries, as he thought it regulated the pressure exerted by the cell sap

Totipotent (L. *totus*, whole + *patens*, being able) capable of development along any of the lines inherently possible to cells of its species

Toxin (L. *toxicum*, poison) a poisonous secretion of a plant or animal

Tracheary element general term for the conducting elements in xylem, a vessel member or tracheid

Tracheid (Gr. *tracheia*, windpipe) an elongated, tapering xylem cell with lignified pitted walls, adapted for conduction and support, and without open end walls

Tracheophytes (Gr. *tracheia*, windpipe + *phyton*, plant) vascular plants

Trait a distinctive definable characteristic; a mark of individuality

Transcription (L. *trans*, across + *scribere*, to write) the process of RNA formation from a DNA code

Transduction recombination that occurs when a virus transfers host DNA from one host organism to another

Transfer cells specialized cells modified by their cell wall projections that may facilitate short-distance transport

Transformation change in the genotype of bacteria by the direct uptake, incorporation, and expression of extracellular DNA

Translation generally, rendering from one language into another; specifically, the use of a sequence of RNA bases to specify a sequence of amino acids in the process of protein synthesis

Translocation (L. *trans*, across + *locare*, to place) the transfer of food materials or products of metabolism; in genetics, the exchange of chromosome segments between nonhomologous chromosomes

Transmission electron microscopy a technique of visualizing an object at high resolution by focusing a beam of electrons through the object onto a fluorescent screen or photographic film

Transmit to pass or convey something from one person, organism, or place to another person, organism, or place

Transpiration (F. *transpirer*, to perspire), the giving off of water vapor from the surface of leaves

Tree (M.E. *treow*) a woody plant having a single trunk

Tricarboxylic acid cycle a system of reactions that contributes to the catabolic breakdown of foods in respiration and that provides building materials for a number of anabolic pathways; also called the Krebs cycle and the citric acid cycle

Trichogyne (Gr. *trichos*, a hair + *gyne*, female) receptive hairlike extension of the female gametangium in the Rhodophyta and Ascomycetes

Trichome (Gr. *trichoma*, a growth of hair) a short filament of one or more cells extending from the epidermis

Triglyceride a lipid compound composed of three fatty acids and glycerol; usually functions as a form of stored carbon and energy

Triose (Gr. *treis*, three + *ose*, suffix indicating a carbohydrate) any three-carbon sugar

Trisomic (Gr. *treis*, three + *soma*, body) a plant containing one additional chromosome, $2n + 1$

Tritium a hydrogen atom, the nucleus of which contains one proton and two neutrons; it is written as ^{3}H; the more common hydrogen nucleus consists only of a proton

Trophic level (Gr. *trophe*, food) a group of organisms sharing the same food source, as producers, herbivores, consumers, or decomposers

Tropism (Gr. *trope*, a turning) movement or curvature caused by an external stimulus that determines the direction of movement

True fungi an informal term for organisms that belong to kingdom Fungi

Tube cell one of the cells of the pollen grain; directs the growth of the pollen tube

Tuber (L. *tuber*, a bump, swelling) a much-enlarged, short, fleshy underground stem

Tundra (Lapp, *tundar,* hill) meadowlike vegetation at low elevation in cold regions that do not experience a single month with average daily maximum temperatures greater than 50°F

Turgid (L. *turgidus,* swollen, inflated) swollen, distended; referring to a cell that is firm because of water uptake

Turgor pressure (L. *turgor,* a swelling) the pressure within a cell resulting from the osmotically generated imbibition of water into the protoplasm and vacuole

Tylose (plural, tyloses) (Gr. *tylos,* a lump or knot) a growth of one cell into the cavity of another

Type specimen an example organism that is placed on file when its species is first formally named

Umbel (L. *umbella,* a sunshade) an inflorescence, the individual pedicles of which all arise from the apex of the peduncle

Unavailable water water held by the soil so strongly that root hairs cannot readily absorb it

Unicell (L. *unus,* one + cell) an organism consisting of a single cell; generally used in describing algae

Uniseriate (L. *unus,* one + M.L. *seriatus,* to arrange in a series) said of a filament having a single row of cells

Unrooted cladogram a cladogram that does not identify the earliest node in the clade (the root), and therefore does not specify the direction of evolution between internal nodes of the cladogram

Uracil a pyrimidine found in RNA but not in DNA

Urediniomycetes a monophyletic class of basidiomycete fungi that lack basidiomata and cause rust diseases in plants

Uredospore (L. *uredo,* a blight + spore) a red, one-celled summer spore in the life cycle of the rust fungi

Ustilagomycetes a monophyletic class of basidiomycete fungi that lack basidiomata and often cause smut diseases in plants

Vacuole (L. dim. of *vacuus,* empty) an organelle of a plant cell, surrounded and differentiated from the rest of the cytoplasm by the tonoplast membrane and containing various substances depending on the type of cell

Vallecular canal an open canal in the cortex of horsetail stems, arranged with others in a circle, alternating with vascular bundles; the function may be to transport air to belowground rhizomes and roots

Valve (L. *valva,* door) openings in bryophyte capsules that allow spores to be dispersed

Variety in taxonomy, a subgroup within a species; also called a cultivar (in cultivated plants), pathovar (in pathogenic organisms), or a race (in animals)

Vascular (L. *vasculum,* a small vessel) referring to a plant tissue (xylem and phloem) or region consisting of or giving rise to conducting tissue; for example, vascular bundle, vascular cambium, phloem ray

Vascular bundle a strand of tissue containing primary xylem and primary phloem (and procambium, if present) and frequently enclosed by a bundle sheath of parenchyma or fibers

Vascular cambium cambium giving rise to secondary phloem and secondary xylem

Vascular cylinder the central portion of a stem or root bounded by and including the pericycle

Vascular tissue system includes the primary and secondary xylem and phloem

Vector (L. *vehere,* to carry) an organism, usually an insect, that carries and transmits disease-causing organisms; a DNA molecule used to transmit genes in a transformation procedure

Vegetation (L. *vegetare,* to quicken) the plant cover that clothes a region; it is formed of the species that make up the flora, but is characterized by the abundance and life form (tree, shrub, herb, evergreen, deciduous plant, and so on) of certain of them

Venation (L. *vena,* a vein) arrangement of veins in a leaf blade

Venter (L. *venter,* the belly) enlarged basal portion of an archegonium in which the egg cell is borne

Ventral canal cell the cell just above the egg cell in the archegonium

Ventral suture (L. *ventralis,* pertaining to the belly) the line of union of the two edges of a carpel

Vernalization (L. *vernalis,* belonging to spring + *izare,* to make) the promotion of flowering by naturally or artificially applied periods of extended low temperature; seeds, bulbs, or entire plants may be so treated

Vesicle (L. *vesicula,* small bladder) a small, membrane-enclosed cavity in the cytoplasm of a cell

Vessel (L. *vasculum,* a small vessel) tube of determinate length composed of vessel members joined end to end by opened perforation plates; the end wall of the terminal vessel member of a vessel are closed

Vessel member a conducting cell in the primary and secondary xylem; typically have a lignified secondary cell wall and open end walls, a perforation plate

Vibrio (L. *vibrare,* to shake, vibrate) a genus of short, rigid motile bacteria having one or more polar flagella, being typically shaped like a comma or an S

Vine (L. *vinea,* of wine) a woody plant with a flexible stem that is supported by trailing along the ground or by climbing upward on a shrub, tree trunk, or some other surface

Virion (L. *virulentus,* full of poison) infectious virus particle as it exists outside a host

Virulence (L. *virulentia,* a stench) the relative infectiousness of a bacteria or virus, or its ability to overcome the resistance of the host metabolism

Virus (L. *virus,* a poisonous or slimy liquid) a disease principle that can be cultivated only in living tissues, or in freshly prepared tissue brei

Vitamins (L. *vita,* life + amine) naturally occurring organic substances necessary in small amounts for the normal metabolism of plants and animals

Volva (L. *volva,* a wrapper) cup at base of stipe or stalk of a basidiomycete fruiting body

Water potential the difference between the activity of water molecules in pure distilled water or water-saturated air and the activity of water molecules in any other system; the activity of water molecules is negative in unsaturated air and in solutions

Weed (A.S. *weod,* used at least since 888 in its present meaning) generally a herbaceous plant or shrub not valued for use or beauty, growing where unwanted, and regarded as using ground or hindering the growth of more desirable plants

Wetland any area that remains wetted throughout the soil profile for a sufficient length of time each year to produce anaerobic conditions in the soil; characterized by the dominance of flood-tolerant plant species

Whisk ferns see Psilophytes

Whorl a circle of flower parts, or of leaves

Whorled referring to bud or leaf arrangement in which there are three or more buds or three or more leaves at a node

Wild type in genetics, the gene normally occurring in the wild population, usually dominant

Wood (M.E. *wode, wude,* a tree) a dense growth of trees, or a piece of a tree, generally the xylem

Wood Age the span of history during which wood was the, or one of the, principal materials for houses, fuel, and vehicles; considered by some to include the Stone Age, Bronze Age, Iron Age, Industrial Age, Nuclear Age, and up to the present

Woodland a vegetation type with two canopy layers, a scattered overstory of trees that collectively shade less than 60% of the understory (but at least 30%), and a nearly continuous understory of forbs and grasses

Xanthophyll (Gr. *xanthos,* yellowish brown + *phyllon,* leaf) a yellow chloroplast pigment

Xanthophyta (Gr. *xanthos,* yellow) a small group of freshwater and (mainly) unicellular algae that contain chlorophylls *a* and *c,* have cellulose walls, and store

carbohydrate as oil droplets; classified in this text with the protists in the heterokont group

Xerophyte (Gr. *xeros,* dry + *phyton,* a plant) a plant resistant to drought or that lives in dry places

Xylem (Gr. *xylon,* wood) a plant tissue consisting of tracheids, vessel members, parenchyma cells, and fibers; wood

Yeast a fungal body that consists of a single, rounded microscopic cell

Zoology (Gr. *zoon,* an animal + logos, speech) the science of animal life

Zoosporangium (Gr. *zoon,* an animal + sporangium) a sporangium bearing zoospores

Zoospore (Gr. *zoon,* an animal + spore) a eukaryotic spore that swims by means of one or more flagella

Zygomorphic (Gr. *zygo,* yoke, pair + *morphe,* form) referring to bilateral symmetry; said of organisms, or a flower, capable of being divided into two symmetrical halves only by a single, longitudinal plane passing through the axis

Zygomycota an artificial (nonmonophyletic) phylum in kingdom Fungi, characterized by formation of zygospores, aseptate hyphae, mitosporangia, and lack of swimming cells

Zygosporangium in zygomycete fungi, a sporangium in which a zygospore is made by fusing two unicellular gametangia

Zygospore (Gr. *zygon,* a yoke + spore) a resistant spore that contains one or more diploid zygote nuclei at some stage in its development

Zygote (Gr. *zygon,* a yoke) a diploid cell resulting from the fusion of gametes or gametangia and their nuclei

Zygotic life cycle a life cycle in which the only diploid phase is the single-celled zygote

Photo Credits

This page constitutes an extension of the copyright page. We have made every effort to trace the ownership of all copyrighted material and to secure permission from copyright holders. In the event of any question arising as to the use of any material, we will be pleased to make the necessary corrections in future printings. Thanks are due to the following authors, publishers, and agents for permission to use the material indicated.

Chapter 1. CO Photodisk/Getty Images **1.1 (a)** Fabrice De Clerck, University of California, Davis **1.1 (b)** Michael G. Barbour **1.1 (c,d)** Terence M. Murphy, University of California, Davis **1.1 (e)** Kathy Garvey, University of California, Davis **1.2 (a)** NASA **1.2 (b)** Michael G. Barbour **1.3 (a–e)** Michael G. Barbour **1.4 (a–c)** D. Brandon, University of California, Davis **1.5** Michael G. Barbour **1.6** Phillip Grime, University of Shefield **Page 6 (figure 1)** Michael G. Barbour

Chapter 2. CO Photodisk/Getty Images **Page 21 (figure 1)** Terence M. Murphy **2.14 (a–b)** Terence M. Murphy **2.17** Terence M. Murphy

Chapter 3. CO Dr. George Wilder/ Visuals Unlimited **3.1 (a)** National Library of Medicine **3.1 (b)** James M. Bell/Photo Researchers, Inc. **3.2 (c–d)** Jeremy Pickett-Heaps, School of Botony, University of Melbourne **3.3 (a)** Art by Raychel Ciemma and Precision Graphics **3.3 (b)** Micrograph M. C. Ledbetter, Brookhaven National Laboratory **3.3 (c)** Brian Gunning **3.4** Art by Raychel Ciemma **3.5** Redrawn from diagram by W. J. Lucas, University of California, Davis **3.6** P. A. Roelofsen **3.7 (d)** Terence M. Murphy **3.8** Don W. Fawcett/Visuals Unlimited **3.9** From Starr Taggart, Wadsworth Publishing **3.10** B. Partels **3.11 (b)** Art by

Robert Demarest after a model by J. Kephart **3.11 (a)** Micrograph, Gary W. Grimes **3.12 (a)** Art by Raychel Ciemma **3.12 (b–e)** L. K. Shunway **3.12 (f)** A. R. Spurr and W. H. Harris, Amer. Journal of Botony, Vol. 55, 12 10. ©1968 By Botanical Society of America **3.13** Art by Raychel Ciemma **3.13 (b)** Micrograph, Keith R. Porter **3.14** Courtesy of B. Liu, University of California, Davis **3.17 (a–j)** Used by permission of Dr. S. Wick. *Journal of Cell Biology* 89, 685–690 (1981) © The Rockefeller University Press, NY **3.19 (a–e)** Andrew S. Bajer, University of Oregon

Chapter 4. CO D. D. Brandon, University of California, Davis **4.2 (a–c)** Thomas L. Rost **4.3 (a–d)** Thomas L. Rost **4.4** Thomas L. Rost **4.5** Used with permission from Gunning, B.E.S, "Transfer cells and their roles in transport of solutes in plants." Science Progress Oxford 977, 64: 539–568. Photo taken by Dr. Alan J. Browning, Australia National University **4.6** Thomas L. Rost **4.7 (a–c)** Thomas L. Rost **4.7 (d)** Leslie Sunell, University of California, Santa Cruz **4.8 (a–b)** Thomas L. Rost **4.10 (c–d)** Steve Mauseth, University of Texas **4.12** Sonia Cook, University of California, Davis **4.14 (a–c)** Thomas L. Rost **4.15 (a–b)** Vincent R. Franceschi, Washington State University **4.16 (a–b)** Richard H. Falk, Virginia Commonwealth University **4.17** Thomas L. Rost **4.18 (a)** Thomas L. Rost **4.19** Thomas L. Rost **Page 62 (figure 1)** David K. Glandish, Miami University, Ohio **Page 63 (figure 3)** David K. Glandish, Miami University, Ohio **4.21** Thomas L. Rost **4.23 (a)** David R. Frazier/Photo Researchers, Inc. **4.23 (b, d)** Thomas L. Rost **4.23 (c)** G. I. Bernard/NHPA

Chapter 5. CO Royalty-Free/Corbis **5.2 (b,c, e,f)** Thomas L. Rost **5.3** Thomas L. Rost **5.7** Todd Jones

5.8 (a–b) Thomas L. Rost **5.10 (a–b)** Thomas L. Rost **5.12 (b)** Thomas L. Rost **5.13 (b–d)** Thomas L. Rost **5.14** Thomas L. Rost **5.15** Thomas L. Rost **5.16** Thomas L. Rost **Page 81 (figures 1–4)** Thomas L. Rost **5.17 (a)** Thomas L. Rost **5.17 (b)** Photo by Dr. Josh Stevenson **5.18 (b–d)** Thomas L. Rost **5.19 (a)** Thomas L. Rost **5.20 (d)** Thomas L. Rost **5.21** Thomas L. Rost **5.22** Thomas L. Rost **5.23** Susan Larson, University of California, Davis **5.26 (a, c)** Thomas L. Rost **5. 26 (b)** D. Cavagnaro/Visuals Unlimited **5.27** Thomas L. Rost **5.28 (a–d)** Thomas L. Rost **5.30** George Hunter/Stone/Getty Images **5.31** Thomas L. Rost

Chapter 6. CO Royalty-Free/Corbis **6.3 (a–c)** T. Elliot Weier **6.4 (a–c)** Thomas L. Rost **6.5 (a)** Art by Raychel Ciemma **6.5 (b)** Photo by Lesley Sunnell, UC Santa Cruz **6.6 (a)** Richard Falk **Page 96 (figure 1, a–e)** Thomas L. Rost **6.7 (a–b)** Richard Falk **6.10 (a–b)** Thomas L. Rost **6.11 (a)** Dr. Joseph Lin **6.13 (a–b)** Thomas L. Rost **6.15 (a–b)** Thomas L. Rost **6.16 (a–c)** Thomas L. Rost **6.17 (a)** Thomas L. Rost **6.17 (b)** William J. Weber/Visuals Unlimited **6.17 (c)** Jerome Wexler/Photo Researchers, Inc. **Page 103 (figure 1)** Tim Metcalf **Page 103 (figure 2–3)** Masters of Linen/USA

Chapter 7. CO Digital Vision/Getty Images **Page 108 (figures 1–3)** Original art of "male and female" mandrake (*Mandragora officinarum*) from Peter Schöffer. *The German Herbarius*, Mencz. 1485 **7.2** Redrawn with permission from Epstein, E., and Bloom, A.J. 2004. *Mineral Nutrition of Plants: Principles and Perspectives, 2nd Edition*. Sunderland, MA: Sinauer Associates, p. 415 **7.5 (a)** J. W. Perry **7.7 (a–b)** Thomas L. Rost **7.9** Professor Margaret McMully,

McMaster University, Carlton, Ontario, Canada **7.10 (a)** J. W. Perry **7. 12 (a)** Art by Raychel Ciemma **7.12 (b)** Thomas L. Rost **7.12 (c–d)** Art by Leonard Morgan **7.13** Zarkowski et al., *Lindleyana 2* (1987): 1–7. Used with the permission of the authors **7.14 (a–d)** Thomas L. Rost **7.15** Thomas L. Rost **7.16 (d–e)** Maude Hinchee **Page 118 (figure 1)** Thomas L. Rost **7.18 (a–b)** Thomas L. Rost **7.19 (a)** Thomas L. Rost **7.19 (b)** Dr. Tom Lanini, University of California, Davis **7.20 (a)** Adrian P. Davies/Bruce Coleman Ltd. **7.21 (a)** Used with permission from J. Parlade, IRTA-Department de Proteccio Begetal, Barcelona, Spain **7.22** Thomas L. Rost **7.23** Dr. Judith Jernstedt, University of California, Davis

Chapter 8. CO Photodisk/Getty Images **8.3** Terence M. Murphy

Chapter 9. CO Jeff Austin, University of California **9.8 (b–c)** Art by Raychel Ciemma

Chapter 10. CO Royalty-Free/Corbis **10.1** Art by Raychel Ciemma **10.1** Photograph by Carlina Biological Supply Company **10.2 (a)** Art by Raychel Ciemma **10.2 (b)** Micrograph by David Fisher **10.10** Thomas L. Rost **10.11** Art by Raychel Ciemma

Chapter 11. CO Royalty-Free/Corbis **11.9** John Troughton and L. A. Donaldson **11.10 (a–h)** E. Epstein, University of California, Davis **11.13 (b)** H. M. Metcalf, University of California, Davis

Chapter 12. CO PhotoDisc/ Getty Images **12.1 (a)** Ronald W. Hoham, Dept. of Biology, Colgate University **12.1 (b)** Ripon Microslides Inc. **12.2** Thomas L. Rost **12.5** Art by Raychel Ciemma **12.7** Art by Raychel Ciemma **12.7** Photo by D. J. Patterson/ Seaphot Limited: Planet Earth Pictures

Chapter 13. CO Jeff Austin, University of California **13.3 (a–b)** T. Elliot Weier **13.4** Rob Preston and Professor Maureen Stanton, University of California, Davis **13.6** Redrawn from Priestley and Scott. *An Introduction to Botany.* © 1949 Longmans Green. Reprinted with permission of the Longmans Group, Ltd. **13.7 (b–h)** Redrawn from Gore, U.R. 1932. *American Journal of Botany* 19:795–807 **13.7 (i)** Modified and redrawn from Jensen, W.A. 1965. *American Journal of Botany* 52:781–797 **13.9 (a–e)** Dr. Joseph Lin **13.11 (a)** T. Elliot Weier **13.11 (b)** Thomas L. Rost **13.12 (a–c)** T. Elliot Weier **Page 206 (figure 1, a–b)** Thomas Eisner, Cornell University **Page 206 (figure 2, a–b)** T. Elliot Weier **Page 206 (figure 3)** Edward S. Ross **Page 206 (figure 4)** Dr. R. Norris, University of California, Davis **13.14** T. Elliot Weier **13.15** T. Elliot Weier **13.16** Michael G. Barbour **13.17 (a–b)** Michael G. Barbour **13.18** Grady L. Webster, University of California, Davis **13.19** Redrawn from Niklas, K.J, Buchmann, S.L. 1985. *American Journal of Botany* 72:530–539

Chapter 14. CO Photodisk/Getty Images **14.1 (a, c–e)** Redrawn from Gore, U.R. 1932. *American Journal of Botany* 19:795–807 **14.2 (a)** Thomas L. Rost **14.3 (a)** Thomas L. Rost **14.5 (a–b)** T. Elliot Weier **14.7 (a–b)** Thomas L. Rost **14.13** T. Elliot Weier **14.14** Michael G. Barbour **14.15 (a–b)** J. W. Perry **14.16** Thomas L. Rost **14.17** Thomas L. Rost **14.18** Michael G. Barbour **14.19** Redrawn from Krosmo, E. 1930. *Unkraut in Ackerbau der Neuzeit. Berlin*: Springer-Verlag. © 1930 by Springer Verlag **14.20** Thomas L. Rost **14.21** Thomas L. Rost **14.24 (a)** Thomas L. Rost **14.25 (a–b)** J. W. Perry **14.26 (b)** Judy Jernstedt **14.27** Thomas L. Rost **14.29** J. W. Perry **14.31 (a)** Inga Spence © 1995/Tom Stack & Assoc. **14.31 (b)** John D. Cunningham/Visuals Unlimited **14.32 (a–b)** Robert S. Boyd, Auburn University **14.33 (a)** Michael G. Barbour

Chapter 15. CO Royalty-Free/Corbis **15.1 (a)** John J. Harada, University of California, Davis **15.1 (b–e)** Terence M. Murphy **15.2 (b)** Terence M. Murphy **15.5** Terence M. Murphy **15.11** Adapted from a model by S.C. Fry, University of Edinburgh, Edinburgh, UK **15.12** Christina McWhorter **15.14** A. Lang **15.15** J. Goeschel and H. Pratt **15.17** Marita Cantwell, University of California, Davis **15.18** Dr. John Ryals and Dr. Eric Ward **15.19 (a)** Terence M. Murphy **15.21 (a)** Jan Zeevart **15.21 (b)** Plant Industry Station, Crops Research Division, Agricultural Research Service, U.S. Department of Agriculture

Chapter 16. CO Digital Vision/Getty Images **16.1 (a–d)** Terence M. Murphy **16.10** Pioneer Hi-Bred International, Inc., © 1996 **16.12** Adapted from information

in Ownbey, M. 1950. *American Journal of Botany* 37:487–499, with permission

Chapter 17. CO Dex Image/Getty Images **17.3** Terence M. Murphy **17.5** Terence M. Murphy **17.8** Keith V. Wood **17.9** Terence M. Murphy **17.10 (a–b)** Monsanto Company **17.10 (c–d)** Calgene, Inc **17.11** Terence M. Murphy **17.12 (a–c)** H.X. Zhang & E. Blumwald, *Nature Biotechnology* 19:765–768, 2001, with permission.

Chapter 18. CO © John Cancalosi/Peter Arnold, Inc. **18.1** Michael G. Barbour **18.5** SEM by John L. Bowman **18.11** Tim Metcalf

Chapter 19. CO PhotoDisc Green/Getty Images **19.1 (a)** L. Santo **19.1 (b)** Palay Beaubois **19.2 (a)** Laurie Feldman, University of California, Davis **19.2 (b)** J. W. Costerton, reproduced with permission from *Annual Review of Microbiology,* 33:459 (1979). Copyright 1979 by Annual Reviews, Inc. **19.3 (b)** Dr. David Dressler and Dr. Huntington Potter, Harvard Medical School **19.4** Charles C. Brinton, Jr. and Judith Carnahan, University of Pittsburg **19.5** Art by Raychel Ciemma **19.6 (a)** J. Pangborn **19.6 (b)** M. L. DePamphilis and Julius Adler **19.8 (a)** W. C. Trentini and the American Society of Microbiology, from *Journal of Bacteriology,* 93:1699 (1967) **19.8 (b)** T. W. Goodwin **19. 8 (c)** Stanley W. Watson, *International Journal of Systematic Bacteriology,* 21:254–270, 1971, used with permission **19.9** T. J. Beveridge, University of Guelph/Biological Photo Service **19.10** Hans Reichenbach **19.11** Richard Cowen, University of California, Davis **19.13** Lawrence Migdale © 1985 **19.14 (a–b)** J. Whattley and R.A. Lewin, *New Phytologist* Vol. 79, pp. 309–313, 1977 **19.16 (a)** Dr. O. Bradfute and Peter Arnold **19.16 (b)** D. L. D. Casper **19.17 (a)** Terence M. Murphy **19.17 (b)** Kenneth Corbett **19.17 (c)** Brecks, Peoria, IL

Chapter 20. CO David M. Dennis/Tom Stack and Associates **20.2** J. W. Perry **20.3** R. M. Thonton **20.5** J. W. Perry **20.6** © Claire Ott/Photo Researchers, Inc. **20.7** J. W. Perry **20.9 (a–c)** R. M. Thonton **20.20** R. M. Thonton **20.24 (a–d)** J. W. Perry

Chapter 21. CO M. I. Walker/NHPA **21.2(a–b)** Robert M. Thornton

21.2 (c) N. Lang 21.3 William Schopf, University of California, Los Angeles 21.6 Terence M. Murphy 21.7 Holt Studios International/Nigel Cattlin/Photo Researchers, Inc. **Page 370 (figure 1)** J. W. Perry **21.10 (c)** J. W. Perry **21.12 (b)** H. Hoops **21.13 (a)** J. W. Perry **21.15** N. Lang

Chapter 22. CO PhotoDisc Green/Getty Images **22.2 (a, e)** Michael G. Barbour **22.2 (b, d, e)** D. K. Canington **22.5 (a)** Reproduced from Scagel, R.F. et al 1984. *Plants: an Evolutionary Survey.* Wadsworth Publishing **22.5 (b)** Reproduced from Janssens, Horton, & Basinger 1979. *Canadian Journal of Botany* 57:2150–2161, with permission **22.8** W. Russell **22.9** J. W. Perry **22.10** J. W. Perry **22.11 (a–b)** J. W. Perry **22.15 (b–c)** J. W. Perry **22.16** Photograph by Jan Burton/ Bruce Coleman Ltd. **22.16** Art by Raychel Ciemma **22.17 (a)** J. W. Perry **22.17 (b–d)** W. Russell and D. Hess

Chapter 23. CO Gregory G. Dimijian/ Photo Researchers, Inc. **23.2 (b)** Dr. Judith Jernstedt, University of California, Davis **Page 404 (figure 1)** Michael G. Barbour **23.5** Field Museum of Natural History **23.6** Art by Raychel Ciemma **23.7 (a–b)** J. W. Perry **23.7 (c)** Doug Sokell/ Visuals Unlimited **23.7 (d)** Michael G. Barbour **23.8 (a)** J. W. Perry **23.13** J. W. Perry **23.18 (a)** J. Michael Elchelberger/ Visuals Unlimited **23.18 (b)** James W. Richardson/Visuals Unlimited **23.20 (b)** J. W. Perry **23.20 (c)** Jack M. Bostrack/Visuals Unlimited **23.20 (d)** John D. Cunningham/ Visuals Unlimited **23.22 (b)** Ken Wagner/ Visuals Unlimited

Chapter 24. CO PhotoDisc Blue/Getty Images **24.2** Michael G. Barbour **24.6** J. W. Perry **24.8** J. W. Perry **24.9** Michael G. Barbour **24.11** Terence M. Murphy **24.12** Michael G. Barbour **24.13** Ameri-can Museum of Natural History **24.14** J. W. Perry **24.15 (a–b)** Michael G. Barbour **Page 437 (figure 1)** Michael G. Barbour **24.16** T. Elliot Weier **24.17** Michael G. Barbour **24.18** Michael G. Barbour **24.20 (a–c)** J. W. Perry **24.21 (a–c)** J. W. Perry **24.22** T. Elliot Weier **24.23 (a)** Michael G. Barbour **24.23 (b–c)** T. Elliot Weier **24.23 (d)** J. W. Perry **24.24 (a)** A. B. Addicot

Chapter 25. CO PhotoDisc Green/Getty Images **25.1 (a–b)** Michael G. Barbour **25.4 (a–c)** Michael G. Barbour **25.5** Adapted from Dorf, E. 1960. *American Scientist* 48:341, with permission **25.6** Redrawn from Dorf, E. 1960. *American Scientist* 48:341, with permission **25.7** Redrawn from Delcourt, H.R., et al. 1983. *Quarterly Science Review* 1:153–175, with permission **25.9** Adapted from Judd, W.S., et al. 2002. *Plant Systematic, 2nd ed.* Sunderland, MA: Sinauer Associates, p. 227, with permission **25.14 (a–b)** Michael G. Barbour **25.15** Electronmicrograph courtesy of J. Ward **25.18** Michael G. Barbour **25.25** Redrawn from Good, R. 1961. *The Geography of the Flowering Plants.* New York: John Wiley, with permission **25.26 (a–b)** G. J. James/Biological Photo Service **25.27** Truman Young, University of California, Davis **Page 473 (figure 1)** David Parker/Science Photo Library/Photo Researchers, Inc.

Chapter 26. CO PhotoDisc Green/Getty Images **26.1** Michael G. Barbour **26.3** Michael G. Barbour **26.4** Redrawn from Fowells, H.A. 1965. Agriculture handbook no. 271. Washington, D.C.: U.S. Department of Agriculture **26.5** Michael G. Barbour **26.6** Michael G. Barbour **26.7** © 1987 Dan Suzio/Photo Researchers, Inc. **26.8** Michael G. Barbour **26.9** Redrawn from Grime, J.P. 1977. *American Naturalist* 11:1169–1194, with permission **26.10** Michael G. Barbour **26.12** Redrawn from Harper, J.L. 1977. *Population Biology of Plants.* New York: Academic Press, with permission **26.13 (a)** Michael G. Barbour **26.13 (b)** John Shaw/NHPA **26.13 (c)** George Ranalli/Photo Researchers, Inc. **26.14** John D. Cunningham/Visuals Unlimited **26.15** Redrawn from Boysen-Jensen, P., Muller, D. 1929. *Jahrbuch fuer Wissenschaftliche Botanische* 70:493, with permission **26.18** Paige Martin/Archbold Biological Station **Page 491 (figure 1, a–b)** Michael G. Barbour **26.19 (a)** Laurie Campbell/ NHPA **26.19 (b)** Steve Krasemann/Photo Researchers, Inc. **26.20 (a–b)** C. Muller

Chapter 27. CO PhotoDisc Blue/Getty Images **27.1** Michael G. Barbour **27.2 (a–b)** Michael G. Barbour **27.6 (a–d)** Norm Christensen, Duke University **27.7** Redrawn from Drury, W.H. Jr. 1956. Contributions from the *Gray Herbarium* 178:1–30, with permission **27.8** Redrawn from Barbour, M.G., Christensen, N. Vegetation of North America. In: *Flora of North America*, Vol. 1. New York: Oxford University Press, with permission **27.9 (a)** James McGraw **27.9 (b)** Michael G. Barbour **27.10** Michael G. Barbour **27.11 (a–b)** Michael G. Barbour **27.12** Michael G. Barbour **27.13** Michael G. Barbour **27.14 (a–b)** Michael G. Barbour **27.15 (a–b)** Michael G. Barbour **27.16** Michael G. Barbour **27.17** Greg Vaughn/Tom Stack & Associates **27.18** Drawing by M. Yuval from Barbour, M., et al. 1993. *California's Changing Landscapes.* Sacramento: California Native Plant Society, with permission **27.19** Michael G. Barbour **27.20** Michael G. Barbour **27.21** Redrawn from M. Foster, with permission **Page 515 (figure 1)** Michael G. Barbour **27.23 (a–b)** From Gruell, G. E. *Fire in Sierra Nevada Forests.* Missoula, MT: Mountain Press, 2001. **27.23 (c)** Michael G. Barbour

Index

Boldface indicates pages on which primary discussion appears.

Ancestral character state, **308**
Androecium, **197**, 198
Andropogon, 490
Andropogon virginicus (broomsedge), 502
Angiopteris evecta, 419f
Angiosperm, **196**, 448, **449**, 450–475
  basal, 459–460, 459f, 460f
  core, 460–468, 460f–469f
    eudiocot, 462–469, 462f–469f
    magnolid, 460, 460f
    monocot, 460–462, 461f–462f
  diversity of, 459–468
  dominance of, 453–458
    in Quaterary period, 455–458, 456f–457f
    in Tertiary period, 455
  enclosed seed in, 449–450, 450f
  families of, 470t
  life cycle of, 458, 458f
  molecular clock and, 451
  origin of, 450–452, 450f, 452f
  plant geography and, 468, 470–473, 470t, 471f–473f
  wood of, 81–83, 82f–83f
Angle of divergence, 74
Animal dispersal of seeds, 229–230
Animalia, 9
Anion, **16**
Annual, **480**
Annual ring, 78f, 79
Anther, **197**
Antheridia, **350, 382**, 383, **388**
Antheridiophore, **393**
Antheridium, of moss, 396
Anthrocyanin, 254
Anthurium, 460
Anticodon, 239
Antiherbivore mechanism, 230–231
Antioxidant, 21
Antipodal cell, **199**
Apical meristem, 64
  flower development in, 201–204, 202f–203f
  root, **111**, 111f
Apicomplexa, **367**
Apium graveolens (celery), 225
Apogamy, **424**
Apomixis, **205**
Apoplast, **34**
Apoplast solution, 166
Apoplastic flow, 167–168, 168f
Apospory, **424**
Apothecium, 352
Apple, 222
  anaerobic respiration in, 138
  controlling respiration of, 139
  fruit of, 226, 227f
  hormones and, 248

Aquatic food chain, 374, 375f, 376
Aquilegia formosa (columbine), 463
  pollination of, 208, 208f
Arabidopsis thaliana (mouse-ear cress)
  growth and development of, 236f
  meristem of, 67f
  mitotubules of, 41f
  mutation of, 248–249
Arachis hypogaea (peanut)
  pod of, 225
  seeds of, 231
Araucariaceae, 438
Araucuria, 438, 438f
Arbuscule, 349, 349f
Archaea, **10, 294, 317**
  characteristics of, 319, 324t
  environments of, 323–325, 324f–325f
  global warming and, 327
Archaeopteris, 428, 428f
Archegoniophore, **393**
Archegonium, **388**, 388f
  of moss, 396
Architecture, flower, 201–203, 202f–203f
Arctic tundra, **505**, 505f
Areca, 462
Arisaema triphylla (jack-in-the-pulpit), 480f
Aristida, 490
Arrow-grass. *See Triglochin maritimum*
Artchaefructus, 451
Arundo donax (giant cane), 62–63, 62f–63f
Asarum (wild ginger), 460
Asci, **350**, 351f
Ascogonia, **350**
Ascoma, **350**, 352, 352f
Ascomycete
  ascoma of, 352, 352f
  dikaryotic stage of, 350–351, 351f
  dimorphic, 353–354
  hyphae and conidia of, 350
  mitosporic, 353–354
  septate hyphae and conidia of, 350
  symbiotic, 352–353, 353f
Ascomycota, **342**, 349–354, 351f–353f
Ascospore, **350**
Ascus, **342**
Aseptate hypha, **344**
Asexual reproduction
  mitosis in, 184–185, 184f–185f
  in mosses, 395
  seed in, 205
Asparagus officinale (asparagus)
  flower of, 202
  root of, 115, 116f
Aspen. *See Populus tremuloides*
Aspergillus, spore of, 338f
Aspergillus flavus, 353–354

Aster pilosus, 502
Asterid clade, **468**
Athyrium felix-femina, 421f
Atom, **14**, 14f, 15
  electrical forces and, 15
Atomic number, 15
ATP. *See* Adenosine triphosphate
ATP synthetase, **143**
Atriplex (saltbush), 97, 512
Attribute, **497**, 501
  biomass and productivity as, 498–499
  nutrient cycling as, 499–501, 499f–501f
  physiognomy as, 498, 498f
  species richness as, 498, 499f
Auricle, 93
Autotroph, **323**
Auxin
  cytokinin and, 244f
  discovery of, **242**–243
    effects of, 241t
  shoot growth and, 245–246, 246f
Avocado. *See Persea americana*
Awn, 231
Axis, of stem, 74
Azolla, 329

B
Bacillus, **319**
Baciullus thuringiensis, 285
Back-crossing, **297**
Backbone of amino acid, **23**
Bacon, Sir Francis, 5
Bacteria, **10, 294, 317**
  biomining and, 326
  characteristics of, 324t
  chemoheterotrophic, 325
  chemosynthesis and, 149
  competing with mycelia, 339–340
  eukaryotic, 294
  global warming and, 327
  nitrogen fixation and, 176
  parasitic, 329–330
  prokaryotic, 294
Baja California, 471
Baker's yeast
cells of, 339f
Bald cypress. *See Taxodium distichum*
Ballistospore, 356, **357**, 357f
Band
  preprophase, 45f
  prophase, **44**
Banksia, 463
Banyan. *See Ficus bengalensis*
Barberry. *See Berberis*
Bark, 83–85, 84f–85f
Barley. *See Hordeum vulgare*
Barrel, wine, 80–81

Barrier, energy, 124–127
Basal angiosperm, **459,** 460, 459f–460f
Base
    hydrogen accepted by, **19**
    nucleotide, 25, 26f
Base sequence of DNA, 258
    gene expression and, 237–238
Basidioma, **354**
Basidiomycetes, saprobic, 356
Basidiomycota, **342,** 354–358,
        355f–357f
Basidiospore, 356–357
Basidium, **342, 354**
Basswood. *See Tilia americana*
Bat-pollinated plant, 208, 208f
Bean. *See Phaseolus vulgaris*
Bee-pollinated plant, 206–208, 206f
Beer, 145
Beet. *See Beta vulgaris*
Begonia, root of, 110
Berberis (barberry), 463
Bermuda grass. *See Cynodon dactylon*
Berry, **225**
    of tomato, 226f
Bertholleda excelsa (Brazil nut), 225
Beta vulgaris (beet), 60f
Betel pepper. *See Piper betle*
Betula papyrifera, 85, 85f
Biennial, **480**
Bigonia capreolata (trumpet flower),
        101f
Binary fission, **320,** 320f
Binomial system, **293**
Bio-organic molecule, **19–20**
Biochemical reaction, 124
Biodiversity, **292**
Biogeographer, plant, **472**
Biogeography, **8**
Biological clock, 252–253
Biological species, **293**
Biology, conservation, 514–516,
        515f–516f
Biomass
    of ecosystem, 476–477, 477f
    productivity and, 498–499
Biomolecule, 20t
Biota, 175
Biotec vector, **229**
Biotechnology, 275–290
    applications of, 283–288
        in medicine and industry, 283
        to produce new plants, 283,
            285–288, 285f–287f
        genetic engineering in, 276–283. *See
            also* Genetic engineering
    plant breeding and, 270
    safety of, 288
Biotic community, 497
Biotic pollen vector, 207–208

Birch. *See Betula papyrifera*
Bird-of-paradise. *See Strelitzia*
Bird-pollinated plant, 208, 208f
Bird's nest fungus, **356**
Bisexual flower, **201**
Blackman, F.F., 150
Blade, **371**
    leaf, **92**
Blancmange, 377
Blight, potato, 368–369
Blood lily. *See Haemantus katherinae*
Blooms, algal, 376, **376**
Blue grama. *See Bouteloua gracilis*
Blue–green algae, **10**
Body
    plant, 50, 51f
    prolamellar, 39
Boehmeria nivea (ramie), 102
    fibers of, 60
Bolting, 247, 247f
Bond
    covalent, **16–18**
    hydrophobic, **19**
    ionic, 15–16
    peptide, **24,** 24f
Bond, William, 452
Bordered pit, **57**
Boreal forest, **505,** 505f, 506
Boron, 172t
Botanical knowledge, 2
Botany, **2**
    careers in, 3f
Botrychium, 416, 416f
Bouteloua gracilis (blue grama), 109
Bracket fungi, **356**
Branch, compressed, 85, 85f
Branch-point enzyme, 128
Brassica campestris (rapid-cycling
        Bassicas), 259t
Brassica napus (canola), 287
Breeding, plant, 269–272, 269f–272f
Brewer's yeast, 340
Bristlecone pine. *See Pinus longaeva*
British India Office, 472
Broadleaf evergreen, **481**
Broomsedge. *See Andropogon virginicus*
Brown algae, **371**
    industrial uses of, 378
    sporic life cycle of, **381,** 381f
Bryophyte, **385,** 386–400
    classification of, 389–390, 390f
    hornworts, 390–391, 390f–391f
    as land plants, 386–389, 386f–388f
    liverworts, 391–394, 392f–393f
    mosses, 394–399. *See also* Moss
Bryum capillare, 396
Buchloe dactyloides (buffalo grass), 109
Buckeye. *See Aesculus*
Buckwheat. *See Fagopyrum*

Bud, 86f
    as compressed branch, 85, 85f
    of moss, 394
Buffalo grass. *See Buchloe dactyloides*
Bulb, **87,** 88
    leaves of, **104**
Bundle, vascular, **56**
Bur clover. *See Medicago denticulata*
Bursera, 471
Butcher's broom. *See Ruscus aculeatus*
Buttercup. *See Ranunculus* sp.
Butterfly-pollinated plant, 207–208
Byrum, 484

C

C. HOPKiNS CaFe mnemonic, 172
C_3 pathway, 156–157, 157f
C_4 pathway, 158–159
C type life history pattern, **483,** 484f
Cactaceae, 301f
Cactus, 101f
    epiphytic, 487f
    flower of, 465f
Cakile maritima (sea rocket), 231, 232f
Calamites, 416f, 417
Calcium, 172t
Calendula officinalis, 54f
California coast redwood forest,
        436–437, 437f
California fan palm, 461
California poppy. *See Eschscholzia
        californica*
Callose, **60,** 60f
Calvatia gigantea, 356
Calvin cycle, 156–157
Calypso (orchid), 206
Calyptra, **396,** 397f
Calyx, **196**
CAM plant, 158
Cambium
    cork, 63f, **67, 83,** 84f
    fascicular, **75**
    interfascicular, **75**
    vascular, **67, 75,** 76f, 117, 117f, 119
Camellia (tea), 468
Camilia japonica (camillia), 170
Camphor. *See Cinnamomum*
Canal
    carinal, **418**
    vascular, **417,** 417f
Cancer, plants to treat, 21
Candida albicans, 350
Cane, giant, 62–63, 62f–63f
Canna, 462
    rhizome of, 87f
Cannabis sativa (marijuana), 55
Canopy, overstory, 506–507
Canopy layer, 497
Cap, **354**

Capillary force, **166,** 167f
Capillary tube, 166
Capsella burse-pastoris, 61f
Capsicum (chili pepper), 468, 473
   chromoplast of, 40f
Capsule, **224**
   of moss sporophyte, 396–397, 398f
Carbohydrate, **22**
   characteristics of, 20t
   phloem transport and, 178–179
Carbon
   covalent bonds of, 16
   isotopes of, 15
   in tricarboxylic acid cycle, 138–140
Carbon cycle, 156–157
Carbon dioxide
   beer and, 145
   in chemical reactions, 124
   compensation point and, 158–159
   light and, 149–150
   in photorespiration, 157–158
   photosynthesis and, 160–161
   radioactive, 156
   in succulents, 158
   transport and, 164
   in tricarboxylic acid cycle, 139–140
Carbon reduction cycle,
      photosynthetic, 156
Carboxylic acid, 23–24
Cardamon. *See Elettaria*
Carex (sedge), 513
   seed dispersal of, 229
Carica papaya (papain), 126f
Carinal canal, **418**
Carnation. *See Dianthus*
Carotenoid, **154**
Carpel, **197**
Carrageenan, **378**
Carrot. *See Daucus carota*
Carthamus (safflower), 468
   mitochondria of, 41f
Caruncle, **216**
Carya (hickory)
   in eastern deciduous forest, 506
   succession and, 502
Caryopsis, **217, 224**
Cascade, signal, 239–240, 240f
Casparian strip, **114,** 114f
Cassava. *See Manihot esculenta*
Castanea (chestnut), 506
   seed of, 225
Catabolic reaction, **128**
Catalpa bignonioides (catalpa), 86f
Catalyst
   enzyme as, 126–127, 126f–127f
   rate of reaction and, **125–126**
Cation, **16**
Cattail. *See Typha*
Caulerpa, 365f, 374

cDNA, 278
Cedrus sp. (cedar), 435
Cell
   albuminous, **60,** 65t
   antipodal, **199**
   as basic unit, 30
   central, **199**
   collenchyma, 54f
   companion, 65t
   cytoskeleton of, 41–43
   determined, **236**
   differentiation of, 236–237
   egg, **199**
   energy metabolism in, 39
   epidermal, **60,** 65t
   epithelial, resin and, 83
   generative, **197**
   gram-negative, 317f, **318**
   gram-positive, 317f, **319**
   guard, **61,** 65t, 168
   hydration of, 143–144
   laticifer, 63, 64f
   meristematic, **64**
   microscopy and, 30–32, 31f–32f
   organelles of, 33f, 36–39, 37f–39f
   palisade parenchyma, **97**
   parenchyma, 51–53, 53f
   phellem, 66t, **83**
   phelloderm, 66t
   plasma membrane of, 34–35, 34f
   procambium, 73
   protein in, 24
   respiration and, 141–143, 142f–143t
   root, 237
      minerals in, 176
   secretory, 66t
   shoot, 237
   sieve, **60,** 65t, 431
   sperm, **197, 383**
      of moss, 396
   spongy parenchyma, **97**
   subsidiary, **61,** 65t
   synergid, **199**
   transfer, 52, 65t
   transport and storage of substances
      in, 41
   tube, **197**
   vacuole of, 39–40, 42
Cell cycle, **43,** 44–48, 44f–46f
   signals regulating, 240–241
Cell differentiation, **64**
Cell theory, **30**
Cell wall, **35,** 36, 35f
   chitinous, 337
   water flow and, 169f
Cellulose, **2, 22,** 35
Cenozoic era, 453, 455
Central cell, **199**
Central placentation, **198**

Centromere, **44**
Ceratocorys aultii, 367f
CH_4 (methane), 16, 17f
Chain
   electron transport, 140–141
   hydrocarbon, 23
   polypeptide, **24**
Chalcone synthase, 239
Channel in plasma membrane, 177f
Chaparral, **509**
Chara (stonewort), 374, 375f
Character
   molecular, **306,** 307
   morphological, **306**
Character matrix, **306,** 306f
Character of species, **292**
Character state, **306**
   ancestral, **308,** 308f
   derived, **308,** 308f
Charophyte, **373,** 374
Chemical
   mutation and, 296
   in plant, 21
Chemical herbivore defense, 493–494
Chemical reaction
   catalyst affecting, 125–126, 125f–126f
   energy barriers and, 124–125, 125f
   enzymes in, 127
   free energy and, 129–131, 129f–131f
   metabolic pathways and, 127–129,
      128f–129f
   respiration as, 135–136, 135f
Chemiosmotic theory, **143**
Chemistry, 14
Chemoautotroph
   characteristics of, 326–327
   oxidation–reduction reactions of,
      325t
Chemoautotrophic organism, 149
Chemoheterotroph
   bacteria as, 325, 325t
   fungi as, 337
   oxidation–reduction reactions of,
      325t
Chemosynthesis, 149
Chemosynthetic fixation, 149
Chemotherapy, plant-derived, 21
Chemotroph, **322–323**
Chemotropism, **338**
Chenopodium quinoa (quinoa), 463
Chestnut. *See Castanea*
Chestnut oat. *See Quercus prinus*
Chiasma, **262**
Chili pepper. *See Capsicum*
Chilopsis (desert willow), 229
Chitin, 337
Chitinous cell wall, 337
Chlamydomonas (green algae), 373
   zygotic life cycle of, 190–191

Chlorella (green alga), 156
Chloride
   function of, 172t
   in hydrochloric acid, 19
Chloride ion, 168
Chlorine, oxidation of, 16f
Chloroflorocarbon, 6
Chlorophyceae, 373
Chlorophyll, **39**
   absorption of light energy by, 153
   light and, 254
Chlorophyte, **373**
Chloroplast, 40f
   endosymbiosis and, 298
   origin of, 364
   in photosynthesis, 150–152,
      151f–152f
Cholesterol, 23
Cholodny, N., 243
Chondrus crispus, 377
Chromatid, **44**
   in meiosis, **260**
      in asexual reproduction, 184
Chromatin, **260**
Chromatography, paper, 156
Chromista, 368
Chromosome, **36**, 37, 37f, **260,**
   260f–261f
   of Escherichia coli, 319, 319f
   of ferns, 422, 424
   genes carried on, 260
   unpaired, 185
Chromosome set, **185**
Chrysophyta, **368**
Chs gene, 239
Chytridiomycota, **342,** 346, 346f
Cichorium sp. (endive, chicory), 40f,
   468
Ciguatera poisoning, 376
Cilia, 42–43
Ciliate, **367**
Cinnamomum (camphor, cinnamon),
   460
Cisterna, **38**
Citrulus vulgaris (watermelon), 226
Citrus, 473
   defining species of, 292–293
   seed of, 225, 226f
Clade, **9, 305**
   asterid, **468**
   rosid, 463, 465
Cladistics, **9, 305**
   cladogram in, 305–308, 305f–308f
   in DNA analysis, 310–313
Cladogram, **9,** 10, **305**
   alternative, 306
   of bryophytes, 386f
   comparison of character states in,
      306–307, 306f

   for conifers, 433f
   of eudicots, 462–463, 463f
   fungal, 343, 343f
   of gymnosperms, 430
   of protists, 363f
   rooted, 307–308, 307f–308f
Cladophyll, **88**
Clamp connection, **355**
Claoxylon sandwicense, 160
Class, **293**
Classification, plant, **9–11,** 10f
Cleared pea. *See Pisum sativum*
Cleistothecia, **352**
Clematis hybrida (clematis), **118**
Climate
   Mediterranean, **508**–509
   in North America, 456f
Climax stage, **502**
Climbing fig. *See Ficus pumilla*
Clock
   biological, 252–253
   molecular, 451
Clone, **277**
   Pteridium aquilinum, 420
Cloning, 277–278, 278f
Closed-cone conifer, **434, 506**
Clostridium tetanus, 322, 323f
Clover. *See Trifolium* sp.
Cluster, flower, 203–204, 204f
Coal, 408–409, 409f
Coal Age, 406, 406f
Coat, seed, **213**
Cobalt, 172t
Coccidioides immitis, 349
Coconut palm. *See Cocos* sp.
Cocos sp. (coconut), 462
   seed dispersal of, 229
Code, genetic, **26**
Codon, 238–239
Coenomycetes, **344,** 344f
Coenzyme, **126**
Coevolution, **301**
Cofactor, definition of, 126
Coffea (coffee), 468, 473
Cohesive molecules, **166**
Coleochaete, 374, 389
Coleoptile, **217,** 220
Coleorhiza, **217,** 220
Coleus, parenchyma cells of, 52
Coleus blumei, meristem of, 67f
Collenchyma, 52–53, 65t
Colocasia, 460
Columbine. *See Aquilegia formosa*
Columella, **391**
Commensal relationship, **494**
Common ancestor, 294
Community, plant, 496–518. *See also*
      Plant community
Companion cell, **58**–60, 65t

Compartmentalization by
      organelles, 41
Compensation depth, **487**
Compensation point, **158**
Competition, **492**
Complete fertilizer, 170
Complete flower, **201**
Compound
   definition of, 14
   high-energy, 131
   hydrocarbon, 17, 17f
   organic, **19–20**
Compound fruit, 227
Compound leaf, **92**
Compound pistil, **198,** 199f
Concentration
   metabolism and, 128–129
   water flow and, 165
Condensation reaction, **130**
Condiospore, **341**
Conduction of water, 164f
Confocal microscopy, **31**
Conidium, **341**
   of *Penicillium,* 350
Conifer, 433–444. *See also Pinus entries*
   cladogram for, 433f
   Cupressaceae, 434–436, 435f
   Pinaceae, 433–434, 433f
   Podocarpaceae, 437
   seedling of, 476f
   Taxaceae, 436–437, 436f
Conifer forest
   montane, **510,** 510f
   Pacific Coast, 509–510, 510f
   upland, 510–511, 510f–511f
Conjugation, **297, 320,** 320f
Conservation, **5, 514**
Conservation biology, 514–516,
   515f–516f
Consumer, **477**
Continuous variation, 271
Control point hypothesis, 43–44, 44f
Convection, **488**
Convergent evolution, **301,** 308–309
Convolvulus arvensis (field bindweed),
   216f
Conyza canadensis (horseweed), 502
Copernicia, 462
Copper, function of, 172t
Coral fungi, **356**
Core angiosperm, **460**–468, 460f–469f
   eudicot, 462–469, 462f–469f
   magnolid, **460,** 460f
   monocot, **460**–462, 461f–462f
Cork cambium, **62,** 63f, **67,** 83, 84f
Cork oak. *See Quercus suber*
Corm, **87**
Corn. *See Zea mays*
Corolla, **197**

Cortex, **52**
  in primary growth, 72–73
Cottonseed, 21
Cottonwood. *See Populus*
Cotyledon, 99, **213**
Coupled reaction, 130
  transfer of energy in, 136
Covalent bond, **16–18**
Cranberry. *See Vaccinium macrocarpon*
Crassula argentea (jade plant), 101, 101f
Crassulacean acid metabolism (CAM),
  158, 508
Cresosote bush, 481f
Cress. *See Arabidopsis thaliana*
Cristae, **39**
Cronquist, Arthur, 459
Cross, test, **264**
Cross-pollination, **204**
Crossing over, **185, 262**
Crotalaria sagittalis (locoweed), 109
Crotonium, 420f
Crozier, **350**
Crustose lichen, **353,** 353f
Crypt, stomatal, **167**
Crysta, 52
Cucumber. *See Cucurbita*
Cucurbita (cucumber)
  phloem of, 59f
  seed of, 225, 226f
Cupressaceae, 434–436, 435f
Cupressus forbesil (Tecate cypress), 479
Curcuma (tumeric), 462
Curve, survivorship, **487–488,** 487f
Cuticle of bryophyte, 389
Cutin, **36**
Cyanobacteria, **10**
Cycad, 432–433, 432f
Cycas revolute, 432
Cycle
  cell, 43–48, 44f–46f
    signals regulating, 240–241
  energy, 135–136, 135f
  glyoxylate, 141
  Krebs, **137**
  life, 182–194. *See also* Life cycle
  nutrient, 499–501, **499–501,**
    499f–501f
  parasexual, **341**
  transpiration, 168
  tricarboxylic acid, **137,** 140f
Cyclic electron transport, **155**
Cyclic photophosphorylation, **155**
Cyclin-dependent protein kinase
  (C-PK), 240–241, 241f
Cyclosis, **41**
Cylinder, vascular, **56**
Cyme, **204**
Cynara (artichoke), 468
Cynodon dactylon (Bermuda grass), 87

Cyperus, 513
Cyrtomium falcatum, 420f
Cytissus scoparius (Scotch broom), 223
Cytochrome, **141**
Cytokinesis, **43,** 44–47, 45f–46f
Cytokinin, **244,** 245, 244f
  effects of, 241t
Cytoplasm, **10, 34**
  of prokaryote, 319
Cytoplasmic organelle, 217
Cytoplasmic streaming, **339**
Cytoskeleton, **41**

D

Daffodil. *See Narcissus*
Dalea (smoke tree), 229
Dandelion. *See Taraxacum vulgare*
Dark reaction, 150
Darwin, Charles, 295
  on climbing plants, 118
  on flowering plants, 450
  growth studies of, 242–243
Darwin, Francis, 242–243
Daucus carota (carrot)
  gibberellic acid and, 247, 247f
  roots of, 109
Dawn redwood. *See Metasequoia*
  glyptotroboides
Dawsonia superba, 386
Day length, 252, 252f
DCIP, 152
de Saussure, Nicholas, 149
Decay, radioactive, 15
Decomposer, **477**
  fungi as, 10–11, 340
    basidiomycetes as, 356
Decomposition, 500
  litter, 501
Defense, herbivore, 493–494
Dehiscent fruit, 221
Delphinium (larkspur), 463
Demography, plant, **484,** 486, 487f
Denatured protein, **24**
Density, **498**
Deoxyribonucleic acid (DNA),
  24–27
  base sequence of, 280, 280f
  chromosomes and, 260
  cladistic analysis of, 307, 310–313
  endosymbiosis and, 298
  gene expression and, 237–238
  genetic information in, 36–39,
    37f–39f
  insertion of, into plant, 279
  in kingdoms, 294
  mutation and, 296
  in nucleoplasm, 37f
  in principle control point
    hypothesis, 44

recombinant, **277,** 277f
  traits and, 258
Dephosphorylation, **241**
Derived character state, **308**
Dermal tissue system, **50**
Descartes, René, 5
Desert ephemeral, 480f
Desert scrub, **507–508**
Determinate growth pattern, **235–236**
Determined cell, **236**
Deuteromycota, **342**
Devonian period, 406, 428
Dianthus (carnation), 463
Diatom, 369, 369f, 371
  life cycle of, **380,** 380f
Diatomaceous earth, 377–378
Diatomite, **377**
Dichlorophenol indophenol (DCIP), 152
Dichotomous venation, **95**
Dichotomously branching rhizome, **403**
Dicot
  cambium of, 117
  growth patterns and, 235
  primary growth in, 75, 75f
  root of, 114f
  stem development of, 71–73, 72f–73f
Dicytosome, **38**
Dietary Reference Intake, 42
Differentially permeable membrane,
  165
Differentiation, 235
  cell, **64**
Diffuse porous pattern, 79
Diffusion
  minerals and, 176
  water flow and, 165
  of water vapor, 166–167
Digestion, **134**
Digitaria sanguinalis (crabgrass), 92
Dikaryomycetes, **344,** 344f
Dikaryon, **344**
  hymenomycetes as, 354–355
Dikaryotic mycelium, **344**
Dikaryotic stage, **344**
  of asmomycetes, 350
Dimorphism, of yeast, **339**
Dinoflagellate, **367,** 367f
  algal blooms and, 376
Dioecious plant, **202, 482**
Dionea muscipula, 96, 96f
Diploid, **185**
Diploid gametophyte, 422, 424
Diploid generation, 192–193
Directional selection, **299,** 299f
Dispersal
  of fruit, 228f
  of population, 493
  seed, 228t, 229–232, 229f, 230f, 231t
Dissolving of substance, **19**

Hydrogen peroxide, 250
Hydroid, **394**–395
Hydrolysis of adenosine triphosphate,
    130–131, 130f
Hydrolysis reaction, **130**
Hydronium, 19
Hydrophilic bond, **19**
Hydrophobic bond, **19**
Hydrophyte, **101**
Hydrostatic pressure, 166
Hymenium, **352**
Hymenomycetes, **354**–358, 355f–357f
Hypanthium, **203**
Hypha, **338**, 338f
    in ascomycetes, 350
    of *Phycomyces blakesleeanus*, 338f
Hypocotyl, **216**
Hypogeal germination, **218**
Hypolimnion, **513**
Hypothesis, **5**
    principal control point, 43–44
    seedling, **452**

I
Ice Age, **455**
Ilex opaca (holly), 246
Imperfect flower, **201**
Incomplete flower, **201**
Incompletely dominant, **263**
Indehiscent fruit, 221
Independent assortment, Mendel's
    law of, 265
Indeterminate growth pattern, **235**–236
Industry
    algae used in, 377–378
    biotechnology in, 283
Infection, 250
    viral, 331–333, 332f–333f
Inflorescence, **203**–204, 204f
Ingen-Housz, Jan, 149
Ingroup, **307**
Inhibition
    of enzymes, 127
    feedback, **129**, 129f
Injection of DNA into plant, 282–283
Integument, **199**, **213**
    of gymnosperm, 429
Intercalary meristem, **94**
Interfascicular cambium, **75**
Intermediary metabolism, **128**
Intermediate in metabolic pathway, **128**
Interphase, **43**
Intertidal ecosystem, rocky, **512**
Intracellular structure of prokaryotic
    cells, 321–322, 321f
Introgression, **297**
Ion
    malate, 168
    transport of, 177–178

Ion pump, **34**
Ionic bond, **16**
Ipomoea (sweet potato), 468
Iris douglasiana (Douglas iris), 461, 461f
Irish moss, 377, 378
Iron, 172t
Irregular symmetry, **202**
Isoetes, 407, 412, 414, 414f
Isolation
    geographic, **302**
    reproductive, **302**
Isomorphic generation, **381**
Isotope, carbon, 15
Iteroparous reproduction, **482**
Ivy. *See Hedera helix*

J
Jack-in-the-pulpit. *See Arisaema triphylla*
Jade. *See Crassula argentea*
Jewel plant. *See Strephthanthus*
John Day Basin, 455
Joshua tree. *See Yucca*
Juglans regia (walnut)
    in eastern-deciduous forest, 506
    flower of, 201
    leaf scars and buds of, 86f
    pollination of, 209
    seed of, 225
    twig of, 85f
Juniper, 435–436, 435f
Juniperus occidentalis (mountain
    juniper), 435f

K
K-selected species, **483**, 483f, 483t
Kalanchoe pinnata (air plant), 101f, 104
    asexual reproduction of, 183, 184f
Karyogamy, **189**
Karyokinesis, **43**
Kauri. *See Agathis*
Kelp, **371**
Kelp forest, 370
Kidnapping of Lindbergh baby, 55
Kinetic energy, **125**
Kinetochore, **44**
Kingdom, **9**, **293**, 294
Krebs cycle, **137**

L
Lactobacillus, 325
Lactuca (lettuce), 468
    hormones and, 247
Laminaria, 377
    industrial uses of, 378
    life cycle of, 382–383, 382f
Land
    bryophytes on, 386–389, 386f–388f
    tracheophytes on, 402–406,
        402f–403f, 405f, 405t

Larix (larch, tamarack), 435
Larkspur. *See Delphinium*
Larrea tridentata (cresosote bush), 481f
Lateral root, 117f
Lateral root primordia, **116**
Latewood, **79**
Lathyrium felix-femina, 421f
Laticifer cell, 63, 64f
Law of independent assortment, 265
Leaf, 91–105
    abscission of, 104, 104f
    adaptation of, 100–101, 101f, 104
    arrangement of cells and, 95,
        95f–98f, 97–98
    compound, **92**
    growth patterns and, 235
    light and, 92–95, 92f–94f
    origin of, 98–99, 99f
    shade, **487**
    shape of, 99–100, 99f, 100f
    simple, **92**
    sun, **487**
    vascular bundles in relation to,
        73–74, 73f
    water flowing through, 167
Leaf primordium, **98**
Leaf scars, 86f
Leafy body, liverwort, 393–394
Legume
    defenses of, 231t
    fertilizer and, 171
    as fruit, **222**–223
Lenticel, **84**
Lepidodendrid group, 412, 414, 414f
Lepidodendron, 407, 407f
Leptoid, **395**
Leptosporangiate fern, **420**
Leptosporangium, 420, **420**
Lettuce. *See Lactuca*
Lettuce, sea, 373–374
Leukoplast, **39**, 40f
Lianas, **481**
Lichen, **340**, 340f
    crustose, **353**, 353f
    fruticose, **353**, 353f
    mutualism and, 494
Life, transport necessary for, 164
Life cycle, 182–194
    alternation of generations in,
        185–193
    of angiosperms, 458, 458f
    diploid generation and, 190–191
    evolutionary trends in, 193f
    gametic, 189–**191**, 191f–192f, **380**, 380f
    genetic information and, 183–185,
        183f–186f
    heterosporic, 189
    homosporous, 189
    of *Lycopodium*, 408–412, 409f–411f

Rumex acetosa (dock), 209
Runner, 88
Ruppia (ditch-grass), 209
Ruscus aculeatus (butcher's broom), 88, 88f
Rust disease, 358
Rye. See Elymus cereale; Secale cereale

S
S type life history pattern, **483,** 484f
Saccharomyces cervisiae, 350
Saccharum (sugarcane), 473
Safety of biotechnology, 288
Safflower. See Carthamus
Sagittaria (arrowhead), 461, 461f
Salicornia virginica (pickleweed), 512
Salicylic acid, 251
Salinity of soil, 175
Salix (willow)
    photosynthesis and, 148
    pollination of, 209f
Salt
    definition of, **16**
    excess of, 171
Salt grass. See Distichlis spicata
Salt marsh, **511,** 511f
Salt tolerant tomato, 287–288, 287f
Saltbush. See Atriplex
Salvia mellifera (sage), 468f
Salvinia, 419f
SAM. See Shoot apical meristem
Samara, **224**–225
Sambucus sp.
    cork cambium of, **84f**
    elder, 84f
    elderberry, 468
Sap, 83
Sapling, solar radiation and, 487
Saprobe, **340**
    basidiomycetes as, 356
Sapwood, **79**
Sarrencia purpurea, 96, 96f
Saturated hydrocarbon chain, **23**
Savanna, **507**
Saw palmetto, 461
Scanning electron microscopy, **32**
Schizocarp, **225**
Schizosaccharomyces pombe, 350
Schüsler, Arthur, 343
Sciadophyton, 404f
Scientific method, **2, 5,** 7–8
Scirpus, 513
Sclereid, **53**–54, 65t
Sclerenchyma, **53,** 54f
Scrub, **507**–508
Scutellum, **217**
Sea grass, 11
Sea lavender. See Limonium
Sea lettuce, 373–374

Seaweed, **373**
    distribution of, 512f
    red, 377
Secale cereale (rye), pollination of, 209
Secondary cell wall, 36
Secondary consumer, **477**
Secondary endosymbiosis, 298
Secondary growth, 75–86. See also
        Growth, secondary
    of root, 117f
Secondary meristem, **64**
Secondary phloem, **58**
    bark composed of, 83
Secondary structure of protein, **24**
Secondary succession, **502**
Secondary xylem, **56**
    wood composed of, 77, 79, 81
Secretory cell, 66t
Secretory structure, **63**
Sedum (stonecrop), flower of, 202–203
Seed, 212–221
    in asexual reproduction, 205
    dispersal of, 228f–230f, 229–232, 231t
    enclosed, 449–450, 450f
    germination of, 217–221, 218f–220f
    of grasses, 217
    of gymnosperm, 426–430, 427f–429f
    as mature ovule, 212
    onion, 217
    pine, 443
    of pine, 442
    reproduction on land and, 428–430
    structure of, 212, 217
Seed coat, **213**
Seed fern, 429f
Seed leaf, 99
Seed plant, **427**
    cladogram of, 427f
Seedless fruit, 228
Seedless vascular plant, 403. See also
        Tracheophyte
Seedling, conifer, 476f
Seedling hypothesis, **452**
Selaginella, 407, 412, 413f, 429
Selection, natural, **295,** 298–301,
        299f–301f
Selective agent, **299**
Selectively permeable membrane, 34
Self-pollination, **204**
Selfing, **204**
Semelparous reproduction, **482**
Seminal root, **109,** 110f
Semitropical forest, 487f
Senebier, Jean, 149
Senescence, ethylene and, 249–250,
        250f
Sepal, **196**
Septal pore, **340**
Septate hypha, in ascomyces, 350

Septation, regular, 345
Septum, **340**
    dolipore, **356,** 356f
Sequence, base, 258
Sequoia sempervirens (redwood)
    characteristics of, 436
    chromosomes of, 260
    forests of, 509–510, 510f
    populations of, 478, 478f
    reproduction of, 183
    wood of, 82, 82f
Sesamum (sesame), 468
Seta, **392**
Setaria lutescens (yellow foxtail), 200
Sexual cycle, heterosporic or
        homosporic, 190
Sexual identity, 482–483
Sexual reproduction
    dikaryotic stage in, 344, 344f
    of ferns, 421–422, 422f
    flower as site of, 196
    of fungi, 341–342, 343f
    genetic difference and, 184
    of gymnosperms, 428–430, 429f
    of Marchantia, 393, 393f
    meiosis in, 185, 186f
    in mosses, 395–397, 396f–397f
    recombination in, **297**
Shade leaf, 100f, **487**
Shasta red fir. See Abis magnifica
Sheath, **93**
Shelf fungi, **356**
Shellfish poisoning, 376
Shepherd's purse. See Capsella burse-
        pastoris
Shoot apical meristem (SAM), **64**
Shoot cell, 237
Shoot system, **50,** 70–105
    function of, 70–71
    growth of, 71–75, 72f–75f
    leaf and, 91–105. See also Leaf
    stem and, 77–90. See also Stem
Shrub, **481**
Side chain, 24
Sieve area, **60**
Sieve cell, **60,** 65t, 431
Sieve plate, **60**
Sieve tube, **58**
    transport of sucrose, 179–180
Sieve-tube member, **58,** 65t, **452**
Signal cascade, 239–240, 240f
Silencing, gene, 284
Silique, **224**
Silurian period, 403, 406
Simple fruit, 222–226, 223f–226f
Simple leaf, **92**
Simple pistil, **198,** 199f
Simple pit, **57**
Simple tissue, 50–53, 53f

Single ovary, 222–226, 223f–226f
Sister chromatid, 185
Sitka spruce. *See Picea sitchensis*
Skunk cabbage. *See Lysichtum americanum*
Slime mold, 362, 363f
Smog, 408
Smooth endoplasmic reticulum, **38**
Smut disease, 358, 358f
Soap weed. *See Yucca* sp.
Sodium, oxidation of, 16f
Softwood tree, **481**
Soil
  environmental importance of, 486–487, 486f
  formation of, 174, 176
  of grassland, 175f
  particles of, 166
  types of, 172, 174–175
  water flow through, 168
Solanum parishii (nightshade), 468f
Solanum (potato), 468
Solanum tuberosum (potato)
  hybrid, 271
  tubers of, 87, 87f–88f
Solar radiation, 487–489, 487f–489f
Solarium nigra (potato)
  epidermis of, 95f
  leaf of, 95f
Solute, **19**
Solution, **19**
  apoplast, 166
Solvent, water as, **19**
Sorbus americana (mountain ash), 497
Soredium, **353**
Soybean. *See Glycine max*
Species, **292**–293, 292f
  biological, **293**
Species name, **293**
Species Plantarum, 458
Species richness, **498**
Specific epithet, **293**
Specimen, type, **292**
Spectrum
  absorption, **153**
  electromagnetic energy, **152**
Sperm cell, **197, 383**
  of moss, 396
Sperm nucleus, 442
Sphagnum, 394–395, 397–398
  succession and, 503
Sphenophyte, **417**–419, 417f–418f
Spike, **204**
Spinacea (spinach), 463
Spindle apparatus, 44
  in asexual reproduction, 184, 184f
Spiral phyllotaxis, 74, 74f
Spirilla, **319**
Spirogyra, 365f

Spongy mesophyll, **97**
Spongy parenchyma cell, **97**
Sporangiophore, **347, 418**
Spore, **2, 186, 338**
  of *Aspergillus*, 338f
  of ferns, 421–422
  natural selection for, 404
Sporic life cycle, 192, **381,** 381f
Sporocyte, **388**
Sporophyll, **410**
Sporophyte, **186**
  evolution of, 387–388, 388f
  moss, 396–397, 398f
  survival on land and, 404
Sporopollenin, 197
Springwood, **79**
Spruce. *See Picea*
Spurge. *See Euphorbia*
Stabilizing selection, **299**
Stamen, **197**
Staminate, **201**
Starch, **22**
  beer and, 145
  transformed into glucose, 135f
State, character, **306**
Stem, 77–90
  economic value of, 88–89, 88f–89f
  growth of, 71–75, 72f–75f
    ground meristem to pith and cortex, 72–73
    monocot versus dicot, 74–75, 75f
    procambium to primary xylem and phloem, 73
    protoderm to epidermis, 71–72, 73f
    vascular bundles in, 73–74, 74f
  modifications of, 87–88, 87f–88f
  monocot, secondary growth of, 85–86
Steppe, **507**
Sterol, **23**
Stigma, **197**
Stipe, **354, 371**
Stobilus, **410**
Stolon, **88**
Stoma, **61,** 61f, **95**
  of bryophyte, 389
  guard cells of, 168
  opening of, 169f
  water vapor diffusion through, 166–167
Stomatal apparatus, **95**
Stomatal crypt, **167**
Stonecrop. *See Sedum*
Stonewort. *See Chara*
Stramenopile, 368
Stratification, **442**
Strawberry. *See Fragaria* sp.
Strelitzia (bird-of-paradise), 462

Strephthanthus (jewel plant), style of, 198f
Stress
  competition and, 492
  environmental, photosynthesis and, 158–159, 158f
Strip, Casparian, **114,** 114f–115f
Strobilum, 418f
  of gnetophyte, 444
Strobilus, **440,** 440f
Stroma, **39**
Stroma lamella, **151**
Style, **197,** 198–199
Subalpine, **511**
Subcellular parasite, 330–333
Suberin, **36**
Subshrub, **481**
Subsidiary cell, **61,** 65t
Substomatal chamber, **97**
Substrate, **124**
  metabolic pathway and, 127–128
Succession, **501**–503, 502f, 503t
  primary, **502**
  progressive, **503**
  retrogressive, **503**
  secondary, **502**
Succulent, **486, 508**
  carbon dioxide and, 158
Sucrose, transport of, 179–180
Sugar, **22**
  maple, 88, 88f
  reduction of carbon dioxide to, 156–157
Sugarcane. *See Saccharum*
Sulfur
  function of, 172t
  oxidation of, 326–327
Summerwood, **79**
Sun leaf, 100, 100f, **487**
Sunflower. *See Helianthus annuus*
Sunshine, 487–489, 487f–489f
Superior ovary, **203**
Survivorship curve, **487**–488, 487f
Suspensor, **213**
Sweet gum. *See Liquidambar styraciflua*
Sycamore. *See Platanus*
Symbiotic relationship, **328**–330
  ascomycetes in, 352–353, 353f
  of fungi, **340**
Symbol, for molecule, 14
Symmetry, regular and irregular, **202**
Symplast, **33**
Symplastic flow, 167–168, 168f
Synapsis, **185,** 260
Synergid cell, **199**
Synthesis, enzyme, 127
Syringa vulgaris (lilac)
  leaf of, 97f
  parenchyma of, 53f

Xylem, **56**
    of lycophytes, 410
    primary, 73
    of roots, water flow through,
        167–168
    secondary, **56**
        wood composed of, 77, 79, 81
    water flow through, 167–168

Y

Yeast, **339**
Yellow foxtail. *See Setaria lutescens*
Yew, 437, 437f
Yucca
    Josua tree, 167f
    secondary growth of, 86
    soap weed, 108

Z

Zamia integrifolia, 432, 432f
Zantedeschia, 460
Zea mays (corn, maize)
    breeding of, 269f
    carbon dioxide and, 158f
    chloroplast of, 40f
    flower of, 201
    germination of, 220, 220f
    hybrid, 271
    leaf of, 92, 98f
    mutations in, 259f, 259t
    parenchyma of, 53f
    photosynthesis in, 159f
    pollination of, 209
    root of, 110f
    roots of, 109
    ropot cap of, 112f
    transpiration in, 165
    Ustilago maydos and, 358, 358f
Zinc, function of, 172t
Zingiber (ginger), 462
Zonation for forest community, 510
Zoospore, **346**
Zostera (eel grass), pollination of, 209
Zosterophyllophyta, 406
Zuckerkandl, Emile, 451
Zygomycetes, 346–348, 347f
Zygomycota, **342, 346–348,** 347f
Zygosporangium, **342, 347**
Zygospore, **347**
Zygote, **185, 200**
    differentiation and, 236
Zygotic life cycle, **190**–191, **378**, 379f